# Understanding Organic Chemistry

# Understanding Organic Chemistry

Edited by
Saul Rowen

Willford Press
www.willfordpress.com

Published by Willford Press,
118-35 Queens Blvd., Suite 400,
Forest Hills, NY 11375, USA



ISBN: 978-1-68285-373-3

**Cataloging-in-Publication Data**

Understanding organic chemistry / edited by Saul Rowen.
p. cm.
Includes bibliographical references and index.
ISBN 978-1-68285-373-3
1. Chemistry, Organic. 2. Chemistry. I. Rowen, Saul.
QD251.3 .U53 2017
547--dc23

For information on all Willford Press publications
visit our website at www.willfordpress.com

WILLFORD PRESS

Printed in China.

# Contents


**Preface** ..... IX

Chapter 1 **Bond cleavage, fragment modification and reassembly in enantioselective three-component reactions** ..... 1
Dan Zhang, Jun Zhou, Fei Xia, Zhenghui Kang, Wenhao Hu

Chapter 2 **Cationic mononuclear ruthenium carboxylates as catalyst prototypes for self-induced hydrogenation of carboxylic acids** ..... 9
Masayuki Naruto, Susumu Saito

Chapter 3 **A molecular nematic liquid crystalline material for high-performance organic photovoltaics** ..... 18
Kuan Sun, Zeyun Xiao, Shirong Lu, Wojciech Zajaczkowski, Wojciech Pisula, Eric Hanssen, Jonathan M. White, Rachel M. Williamson, Jegadesan Subbiah, Jianyong Ouyang, Andrew B. Holmes, Wallace W.H. Wong, David J. Jones

Chapter 4 **One-shot K-region-selective annulative π-extension for nanographene synthesis and functionalization** ..... 27
Kyohei Ozaki, Katsuaki Kawasumi, Mari Shibata, Hideto Ito, Kenichiro Itami

Chapter 5 **Carbon-catalysed reductive hydrogen atom transfer reactions** ..... 34
Huimin Yang, Xinjiang Cui, Xingchao Dai, Youquan Deng, Feng Shi

Chapter 6 **Using carbon dioxide as a building block in organic synthesis** ..... 45
Qiang Liu, Lipeng Wu, Ralf Jackstell, Matthias Beller

Chapter 7 **Computational ligand design in enantio- and diastereoselective ynamide [5+2] cycloisomerization** ..... 60
R.N. Straker, Q. Peng, A. Mekareeya, R.S. Paton, E.A. Anderson

Chapter 8 **Direct and selective hydrogenolysis of arenols and aryl methyl ethers** ..... 69
Shuhei Kusumoto, Kyoko Nozaki

Chapter 9 **A simple and versatile design concept for fluorophore derivatives with intramolecular photostabilization** ..... 76
Jasper H. M. van der Velde, Jens Oelerich, Jingyi Huang, Jochem H. Smit, Atieh Aminian Jazi, Silvia Galiani, Kirill Kolmakov, Giorgos Guoridis, Christian Eggeling, Andreas Herrmann, Gerard Roelfes, Thorben Cordes

Chapter 10 **Design concept for α-hydrogen-substituted nitroxides** ..... 91
Michal Amar, Sukanta Bar, Mark A. Iron, Hila Toledo, Boris Tumanskii, Linda J.W. Shimon, Mark Botoshansky, Natalia Fridman, Alex M. Szpilman

Chapter 11 **Total synthesis of periploside A, a unique pregnane hexasaccharide with potent immunosuppressive effects** ..... 100
Xiaheng Zhang, Yu Zhou, Jianping Zuo, Biao Yu

Chapter 12 **Nucleophilic arylation with tetraarylphosphonium salts** ..... 110
Zuyong Deng, Jin-Hong Lin, Ji-Chang Xiao

Chapter 13 **Total synthesis and antiviral activity of indolosesquiterpenoids from the xiamycin and oridamycin families** ..... 118
Zhanchao Meng, Haixin Yu, Li Li, Wanyin Tao, Hao Chen, Ming Wan, Peng Yang, David J. Edmonds, Jin Zhong, Ang Li

Chapter 14 **Late-stage C–H functionalization of complex alkaloids and drug molecules via intermolecular rhodium-carbenoid insertion** ..... 126
Jing He, Lawrence G. Hamann, Huw M.L. Davies, Rohan E.J. Beckwith

Chapter 15 **Dynamic control of chirality in phosphine ligands for enantioselective catalysis** ..... 135
Depeng Zhao, Thomas M. Neubauer, Ben L. Feringa

Chapter 16 **The evolution of cyclopropenium ions into functional polyelectrolytes** ..... 142
Yivan Jiang, Jessica L. Freyer, Pepa Cotanda, Spencer D. Brucks, Kato L. Killops, Jeffrey S. Bandar, Christopher Torsitano, Nitash P. Balsara, Tristan H. Lambert, Luis M. Campos

Chapter 17 **Pd(II)-catalysed *meta*-C–H functionalizations of benzoic acid derivatives** ..... 149
Shangda Li, Lei Cai, Huafang Ji, Long Yang, Gang Li

Chapter 18 **Enantioselective acyl transfer catalysis by a combination of common catalytic motifs and electrostatic interactions** ..... 157
Hiroki Mandai, Kazuki Fujii, Hiroshi Yasuhara, Kenko Abe, Koichi Mitsudo, Toshinobu Korenaga, Seiji Suga

Chapter 19 **The limits of precision monomer placement in chain growth polymerization** ..... 168
Guillaume Gody, Per B. Zetterlund, Sébastien Perrier, Simon Harrisson

Chapter 20 **Luminescent platinum(II) complexes with functionalized N-heterocyclic carbene or diphosphine selectively probe mismatched and abasic DNA** ..... 176
Sin Ki Fung, Taotao Zou, Bei Cao, Tianfeng Chen, Wai-Pong To, Chen Yang, Chun-Nam Lok, Chi-Ming Che

Chapter 21 **A competitive and reversible deactivation approach to catalysis-based quantitative assays** ..... 185
Kazunori Koide, Matthew P. Tracey, Xiaodong Bu, Junyong Jo, Michael J. Williams, Christopher J. Welch

Chapter 22 **Homochiral $D_4$-symmetric metal–organic cages from stereogenic Ru(II) metalloligands for effective enantioseparation of atropisomeric molecules** ..... 192
Kai Wu, Kang Li, Ya-Jun Hou, Mei Pan, Lu-Yin Zhang, Ling Chen, Cheng-Yong Su

Chapter 23 **Selective catalytic two-step process for ethylene glycol from carbon monoxide** ..... 202
Kaiwu Dong, Saravanakumar Elangovan, Rui Sang, Anke Spannenberg, Ralf Jackstell, Kathrin Junge, Yuehui Li, Matthias Beller

Chapter 24 **Orthogonal ring-closing alkyne and olefin metathesis for the synthesis of small GTPase-targeting bicyclic peptides** ..... 209
Philipp M. Cromm, Sebastian Schaubach, Jochen Spiegel, Alois Fürstner, Tom N. Grossmann, Herbert Waldmann

Chapter 25 **Total synthesis of atropurpuran.......216**
Jing Gong, Huan Chen, Xiao-Yu Liu, Zhi-Xiu Wang, Wei Nie, Yong Qin

**Permissions**

**List of Contributors**

**Index**

# Preface

For all readers who are interested in organic chemistry, this book will serve as an excellent guide to develop a comprehensive understanding. Organic chemistry delves into the structure, properties, composition, reactions and preparation of carbon-containing compounds, which include not only hydrocarbons but also compounds with any number of other elements, including nitrogen, oxygen, etc. This book is a valuable compilation of topics, ranging from the basic to the most complex advancements in the field of organic chemistry. Some of the diverse topics covered herein address the varied branches that fall under this category. Coherent flow of topics, student-friendly language and extensive use of examples make this book an invaluable source of knowledge.

This book is a result of research of several months to collate the most relevant data in the field.

When I was approached with the idea of this book and the proposal to edit it, I was overwhelmed. It gave me an opportunity to reach out to all those who share a common interest with me in this field. I had 3 main parameters for editing this text:

1. Accuracy - The data and information provided in this book should be up-to-date and valuable to the readers.

2. Structure - The data must be presented in a structured format for easy understanding and better grasping of the readers.

3. Universal Approach - This book not only targets students but also experts and innovators in the field, thus my aim was to present topics which are of use to all.

Thus, it took me a couple of months to finish the editing of this book.

I would like to make a special mention of my publisher who considered me worthy of this opportunity and also supported me throughout the editing process. I would also like to thank the editing team at the back-end who extended their help whenever required.

**Editor**

# 1

# Bond cleavage, fragment modification and reassembly in enantioselective three-component reactions

Dan Zhang[1], Jun Zhou[1], Fei Xia[1], Zhenghui Kang[1] & Wenhao Hu[1]

Chemical bond cleavage and reconstruction are common processes in traditional rearrangement reactions. In contrast, the process that involves bond cleavage, fragment modification and then reconstruction of the modified fragment provides an efficient way to build structurally diversified molecules. Here, we report a palladium(II)/chiral phosphoric acid catalysed three-component reaction of aryldiazoacetates, enamines and imines to afford α-amino-δ-oxo pentanoic acid derivatives in good yields with excellent diastereoselectivities and high enantioselectivities. The stereoselective reaction went through a unique process that involves cleavage of a C–N bond, modification of the resulting amino fragment and selective reassembly of the modified fragment. This innovative multi-component process represents a highly efficient way to build structurally diversified polyfunctional molecules in an atom and step economic fashion. A keto-iminium is proposed as a key intermediate and a chiral palladium/phosphate complex is proposed as an active catalyst.

[1] Shanghai Engineering Research Center of Molecular Therapeutics and New Drug Development, and Department of Chemistry, East China Normal University, Shanghai 200062, China. Correspondence and requests for materials should be addressed to W.H. (email: whu@chem.ecnu.edu.cn).

Innovative processes involving new modes of chemical bond formation are highly desirable to achieve challenging synthetic tasks that existing ones cannot easily accomplish[1]. Rearrangement reactions, for example, have been intensively investigated to provide a unique way to build one molecule from its mother molecule in a 100% atom-efficient fashion[2]. Bond cleavage and reconstruction involved in the rearrangement reactions occur in the same molecule[3–5] and typically in a concerted fashion (Fig. 1 path a)[2]. In addition, bond cleavage and reassembly of fragments in traditional two-component reactions have been reported[6–10], but in most cases, these reactions resulted in either loss of components[8] or random combination of the fragments such as radical fragments without any selectivity[9,10]. We communicate here a multi-component process that involves chemical bond cleavage, fragment modification and specific reassembly of the modified fragment to afford functional group-enriched molecules (Fig. 1 path b). This innovative process provides a unique way to build diversified molecules in an efficient and atom-economic fashion.

We have recently reported an efficient trapping process in which zwitterionic intermediates were trapped by electrophiles to produce polyfunctional molecules in a single synthetic step[11,12]. In the reported reactions, for example, indoles and pyrroles containing a cyclic enamine as a key structural motif were successfully employed as substrates (Fig. 2a). In continuation with this study, we envisioned that using an acyclic enamine should undergo a similar transformation to afford a zwitterionic intermediate trapping product (Fig. 2b)[13]. Thus, the three-component reaction of methyl phenyldiazoacetate, an acyclic enamine and an imine was conducted. However, the proposed process did not occur as expected, but we discovered a novel process that involves cleavage of a C–N bond, modification of the resulting amino fragment by forming an ammonium ylide and regio- and stereoselective reassembly of the modified fragment back to another iminium fragment to produce a chiral α-amino-δ-oxo pentanoic acid derivative (Fig. 2c). In this three-component reaction, one bond cleavage and three bonds formation take place in one step to produce a new molecule bearing two stereogenic centres including one quaternary carbon centre in excellent diastereoselectivity and high enantioselectivity. This unique process provides an easy access for highly efficient construction of chiral α-amino-δ-oxo pentanoic acid derivatives[14–17] from simple substrates under mild reaction conditions.

## Results

**Preliminary optimization**. Initial reaction of methyl phenyldiazoacetate **2a**, enamine **3a** and imine **4a** was carried out in dichloromethane ($CH_2Cl_2$) at ambient temperature under the catalysis of $[PdCl(\eta^3\text{-}C_3H_5)]_2$ and Brønsted acid *rac*-**1a**. After hydrolysis, product **5a** was obtained in just 15% isolated yield with excellent diastereoselectivity (>95:5 diastereomeric ratio, dr) accompanying with 35% of side product 1,5-diketo **6** (Table 1, entry 1). The structural identity of **5a** was later confirmed by single-crystal X-ray analysis (see Supplementary Fig. 63). The interesting process for the formation of **5a** attracted us for further studies. To suppress the formation of the side product **6**, minimizing the concentration of **3a** in the reaction mixture was thought to be critical. Thus, a solution of **2a** and **3a** was slowly introduced to the reaction mixture over 1 h. As a result, the yield of **5a** was increased to 45% (Table 1, entry 2). The product yield of **5a** was further enhanced to 63% and side product **6** was decreased to 6% yield by performing the reaction in an ice-bath (Table 1, entry 3). Lowering the reaction temperature to −20 °C resulted in a very slow process for the diazo decomposition and no improved results were obtained (Table 1, entry 4).

**Enantioselective reaction**. We then explored the enantioselective control of the novel process. A series of chiral 3,3'-disubstituted BINOL-derived phosphoric acids (PPAs)[11,12] were tested and the results were summarized in Table 2. Bulky substituted phosphoric acids **1b** and **1c** were found to be less effective to promote this reaction, as only trace amount of product was observed (Table 2, entries 1–2). Moderate-to-good enantioselectivities were obtained by using PPAs **1d**–**1k** (upto 80:20 enantiomeric ratio (er), entries 3–10, Table 2). Other solvents did not improve the er further in this case (see Supplementary Table 1). To further improve the enantioselectivity, a more sterically hindered diazo compound *tert*-butyl phenyldiazoacetate **2b** was used. While a sharp drop in enantiomeric excess (ee) value was observed with PPA **1e**, less bulkier PPA **1i** gave the product with increased er to 82:18 (Table 2, entry 11 versus 12). Additional reaction condition optimization increased the er to 92:8 when chloroform ($CHCl_3$) was used as the solvent at a reaction temperature of −20 °C (Table 2, entry 14). When the extremely bulky diazo compound **2c** was used, the reaction proceeded in $CHCl_3$ at −10 °C to give **5a** in 69% yield, >95:5 dr and 94:6 er (Table 2, entry 16).

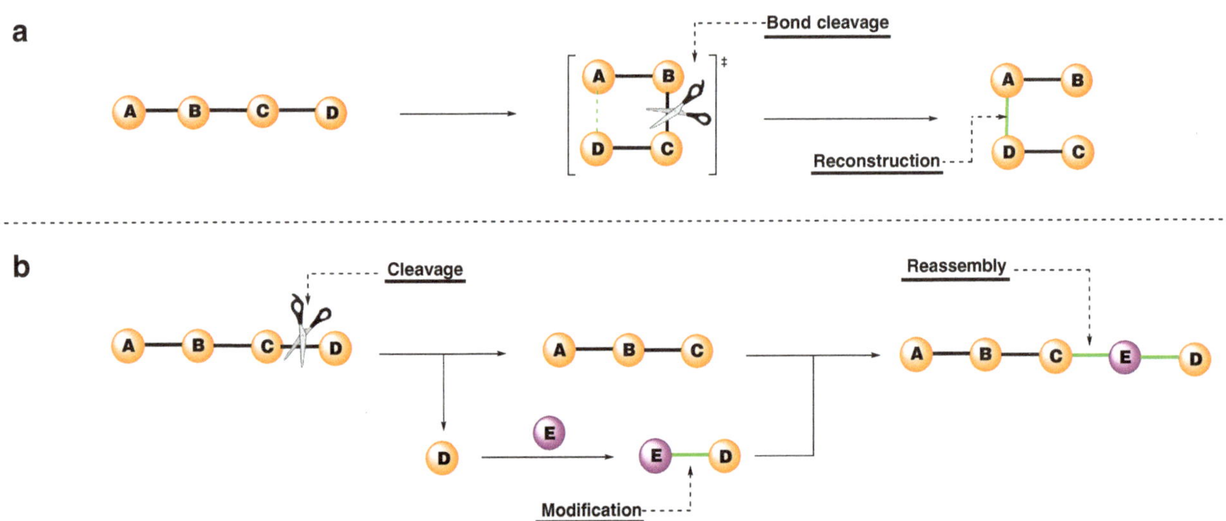


**Figure 1 | Conceptual framework.** (**a**) Intramolecular rearrangement: bond cleavage and reconstruction. (**b**) Concept of bond cleavage, fragment modification and reassembly.

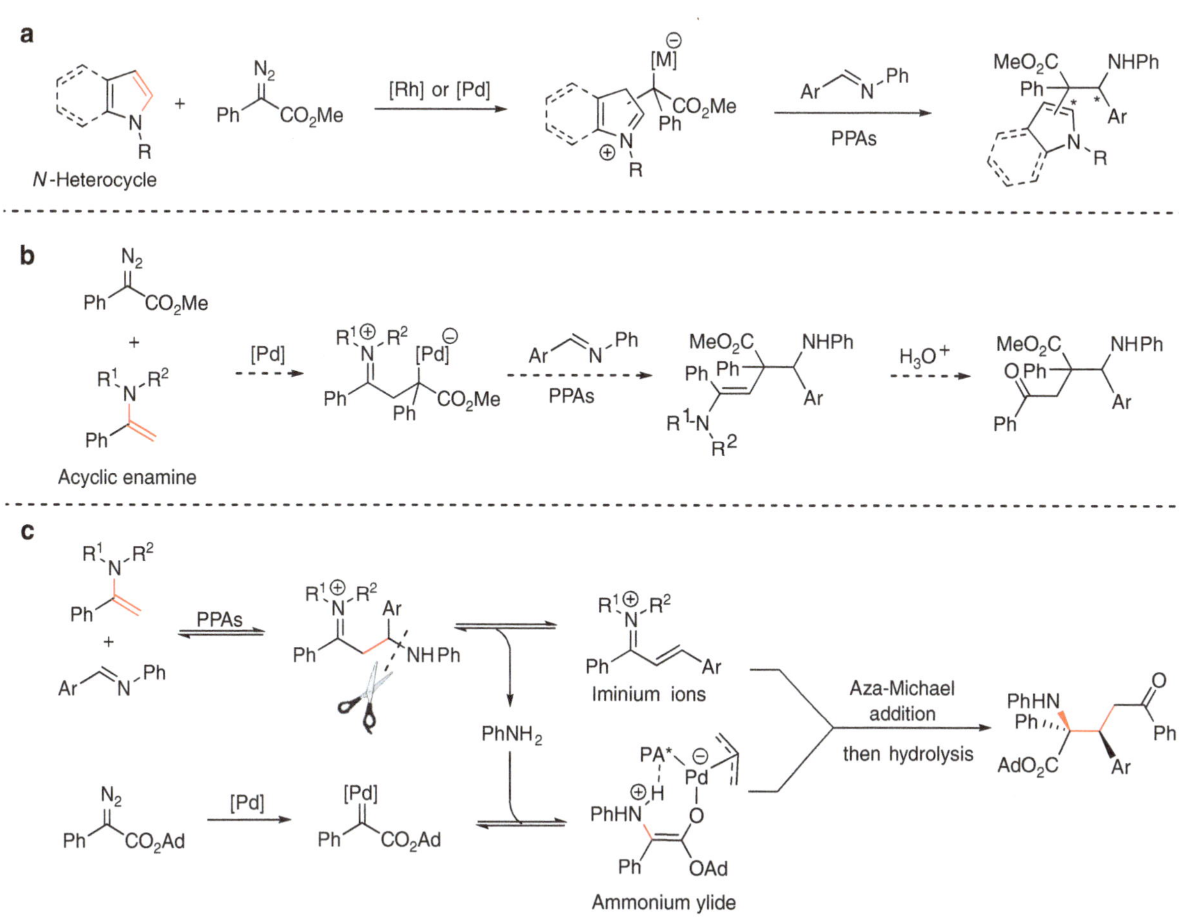


**Figure 2 | Discovery of the novel process.** (**a**) Prior works for the enantioselective trapping of zwitterionic intermediates. (**b**) Hypothetical pathway for the enamine-participated multi-component reaction. (**c**) Actual process: bond cleavage, fragment modification and reassembly.

**Table 1 | Preliminary results for the title three-component reaction*.**

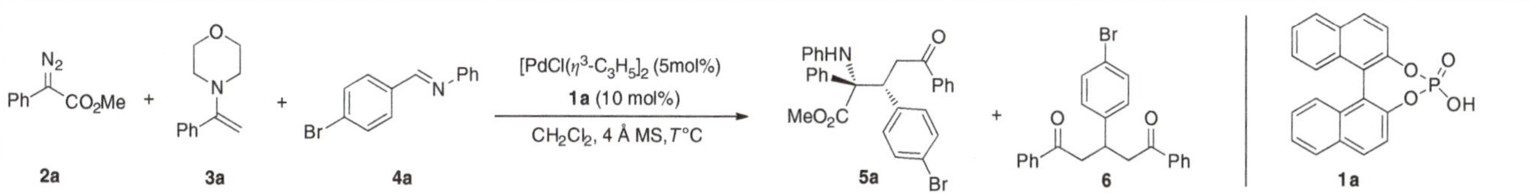


| Entry | T (°C) | 5a, Yield (%)† | 6, Yield (%)† | Dr (syn:anti)‡ |
|---|---|---|---|---|
| 1§ | 25 | 15 | 35 | >95:5 |
| 2‖ | 25 | 45 | 20 | >95:5 |
| 3‖ | 0 | 63 | 6 | >95:5 |
| 4‖,¶ | −20 | 41 | 25 | >95:5 |

Dr, diastereomeric ratio.
*$[PdCl(\eta^3\text{-}C_3H_5)]_2$/**1**/**2a**/**3a**/**4a** = 0.05:0.1:1.5:1.1:1.
†Isolated yields.
‡Dr of **5a**, determined by $^1H$ NMR analysis of the crude mixture.
§**3a** was added to the system in one portion.
‖**2a** and **3a** were dissolved in 1 ml $CH_2Cl_2$ and the solution was added to the system in 1 h.
¶The reaction ran over 10 h.

**Substrate scope.** With optimized reaction conditions identified, the substrate scope of the reaction was explored. With respect to the imine substrate, the reactions gave good yields with high-to-excellent dr and er (Table 3, entries 1–10). The best result came from electron-deficient imine **4b**, which gave product **5d** in 65% yield with 95:5 dr and 96:4 er (Table 3, entry 2). The reaction was extended to other aryldiazoacetates (Table 3, entries 11–15) and enamines (Table 3, entries 16–17). All the reactions gave comparably good results except for diazo compound **2f**, which bears three electron-donating methoxyl groups, resulting in product **5o** with decreased er to 73:27 (Table 3, entry 13). For a representative example, with 4-fluoro substituted enamine **3c**, the reaction gave corresponding product **5s** in 64% isolated yield with 96:4 dr and 97:3 er (Table 3, entry 17).

## Discussion

To understand the reaction pathway for this new process, several control experiments were initially performed. Aziridination

**Table 2 | Optimization for the enantioselective three-component reaction of 2 and 3a with 4a*.**

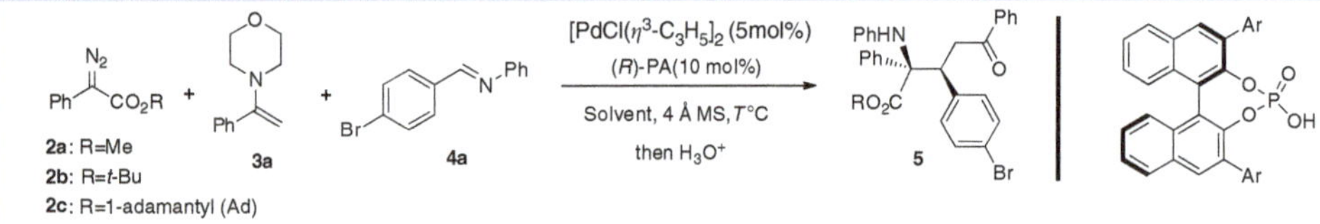


1b, Ar=2.4.6-$^{i}$Pr$_3$-C$_6$H$_2$
1c, Ar=SiPh$_3$
1d, Ar=3,5-Cl$_2$-C$_6$H$_3$
1e, Ar=9-phenanthryl
1f, Ar=Ph
1g, Ar=p-MeO-C$_6$H$_4$
1h, Ar=p-CF$_3$-C$_6$H$_4$
1i, Ar=3,5-(CF$_3$)$_2$-C$_6$H$_3$
1j, Ar=biphenyl
1k, Ar=β-naphthyl

| Entry | 2 | Solvent | PA | T (°C) | Yield (%)† | Dr (syn:anti)‡ | Er§ |
|---|---|---|---|---|---|---|---|
| 1 | **2a** | $CH_2Cl_2$ | **1b** | 0 | Trace | — | — |
| 2 | **2a** | $CH_2Cl_2$ | **1c** | 0 | Trace | — | — |
| 3 | **2a** | $CH_2Cl_2$ | **1d** | 0 | 41 | >95:5 | 53:47 |
| 4 | **2a** | $CH_2Cl_2$ | **1e** | 0 | 58 | >95:5 | 80:20 |
| 5 | **2a** | $CH_2Cl_2$ | **1f** | 0 | 50 | >95:5 | 68:32 |
| 6 | **2a** | $CH_2Cl_2$ | **1g** | 0 | 53 | >95:5 | 61:39 |
| 7 | **2a** | $CH_2Cl_2$ | **1h** | 0 | 38 | >95:5 | 65:35 |
| 8 | **2a** | $CH_2Cl_2$ | **1i** | 0 | 49 | >95:5 | 67:33 |
| 9 | **2a** | $CH_2Cl_2$ | **1j** | 0 | 42 | >95:5 | 65:35 |
| 10 | **2a** | $CH_2Cl_2$ | **1k** | 0 | 50 | >95:5 | 67:33 |
| 11 | **2b** | $CH_2Cl_2$ | **1e** | 0 | 55 | >95:5 | 53:47 |
| 12 | **2b** | $CH_2Cl_2$ | **1i** | 0 | 51 | >95:5 | 82:18 |
| 13 | **2b** | $CHCl_3$ | **1i** | 0 | 62 | >95:5 | 90:10 |
| 14 | **2b** | $CHCl_3$ | **1i** | – 20 | 48 | >95:5 | 92:8 |
| 15 | **2c** | $CHCl_3$ | **1i** | 0 | 61 | >95:5 | 91:9 |
| 16 | **2c** | $CHCl_3$ | **1i** | – 10 | 69 | >95:5 | 94:6 |

Dr, diastereomeric ratio; Er, enantiomeric ratio; PA, phosphoric acid.
*[PdCl($\eta^3$-C$_3$H$_5$)]$_2$/**1**/**2**/**3a**/**4a** = 0.05:0.1:1.5:1.1:1.
†Isolated yields.
‡Determined by $^1$H NMR analysis of the crude mixture.
§Determined by HPLC analysis using a chiral stationary phase. Value is that of the major isomer.

**Table 3 | Substrate scope of enantioselective three-component reaction of 2 and 3 with 4*.**

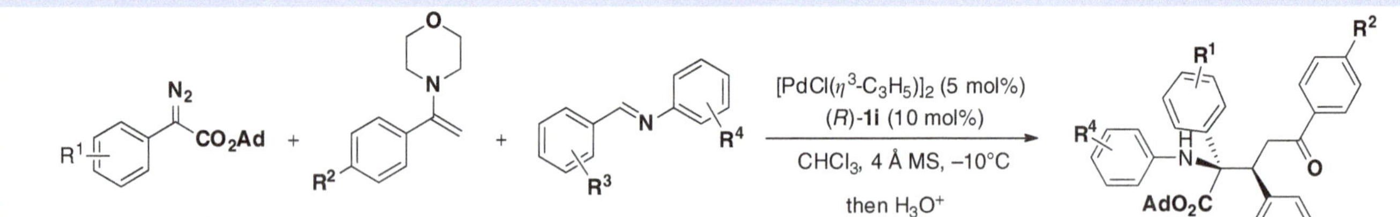


2c, R$^1$ = H; 2d, R$^1$ = 4-Me;
2e, R$^1$ = 4-MeO;
2f, R$^1$ = 3,4,5-(MeO)$_3$;
2g, R$^1$ = 4-F; 2h, R$^1$ = 4-CF$_3$.

3a, R$^2$ = H;
3b, R$^2$ = Me;
3c, R$^2$ = F.

| Entry | 2/3 | 4: R³/R⁴ | 5: Yield (%)† | Dr‡ | Er§ |
|---|---|---|---|---|---|
| 1 | **2c/3a** | **4a**: 4-Br/H | **5c**: 69 | 95:5 | 94:6 |
| 2 | **2c/3a** | **4b**: 4-CF$_3$/H | **5d**: 65 | 95:5 | 96:4 |
| 3 | **2c/3a** | **4c**: 4-CN/H | **5e**: 70 | 88:12 | 91:9 |
| 4 | **2c/3a** | **4d**: 4-NO$_2$/H | **5f**: 65 | 93:7 | 95:5 |
| 5 | **2c/3a** | **4e**: 3-NO$_2$/H | **5g**: 70 | 95:5 | 93:7 |
| 6 | **2c/3a** | **4f**: 2,4-F$_2$/H | **5h**: 63 | 93:7 | 88:12 |
| 7 | **2c/3a** | **4g**: 4-Me/H | **5i**: 61 | 95:5 | 89:11 |
| 8 | **2c/3a** | **4h**: 4-MeO/H | **5j**: 60 | 95:5 | 81:19 |
| 9 | **2c/3a** | **4i**: 4-NO$_2$/4-Br | **5k**: 61 | 90:10 | 90:10 |
| 10 | **2c/3a** | **4j**: 4-NO$_2$/4-Me | **5l**: 65 | 92:8 | 95:5 |
| 11 | **2d/3a** | **4k**: 4-NO$_2$/H | **5m**: 55 | 95:5 | 94:6 |
| 12 | **2e/3a** | **4l**: 4-NO$_2$/H | **5n**: 71 | 95:5 | 96:4 |
| 13 | **2f/3a** | **4m**: 4-NO$_2$/H | **5o**: 55 | 74:26 | 73:27 |
| 14 | **2g/3a** | **4n**: 4-NO$_2$/H | **5p**: 74 | 96:4 | 96:4 |
| 15 | **2h/3a** | **4o**: 4-NO$_2$/H | **5q**: 45 | 95:5 | 91:9 |
| 16 | **2c/3b** | **4p**: 4-NO$_2$/H | **5r**: 59 | 94:6 | 97:3 |
| 17 | **2c/3c** | **4q**: 4-NO$_2$/H | **5s**: 64 | 96:4 | 97:3 |

Dr, diastereomeric ratio; er, enantiomeric ratio.
*[PdCl($\eta^3$-C$_3$H$_5$)]$_2$/**1i**/**2**/**3**/**4** = 0.05:0.1:1.5:1.1:1.
†Isolated yields.
‡*Syn/anti*, determined by $^1$H NMR analysis of the crude mixture or by HPLC (entries 13, 14 and 17).
§Determined by HPLC analysis using a chiral stationary phase. Value is that of the *syn* isomer.

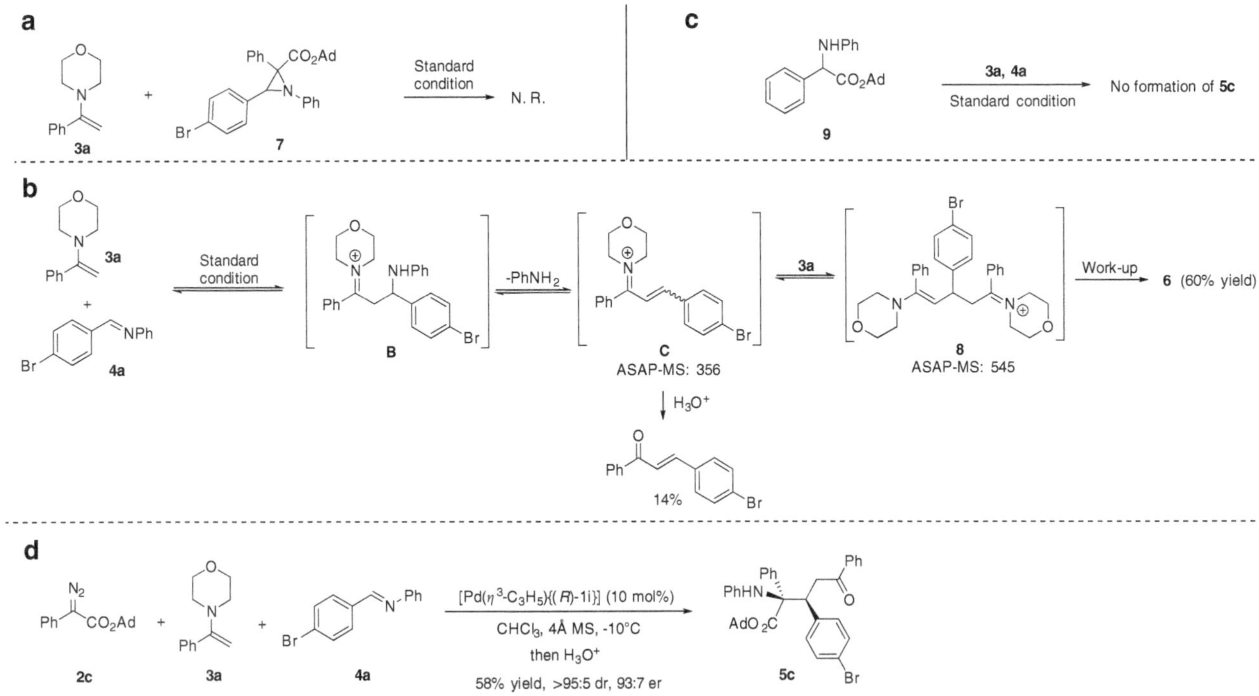


**Figure 3 | Discussion on the reaction pathway.** (**a**) Control reaction to exclude the formation of **5c** from aziridine **7**. (**b**) Proposed reaction pathway for the formation of **6** via iminium intermediate **C**. (**c**) Control reaction to exclude the formation of **5c** from N-H insertion product **9**. (**d**) Reaction under the catalysis of a premade palladium/phosphate complex to confirm the active catalyst.

product **7** was first isolated from the reaction of diazo compound **2c** and imine **4a**. Then treatment of aziridine **7** with enamine **3a** did not give any ring opening product (Fig. 3a). This experiment excluded the possibility that the product **5c** was resulted from a stepwise reaction pathway via intermediate **7**. In addition, in the absence of the diazo compound, the reaction gave side product **6** in a 60% yield together with a 14% yield of chalcone (chalcone is inactive in trapping of ammonium ylide under current conditions, see Supplementary Figure 65). Process of the above Mannich-type reaction from the addition of **3a** to **4a** was monitored by $^1$H NMR, and it was found that concentration of liberated free aniline increased along with the reaction proceeding, which indicates that a retro aza-Michael addition to cleave the C–N bond does occur (see Supplementary Fig. 67). The cleavage of the C–N bond is also consistent with the formation of side product **6** as proposed in the following reaction pathway (Fig. 3b): (1) a Mannich-type addition of enamine **3a** to imine **4a** gives its adduct **B**; (2) a C–N cleavage occurs through a retro aza-Michael addition to give a keto-iminium intermediate[18–21] **C** and an aniline fragment; (3) a conjugated addition of another enamine **3a** to the keto-iminium produces the intermediate **8**; (4) hydrolysis work-up of **8** produces the side product **6**. In addition, the keto-iminium intermediate was detected by Atmospheric Solids Analysis Probe Mass Spectrometry (ASAP-MS) (see Supplementary Fig. 66). All the information shown above indicates that the C–N bond cleavage does not need any assistance of a metal carbene. To rule out the possibility that product **5c** was resulted from stepwise reactions of an N–H insertion of the free aniline and then a Michael-type addition to iminium intermediate C, an additional control experiment by addition of the premade N–H insertion product **9** to the reaction mixture was conducted, and no **5c** was observed (Fig. 3c). The concerted reaction pathway is in agreement with our previous studies in the trapping of active ylides by other electrophiles[22]. Finally, to get information regarding active catalytic species, the crude reaction mixture was subjected to the high-resolution mass spectrum (HRMS). The palladium/phosphate complex [Pd(allyl)(**1i**)] ($[M+1]^+$: 918.8259, see Supplementary Fig. 72) was detected in the HRMS analysis. To further confirm the critical role of the chiral palladium/phosphate complex, [Pd(allyl)(**1i**)] (Supplementary Figure 73 and Supplementary method 3) was prepared[12] and used in the reaction of **2c**, **3a** and **4a**. Product **5c** was obtained in 58% yield with exactly the same stereoselective control of 95:5 dr and 93:7 er (Fig. 3d versus entry 6 in Table 3), indicating that the palladium/phosphate complex is the real catalyst in the catalytic cycles.

On the basis of the information above, a plausible mechanism is proposed in Fig. 4. First, reaction of an imine, a phosphoric acid and allylpalladium(II) chloride dimer gives iminium **A** and generates palladium/phosphate complex **[Pd(allyl)PA*]** as an active catalyst. Nucleophilic addition of an enamine **3** to the iminium **A** affords a Mannich-type adduct **B** as a 'mother' molecule. We propose that a retro aza-Micheal addition occurs here to cleave the C–N bond and produce an aniline fragment $Ar^3NH_2$ and conjugated iminium intermediate **C** as a 'mother' fragment (cycle I). In the meantime, in the catalytic cycle II, **[Pd(allyl)PA*]** catalyzes diazo decomposition of **2** generating a palladium carbenoid **D** by the extrusion of dinitrogen. Modification of the aniline fragment by reacting with the carbenoid **D** gives a proposed Pd-associated ammonium ylide **E** or its enolate form **F**. Then, reassembly of the ammonium ylide back to the 'mother' fragment **C** via an enantioselective Michael-type addition produces an enamine intermediate **G** and regenerates **[Pd(allyl)PA*]** catalyst. Finally, hydrolysis of **G** during work-up affords the final products **5**.

To further confirm the C–N bond cleavage of the mother molecule **B** producing iminium intermediates **C** and eliminating the aniline fragment, *p*-bromoaniline was introduced to the reaction and the control experiment gave the cross-coupling product **10** in 17% yield (Fig. 5a), supporting the existence

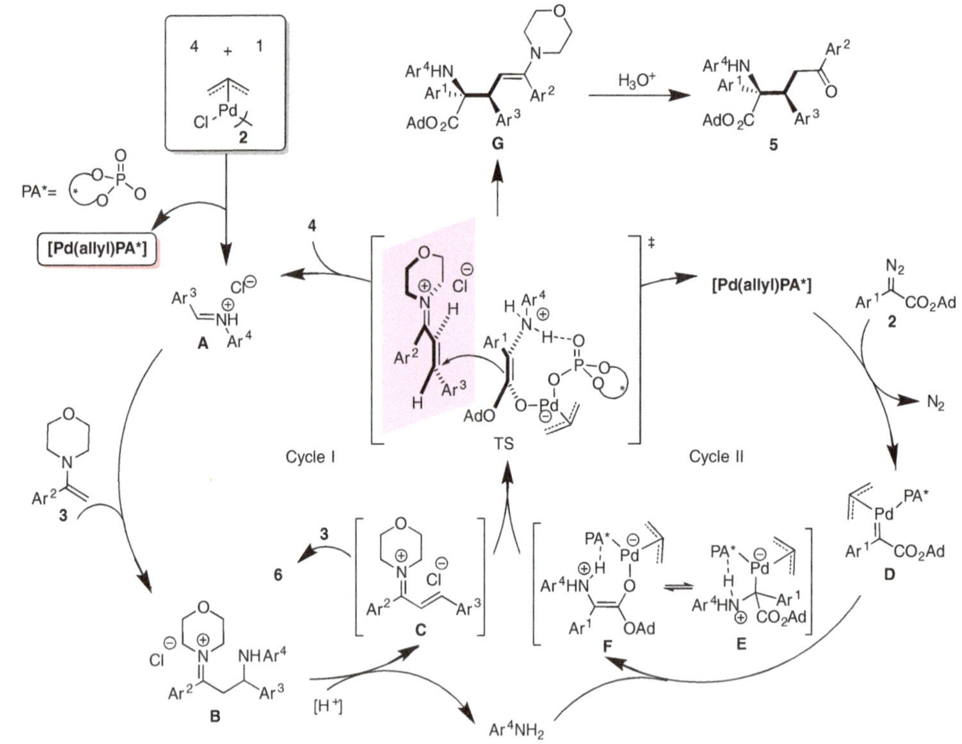


**Figure 4 | Putative reaction pathways.** Cycle I: formation of iminium salt C; cycle II: formation of ammonium ylides.

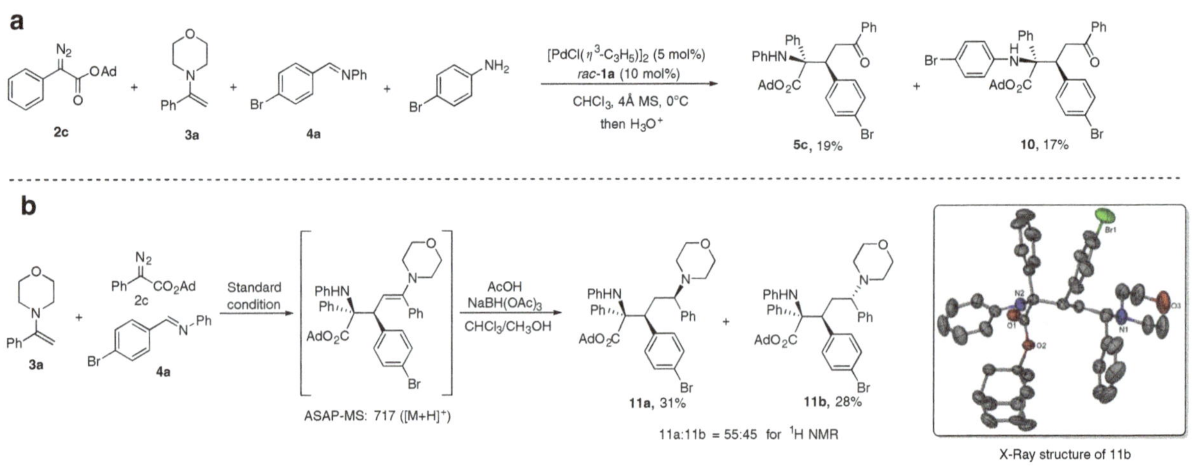


**Figure 5 | The additional evidence and structure determination.** (**a**) Cross-coupling control experiment reaction to confirm the C-N bond cleavage. (**b**) Reduction of enamine intermediate **G** to diamines **11a** and **11b**, and the structure determination of **11b** by single-crystal X-ray diffraction analysis.

of fragments **C** (for other cross-coupling reactions, see Supplementary Figs 68–70). The existence of enamine intermediate **G** can be confirmed by direct detection by ASAP-MS ($[M+H]^+$: 717, see Supplementary Fig. 71) and by reduction of the crude reaction mixture to corresponding amine derivatives. Treatment of the crude product mixture with $NaBH(OAc)_3$/AcOH afforded α,δ-diamino acid derivatives **11a** and **11b** in 59% yield with 55:45 dr (Fig. 5b). The absolute configuration of **11b** was confirmed as (2*R*, 3*S*, 5*R*) by X-ray single-crystal analysis (see Supplementary Fig. 64) and that of **5** was determined as (2*R*, 3*S*) by analogy.

Since the retro aza-Michael addition is considered to be a reversible process, it is still possible that the product formation is directly through the Mannich product **B** via a ylide formation from **B** and followed by a 1,2-C shift. To evaluate this possibility, density functional theory calculation was carried out. A reaction pathway involving the ylide formation and the 1,2-C shift is exothermic by 24.3 kcal mol$^{-1}$. However, this pathway has to proceed through a transition state with a higher energy than that of the **B** by 21.9 kcal mol$^{-1}$ and surmounts an energy barrier of 29.4 kcal mol$^{-1}$, which is considered to be unfavourable in thermodynamics (see Supplementary Figs 74–77).

In conclusion, we report an enantioselective three-component reaction of diazo compounds, enamines and imines. This reaction proceeds through an innovative process that involves C–N cleavage, modification of the amino fragment and selective reassembly of the modified fragment. This unique process provides a convenient pathway to chiral α-amino-δ-oxo pentanoic acid derivatives in good yields and high stereoselectivities. Mechanistic study reveals that a keto-iminium fragment is the key intermediate and a chiral palladium/phosphate complex is the real catalyst. This innovative mode of bond formation provides an atom-efficient and step-economic way to build functional group-enriched molecules that existing methods cannot easily accomplish, and could inspire more discoveries in the efficient construction of unique molecules.

## Methods

**General.** $^1$H NMR (400 MHz), $^{13}$C NMR (100 MHz) and $^{19}$F NMR (376 MHz) spectra were recorded on Brucker Asend 400 spectrometers. HRMS was performed on Waters Micromass Q-TOF micro Synapt High Definition Mass Spectrometer. HPLC analysis using a chiral stationary phase was performed on Shimadzu or Dalian Elite (UV230+ UV/Vis Detector and P230P High Pressure Pump). CHIRALCEL OD-H or CHIRALPAK IC column was purchased from Daicel Chemical Industries, Ltd. Single-crystal X-ray diffraction data were recorded on Bruker-AXS SMART APEX II single-crystal X-ray diffractometer. See Supplementary Figs 1–62 for the NMR spectra and HPLC chromatograms, Supplementary Table 1 for the optimization of reaction, Supplementary Discussion for the mechanism studies and computational details and Supplementary Methods for the characterization data of compounds not listed in this part.

**Materials.** THF and toluene were distilled over sodium. $CH_2Cl_2$ and $CHCl_3$ were distilled over calcium hydride. Enamines **3** were synthesized from acetophenone and morpholine[13]. Diazo compounds were prepared according to the literature[12,23]. Aldimines **4** were prepared from condensation of the corresponding aldehydes with arylamines according to the reported methods[24]. Chiral BINOL-derived phosphoric acids (**1a–1k**) were prepared according to the literature procedures[11,12].

**General procedure for enantioselective three-component reactions.** A suspension of $[PdCl(\eta^3\text{-}C_3H_5)]_2$ (3.6 mg, 5.0 mol%), chiral phosphoric acid (*R*)-**1i** (15 mg, 10 mol%), imine **4** (0.2 mmol, 1.0 equivalent (eq)) and 4 Å molecular sieve (0.2–0.3 g) was stirred in 1.5 ml of $CHCl_3$ in a test tube equipped with plug at − 10 °C. The mixture of diazo compound **2** (0.3 mmol, 1.5 eq) and enamine **3** (0.22 mmol, 1.1 eq) in 1.0 ml $CHCl_3$ was introduced to the suspension over 1 h via a syringe pump. After completion of the addition, the mixture was stirred for another 5 min and then was hydrolyzed by elution through short column chromatography on silica gel (eluent: EtOAc/petroleum ether = 1/10 ~ 1/5) to give a crude product, which was subjected to $^1$H NMR spectroscopy analysis for the determination of diastereoselectivity and to chiral HPLC analysis for the ee values. Purification of the crude products by flash chromatography on silica gel (eluent: EtOAc/petroleum ether = 1/80 ~ 1/20) afforded pure products.

**(2*R*,3*S*)-adamantan-1-yl 3-(4-bromophenyl)-5-oxo-2,5-diphenyl-2-(phenylamino)pentanoate (5c).** Obtained following the general procedure in 69% yield (89 mg). $^1$H NMR (400 MHz, $CDCl_3$) $\delta$ 7.75–7.66 (m, 2H), 7.64–7.56 (m, 2H), 7.48 (dd, *J* = 10.5, 4.3 Hz, 1H), 7.39–7.29 (m, 7H), 7.09–7.02 ( m, 1H), 6.68 (m, 1H), 6.41 (d, *J* = 7.8 Hz, 1H), 4.88 (s, 1H), 4.62 (dd, *J* = 11.4, 2.2 Hz, 1H), 3.61 (dd, *J* = 17.4, 11.5 Hz, 1H), 3.41 (dd, *J* = 17.4, 2.4 Hz, 1H), 2.04 (s, 3H), 1.76 (q, *J* = 11.4 Hz, 6H), 1.62–1.41 (m, 6H); $^{13}$C NMR (100 MHz, $CDCl_3$) $\delta$ 197.5, 171.0, 145.7, 137.8, 137.3, 136.7, 133.2, 131.6, 131.0, 128.6, 128.5, 128.3, 127.9, 127.7, 121.4, 117.8, 115.9, 82.8, 70.1, 48.3, 40.9, 40.8, 36.0, 30.7; FT-IR (KBr) $\tilde{\nu}$ 3,404; 3,052; 2,912; 2,852; 1,728; 1,687; 1,600; 1,502; 1,447; 1,354; 1,322; 1,259; 1,235; 1,176; 1,073; 1,053; 965, 747, 696, 684 $cm^{-1}$. HRMS(ESI) *m/z*: calculated (calcd.) for $C_{39}H_{38}NO_3NaBr$ $(M+Na)^+$ 670.1933, found 670.1929. HPLC (CHIRALCEL OD-H, $\lambda$ = 254 nm, hexane/2-propanol = 99/1, flow rate = 1.0 ml $min^{-1}$), 88–90% ee, $t_{major}$ = 5.74 min, $t_{minor}$ = 8.47 min.

**Procedure for the preparation of 11a and 11b.** A suspension of $[PdCl(\eta^3\text{-}C_3H_5)]_2$ (7.2 mg, 5.0 mol%), chiral phosphoric acid (*R*)-**1i** (30 mg, 10 mol%), imine **4a** (105 mg, 0.4 mmol, 1.0 eq) and 4 Å molecular sieve (0.1 g) was stirred in 2 ml of $CHCl_3$ at − 10 °C, and then the mixture of diazo compound **2c** (199 mg, 0.6 mmol, 1.5 eq) and enamine **3a** (84 mg, 0.44 mmol, 1.1 eq) in 1.5 ml $CHCl_3$ was introduced to the suspension over 1 h via a syringe pump. After completion of the addition, the mixture was stirred for another 10 min. To the mixture, $NaBH(OAc)_3$ (5.0 eq.), AcOH (0.4 ml) and $CH_3OH$ (3 ml) were added at − 10 °C and stirred at room temperature. After the reaction completed (monitored by TLC), the mixture was filtered through Celite. The filtrate was concentrated down to give a residue, which was subjected to $^1$H NMR for determination of dr and was purified by flash chromatography on silica gel (eluent: EtOAc/petroleum ether = 1/80 ~ 1/10) to afford pure **11a** and **11b**.

**(2*R*,3*S*,5*R*)-adamantan-1-yl 3-(4-bromophenyl)-5-morpholino-2,5-diphenyl-2-(phenylamino)pentanoate (11a).** A total 32% yield (92 mg); $[\alpha]20_D = -43.0$ (*c* = 0.348, $CHCl_3$); $^1$H NMR (400 MHz, $CDCl_3$) $\delta$ 7.67 (d, *J* = 3.9 Hz, 2H), 7.35 (d, *J* = 8.3 Hz, 2H), 7.32–7.26 (m, 3H), 7.22 (dt, *J* = 10.7, 4.0 Hz, 3H), 7.03–6.95 (m, 2H), 6.92–6.78 (m, 4H), 6.68–6.57 (m, 1H), 6.37 (d, *J* = 8.0 Hz, 2H), 4.61 (s, 1H), 4.16 (d, *J* = 11.0 Hz, 1H), 3.67 (t, *J* = 4.4 Hz, 4H), 2.87 (dd, *J* = 11.7, 3.1 Hz, 1H), 2.54 (t, *J* = 12.0 Hz, 1H), 2.35–2.20 (m, 2H), 2.03 (s, 3H), 1.98–1.89 (m, 2H), 1.83 (dd, *J* = 19.0, 7.1 Hz, 1H), 1.76–1.65 (m, 6H), 1.58–1.38 (m, 7H); $^{13}$C NMR (100 MHz, $CDCl_3$) $\delta$ 171.5, 145.9, 138.1, 137.5, 137.2, 131.9, 131.0, 129.6, 128.5, 128.4, 127.9, 127.5, 127.4, 121.2, 117.9, 115.7, 82.2, 77.3, 77.0, 76.7, 70.7, 67.6, 65.9, 52.1, 49.2, 40.8, 36.0, 32.0, 30.7. FT-IR (KBr) $\tilde{\nu}$3,416; 3,056; 3,026; 1,725; 1,703; 1,604; 1,501; 1,452; 1,355; 1,317; 1,254; 1,230; 1,182; 1,118; 1,051; 1,108; 967, 822, 751, 703, 553 $cm^{-1}$; HRMS(ESI) *m/z*: calcd. for $C_{43}H_{48}N_2O_3Br$ $(M+H)^+$ 719.2848, found 719.2876.

**(2*R*,3*S*,5*S*)-adamantan-1-yl 3-(4-bromophenyl)-5-morpholino-2,5-diphenyl-2-(phenylamino)pentanoate (11b).** A total 28% yield (80 mg); $^1$H NMR (400 MHz, $CDCl_3$) $\delta$ 7.41–7.31 (m, 6H), 7.31–7.19 (m, 4H), 7.03–6.91 (m, 4H), 6.74–6.53 (m, 3H), 6.28 (d, *J* = 7.8 Hz, 2H), 4.44 (s, 1H), 3.61–3.44 (m, 1H), 3.14 (d, *J* = 10.5 Hz, 1H), 2.58 (dd, *J* = 11.2, 3.2 Hz, 1H), 2.41 (td, *J* = 12.9, 3.3 Hz, 1H), 2.26–2.12 (m, 3H), 2.11–2.02 (m, 2H), 1.99 (s, 3H), 1.66–1.56 (m, 6H), 1.56–1.43 (m, 6H); $^{13}$C NMR (100 MHz, $CDCl_3$) $\delta$ 171.0, 145.7, 139.4, 137.1, 136.5, 131.9, 131.0, 129.6, 129.1, 128.4, 128.3, 127.67, 127.5, 127.4, 121.5, 117.8, 115.7, 82.2, 77.3, 77.0, 76.7, 70.5, 68.6, 67.0, 51.4, 40.6, 36.0, 33.6, 30.7; FT-IR (KBr) $\tilde{\nu}$3,390; 3,059; 3,026; 2,955; 2,911; 2,852; 2,812; 1,713; 1,604; 1,500; 1,450; 1,258; 1,233; 1,181; 1,119; 1,054; 1,010; 876, 821, 759, 703, 547 $cm^{-1}$; HRMS(ESI) *m/z*: calcd. for $C_{43}H_{48}N_2O_3Br$ $(M+H)^+$ 719.2848, found 719.2833.

## References

1. Brinker, U. H. *Advances in Carbene Chemistry* Vol. 2 (JAI, 1998).
2. Carey, F. A. & Sundberg, R. J. *Advanced Organic Chemistry. Concerted Cycloadditions, Unimolecular Rearrangements and Thermal Eliminations* (Springer, 2007).
3. Yen, Y. S. *et al.* Intramolecular metathesis of a vinyl group with vinylidene C = C double bond in Ru complexes. *J. Am. Chem. Soc.* **127,** 18037–18045 (2005).
4. Cheng, C. W. *et al.* Intramolecular cyclization of ruthenium vinylidene complexes with a tethering vinyl group: facile cleavage and reconstruction of the C-C double bond. *J. Am. Chem. Soc.* **129,** 14974–14980 (2007).
5. Schultz, A. G. & Berger, M. H. Fragmentation-recombination of methoxymethyl ester enolates. A novel method for preparation of hydracrylic and glycidic esters. *J. Org. Chem.* **41,** 585–586 (1976).
6. Terada, M., Machioka, K. & Sorimachi, K. Activation of hemiaminal ethers by chiral brønsted acids for facile access to enantioselective two-carbon homologation using enecarbamates. *Angew. Chem. Int. Ed.* **48,** 2553–2556 (2009).
7. Linghu, X., Bausch, C. C. & Johnson, J. S. Mechanism and scope of the cyanide-catalyzed cross silyl benzoin reaction. *J. Am. Chem. Soc.* **127,** 1833–18409 (2005).
8. Grubbs, R. H. *Handbook of Metathesis* (Wiley-VCH, 2005).
9. Heine, H. G. *et al.* Photochemical α cleavage and free-radical reactions of some deoxybenzoins. *J. Org. Chem.* **39,** 691–698 (1974).
10. Step, E. N., Buchachenko, A. L. & Turro, N. J. The cage effect in the photolysis of (*S*)-( + )-α-methyldeoxybenzoin: can triplet radical pairs undergo geminate recombination in nonviscous homogeneous solution? *J. Org. Chem.* **57,** 7018–7024 (1992).
11. Qiu, H. *et al.* Highly enantioselective trapping of zwitterionic intermediates by imines. *Nat. Chem.* **4,** 733–738 (2012).
12. Zhang, D. *et al.* Enantioselective palladium(II) phosphate catalyzed three-component reactions of pyrrole, diazoesters and imines. *Angew. Chem. Int. Ed.* **52,** 13356–13360 (2013).
13. Zhao, W. J., Yan, M., Huang, D. & Ji, S. J. New reaction of enamines with aryldiazoacetates catalyzed by transition metal complexes. *Tetrahedron* **61,** 5585–5593 (2005).
14. Yu, G., Wang, S. Z., Wang, K., Hu, Y. F. & Hu, H. W. A novel approach to 1,2-dihydro-2-oxo-3-pyridinecarboxylic ester via aromatization induced by deamidation. *Synthesis* **7,** 1021–1028 (2004).
15. Ye, Y., Wang, H. & Fan, R. H. Stereoselective construction of highly functionalized azetidines via a [2 + 2]-cycloaddition. *Org. Lett.* **12,** 2802–2805 (2010).
16. Truong, T. K. T. & Vo-Thanh, G. Synthesis of functionalized chiral ammonium, imidazolium and pyridiniumbased ionic liquids derived from (-)-ephedrine using solvent-free microwave activation. Applications for the asymmetric Michael addition. *Tetrahedron* **66,** 5277–5282 (2010).
17. Miao, C. B. *et al.* Highly stereoselective, one-pot synthesis of azetidines and 2,4-dioxo-1,3-diazabicyclo[3.2.0] compounds mediated by $I_2$. *J. Org. Chem.* **78,** 4329–4340 (2013).
18. Mi, X. L., Luo, S. Z., He, J. Q. & Cheng, J. P. Dy(OTf)$_3$ in ionic liquid: an efficient catalytic system for reactions of indole with aldehydes/ketones or imines. *Tetrahedron Lett.* **45,** 4567–4570 (2004).
19. Esquivias, J., Gómez-Arrayás, R. & Carretero, J. C. A copper(II)-catalyzed aza-Friedel–Crafts reaction of *N*-(2-pyridyl)sulfonyl aldimines: synthesis of unsymmetrical diaryl amines and triaryl methanes. *Angew. Chem. Int. Ed.* **45,** 629–633 (2006).

20. Alonso, I., Esquivias, J., Gómez-Arrayás, R. & Carretero, J. C. Understanding the behavior of N-tosyl and N-2-pyridylsulfonyl imines in $Cu^{II}$-catalyzed aza-Friedel-Crafts reactions. *J. Org. Chem.* **73,** 6401–6404 (2008).
21. Barbero, M., Cadamuro, S., Dughera, S., Magistris, C. & Venturello, P. A new practical synthesis of triaryl and trisindolylmethanes under solvent-free reaction conditions. *Org. Biomol. Chem.* **9,** 8393–8399 (2011).
22. Guo, X & Hu, W. H. Novel multicomponent reactions via trapping of protic onium ylides with electrophiles. *Acc. Chem. Res.* **46,** 2427–2440 (2013).
23. Harker, W. R. R., Carswell, E. L. & Carbery, D. R. A practical protocol for the highly E-selective formation of aryl-substituted silylketene acetals. *Org. Lett.* **12,** 3712–3715 (2010).
24. Shiino, M., Watanabe, Y. & Umezawa, K. Synthesis of *N*-substituted *N*-nitrosohydroxylamine as inhibitor of mushroom Tyrosinase. *Bioorg. Med. Chem.* **9,** 1233–1240 (2001).

## Acknowledgements

We are grateful for the financial support from NSFC (21125209 and 21332003), the Ministry of Science and Technology (MOST) of China (2011CB808600), STCSM of Shanghai (12JC1403800) and Specialized Research Fund for the Doctoral Program of Higher Education (20100076110005). F.X. thanks the National Natural Science Foundation of China (Grant Nos. 21433004 and 21473056) and the Shanghai Natural Science Foundation (No. 14ZR1411800).

## Author contributions

W.H. directed the project and developed the strategy. D.Z. planned, conducted and analyzed the experiments. J.Z. assisted with some experiments and analysed some NMR spectra. F.X. performed the density functional theory calculations. Z.K. performed some experiments. W.H. and D.Z. wrote the manuscript.

## Additional information

**Competing financial interests:** The authors declare no competing financial interests.

2

# Cationic mononuclear ruthenium carboxylates as catalyst prototypes for self-induced hydrogenation of carboxylic acids

Masayuki Naruto[1] & Susumu Saito[1,2,3]

Carboxylic acids are ubiquitous in bio-renewable and petrochemical sources of carbon. Hydrogenation of carboxylic acids to yield alcohols produces water as the only byproduct, and thus represents a possible next generation, sustainable method for the production of these alternative energy carriers/platform chemicals on a large scale. Reported herein are molecular insights into cationic mononuclear ruthenium carboxylates ($[Ru(OCOR)]^+$) as prototypical catalysts for the hydrogenation of carboxylic acids. The substrate-derived coordinated carboxylate was found to function initially as a proton acceptor for the heterolytic cleavage of dihydrogen, and subsequently also as an acceptor for the hydride from $[Ru\text{-}H]^+$, which was generated in the first step (self-induced catalysis). The hydrogenation proceeded selectively and at high levels of functional group tolerance, a feature that is challenging to achieve with existing heterogeneous/homogeneous catalyst systems. These fundamental insights are expected to significantly benefit the future development of metal carboxylate-catalysed hydrogenation processes of bio-renewable resources.

[1]Graduate School of Science, Nagoya University, Chikusa, Nagoya 464-8602, Japan. [2]Institute for Advanced Research, Nagoya University, Chikusa, Nagoya 464-8601, Japan. [3]ACT-C, JST, Chikusa, Nagoya 464-8602, Japan. Correspondence and requests for materials should be addressed to S.S. (email: saito.susumu@f.mbox.nagoya-u.ac.jp).

A great variety of carboxylic acids (CAs) is abundantly available from both the petrochemical industry and natural sources. The hydrogenation of CAs is an important research subject with respect to using the resulting alcohols as alternative organic energy ($H_2$) carriers, or as platform chemicals[1]. According to a 2004 report from the US Department of Energy, the vast majority of high value-added chemicals from biomass at that time were CAs[2]. While this number has been revised since, due to technological advancements, CAs remain prominent biomass products on this continuously changing list[3]. Alternatively, CAs can be generated synthetically, and an elegant synthesis was recently reported using $CO_2$, $H_2$ and olefins[4]. Furthermore, formic acid has been produced by hydrogenation of $CO_2$ (refs 5–8), or by the photo-reduction of $CO_2$ with $H_2O$ using solar energy[9]. The availability of carbon-neutral alternatives to fossil fuels presents a critical challenge for the scientific community on the way to a more sustainable society. When CAs are sourced from biomass and/or produced from $CO_2$, they represent a potential renewable resource. Finally, recent progress in C–H bond functionalization chemistry has demonstrated the versatility and utility of CAs as effective directing groups in organic synthesis[10].

The hydrogenation of CAs is an ideal method for the bulk production of alcohols, and may thus benefit the 'methanol economy', that is, the anthropogenic chemical carbon cycle[11,12], since water is the only byproduct and salt waste is not formed. A hydrogenation method that is widely applicable to a broad variety of CAs and that selectively produces alcohols is therefore highly desirable. Unfortunately, however, simple molecular hydrogenation catalysts that enable such conversions still remain elusive. This is predominantly due to the lack of a rational design strategy for single-active-site catalysts that effectively hydrogenate the thermodynamically stable and kinetically inert COOH group in the absence of undesirable side reactions involving the COOH or other potentially present functional groups. In addition, the catalyst should be able to operate under acidic conditions, given the *ex vi termini* acidity of CAs. This is an important issue, since the catalytic hydrogenation of CA derivatives such as esters[13–15] and amides[14–16] proceeds well under basic to neutral conditions, but has seldomly been investigated when using a CA as the acidic reaction medium[15]. Even though conventional approaches to the hydrogenation of CAs with heterogeneous or homogeneous catalysts allow control over a number of reaction parameters, they usually proceed under fair to vigorous conditions ($P_{H2} = 2.5$–$13$ MPa and/or $T = 100$–$500\,°C$), and they frequently suffer from low alcohol selectivity/productivity[17–22] as a result of esterification ($Ru^0/Ru^{II}$)[17–19], partial reduction to afford aldehydes ($Cr_2O_3$; ref. 20), over reduction ($Re^{VII}$–$Os^{VIII}$; ref. 21), and/or dearomatic hydrogenation ($Rh^0$–$Mo^0$; ref. 22). Even under milder conditions (typically: $P_{H2} = 2$–$3$ MPa, $T = 120$–$130\,°C$), decarboxylation (Pt(–Re)/$TiO_2$; ref. 23) and esterification ($Ir^{III}$ or $Rh^{III}$; ref. 24) of some CAs represent significant shortcomings. Moreover, studies on functional group tolerance, which may provide useful information for organic synthesis have not yet been reported. In the presence of catalytic Brønsted acids and Ru–triphos complexes[25–32] (triphos = $CH_3C[CH_2PPh_2]_3$; Ph = $C_6H_5$), a group of 4-oxo- and 4-hydroxy-CAs, including levulinic acid and succinic acid, undergoes hydrogenation under harsher conditions ($P_{H2} = \sim 10$ MPa, $T = \sim 200\,°C$) to afford alcohols[27–30]. Very recently, Leitner and co-workers[31] reported the [Ru(triphos)(TMM)]-catalysed (TMM = trimethylene methane) hydrogenation of acetic acid (AcOH; $P_{H2} = 6$ MPa, $T = 180\,°C$), hexanoic acid and benzoic acid ($P_{H2} = 5$ MPa, $T = \sim 220\,°C$) (ref. 32) in the presence or absence of catalytic amounts of $HN(SO_2CF_3)_2$ under relatively strenuous conditions. The corresponding alcohols were obtained in moderate to excellent yields (conversion of CAs: 50–99%). However, in many cases, this should be the result of *in situ*-generated esters (lactones) or anhydrides that serve as intermediates, which are able to undergo hydrogenation more effectively[26–34]. For instance, Ru–triphos complexes hydrogenate esters such as methyl benzoate and alkyl formates under milder conditions (for example, $P_{H2} = 3$–$5$ MPa, $T = 140\,°C$) (refs 28,32). In addition, Goldberg and co-workers[24] demonstrated experimentally that CAs with shorter aliphatic carbon chains such as AcOH react much more rapidly, which is consistent with previous observations[17–19]. When the size of the CA increases from $C_1$ to $C_4$, the corresponding carboxylate carbon atom becomes more electron rich and thus less susceptible to nucleophilic attack from transition metal hydrides.

Herein, we would like to report a molecular prototype that benefits from the rational design for a single-active-site Ru catalyst for the hydrogenation of CAs[35]. The CA-derived carboxylate coordinated to the Ru centre was found to function initially as a proton acceptor in the heterolytic cleavage of dihydrogen, and subsequently also as a acceptor for a hydride from $[Ru–H]^+$, which was generated in the first step. This catalytic cycle thus represents a CA self-induced CA hydrogenation. This catalyst affords the corresponding alcohols selectively and exhibits a high functional group tolerance. Conversely, the hydrogenation of esters is not promoted, and this catalytic system thus constitutes a benchmark achievement that may be potentially useful for selective organic synthesis and for the development of future catalytic hydrogenation methods for feedstock derived from biomass or $CO_2$.

## Results

**Ruthenium carboxylates as catalyst prototypes.** Catalyst prototypes (Fig. 1a) for CA self-induced CA hydrogenation, and the characteristics of the present work that differ from earlier studies (Fig. 1b) are shown as the outline of this study.

**Hydrogenation using Ru complex with monodentate phosphine.** Since the milestone discovery of the Wilkinson-type ruthenium complex $RuCl_2(PPh_3)_3$ (Ru-**a**) for the hydrogenation of olefins in the 1960s (refs 36,37), custom-tailored molecular single-active-site catalysts that hydrogenate CAs effectively have scarcely been investigated systematically[24,32]. The hydrogenation of 3-phenylpropionic acid (CA-**a**; Table 1, entry 1) in toluene using a combination of Ru-**a** (2 mol%: $[Ru]_0 = 6.7$ mM), $NaBPh_4$ (10 mol%), and high pressure of $H_2$ ($P_{H2} = 8$ MPa, $T = 160\,°C$, $t = 24$ h) resulted in the formation of alcohol AL-**a** and the ester $Ph(CH_2)_2CO_2(CH_2)_3Ph$ (ES-**a**) in 58% and 16% yield, respectively (overall conversion of CA-**a**: 92%). In the absence of $NaBPh_4$, Ru-**a** was unable to promote the hydrogenation, suggesting that a cationic Ru complex $[L_nRu]^+$ should be of critical importance for a successful hydrogenation. Instead of $NaBPh_4$, other similar additives were tested, for example, $NaB[3,5\text{-}(CF_3)_2C_6H_3]_4$, $NaBF_4$, $NaPF_6$, NaOTs, NaOTf and $NaNTf_2$, but the results were unsatisfactory (AL-**a**: 0–17%). Encouraged by these results, several commercial derivatives of Ru-**a**, including $RuCl_2(PPh_3)_4$, $RuHCl(CO)(PPh_3)_3$, $RuCl_2(CO)_2(PPh_3)_2$, Ru(*p*-cymene)$Cl_2$ $[P(C_6H_{11})_3]$, $Ru(C_5H_5)Cl(PPh_3)_2$, *cis*-$RuCl_2(DMSO)_4$, and $Ru(C_5H_5)Cl$(dppm) (DMSO = dimethylsulfoxide; dppm = 1,1-bis(diphenylphosphino)methane) were screened in the presence of $NaBPh_4$ (for details, see: Supplementary Table 1). However, the obtained yield of AL-**a** and ES-**a** was consistently below 10%. In contrast, the acetato complex $RuCl(OAc)(PPh_3)_3$ (Ru-**b**; OAc = $CH_3CO_2^-$) afforded AL-**a** (55%) and ES-**a** (17%) in yields that were similar to those of Ru-**a**. Since a catalytic amount of the Ru-acetato complex Ru-**b** can readily undergo ligand

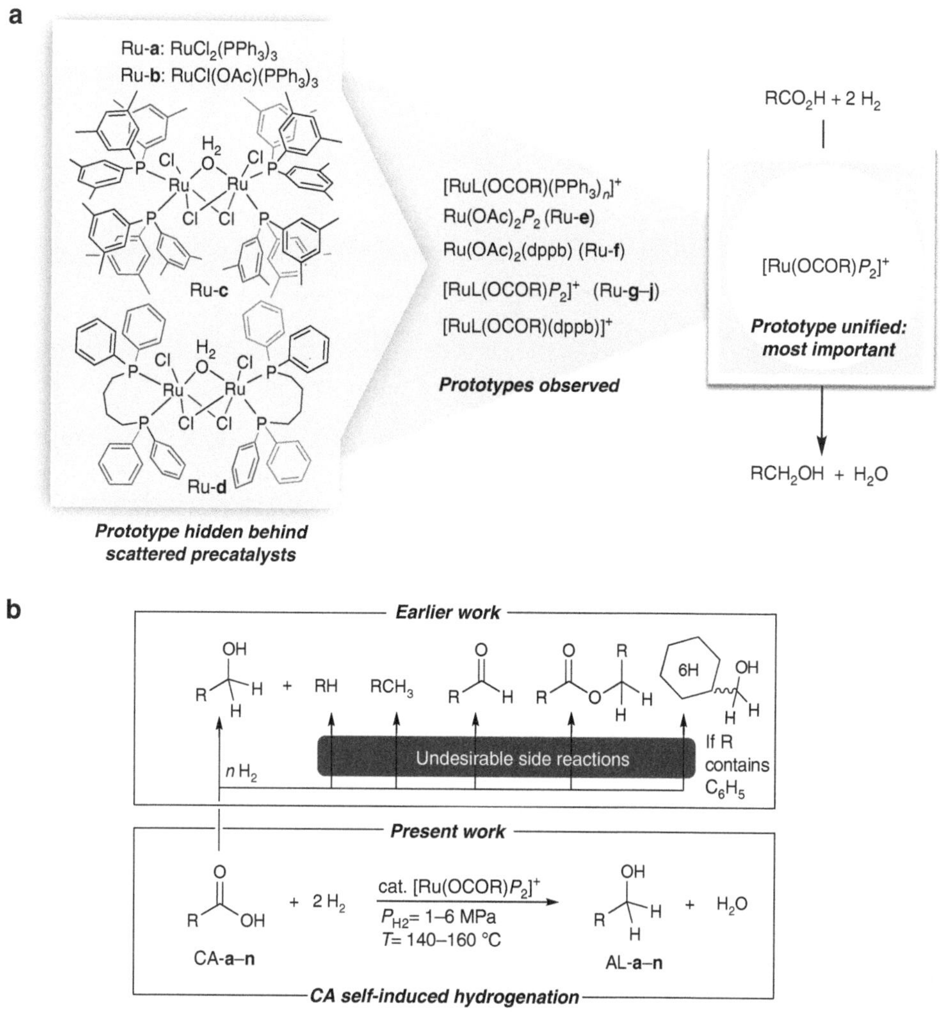


**Figure 1 | Outline of this study.** (**a**) The most favourable R is a saturated aliphatic group. $P_2$ denotes two monodentate or a bidentate phosphine ligand(s) and the search for optimal $P_2$ still remains the subject of further optimization. (**b**) CA self-induced CA hydrogenation (present work), in contrast to the shortcomings of earlier CA hydrogenation methods.

exchange with excess CA-**a** in the reaction mixture, the general formula '[Ru(OCOR)]$^+$' (R = aliphatic group) is assigned to the critical structure of the catalyst. Although H–H bond scission is known to be promoted by [Ru(OAc)] species[25,38–40], investigations of their potential use have so far been focused more on the hydrogenation of unsaturated carbon–carbon bonds.

To determine the best molar amount of $PPh_3$ relative to Ru sufficient to catalyse the hydrogenation of CA-**a**, the precatalyst was switched from Ru-**a** to $RuCl_2(DMSO)_4$ (2 mol%, $[Ru]_0 = 6.7$ mM) and varying molar amounts of the phosphine ligand was used in the presence of 10 mol% $NaBPh_4$. The best results, which provided similar hydrogenation rates, were observed for a 2:1 and 3:1 ratio of $PPh_3$ and $RuCl_2(DMSO)_4$, while a 1:1 ratio significantly diminished the reaction rate. From the thirteen different monodentate phosphines tested (for details, see: Supplementary Table 2), $P(3,5\text{-}(CH_3)_2(C_6H_3))_3$ (P(3,5-xylyl)$_3$) afforded the best yield (AL-**a** (ES-**a**): 49% (14%)).

To gain insight into the structural differences between the [Ru – P(3,5-xylyl)$_3$] complex and Ru-**a**, the synthesis of the corresponding Ru complex was attempted by treatment of $RuCl_3 \cdot nH_2O$ with P(3,5-xylyl)$_3$ in a molar ratio of 1:6, following the experimental procedure reported for Ru-**a**[36] (Supplementary Methods). However, the expected Wilkinson-type 16e$^-$ complex $Ru^{II}Cl_2$(P(3,5-xylyl)$_3$)$_3$ was not obtained. Instead, the binuclear 18e$^-$ Ru complex $Ru^{II}{}_2Cl_2(\mu\text{-}Cl)_2(\mu\text{-}OH_2)$(P(3,5-xylyl)$_3$)$_4$ (Ru-**c**; see Figs 1a and 2, wherein hydrogen atoms are omitted for clarity) was isolated as a reddish brown precipitate in 83% yield (for the X-ray crystallographic analysis, metric parameters, and an ORTEP structure of Ru-**c**, see: Supplementary Fig. 1). As a solid, Ru-**c** can be easily stored and handled, even under atmospheric conditions. Addition of a third phosphine ligand to the Ru centre, analogous to the formation of Ru-**a** proved to be impossible, presumably due to the sterically congested environment of Ru-**c**. This structural preference is consistent with the observation that a 2:1 ratio between the monodentate phosphine and Ru secured efficient catalysis. When using Ru-**c** (1 mol%) with $NaBPh_4$ (10 mol%), hydrogenation of CA-**a** proceeded more effectively, even under a lower hydrogen pressure ($P_{H2} = 4$ MPa), affording AL-**a** and ES-**a** in 65 and 12% yield, respectively ($T = 160$ °C, $t = 24$ h; conversion of CA-**a**: 92%). Replacing $NaBPh_4$ with NaOAc and Na(acac; acac = acetylacetonate) resulted in comparable effectiveness, furnishing AL-**a** (ES-**a**) in 62% (15%) and 64% (14%) yield, respectively (Supplementary Table 3).

The reaction conditions were further optimized by slightly increasing the load of Ru-**c** to 1.5 mol% so that the hydrogenation was accelerated relative to the simultaneously occurring *in situ* esterification. Furthermore, the more atom-economical NaOAc (10 mol%) was used as the additive for the hydrogenation of various CAs ($P_{H2} = 2$–4 MPa, $T = 140$–160 °C). The results are

**Table 1 | Hydrogenation of CAs using catalytic amounts of Ru-c, Ru-d or Ru-e.***

| Entry | Carboxylic acid (CA) | Conditions | $P_{H2}$ (MPa), $T$ (°C) | Product alcohol (AL) | AL Yield (%)†,‡ (ester (ES)) yield (%)†,‡ |
|---|---|---|---|---|---|
| 1 | 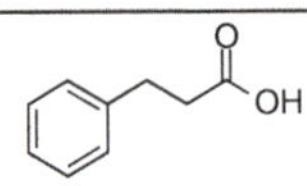 **CA-a** | A | 4,160 | 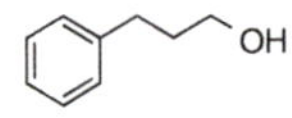 **AL-a** | 89 ± 3 (6 ± 1) |
| 2 | $CH_3(CH_2)_{16}CO_2H$ **CA-b** | A | 4,160 | $CH_3(CH_2)_{16}CH_2OH$ **AL-b** | 70 (9)[§] |
| 3 | | B | 4,160 | | 90 ± 1 (5 ± 0.5), 85[‖] |
| 4 | 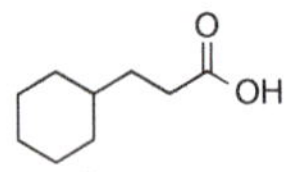 **CA-c** | A | 4,160 | 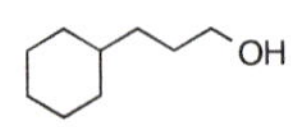 **AL-c** | 78 ± 3 (12 ± 4) |
| 5 | 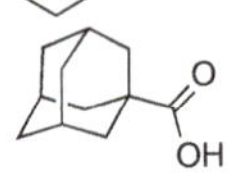 **CA-d** | A | 4,160 | 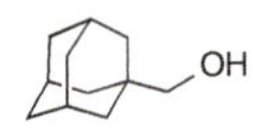 **AL-d** | 96 ± 4 (–), 95[‖] |
| 6 | 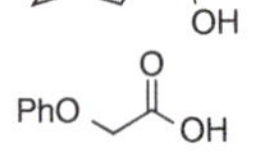 **CA-e** | A | 2,140 |  **AL-e** | 91 ± 1 (5 ± 1) |
| 7 | 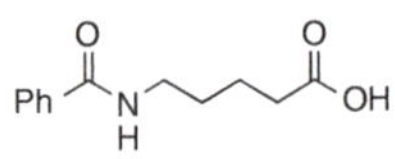 **CA-f** | A | 4,160 | 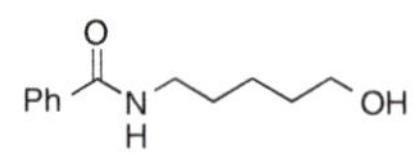 **AL-f** | 45 (17)[§] |
| 8 | | C[¶] | 4,160 | | 55 ± 1 (7 ± 2) |
| 9 | 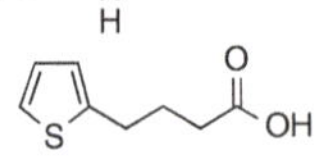 **CA-g** | A | 4,160 | 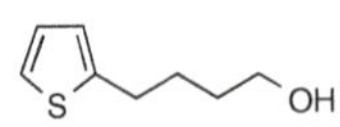 **AL-g** | 24 (18)[§] |
| 10 | 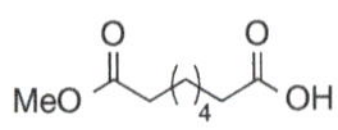 **CA-h** | A | 4,160 | 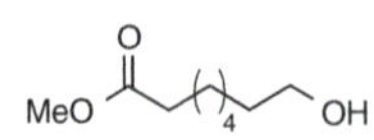 **AL-h** | 69 (16)[§], 58[‖] |
| 11 | | B | 6,160 | | 80 ± 2 (10 ± 2) |
| 12 | 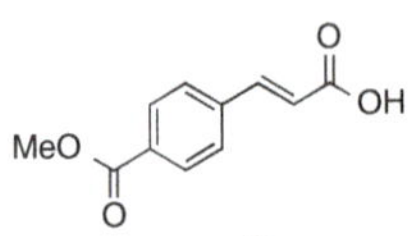 **CA-i** | A | 4,160 | 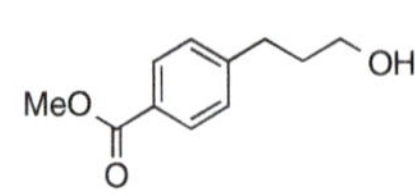 **AL-i** | 86 ± 3 (7 ± 1), 78[‖] |
| 13 | 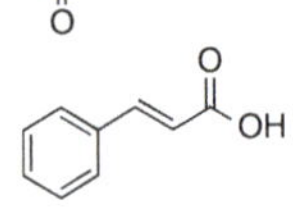 **CA-j** | A | 4,160 | **AL-a** | 86 ± 2 (7 ± 1) |
| 14 | 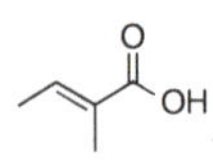 **CA-k** | C | 4,160 | 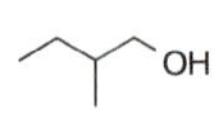 **AL-k** | 65 ± 6 (5 ± 3) |
| 15 | 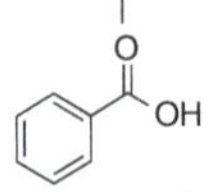 **CA-l** | A | 4,160 | 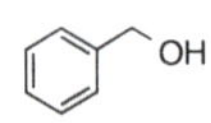 **AL-l** | 34 (1)[§] |
| 16 | | B[#] | 6,160 | | 93 ± 3 (1 ± 0.5) |
| 17 | 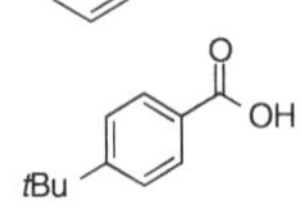 **CA-m** | B[#] | 6,160 | 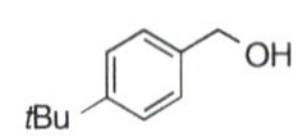 **AL-m** | 93 ± 1 (4 ± 1) |
| 18 | 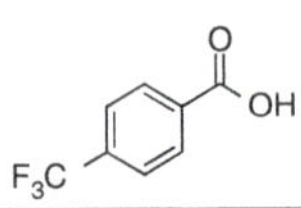 **CA-n** | B[#] | 6,160 | 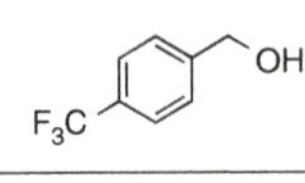 **AL-n** | 87 ± 7 (5 ± 3) |

AL, alcohol; CA, carboxylic acid; ES, ester; *T*, temperature.
*Unless otherwise specified, reactions were carried out over a period of 24 h ($[Ru]_0 = 10$ mM; $[CA]_0 = 333$ mM; in toluene). Conditions A: Ru-**c**:NaOAc:CA = 1.5:10:100 mol%; conditions B: Ru-**d**:Na(acac)·$(H_2O)_n$:CA = 1.5:10:100 mol%; conditions C: 3 mol% Ru-**e** without NaOAc.
†NMR yields using an internal standard (anisole, mesitylene or 1,1,2,2-tetrachloroethane).
‡Average of three runs with calculated s.d.
§Average of two runs.
‖Isolated yield.
¶In dioxane/toluene (v/v = 1:3.5).
#*t* = 48 h.

given in Table 1 (Supplementary Methods for representative methods).

Aliphatic acids CA-**b** and CA-**c** were hydrogenated chemoselectively, exclusively producing alcohols AL-**b** and AL-**c** (entries 2 and 4). This contrasts with the heterogeneous Pt–Re/$TiO_2$-catalysed hydrogenation system, in which the decarboxylation of CA-**c** predominates, affording the by one carbon atom diminished alkane[23]. As CA-**d**, was sterically the most demanding CA, we expected it to be kinetically the most inert. Indeed for the hydrogenation of CA-**d**, a slower reaction rate was observed, but AL-**d** was generated cleanly and esterification was not observed (entry 5). CA-**e** (α-phenoxy acid) was one of the most reactive CAs tested, and hydrogenation proceeded smoothly even under relatively mild conditions ($P_{H2} = 2$ MPa, $T = 140$ °C; entry 6). When the hydrogenation of CA-**e** with Ru-**c** (1.5 mol%) and NaOAc (10 mol%) was stopped after 6 h (AL-**e**: 50%), more than 90% of free AcOH (based on added NaOAc) were detected by $^1H$ NMR. This result suggests the exclusive formation of a $[Ru(OCOCH_2OPh)]^+$ species, which does not promote the hydrogenation of AcOH, but should be responsible for hydrogenation of CA-**e** (that is, a CA-**e** self-induced CA-**e** hydrogenation). This is in agreement with the previously discussed result, which proposed a $[Ru(OCOR)]^+$ complex derived from Ru-**b** as the active catalyst. In contrast, benzoic acid

CA-l was one of the least reactive substrates (entry 15). And indeed, with the exception of one very recent example[32], the chemoselective hydrogenation of arenyl (Ar) CAs ($ArCO_2H$) using molecular catalysts has so far failed. For instance, both the benzene ring and the carboxyl group of CA-l are fully hydrogenated using $Rh/Al_2O_3$–$Mo(CO)_6$ as the catalyst system[22]. Relative to aliphatic carboxylates, arenyl carboxylates ($^-OCOAr$) such as those derived from CA-l should be more strongly coordinated to Ru complexes in a $\kappa^2$-manner[41,42]. The nature of the more favourable $\kappa^2$-complexation prevents $[Ru(OCOAr)]^+$ from forming a $[Ru(\eta^2\text{-}H_2)]^+$ species, since the coordination sites of Ru are effectively occupied, rendering the electrophilicity of the Ru centre too low to trap the $\sigma_{H-H}$ bond. Subsequent formation of a ruthenium hydride, bearing the protonated carboxylate, $[RuH(HOCOR)]^+$, is also more favourable with an aliphatic group than with aryl group, that is, the intramolecular deprotonation of $\eta^2$-$H_2$ should proceed more readily in the $\kappa^1$-carboxylate structure $[Ru(\eta^2\text{-}H_2)(\kappa^1\text{-}OCOR)]^+$ than in the $\kappa^2$-carboxylate, where the carboxylate ($R \neq Ar$) is more basic and thus a less innocent ligand[38–44]. Indeed, the hydrogenation of CA-**a** in three different initial concentrations ($[\text{CA-}\mathbf{a}]_0 = 200$, 333 and 500 mM) was consistently more sluggish in the presence of a small amount of CA-l ($[\text{CA-l}]_0 = 67$ mM; CA-l:Ru-**c** = 13:1) than in the absence of CA-l, and it also afforded AL-**a** consistently in lower yields. However, irrespective of the presence ($t = 2.5$ h) or absence ($t = 1.5$ or 2.5 h) of CA-l, the reaction rates obtained with the three different initial concentrations $[\text{CA-}\mathbf{a}]_0$ in each series were roughly equivalent, providing approximately the same $[\text{AL-}\mathbf{a}]_{t=2.5}$ (0.06–0.09 M with CA-l versus 0.08–0.11 M without CA-l. For the dependence of different $[\text{CA-}\mathbf{a}]_0$ with the same $[\text{CA-l}]_0$ on yields of AL-**a** and AL-l, see: Supplementary Fig. 2). The hydrogenation rate was observed to be virtually independent of $[\text{CA-}\mathbf{a}]_0$, whereby the initially added CA-l was recovered effectively (CA-l: 78–99%; AL-l: 5–10%) in all three runs. These results suggested that the catalytic hydrogenation giving AL-l promoted by $[Ru(OCOPh)]^+$ via an inner-sphere mechanism was barely involved. The catalysis facilitated by the more reactive $[Ru(OCO(CH_2)_2Ph)]^+$ via an outer-sphere mechanism (cleavage of H–H bond, followed by an intermolecular hydride transfer from the resulting $[RuH(HOCO(CH_2)_2Ph)]^+$ to a free CA-l) was also negligible. These processes scarcely interfere with the faster catalysis mediated by $[Ru(OCO(CH_2)_2Ph)]^+$ or $[RuH(HOCO(CH_2)_2Ph)]^+$ that affords AL-**a** (CA-**a** self-induced CA-**a** hydrogenation in the presence of CA-l). To reconfirm that indeed the suspected 'CA self-induced CA hydrogenation' was observed here, the hydrogenation of CA-l was also carried out using three different initial concentrations ($t = 6$ h). As in the previous case, a zeroth order rate with respect to [CA-l] corroborates a CA-l self-induced CA-l hydrogenation (Supplementary Fig. 3). These preliminary kinetic experiments indicated that the apparent reaction rate of CA hydrogenation with Ru-**c** and NaOAc at the same parameters for $[H_2]$ and $T$ is independent of $[CA]_t$ and almost constant, assuming a constant $[[Ru(OCOR)]^+]_t$. In other words, the velocity of the rate-determining step should change only on varying the concentration of $[Ru(OCOR)]^+$. It is therefore suggested that CA-**a** and CA-l should be involved not only as integral 'carboxylates' of the catalysts for cleaving the H–H bond, but also as 'protonated carboxylates' that are spontaneously activated and reduced by the resulting $[Ru\text{–}H]^+$, which is generated in the first step of the CA self-induced CA hydrogenation.

The $CO_2H$ groups of CA-**h** and CA-**i** were hydrogenated more rapidly than the interior aliphatic and aromatic esters, furnishing AL-**h** and AL-**i** chemoselectively, under preservation of the methyl ester moieties (entries 10 and 12). In addition, when a 1:1 molar mixture of CA-**d** and ethyl stearate ($CH_3(CH_2)_{16}CO_2CH_2CH_3$) was hydrogenated using Ru-**c** (1.5 mol%) and NaOAc (10 mol%; $P_{H2} = 4$ MPa, $T = 160$ °C, $t = 24$ h), AL-**d** was obtained in 99% yield, while the ester was recovered unchanged. An external ester did not inhibit the catalysis and did not engage with the $[RuH(HOCOR)]^+$ species (R = 1-adamantyl). The $[RuH(HOCOR)]^+$ complex should be accessible only by a CA covalently attached to the Ru centre. In contrast, esters that cannot covalently bind to the Ru complex may have little chance to participate in the catalytic cycle as an integral structure of the catalyst. The applied conditions are most likely not basic enough to effectively promote the homocoupling of alcohols, and therefore the occurrence of a Tishchenko-type reaction giving esters can be excluded (Supplementary Methods). Moreover, it seems that the condensation of the CA with the alcohol to give the corresponding ester is also effectively suppressed in the present system, since the catalyst is unable to hydrogenate esters.

The benzene rings in close proximity to or spatially removed from the CA groups of CA-**a**, **e**, **f**, **i**, **j** and CA-**l–n** were well tolerated (entries 1, 6–8, 12, 13 and 15–18), whereas the amide moiety of CA-**f** (entry 7) and the thiophene moiety of CA-**g** (entry 9) inhibited the catalysis. Despite the fact that the olefin moieties of CA-**i–k** were hydrogenated (entries 12–14), the high chemoselectivity and concurrent compatibility of aromatic rings and ester moieties achieved here is virtually unprecedented and rarely achieved with typical, strong, stoichiometric agents such as $LiAlH_4$, $LiBH_4$, $BH_3$ or $LiBEt_3H$ that reduce not only CAs, but also esters[45].

**Hydrogenation using Ru complex with bidentate phosphine.** To improve the catalytic activity, the combination of a multidentate ligand (2 mol%), $RuCl_2(DMSO)_4$ (2 mol%) and $NaBPh_4$ (10 mol%) was examined in detail. From a variety of multidentate ligands screened (Supplementary Table 2), 1,4-(diphenylphosphino)butane (dppb) proved to be one of the best (AL-**a**: 52%, ES-**a**: 14%; $P_{H2} = 8$ MPa, $T = 160$ °C, $t = 24$ h). Another elegant approach that differs from ours was reported by Leitner and Klankermayer, who proposed $[Ru(triphos)(OCOH)]^+$ and Ru(triphos)(TMM) species as the catalytically active species for the hydrogenation of $CO_2$ ($P_{H2} = 5$ MPa, $T = 140$ °C) and CA-l ($P_{H2} = 50$ bar, $T = 220$ °C) (ref. 32), giving $CH_3OH$ (ref. 31) and AL-l, respectively. Nevertheless, $RuCl_2(DMSO)_4$/triphos (2 mol% each) was observed to be less effective for CAs with a longer carbon chain (AL-**a**: 22%, ES-**a**: 13%). Moreover, the catalyst systems $Ru(acac)_3$/triphos (2 mol% each)[26,30] and [Ru(triphos)(TMM)] (refs 31,32) (2 mol%) were tested separately ($P_{H2} = 8$ MPa, $T = 160$ °C, $t = 24$ h), but the observed reactivity was consistently low (AL-**a**: 22% and 10%, ES-**a**: 10% and 8%, respectively). In contrast, it became clear that a [Ru(dppb)] complex could adopt a structure Ru-**d**[46], which is similar to Ru-**c**. Hence, Ru-**d** (1 mol%) with Na(acac) (10 mol%) was tested for the hydrogenation of CA-**a** ($P_{H2} = 4$ MPa, $T = 160$ °C, $t = 24$ h), which furnished an improved yield of 78 and 10% for AL-**a** and ES-**a**, respectively, under almost quantitative conversion of CA-**a**. Using Ru-**d**, it was also possible to hydrogenate the relatively unreactive substrate CA-l under much milder conditions than those reported by Leitner[32], affording AL-l in 93% yield (Table 1, entry 16). The increased catalytic activity of Ru-**d** relative to that of Ru-**c** could also be demonstrated by the following experiment: Ru-**c** and Ru-**d** (1.5 mol% each) were used separately with Na(acac) (10 mol%) for the hydrogenation of CA-l ($P_{H2} = 4$ MPa, $T = 160$ °C, $t = 48$ h), giving AL-l in 26% and 56%, respectively (benzyl benzoate: 2% and 1%, respectively). Substrates CA-**m** and CA-**n**, carrying electron-donating and -withdrawing groups, respectively, were also compatible with these conditions (Table 1, entries 17 and 18). Moreover, the hydrogenation of other substrates that were relatively inert to the Ru-**c**/NaOAc system also proceeded more readily when using Ru-**d** and Na(acac; Table 1, entries 3 and 11). Another advantage of Ru-**d** over Ru-**c** is the structural robustness of the catalyst derived from Ru-**d** under aqueous conditions (Supplementary Table 4).

**Insight into catalytic $[Ru(OCOR)P_2]^+$.** The prospective resting state of these catalysts, a $[Ru(OCOR)P_2]^+$ species, was elucidated by electrospray ionization-mass spectroscopy (ESI–MS) analysis (Fig. 2 and Supplementary Figs 4 and 5). After a toluene solution of a 1:6.7 mixture of Ru-**c** and NaOAc was heated to 160 °C for 3 h, a sample thereof was dissolved in MeCN and examined by ESI–MS, which revealed two primary signals ($m/z$ = 853.2856 and 894.3137) corresponding to Ru-**g** and Ru-**h**, respectively. Presumably, the catalytically innocent species $[RuH(CO)[P(3,5\text{-xylyl})_3]_2]^+$ ($m/z$ = 823.2752, exact mass: 823.2766) was observed, as its parent compound $Ru(OAc)H(CO)[P(3,5\text{-xylyl})_3]_2$ could be detected by $^{31}P\{^1H\}$ NMR in $CDCl_3$ (δ 45.8 (s, *P*RuP) and 45.9 (s, PRu*P*) p.p.m. (ref. 47). When a toluene solution of a 1:6.7:33 mixture of Ru-**c**, NaOAc and CA-**a** was heated to 160 °C for 3 h, Ru-**i**, Ru-**j** and Ru-**h** were detected as the primary signals ($m/z$ = 943.3291, 984.3645 and 894.3083, respectively). In both ESI–MS measurements, binuclear structures (involving two Ru centres) could not be detected. The catalytically important structure '$[Ru(OCOR)P_2]^+$', consistent with Ru-**i** and Ru-**j**, could not be obtained in significant amounts when $NaB[3,5\text{-}(CF_3)_2C_6H_3]_4$, $NaBF_4$, NaOTf or $NaNTf_2$ were used as additives instead of NaOAc or $NaBPh_4$ (Supplementary Figs 6 and 7). One of the original $Cl^-$ groups of Ru-**c** remained unaffected, or was replaced by CO.

In addition, after a $(CH_3)_3COH$ solution of a 1:20 mixture of Ru-**c** and NaOAc was heated to 90 °C for 2 h, Ru-**e** was isolated in 49% yield (Fig. 2; Supplementary Methods). The hydrogenation of CA-**a** with Ru-**e** (3 mol%, $[Ru]_0$ = 10 mM) proceeded effectively ($P_{H2}$ = 4 MPa, $T$ = 160 °C, $t$ = 24 h) in the absence of NaOAc, giving AL-**a** (ES-**a**) in 87% (6%) yield, which is comparable to previously obtained results. In the absence of NaOAc (under less basic conditions) and using a small amount of dioxane (10 equiv with respect to CA-**a**), esterification was more effectively suppressed (Table 1, entry 8). Using Ru-**e**, the α,β-unsaturated aliphatic CA tiglic acid (CA-**k**) was hydrogenated at both the olefin and the $CO_2H$ groups, giving 2-methyl-1-butanol (AL-**k**) in ca. 65% yield (Table 1, entry 14). We would like to use this opportunity to underline the unique reactivity of our system relative to previously described systems: using the comparable Ru-acetato complex $Ru(CO)_2(OCOCH_3)_2(PBu_3)_2$ (Bu = $CH_3(CH_2)_3$) did not lead to the hydrogenation of the $CO_2H$ groups of CA-**k** or 2-methylbutanoic acid ($P_{H2}$ = ca. 13 MPa, $T$ = 100 °C)[19]. Here the CA self-induced CA hydrogenation via the cationic mononuclear complexes $[Ru(OCOR)]^+$ was not observed and the hydrogenation of AcOH was ascribed to 'unknown catalytic species', affording the ester $AcOCH_2CH_3$. A higher catalytic activity of binuclear Ru-carboxylato complexes relative to that of mononuclear derivatives[18,19] and the importance of intermediary Ru(acyl)(alkoxy) complexes were briefly discussed[18].

The Ru-diacetato complex Ru-**f** was synthesized from Ru-**d** (Fig. 2)[48] and was one of the most active Ru catalysts for the hydrogenation of CA-**a** under relatively mild conditions ($P_{H2}$ = 2 MPa; Table 2), even though a slightly decreased yield of AL-**a** was observed under lower hydrogen pressure ($P_{H2}$ = 1 MPa; entry 5). In contrast, hydrogenation with Ru-**e** proceeded comparably, but more selectively using a lower $[CA\text{-}a]_0$ (entry 3 versus 4). The high alcohol selectivity should be ascribed, at least partially, to the low $[CA]_0$.

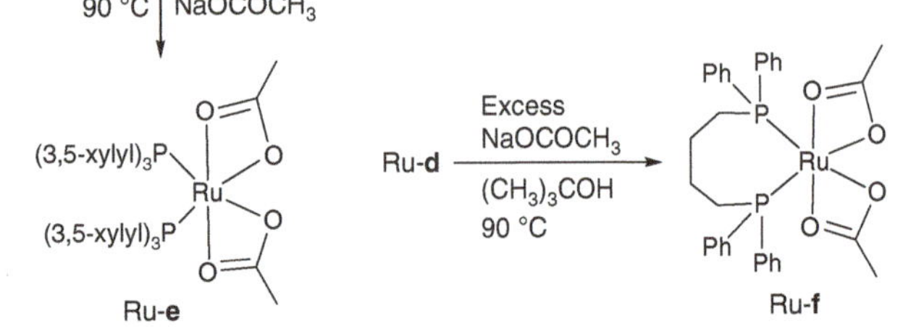


**Figure 2 | Importance of '$[Ru(OCOR)]^+$' and its precursors for CA hydrogenation.** Single crystal X-ray structure of Ru-**c** (containing ethanol). Colour code: Ru (purple), Cl (green), P (orange), O (red) and C (grey). Calculated exact masses: Ru-**g** (853.2872), Ru-**h** (894.3137), Ru-**i** (943.3341) and Ru-**j** (984.3607).

## Discussion

On the basis of these observations, we would like to propose a catalytic cycle (Fig. 3).

Important points that corroborate the catalytic cycle can be summarized as: (1) Ru-dicarboxylates such as Ru-**e** and Ru-**f** are stable enough to be isolated; (2) $NaBPh_4$ (replacing $Cl^-$ in Ru-**c** or Ru-**d** to generate the catalytically active cationic Ru complex) as well as NaOAc effective induced the catalyst; (3) the rate of the hydrogenation was effectively decreased when using Ru-**e** ($P_{H2}$ = 4 MPa, $T$ = 160 °C, $t$ = 24 h) in dioxane as a solvent (AL-**a**: 60%; ES-**a**: 2%), which strongly coordinates to the cationic Ru centre; (4) although Ru(CO) complexes were detected by ESI–MS (*vide supra*), the attachment of CO to the Ru centres should be excluded from the catalytic cycle, since such a coordination of CO would be detrimental to the catalytic activity of the CA hydrogenation under mild conditions; and (5) the intermediate '$[RuH(dppb)]^+$', corresponding to $\mathbf{I_D}$ or $\mathbf{I_E}$ stabilized by toluene ($m/z$ found: 621.1425, calcd: 621.1409) or by $CH_3CN$ ($m/z$ found: 570.1047, calcd: 570.1048) was observed in addition to a $[Ru(OCOR)(dppb)]^+$ species analogous to Ru-**i** and Ru-**j**. Those species were generated by treatment of a toluene solution of Ru-**f** with CA-**a** and $H_2$ (Ru-**f**:CA-**a** (mol%) = 1:33, $[CA\text{-}a]_0$ = 333 mM, $P_{H2}$ = 4 MPa, $T$ = 160 °C, $t$ = 3 h) and were detected in $CH_3CN$ solution by ESI–MS analysis (for all Ru species observed, see: Supplementary Fig. 8). Another critical intermediate identified was 18e$^-$ $\eta^6$-toluene-stabilized $\mathbf{I_B}[Ru(OCH(OH)(CH_2)_2Ph)(dppb)(toluene)]^+$ ($m/z$ found: 771.2001, calcd: 771.2089). Moreover, $[Ru[OCO(CH_2)_2Ph](dppb)(CH_3CN)]^+$ [$m/z$ found: 718.1601 (the strongest intensity), calcd: 718.1572] as stabilized $\mathbf{cat_A}$ is consistent with the unified 'prototype $[Ru(OCOR)P_2]^+$' in the catalytic cycle involving the CA self-induced CA hydrogenation ($\mathbf{I_A} \rightarrow \mathbf{TS_{in}} \rightarrow \mathbf{I_B}$). By taking these facts into account, a cationic ruthenium complex $\mathbf{cat_A}$ incorporating one carboxylate is likely the most critical and active species, although catalysis involving the two carboxylate-bearing $\mathbf{cat_B}$ can not be fully ruled out. Since the hydrogenation rate using Ru-**e** was affected by the $H_2$ pressure (1–4 MPa; Supplementary Fig. 9), the formation of Ru–$\eta^1$-H ($\mathbf{TS_{in}}$) via Ru–$\eta^2$-$H_2$ complex $\mathbf{I_A}$ should be the turnover-limiting step in this $P_{H2}$ range. A multifunctional

**Table 2 | Comparison experiments using the $Ru(OAc)_2P_2$ complexes Ru-c and Ru-d under lower hydrogen pressure ($P_{H2}$ = 1-2 MPa) for the hydrogenation of CA-a.***

| Entry | Ru complex, mol% | Additive, mol% | Yield (%)†,‡ | |
|---|---|---|---|---|
| | | | AL-a | ES-a |
| 1 | Ru-**c**, 1.5 | Na(OAc), 10 | 69 | 13 |
| 2 | Ru-**d**, 1.5 | Na(acac), 10 | 69 | 11 |
| 3 | Ru-**e**, 3 | none | 78 ± 3§ (85)‖ | 10 ± 1§ (5)‖ |
| 4 | Ru-**f**, 3 | none | 76 ± 3§ (80)‖ | 10 ± 1§ (5)‖ |
| 5¶ | Ru-**f**, 3 | none | 53 | 11 |
| 6 | $Ru(OAc)_2$[(*R*)-BINAP], 3 | none | 15 | 24 |
| 7 | $Ru(OAc)_2$[(*S*)-DMBINAP], 3 | none | 30 | 23 |

AL, alcohol; ES, ester.
*Unless otherwise specified, reactions were carried out with CA-**a**:Ru (mol%) = 100:3, $P_{H2}$ = 2 MPa, *T* = 160 °C for 24 h ($[Ru]_0$ = 10 mM; $[CA\text{-}\mathbf{a}]_0$ = 333 mM; in toluene).
†NMR yields are based on the internal standard mesitylene.
‡Average of two runs.
§Average of three runs with calculated s.d.
‖($[Ru]_0$ = 5 mM; $[CA\text{-}\mathbf{a}]_0$ = 167 mM). BINAP = 2,2′-bis(diphenylphosphanyl)-1,1′-binaphthalene; DMBINAP = 2,2′-bis(bis(3,5-dimethylphenyl) phosphanyl)-1,1′-binaphthalene.
¶$P_{H2}$ = 1 MPa, *t* = 48 h; in dioxane/toluene (v/v = 1:3.5).

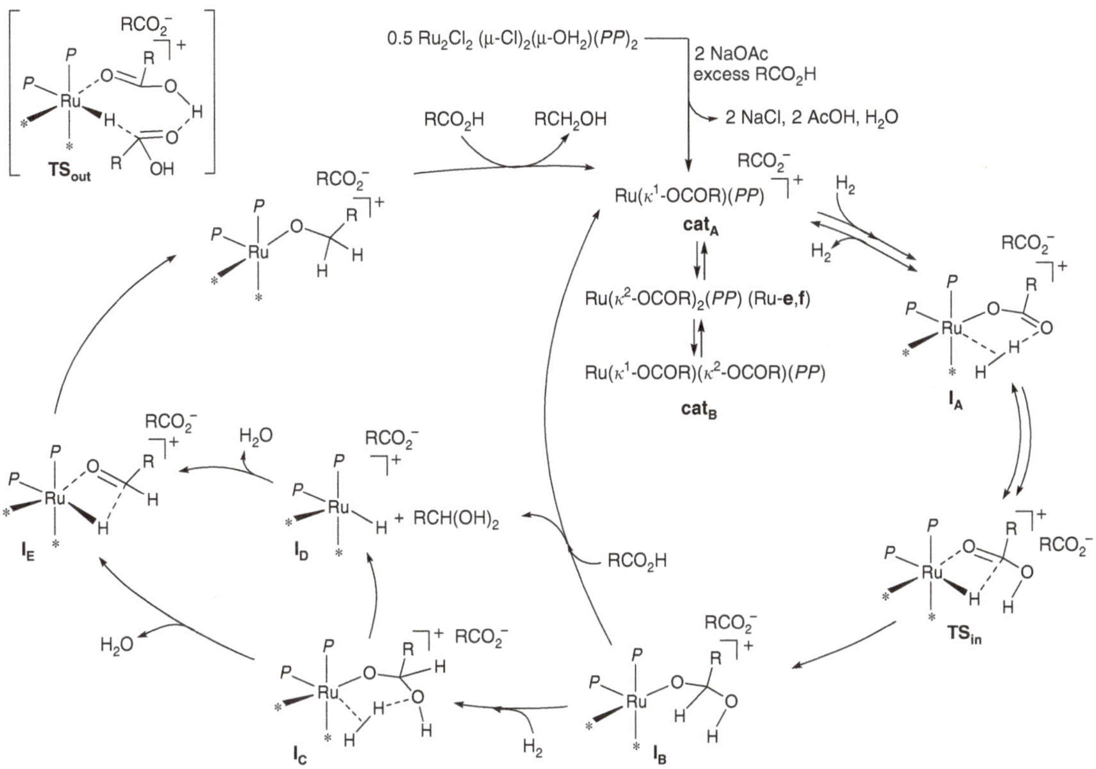


**Figure 3 | Plausible catalytic cycle involving '$[Ru(OCOR)P_2]^+$'.** $P_2$ denotes $2P(3,5\text{-xylyl})_3$ or dppb. The asterisk (*) denotes various neutral ligands including solvents such as toluene or dioxane.

(outer sphere) mechanism (**TS$_{out}$**; Fig. 3) involving a $RCO_2^-$ moiety[49] and a $RCO_2H$ molecule (with a cooperative role for RuH(δ$^-$) and Ru[H(δ$^+$)OCOR] for hydrogen transfer to another $RCO_2H$), as an alternative to a **TS$_{in}$** (inner-sphere mechanism involving direct interaction of the CA with a Ru centre; Fig. 3), might be also possible, although an unambiguous assignment requires a further detailed kinetic and theoretical analysis. Even though free aldehydes were not detected, the release of free hemiacetal from **I$_C$** via a four-centered σ-bond metathesis giving **I$_D$**, or through the hemiacetal-CA exchange of **I$_B$** that affords **cat$_A$**, followed by dehydrative formation of free aldehyde that re-enters the catalytic cycle to give **I$_E$**, is another possible scenario.

In any case, further refinement of the molecular design of CA hydrogenation catalysts based on the catalytic cycle proposed involves two key factors: (1) the formation of a Ru-$\eta^1$-H species through deprotonation of the Ru-$\eta^2$-$H_2$ is facilitated by the participation of a cationic $[Ru(OCOR)]^+$ (R = aliphatic group) species. This step may involve a pathway analogous to the concerted metalation-deprotonation mechanism proposed for the cleavage of either H-H (refs 25,38-40,49) or C-H bonds[41-44]. (2) The CA reformed via the protonation of Ru-carboxylate remains coordinated to the cationic Ru centre of Ru-H species (**I$_A$** → **TS$_{in}$**). This activated CA in **TS$_{in}$** is more electrophilic and should be able to facilitate a smooth hydride transfer from the Ru-H species to the carbonyl carbon atom.

In summary, the results obtained from the systematic study of these cationic, mononuclear ruthenium mono-carboxylate catalyst prototypes enabled a rational approach to the design of hydrogenation catalysts for CAs. The results demonstrate that the CAs should act not only as integral carboxylates for the catalysts to cleave the H–H bond, but also as protonated carboxylates that are simultaneously activated and reduced by the resulting $[Ru\text{-}H]^+$, which is generated in the first step (CA-induced CA hydrogenation). As the availability of bidentate phosphines is virtually infinite, further optimization of the ligand for better catalyst performance under milder conditions is ongoing, together with an extension of the present fundamental research, which has provided this milestone discovery. Identification of ideal metal carboxylates that hydrogenate bio-renewable resources in high oxidation states will significantly benefit the future development of chemical processes directed towards a sustainable society.

## References


1. Teichmann, D., Arlt, W., Wasserscheid, P. & Freymann, R. A future energy supply based on liquid organic hydrogen carriers (LOHC). *Energy Environ. Sci.* **4,** 2767–2773 (2011).
2. Werpy, T. & Petersen, G. Top value added chemicals from biomass volume I-results of screening for potential candidates from sugars and synthesis gas http://energy.gov/eere/bioenergy/downloads/top-value-added-chemicals-biomass-volume-i-results-screening-potential (2004).
3. Bozell, J. J. & Petersen, G. R. Technology development for the production of biobased products from biorefinery carbohydrates—the US Department of Energy's 'Top 10' revisited. *Green Chem.* **12,** 539–554 (2010).
4. Ostapowicz, T. G., Schmitz, M., Krystof, M., Klankermayer, J. & Leitner, W. Carbon dioxide as a C1 building block for the formation of carboxylic acids by formal catalytic hydrocarboxylation. *Angew. Chem. Int. Ed.* **52,** 12119–12123 (2013).
5. Jessop, P. G., Ikariya, T. & Noyori, R. Homogeneous catalytic hydrogenation of supercritical carbon dioxide. *Nature* **368,** 231–233 (1994).
6. Jessop, P. G., Ikariya, T. & Noyori, R. Homogeneous hydrogenation of carbon dioxide. *Chem. Rev.* **95,** 259–272 (1995).
7. Hull, J. F. *et al.* Reversible hydrogen storage using $CO_2$ and a proton-switchable iridium catalyst in aqueous media under mild temperatures and pressures. *Nat. Chem.* **4,** 383–388 (2012).
8. Tanaka, R., Yamashita, M. & Nozaki, K. Catalytic hydrogenation of carbon dioxide using Ir(III)-pincer complexes. *J. Am. Chem. Soc.* **131,** 14168–14169 (2009).
9. Sato, S. *et al.* Selective $CO_2$ conversion to formate conjugated with $H_2O$ oxidation utilizing semiconductor/complex hybrid photocatalysts. *J. Am. Chem. Soc.* **133,** 15240–15243 (2011).
10. Giri, R. *et al.* Palladium-catalyzed methylation and arylation of $sp^2$ and $sp^3$ C-H bonds in simple carboxylic acids. *J. Am. Chem. Soc.* **129,** 3510–3511 (2007).
11. Olah, G. A. Towards oil independence through renewable methanol chemistry. *Angew. Chem. Int. Ed.* **52,** 104–107 (2013).
12. Olah, G. A., Prakash, G. K. S. & Goeppert, A. Anthropogenic chemical carbon cycle for a sustainable future. *J. Am. Chem. Soc.* **133,** 12881–12898 (2011).
13. Clarke, M. L. Recent developments in the homogeneous hydrogenation of carboxylic acid esters. *Catal. Sci. Technol.* **2,** 2418–2423 (2012).
14. Dub, P. A. & Ikariya, T. Catalytic reductive transformation of carboxylic and carbonic acid derivatives using molecular hydrogen. *ACS Catal.* **2,** 1718–1741 (2012).
15. Werkmeister, S., Junge, K. & Beller, M. Catalytic hydrogenation of carboxylic acid esters, amides, and nitriles with homogeneous catalysts. *Org. Process Res. Dev.* **18,** 289–302 (2014).
16. Smith, A. M. & Whyman, R. Review of methods for the catalytic hydrogenation of carboxamides. *Chem. Rev.* **114,** 5477–5510 (2014).
17. Bianchi, M. *et al.* Homogeneous catalytic hydrogenation of free carboxylic acids in the presence of cluster ruthenium carbonyl hydrides. *J. Organomet. Chem.* **188,** 109–119 (1980).
18. Salvini, A. *et al.* The role of functionalized phosphines in the hydrogenation of carboxylic acids in the presence of phosphine substituted hydrido ruthenium complexes. *J. Organomet. Chem.* **582,** 218–228 (1999).
19. Salvini, A., Frediani, P., Giannelli, C. & Rosi, L. On the behaviour of Ru(I) and Ru(II) carbonyl acetates in the presence of $H_2$ and/or acetic acid and their role in the catalytic hydrogenation of acetic acid. *J. Organomet. Chem.* **690,** 371–382 (2005).
20. Yokoyama, T. & Yamagata, N. Hydrogenation of carboxylic acids to the corresponding aldehydes. *Appl. Catal. A: Gen.* **221,** 227–239 (2001).
21. Yoshino, K., Kajiwara, Y., Takaishi, N., Inamoto, Y. & Tsuji, J. Hydrogenation of carboxylic acids by rhenium-osmium bimetallic catalyst. *J. Am. Oil Chem. Soc.* **67,** 21–24 (1990).
22. He, D.-H., Wakasa, N. & Fuchikami, T. Hydrogenation of carboxylic acids using bimetallic catalysts consisting of group 8 to 10, and group 6 or 7 metals. *Tetrahedron Lett.* **36,** 1059–1062 (1995).
23. Manyar, H. G. *et al.* Highly selective and efficient hydrogenation of carboxylic acids to alcohols using titania supported Pt catalysts. *Chem. Commun.* **46,** 6279–6281 (2010).
24. Brewster, T. P., Miller, A. J. M., Heinekey, D. M. & Goldberg, K. I. Hydrogenation of carboxylic acids catalyzed by half-sandwich complexes of iridium and rhodium. *J. Am. Chem. Soc.* **135,** 16022–16025 (2013).
25. Chaplin, A. B. & Dyson, P. J. Reactivity and catalytic activity of a robust ruthenium(II)-triphos complex. *Inorg. Chem.* **47,** 381–390 (2008).
26. Geilen, F. M. A. *et al.* Selective and flexible transformation of biomass-derived platform chemicals by a multifunctional catalytic system. *Angew. Chem. Int. Ed.* **49,** 5510–5514 (2010).
27. Geilen, F. M. A., Engendahl, B., Hölscher, M., Klankermayer, J. & Leitner, W. Selective homogeneous hydrogenation of biogenic carboxylic acids with $[Ru(TriPhos)H]^+$: A mechanistic study. *J. Am. Chem. Soc.* **133,** 14349–14358 (2011).
28. Wesselbaum, S., Stein, T. V., Klankermayer, J. & Leitner, W. Hydrogenation of carbon dioxide to methanol by using a homogeneous ruthenium-phosphine catalyst. *Angew. Chem. Int. Ed.* **51,** 7499–7502 (2012).
29. Rosi, L., Frediani, M. & Frediani, P. Isotopomeric diols by "one-pot' Ru-catalyzed homogeneous hydrogenation of dicarboxylic acids. *J. Organomet. Chem.* **695,** 1314–1322 (2010).
30. Mehdi, H. *et al.* Integration of homogeneous and heterogeneous catalytic processes for a multi-step conversion of biomass: from sucrose to levulinic acid, γ-valerolactone, 1,4-pentanediol, 2-methyl-tetrahydrofuran, and alkanes. *Top. Catal.* **48,** 49–54 (2008).
31. Wesselbaum, S. *et al.* Hydrogenation of carbon dioxide to methanol using a homogeneous ruthenium-triphos catalyst: from mechanistic investigations to multiphase catalysis. *Chem. Sci.* **6,** 693–704 (2015).
32. Stein, T. V. *et al.* Highly versatile catalytic hydrogenation of carboxylic and carbonic acid derivatives using a Ru-triphos complex: molecular control over selectivity and substrate scope. *J. Am. Chem. Soc.* **136,** 13217–13225 (2014).
33. Huff, C. A. & Sanford, M. S. Cascade catalysis for the homogeneous hydrogenation of $CO_2$ to methanol. *J. Am. Chem. Soc.* **133,** 18122–18125 (2011).
34. Delhomme, C., Weuster-Botz, D. & Kühn, F. E. Succinic acid from renewable resources as a $C_4$ building-block chemical–a review of the catalytic possibilities in aqueous media. *Green Chem.* **11,** 13–26 (2009).
35. Preliminary results of this report (including Ru complexes Ru-**c,d**) were used in a patent application, see: Saito, S., Noyori, R. & Naruto, M. *JP patent*, Appl. #2013-267866, Filed: Dec 25, 2013.
36. Stephenson, T. A. & Wilkinson, G. New complexes of ruthenium (II) and (III) with triphenylphosphine, triphenylarsine, trichlorostannate, pyridine and other ligands. *J. Inorg. Nucl. Chem.* **28,** 945–956 (1966).
37. Brown, J. M. Directed homogeneous hydrogenation. *Angew. Chem. Int. Ed.* **26,** 190–203 (1987).
38. Kitamura, M. *et al.* Mechanism of asymmetric hydrogenation of α-(acylamino)acrylic esters catalyzed by BINAP-ruthenium(II) diacetate. *J. Am. Chem. Soc.* **124,** 6649–6667 (2002).
39. Ashby, M. T. & Halpern, J. Kinetics and mechanism of catalysis of the asymmetric hydrogenation of α, β-unsaturated carboxylic acids by bis(carboxylato) {2,2'-bis(diphenylphosphino)-1,1'-binaphthyl}ruthenium(II), $[Ru^{II}(BINAP)(O_2CR)_2]$. *J. Am. Chem. Soc.* **113,** 589–594 (1991).
40. Fries, J. G. D. & Elsevier, C. J. *Handbook of Homogeneous Hydrogenation: 3 Volumes* Vol. 1, 45 (Wiley VCH, 2007).
41. Lynam, J. M., Welby, C. E. & Whitwood, A. C. Exploitation of a chemically non-innocent acetate ligand in the synthesis and reactivity of ruthenium vinylidene complexes. *Organometallics* **28,** 1320–1328 (2009).
42. Hiett, N. P., Lynam, J. M., Welby, C. E. & Whitwood, A. C. Ruthenium carboxylate complexes as easily prepared and efficient catalysts for the synthesis of β-oxopropyl esters. *J. Organomet. Chem.* **696,** 378–387 (2011).
43. Ackermann, L. Carboxylate-assisted transition-metal-catalyzed C – H bond functionalizations: mechanism and scope. *Chem. Rev.* **111,** 1315–1345 (2011).
44. Arockiam, P. B., Bruneau, C. & Dixneuf, P. H. Ruthenium(II)-catalyzed C – H bond activation and functionalization. *Chem. Rev.* **112,** 5879–5918 (2012).
45. Brown, H. C. From little acorns to tall oaks: from boranes through organoboranes. *Science* **210,** 485–492 (1980).
46. MacFarlane, K. S. *et al.* Triply-bridged diruthenium(II) 1,4-bis(diphenylphosphino) butane(dppb) and (*R*)-2,2'-bis(diphenylphosphino)-l,l'-binaphthyl (binap) complexes, including structural characterisation of $[(dppb)ClRu(\mu\text{-}D_2O)(\mu\text{-}Cl)_2RuCl(dppb)]$, $[\eta^2\text{-}H_2](dppb)Ru(\mu\text{-}Cl)_3RuCl(dppb)]$ and the $[(dppb)ClRu(\mu\text{-}Cl)_3RuCl(dppb)]^-$ anion. *Inorg. Chim. Acta* **270,** 130–144 (1998).

47. Rosales, M. *et al.* A general route for the synthesis of hydrido-carboxylate complexes of the type $MH(CO)(\kappa^3\text{-}OCOR)(PPh_3)_2$ [M = Ru, Os; R = $CH_3$, $CH_2Cl$, $C_6H_5$, $CH(CH_3)_2$] and their use as precatalysts for hydrogenation and hydroformylation reactions. *Polyhedron* **27,** 530–536 (2008).
48. Wong, W.-K. *et al.* Reactivity of $Ru(OAc)_2(Ph_3P)_2$ toward chelating diphosphine ligands. X-ray crystal structures of *fac*-$Ru(OAc)_2(Ph_3P)$(dppm) and *trans*-$Ru(OAc)_2(P_2N_2H_4)$. *Polyhedron* **13,** 2751–2762 (1994).
49. Duan, Y. *et al.* Homogenous Pd-catalyzed asymmetric hydrogenation of unprotected indoles: scope and mechanistic studies. *J. Am. Chem. Soc.* **136,** 7688–7700 (2014).

## Acknowledgements

This work was supported by the research area of Japan Science and Technology Corporation (JST), 'Creation of Advanced Catalytic Transformation for the Sustainable Manufacturing at Low Energy, Low Environmental Load (ACT-C)'. M.N. acknowledges support from a IGER (NU) fellowship, as well as from Iue memorial and Iwadare fellowships. The authors wish to thank Professor Shohei Saito (NU) for technical support with the single crystal X-ray analysis and Professor R. Noyori (NU, JST and RIKEN) for fruitful discussions.

## Author contributions

M.N. made the initial discovery and carried out the experiments. S.S. directed the project and wrote the manuscript.

## Additional information

**Competing financial interests:** The authors declare no competing financial interests.

# 3

# A molecular nematic liquid crystalline material for high-performance organic photovoltaics

Kuan Sun[1,2,3,*], Zeyun Xiao[1,*], Shirong Lu[1,*], Wojciech Zajaczkowski[4], Wojciech Pisula[4], Eric Hanssen[5], Jonathan M. White[1], Rachel M. Williamson[6], Jegadesan Subbiah[1], Jianyong Ouyang[2], Andrew B. Holmes[1], Wallace W.H. Wong[1] & David J. Jones[1]

Solution-processed organic photovoltaic cells (OPVs) hold great promise to enable roll-to-roll printing of environmentally friendly, mechanically flexible and cost-effective photovoltaic devices. Nevertheless, many high-performing systems show best power conversion efficiencies (PCEs) with a thin active layer (thickness is ~100 nm) that is difficult to translate to roll-to-roll processing with high reproducibility. Here we report a new molecular donor, benzodithiophene terthiophene rhodanine (BTR), which exhibits good processability, nematic liquid crystalline behaviour and excellent optoelectronic properties. A maximum PCE of 9.3% is achieved under AM 1.5G solar irradiation, with fill factor reaching 77%, rarely achieved in solution-processed OPVs. Particularly promising is the fact that BTR-based devices with active layer thicknesses up to 400 nm can still afford high fill factor of ~70% and high PCE of ~8%. Together, the results suggest, with better device architectures for longer device lifetime, BTR is an ideal candidate for mass production of OPVs.

[1] School of Chemistry, Bio21 Institute, The University of Melbourne, 30 Flemington Road, Parkville, Victoria 3010, Australia. [2] Department of Materials Science and Engineering, National University of Singapore, 7 Engineering Drive 1, Singapore 117574, Singapore. [3] Department of Renewable Energy, School of Power Engineering, Chongqing University, 174 Shazhengjie, Shapingba, Chongqing 400044, China. [4] Max Planck Institute for Polymer Research, Ackermannweg 10, 55128 Mainz, Germany. [5] Advanced Microscopy Facility, Bio21 Institute, The University of Melbourne, 30 Flemington Road, Parkville, Victoria 3010, Australia. [6] MX Beamlines, Australian Synchrotron, 800 Blackburn Road, Clayton, Victoria 3168, Australia. * These authors contributed equally to this work. Correspondence and requests for materials should be addressed to J.O. (email: mseoj@nus.edu.sg) or to W.W.H.W. (email: wwhwong@unimelb.edu.au) or to D.J.J. (email: djjones@unimelb.edu.au).

Despite recent developments in solid-state photovoltaic devices[1–3], bulk-heterojunction (BHJ) organic photovoltaics (OPVs)[4] continue to be a promising low-cost renewable energy technology. The reasons for this outlook include the versatility of organic semiconducting materials and simple device architectures that can be constructed from a variety of printing techniques. The development of the BHJ OPVs has been rapid in recent years driven by a combination of organic material design, interface engineering and improvements in device geometry. The reported power conversion efficiency (PCE) of single-junction small-area devices is now routinely in the 6–8% range[5–7].

Published reports of single-junction BHJ OPVs over 9% PCE are still rare. A handful of polymeric electron donor materials and only one molecular donor have been reported in devices that reached this benchmark[8–13]. Molecular OPVs are an attractive alternative to polymer-based OPVs. Higher material purity can be achieved with the well-defined discrete structure of molecules and this should ensure greater reproducibility in devices[14–16]. To achieve high photovoltaic conversion efficiency, the material should be capable of forming good films with high molecular order. This can be achieved by smart molecular design and control over crystallization processes. For molecular semiconductors, conjugated flat and rigid backbones are preferred for easy packing via π–π interactions. Good solubility is conferred through employment of the appropriate number, type and length of side chains, without hindering the packing of the backbones. The desirable donor phase in BHJ OPVs needs well-ordered nanocrystals with sizes comparable to the exciton diffusion length for efficient charge generation. Strategies simultaneously to enhance molecular order and restrict crystal size have been reported, including thermal annealing, solvent additives, solid additives and solvent vapour annealing (SVA)[7,17–22]. Recent reports showed that the rapid SVA treatment was particularly useful in achieving high fill factor (FF) and PCE in molecular OPVs[23,24]. Solvent selection rules for SVA treatment were identified in our previous study[23].

Despite the important progress achieved in small-area OPV devices fabricated in laboratories, the successful commercialization of OPV technology relies on the application of solution-processed roll-to-roll techniques for large-scale printing[25]. One of the challenges in printing OPV devices is in printing the optimal active layer thickness of 80–120 nm that many of the high-performing material systems require, while obtaining pinhole-free thin films reproducibly at high printing speed. This problem can be relieved by printing thick films with thickness over 200 nm[26]. Unfortunately, due to limited charge diffusion length, thick-film OPV devices often experience severe bimolecular recombination and space charge effect, leading to reduced FF and PCE[27]. So far only a few studies have achieved high photovoltaic performance on polymer-based OPVs, with active layers above 200 nm[26,28–30]. No report has been found for molecular OPVs.

In this work, a new molecular electron donor material, benzodithiophene terthiophene rhodanine (BTR), with a benzo[1,2-*b*:4,5-*b*′]dithiophene (BDT) core and rhodanine peripheral units was developed and used in OPV devices giving PCEs >9%. While its π-conjugated structure is analogous to a high-performance compound reported previously[7,31], the strategic placement of the side chains provided BTR with strong intermolecular interactions, as evidenced by its liquid crystalline (LC) behaviour. Such interactions translated successfully into excellent hole transport properties; hole mobilities up to 0.1 and $1.6 \times 10^{-3}$ cm$^2$ V$^{-1}$ s$^{-1}$ were recorded by organic field-effect transistor (OFET) and space-charge-limited current (SCLC) methods, respectively. Thus, BTR-based OPVs with thick active layers (300–400 nm) could still afford PCEs of over 8% with high FF of ~70%. Normal cell architecture employed in this study showed a moderate device lifetime. With better cell architectures or proper encapsulation for longer device lifetime, it is believed that BTR is a very attractive candidate for roll-to-roll printed OPV modules.

## Results

**Physical properties of BTR.** The BTR molecule was synthesized in two steps from known precursors in a good yield (Supplementary Fig. 1). The chemical structure of BTR is shown in Fig. 1a. The backbone consisting of the BDT unit, two terthiophenes and two rhodanine groups formed a coplanar structure. In comparison with analogous structures in the literature[7,31], the side chains of BTR were shortened and positioned at the terthiophene building blocks in a regioregular manner to facilitate side-chain interdigitation[32,33]. In combination with the additional hexyl group on the thienyl-BDT unit, the side chains of BTR imparted LC behaviour (vide infra) that was not observed in previous reports.

BTR shows an excellent solubility of 211 mg ml$^{-1}$ in chloroform, as derived from concentration and absorption data (Supplementary Fig. 2). BTR in solution displays an absorption maximum ($\lambda_{max}$) at 523 nm, with an extinction coefficient ($\varepsilon$) of $1.10 \times 10^5$ M$^{-1}$ cm$^{-1}$ (Fig. 1b; Supplementary Table 1). The high $\varepsilon$ is attributed to the planarity of its backbone. The BTR solid film exhibits a red-shift of the $\lambda_{max}$ to 572 nm relative to that in solution. Furthermore, an additional absorption peak at 620 nm appears in the absorption spectrum of a thin film. The red-shift and new absorption peak of the BTR film suggest the presence of strong intermolecular interaction and aggregation in the solid film. The absorption onset of the BTR film is at 681 nm, equivalent to an optical frontier orbital energy gap of 1.82 eV. Determined by cyclic voltammetry (CV) (Supplementary Fig. 3), the highest occupied molecular orbital (HOMO) and lowest unoccupied molecular orbital (LUMO) of BTR are −5.34 and −3.52 eV, respectively. The HOMO–LUMO gap of BTR is 1.82 eV, which is in good agreement with the optical energy gap. Because the open-circuit voltage ($V_{oc}$) is largely determined by the HOMO–LUMO gap of the donor and acceptor, a deep-lying HOMO of BTR can potentially support a high $V_{oc}$. In combination with fullerene acceptor [6,6]-phenyl $C_{71}$ butyric acid methyl ester ($PC_{71}BM$), whose LUMO level is around −4.0 eV, the LUMO energy offset of 0.48 eV between BTR and $PC_{71}BM$ should provide enough driving force for exciton dissociation[34].

The BTR molecule has good thermal stability with a decomposition temperature of 405 °C in nitrogen (5% weight loss in thermogravimetric analysis, Supplementary Fig. 4). BTR exhibits a sharp differential scanning calorimetry (DSC) peak at 175 °C (Fig. 1c), which is assigned to secondary crystalline phase transition by means of a structural analysis (vide infra). Furthermore, a melting temperature of 186 °C into a LC phase and a clearing temperature at 196 °C of small enthalpy into an isotropic melt were observed. Upon cooling, three exothermic peaks at 193, 181 and 133 °C were recorded. The first minor transition was attributed to the LC phase transition, while the two major ones were related to the crystallization process of the two crystalline phases. To observe directly the LC transition and precisely assign the phases, BTR powder was sandwiched in between two glass slides, heated and examined under a polarized optical microscope (POM). The BTR molecule was highly crystalline below a stage temperature of 185 °C (Fig. 1d). Between 185 and 195 °C, the crystalline solid was replaced by a liquid crystal nematic texture (Fig. 1e). The nematic phase suggests that BTR molecules have a rigid rod-like shape, which can maintain a long-range directional order with their long axes parallelly

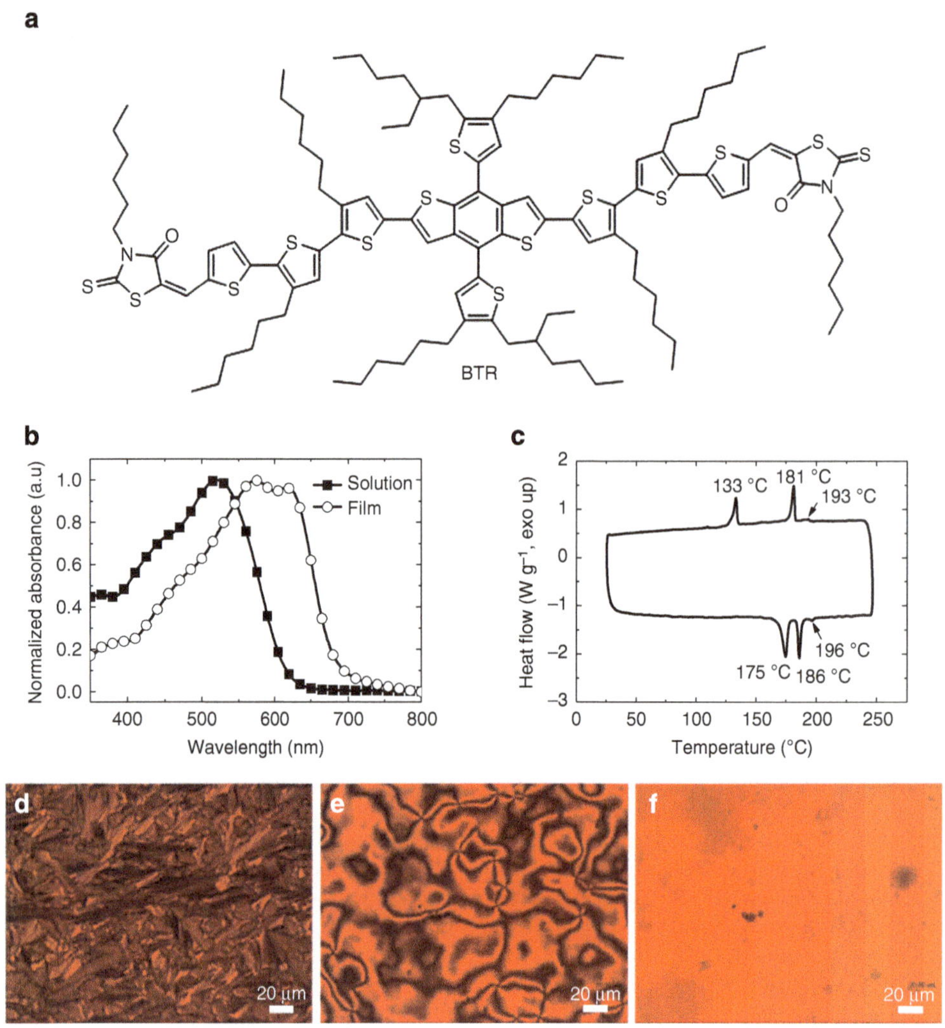


**Figure 1 | BTR chemical structure and physical properties.** (**a**) Chemical structure of BTR. (**b**) Normalized UV-vis absorption spectra of BTR in chloroform (5 mg ml$^{-1}$) and in a spin-cast film. (**c**) DSC thermogram of BTR in nitrogen at a ramp rate of 10 °C min$^{-1}$. The lower trace is from the heating cycle and upper trace from the cooling cycle. (**d**) BTR thin film sandwiched in between two glass slides observed under a polarized optical microscope (POM) at a stage temperature of 185 °C. (**e**) The POM image of the same BTR thin film at the same settings when the stage temperature rises to 195 °C. (**f**) The POM image taken at a stage temperature of 197 °C.

aligned. They can thus have high crystallinity in solid state[17,35–37]. The liquid crystal transformed above 196 °C into an isotropic melt, leaving no prominent feature under the POM (Fig. 1f). Thereby, the small transition enthalpy determined by DSC is in agreement with the low-ordered nematic phase. The nematic LC behaviour is an important feature of the BTR molecule, implying strong intermolecular interaction resulted from side-chain modifications, and potentially high charge carrier mobility due to three-dimensional (3D) charge transport[38].

**Crystal packing of BTR molecules.** To obtain a better understanding of the packing of BTR molecules in the solid, X-ray quality single crystals of BTR were grown from a mixed solution of 2-propanol and dichloromethane by slow evaporation. The single crystal structure was solved using data from the MX2 beamline at the Australian Synchrotron[39] (Fig. 2a; Supplementary Figs 5–8). The X-ray crystal structure of the BTR molecule revealed a coplanar structure of the conjugated backbone, which should facilitate light absorption and also crystal stacking. The crystal packing is dominated by π-stacking between the individual BTR backbones that arrange themselves into π-stacked centrosymmetric dimers with an average inter-plane separation of ca 3.60 Å (Fig. 2a). These individual dimers aggregate together by π-stacking, with an average interplaner separation of 3.62 Å (Supplementary Fig. 6). This type of packing is consistent with the bathrochromic shifting of the absorption from solution to the solid film (*J*-aggregate).

The solid-state structure was also examined using two-dimensional wide-angle X-ray scattering (2D-WAXS) on neat BTR filaments. The samples were prepared by filament extrusion[40], which imparted bulk orientation on the crystalline material. The 2D-WAXS pattern suggests a crystalline character of BTR in the low-temperature phase as evident by the high number of distinct reflections (Fig. 2b). The molecules are organized in a layered structure that is aligned in the direction of the fibre axis. An interlayer distance of 18.3 Å is determined from reflections located in the equatorial small-angle range. On the same plane of the pattern, two π-stacking peaks appear that are related to distances of 3.70 and 3.65 Å of stacked BTR dimers. These values are in the same range as found for the single crystal. Further meridional reflections are originated from intramolecular correlations along the extended conjugated BTR backbone. At 179 °C, the sample maintains a crystalline phase, however, with a slightly smaller degree of order (Supplementary Fig. 9). The interlayer spacing remains identical at 18.3 Å, while only one and a little larger stacking distance was observed at 3.76 Å.

In a thin solid film, BTR organizes in two different molecular arrangements as indicated by the grazing incidence wide-angle X-ray scattering (GIWAXS) pattern in Fig. 2c. Reflections in the meridional plane (along $q_{xy} = 0\,Å^{-1}$) in the small- and middle-range scattering region are related to the formation of a layered structure with an interlayer distance of 18.7 Å. In addition, 3rd-order reflections are visible typical for a long-range order, while their position on the meridional plane of the pattern is characteristic for an edge-on molecular organization. In this arrangement, the backbone plane is aligned perpendicular to the surface. However, the corresponding equatorial π-stacking peak of the edge-on arranged molecules assembled in the layered structure is too weak to be detected. Instead, the π-stacking reflection related to a single distance of 3.70 Å is located also on the meridional plane, which is typical for a face-on arrangement. These results imply two distinct surface organizations. In the first phase, the molecules are π-stacked and face-on arranged, but do not organize in a layered structure. In the second phase, the molecules are edge-on aligned with respect to the substrate, but are disordered within the layer organization. Because charge transport in organic semiconductors is mainly via hopping between adjacent molecules, the co-existence of edge-on and face-on orientations can potentially form a 3D network for hopping, thus beneficial to charge transport[41].

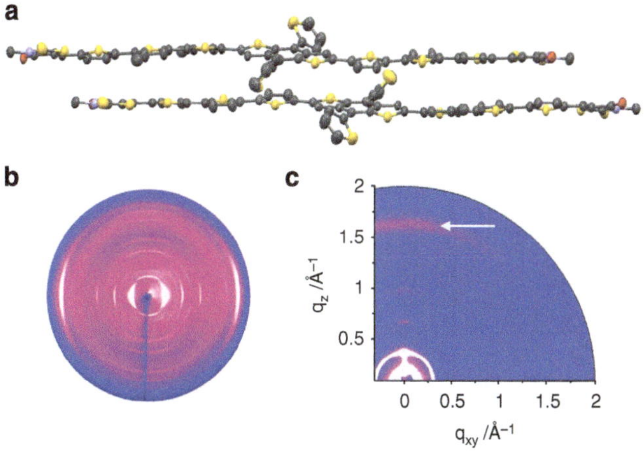


**Figure 2 | Crystal packing resolved by X-ray techniques.** (**a**) Centrosymmetric π-stacked dimers of BTR molecules in its single crystal, the alkyl side chains have been omitted for clarity. (**b**) 2D-WAXS of BTR filament measured at 30 °C. (**c**) GIWAXS of the as-cast BTR thin film on silicon wafer via spin coating (π-stacking reflection is indicated by an arrow).

**OFET mobility**. To study the charge carrier transport, OFETs using different procedures were built. For top-contact devices, BTR was spin-coated from a 4.5-mg ml$^{-1}$ toluene solutionand subsequently annealed at 179 °C. These transistors delivered hole mobilities up to 0.01 cm$^2$ V$^{-1}$ s$^{-1}$ (Supplementary Fig. 10). Bottom-contact OFET devices with the BTR molecules deposited by drop-casting gave mobility values as high as 0.1 cm$^2$ V$^{-1}$ s$^{-1}$ (Supplementary Fig. 11). It should be noted that the OFET devices were not intensively optimized. The primary purpose of the OFET experiments was to show the potential of the BTR material as a semiconductor.

**Photovoltaic performances**. The excellent solubility, strong intermolecular interaction, suitable absorption profile and energy levels, as well as encouraging semiconducting properties prompted us to explore the photovoltaic performance of the BTR

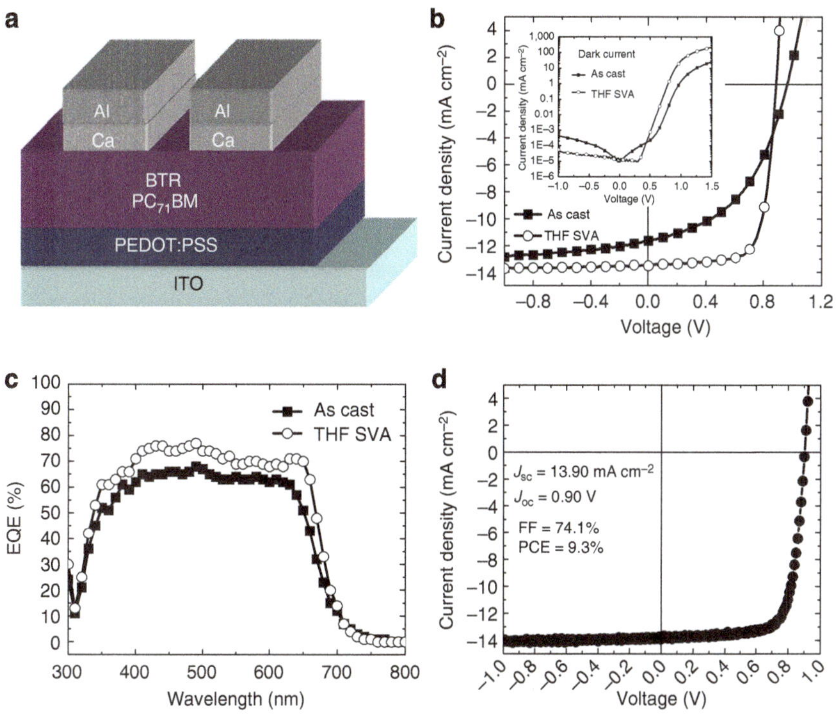


**Figure 3 | Device architecture and photovoltaic performances.** (**a**) Schematic diagram of a normal cell architecture used in this study. (**b**) *J*–*V* characteristics of BTR:$PC_{71}BM$ BHJ solar cells with or without THF solvent vapour annealing tested in air under 98 mW cm$^{-2}$ AM1.5G illumination. Inset: dark current plotted in a semi-log scale of the two solar cells. (**c**) EQE spectra of optimized BTR-based solar cells with or without THF SVA treatment. (**d**) *J*–*V* curve of the most efficient BTR:$PC_{71}BM$ BHJ solar cell after 15 s of THF SVA measured by an independent research institute in nitrogen atmosphere under an illumination of 100 mW cm$^{-2}$.

molecule. The OPV cells adopted a simple normal architecture, with the BTR:$PC_{71}BM$ blend film sandwiched between a poly(3,4-ethylenedioxythiophene):poly(styrenesulfonate) (PEDOT:PSS)-coated indium tin oxide (ITO) transparent anode and a Ca/Al back cathode (Fig. 3a). We further treated the active layer with SVA, which has been shown to be effective in enhancing the performance of molecular OPVs[22–24]. The SVA treatment was carried out by exposing the as-cast active layer to solvent vapours. According to the solvent selection rules previously identified[23], tetrahydrofuran (THF) was chosen for SVA owing to the moderate solubility of BTR in THF ($89\,mg\,ml^{-1}$).

The BTR-based OPVs with an optimal active layer thickness of 250 nm were encapsulated and tested in air. The current density ($J$)–voltage ($V$) curves of the best devices are shown in Fig. 3b, with the photovoltaic parameters summarized in Table 1. Without SVA treatment, the highest performance for the as-cast OPVs showed short-circuit current density ($J_{sc}$) = $11.64\,mA\,cm^{-2}$, $V_{oc} = 0.96\,V$, FF = 47% and PCE = 5.2%. SVA treatment significantly enhanced the photovoltaic performance. OPVs with 15 s of THF SVA exhibited $J_{sc} = 13.52\,mA\,cm^{-2}$, $V_{oc} = 0.89\,V$, FF = 73% and PCE = 8.7%. Device assembly was reproducible with around 60 SVA-treated OPV devices having an average PCE of 8.3 ± 0.2%. Thermal annealing was found to diminish the device performance, due to the overgrowth of the phases (Supplementary Fig. 12).

The causes for the enhanced FF after SVA treatment were investigated by measuring dark currents (inset of Fig. 3b). Compared with an as-cast molecular OPV, the SVA-treated sample displayed notably higher current density under positive bias. In great contrast, the current density was one order of magnitude smaller in reverse bias. To further understand the SVA treatment effect, series resistance ($R_s$) and shunt resistance ($R_{sh}$) were extracted at 1.5 and 0 V of the dark curves (Table 1). Without SVA treatment, the OPV had a $R_s$ of $14.0\,\Omega\,cm^2$ and a $R_{sh}$ of $5.5\,M\Omega\,cm^2$. SVA treatment led to a reduction of $R_s$ by six times and a slight increase of $R_{sh}$. Together, the results suggest the SVA treatment can suppress leakage current and improve the diode behaviour.

The slight improvement in $J_{sc}$ after SVA treatment was monitored by external quantum efficiency (EQE) measurement (Fig. 3c). A high EQE of over 60% was measured in the visible region from 400 to 650 nm for the non-annealed OPV. The $J_{sc}$ calculated by integrating the product of photon flux and EQE at each wavelength was $11.70\,mA\,cm^{-2}$, which was in good agreement with the measured $J_{sc}$ ($11.64\,mA\,cm^{-2}$). The SVA treatment lifted the EQE in the entire absorption range. In particular, the EQE stayed above 70% between 400 and 650 nm, and a shoulder was found at 640 nm. As a result, the calculated $J_{sc}$ increased to $13.53\,mA\,cm^{-2}$. The EQE result clearly indicates SVA treatment plays a positive role in charge generation, transport and/or collection.

Bearing in mind that OPVs with normal cell architecture are not stable in air, we fabricated a batch of 20 devices in Singapore and 8 devices in Australia and tested them under inert atmosphere using the facilities at Solar Energy Research Institute of Singapore and the Commonwealth Scientific and Industrial Research Organisation, respectively. The best BTR-based OPV fabricated in Singapore exhibited a record efficiency of 9.3%, with $J_{sc} = 13.90\,mA\,cm^{-2}$, $V_{oc} = 0.90\,V$ and FF = 74.1% (Fig. 3d; Table 1). The results were highly reproducible. The same PCE of 9.3% with a $J_{sc}$ of $13.40\,mA\,cm^{-2}$, a $V_{oc}$ of 0.90 V and an extremely high FF of 77.0% was achieved in Australia (Table 1). This result demonstrates molecular OPVs can achieve comparable efficiencies attainable by polymer-based OPVs[8–11]. It is worth noting that the FF of 77.0% is among the highest FF value reported in the literature for solution-processed molecular OPVs[12,42]. The average photovoltaic parameters for the 28 devices were $J_{sc} = 13.49 \pm 0.28\,mA\,cm^{-2}$, $V_{oc} = 0.89 \pm 0.01\,V$, FF = 74 ± 1% and PCE = 8.9 ± 0.2% (Table 1).

**OPVs of a thick active layer.** The high FF values suggest that the BTR-based OPVs can accommodate a greater range of active layer thicknesses. This is particularly important in roll-to-roll printing of very thin films, which are difficult to be precisely controlled, and pinholes are often found in thin-film devices. We were motivated to explore the thickness-dependent solar cell performance using the BTR molecule. Active layers with different thicknesses ranging from 80 to 400 nm were fabricated by tuning the solution concentrations and spin rates. Figure 4 and Supplementary Table 2 show that BTR-based OPVs maintain a nearly constant $V_{oc}$ between 0.87 and 0.90 V. The average $J_{sc}$ increases from ~10 to ~$13\,mA\,cm^{-2}$ as the active layer thickness increases from 80 to 250 nm and then it saturates around $13\,mA\,cm^{-2}$ when the thickness further increases to 400 nm. Surprisingly, the FF values for BTR-based OPVs remain high and close to 70% even at

**Table 1 | Photovoltaic parameters of BTR:$PC_{71}BM$ BHJ solar cells fabricated and tested under different conditions**

| $J_{sc}$ ($mA\,cm^{-2}$) | $V_{oc}$ (V) | FF (%) | PCE (%) | $R_s$* ($\Omega\,cm^2$) | $R_{sh}$† ($M\Omega\,cm^2$) | $\mu_h$‡ ($cm^2\,V^{-1}\,s^{-1}$) | $\mu_e$‡ ($cm^2\,V^{-1}\,s^{-1}$) |
|---|---|---|---|---|---|---|---|
| *As-cast*§ | | | | | | | |
| 11.64 | 0.96 | 47 | 5.2 | 14.0 | 5.5 | $2.2 \times 10^{-4}$ | $3.5 \times 10^{-4}$ |
| (11.20 ± 0.51) | (0.96 ± 0.01) | (42 ± 3) | (4.5 ± 0.4) | — | — | | |
| *THF SVA*§ | | | | | | | |
| 13.52 | 0.89 | 73 | 8.7 | 2.4 | 42 | | |
| [12.16] | [0.90] | [76] | [8.3] | 2.2 | 6.4 | | |
| (13.17 ± 0.43) | (0.89 ± 0.01) | (71 ± 2) | (8.3 ± 0.2) | — | — | $1.6 \times 10^{-3}$ | $9.6 \times 10^{-3}$ |
| *THF SVA*‖ | | | | | | | |
| 13.90 | 0.90 | 74 | 9.3 | 2.7 | 15 | | |
| [13.40] | [0.90] | [77] | [9.3] | 1.9 | 20 | | |
| (13.61 ± 0.16) | (0.89 ± 0.01) | (74 ± 1) | (8.9 ± 0.2) | — | — | | |

FF, fill factor; PCE, power conversion efficiency; SCLC, space-charge-limited current; SVA, solvent vapour annealing; THF, tetrahydrofuran.
Data in brackets show the cell with the highest FF. Data in parentheses are average results out of 40 + encapsulated devices tested in air or 28 non-encapsulated devices tested in nitrogen.
*Series resistance of the device.
†Shunt resistance of the device.
‡Charge mobility values obtained from SCLC experiments.
§Cells were tested in air with encapsulation, illumination intensity was $98\,mW\,cm^{-2}$.
‖Cells without encapsulation were tested in a glove box filled with dry nitrogen; illumination intensity was $100\,mW\,cm^{-2}$.

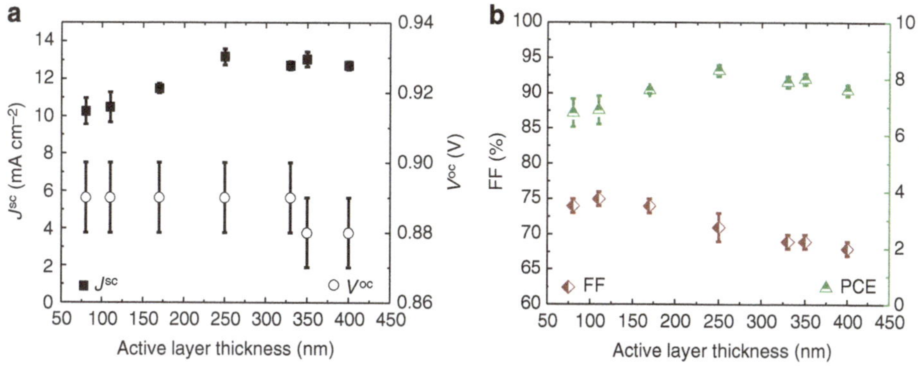


**Figure 4 | Active layer thickness-dependent variation of photovoltaic performances.** (**a**) Plots of $J_{sc}$ or $V_{oc}$ vs active layer thickness ranging from 80 to 400 nm. (**b**) Plots of FF or PCE against active layer thickness. The results are an average value of >8 devices. The error bars represent the standard deviation from >8 devices.

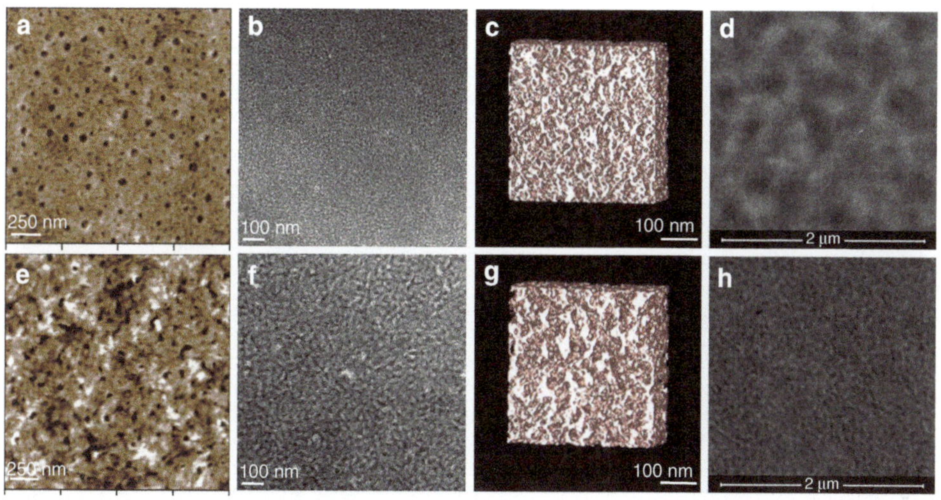


**Figure 5 | Solvent vapor annealing induced morphological changes.** (**a**) AFM image shows the topography of an as-cast BTR:$PC_{71}BM$ (1:1 weight ratio) blend film. (**b**) TEM bright-field image of the as-cast film taken at a defocusing range of 3 μm. (**c**) Computer model generated from the TEM tomogram of the as-cast film. (**d**) Low-energy HAADF STEM image of the as-cast film at focus using a beam energy of 15 keV. (**e**) AFM image of the BTR:$PC_{71}BM$ blend film after THF SVA for 15 s. (**f**) TEM bright-field image of the SVA-treated film at a defocusing range of 3 μm. (**g**) Computer model of the THF SVA film. (**h**) HAADF STEM image of the blend film after SVA treatment.

thicknesses up to 400 nm. This is not commonly observed in thick-film OPVs, whether it is a molecular OPV or a polymer-based solar cell[26–29,43]. As a result, the overall PCEs formed a flat bell curve with a minimum average value of 6.8% and maximum average value of 8.3% at an active layer thickness of 250 nm. The large tolerance for the active layer thickness makes the BTR molecule a strong candidate for printed OPVs.

**Solvent vapour annealing**. To understand the effect of SVA treatment on the photovoltaic performance of BTR-based OPVs, we carried out studies on active layer morphology and the optoelectronic properties. The surface topography of the active layer was recorded by atomic force microscopy (AFM) operated in the tapping mode. Before the SVA treatment, Fig. 5a depicts a rather smooth surface, with root-mean-square roughness ($R_{rms}$) of 0.61 nm. Fine crystal domains co-exist with random pinholes, which are believed to be related with the escaping of processing solvent. After a short THF SVA treatment of 15 s, the active layer exhibits a coarser surface (Fig. 5e). The $R_{rms}$ value almost doubles to 1.04 nm. Transmission electron microscopy (TEM) is able to provide morphological information inside the active layer. The bright-field TEM images (Fig. 5b,f) suggest THF SVA treatment leads to larger and more well-defined domains. Because of the sharp contrast in the TEM images, we were able to obtain TEM tomograms and computer models to view the morphological change in 3D (Fig. 5c,g; Supplementary Movies). Both the TEM tomograms and their computer models show that fine-sized domains in the as-cast active layer (Supplementary Movies 1 and 2) evolve into larger domains that are inter-connected to form networks throughout the entire active layer after THF SVA for 15 s (Supplementary Movies 3 and 4). Such networks resemble '3D charge highways' that are beneficial to fast charge transport. The feature size on TEM images is verified by low-energy high-angle angular dark-field scanning TEM (HAADF STEM) images (Fig. 5d,h).

The SVA treatment can be monitored by colour change of the active layer. The inset of Supplementary Fig. 13 is a digital image of the active layer before and after the THF SVA treatment. The colour of the film changed from maroon to purple upon annealing by THF vapour. Such a colour change was reflected by the change in absorption profile (Supplementary Fig. 13). There was a slight red-shift of the absorption maximum from 555 to 565 nm. Besides, the shoulder at 620 nm became more prominent, suggesting good alignment of the rod-like molecules. The absorption enhancement at 620 nm directly translated to increased photocurrent, as suggested by the EQE plot (Fig. 3c).

GIWAXS measurements were performed to understand the organization of BTR in the active layer before and after SVA. In comparison with the BTR neat film, the edge-on layered organization remains unchanged in the as-cast BTR:$PC_{71}BM$ blend film, while the π-stacking distance slightly increases to 3.80 Å and becomes randomly distributed towards the surface, as confirmed by the isotropic intensity of the corresponding peak (Supplementary Fig. 14a). The amorphous halo at $q$-range of ca. 1.25 $Å^{-1}$ is attributed to $PC_{71}BM$ domains. SVA improves the crystallinity and surface ordering of BTR. The interlayer distance is reduced to 17.75 Å, while the π-stacking distance decreases to 3.60 Å. The random orientation of π-stacking evolves into the co-existence of both edge-on and face-on arrangements after SVA, evidenced from the π-stacking reflections at *ca.* 1.7 $Å^{-1}$ in both $q_{xy}$ and $q_z$ directions (Supplementary Fig. 14b). Such a molecular arrangement is beneficial to 3D charge transport.

**SCLC mobilities.** The hole mobility was measured using the SCLC method with a cell architecture: ITO/PEDOT:PSS/BTR:$PC_{71}BM$/Au. Without SVA treatment, the blend film exhibited a relatively high hole mobility of $2.2 \times 10^{-4} cm^2 V^{-1} s^{-1}$ (Supplementary Fig. 15; Table 1). The rigid and planar backbone facilitates easy stacking and strong intermolecular interaction due to side-chain modifications. The SVA treatment substantially enhanced the mobility by one order of magnitude to $1.6 \times 10^{-3} cm^2 V^{-1} s^{-1}$, which is comparable to or greater than those reported for high-performing donor:acceptor blend systems[7,31,41,44,45]. The electron mobility derived from the SCLC method was also improved by one order of magnitude after SVA treatment (Supplementary Fig. 16; Table 1). Such an enhancement can be attributed to larger and more structured domains, as well as better molecular arrangement. The extremely high mobility would partially account for the high FF observed in BTR-based OPVs. However, we do not exclude other possible factors including vertical phase separation or removal of recombination centres and so on[33].

**Solar cell stability.** For practical application, solar cell-stability experiments were carried out in both air and nitrogen environments. Due to the use of active metal-like calcium for the top electrode, the unencapsulated OPVs of the thick active layer (~400 nm) degraded to almost zero efficiency within three days of storage in air (Supplementary Fig. 17; Supplementary Table 3). However, a simple encapsulation with ultraviolet (UV)-curable epoxy and thin glass slides could greatly improve the device stability in air. The OPVs degraded three times slower than that without encapsulation (Supplementary Fig. 17; Supplementary Table 4). To minimize the degradation factor due to the oxidation of electrode and further explore the stability of the active layer, one BTR-based solar cell device was stored in a glove box filled with dry nitrogen and was monitored over a time span of 30 days. The cell retained 86% of initial PCE after 7 days, and it exhibited >50% of original PCE after 30 days of storage (Supplementary Fig. 18a,b; Supplementary Table 5). Further enhancement of device stability could be achieved by improved device architecture. Average of 10 OPV cells of thick active layer of 400 nm and an additional 30-nm-thick silver top electrode/protection layer with or without encapsulation retained 92 and 86% of initial average PCE after 30 days of storage in a glove box (Supplementary Fig. 18c; Supplementary Tables 6 and 7). We believe even better stability can be obtained if the cells were properly encapsulated or inverted cell architecture was employed.

## Discussion

In summary, we present a new molecular donor, BTR, which possesses a rigid and flat backbone and a large number of flexible side chains that could work synergistically to provide excellent processability, nematic liquid crystal behaviour and optoelectronic properties. The neat BTR film exhibited hole mobilities up to 0.1 $cm^2 V^{-1} s^{-1}$ in OFET devices. The solution-processed single-junction BHJ solar cells based on BTR and $PC_{71}BM$ demonstrated a reproducible record efficiency of 9.3%. The blend film also supported a high FF of 77% and a high SCLC hole mobility of $1.6 \times 10^{-3} cm^2 V^{-1} s^{-1}$ after SVA with THF. Thick-film molecular solar cells with an active layer thickness up to 400 nm were demonstrated, showing a low thickness dependence of photovoltaic performance. Together, the results suggest BTR is an ideal candidate for printed OPVs. Moreover, enhancing the intermolecular interaction through side-chain modification is a viable way further to enhance the efficiency of molecular solar cells in excess of 10%.

## Methods

**Materials.** Unless noted, all materials were reagent grade and used as received without further purification. Anhydrous solvents were prepared by drying HPLC-grade solvents using freshly activated molecular sieves.

**Synthesis of aldehyde compound 3.** Synthetic route could be found in Supplementary Fig. 1. Precursor **1** and **2** were prepared using literature methods[46,47]. In protection of $N_2$, to a dry 250-ml flask were added compound **1** (1.07 g, 1.0 mmol), compound **2** (1.05 g, 2.0 mmol), $Pd_2(dba)_3$ (46 mg, 0.04 mmol), $P(o\text{-tolyl})_3$ (97 mg, 0.32 mmol) and 50 ml of toluene. The reaction mixture was refluxed at 125 °C for 12 h. The reaction mixture was filtered and the crude product was purified by silica gel column chromatography (petroleum spirit 40–60 °C: dichloromethane = 1:1.5, $R_f$ 0.5) to give the product as yellow solid (830 mg, 53%). mp 146–147 °C; infrared (IR) (neat) ν 2,955, 2,923, 2,855, 1,656, 1,432, 1,225, 1,058, 820, 786 and 663 $cm^{-1}$; $^1H$ NMR (δ, $CDCl_3$) 9.89 (s, 2 H), 7.72 (d, $J$ = 4.0 Hz, 2 H), 7.68 (s, 2 H), 7.23 (d, $J$ = 4.0 Hz, 2 H), 7.12 (s, 2 H), 7.01 (s, 2 H), 2.82 (m, 12 H), 2.65 (m, 4 H), 1.60–1.75 (m, 14 H), 1.26–1.45 (m, 52 H), 0.88 (m, 30 H); $^{13}C$ NMR (δ, $CDCl_3$) 182.6, 146.1, 142.6, 142.2, 141.2, 139.6, 139.0, 138.6, 137.3, 137.2, 136.9, 136.0, 135.8, 134.8, 130.3, 129.8, 129.5, 129.1, 128.3, 125.9, 123.5, 119.5, 41.7, 32.7, 32.3, 31.8, 31.6, 30.8, 30.4, 30.2, 29.8, 29.6, 29.3, 29.2, 29.1, 28.9, 28.4, 26.0, 23.1, 22.7, 22.6, 14.2 and 13.1; mass spectrometry (MS) (matrix-assisted laser desorption/ionization (MALDI)) $m/z$ 1,630 [M].

**Synthesis of BTR molecule.** To the solution of compound **3** (327 mg, 0.2 mmol) in dry chloroform (20 ml) was added rhodanine **4** (434 mg, 2.0 mmol) followed by one drop of 1,8-Diazabicyclo[5.4.0]undec-7-ene (DBU). The reaction mixture was then placed under continuous stirring at room temperature for 3 h. Reaction solvent was removed and the crude product was purified by silica gel column chromatography (petroleum spirit 40–60 °C: dichloromethane = 1:1, $R_f$ 0.8). The desired fractions were collected and the product was obtained as a purple solid (310 mg, 76%) after washing with acetone. mp 178–180 °C; IR (neat) ν 2,924, 2,855, 1,699, 1,575, 1,423, 1,327, 1,238, 1,180 and 820 $cm^{-1}$; $^1H$ NMR (δ, $CDCl_3$) 7.83 (s, 2 H), 7.65 (s, 2 H), 7.35 (d, $J$ = 4.0 Hz, 2 H), 7.23 (s, 2 H), 7.20 (d, $J$ = 4.0 Hz, 2 H), 7.10 (s, 2 H), 7.00 (s, 2 H), 4.09 (t, $J$ = 7.6 Hz, 4 H), 2.81 (m, 12 H), 2.78 (m, 4 H), 1.60–1.75 (m, 18 H), 1.26–1.45 (m, 64 H), 0.88 (m, 36 H); $^{13}C$ NMR (δ, $CDCl_3$) 192.2, 167.5, 144.3, 141.9, 141.0, 139.5, 139.0, 138.6, 137.3, 137.2, 137.1, 135.7, 135.6, 134.8, 134.6, 130.4, 129.8, 129.7, 129.0, 128.3, 126.5, 125.0, 123.5, 120.2, 119.5, 44.9, 41.7, 32.7, 32.4, 31.8, 31.7, 31.6, 31,3, 30.8, 30.4, 30.3, 29.8, 29.7, 29.3, 29.2, 29.1, 28.9, 28.4, 27.0, 26.4, 26.0, 23.1, 22.7, 22.6, 22.5, 14.2, 13.1; MS (MALDI) $m/z$ 2,028.7 $[M]^+$. Elemental analyses calcd (%) for C114H152N2O2S14 (BTR): C, 67.41; H, 7.54; N, 1.38; O, 1.58; S, 22.10; found: C, 67.64; H, 7.65; N, 1.24; O, 1.33; S, 21.87.

**Material characterizations.** IR spectra were obtained on a Perkin-Elmer Spectrum One Fourier transform infrared spectrometer and UV–visible spectra were recorded using a Cary 50 UV–visible spectrometer. Photoluminescence was measured with a Varian Cary Eclipse fluorimeter. Melting points were determined on a Büchi 510 melting point apparatus. $^1H$ NMR and $^{13}C$ NMR spectra were carried out on a 400-MHz spectrometer. All NMR data were referenced to the chloroform signal and peak multiplicity was reported as follows: s = singlet, d = doublet, t = triplet, q = quartet, p = pentet, dd = doublets of doublets, m = multiplet and br = broad). MALDI-time-of-flight MS was performed on a Bruker microflex instrument, using chloroform as solvent and dithranol as the assisted matrix. Elemental analyses were obtained commercially through Chemical & Analytical Services Pty Ltd. (Australia) an Exeter Analytical CE-440 elemental analyzer. Thermal gravimetric analysis experiments were carried out with a Mettler Toledo TGA/SDTA851e, and DSC experiments were performed on a Perkin-Elmer Sapphire DSC. CV experiments were performed at a sweep rate of 100 $mV s^{-1}$. CVs were carried out in a three-electrode cell consisting of a glassy carbon working electrode, a platinum wire auxiliary electrode and a $Ag/Ag^+$ pseudo-reference electrode. The supporting electrolyte was 0.10 M tetrabutylammonium hexafluorophosphate ($Bu_4NPF_6$) in $CH_3CN$. The solutions were deoxygenated by sparging with argon prior to each scan and blanketed with argon during the scans.

The glassy carbon working electrode was prepared by polishing with 5 mm alumina and washed and dried before the polymer was drop-casted on the electrode from chlorobenzene solution to form a film. Ferrocene/ferrocenium redox couple was used as the internal standard. The HOMO energy level was calculated from the onset of the oxidation potential of the polymer using the following: $E_{HOMO} = -(4.8 + E_{ox}$ onset) eV.

**Crystal structure determination.** Single-crystal data were collected on the MX2 beamline at The Australian Synchrotron at 100 K using a 360° rotation in steps of 1° with 1-s exposure time per step. The beamline was set to the Mo-K-α edge (17.444 keV or 0.71073 Å). Crystal data for BTR: C112 H150 N2 O2 S14, $M = 2{,}005.17$, $T = 100.0(2)$ K, $\lambda = 0.71073$ Å, Triclinic, space group *P-1*, $a = 314.257(3)$, $b = 20.519(4)$, $c = 21.795(4)$ Å, $\alpha = 114.76(3)°$, $\beta = 98.08(3)°$, $\gamma = 02.00(3)°$, $V = 5{,}474(2)$ Å$^3$, $Z = 2$, $D_c = 1.217$ Mg M$^{-3}$, $\mu = 0.326$ mm$^{-1}$, $F(000) = 2{,}152$, crystal size 0.10 × 0.05 × 0.05 mm. $\theta_{max} = 27.47°$, 88,775 reflections measured, 22,889 independent reflections ($R_{int} = 0.055$) the final $R = 0.11$ ($I > 2\sigma(I)$, 14,445 data) and $wR(F^2) = 0.3696$ (all data) GOOF = 1.371.

**Thin-film characterizations.** To study the liquid crystal property of BTR, we used a Nikon Eclipse LV100 POM equipped with Linkam LTS350 heating and cooling stage connected to a Linkam TMS 94 temperature programmer. GIWAXS samples were prepared by spin-coating chloroform solution of BTR onto PEDOT:PSS-coated silicon substrates. The measurements were performed by means of a solid anode X-ray tube (Siemens Kristalloflex X-ray source, copper anode X-ray tube operated at 35 kV and 40 mA), Osmic confocal MaxFlux optics, X-ray beam with pinhole collimation and a MAR345 image plate detector. The beam size was 0.5 × 0.5 mm and samples were irradiated just below the critical angle for total reflection with respect to the incoming X-ray beam (~0.18°). The scattering intensity was detected on a 2D image plate (MAR345) with a pixel size of 100 μm (3,450 × 3,450 pixels). Data analysis was performed using the Datasqueeze 3.0.0 software. Atomic force microscopy images were acquired using an Asylum Research Cypher scanning probe microscope operated in tapping mode. Samples for electron microscopy were prepared by dissolving the PEDOT:PSS layer using water and transferring the floating active layer to TEM grids. TEM bright-field images were obtained by a FEI Tecnai TF30 TEM equipped a with beam-blank function. For TEM tomography, tilt series were acquired using the Xplore 3D software (FEI Company). Tomograms were recorded between −65 and +65 degrees at 2° intervals and aligned with IMOD. A 3D model rendering employed 3d mod software. Each model was generated from the aligned tomogram. The BTR donor phase was rendered in pink colour, while the $PC_{71}BM$ acceptor phase was represented by empty space in model movies. The scale bar was 100 nm in both the model movies and the tomogram movies. HAADF STEM was performed at a primary electron energy of 15 keV with a FEI Quanta 3D Microscope equipped with a HAADF STEM detector. To obtain hole and electron mobilities using SCLCs, hole-only devices (ITO/PEDOT:PSS/BTR:$PC_{71}BM$/Au) and electron-only devices (ITO/Al/ BTR:$PC_{71}BM$/Al) were constructed. Their dark currents were recorded by a computer-programmed Keithley 2400 source meter and then fitted by the Mott–Gurney equation. Film thickness was determined by a Veeco Dektak 150+ Surface Profiler.

**Solvent vapour annealing.** SVA was conducted in a glove box filled with dried nitrogen. THF (1 ml) were injected into a 30-mm glass Petri dish. The Petri dish was closed for 1 min to let the vapour saturate the treatment chamber. Then as-cast films were attached on the backside of the Petri dish lid, which was quickly swapped with the lid covering the solvent-containing Petri dish. The film was about 1 cm above the solvent level during the SVA. After certain duration, the film was removed from the treatment chamber. The optimal duration in this study was 15 s.

**OFET fabrication and measurement.** OFETs were fabricated employing two types of device configurations: bottom-gate, bottom-contact and bottom-gate, top-contact on the 300-nm-thick silicon-dioxide dielectric covering the highly doped silicon that acted as the gate electrode. The source–drain channel length ($L$) to channel width ($W$) ratio was 1:70 for bottom-contact and 1:33 for top-contact configurations. The substrates were first cleaned by sonication in acetone and isopropanol for 20 min each, then dried under nitrogen flow and dipped into a freshly prepared piranha solution (7:3 v/v $H_2SO_4$/30% $H_2O_2$) at 90 °C for 60 min. To remove residues of the piranha solution, substrates were rinsed with copious deionized water and dried under nitrogen flow. The BTR thin film was deposited via spin-coating (for top-contact OFETs) or drop-casting (bottom-contact OFETs) using 4.5 mg ml$^{-1}$ toluene with subsequent annealing at 60 °C for 60 min to remove residual solvent. In addition, the BTR films were measured after thermal annealing at 179 and 190 °C. All the electrical measurements are performed in a glove box under nitrogen atmosphere by means of a Keithley 4200 SCS.

**Solar cell fabrication and characterization.** Patterned ITO glasses were washed sequentially by detergent, deionized water, acetone and 2-propanol in an ultra-sonication bath and UV/ozone-treated. PEDOT:PSS (Clevios P VP AI 4083) was spin-coated at 6.000 r.p.m. and then baked at 140 °C for 10 min in air. The substrates were transferred to a glove box filled with dried nitrogen, where a chloroform solution of the donor–acceptor blend was spin coated at various spin rates on top of the substrate. The best PCE was achieved by spin coating 20 mg BTR and 20 mg $PC_{71}BM$ in 1 ml of chloroform at 1,000 r.p.m. The resulting film thickness was around 250 nm. Thicker or thinner films were produced by varying the solution concentration and spin rate from 1.6 to 2.0% and from 600 to 2,000 r.p.m., respectively. To obtain high photovoltaic performance, the films were solvent vapour annealed by THF. Then they were transferred to a thermal evaporator where 40 nm Ca and 100 nm aluminium were deposited through a shadow mask (active area was 0.11 cm$^2$) at a base pressure of $4 \times 10^{-7}$ mbar. The solar cells were encapsulated under nitrogen by UV-curable epoxy (Epotek OG112-6 by Epoxy Technology Inc.) and cover glass. They were tested in air with a computer-programmed Keithley 2400 source meter under a Newport Oriel class A solar similar, which simulated the AM1.5 solar irradiance with energy density of 98 mW cm$^{-2}$. For accurate measurement and to avoid oxidation during testing, a batch of optimized solar cells were brought to a third-party research institute, Solar Energy Research Institute of Singapore, where the cells were tested under nitrogen and 1 SUN (100 mW cm$^{-2}$) condition provided by a 1 kW solar simulator (Sun 2000 Solar Simulator by ABET Technologies) that was calibrated by a silicon reference cell (Fraunhofer ISE) and a Schott visible-colour glass-filtered (KG5 colour filtered) Si diode (Hamamatsu S1133). OPVs made in Australia followed the same fabrication procedures, and were tested at Clayton Laboratories of Commonwealth Scientific and Industrial Research Organisation, where a 1-kW Oriel solar simulator with an AM 1.5G filter was used as the light source.

## References


1. Burschka, J. *et al.* Sequential deposition as a route to high-performance perovskite-sensitized solar cells. *Nature* **499**, 316–319 (2013).
2. Liu, M., Johnston, M. B. & Snaith, H. J. Efficient planar heterojunction perovskite solar cells by vapour deposition. *Nature* **501**, 395–398 (2013).
3. Zhou, H. *et al.* Interface engineering of highly efficient perovskite solar cells. *Science* **345**, 542–546 (2014).
4. Yu, G., Gao, J., Hummelen, J. C., Wudl, F. & Heeger, A. J. Polymer photovoltaic cells: enhanced efficiencies via a network of internal donor-acceptor heterojunctions. *Science* **270**, 1789–1791 (1995).
5. Chen, H. Y. *et al.* Polymer solar cells with enhanced open-circuit voltage and efficiency. *Nat. Photon.* **3**, 649–653 (2009).
6. Small, C. E. *et al.* High-efficiency inverted dithienogermole-thienopyrrolodione-based polymer solar cells. *Nat. Photon.* **6**, 115–120 (2012).
7. Liu, Y. *et al.* Solution-processed small-molecule solar cells: breaking the 10% power conversion efficiency. *Sci. Rep.* **3**, 3356 (2013).
8. He, Z. *et al.* Enhanced power-conversion efficiency in polymer solar cells using an inverted device structure. *Nat. Photon.* **6**, 591–595 (2012).
9. Ye, L., Zhang, S., Zhao, W., Yao, H. & Hou, J. Highly efficient 2D-conjugated benzodithiophene-based photovoltaic polymer with linear alkylthio side chain. *Chem. Mater.* **26**, 3603–3605 (2014).
10. Liao, S.-H., Jhuo, H.-J., Cheng, Y.-S. & Chen, S.-A. Fullerene derivative-doped zinc oxide nanofilm as the cathode of inverted polymer solar cells with low-bandgap polymer (PTB7-Th) for high performance. *Adv. Mater.* **25**, 4766–4771 (2013).
11. Guo, X. *et al.* Enhanced photovoltaic performance by modulating surface composition in bulk heterojunction polymer solar cells based on PBDTTT-C-T/$PC_{71}BM$. *Adv. Mater.* **26**, 4043–4049 (2014).
12. Gupta, V. *et al.* Barium: an efficient cathode layer for bulk-heterojunction solar cells. *Sci. Rep.* **3**, 1965 (2013).
13. Li, C.-Z. *et al.* Suppressed charge recombination in inverted organic photovoltaics via enhanced charge extraction by using a conductive fullerene electron transport layer. *Adv. Mater.* **26**, 6262–6267 (2014).
14. Walker, B., Kim, C. & Nguyen, T.-Q. Small molecule solution-processed bulk heterojunction solar cells. *Chem. Mater.* **23**, 470–482 (2010).
15. Mishra, A. & Bäuerle, P. Small molecule organic semiconductors on the move: promises for future solar energy technology. *Angew. Chem. Int. Ed.* **51**, 2020–2067 (2012).
16. Lin, Y., Li, Y. & Zhan, X. Small molecule semiconductors for high-efficiency organic photovoltaics. *Chem. Soc. Rev.* **41**, 4245–4272 (2012).
17. Shin, W., Yasuda, T., Watanabe, G., Yang, Y. S. & Adachi, C. Self-organizing mesomorphic diketopyrrolopyrrole derivatives for efficient solution-processed organic solar cells. *Chem. Mater.* **25**, 2549–2556 (2013).
18. Graham, K. R. *et al.* Improved performance of molecular bulk-heterojunction photovoltaic cells through predictable selection of solvent additives. *Adv. Funct. Mater.* **22**, 4801–4813 (2012).
19. Wang, H. *et al.* The role of additive in diketopyrrolopyrrole-based small molecular bulk heterojunction solar cells. *Adv. Mater.* **25**, 6519–6525 (2013).
20. Graham, K. R. *et al.* Tailor-made additives for morphology control in molecular bulk-heterojunction photovoltaics. *ACS Appl. Mater. Interfaces* **5**, 63–71 (2013).
21. Zhou, J. *et al.* Small molecules based on benzo[1,2-b:4,5-b′]dithiophene unit for high-performance solution-processed organic solar cells. *J. Am. Chem. Soc.* **134**, 16345–16351 (2012).

22. Wei, G., Wang, S., Sun, K., Thompson, M. E. & Forrest, S. R. Solvent-annealed crystalline squaraine: $PC_{70}BM$ (1:6) solar cells. *Adv. Energy Mater.* **1,** 184–187 (2011).
23. Sun, K. *et al.* The role of solvent vapor annealing in highly efficient air-processed small molecule solar cells. *J. Mater. Chem. A* **2,** 9048–9054 (2014).
24. Wessendorf, C. D. *et al.* Efficiency improvement of solution-processed dithienopyrrole-based A-D-A oligothiophene bulk-heterojunction solar cells by solvent vapor annealing. *Adv. Energy Mater.* **4,** 1400266 (2014).
25. Krebs, F. C. Fabrication and processing of polymer solar cells: a review of printing and coating techniques. *Sol. Energy Mater. Sol. Cells* **93,** 394–412 (2009).
26. Li, W. *et al.* Efficient small bandgap polymer solar cells with high fill factors for 300 nm thick films. *Adv. Mater.* **25,** 3182–3186 (2013).
27. Li, G., Shrotriya, V., Yao, Y. & Yang, Y. Investigation of annealing effects and film thickness dependence of polymer solar cells based on poly(3-hexylthiophene). *J. Appl. Phys.* **98,** 043704 (2005).
28. Murphy, L., Hong, W., Aziz, H. & Li, Y. Organic photovoltaics with thick active layers (~800 nm) using a high mobility polymer donor. *Sol. Energy Mater. Sol. Cells* **114,** 71–81 (2013).
29. Zeng, L., Tang, C. W. & Chen, S. H. Effects of active layer thickness and thermal annealing on polythiophene: fullerene bulk heterojunction photovoltaic devices. *Appl. Phys. Lett.* **97,** 053305 (2010).
30. Wang, N., Chen, Z., Wei, W. & Jiang, Z. Fluorinated benzothiadiazole-based conjugated polymers for high-performance polymer solar cells without any processing additives or post-treatments. *J. Am. Chem. Soc.* **135,** 17060–17068 (2013).
31. Zhou, J. *et al.* Solution-processed and high-performance organic solar cells using small molecules with a benzodithiophene unit. *J. Am. Chem. Soc.* **135,** 8484–8487 (2013).
32. Kline, R. J. *et al.* Critical role of side-chain attachment density on the order and device performance of polythiophenes. *Macromolecules* **40,** 7960–7965 (2007).
33. Kim, M. S., Kim, B. G. & Kim, J. Effective variables to control the fill factor of organic photovoltaic cells. *ACS Appl. Mater. Interfaces* **1,** 1264–1269 (2009).
34. Bredas, J. L., Norton, J. E., Cornil, J. & Coropceanu, V. Molecular understanding of organic solar cells: the challenges. *Acc. Chem. Res.* **42,** 1691–1699 (2009).
35. Schmidt-Mende, L. *et al.* Self-organized discotic liquid crystals for high-efficiency organic photovoltaics. *Science* **293,** 1119–1122 (2001).
36. Goel, M. & Jayakannan, M. Supramolecular liquid crystalline π-conjugates: the role of aromatic π-stacking and van der Waals forces on the molecular self-assembly of oligophenylenevinylenes. *J. Phys. Chem. B* **114,** 12508–12519 (2010).
37. Kim, B.-G. *et al.* A molecular design principle of lyotropic liquid-crystalline conjugated polymers with directed alignment capability for plastic electronics. *Nat. Mater.* **12,** 659–664 (2013).
38. Hanna, J.-i., Ohno, A. & Iino, H. Charge carrier transport in liquid crystals. *Thin Solid Films* **554,** 58–63 (2014).
39. McPhillips, T. M. *et al.* Blu-Ice and the distributed control system: software for data acquisition and instrument control at macromolecular crystallography beamlines. *J. Synchrotron Radiat.* **9,** 401–406 (2002).
40. Guo, X., Puniredd, S. R., Baumgarten, M., Pisula, W. & Müllen, K. Benzotrithiophene-based donor-acceptor copolymers with distinct supramolecular organizations. *J. Am. Chem. Soc.* **134,** 8404–8407 (2012).
41. Guo, X. *et al.* Polymer solar cells with enhanced fill factors. *Nat. Photon.* **7,** 825–833 (2013).
42. Lin, Y. *et al.* Small-molecule solar cells with fill factors up to 0.75 via a layer-by-layer solution process. *Adv. Energy Mater.* **4,** 1300626 (2014).
43. Murphy, L., Hong, W., Aziz, H. & Li, Y. Influences of using a high mobility donor polymer on solar cell performance. *Org. Electron.* **14,** 3484–3492 (2013).
44. Liang, Y. *et al.* For the bright future: bulk heterojunction polymer solar cells with power conversion efficiency of 7.4%. *Adv. Mater.* **22,** E135–E138 (2010).
45. Kyaw, A. K. K. *et al.* Intensity dependence of current–voltage characteristics and recombination in high-efficiency solution-processed small-molecule solar cells. *ACS Nano* **7,** 4569–4577 (2013).
46. Qin, T. *et al.* Tailored donor–acceptor polymers with an A–D1–A–D2 structure: controlling intermolecular interactions to enable enhanced polymer photovoltaic devices. *J. Am. Chem. Soc.* **136,** 6049–6055 (2014).
47. Kumar, R. J., MacDonald, J. M., Singh, T. B., Waddington, L. J. & Holmes, A. B. Hierarchical self-assembly of semiconductor functionalized peptide α-helices and optoelectronic properties. *J. Am. Chem. Soc.* **133,** 8564–8573 (2011).

## Acknowledgements

This research was supported by the Victorian Organic Solar Cell Consortium with funding from the Victorian Government (DSDBI) and the Australian Renewable Energy Agency (ARENA) through grants 2-A018, 1-GER001 and the ARENA funded Australian Centre for Advanced Photovoltaics. The research was also supported by the Ministry of Education, Singapore (R-284-000-113-112). W.W.H.W. is supported by an ARC Future Fellowship (FT130100500). We acknowledge the Monash Centre for Electron Microscopy (MCEM) at Monash University, the MX2 beamline at the Australian Synchrotron, CSIRO, Department of Materials Science & Engineering (DMSE) at National University of Singapore (NUS) and Solar Energy Research Institute of Singapore (SERIS) for sharing their facilities with us. We acknowledge Dr Doojin Vak at CSIRO, Dr Xi-Ya Fang at MCEM, Henche Kuan, Yeow Koon Liew, Drs Lei Chen and Pengcheng Li at DMSE NUS and Fang Jeng Lim at SERIS for experiment assistance and in-depth discussions.

## Author contributions

K.S. fabricated the solar cell device and characterized the film and device properties using POM, XRD, AFM, TEM, HAADF STEM and SCLC. Z.X. designed and performed materials synthesis and characterization including NMR, MS, CV and UV–vis absorption spectra for solution and thin film. S.L. scaled up the synthesis, grew single crystals and measured DSC, TGA, PL and solubility. W.Z. and W.P. fabricated OFET, measured and analysed OFET, 2D-XRD and GIWAXS data. E.H. constructed TEM tomograms and computer models. J.M.W. solved the crystal structure using XRD acquired by R.M.W. at Australian Synchrotron. J.S. reproduced the OPV results. K.S., Z.X. and S.L. prepared the manuscript. All authors discussed the results and commented on the manuscript. J.O., W.W.H.W., A.B.H. and D.J.J. supervised the project and revised the manuscript.

## Additional information

**Competing financial interests**: The authors declare no competing financial interests.

# 4

# One-shot K-region-selective annulative $\pi$-extension for nanographene synthesis and functionalization

Kyohei Ozaki[1], Katsuaki Kawasumi[1], Mari Shibata[1], Hideto Ito[1] & Kenichiro Itami[1,2]

The optoelectronic nature of two-dimensional sheets of $sp^2$-hydridized carbons (for example, graphenes and nanographenes) can be dramatically altered and tuned by altering the degree of $\pi$-extension, shape, width and edge topology. Among various approaches to synthesize nanographenes with atom-by-atom precision, one-shot annulative $\pi$-extension (APEX) reactions of polycyclic aromatic hydrocarbons hold significant potential not only to achieve a 'growth from template' synthesis of nanographenes, but also to fine-tune the properties of nanographenes. Here we describe one-shot APEX reactions that occur at the K-region (convex armchair edge) of polycyclic aromatic hydrocarbons by the $Pd(CH_3CN)_4(SbF_6)_2$/*o*-chloranil catalytic system with silicon-bridged aromatics as $\pi$-extending agents. Density functional theory calculations suggest that the complete K-region selectivity stems from the olefinic (decreased aromatic) character of the K-region. The protocol is applicable to multiple APEX and sequential APEX reactions, to construct various nanographene structures in a rapid and programmable manner.

[1] Institute of Transformative Bio-Molecules (WPI-ITbM) and Graduate School of Science, Nagoya University, Chikusa, Nagoya 464-8602, Japan.
[2] JST-ERATO, Itami Molecular Nanocarbon Project, Nagoya University, Chikusa, Nagoya 464-8602, Japan. Correspondence and requests for materials should be addressed to K.I. (email: itami@chem.nagoya-u.ac.jp).

Nanographenes, which are nanometre-size subunits of graphenes (single-layer two-dimensional $sp^2$-hybridized carbon sheets)[1–5] with a tunable bandgap, have become hot molecular entities in the field of nanocarbon materials science[6]. As the properties of nanographenes depend heavily on the degree of π-extension, shape, width and edge topology, a novel bottom-up methodology for the precisely controlled synthesis of structurally uniform nanographenes is highly desirable[6]. Among various approaches to synthesize nanographenes with atom-by-atom precision[6], one-shot annulative π-extension (APEX) reactions of polycyclic aromatic hydrocarbons (PAHs) hold significant potential not only to achieve a 'growth from template' synthesis of nanographenes, but also to fine-tune the properties of nanographenes.

In the last two decades, various bottom-up organic synthesis methods have been established for the controlled synthesis of large π-extended PAHs and nanographenes, as exemplified by the groundbreaking achievements of Müllen and colleagues[7–10], Scott and colleagues[11,12], Fasel and colleagues[13,14] and others[15] (Fig. 1a). In essence, most of the reported nanographene syntheses rely on a two-step sequence of (i) component assembly of small π-components, using reactions such as Diels–Alder reactions, Suzuki–Miyaura couplings and C–H activation reactions, to synthesize soluble nanographene

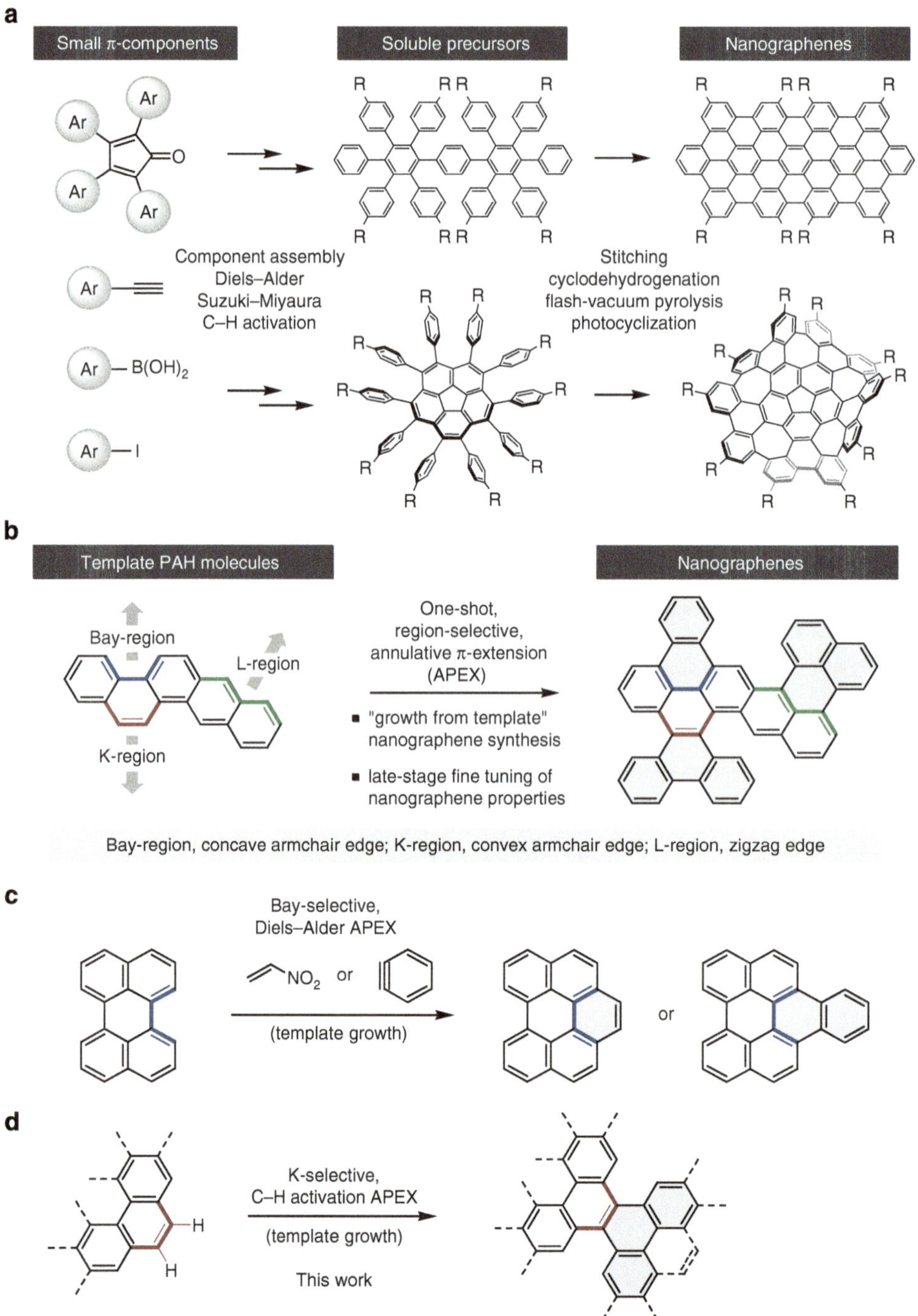


**Figure 1 | Organic synthesis approaches for structurally uniform nanographenes.** (**a**) The well-established two-step synthesis of nanographenes through π-component-assembling reaction and stitching (graphenization). (**b**) One-shot, region-selective APEX as alternative synthesis and functionalization of nanographenes through 'growth from template'. (**c**) One-shot, bay-region-selective APEX by Diels-Alder reaction. (**d**) One-shot, K-region-selective APEX by double C-H activation (this work).

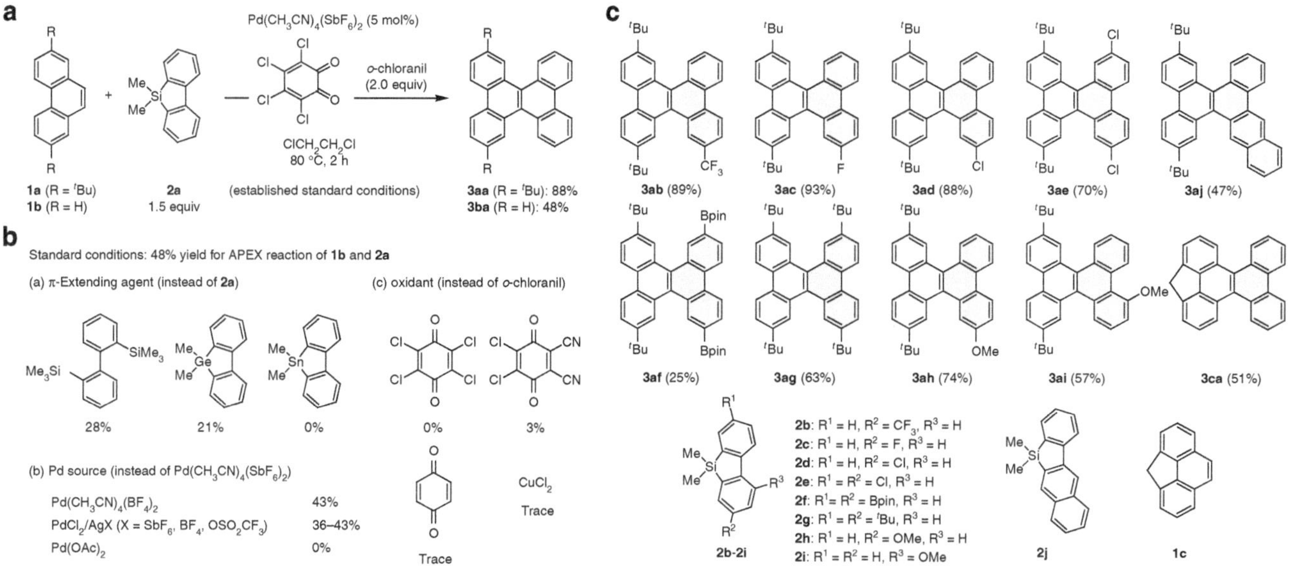


**Figure 2 | One-shot K-region-selective C-H activation APEX.** (**a**) The established reaction conditions. (**b**) Deviation from standard conditions (effect of reaction parameters) in the reaction of **1b** and **2a**. (**c**) Scope of one-shot, K-region-selective, C-H activation APEX. Reaction conditions: polycyclic aromatic compounds **1** (0.2 mmol), silicon-bridged aromatics **2** (1.5 equiv), $Pd(CH_3CN)_4(SbF_6)_2$ (5 mol%), *o*-chloranil (2.0 equiv), 1,2-dichloroethane, 80 °C, 2 h. Bpin, pinacolatoboryl.

precursors, followed by (ii) stitching (graphenization[6]) of soluble polyphenylene precursors by cyclodehydrogenation, flash-vacuum pyrolysis or photocyclization, to yield the target nanographenes (Fig. 1a)[6,16]. Although this state-of-the-art methodology has contributed significantly to the rapid progress of nanographene materials science, it has been well documented that the final and vital step of stitching (graphenization) is usually problematic[16]. For example, intramolecular oxidative cyclodehydrogenation by Lewis acids and oxidants (Scholl-type reactions) often suffer from problems such as incomplete stitching, lack of regioselectivity and undesirable rearrangements.

The continuing evolution of nanographene science is heavily dependent on the discovery of new reactions and strategies that allow rapid and predictable synthesis, and functionalization of nanographenes. Here we illustrate the significant potential of one-shot APEX reactions of template PAH molecules with π-extending agents (Fig. 1b) as an enabling concept that is complementary to the aforementioned two-step methodology. By devising APEX reactions that are selective to the specific regions of PAH structures (bay region, K-region and L-region, see Fig. 1b)[17], the direction-controlled growth or polymerization of a template PAH producing structurally uniform nanographenes would become possible. These regions correspond to the concave armchair edge (bay region), convex armchair edge (K-region) and zigzag edge (L-region) in graphene nomenclature. APEX technology should also be useful for the late-stage fine-tuning of nanographene properties by subtly modifying their edge structures. Moreover, as a result of significant recent progress in surface-catalysed cyclodehydrogenation and subsequent epitaxial elongation reactions of PAHs by the group of Fasel and colleagues[18], a number of π-extended PAHs provided by the APEX technology could also be converted into structurally uniform carbon nanotubes[18,19].

Although the full potential of APEX reactions in nanographene chemistry is yet to be realized partly because the concept had not been formulated, significant experimental[20–25] and theoretical[26] advances have appeared, describing methods for bay-region-specific π-extension in PAHs. For example, Scott has revisited the finding of the Diels–Alder reactivity at the PAH bay regions by Clar *et al.*[27] and proposed its use for metal-free growth of single-chirality carbon nanotubes[20–26] (Fig. 1c). This represents a bay-region-selective APEX reaction in our definition. However, no APEX reactions at other PAH regions (L-region and K-region) have been developed up until now. Although the dimerization of perylene under Scholl-type coupling to furnish quaterrylene represents the closest example of an L-region-selective APEX reaction, this can be categorized as the two-step method having a serious problem at the stitching (cyclodehydrogenation) stage[28]. The reactions at K-regions (convex armchair edges) are considered to be particularly difficult, as K-region bonds tend to have relatively high bond orders, and most aromatic substitution reactions occur preferentially at other aromatic C–H bonds[17].

Our APEX campaign[29–34] began when we serendipitously discovered $Pd(OAc)_2$/*o*-chloranil as the first-generation C–H activation catalyst for PAHs in 2011 (ref. 35). This catalyst uniquely and effectively promotes the C–H arylation of non-functionalized PAHs with arylboroxines, with complete K-region selectivity. When coupled with the Scholl-type cyclodehydrogenation of the thus-formed arylated PAHs, a number of structurally intriguing nanographenes such as warped nanographenes[34] (Fig. 1a) have been synthesized. Although this was a two-step nanographene synthesis at the time, we felt that the observed C–H activation reactivity and selectivity might be translated into an APEX reaction when the K-region is activated and annulated with properly arranged 1,4-dimetal π-units in a [2 + 4] annulation manner. Here, in we describe a one-shot APEX reaction that occurs selectively at the K-region of PAHs via double C–H activation[36–39] (Fig. 1d). This protocol is applicable to multiple APEX and sequential APEX reactions to construct various nanographene structures in a rapid and programmable manner.

## Results

**Development of catalyst.** We began our study by examining various palladium salts, ligands, oxidants and π-extending agents for the APEX reaction of phenanthrenes **1a** and **1b**. After extensive screening, we determined that $Pd(CH_3CN)_4(SbF_6)_2$ and *o*-chloranil serve as an efficient pre-catalyst and oxidant

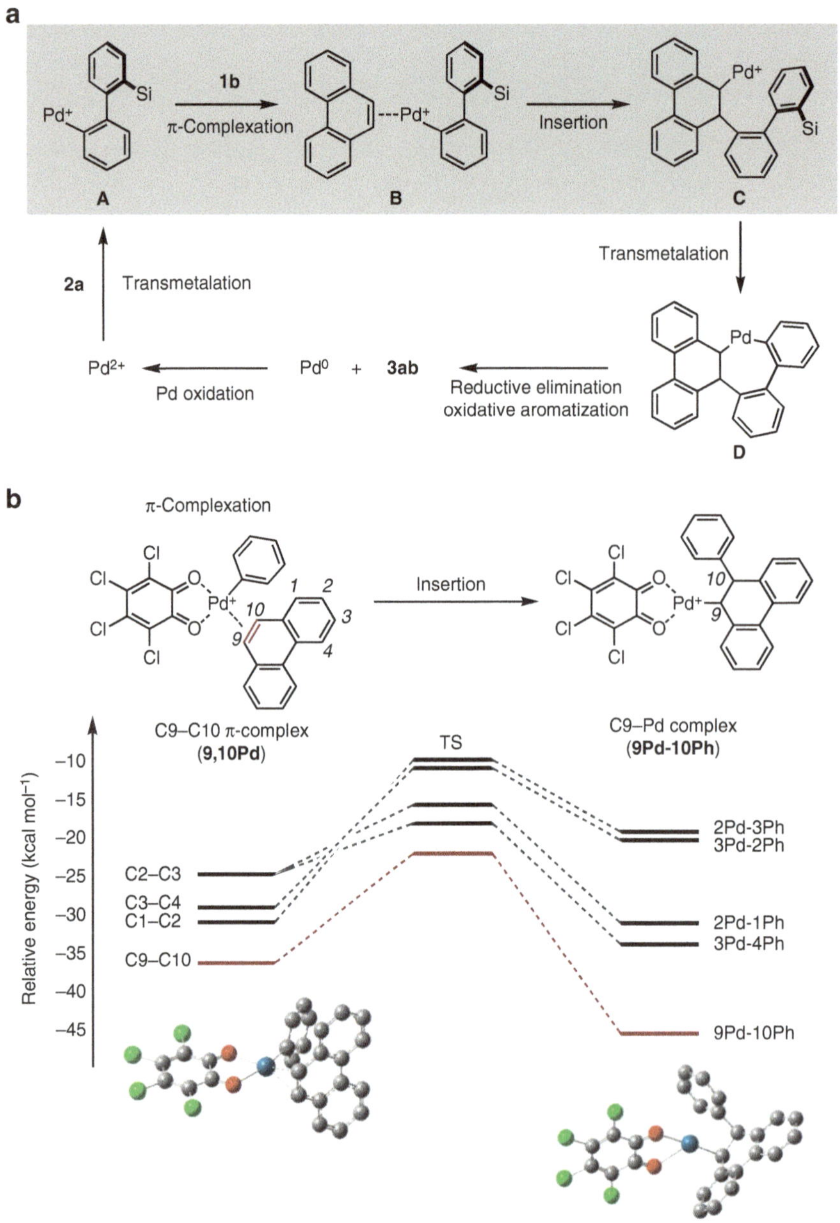


**Figure 3 | Mechanistic considerations.** (**a**) A possible mechanism of K-region-selective C-H activation APEX. (**b**) Theoretical calculations of π-complexation and insertion steps. Structures were optimized by the density functional theory (DFT[44]) calculations using B3PW91 hybrid functional[45,46] (hydrogen atoms are omitted for clarity). Reaction pathways were followed by intrinsic reaction coordinate (IRC[47,48]) computations, and high-accuracy, single-point energy calculations of DFT-optimized structures were performed with Møller–Plesset perturbation theory (MP2) (ref. 49). Energies are relative to that of *o*-chloranil-bound cationic phenylpalladium species.

respectively, while silicon-bridged aromatics **2** are optimal π-extending agents (Fig. 2a). For example, when 2,7-di-*tert*-butyl-phenanthrene (**1a**: 1.0 equiv) was treated with dimethyldibenzosilole[40] (**2a**: 1.5 equiv) in 1,2-dichloroethane at 80 °C for 2 h in the presence of $Pd(CH_3CN)_4(SbF_6)_2$ (5 mol%) and *o*-chloranil (2.0 equiv), the corresponding K-region-annulated APEX product **3aa** was obtained in 88% isolated yield, representing our standard APEX conditions. Unfunctionalized phenanthrene (**1b**) also underwent an APEX reaction with **2a** to yield dibenzo[*g,p*]chrysene (**3ba**). It should be noted that under these conditions, we observed the annulation exclusively at the K-region of phenanthrene **1a** or **1b**.

Listed in Fig. 2b are the effects of variations from the standard APEX conditions (**1b** + **2a** → **3ba**: 48% yield). For full lists of the effects of reaction parameters, see Supplementary Tables 2–4. As for the π-extension agents, we found that other dimetallobiphenyl derivatives such as 2,2′-bis(trimethylsilyl)-1,1′-biphenyl and dimethyldibenzogermole also reacted but led to the desired product in much lower yield. Dimethyldibenzostannole did not yield the product but generated a considerable amount of tetraphenylene and quaterphenyl by homodimerization. Changing the methyl groups of **2a** only had detrimental effect on reaction efficiency. The commercially available cationic palladium complex $Pd(CH_3CN)_4(BF_4)_2$ showed APEX activity comparable to $Pd(CH_3CN)_4(SbF_6)_2$. Similar catalytic activities were also observed with the combined use of $PdCl_2$ and silver salts such as $AgOSO_2CF_3$, $AgBF_4$ and $AgSbF_6$. On the other hand, neutral $Pd(OAc)_2$ (our previous palladium pre-catalyst)[29] did not promote the APEX reaction at all. These results clearly indicate the importance of a cationic palladium species for the APEX

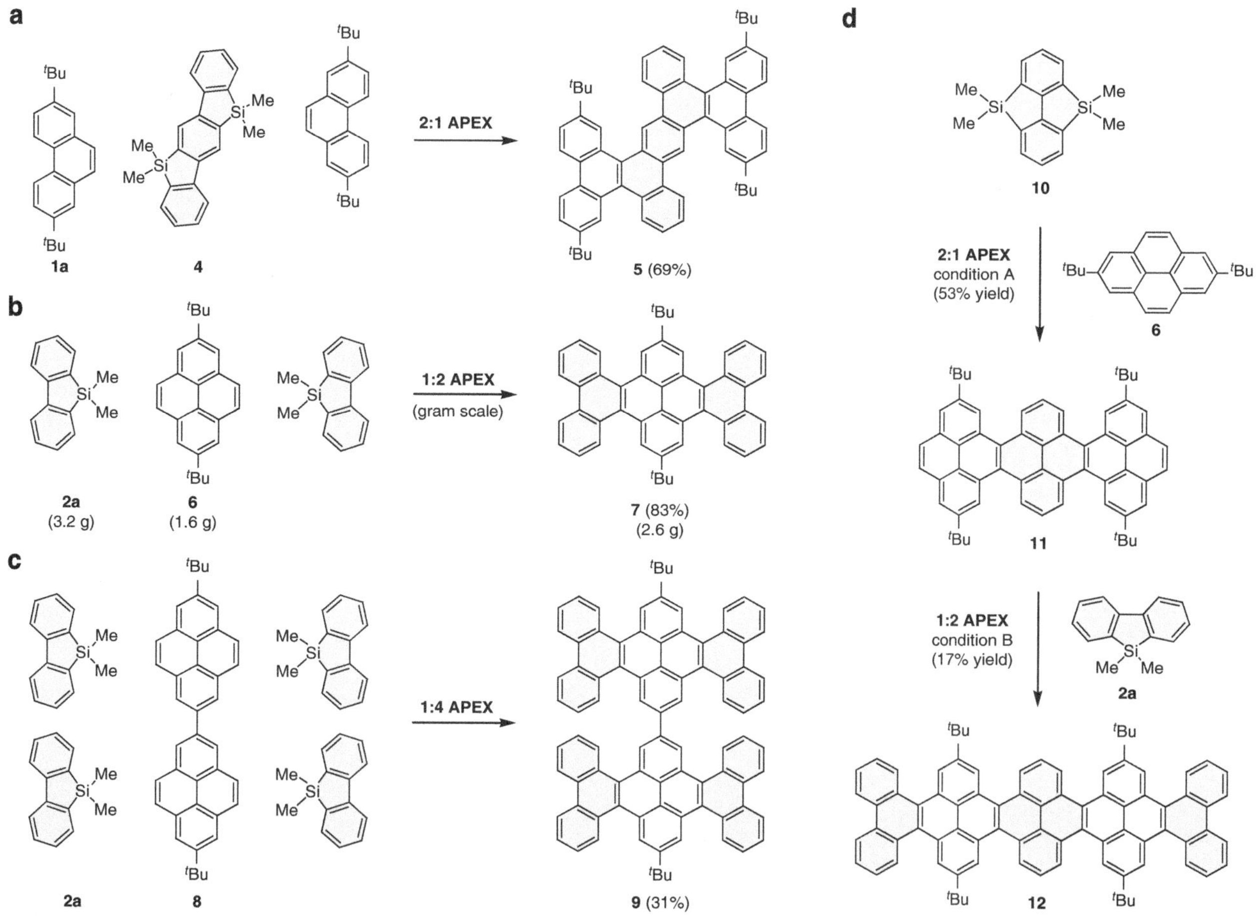


**Figure 4 | Multiple APEX and sequential APEX.** Multiple APEX (**a–c**). (**a**) 2:1 APEX. Reaction conditions: **4** (1 equiv), **1a** (5 equiv), $Pd(CH_3CN)_4(SbF_6)_2$ (5 mol%), *o*-chloranil (4 equiv), 1,2-dichloroethane, 80 °C, 2 h. (**b**) 1:2 APEX. Reaction conditions: **6** (1 equiv), **2a** (3 equiv), $Pd(CH_3CN)_4(SbF_6)_2$ (5 mol%), *o*-chloranil (4 equiv), 1,2-dichloroethane, 80 °C, 8 h. (**c**) Reaction conditions: **8** (1 equiv), **2a** (10 equiv), $Pd(CH_3CN)_4(SbF_6)_2$ (15 mol%), *o*-chloranil (10 equiv), 1,2-dichloroethane, 80 °C, 1 h. (**d**) Sequential APEX. 2:1 APEX conditions (A): **6** (3 equiv), **10** (1 equiv), $Pd(CH_3CN)_4(SbF_6)_2$ (5 mol%), *o*-chloranil (4 equiv), 1,2-dichloroethane, 80 °C, 6 h. 1:2 APEX conditions (B): **11** (1 equiv), **2a** (4 equiv), $Pd(CH_3CN)_4(SbF_6)_2$ (15 mol%), *o*-chloranil (5 equiv), 1,2-dichloroethane, 80 °C, 4 h.

reaction to occur. In the investigation of oxidants, commonly used oxidants such as benzoquinone, *p*-chloranil, DDQ and $CuCl_2$ (ref. 35) displayed virtually no APEX-type activity. We assume that the high reactivity of *o*-chloranil stems not only from its high oxidation aptitude but also from its unique *o*-quinone structure, which can bind to palladium in a bidentate manner and modulate the redox property effectively[41,42]. Our preferred solvent is 1,2-dichloroethane, but aromatic solvents such as toluene, chlorobenzene, 1,2-dichlorobenzene, fluorobenzene and trifluoromethylbenzene can also be used for the present APEX reactions.

**Scope of K-region-selective APEX.** As shown in Fig. 2c, various structurally and electronically diverse silicon-bridged aromatics **2** were found to react with 2,7-di-*tert*-butylphenanthrene (**1a**), providing the corresponding dibenzo[*g,p*]chrysenes (**3ab–3aj**) in good-to-excellent yield with virtually complete K-region selectivity. In particular, dibenzosiloles having electron-deficient substituents (**2b–2e**)[40,43] showed excellent reactivity. The tribenzo[*a,c,f*]tetraphene framework (**3aj**) can be readily constructed by the APEX reaction of **1a** and benzonaphthosilole **2j**[43]. Notably, dichloro- and diboryl-substituted dibenzosiloles underwent the APEX reaction smoothly, leaving C–Cl and C–B bonds intact (**3ad**, **3ae** and **3af**). The tolerance of the reaction for these bonds makes it attractive for further π-extension and functionalization, using well-established cross-coupling chemistry. Furthermore, methylene-bridged phenanthrene **1c** reacted with **2a** to afford benzoindenochrysene **3ca**, which can potentially lead to soluble nanographenes by facile substitution at the methylene moiety. Ease of post-functionalization is particularly advantageous for controlled surface alignment of nanographenes for device applications.

**Mechanistic considerations of K-region-selective APEX.** Although the exact mechanism of the present APEX reaction remains unclear, our current assumption is shown in Fig. 3a. A palladium(II) species undergoes the first transmetalation with silicon-bridged aromatic **2a**, followed by coordination of phenanthrene (**1b**) to arylpalladium intermediate **A** at the K-region, yielding π-complex **B**. Coordination-induced insertion at the K-region C=C bond, followed by a second transmetalation forms palladacycle intermediate **D**. Reductive elimination and oxidative aromatization yields APEX product **3ab** and palladium(0), the latter of which is oxidized to the active palladium(II) species by the action of *o*-chloranil.

We assume that the K-region selectivity in the present APEX reaction stems from the preferential palladium π-complexation at the K-region of PAHs before the insertion step, as depicted in Fig. 3a. To prove this hypothesis, we conducted density functional theory[44] calculations (using B3PW91 hybrid functional[45,46]) for the $\pi$-complexation and insertion steps on a model reaction of *o*-chloranil-bound cationic phenylpalladium species with phenanthrene yielding the alkylpalladium species shown in Fig. 3b (a model reaction relevant to **A**→**B**→**C** in Fig. 3a). Reaction pathways were followed by intrinsic reaction coordinate[47,48] computations, and high-accuracy, single-point energy calculations of density functional theory-optimized structures were performed with Møller–Plesset perturbation theory[49]. Among possible $\pi$-coordination complexes at C1–C2, C2–C3, C3–C4 and C9–C10 bonds, the $\pi$-complex at C9–C10 (K-region) was found to be most stable (**9,10Pd**). This may be due to the tendency of this bond to have the most olefinic (less aromatic) character[17,50]. We also calculated all possible transition states of insertion from these $\pi$-complexes to give alkylpalladium complexes (Fig. 3b). The formation of C9–Pd complex (**9Pd–10Ph**), which leads to APEX at the K-region, was found to be most favourable both kinetically and thermodynamically. Thus, the basis of K-region selectivity in our new APEX reaction has been supported by computational theory. Detailed computational studies of Pd/*o*-chloranil-catalysed C–H activation, including the present APEX reactions, will be reported in due course. Based on these calculations, we predict the most olefinic (least aromatic) K-region $\pi$-bond to be the first APEX reaction site in future functionalizations of related $\pi$-extended PAHs and nanographenes.

**Multiple APEX and sequential APEX.** To showcase the utility of our APEX methodology in accessing a variety of nanographene ($\pi$-extended PAH) structures in a rapid and programmable manner, we examined several types of multiple APEX reactions (Fig. 4a–c). For example, a 2:1 APEX reaction occurs when treating ladder-type *bis*-silicon-bridged *p*-terphenyl **4** (ref. 51) with an excess of 2,7-di-*tert*-butylphenanthrene (**1a**) in the presence of $Pd(CH_3CN)_4(SbF_6)_2$/*o*-chloranil, to construct the dibenzodiphenanthroanthracene framework **5** in 69% yield (Fig. 4a). Other isomers were not observed in the reaction, highlighting the fidelity of the present method to specific reaction sites. An alternative mode of the double APEX reaction (1:2 APEX) was also possible by the reaction of 2,7-di-*tert*-butylpyrene **6** (1.6 g, 1.0 equiv) and dibenzosilole **2a** (3.2 g, 3.0 equiv) in the presence of $Pd(CH_3CN)_4(SbF_6)_2$ (5 mol%) and *o*-chloranil (4.0 equiv) (Fig. 4b). It is noteworthy that the reaction could be conducted on a gram scale to yield di-*tert*-butylhexabenzotetracene **7** in 2.6 g (83% yield). The gram-scale synthesis clearly underscores the high capability of the present reaction conditions for multiple APEX reactions. Furthermore, the $Pd(CH_3CN)_4(SbF_6)_2$/*o*-chloranil system effectively promoted the one-shot fourfold APEX reaction of 7,7′-di-*tert*-butyl-2,2′-bipyrene **8** (1.0 equiv) with dibenzosilole **2a** (10 equiv), to provide bihexabenzotetracene **9** in 31% yield (Fig. 4c).

To further examine the applicability of APEX technology for the construction of larger molecules, sequential APEX reactions were investigated (Fig. 4d). Pleasingly, the 2:1 APEX reaction of 2,7-di-*tert*-butylpyrene **6** (3.0 equiv) with *bis*-silicon bridged biphenyl **10** (ref. 51) (1.0 equiv) under $Pd(CH_3CN)_4(SbF_6)_2$/*o*-chloranil conditions afforded tetra-*tert*-butylhexabenzohexacene **11** in 53% yield. The follow-up 1:2 APEX reaction of **11** (1.0 equiv) with dibenzosilole **2a** (4.0 equiv) also took place to furnish the target decabenzooctacene framework **12** in 17% yield. In this particular reaction, we recovered a considerable amount of starting material likely to be attributed to poor solubility in the reaction media. Despite being relatively small in molecular size, the electronic structures of these nanographenes can be systematically altered. With the increase in molecular length, the HOMO-LUMO gap becomes smaller. In line with this trend, we observed decent red shift in both absorption and fluorescence (see Supplementary Fig. 41 for details).

## Discussion

It should also be mentioned that we observed the first sign of the possibility of employing APEX in an oligomerization/polymerization manifold when we detected oligomers by matrix-assisted laser desorption/ionization time-of-flight mass spectrometry analysis in the first 2:1 APEX reaction shown in Fig. 4d. Although we have not yet investigated this possible mode of reaction extensively, we envisage that the present APEX reaction could be applied to even larger molecules through judicious choice of solubilizing substituents on the substrates. Thus, the present result bodes well for the potential application of our APEX methodology to the bottom-up synthesis of graphene nanoribbons with controlled edge structures.

The present APEX methodology is complementary to the state-of-the-art two-step nanographene synthetic methods, thereby finding significant use in the 'growth from template' nanographene synthesis and in the late-stage fine-tuning of nanographene properties. Moreover, our APEX technology is not limited to the synthesis and functionalization of $\pi$-extended PAHs and nanographenes. One of the most significant features of the present APEX reaction is that unfunctionalized PAHs can be directly used for $\pi$-component assembly and $\pi$-extension without any pre-functionalization. Thus, various $\pi$-conjugated molecules made by many research groups (for many different purposes) will be suitable substrates for our APEX reaction, furnishing even more exciting classes of $\pi$-conjugated systems. The realization of APEX polymerization, the development of new APEX reactions for other PAH regions and acquisition of the first structure–property relationships for various nanographene structures are now ongoing in our laboratory.

## Methods

**Materials and characterization data.** For the synthesis of all starting materials, siloles and $\pi$-extended PAHs, and their characterization, see Supplementary Methods. $^1H$, $^{13}C$ and $^{19}F$ NMR spectra were obtained for all compounds, see Supplementary Fig. 1–40. Ultraviolet–visible absorption and fluorescence spectra of **6**, **11** and **12** are provided in Supplementary Fig. 41. Calculated energy surface of $\pi$-complexation and insertion are provided in Supplementary Fig. 42. Calculated energies of stationary points and their Cartesian coordinates are summarized in Supplementary Table 1. Effects of reaction parameters are provided in Supplementary Tables 2–4.

**Typical procedure of K-region-selective APEX reaction.** 2,7-Di-*tert*-butylphenanthrene (**1a**: 58 mg, 0.2 mmol, 1.0 equiv), dimethyldibenzosilole (**2a**: 63 mg, 0.3 mmol, 1.5 equiv), $Pd(CH_3CN)_4(SbF_6)_2$ (7.4 mg, 10 μmol, 5 mol%), *o*-chloranil (98 mg, 0.4 mmol, 2.0 equiv) and a stirring bar were placed in a screw cap test tube. The tube was sealed with a perforated plastic cap and air was removed though a needle under reduced pressure. After back filling with nitrogen or argon, 1,2-dichloroethane (2 ml) was added and the needle was removed to seal up the test tube. The tube was fixed in an aluminum heating block and the mixture was stirred at 80 °C. After 2 h, the reaction mixture was cooled to room temperature and then passed through a short pad of silica gel (eluent: $CH_2Cl_2$). After the organic solvents were removed under reduced pressure, the residue was purified by silica-gel chromatography (eluent: hexane) to afford 3,14-di-*tert*-butyldibenzo[*g,p*]chrysene (**3aa**) in 88% yield (77.0 mg) as a white powder.

## References

1. Novoselov, K. S. *et al.* Electric field effect in atomically thin carbon films. *Science* **306**, 666–669 (2004).
2. Allen, M. J., Tung, V. C. & Kaner, R. B. Honeycomb carbon: a review of graphene. *Chem. Rev.* **110**, 132–145 (2010).
3. Geim, A. K. Graphene: status and prospects. *Science* **324**, 1530–1534 (2009).
4. Wu, J., Psula, W. & Müllen, K. Graphenes as potential material for electronics. *Chem. Rev.* **107**, 718–747 (2007).
5. Dreyer, D. R., Ruoff, R. S. & Bielawski, C. W. From conception to realization: an historical account of graphene and some perspectives for its future. *Angew. Chem. Int. Ed.* **49**, 9336–9344 (2010).

6. Chen, L., Hernandez, Y., Feng, X. & Müllen, K. From nanographene and graphene nanoribbons to graphene sheets: chemical synthesis. *Angew. Chem. Int. Ed.* **51,** 7640–7654 (2012).
7. Müllen, K. Evolution of graphene molecules: structural and functional complexity as driving forces behind nanoscience. *ACS Nano* **8,** 6531–6541 (2014).
8. Watson, M. D., Fechtenkoetter, A. & Müllen, K. Big is beautiful – "aromaticity" revisited from the viewpoint of macromolecular and supramolecular benzene chemistry. *Chem. Rev.* **101,** 1267–1300 (2001).
9. Narita, A. *et al.* Synthesis of structurally well-defined and liquid-phase-processable graphene nanoribbons. *Nat. Chem.* **6,** 126–132 (2014).
10. Schwab, M. G. *et al.* Structurally defined graphene nanoribbons with high lateral extension. *J. Am. Chem. Soc.* **134,** 18169–18172 (2012).
11. Scott, L. T. *et al.* Geodesic polyarenes with exposed concave surfaces. *Pure Appl. Chem.* **71,** 209–219 (1999).
12. Tsefrikas, V. M. & Scott, L. T. Geodesic polyarenes by flash vacuum pyrolysis. *Chem. Rev.* **106,** 4868–4884 (2006).
13. Cai, J. M. *et al.* Atomically precise bottom-up fabrication of graphene nanoribbons. *Nature* **466,** 470–473 (2010).
14. Treier, M. *et al.* Surface-assisted cyclodehydrogenation provides a synthetic route towards easily processable and chemically tailored nanographenes. *Nat. Chem.* **3,** 61–67 (2011).
15. Vo, T. H. *et al.* Large-scale solution synthesis of narrow graphene nanoribbons. *Nat. Commun.* **5,** 3189 (2014).
16. Feng, X., Pisula, W. & Müllen, K. Large polycyclic aromatic hydrocarbons: synthesis and discotic organization. *Pure Appl. Chem.* **81,** 2203–2224 (2009).
17. Harvey, R. G. *Polycyclic Aromatic Hydrocarbons* (Wiley-VCH, 1997).
18. Sanchez-Valencia, J. R. *et al.* Controlled synthesis of single-chirality carbon nanotubes. *Nature* **512,** 61–64 (2014).
19. Omachi, H., Nakayama, T., Takahashi, E., Segawa, Y. & Itami, K. Initiation of carbon nanotube growth by well-defined carbon nanorings. *Nat. Chem.* **5,** 572–576 (2013).
20. Fort, E. H., Donovan, P. M. & Scott, L. T. Diels–Alder reactivity of polycyclic aromatic hydrocarbon bay regions: implications for metal-free growth of single-chirality carbon nanotubes. *J. Am. Chem. Soc.* **131,** 16006–16007 (2009).
21. Fort, E. H. & Scott, L. T. One-step conversion of aromatic hydrocarbon bay regions into new unsubstituted benzene rings. A reagent for the low-temperature, metal-free growth of single-chirality carbon nanotubes. *Angew. Chem. Int. Ed.* **49,** 6626–6628 (2010).
22. Fort, E. H. & Scott, L. T. Gas-phase Diels–Alder cycloaddition of benzyne to an aromatic hydrocarbon bay region. Groundwork for the selective solvent-free growth of armchair carbon nanotubes. *Tetrahedron Lett.* **52,** 2051–2053 (2011).
23. Li, J., Jiao, C., Huang, K. W. & Wu, J. Lateral extension of π conjugation along the bay regions of bisanthene through a Diels–Alder cycloaddition reaction. *Chem. Eur. J.* **17,** 14672–14680 (2011).
24. Konishi, A., Hirano, Y., Matsumoto, K., Kurata, H. & Kubo, T. Facile synthesis and lateral π-expansion of bisanthenes. *Chem. Lett.* **42,** 592–594 (2013).
25. Schuler, B. *et al.* From perylene to a 22-ring aromatic hydrocarbon in one-pot. *Angew. Chem. Int. Ed.* **53,** 9004–9006 (2014).
26. Fort, E. H. & Scott, L. T. Carbon nanotubes from short hydrocarbon templates. Energy analysis of the Diels–Alder cycloaddition/rearomatization growth strategy. *J. Mater. Chem.* **21,** 1373–1381 (2011).
27. Clar, E. & Zander, M. Syntheses of coronene and 1: 2-7: 8-dibenzocoronene. *J. Chem. Soc.* 4616–4625 (1957).
28. Thamatam, R., Skraba, S. L. & Johnson, R. P. Scalable synthesis of quaterrylene: solution-phase $^1$H NMR spectroscopy of its oxidative dication. *Chem. Commun.* **49,** 9122–9124 (2013).
29. Mochida, K., Kawasumi, K., Segawa, Y. & Itami, K. Direct arylation of polycyclic aromatic hydrocarbons through palladium catalysis. *J. Am. Chem. Soc.* **133,** 10716–10719 (2011).
30. Kawasumi, K., Mochida, K., Kajino, T., Segawa, Y. & Itami, K. $Pd(OAc)_2$/*o*-chloranil/$M(OTf)_n$: a catalyst for the direct C–H arylation of polycyclic aromatic hydrocarbons with boryl-, silyl-, and unfunctionalized arenes. *Org. Lett.* **14,** 418–421 (2012).
31. Zhang, Q., Kawasumi, K., Segawa, Y., Itami, K. & Scott, L. T. Palladium-catalyzed C–H activation taken to the limit. Flattening an aromatic bowl by total arylation. *J. Am. Chem. Soc.* **134,** 15664–15667 (2012).
32. Kawasumi, K., Mochida, K., Segawa, Y. & Itami, K. Palladium-catalyzed direct phenylation of perylene: structural and optical properties of 3,4,9-triphenylperylene and 3,4,9,10-tetraphenylperylene. *Tetrahedron* **69,** 4371–4374 (2013).
33. Ozaki, K., Zhang, H., Ito, H., Lei, A. & Itami, K. One-shot indole-to-carbazole π-extension by Pd-Cu-Ag trimetallic system. *Chem. Sci.* **4,** 3416–3420 (2013).
34. Kawasumi, K., Zhang, Q., Segawa, Y., Scott, L. T. & Itami, K. A grossly warped nanographene and the consequences of multiple odd-membered-ring defects. *Nat. Chem.* **5,** 739–744 (2013).
35. Funaki, K., Kawai, H., Sato, T. & Oi, S. Palladium-catalyzed direct C–H bond arylation of simple arenes with aryltrimethylsilanes. *Chem. Lett.* **40,** 1050–1052 (2011).
36. Okamoto, K., Zhang, J., Housekeeper, J. B., Marder, S. R. & Luscombe, C. K. C–H arylation reaction: atom efficient and greener syntheses of π-conjugated small molecules and macromolecules for organic electronic materials. *Macromolecules* **46,** 8059–8078 (2013).
37. Segawa, Y., Maekawa, T. & Itami, K. Synthesis of extended π-systems through C–H activation. *Angew. Chem. Int. Ed.* **54,** 66–81 (2015).
38. Yamaguchi, J., Yamaguchi, A. D. & Itami, K. C–H bond functionalization: emerging synthetic tools for natural products and pharmaceuticals. *Angew. Chem. Int. Ed.* **51,** 8960–9009 (2012).
39. Wencel-Delord, J. & Glorius, F. C–H bond activation enables the rapid construction and late-stage diversification of functional molecules. *Nat. Chem.* **5,** 369–375 (2013).
40. Neugebauer, W., Kos, A. J. & Schleyer, P. v. R. Regioselektive dimetallierung von aromaten. Bequemer zugang zu 2,2′-disubstituierten biphenylderivaten. *J. Organomet. Chem.* **228,** 107–118 (1982).
41. Balch, A. L. Oxidation of *o*-quinone adducts of transition metal complexes. *J. Am. Chem. Soc.* **95,** 2723–2724 (1973).
42. Fox, G. A. & Pierpont, C. G. Periodic trends in charge distribution for transition metal complexes containing catecholate and semiquinone ligands. Metal-mediated spin coupling in the bis(3,5-di-*tert*-butyl-1,2-semiquinone) complexes of palladium(II) and platinum(II) $M(DBSQ)_2$ (M = Pd, Pt). *Inorg. Chem.* **31,** 3718–3723 (1992).
43. Ureshino, T., Yoshida, T., Kuninobu, Y. & Takai, K. Rhodium-catalyzed synthesis of silafluorene derivatives via cleavage of silicon – hydrogen and carbon – hydrogen bonds. *J. Am. Chem. Soc.* **132,** 14324–14326 (2010).
44. Frisch, M. J. *Gaussian09, Revision E.01* (Gaussian, Inc., Wallingford CT, 2004).
45. Becke, A. D. Density functional thermochemistry. III. The role of exact exchange. *J. Chem. Phys.* **98,** 5648–5652 (1993).
46. Perdew, J. P. *Unified Theory of Exchange and Correlation Beyond the Local Density Approximation/Electronic Structure of Solids.* (Eds Ziesche, P. & Eschrig, H.) 11–20 (Akademie Verlag, 1991).
47. Dunning, Jr T. H. & Hay, P. J. *Methods of Electronic Structure Theory/Modern Theoretical Chemistry.* (Ed Schaefer, III H. F.3, 1–27 (Springer, 1977).
48. Hay, P. J. & Wadt, W. R. *Ab initio* effective core potentials for molecular calculations. Potentials for the transition metal atoms Sc to Hg. *J. Chem. Phys.* **82,** 270–283 (1985).
49. Grimme, S. Improved second-order Møller–Plesset perturbation theory by separate scaling of parallel- and antiparallel-spin pair correlation energies. *J. Chem. Phys.* **118,** 9095–9102 (2003).
50. Hasegawa, T., Sekine, M., Schaefer, W. P. & Taube, H. Preparation and characterization of a bis(pentaammineosmium(II)) pyrene complex. *Inorg. Chem.* **30,** 449–452 (1991).
51. Shimizu, M., Tatsumi, H., Mochida, K., Oda, K. & Hiyama, T. Silicon-bridge effects on photophysical properties of silafluorenes. *Chem. Asian J.* **3,** 1238–1247 (2008).

## Acknowledgements

This work was supported by the ERATO programme from JST (K.I.), the Funding Program for Next Generation World-Leading Researchers from JSPS (220GR049 to K.I.) and KAKENHI from MEXT (26810057 to H.I.). K.O. and K.K. thank the Japan Society for the Promotion of Science (JSPS) for fellowships. We thank Dr Yasutomo Segawa (Nagoya University), Professor Cathleen M. Crudden (Queen's University) and Dr Ayako Miyazaki (Nagoya University) for fruitful discussion and critical comments. Dr Keiko Kuwata (Nagoya University) is acknowledged for assistance in mass spectroscopy. Calculations were performed using the resources of the Research Center for Computational Science, Okazaki, Japan. ITbM is supported by the World Premier International Research Center (WPI) Initiative, Japan.

## Author contributions

K.I. conceived the concept. K.O., K.K. and H.I. conducted experiments. M.S. performed theoretical calculations. K.I and H.I. prepared the manuscript, with feedback from other authors.

## Additional information

**Competing financial interests:** The authors declare no competing financial interests.

5

# Carbon-catalysed reductive hydrogen atom transfer reactions

Huimin Yang[1,2], Xinjiang Cui[1], Xingchao Dai[1,2], Youquan Deng[1] & Feng Shi[1]

Generally, transition metal catalysts are essential for the reductive hydrogen atom transfer reaction, which is also known as the transfer hydrogenation reaction or the borrowing-hydrogen reaction. It has been reported that graphene can be an active catalyst in ethylene and nitrobenzene reductions, but no report has described carbon-based materials as catalysts for alcohol amination via the borrowing-hydrogen reaction mechanism. Here we show the results from the preparation, characterization and catalytic performance investigation of carbon catalysts in transition metal-free borrowing-hydrogen reactions using alcohol amination and nitro compound/ketone reduction as model reactions. XPS, XRD, SEM, FT-IR and $N_2$ adsorption–desorption studies revealed that C=O group in the carbon catalysts may be a possible catalytically active site, and high surface area is important for gaining high activity. The activity of the carbon catalyst remained unchanged after reuse. This study provides an attractive and useful methodology for a wider range of applications.

[1] State Key Laboratory for Oxo Synthesis and Selective Oxidation, Centre for Green Chemistry and Catalysis, Lanzhou Institute of Chemical Physics, Chinese Academy of Sciences, No.18, Tianshui Middle Road, Lanzhou 730000, China. [2] University of Chinese Academy of Sciences, No. 19A, Yuquanlu, Beijing 100049, China. Correspondence and requests for materials should be addressed to F.S. (email: fshi@licp.cas.cn).

Hydrogen atom transfer is the most common reaction and is the fundamental step in many processes, ranging from combustion, aerobic oxidation and reduction to enzymatic catalysis and the destructive effects of reactive oxygen species[1]. In practical transformations, two types of reactions are involved in the hydrogen atom transfer reaction (that is, oxidation and reduction[2]). In oxidation reactions, the hydrogen atom transfer initially occurs between the starting material/hydrogen donor and the catalyst, and the hydrogen atom is then removed by oxygen or other oxidants[3]. The classical reactions with this mechanism involve the oxidation of an alkane and an alcohol in the presence or absence of transition metal catalysts. The reductive hydrogen atom transfer reaction, which is also known as the transfer hydrogenation reaction or the borrowing-hydrogen reaction, is often observed in the reduction of ketones, nitro compounds and imines using isopropanol as the hydrogen donor or in the alcohol amination reaction[4,5]. Generally, this reaction has been observed in biotransformations[6], and the presence of transition metals is essential to achieve this transformation[4,5,7,8].

Carbon materials, including amorphous carbon, ordered mesoporous carbon, graphite/graphene (oxide) and carbon nanotubes, are widely applied in many catalytic transformations in modern organic chemistry[9–11], and good performance has been obtained in the oxidation of an alkane[12–15], alcohol[16–18], amine[19], thiol and sulphide[20,21]. These reactions encompass the oxidative hydrogen atom transfer reactions. In addition, the O-rich carbon materials have been active catalysts in various reactions (that is, nucleophilic addition of alcohol to an epoxide[22], aldehyde acetalization or esterification[23,24], Michael addition reactions[25,26], F-C addition of indoles to α,β-unsaturated ketones[27] and synthesis of dipyrromethane[28], among others[29,30]). In these studies, nearly all of the organic transformations have been focused on oxidative hydrogen atom transfer reactions or acid-catalysed reactions with heteroatom-modified carbon materials (Fig. 1). It has been reported that graphene or phosphonium salt can be an active catalyst in olefin[31,32] and nitrobenzene[33] reductions, but no report has described carbon-based materials as catalysts for alcohol amination via the borrowing-hydrogen reaction mechanism.

In this study, we report new results on the preparation, characterization and catalytic performance of carbon materials in reductive hydrogen transfer reactions (that is, the alcohol amination reaction and ketone and nitrobenzene reduction with isopropanol as the hydrogen donor). The results suggest that the above-mentioned transformations are efficiently catalysed by carbon materials without the addition of transition metals.

## Results

**Catalyst preparation.** The carbon materials were prepared via sol-gel polymerization of resorcinol and formaldehyde with $Na_2CO_3$ as a catalyst (Fig. 2). First, a wet RF gel was prepared by polymerization of resorcinol and formaldehyde using a hydrothermal method. Then, the wet RF gel was mixed with KOH or another base and heated at 800 °C under a nitrogen flow. Next the carbonized sample was washed with deionized water to remove the base, and the carbon material was obtained. A series of carbon materials was prepared using this method by varying the amount and type of base. The carbon materials that were not treated with a base are denoted C-0. C-1, C-2, C-3 and C-4 were prepared using various bases (that is, KOH, NaOH, $K_2CO_3$ and $Na_2CO_3$, respectively) with a fixed ratio of wet RF gel to base (that is, 1:1). Catalysts C-5, C-6 and C-7 were prepared by varying the mass ratio of the wet RF gel to KOH (that is, 1:0.25, 1:0.5 and 1:1.25, respectively).

**Catalyst screening and optimization of the reaction conditions.** Alcohol amination is one of the classical reactions using the borrowing-hydrogen reaction mechanism[7,8]. In addition, alcohol amination is a potential approach for green and economic *N*-alkyl amine synthesis because alcohol is readily available, and water is generated as the sole byproduct[7,8]. Over the past few decades, various transition metals, such as Ru[34–36], Ir[37,38], Rh[39], Pd[40–42], Pt[43], Au[41,44], Ag[45,46], Ni[47–50], Mn[51,52], Cu[53–58] and Fe[59–61], have been employed as homogeneous or heterogeneous catalysts in the alcohol amination reaction and have yielded good results. Because it is challenging to involve the carbon catalyst in the hydride formation, carbon catalysts have not been applied to this transformation. Thus, alcohol amination was chosen as the model reaction to explore the reactivity of the carbon catalysts. The application of carbon catalysts in this reaction can offer a novel catalyst for this transformation and can aid our understanding of the hydrogen atom transfer mechanism in the

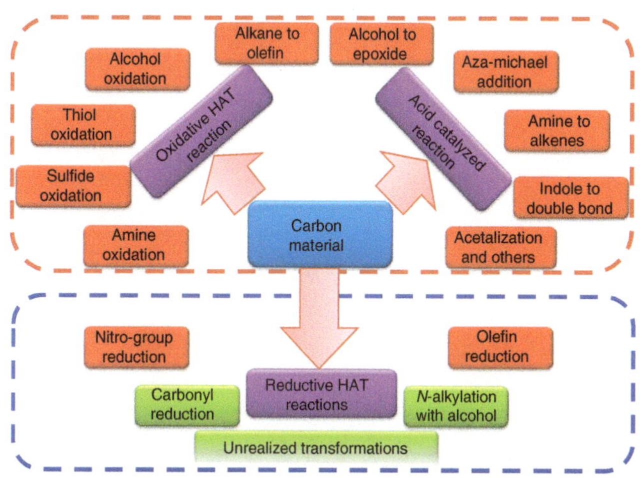


**Figure 1 | Organic transformations catalysed by carbon materials.** Carbon materials are widely applied in many catalytic transformations, including selective oxidation, nucleophilic addition and olefin/nitrobenzene reduction, but no report has described carbon-based materials as catalysts for alcohol amination via the borrowing-hydrogen reaction mechanism.

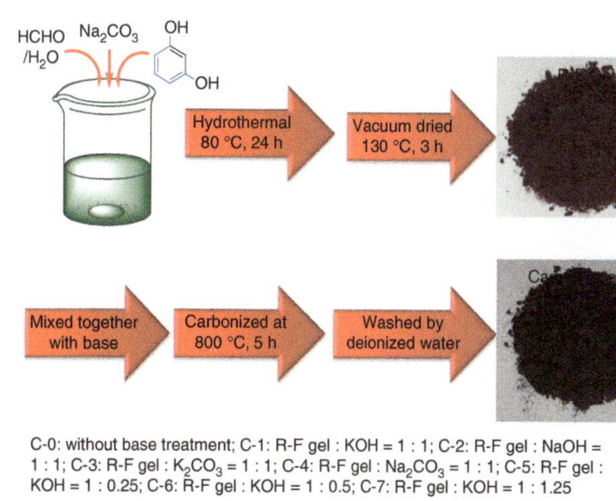


**Figure 2 | An illustration of the carbon catalyst preparation.** (1) Sol-gel polymerization of resorcinol and formaldehyde using the hydrothermal method. (2) The wet RF gel was mixed with base and heated at 800 °C under a nitrogen flow. (3) The carbonized sample was washed with deionized water to remove the base.

absence of transition metals. The catalytic performance of the carbon materials was first explored in the benzyl alcohol amination reaction (Table 1). Only 60% conversion of aniline and a 42% yield of *N*-benzyl aniline were obtained when C-0 was directly used as the catalyst (entry 1). Generally, the modification of the carbon material with bases, such as KOK, NaOH and $K_2CO_3$, can significantly improve the catalytic performance, and an *N*-benzyl aniline yield of >99% was obtained with the C-1 catalyst, which was modified with KOH (entries 2–4). In this case, there were no imines or tertiary amines observed by gas chromatography-flame ionization detector (GC-FID) and —gas chromatography–mass spectrometry. Interestingly, the modification of the carbon with $Na_2CO_3$ resulted in poor activity, and the yield of *N*-benzyl aniline was <5% (entry 5). The influence of the amount of base on the carbon material modification was further studied by varying the mass ratio of the RF gel to KOH (that is, 1:0.25, 1:0.5 and 1:1.25). A lower activity would be obtained if the mass ratio of RF gel to KOH was <1:1. The application of additional KOH had a smaller influence on the catalytic performance of the final carbon catalyst (entries 6–8). In addition, catalyst loading was explored, and 99% *N*-benzyl aniline was maintained, even when 50 mg of the carbon catalyst was employed (entries 9–11). The ratio of aniline to benzyl alcohol was reduced to 1:1.2, and the yield of *N*-benzyl aniline was 98%. However, when the reaction was performed at 120 or 100 °C, the yield of *N*-benzyl aniline decreased to 89% or 21%, respectively. The amount of base added during the reaction was further optimized. Nearly no desired product was formed when no base was added, and the yield of *N*-benzyl aniline was only 7%, even when 30% KOH was added (entries 16 and 17). However, the yield was substantially increased to 87 and >99% when 40 and 50% KOH were added as a co-catalyst, respectively (entries 2 and 18). When other bases, including $Na_2CO_3$, NaOH, $K_2CO_3$ and KO-t-Bu, were used as catalysts, different results were obtained. Yields of 68 and 89% were obtained when NaOH and KO-t-Bu were applied, respectively. In addition, the yields were <5% when $Na_2CO_3$ or $K_2CO_3$ was employed (entries 19–22). The difference in activity may be due to the different strengths of the bases. Finally, to exclude contributions of contaminating transition metals to the reaction, metal loadings of Pd, Ni and Cu in catalyst C-1 were investigated by ICP-AES. The results indicated that all of the metal loadings were <0.001 wt%, but no accurate results could be obtained due to the low loadings. In addition, two catalysts (that is, Ni/C-1 and Pd/C-1 containing ~0.18 wt% Ni

**Table 1 | Catalyst screening and optimization of the reaction conditions*.**

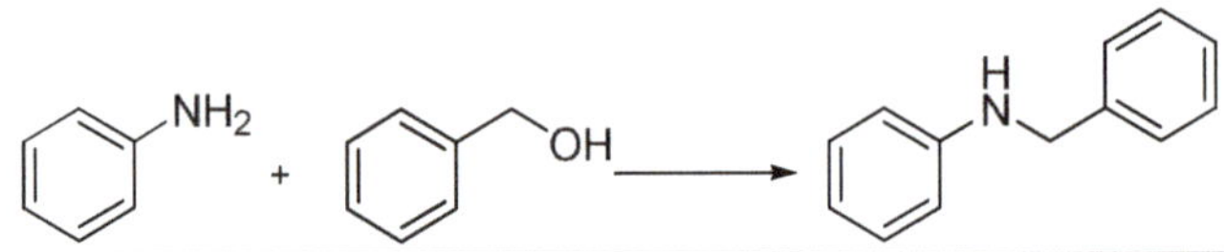


| Entry | Catalyst/mg | base/mmol | Conversion (%) | Yield (%)† |
|---|---|---|---|---|
| 1 | C-0/100 | KOH/0.5 | 60 | 42 |
| 2 | C-1/100 | KOH/0.5 | >99 | >99 (99‡) |
| 3 | C-2/100 | KOH/0.5 | 80 | 68 |
| 4 | C-3/100 | KOH/0.5 | 80 | 74 |
| 5 | C-4/100 | KOH/0.5 | 14 | 3 |
| 6 | C-5/100 | KOH/0.5 | 70 | 50 |
| 7 | C-6/100 | KOH/0.5 | 80 | 71 |
| 8 | C-7/100 | KOH/0.5 | >99 | >99 |
| 9 | C-1/75 | KOH/0.5 | >99 | >99 |
| 10 | C-1/50 | KOH/0.5 | >99 | 99 |
| 11§ | C-1/25 | KOH/0.5 | 42 | 33 |
| 12‖ | C-1/100 | KOH/0.5 | 99 | 98 |
| 13¶ | C-1/100 | KOH/0.5 | 99 | 98 |
| 14# | C-1/100 | KOH/0.5 | 97 | 89 |
| 15** | C-1/100 | KOH/0.5 | 45 | 21 |
| 16 | C-1/100 | — | 15 | 0 |
| 17 | C-1/100 | KOH/0.3 | 32 | 7 |
| 18 | C-1/100 | KOH/0.4 | >99 | 87 |
| 19 | C-1/100 | $Na_2CO_3$/0.5 | 14 | 3 |
| 20 | C-1/100 | NaOH/0.5 | 80 | 68 |
| 21 | C-1/100 | $K_2CO_3$/0.5 | 31 | 2 |
| 22 | C-1/100 | K-O-t-Bu/0.5 | 85 | 89 |
| 23§ | 0.18 wt% Ni/C-1/25 | KOH/0.5 | 28 | 23 |
| 24§ | 0.07% wt Pd/C-1/25 | KOH/0.5 | 22 | 15 |
| 25 | graphene/100 | KOH/0.5 | 22 | 16 |
| 26 | graphite/100 | KOH/0.5 | 50 | 39 |
| 27 | carbon nanotubes/100 | KOH/0.5 | 29 | 20 |
| 28 | AC-Vulcan XC-72/100 | KOH/0.5 | 45 | 35 |
| 29 | AC-KB/100 | KOH/0.5 | 48 | 14 |

*Reaction conditions: 1 mmol aniline, 2 mmol benzyl alcohol, 100 mg catalyst, 0.5 mmol KOH, 2 ml toluene, argon atmosphere, 130 °C and 24 h.
†Determined by GC-FID using biphenyl as the external standard material.
‡C-1 was used in the fifth run.
§The reactions were repeated two times and the average results are given in the table.
‖1.5 equiv., benzyl alcohol.
¶1.2 equiv., benzyl alcohol.
#120 °C.
**100 °C.

and 0.07 wt% Pd, respectively) were prepared as control reactions. The results were even less desirable than that obtained with C-1 (entries 23 and 24, for the explanation of the results, see the discussions below). Therefore, the catalytic activity was not derived from metal contamination. When other carbon materials, such as commercially available multi-wall carbon nanotubes, graphene, graphite and activated carbon, were employed in the reaction, the results were less desirable than that obtained using C-0 as a catalyst (entries 25–29). Therefore, the catalytic activity of our catalyst is unique for the alcohol amination reactions. In addition, note that air/oxygen should be excluded because it is detrimental to borrowing-hydrogen reactions. For example, the conversions of benzyl alcohol were 75–95% with 0–14% *N*-benzyl aniline yields when the reactions were performed under $O_2$ with C-0 to C-6 as catalysts (Supplementary Table 1). The major byproduct was *N*-benzylideneaniline. The optimized reaction conditions for benzyl alcohol amination with aniline are as follows: 1 mmol aniline, 1.2 equiv. benzyl alcohol, 50 mg catalyst, 50 mol% KOH, 2 ml toluene, Ar, 130 °C and 24 h.

Note that the carbon catalyst can be easily recovered and reused. After the reaction, the carbon was recovered, washed with acetone, separated by centrifugation and dried for use in subsequent applications. The reaction was studied with aniline and benzyl alcohol in a molar ratio of 1:2, and the carbon catalyst exhibited stability for up to five runs without any deactivation (entry 2).

**Amine *N*-alkylation with benzyl alcohol.** The catalytic performance of the carbon catalyst in the alcohol amination reactions with respect to amine and alcohol derivatives was explored. First, the amination reactions of alcohol with different amines were explored (Table 2). Various structurally diverse amines, regardless of the presence of electron-withdrawing or electron-donating functional groups, were monoalkylated with benzyl alcohol to yield the corresponding secondary amines in excellent yields. For example, *N*-benzyl aniline was synthesized in 99% yield at 130 °C (entry 1). The substituents at different positions on the aniline significantly affected the reaction rate. For example, a lower yield was obtained when *o*-methylaniline was used as a substrate compared with *p*-methylaniline and *m*-methylaniline (entries 2–4). The reaction tolerated the presence of halogens (entries 5–7), and 90–99% yields of the desired products were observed when *o*-, *m*- or *p*-chloroanilines were used as starting materials. However, the lower reaction rate of *o*-chloronitrobenzene relative to those of the para- and meta-analogues indicated a steric effect. The reaction of 2-aminobiphenyl and benzyl alcohol resulted in a 92% yield using a prolonged reaction time of 36 h (entry 8). Aniline derivatives, heteroaromatic amines, substituted benzyl amines, aliphatic secondary cyclic amines and aliphatic primary amines produced high yields of the corresponding secondary amines, demonstrating the high versatility of the current methodology for secondary amine synthesis (entries 9–16).

**Aniline *N*-alkylation with alcohol derivatives.** The reaction also proceeded successfully with other structurally and electronically diverse alcohols (Table 3). Benzyl alcohol with moderately activating groups reacted smoothly and furnished excellent yields of the respective mono-*N*-alkylated anilines (entries 1–6). The substituent at the ortho position had a substantial effect on the reaction. Harsher conditions (that is, higher temperatures and increased time) were required to obtain high yields when *o*-chlorobenzyl alcohol, *o*-methoxybenzyl alcohol and *o*-methylbenzyl alcohol were used (entries 2, 5, 8). Alcohols with a chlorine substituent reacted with aniline to afford the corresponding products with 100% conversion but with a low yield due to the

**Table 2 | Benzyl alcohol amination with amines*.**

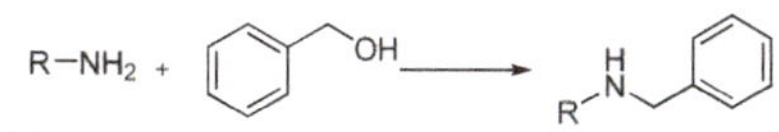

| Entry | Amine | Product | Yield (%)† |
|---|---|---|---|
| 1 | 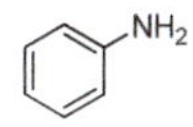 | 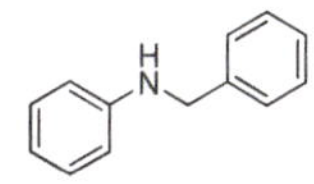 | >99%‡ |
| 2 | 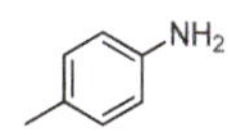 | 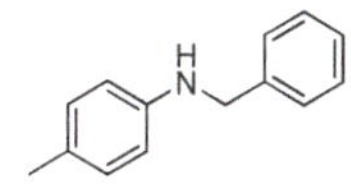 | 97 |
| 3 | 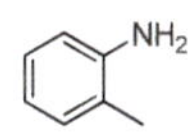 | 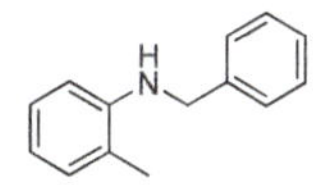 | 80 |
| 4 | 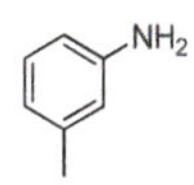 | 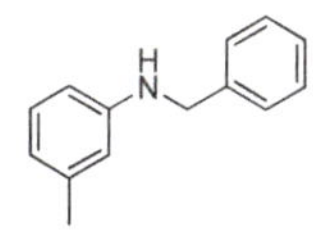 | 91 |
| 5 | 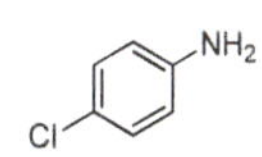 | 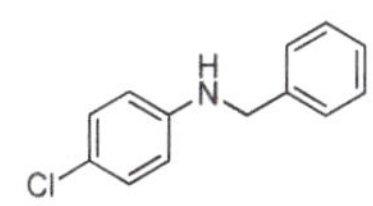 | 99 |
| 6§ | 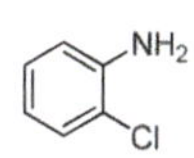 | 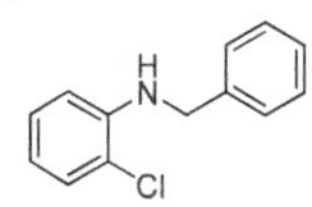 | 90 |
| 7 | 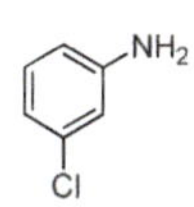 | 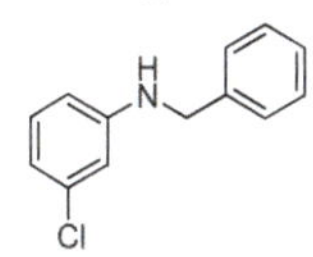 | 95 |
| 8‖,¶ | 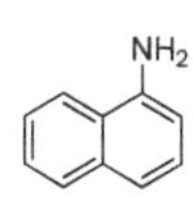 | 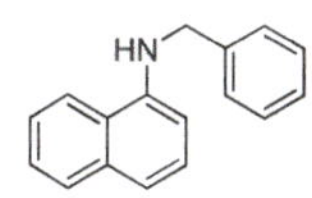 | 92 |
| 9§,¶ | 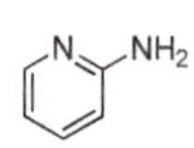 | 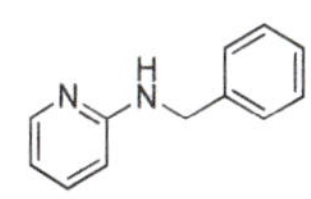 | 90 |
| 10 | 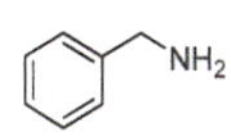 | 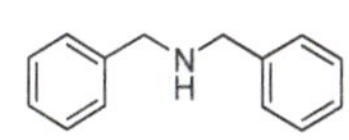 | 85‡ |
| 11 | | | 87‡ |
| 12 | 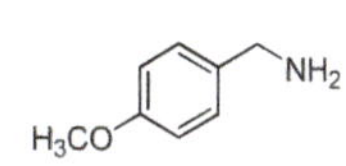 | 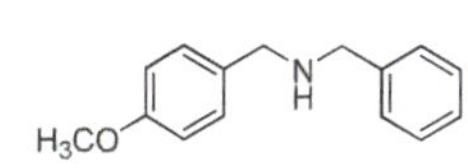 | 74‡ |
| 13 | 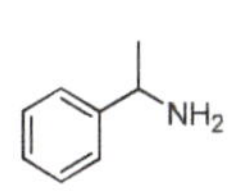 | 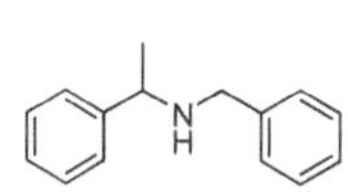 | 81‡ |
| 14 | 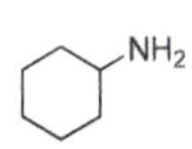 | 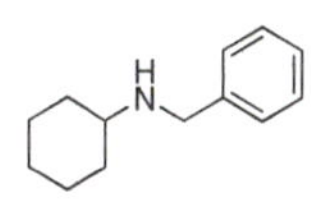 | 85‡ |
| 15 | 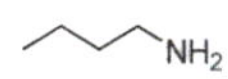 | 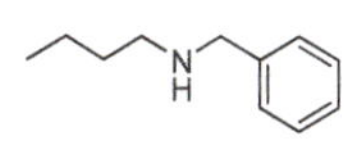 | 80‡ |
| 16‖,# | 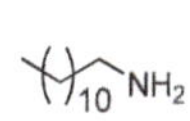 | 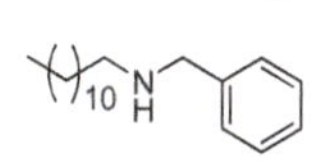 | 73‡ |

*Reaction conditions: 1 mmol amine, 2 mmol benzyl alcohol, 100 mg catalyst C-1, 0.5 mmol KOH, 2 ml toluene, argon atmosphere, 130 °C and 24 h.
†Isolated yield.
‡Determined by GC-FID using biphenyl as the internal standard.
§The reactions were performed at 140 °C.
‖1 equiv. KOH was used.
¶The reaction time was 36 h.
#The reaction time was 48 h.

 

**Table 3 | Results of the amination reaction of alcohols*.**

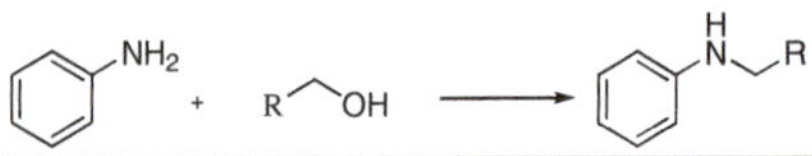


| Entry | Alcohol | Product | Yield (%)† |
|---|---|---|---|
| 1 | 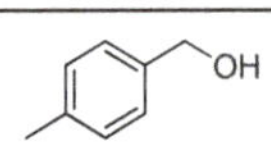 | 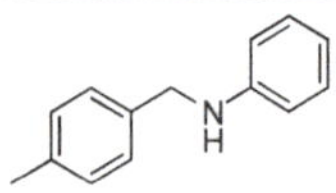 | 97 |
| 2 | 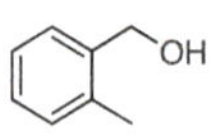 | 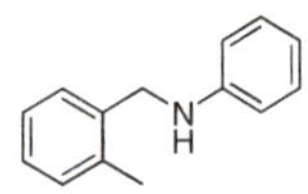 | 93 |
| 3 | 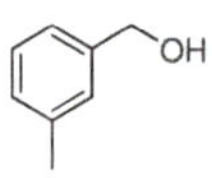 | 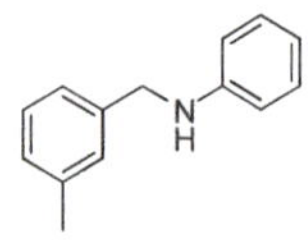 | 88 |
| 4 | 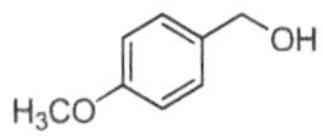 | 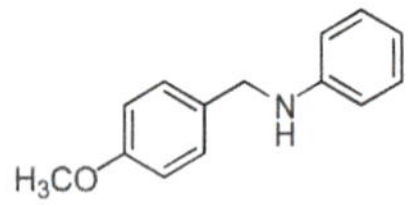 | 95 |
| 5 | 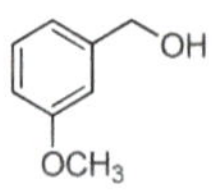 | 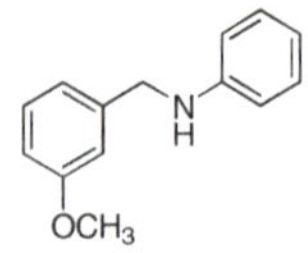 | 85 |
| 6 | 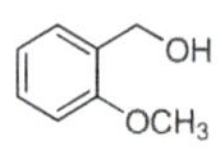 | 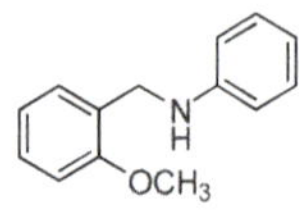 | 81 |
| 7 | 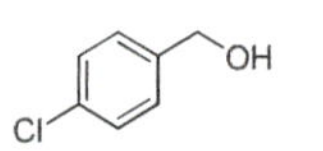 | 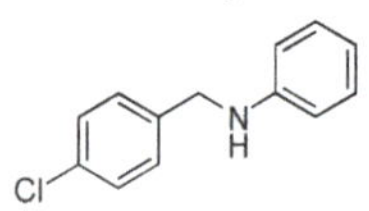 | 65‡ |
| 8 | 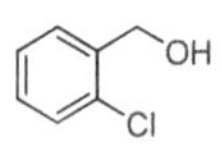 | 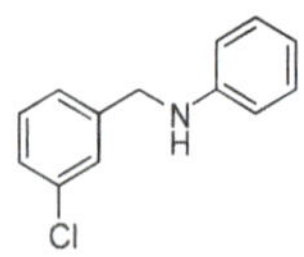 | 47‡ |
| 9 | 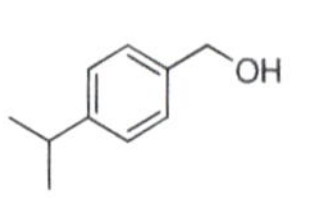 | 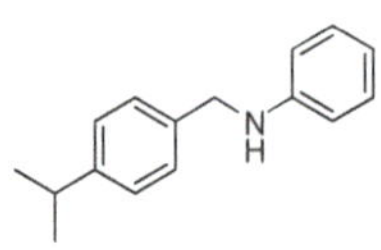 | 96 |
| 10§ | 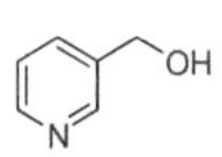 | 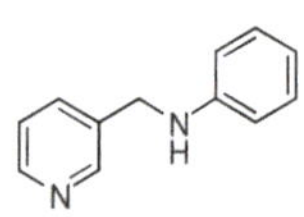 | 93 |
| 11§ | 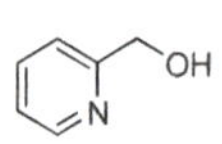 | 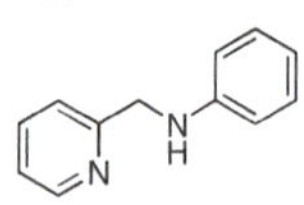 | 74 |
| 12§ | 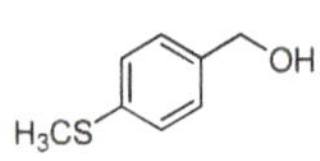 | 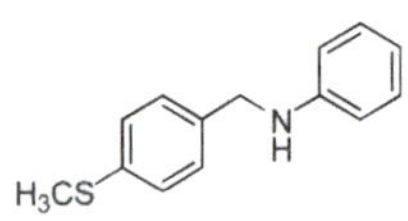 | 92 |
| 13§ | 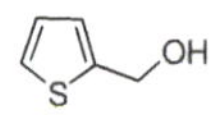 | 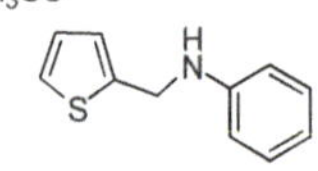 | 80 |
| 14§ | 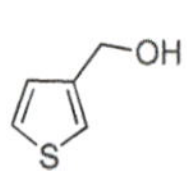 | 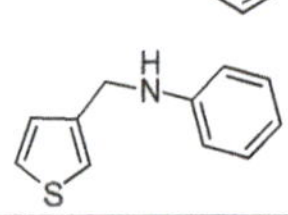 | 71 |

*Reaction conditions: 1 mmol aniline, 2 mmol alcohol, 100 mg catalyst, 0.5 mmol KOH, 2 ml toluene, argon atmosphere, 130 °C and 24 h.
†Isolated yield.
‡Determined by GC-FID using biphenyl as the internal standard.
§1.0 mmol KOH.

dechlorination reaction (entries 7–8). Unfortunately, the reactions with a fluorine and bromine substituent were unsuccessful. Nearly no reaction was detectable and the results are not listed in the table. A prolonged reaction time of 36 h was required to produce a high yield of the desired product when *p*-isopropylbenzyl alcohol was used as the substrate (entry 9). The C-1 catalyst also exhibits promise for direct coupling of aniline with simple electron-poor heteroaromatic alcohols under similar conditions in moderate yields (entries 10–11, 13–14). *p*-Methylthiobenzyl alcohol was converted to the respective secondary amine in high yield. Unfortunately, aliphatic primary alcohols and aliphatic secondary cyclic alcohols failed to yield any product under similar conditions.

**Nitrobenzene reduction with isopropanol.** As discussed above, in the reductive hydrogen atom transfer reactions, there are two classical reactions, that is, transfer hydrogenation of nitro compounds[62–64] and carbonyl group reduction[65–67]. In addition, it has been shown that carbon can function as an active catalyst for nitrobenzene reduction with hydrazine as the hydrogen source[33]. Therefore, the catalytic activity of catalyst C-1 in nitro compound reductions was further tested, and the results are shown in Table 4. For the catalytic reduction of nitrobenzene and its derivatives with a methyl group at various positions on the aromatic ring, the corresponding aniline derivatives were synthesized with up to an 88% yield (entries 1–5). The major byproducts were azobenzene and azoxybenzene, based on the gas chromatography–mass spectrometry and GC-FID analyses. On the basis of the appearance of the reaction mixture, it is possible that a small amount of high molecular weight material was formed, but this possibility was difficult to determine. The presence of other functional groups, such as acetyl, halogen and phenyl groups, only had a small influence on the reaction, and similar results were obtained with a prolonged reaction time (entries 6–9). A good yield was also obtained when 1-nitronaphthalene was employed as a starting material (entry 10). Therefore, the carbon catalyst exhibited good activity for the reduction of nitrobenzene derivatives using isopropanol as a hydrogen donor.

**Ketone reduction with isopropanol.** The catalytic reduction of ketones with isopropanol as a hydrogen donor was performed using C-1 as a catalyst, and the results are listed in Table 5. Acetophenone derivatives containing different substituents were efficiently transformed into the corresponding alcohols in up to 99% yields (entries 1–7). Although no dechlorination reactions were observed for the reduction of *p*-Cl-acetophenone, debromination and dechlorination reactions occurred when 2,4-dichloroacetophenone and *p*-Br-acetophenone were applied as starting materials. In addition, 1-(naphthalenyl)-2-ethanone, propiophenone and benzophenone could be reduced smoothly, and the yields of the desired products were 98–99% (entries 8–10). The catalytic reduction of cyclic ketones, such as cyclohexanone and cyclooctanone, was also achieved in 91–99% yields (entries 11–13). In addition, linear ketones with different alkyl chain lengths could be used as starting materials, and these compounds afforded the desired products in 98% yields (entries 14–15). Therefore, our carbon catalyst system can be employed with ketone derivatives with versatile structures.

**Catalyst structure and reaction mechanism.** Next the carbon catalysts were characterized by X-ray photoelectron spectroscopy (XPS), Brunauer–Emmett–Teller (BET), X-ray diffraction (XRD), scanning electron microscopy (SEM) and Fourier transform infrared spectroscopy (FT-IR) to explore the structure of the

**Table 4 | Reduction of nitro compounds with isopropanol as the hydrogen donor*.**

| Entry | Nitrobenzene | Conversion (%) | Yield (%)† |
|---|---|---|---|
| 1 | 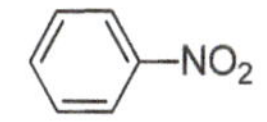 | >99 | 80‡ |
| 2 | 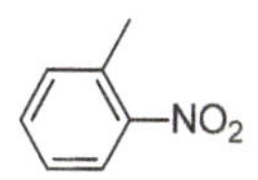 | >99 | 81‡ |
| 3 | 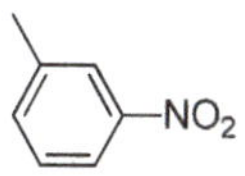 | >99 | 75 |
| 4 | 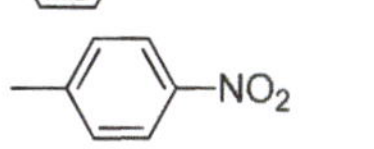 | >99 | 60‡ |
| 5 | 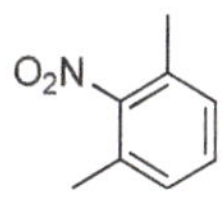 | >99 | 88‡ |
| 6 | 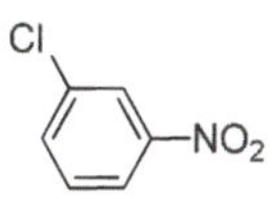 | >99 | 61 |
| 7 | 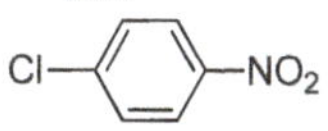 | >99 | 72 |
| 8 | 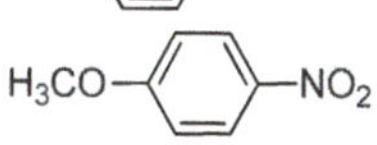 | >99 | 79 |
| 9 | 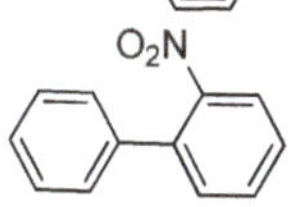 | >99 | 85 |
| 10 | 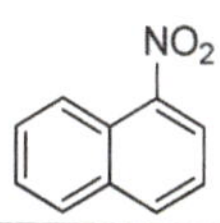 | >99 | 68 |

*Reaction conditions: 1 mmol nitro compound, 100 mg catalyst C-1, 3 mmol KOH, 2 ml isopropanol, argon atmosphere, 100 °C and 24 h (entries 1-6) or 48 h (entries 7-10).
†Isolated yields.
‡Determined by GC-FID using biphenyl as the internal standard. The primary byproducts observed by gas chromatography-mass spectrometry and GC-FID were azobenzene and azoxybenzene.

**Table 5 | Ketone reduction with isopropanol as the hydrogen donor*.**

| Entry | Ketone | Conversion (%) | Yield (%)† |
|---|---|---|---|
| 1 | 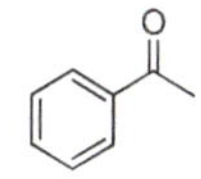 | >99 | >99 |
| 2 | 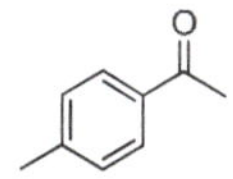 | >99 | 94 |
| 3 | 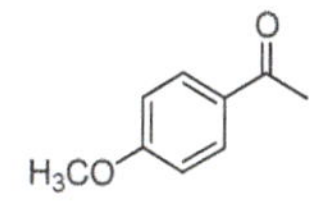 | >99 | 94‡ |
| 4 | 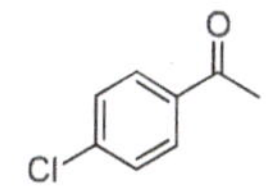 | >99 | 99 |
| 5 | 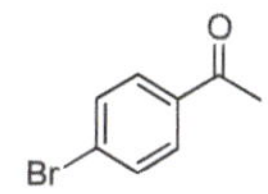 | >99 | 75‡ |
| 6 | 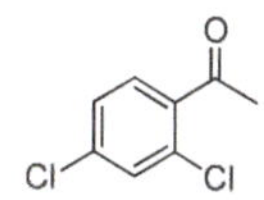 | >99 | 99‡ |
| 7 | 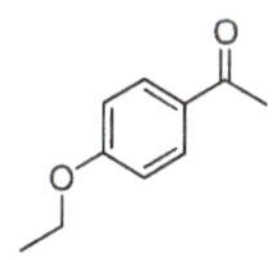 | >99 | 92‡ |
| 8 | 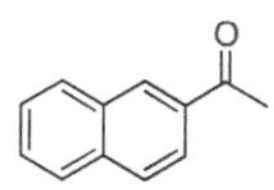 | >99 | 98‡ |
| 9 | 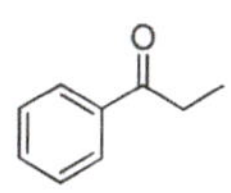 | >99 | 95 |
| 10 | 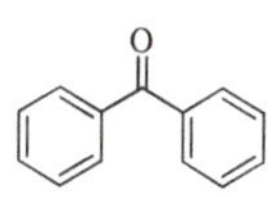 | >99 | 99 |
| 11 | 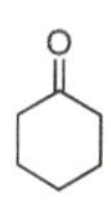 | >99 | >99‡ |
| 12 | 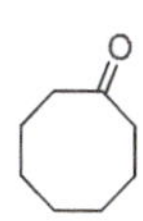 | >99 | 91‡,‖ |
| 13 | 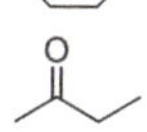 | >99 | 95§ |
| 14 | 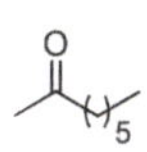 | >99 | 96§ |
| 15 | 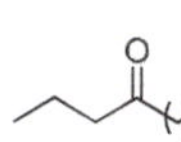 | >99 | 98‡,‖ |

*Reaction conditions: 1 mmol ketone, 100 mg catalyst C-1, 1 mmol KOH, 2 ml isopropanol, argon atmosphere, 130 °C and 24 h (entries 1-4,6,9,12,14) or 36 h (entries 5,8,15) or 48 h (entries (7,10-11,13,16).
†Isolated yield.
‡Determined by GC-FID using biphenyl as the internal standard.
§2 mmol KOH was used.
‖$^1$H NMR yield.

active catalyst. The C1*s* spectra suggested that catalysts C-1 to C-5 possess similar surface carbon species. In addition, the O1*s* spectra of the different carbon materials were different (Supplementary Fig. 1). Active catalysts C-1 and C-3 exhibited a similar activity, although their O1*s* spectra were different. The less active catalyst (C-4) possessed an O1*s* spectrum that was similar to that of the most active catalyst (C-1). Similar phenomena were observed from the surface C:O ratios. The surface C:O ratios of C-0, C-1, C-3, C-4 and C-5 were over the range of 7:1 to 8:1. However, the surface C:O ratio of C-2 was 16:1. Therefore, it is difficult to explain the variations in catalytic activity based on the XPS characterization results. The catalysts were also characterized by XRD. Although the diffraction patterns are not shown, similar results were obtained from all of the catalysts, and nearly all of the carbon materials were amorphous.

Next the carbon catalysts were characterized using s.e.m. and the nitrogen adsorption–desorption method. The SEM images of the different catalysts indicated that the treatment of the RF gel with base before carbonization resulted in the formation of a different porous structure (Fig. 3). The porosity of the carbon materials was consistent with their BET surface areas. The surface area of C-0, which was not treated with a base, was only 220.5 $m^2 g^{-1}$, and the surface areas were 1,220.9, 408.1, 841.8, 407.7 and 582.0 $m^2 g^{-1}$ for catalysts C-1 to C-5, respectively. In

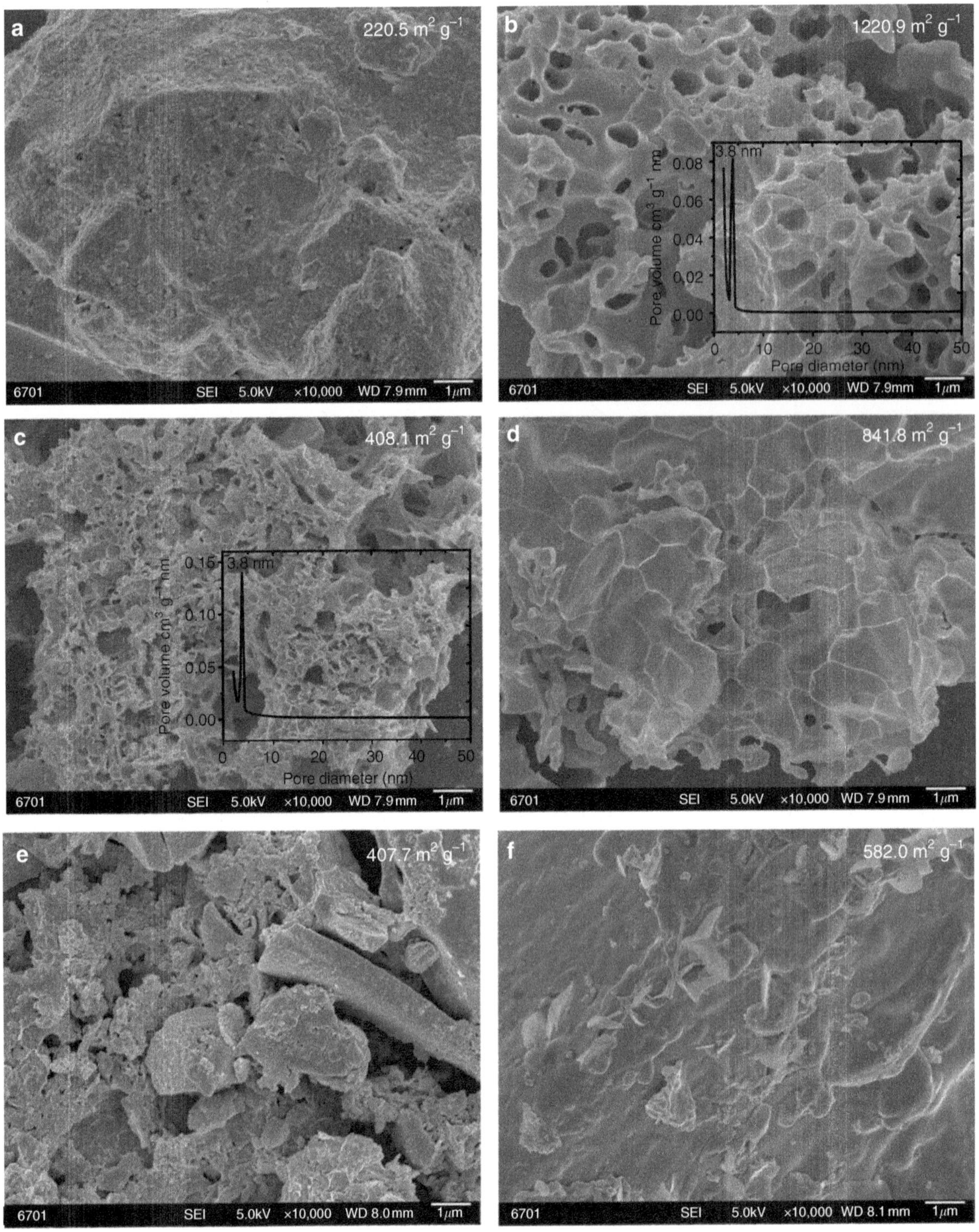


**Figure 3 | SEM images of the carbon materials.** (**a**) C-0, (**b**) C-1, (**c**) C-2, (**d**) C-3, (**e**) C-4 and (**f**) C-5 (scale bar, 1 μm).

addition, mesoporous structures were observed in samples C-1 and C-2. However, the surface area contributed less to the catalytic performance because a similar activity was obtained with C-3. The high BET surface area, along with the mesoporous structure, may be responsible for the high catalytic activity. In addition, the BET surface areas of Ni/C-1 and Pd/C-1 were 1,168 and 1,268 $m^2 g^{-1}$, respectively, and may not be responsible for their relatively low activities compared with C-1.

To determine the possible active site on the carbon catalyst, FT-IR characterization was performed, and the band associated with C = C and the strongly conjugated C = O was observed[68,69]. Interestingly, the variations in catalytic performance were related to the density of the C = C and C = O groups at ~1,570 $cm^{-1}$ in the different catalyst samples, and the strongest adsorptions were observed in samples C-1 and C-2 (Fig. 4a–f). It is illogical to attribute the reactivity to the C = C group because the reverse reaction, that is, alkane dehydrogenation, is less feasible. Thus, the catalytically active site might be the C = O group. For catalysts, 0.18 wt% Ni/C-1 and 0.07 wt% Pd/C-1, the FT-IR adsorption peaks at 1,570 $cm^{-1}$ were weaker than those observed in the FT-IR spectra of C-1 (Fig. 4g,h). Thus, the lower C = O concentrations might have caused their lower activity. To clarify this result, catalyst C-1 was treated with molecular oxygen at 150 °C to verify the variation in the FT-IR spectra

(Fig. 4a′–d′). The C=O adsorption band increased in intensity after the treatment of C-1 with oxygen. The intensity of this band decreased again when it was further treated with isopropanol, and acetone was formed simultaneously. These results suggested that recycling of the C=O group occurred in the transfer hydrogenation reactions. The FT-IR adsorption peaks of multi-wall carbon nanotubes, graphene, graphite and activated carbon were weak at 1,570 cm$^{-1}$, which might explain their low activities (Supplementary Fig. 2 and Table 1, entries 25–29). It should be mentioned that a high surface area may be important to achieve high activity as well.

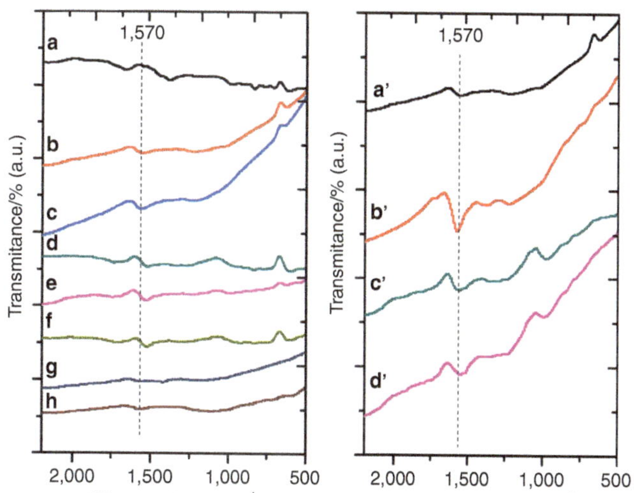


**Figure 4 | FT-IR spectra.** FT-IR spectra of C-0 to C-5 (**a**-**f**), 0.18 wt% Ni/C-1 (**g**), 0.07 wt% Pd/C-1 (**h**) and C-1 treated with oxygen and alcohol (**a′**: C-1; **b′**: C-1-O and C-1/KOH were treated with oxygen at 150 °C for 24 h; **c′**: C-1-O-R and C-1-O/KOH were treated with isopropanol at 150 °C for 24 h; **d′**: C-1-U and C-1 were reused for five runs in the coupling reaction of aniline and benzyl alcohol).

Subsequently, benzophenone and tetraphenyl ethylene were used as the model substrates to explore the potential roles of ketone and olefin in the transfer hydrogenation reaction (Fig. 5 and Supplementary Fig. 3). First, the reactions of benzophenone and tetraphenyl ethylene with benzyl alcohol were performed (**a**). As expected, 38.5% of the benzophenone was converted to benzohydrol, and benzaldehyde was simultaneously generated. For the reaction of tetraphenyl ethylene and benzyl alcohol, no tetraphenyl ethane was observed (**b**). Then, the benzophenone-catalysed amination reaction of aniline and benzyl alcohol was performed (**c**). Although this catalyst is not as active as our carbon catalyst, 85% aniline conversion was obtained and the selectivity to *N*-benzyl aniline was 34.5%. The byproduct was *N*-benzylideneaniline with 65.5% selectivity. The results suggested that benzophenone was an active catalyst for the dehydrogenation of benzyl alcohol to generate benzaldehyde, but it was less active in the transfer hydrogenation of *N*-benzylideneaniline to produce *N*-benzyl aniline. Therefore, these results support the catalytic recycling between C=O and CHOH. In addition, the alcohol amination reaction of aniline with benzyl alcohol was performed with the addition of 5 mol% butylated hydroxyl toluene or 2,2,6,6-tetramethylpiperidine-1-oxyl to determine whether the reaction is radical-based (for the reaction conditions, see Table 1, entry 2). The yields of *N*-benzyl aniline were 98 and 99%, which were nearly the same as the C-1/KOH catalyst system. So, the incorporation of a radical-based mechanism in the carbon-catalysed reductive hydrogen atom transfer reactions can be excluded. On the basis of these discussions, a reaction mechanism was proposed (**d**). Generally, in the alcohol amination reaction, (1) the alcohol was transformed to an aldehyde via hydrogen transfer from the alcohol to the C=O groups of the carbon catalyst; (2) the aldehyde reacted with the amine to generate an imine; and (3) the hydrogen was transferred from the carbon

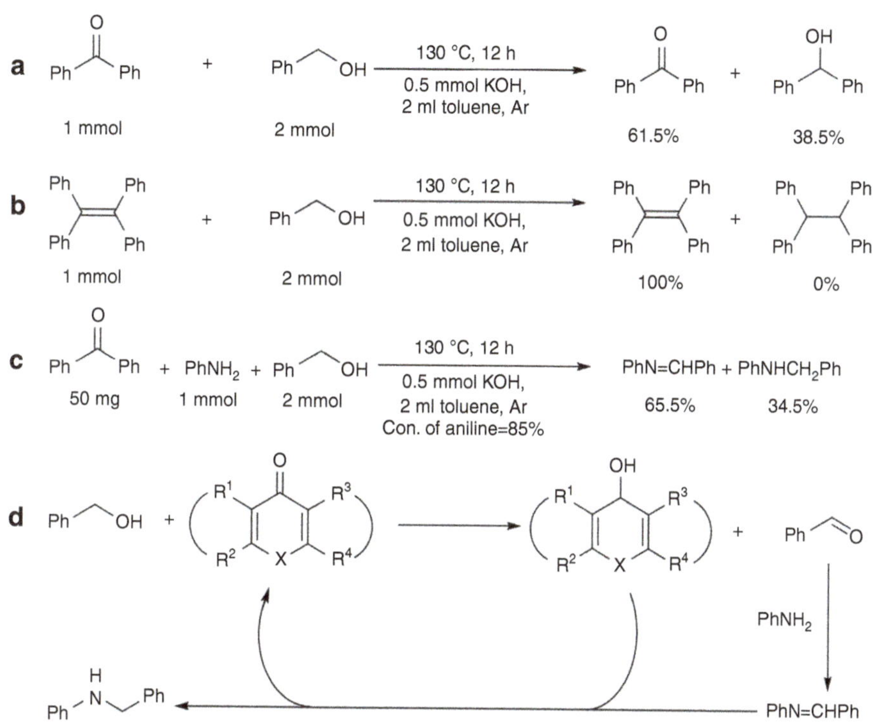


**Figure 5 | Reaction mechanism exploration.** (**a**) Reaction of benzophenone and benzyl alcohol; (**b**) Reaction of tetraphenyl ethylene with benzyl alcohol; (**c**) Benzophenone catalyzed benzyl alcohol amination with aniline and (**d**) Possible reaction mechanism. All the results were determined by gas chromatography-mass spectrometry. Con., conversion.

catalyst to the imine, resulting in the formation of an *N*-alkyl amine and the regeneration of the C=O group.

## Discussion

In summary, the transition metal-free reductive hydrogen atom transfer transformation was realised using a carbon material as the catalyst with alcohol amination as well as nitro compound and ketone reduction as model reactions. The carbon catalyst exhibited excellent catalytic performance and good generality in the above reactions. The catalyst/product separation was easily performed, and the carbon catalyst was recyclable for several runs without observable deactivation. The results suggested that the C=O group may be the catalytically active site in this catalyst, while a high surface area of the carbon catalyst is important for achieving high activity. The activity of the base-activated carbon catalyst remained unchanged during the reuse process due to its stable structure. This new catalytic system may provide an attractive and useful methodology for a wider range of applications for industrially important compounds and intermediates.

## Methods

**General.** XRD measurements were conducted on an X'Pert PRO (PANalytical) diffractometer. The XRD diffraction patterns were scanned in the $2\theta$ range of 10 to 80°. The XPS measurements were performed using a Thermo Scientific ESCALAB 250 instrument with a dual Mg/Al anode X-ray source, a hemispherical capacitor analyser and a 5 keV $Ar^+$ ion-gun. All of the spectra were recorded using non-monochromatic Mg K$\alpha$ (1,253.6 eV) radiation. Nitrogen adsorption–desorption isotherms were measured at 77 K using a Micromeritics 2010 instrument. The pore size distribution was calculated using the Barrett, Joyner and Halenda method from the desorption isotherm. FT-IR spectroscopy characterizations were performed on a Nicolet 5700 spectrometer. The sample was prepared by mixing 0.5 mg carbon material with 100 mg KBr. SEM was performed with a JEOL JSM-6701F equipped with a cold FEG (Field Emission Gun). The metal loadings of the catalysts were measured with an inductively coupled plasma-atomic emission spectrometry (ICP-AES) instrument (ARL Co. USA, 3520). NMR spectra were measured using a Varian NMR system at 400.1 MHz ($^1H$) and 100.6 MHz ($^{13}C$). All spectra were recorded in $CDCl_3$ and chemical shifts ($\delta$) are reported in p.p.m. relative to tetramethylsilane referenced to the residual solvent peaks.

**Typical procedure for the preparation of the carbon catalysts.** The carbon materials were prepared via sol-gel polymerization of resorcinol and formaldehyde with $Na_2CO_3$ as a catalyst. Generally, resorcinol (R, 2.20 g, 20 mmol), formaldehyde (F, 3.25 g, 40 mmol, 36.5% in water, methanol stabilized) and 9 ml of distilled water were added into a 100- ml PTFE autoclave. Then, 22.1 mg of $Na_2CO_3$ (1 mol% to resorcinol) was added to the mixture. After vigorous stirring for 1 h, the autoclave was sealed and maintained at 80 °C for 1 day. Then, the autoclave was cooled to room temperature. The wet gels were placed in a round-bottom flask and vacuum dried at 130 °C for 3 h. Next the RF gel and KOH were mixed with a mass ratio of 1:1 (1.5 g/1.5 g) and carbonized at 800 °C for 5 h under a nitrogen flow (20 ml $min^{-1}$). The resulting carbon material was ultrasonically washed with distilled water until neutral (100 ml × 5) and dried at 80 °C for 6 h. Finally, ~730 mg of the carbon material was obtained and denoted C-1. Several carbon materials were prepared using the same procedure involving treatment of the RF gel and different bases (that is, NaOH (1.5 g/1.5 g), $K_2CO_3$ (1.5 g/1.5 g) and $Na_2CO_3$ (1.5 g/1.5 g)), and the resulting carbon materials are referred to as C-2, C-3 and C-4, respectively. By varying the amount of KOH (that is, 0.375 g, 0.75 g and 1.88 g), carbon materials C-5, C-6 and C-7 were obtained, respectively. For comparison, the RF gel itself was also carbonized at 800 °C for 5 h under a nitrogen flow (20 ml $min^{-1}$), and the final sample was called C-0. C-1-O was prepared by treating C-1 (400 mg C-1, 112 mg KOH, and 4 ml toluene) under 1 MPa oxygen atmosphere at 150 °C for 24 h. C-1-O-R was obtained by treating C-1-O (400 mg C-1-O, 112 mg KOH, and 2 ml isopropanol) with isopropanol at 150 °C for 24 h. C-1-U refers to catalyst C-1 being used for five runs in the coupling reaction of aniline and benzyl alcohol, as shown in Table 1, entry 2 and Supplementary Fig. 1.

Multi-wall carbon nanotubes (Shenzhen Nanotech Port Co., surface area: 40–300 $m^2 g^{-1}$), graphite (Sinopharm Chemical Reagent Co.), graphene (prepared using the arc-discharge method by the Laboratory of Clean Energy Chemistry and Materials, LICP, CAS, surface area: 30–100 $m^2 g^{-1}$), AC-Vulcan XC-72 (Vulcan XC-72 carbon, Cabot, USA, surface area: ~200 $m^2 g^{-1}$) and AC-KB (Henan Huaxian Activated Carbon Company, surface area: ~2,000 $m^2 g^{-1}$) were used for control reactions.

**Typical procedure for benzyl alcohol amination with aniline.** All of the reactions were performed in a pressure tube. Aniline (1 mmol), 2 mmol benzyl alcohol, 100 mg of catalyst C-1, 0.5 mmol KOH and 2 ml of toluene were added to a pressure tube followed by exchange with Ar. The tube was sealed and maintained at 130 °C for 24 h. After completion of the reaction, the tube was cooled to room temperature, and 1 mmol of biphenyl and 10 ml of EtOH were added for quantitative analysis with a GC-FID (Agilent 7890 A). The crude mixture was purified by column chromatography. The $^1H$ and $^{13}C$ NMR spectra and characterization data of the isolated compounds were given in Supplementary Figs 4–23 and Supplementary Methods.

**Typical procedure for nitrobenzene reduction.** All of the reactions were performed in a Shrek tube. Nitrobenzene (1 mmol), 100 mg of catalyst C-1, 3 mmol KOH and 2 ml of isopropanol were added to a Shrek tube followed by exchange with $H_2$. Then, the tube was sealed and maintained under a $H_2$ atmosphere at 100 °C for 24 h ($H_2$ balloon). After completion of the reaction, the tube was cooled to room temperature, and 1 mmol of biphenyl and 10 ml of EtOH were added for quantitative analysis with a GC-FID (Agilent 7890 A). The crude mixture was purified by the following steps. First, the isopropanol was removed on a Rot-Vap. Then, the aniline was extracted by $CH_2Cl_2$, and pure aniline was obtained after removing the $CH_2Cl_2$ and vacuum drying. The $^1H$ and $^{13}C$ NMR spectra and characterization data of the isolated compounds were given in Supplementary Figs 24–29 and Supplementary Methods.

**Typical procedure for acetophenone reduction.** All of the reactions were performed in a Shrek tube. Acetophenone (1 mmol), 100 mg of catalyst C-1, 1 mmol KOH and 2 ml of isopropanol were added to a Shrek tube followed by exchange with $H_2$. Then, the tube was sealed and maintained under a $H_2$ atmosphere at 100 °C for 24 h ($H_2$ balloon). After completion of the reaction, the tube was cooled to room temperature, and 1 mmol of biphenyl and 10 ml of EtOH were added for quantitative analysis with a GC-FID (Agilent 7890 A). The crude mixture was purified by the following steps. First, the catalyst was removed by filtration. Then, the pure product was obtained after removing the organic solvent and vacuum drying. The $^1H$ and $^{13}C$ NMR spectra and characterization data of the isolated compounds were given in Supplementary Figs 30–34 and Supplementary Methods.

## References

1. Zewail, A. H. *The Remarkable Phenomena of Hydrogen Transfer* (WILEY-VCH, 2007).
2. Samec, J. S. M., Baeckvall, J.-E., Andersson, P. G. & Brandt, P. Mechanistic aspects of transition metal-catalyzed hydrogen transfer reactions. *Chem. Soc. Rev.* **35,** 237–248 (2006).
3. Mayer, J. M. Understanding hydrogen atom transfer: from bond strengths to Marcus theory. *Acc. Chem. Res.* **44,** 36–46 (2010).
4. Alonso, F., Riente, P. & Yus, M. Nickel nanoparticles in hydrogen transfer reactions. *Acc. Chem. Res.* **44,** 379–391 (2011).
5. Dobereiner, G. E. & Crabtree, R. H. Dehydrogenation as a substrate-activating strategy in homogeneous transition-metal catalysis. *Chem. Rev.* **110,** 681–703 (2010).
6. Blomberg, M. R. A. & Siegbahn, P. E. M. Different types of biological proton transfer reactions studied by quantum chemical methods. *Biochim. Biophys. Acta* **1757,** 969–980 (2006).
7. Allen, C. L. & Williams, J. M. Metal-catalysed approaches to amide bond formation. *Chem. Soc. Rev.* **40,** 3405–3415 (2011).
8. Bähn, S. *et al.* The catalytic amination of alcohols. *ChemCatChem* **3,** 1853–1864 (2011).
9. Haag, D. & Kung, H. H. Metal free graphene based catalysts: a review. *Top. Catal.* **57,** 762–773 (2014).
10. Shearer, C. J., Cherevan, A. & Eder, D. Application and future challenges of functional nanocarbon hybrids. *Adv. Mater.* **26,** 2295–2318 (2014).
11. Kong, X. K., Chen, C. L. & Chen, Q. W. Doped graphene for metal-free Catalysis. *Chem. Soc. Rev.* **43,** 2841–2857 (2014).
12. Zhang, J. *et al.* Surface-modified carbon nanotubes catalyze oxidative dehydrogenation of n-butane. *Science* **322,** 73–77 (2008).
13. Frank, B., Morassutto, M., Schomaecker, R., Schloegl, R. & Su, D. S. Oxidative dehydrogenation of ethane over multiwalled carbon nanotubes. *ChemCatChem* **2,** 644–648 (2010).
14. McGregor, J. *et al.* Active coke: carbonaceous materials as catalysts for alkane dehydrogenation. *J. Catal.* **269,** 329–339 (2010).
15. Rinaldi, A. *et al.* Oxidative purification of carbon nanotubes and its impact on catalytic performance in oxidative dehydrogenation reactions. *ChemSusChem* **3,** 254–260 (2010).
16. Dreyer, D. R., Jia, H. P. & Bielawski, C. W. Graphene oxide: a convenient carbocatalyst for facilitating oxidation and hydration reactions. *Angew. Chem. Int. Ed.* **49,** 6813–6816 (2010).
17. Zhang, P., Gong, Y., Li, H., Chen, Z. & Wang, Y. Solvent-free aerobic oxidation of hydrocarbons and alcohols with Pd@N-doped carbon from glucose. *Nat. Commun.* **4,** 1593 (2013).

18. Chen, Y., Zhang, J., Zhang, M. & Wang, X. Molecular and textural engineering of conjugated carbon nitride catalysts for selective oxidation of alcohols with visible light. *Chem. Sci.* **4,** 3244–3248 (2013).
19. Huang, H. *et al.* Graphite oxide as an efficient and durable metal-free catalyst for aerobic oxidative coupling of amines to imines. *Green Chem.* **14,** 930–934 (2012).
20. Dreyer, D. R., Jia, H.-P., Todd, A. D., Geng, J. & Bielawski, C. W. Graphite oxide: a selective and highly efficient oxidant of thiols and sulfides. *Org. Biomol. Chem.* **9,** 7292–7295 (2011).
21. Chen, H. *et al.* Metal-free direct amidation of peptidyl thiol esters with α-amino acid esters. *Green Chem.* **13,** 2723–2726 (2011).
22. Dhakshinamoorthy, A., Alvaro, M., Concepcion, P., Fornes, V. & Garcia, H. Graphene oxide as an acid catalyst for the room temperature ring opening of epoxides. *Chem. Commun.* **48,** 5443–5445 (2012).
23. Qi, J., Xu, Y., Ma, N. & Sun, F. Graphite oxide-catalyzed esterification and transesterification. *Chinese J. Org. Chem.* **33,** 1839–1846 (2013).
24. Dhakshinamoorthy, A., Alvaro, M., Puche, M., Fornes, V. & Garcia, H. Graphene oxide as catalyst for the acetalization of aldehydes at room temperature. *ChemCatChem* **4,** 2026–2030 (2012).
25. Verma, S. *et al.* Graphene oxide: an efficient and reusable carbocatalyst for aza-Michael addition of amines to activated alkenes. *Chem. Commun.* **47,** 12673–12675 (2011).
26. Acocella, M. R., Mauro, M., Falivene, L., Cavallo, L. & Guerra, G. Inverting the diastereoselectivity of the Mukaiyama-Michael addition with graphite-based catalysts. *ACS Catal.* **4,** 492–496 (2014).
27. Kumar, A. V. & Rao, K. R. Recyclable graphite oxide catalyzed friedel-crafts addition of indoles to alpha, beta-unsaturated ketones. *Tetrahedron Lett.* **52,** 5188–5191 (2011).
28. Chauhan, S. M. S. & Mishra, S. Use of graphite oxide and graphene oxide as catalysts in the synthesis of dipyrromethane and Calix 4 pyrrole. *Molecules* **16,** 7256–7266 (2011).
29. Wang, H. *et al.* Graphene oxide as a facile acid catalyst for the one-pot conversion of carbohydrates into 5-ethoxymethylfurfural. *Green Chem.* **15,** 2379–2383 (2013).
30. Basu, B., Kundu, S. & Sengupta, D. Graphene oxide as a carbocatalyst: the first example of a one-pot sequential dehydration-hydrothiolation of secondary aryl alcohols. *RSC Adv.* **3,** 22130–22134 (2013).
31. Perhun, T. I., Bychko, I. B., Trypolsky, A. I. & Strizhak, P. E. Catalytic properties of graphene material in the hydrogenation of ethylene. *Theor. Exp. Chem.* **48,** 367–370 (2013).
32. Pérez, M., Caputo, C. B., Dobrovetsky, R. & Stephan, D. W. Metal-free transfer hydrogenation of olefins via dehydrocoupling catalysis. *Proc. Natl Acad. Sci. USA* **111,** 10917–10921 (2014).
33. Gao, Y., Ma, D., Wang, C., Guan, J. & Bao, X. Reduced graphene oxide as a catalyst for hydrogenation of nitrobenzene at room temperature. *Chem. Commun.* **47,** 2432–2434 (2011).
34. Grigg, R., Mitchell, T. R. B., Sutthivaiyakit, S. & Tongpenyai, N. Transition metal-catalysed N -alkylation of amines by alcohols. *J. Chem. Soc. Chem. Soc.* 611–612 (1981).
35. Watanabe, Y., Tsuji, Y., Ige, H., Ohsugi, Y. & Ohta, T. Ruthenium-catalyzed N-alkylation and N-benzylation of aminoarenes with alcohols. *J. Org. Chem.* **49,** 3359–3363 (1984).
36. Hamid, M. H. S. A. C. *et al.* Ruthenium-catalyzed N-alkylation of amines and sulfonamides using borrowing hydrogen methodology. *J. Am. Chem. Soc.* **131,** 1766–1774 (2009).
37. Saidi, O. & Williams, J. M. J. Iridium-catalyzed hydrogen transfer reactions. *Top. Organomet. Chem.* **34,** 77–106 (2011).
38. Kawahara, R., Fujita, K.-i. & Yamaguchi, R. N-alkylation of amines with alcohols catalyzed by a water-soluble Cp*iridium complex: an efficient method for the synthesis of amines in aqueous media. *Adv. Synth. Catal.* **353,** 1161–1168 (2011).
39. Lee, S. A., Kwak, S. H. & Lee, K. I. Highly enantioselective synthesis of cyclic sulfamidates and sulfamides via rhodium-catalyzed transfer hydrogenation. *Chem. Commun.* **47,** 2372–2374 (2011).
40. Ghosh, R. & Sarkar, A. Palladium-catalyzed amination of allyl alcohols. *J. Org. Chem.* **76,** 8508–8512 (2011).
41. Zhang, Y., Qi, X., Cui, X., Shi, F. & Deng, Y. Palladium catalyzed N-alkylation of amines with alcohols. *Tetrahedron Lett.* **52,** 1334–1338 (2011).
42. Nagano, T. & Kobayashi, S. Palladium-catalyzed allylic amination using aqueous ammonia for the aynthesis of primary amines. *J. Am. Chem. Soc.* **131,** 4200–4201 (2009).
43. He, W. *et al.* Pt-Sn/gamma-$Al_2O_3$-catalyzed highly efficient direct synthesis of secondary and tertiary amines and imines. *Chem. Eur. J.* **17,** 13308–13317 (2011).
44. He, L. *et al.* Efficient and clean gold-catalyzed one-pot selective N-alkylation of amines with alcohols. *Chem. Eur. J.* **16,** 13965–13969 (2010).
45. Shimizu, K.-I., Shimura, K., Nishimura, M. & Satsuma, A. Silver cluster-promoted heterogeneous copper catalyst for N-alkylation of amines with alcohols. *RSC Adv.* **1,** 1310–1317 (2011).
46. Cui, X., Zhang, C., Shi, F. & Deng, Y. Au/Ag-Mo nano-rods catalyzed reductive coupling of nitrobenzenes and alcohols using glycerol as the hydrogen source. *Chem. Commun.* **48,** 9391–9393 (2012).
47. Winans, C. F. & Adkins, H. The alkylation of amine as catalyzed by nickel. *J. Am. Chem. Soc.* **54,** 306–312 (1932).
48. Shimizu, K.-i., Kon, K., Onodera, W., Yamazaki, H. & Kondo, J. N. Heterogeneous Ni catalyst for direct synthesis of primary amines from alcohols and ammonia. *ACS Catal.* **3,** 112–117 (2013).
49. Shimizu, K.-i., Imaiida, N., Kon, K., Hakim Siddiki, S. M. A. & Satsuma, A. Heterogeneous Ni catalysts for N-alkylation of amines with alcohols. *ACS Catal.* **3,** 998–1005 (2013).
50. Cui, X., Dai, X., Deng, Y. & Shi, F. Development of a general non-noble metal catalyst for the benign amination of alcohols with amines and ammonia. *Chem. Eur. J.* **19,** 3665–3675 (2013).
51. Yamaguchi, K., Kobayashi, H., Oishi, T. & Mizuno, N. Heterogeneously catalyzed synthesis of primary amides directly from primary alcohols and aqueous ammonia. *Angew. Chem. Int. Ed.* **51,** 544–547 (2012).
52. Yu, X., Liu, C., Jiang, L. & Xu, Q. Manganese dioxide catalyzed N-alkylation of sulfonamides and amines with alcohols under air. *Org. Lett.* **13,** 6184–6187 (2011).
53. Kim, I., Itagaki, S., Jin, X., Yamaguchi, K. & Mizuno, N. Heterogeneously catalyzed self-condensation of primary amines to secondary amines by supported copper catalysts. *Catal. Sci. Technol.* **3,** 2397–2403 (2013).
54. Martínez-Asencio, A., Ramón, D. J. & Yus, M. N-alkylation of poor nucleophilic amine and sulfonamide derivatives with alcohols by a hydrogen autotransfer process catalyzed by Copper(II) acetate. *Tetrahedron Lett.* **51,** 325–327 (2010).
55. Jammi, S. *et al.* CuO nanoparticles catalyzed C-N, C-O, and C-S cross-coupling reactions: scope and mechanism. *J. Org. Chem.* **74,** 1971–1976 (2009).
56. Cui, X. *et al.* Copper-catalyzed N-alkylation of sulfonamides with benzylic alcohols: catalysis and mechanistic studies. *Adv. Synth. Catal.* **351,** 2949–2958 (2009).
57. Likhar, P. R., Arundhathi, R., Kantam, M. L. & Prathima, P. S. Amination of alcohols catalyzed by copper-aluminium hydrotalcite: a green synthesis of amines. *Eur. J. Org. Chem.* 5383–5389 (2009).
58. Shi, F. *et al.* Copper-catalyzed alkylation of sulfonamides with alcohols. *Angew. Chem. Int. Ed.* **48,** 5912–5915 (2009).
59. Martinez, R., Ramon, D. J. & Yus, M. Selective N-monoalkylation of aromatic amines with benzylic alcohols by a hydrogen autotransfer process catalyzed by unmodified magnetite. *Org. Biomol. Chem.* **7,** 2176–2181 (2009).
60. Zhao, Y., Foo, S. W. & Saito, S. Iron/amino acid catalyzed direct N-alkylation of amines with alcohols. *Angew. Chem. Int. Ed.* **50,** 3006–3009 (2011).
61. Cui, X., Shi, F., Zhang, Y. & Deng, Y. Fe(II)-catalyzed N-alkylation of sulfonamides with benzylic alcohols. *Tetrahedron Lett.* **51,** 2048–2051 (2010).
62. Sabater, S., Mata, J. A. & Peris, E. Dual catalysis with an IrIII-AuI heterodimetallic complex: reduction of nitroarenes by transfer hydrogenation using primary alcohols. *Chem. Eur. J.* **18,** 6380–6385 (2012).
63. Sorribes, I. *et al.* Chemoselective transfer hydrogenation to nitroarenes mediated by cubane-type $Mo_3S_4$ cluster catalysts. *Angew. Chem. Int. Ed.* **51,** 7794–7798 (2012).
64. Verho, O., Nagendiran, A., Tai, C.-W., Johnston, E. V. & Backvall, J.-E. Nanopalladium on amino-functionalized mesocellular foam as an efficient and recyclable catalyst for the selective transfer hydrogenation of nitroarenes to anilines. *ChemCatChem* **6,** 205–211 (2014).
65. Hsu, S.-F. & Plietker, B. Selective transfer hydrogenation and hydrogenation of ketones using a defined monofunctional (P boolean and N(Bn)boolean and N(Bn)boolean and P)-Ru-II complex. *Chem. Eur. J.* **20,** 4242–4245 (2014).
66. Wei, J. *et al.* Facile synthesis of hybrid core-shell nanospheres for the asymmetric transfer hydrogenation of aromatic ketones. *ChemCatChem* **6,** 1368–1374 (2014).
67. Zell, T., Ben-David, Y. & Milstein, D. Unprecedented iron- catalyzed ester hydrogenation. mild, selective, and efficient hydrogenation of trifluoroacetic esters to alcohols catalyzed by an iron pincer complex. *Angew. Chem. Int. Ed.* **53,** 4685–4689 (2014).
68. Larkin, P. *Infrared and Raman Spectroscopy: Principles and Spectral Interpretation* 126–127 (Elsevier, 2011).
69. Xuan, X. P., Wang, X. S. & Wang, N. Theoretical study of molecular structure and vibrational spectra of 1,4-dihydroxyanthraquinone. *Spectrochim. Acta Part A* **79,** 1091–1098 (2011).



## Acknowledgements

We thank the National Natural Science Foundation of China (21303228 and 21073208) and the CAS for financial support. We thank Professor Weiqiang Zhang for helpful discussions.

## Author contributions

F.S., X.C. and Y.D. designed the project. H.Y. and X.D. performed the experiments. H.Y. and F.S. authored the manuscript.

## Additional information

**Competing financial interests:** The authors declare no competing financial interests.

# 6

# Using carbon dioxide as a building block in organic synthesis

Qiang Liu[1], Lipeng Wu[1], Ralf Jackstell[1] & Matthias Beller[1]

Carbon dioxide exits in the atmosphere and is produced by the combustion of fossil fuels, the fermentation of sugars and the respiration of all living organisms. An active goal in organic synthesis is to take this carbon—trapped in a waste product—and re-use it to build useful chemicals. Recent advances in organometallic chemistry and catalysis provide effective means for the chemical transformation of $CO_2$ and its incorporation into synthetic organic molecules under mild conditions. Such a use of carbon dioxide as a renewable one-carbon (C1) building block in organic synthesis could contribute to a more sustainable use of resources.

A more sensible resource management is the prerequisite for the sustainable development of future generations. However, when dealing with the feedstock of the chemical industry, the level of sustainability is still far from satisfactory. Until now, the vast majority of carbon resources are based on crude oil, natural gas and coal. In addition to biomass, $CO_2$ offers the possibility to create a renewable carbon economy. Since pre-industrial times, the amount of $CO_2$ has steadily increased and nowadays $CO_2$ is a component of greenhouse gases, which are primarily responsible for the rise in atmospheric temperature and probably abnormal changes in the global climate. This increase in $CO_2$ concentration is largely due to the combustion of fossil fuels, which are required to meet the world's energy demand[1]. Obviously, there is an urgent need to control $CO_2$ emissions and develop efficient carbon capture systems. Although the extensive use of carbon dioxide for chemical production cannot solve this problem alone, $CO_2$ is a useful one-carbon (C1) building block in organic synthesis due to its abundance, availability, nontoxicity and recyclability. As a result, valorization of $CO_2$ is currently receiving considerable and ever increasing attention by the scientific community[2–4]. However, activation and utilization of $CO_2$ is still problematic due to the fact that it is the most oxidized form of carbon, which is also thermodynamically stable and/or kinetically inert in certain desired transformations. Consequently, most of the known studies used highly reactive substrates and/or severe reaction conditions to activate $CO_2$, limiting the application of such methods. In particular, the catalytic coupling of $CO_2$ with energy-rich substrates, such as epoxides and aziridines, to generate polycarbonates/polycarbamates and/or cyclic carbonates/carbamates has drawn significant attention over the past decades. To create C-C bonds with $CO_2$, the use of carbon nucleophiles is specifically limited to strong nucleophilic organolithiums and Grignard reagents, as well as phenolates.

[1] Leibniz-Institut für Katalyse an der Universität Rostock, Albert-Einstein-Strasse 29a, 18059 Rostock, Germany. Correspondence should be addressed to M.B. (email: matthias.beller@catalysis.de).

Along with the rapid development of organometallic chemistry and catalysis (see Box 1), various types of efficient $CO_2$ transformations in organic synthesis have been discovered in the past decades, greatly improving its efficiency and applicability.

This review will describe the most recent advances made in the area of $CO_2$ valorization—turning $CO_2$ into a useful chemical feedstock—under mild conditions. A special focus is given on the reaction modes for the $CO_2$ activation and its application as C1 building block in organic synthesis. While classical methods for $CO_2$ hydrogenation have been reviewed elsewhere and will not be included here[5–7], the following subjects will be addressed: (1) novel transformations using carbon dioxide (briefly summarized in Fig. 1); (2) different reaction modes for $CO_2$ activation (main focus of this review); and (3) potential new applications of $CO_2$ valorization.

## Activation modes for catalytic valorization of $CO_2$

$CO_2$ is the most stable form of oxidized carbon compounds. Nevertheless, it reacts easily with strong nucleophiles to form new C-C or C-H bonds. However, to use carbon dioxide in a more benign and practical manner, efficient transformations with less-activated substrates under mild conditions have to be developed. Obviously, reactions of $CO_2$ that require a high-energy input are not benign because in general this energy leads also to formation of $CO_2$.

**Cooperative effects for $CO_2$ activation.** To achieve low-energy-demanding synthetic protocols using $CO_2$ as C1 building block, efficient interactions with the specific substrate is required. In this respect, cooperative activation of carbon dioxide with such substrates often in the presence of a suitable catalyst is of fundamental importance. Among the $CO_2$ functionalization methodologies, traditional examples that make use of cooperative effects are the catalytic valorization of energy-rich substrates (for example, strong nucleophilic organometallic reagents, epoxides, aziridines and so on). They provide a straightforward way for the thermodynamic driving force.

One of the more successful processes for $CO_2$ utilization for material synthesis is the catalytic productions of cyclic carbonates and polycarbonates from epoxides[8], which has been industrialized as well[9]. Nowadays, cyclic carbonates are widely used as organic synthetic intermediates, aprotic polar solvents, precursors for biomedical applications and raw materials for engineering plastics. Although this 100% atom economical reaction has been commercialized for the synthesis of ethylene and propylene carbonates for over 50 years[10], recently a variety of highly selective and reactive catalysts were developed. Representative systems are summarized in Fig. 2 (ref. 11). Notably, in 1983, Inoue and co-workers described in their seminal work that a (TPP)AlOMe complex can trap carbon dioxide readily and reversibly at room temperature. The trapped carbon dioxide is active enough to react with epoxide at ambient conditions to afford the corresponding cyclic carbonates[12]. Since then, various metalloprophyrin catalysts were studied in this transformation. For example, in 2007, Jing *et al.*[13] reported a Co(TPP)(OAc)/PTAT (phenyltrimethylammonium tribromide) catalyst system to achieve this $CO_2$ fixation at room temperature and low pressure. More recently, Sakai *et al.* developed a highly active bifunctional porphyrin catalyst, which combines an organocatalytic group (quaternary ammonium bromide) and a metal ion ($Mg^{II}$). It shows a high turnover number (TON = 103,000) for the synthesis of cyclic carbonates from $CO_2$ and epoxides under solvent-free conditions[14]. Related salen-based complexes represent another kind of effective catalysts for the electrophilic addition of $CO_2$ to epoxides. Already in 2001, air stable and commercially available Cr(III) salen complexes were presented to be highly efficient catalysts in the coupling of carbon dioxide with epoxides at mild temperatures and pressures by the group of Nguyen[15]. While most of the work in this area made use of pressurized carbon dioxide, Kleij and co-workers[16] showed that inexpensive and structurally simple Zn(salphen) catalysts can be employed as efficient catalysts for the synthesis of cyclic carbonates from $CO_2$ and terminal epoxides under mild conditions (2 bar $CO_2$). Noteworthy, the group of North[17] communicated the formation of cyclic carbonates even at 25 °C

**Box 1 | Coordination modes of $CO_2$ with transition metals.**

A key issue for valorization of $CO_2$ to valuable chemical intermediates and further on to consumer products relies on efficient catalysts. As a basis, an insightful understanding of the detailed organometallic chemistry of $CO_2$ is of great importance for the further design of the catalytic processes. In fact, the chelating mode between transition metal centres and $CO_2$ has been investigated intensively via stoichiometric experimental studies[46,47].

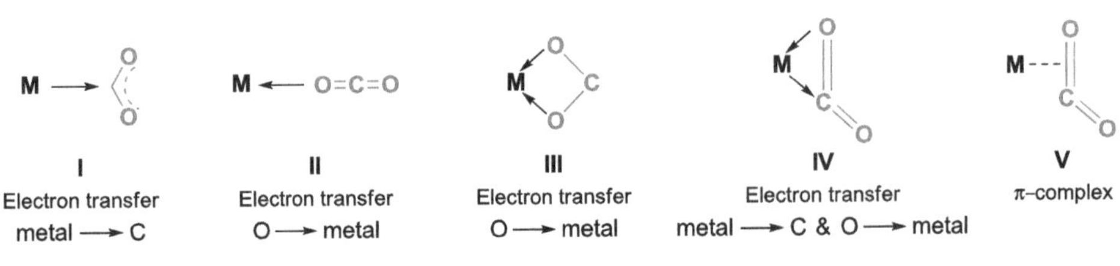


As shown above, $CO_2$ has multiple reactive sites: the carbon atom is an electrophilic Lewis acid centre and the oxygen atoms act as weak nucleophilic Lewis base. In its ground state, carbon dioxide possesses two equivalent C-O bonds that could both coordinate to a transition metal centre. As a result, a series of transition metal $CO_2$ complexes are known. If a metal centre reacts with one molecule of $CO_2$, there are five different chelating modes possible. More specifically, complex **I** with an M-C bond is sometimes termed as a metallacarboxylate. Electron-rich metal centres are more feasible to form these types of complexes via electron transfer from metal centre to carbon atom. Adduct **II** (end-on type) seems less plausible due to the weak interaction between the lone pair of only one oxygen atom and the metal centre. Meanwhile, complex **III** should be more stable, since $CO_2$ acts as a bidentate ligand with two oxygen atoms. In this case, a more electron-deficient metal centre is favoured for the electron transfer from oxygen atom to transition metal. A combination of the above two mentioned electron transfer processed affords the three-membered metallacycle complex **IV**. Moreover, the side-on-bonding π-complex **V** can also be formed in a similar spatial arrangement of atoms through the coordination of the C-O double bond to the central metal.

On the basis of these different coordination and activation modes, well-designed transition metal complex catalysts are able to promote the reactivity and also control the selectivity of $CO_2$ fixation reactions in the organic synthesis. This promising research area still has many novelties to be discovered.

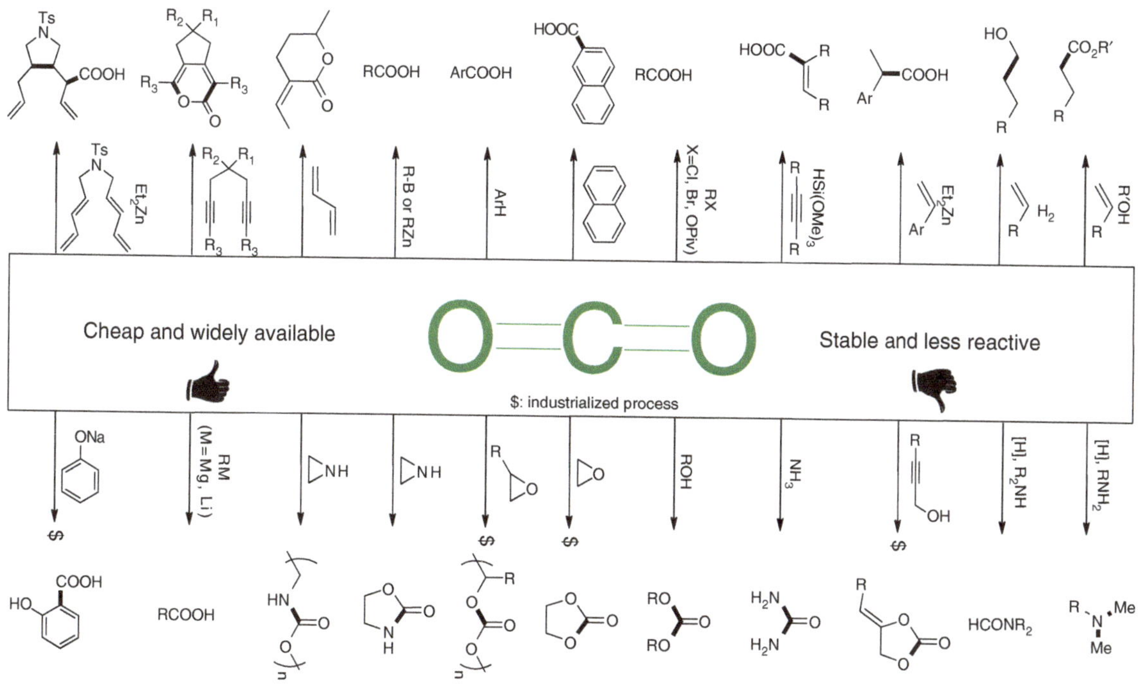


**Figure 1 | Representative examples using $CO_2$ as C1 building block in organic synthesis.** Along with the rapid development of organometallic chemistry and catalysis, various types of efficient $CO_2$ transformations were have been discovered in the past decades. However, in general, the substrates' scope and efficiency of these reactions are still limited due to the requirement of reactive agents for $CO_2$ activation. As a result, only a few processes have been industrialized until now (marked by $).

and 1 bar $CO_2$ using bimetallic $Al^{III}$-salen complexes. More recently, the Kleij group[18] presented an easily accessible amino triphenolate complex based on $Al^{III}$ metal centre. This complex is demonstrated to be a highly active and versatile catalyst for organic carbonate formation. Remarkably, initial turnover frequencies as high as 36,000 per hour during the cycloaddition of $CO_2$ to epoxides were obtained with TONs exceeding 100,000. Except for the above-mentioned sophisticated metal complexes, simple onium salts can also acts as efficient catalysts. For instance, a combination of $ZnCl_2$ and [Bmim]Br reported by Xia's group[19] is a really practical, highly efficient and environmentally benign catalyst system for the chemical fixation of $CO_2$ with epoxides. Finally, it is worth mentioning that Jiang, Han and co-workers[20] in 2011 found that cellulose/KI was a very active, selective and recyclable catalyst for this reaction without the assistance of any transition metal catalyst.

The direct polymerization of epoxides with $CO_2$ to produce polycarbonates has been accomplished too. In this respect, the contributions from Coates and co-workers[21,22] described very active zinc β-diiminate catalysts for the synthesis of monodispersed, highly alternating carbon dioxide/epoxide copolymers with high molecular weight. In addition, the group of Williams[23] demonstrated a zinc-based macrocyclic bimetallic catalyst for epoxide/$CO_2$ copolymerization, which shows remarkable activity at only 1 bar of $CO_2$. Very recently, iron – corrole complexes were synthesized by Nozaki and co-workers[24] to copolymerize epoxides with $CO_2$, providing the first example of iron-catalysed PO (propylene oxide)/$CO_2$ and GPE (glycidyl phenyl ether)/$CO_2$ copolymerization. The same group also demonstrated the stereoselective alternating copolymerization of cyclohexene oxide and $CO_2$ with dimeric zinc complexes for the first time[25]. Besides, improved catalysts are still desirable to better control the properties of the resulting materials[26]. Similarly, many efforts are also devoted to designing catalysts able to promote the reaction of $CO_2$ with less-reactive oxiranes, which could also provide novel materials[27].

Apart from the utilization of energy-rich substrates (for example, epoxides), catalytic reductive formylation and methylation of amines represent important modern examples for cooperative effects in the activation of $CO_2$. Here the activation energy for reductive deoxygenation of $CO_2$ is significantly decreased by the coordination of amines to $CO_2$ and the formation of N-C bonds (in a carbamate or urethane intermediate), which thus facilitates the reduction of $CO_2$. Likely, the other advantage of amines is to increase $CO_2$ concentration in the condensed phase[28]. These transformations represent diagonal approaches to the chemical recycling of $CO_2$, which combine both reduction of $CO_2$ and C-N bond formation processes (Fig. 3)[29]. More specifically, Cantat and co-workers reported an unprecedented organocatalytic synthesis of formamides from $CO_2$. Mechanistic studies uncovered the synergistic effect of the organocatalyst TBD (triazabicyclodecene) and amine substrates in the promotion of $CO_2$ activation (Fig. 3, equation a)[30]. Shortly afterwards, a highly active *N*-heterocyclic carbene organocatalyst was designed by the same group for the formylation of N-H bonds in various amines and heterocycles under very mild conditions using $CO_2$ and polymethylhydrosiloxane (PMHS; Fig. 3, equation b)[31].

Meanwhile, the methylation of amines via six-electron reduction of $CO_2$ remains unknown until 2013. The market value of methylamines such as $MeNH_2$, $Me_2NH$ and $Me_3N$ exceeds 4,000 Euro per ton. Therefore, the reductive methylation of amines with $CO_2$ can create additional value. In early 2013, our group developed a Ru catalyst system that was able to convert carbon dioxide and amines into various kinds of *N*-methylated products in the presence of hydrosilanes ($PhSiH_3$). Notably, diverse functional groups were well tolerated under these conditions (Fig. 3, equation c)[32]. In the same year, Cantat *et al.*[33] reported zinc catalysts able to perform the methylation of amines with $CO_2$ and hydrosilanes at low pressure. Furthermore, the selective reduction of ureas was possible under similar reaction conditions. Although some silanes are considered to be

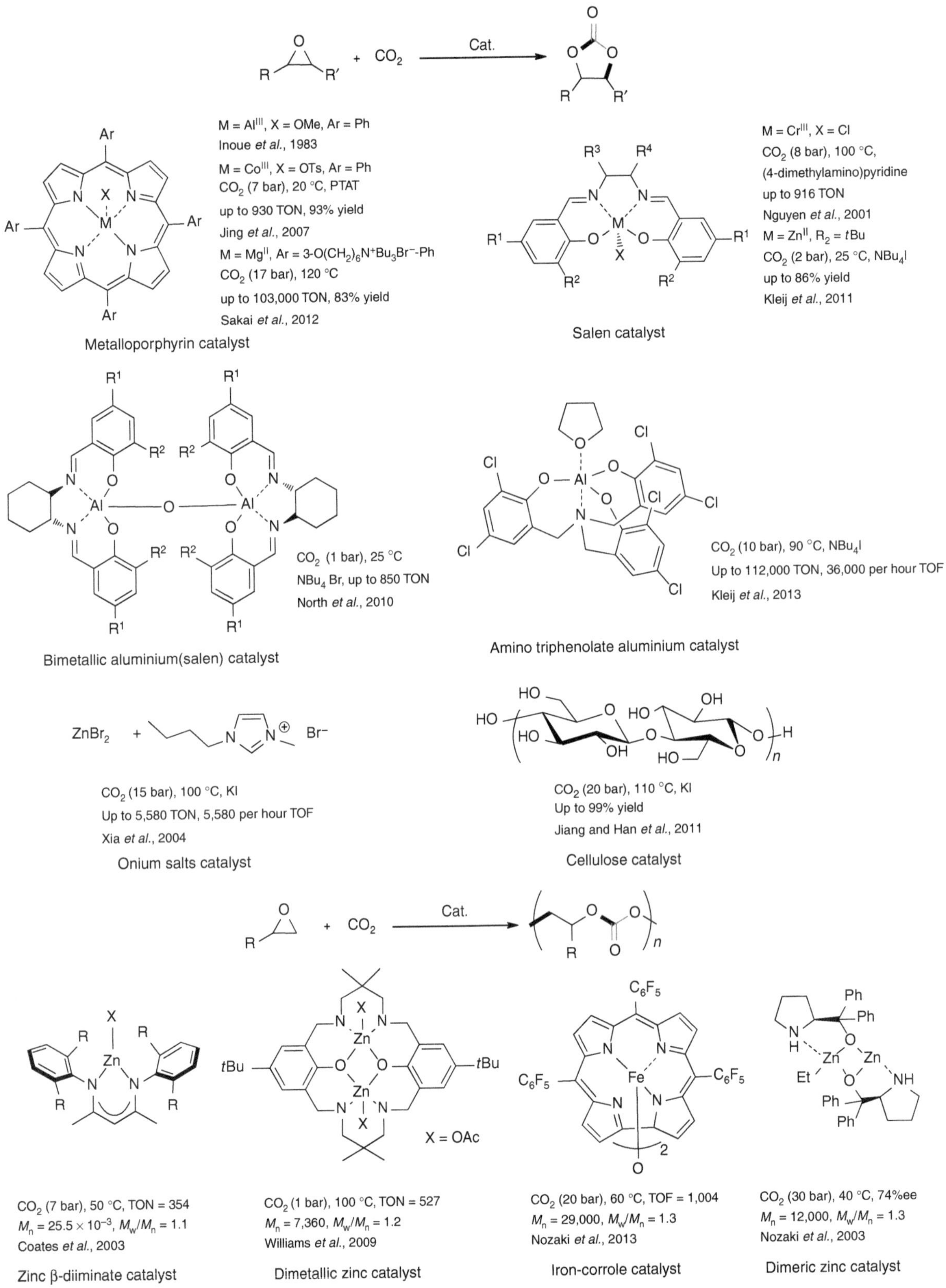


**Figure 2 | Established catalysts and reaction conditions for the catalytic productions of cyclic carbonate and polycarbonate from oxiranes using $CO_2$ as the feedstock.** Examples include metalloporphyrin, salen, metal-corrole, dimetallic as well as metal-free catalysts.

waste products of the silicone industry, all these methodologies are limited by the accessibility of the hydrosilanes and an additional workup step to remove siloxane by-products. Obviously, catalytic methylations using $CO_2$ and $H_2$ represent a greener method with $H_2O$ as the only by-product. In this respect, two research groups independently reported similar catalytic systems shortly afterwards to convert amines into methylamines in the presence of $CO_2$ and $H_2$. On one hand, Klankermayer and co-workers[34] presented the use of a molecularly defined ruthenium complex [Ru(triphos)(tmm)] together with readily available organic acids as co-catalysts to afford the methylation of aryl amines in good yields (Fig. 3, equation d). Meanwhile, the

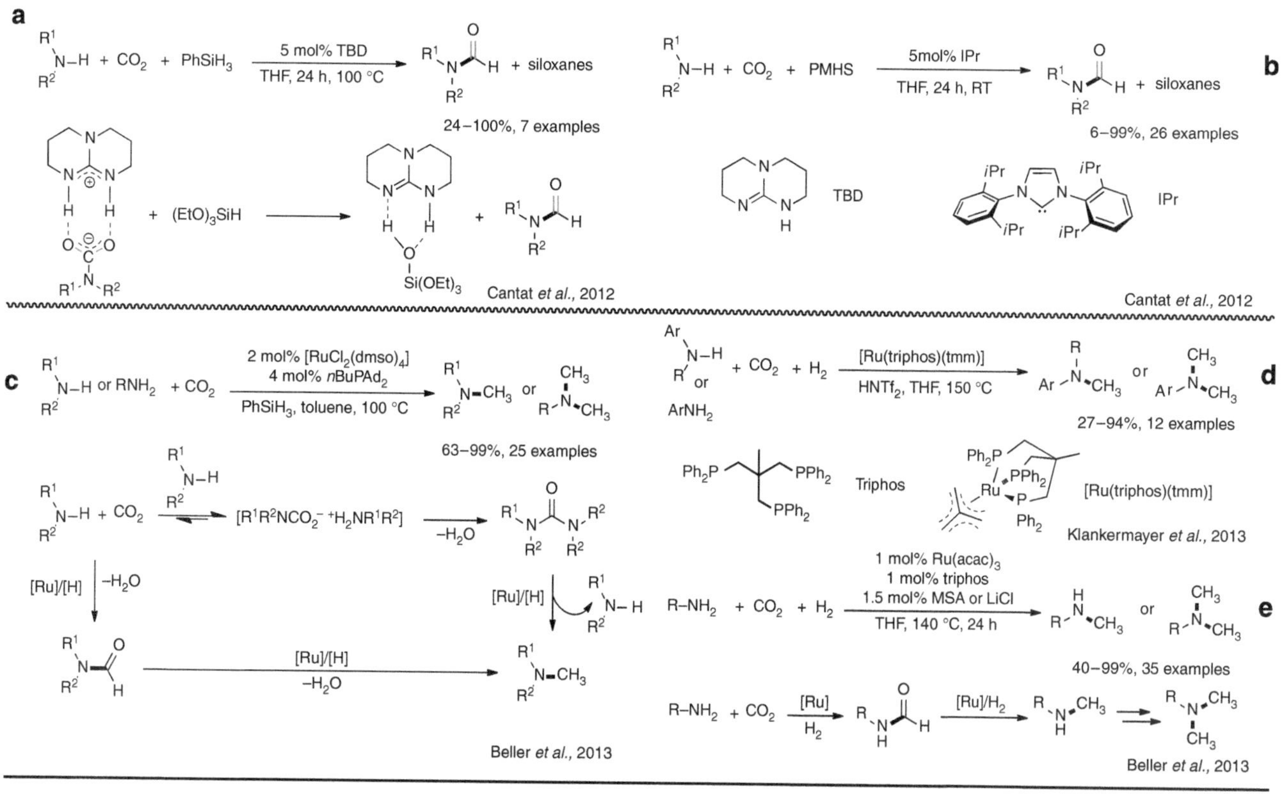


**Figure 3 | Catalytic reductive formylation and methylation of amines with $CO_2$ as C1 source.** (**a**,**b**) Organocatalytic formylation of amines with $CO_2$ using silanes as the reductants. (**c**) Ru-catalysed methylation of amines with $CO_2$ using silanes as the reductants. (**d**,**e**) Ru-catalysed methylation of amines with $CO_2$ using $H_2$ as the reductant.

group of Beller demonstrated the efficient *N*-methylation of both aromatic and aliphatic amines using $CO_2/H_2$ as the methylation reagent. Applying an *in situ* combination of $Ru(acac)_3$, triphos and either acid additives or LiCl, the desired methylated amines were obtained with high efficiency. Notably, selective monomethylation of primary amines as well as the methylation of more challenging aliphatic amines proceeded smoothly under these conditions (Fig. 3, equation e)[35]. Interestingly, Shi *et al.*[36] described in 2014 heterogeneous copper catalysts for the same transformation with $CO_2$ and $H_2$. In general, formamides are the common reaction intermediates for the above-mentioned methylation reactions. Again, this indicates the advantageous activation effect of amine substrates for the $CO_2$ reduction step.

**$CO_2$ fixation via oxidative cyclometallation.** Pioneered by the synthesis and isolation of Aresta's complex $[Ni(CO_2)(PCy_3)_2]$ in 1975 (ref. 37), the development of low-valent metal (nickel or palladium) complexes that mediate the oxidative couplings of $CO_2$ with various unsaturated hydrocarbons provides a facile approach to overcome the obstacles of $CO_2$ fixation. While it permitted the use of $CO_2$ under atmospheric pressure, stoichiometric amounts of metal were necessary to generate various metallacycle complexes via the oxidative cyclometallation of Ni(0) complexes with $CO_2$ and olefins or alkynes (Fig. 4, equations a and b)[38,39]. The nickel-mediated stoichiometric fixation of carbon dioxide with alkenes[40] has been known for over 20 years largely due to the works from Hoberg and co-workers. However, the limited substrate scope and/or relatively harsh conditions still remain concerns for this $CO_2$ activation strategy. Meanwhile, only a few catalytic systems based on this strategy have been developed. In 2002, Louie *et al.*[41] communicated an efficient nickel-catalysed [2+2+2] cycloaddition of $CO_2$ and diynes in the presence of catalytic amounts of Ni(0) and specific carbene ligands (Fig. 4, equation c). In the same year, the group of Mori[42,43] developed a nickel-catalysed regio- and stereoselective ring-closing carboxylation of bis-1,3-dienes, which proceeds via insertion of $CO_2$ into a bis-$\pi$-allylnickel intermediate followed by a transmetalation process of the resulting cyclic nickel carboxylate with a sensitive dialkylzinc reagent (Fig. 4, equation d). Except for nickel-catalysed cyclization reactions, the Pd-catalysed telomerization of 1,3-butadiene with $CO_2$ is one of the first successful examples of the catalytic formation of a new C-C bond between $CO_2$ and an organic compound. Thirty years ago, Behr *et al.* found that the cyclization product is selectively produced employing $Pd(acac)_3$ and $PiPr_3$ as the catalyst. The resulting δ-lactone is formed via an allylic palladium carboxylate complex as a key intermediate (Fig. 4, equation e)[44]. Generally, the $CO_2$ transformations via oxidative cyclometallation pathways are restricted to substrates with extensive $\pi$ systems (for example, dienes and diynes). Interestingly, in 2008, the group of Dong discovered that Aresta's complex catalysed the cross-coupling of organozinc reagents with $CO_2$. By replacing nickel catalysts with palladium, the latter complexes were found to be catalytically active as well for the reaction of aryl zinc bromides (Fig. 4, equation f). The proposed mechanism consisted of an oxidative addition of Ni(0) to $CO_2$ to generate Aresta's complex; transmetalation with the organozinc reagent and reductive elimination affording the corresponding zinc carboxylate that was finally hydrolysed by acidic workup (Fig. 4, equation g)[45].

**Carboxylation reactions via $CO_2$ insertion into C-M bonds.** The formation of a thermodynamically and kinetically stable C-C bond is the most desirable form of $CO_2$ fixation. Since carbon

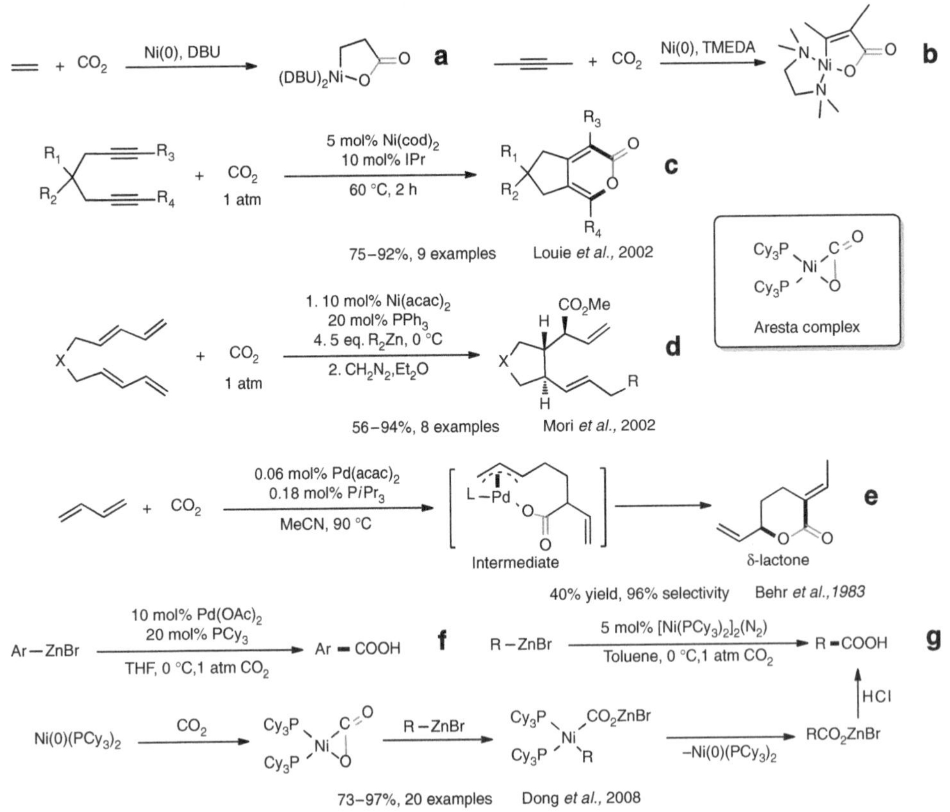


**Figure 4 | Catalytic $CO_2$ fixation via oxidative cyclometallation with low-valent Ni and Pd complexes.** (**a**,**b**) The generation of metallacycle complexes via the oxidative cyclometallation of Ni(0) complexes with $CO_2$ and unsaturated bonds. (**c**) Ni-catalysed [2+2+2] cycloaddition of $CO_2$ and diynes. (**e**) Pd-catalysed telomerization of 1,3-butadiene with $CO_2$. (**f**,**g**) Aresta's complex catalysed the carboxylation of organozinc reagents with $CO_2$.

dioxide is an inert electrophile, highly reactive organolithium and Grignard reagents are typically utilized as strong nucleophiles to react with $CO_2$ directly to construct C-C bonds and furnish valuable carboxylic acids and their derivatives. In fact, the current industrial synthesis of salicylic acid derivatives is a prime example for this type of reaction. In general, nitrogen or oxygen nucleophiles exhibit much better reactivity than carbon nucleophiles for $CO_2$ activation. As a result, catalytic C-C bond formation using $CO_2$ as C1 building block with less-reactive substrates has been and constitutes still a major challenging task for organic synthesis and catalysis research[46]. This goal can be achieved with the aid of transition metal catalysts that lower the activation energy and thus considerably increase the reaction rate for the C-C bond-forming step. Particularly, the $CO_2$ insertion reactions into various C-M bonds to form carboxylato complex via the formation of a new C-C bond are well known[47]. Moreover, transition metal complexes can react with less-nucleophilic organometallic reagents, for example, organoboron compounds, as well as organohalides, unsaturated bonds even activated C-H bonds to form C-M bonds efficiently, similar to their performance in the classic cross-coupling reactions[48]. Furthermore, the $CO_2$ insertion intermediate (RCOO-M) might be quenched in a suitable way to regenerate the catalytically active species. This provides numerous possibilities for catalytic carboxylation reactions with all the above-mentioned substrates[49–52]. Representative examples for such processes are listed in Table 1 including a summary of substrates, catalyst precursors, ligands, additives as well as other conditions. In addition, a comparison of the postulated mechanisms for carboxylation and hydro(bora)carboxylation reactions are presented in Fig. 5.

Pioneered by the work of Iwasawa *et al.* in 2006 (Table 1, entry 1) the catalytic carboxylation of organoboron reagents with $CO_2$ in the presence of Rh[53], Cu[54–57] and Ag (Table 1, entry 4)[58] catalysts has been developed. These coupling reagents exhibit milder nucleophilicity, better stability to air and moisture as well as higher compatibility with sensitive functional groups. Among these results, the combination of IPrCu-NHC complex and alkoxide base (M′OR′) developed by the group of Hou[55,57] since 2008 was found to be the most active catalyst system for the carboxylation of aryl-, alkenyl- and also alkylboronic acid derivatives (Table 1, entries 3 and 6). A general synthesis of various carboxylic acids with $CO_2$ was accomplished, and a wide range of functional groups were well tolerated. On the basis of stoichiometric experiments (taken the reactions of ArBPin, for example, as shown in Fig. 5a), a general mechanism for such reactions was also proposed: the metathesis reaction between [(IPr)CuCl] and alkoxide generates the corresponding alkoxide complex [(IPr)Cu(OR′)], which could coordinate with organoboron reagents (R-$BX_2$) through interaction between the alkoxy group and the boron atom. Subsequent transmetalation to the copper atom generates the organocopper complex [(IPr)CuR]. Then, $CO_2$ insertion into the C-Cu bond yields the carboxylate [(IPr)Cu(OCOR)], which on metathesis with alkoxide base M′OR′ regenerates [(IPr)Cu(OR′)] and releases the carboxylate salt $RCO_2M'$. Finally, the latter yields the carboxylic acid $RCO_2H$ via hydrolysis. Some key reaction intermediates, such as the copper(I) alkoxide complexes [(IPr)Cu-(OR′)], organocopper species [(IPr)CuR] and the carboxylate copper intermediates [(IPr)Cu(OCOR)], were successfully isolated and structurally characterized, which were also confirmed to be catalytically active species in these transformations.

**Table 1 | A summary of carboxylation reactions of C-B, C-X and C-H bonds with $CO_2$: catalytic systems and reaction conditions.**

$$\text{R–X} + CO_2 \xrightarrow[\text{[H}^+\text{]}]{\text{[Rh], [Pd], [Ag], [Ni], [Cu], } metal\ free} \text{R–COOH}$$

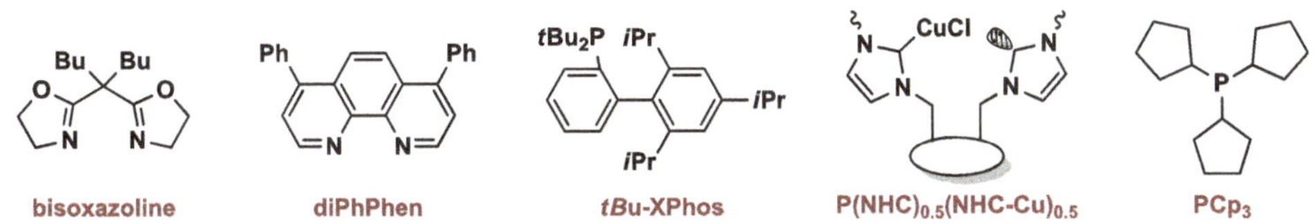


| Entry | R | X | Catalyst precursor/ligand | Additive | Condition | Ref. |
|---|---|---|---|---|---|---|
| 1 | Aryl, alkenyl | $B(OR')_2$ | 3 mol% $[Rh(OH)(cod)]_2$/7 mol% dppp | 3 equiv. CsF | 1,4-dioxane, 60 °C | 53 |
| 2 | Aryl, alkenyl | $B(OR')_2$ | 5 mol% CuI/6 mol% bisoxazoline | 3 equiv. CsF | DMF, 60 °C | 54 |
| 3 | Aryl, alkenyl | $B(OR')_2$ | 1 mol% [(IPr)CuCl] | 1 equiv. *t*BuOK | THF, 70 °C | 55 |
| 4 | Aryl, alkenyl | $B(OR')_2$ | 10 mol% AgOAc/10 mol% $PPh_3$ | 2 equiv. *t*BuOK | THF, 70 °C | 58 |
| 5 | Alkyl | $B(OR')_2$ | 10 mol% CuOAc/1,10-Phen | 1 equiv. *t*BuOK | Toluene, 100 °C | 56 |
| 6 | Alkyl | $B(OR')_2$ | 3 mol% [(IPr)CuCl] | 1.05 equiv. MeOLi | THF, 70 °C | 57 |
| 7 | Akynyl | H | 10 mol% $Cu(diPhPhen)(PPh_3)_2NO_3$ | 1.2 equiv. $Cs_2CO_3$ | DMF, 35-50 °C | 62 |
| 8 | Akynyl | H | 5 mol% $P(NHC)_{0.5}(NHC\text{-}Cu)_{0.5}$ | 1.2 equiv. $Cs_2CO_3$ | DMF, RT | 61 |
| 9 | Akynyl | H | 1 mol% AgI | 1.5 equiv. $Cs_2CO_3$ | DMF, 50 °C | 59 |
| 10 | Akynyl | H | None | 1.2 equiv. $Cs_2CO_3$ | DMF, 120 °C | 60 |
| 11 | Aryl | H (pKa(C-H) < 30.3) | 1.5 mol% [(IPr)AuOH] | 1.05 equiv. KOH | THF, RT | 65 |
| 12 | Aryl | H (pKa(C-H) < 27.3) | 3 mol% [(IPr)CuOH] | 1.1 equiv. CsOH | THF, 40 °C | 66 |
| 13 | Aryl | H (pKa(C-H) < 32.5) | 5 mol% [(IPr)CuCl] | 1.1 equiv. *t*BuOK | THF, 80 °C | 63 |
| 14 | Aryl | H (pKa(C-H) < 27.3) | None | 1.2 equiv. $Cs_2CO_3$ | DMF, 125 °C | 64 |
| 15 | Aryl | Br | 5 mol% $Pd(OAc)_2$/10 mol% *t*Bu-XPhos | 2 equiv. $Et_2Zn$ | DMA, 40 °C | 69 |
| 16 | Aryl, alkenyl | Cl | 5 mol% $NiCl_2(PPh_3)_2$/10 mol% $PPh_3$/10 mol% $Et_4NI$ | 2 equiv. Mn | DMI, 25 °C | 70 |
| 17 | Benzyl | Cl | 10 mol%$NiCl_2 \cdot$glyme/$PCp_3 \cdot HBF_4$ | 5 equiv. Zn, 2 equiv. $MgCl_2$ | DMA, RT | 71 |
| 18 | Aryl or benzyl | OPiv | 7 mol% $NiCl_2(dppf)$/10 mol% $NiCl_2(PMe_3)$ | 1 or 2 equiv. Mn | DMA, RT or DMF, 80 °C | 72 |

COD, cyclooctadiene; DMA, dimethylacetamide; DMF, dimethylformamide; DMI, 1,3-dimethyl-2-imidazolidinone; dppp, 1,3-bis(diphenylphosphino)propane; equiv., equivalent; THF, tetrahydrofuran.

Apart from organoboron reagents, acidic C-H bonds can also undergo deprotonative metallation with specific transition metal complexes to form C-M bonds with the assistance of strong bases. Accordingly, the catalytic carboxylation of terminal alkynes (Table 1, entries 7–10)[59–62] and electron-deficient (hetero)arenes (Table 1, entries 11–14)[63–66] (pKa(C-H) < 32.5) is achieved through the following $CO_2$ insertion and metathesis of metal carboxylate intermediates with base. Except the active copper and gold *N*-heterocyclic carbene complexes, $Cu(diPhPhen)(PPh_3)_2NO_3$ (Table 1, entry 7) and AgI (Table 1, entry 9) were also demonstrated to be the effectual catalysts for the carboxylation of terminal alkynes by the group of Goossen[62] and Lu[59], respectively. Interestingly, at higher temperature (> 120 °C), the $Cs_2CO_3$-mediated carboxylation of heteroarenes (Table 1, entry 14)[64] and terminal alkynes (Table 1, entry 10)[60] with $CO_2$ were reported by Hu and Zhang's groups independently as well.

Later on, the group of Iwasawa showed the formation of C-M bonds by a directing group-assisted C-H activation process, which was used for subsequent carboxylation reactions with $CO_2$. In fact, treatment of 2-hydroxystyrenes bearing a hydroxy group as the chelating group with catalytic amount of $Pd(OAc)_2$ in the presence of $Cs_2CO_3$ and $CO_2$ afforded the corresponding coumarins via carboxylation/condensation (Fig. 5c)[67]. It is proposed that first the six-membered alkenyl palladium intermediate is generated by chelation-assisted alkenyl C-H bond cleavage of 2-hydroxystyrene with $Pd(OAc)_2$ along with coordination of the second molecule of substrate. Subsequently, the alkenyl palladium(II) species undergoes nucleophilic carboxylation to afford a palladium carboxylate intermediate, which reacts with 2-hydroxystyrene and base to give lactone with regeneration of the cyclometalated intermediate. The same group also reported another approach for the direct carboxylation reaction of aromatic C-H bonds: namely, the Rh(I)-catalysed carboxylation of aromatic compounds via chelation-assisted ortho-metalation (Fig. 5d)[68]. The key intermediate is proposed to be the methylrhodium(I) complex **F**, which undergoes chelation-assisted oxidative addition with a suitable C-H bond to afford rhodium(III) complex **G**, followed by reductive elimination of methane to give the nucleophilic arylrhodium(I) complex **H**. Consequently, nucleophilic carboxylation of intermediate **H** affords rhodium carboxylate **I**, which undergoes transmetalation with a methylaluminium reagent to give aluminium carboxylate **J** and regenerate the active methylrhodium(I) complex **F**.

The oxidative addition of organo(pseudo)halides with low-valent transition metals constitutes another way to form C-M bonds. In this respect, in 2009, Martin and co-workers[69] presented a palladium-catalysed direct carboxylation of aryl bromides with $CO_2$ (Table 1, entry 15). In contrast to the reactions vide supra, this is the first example for carboxylation reactions of electrophiles instead of nucleophiles. Therefore, additional reductant (here $Et_2Zn$) is needed to regenerate the

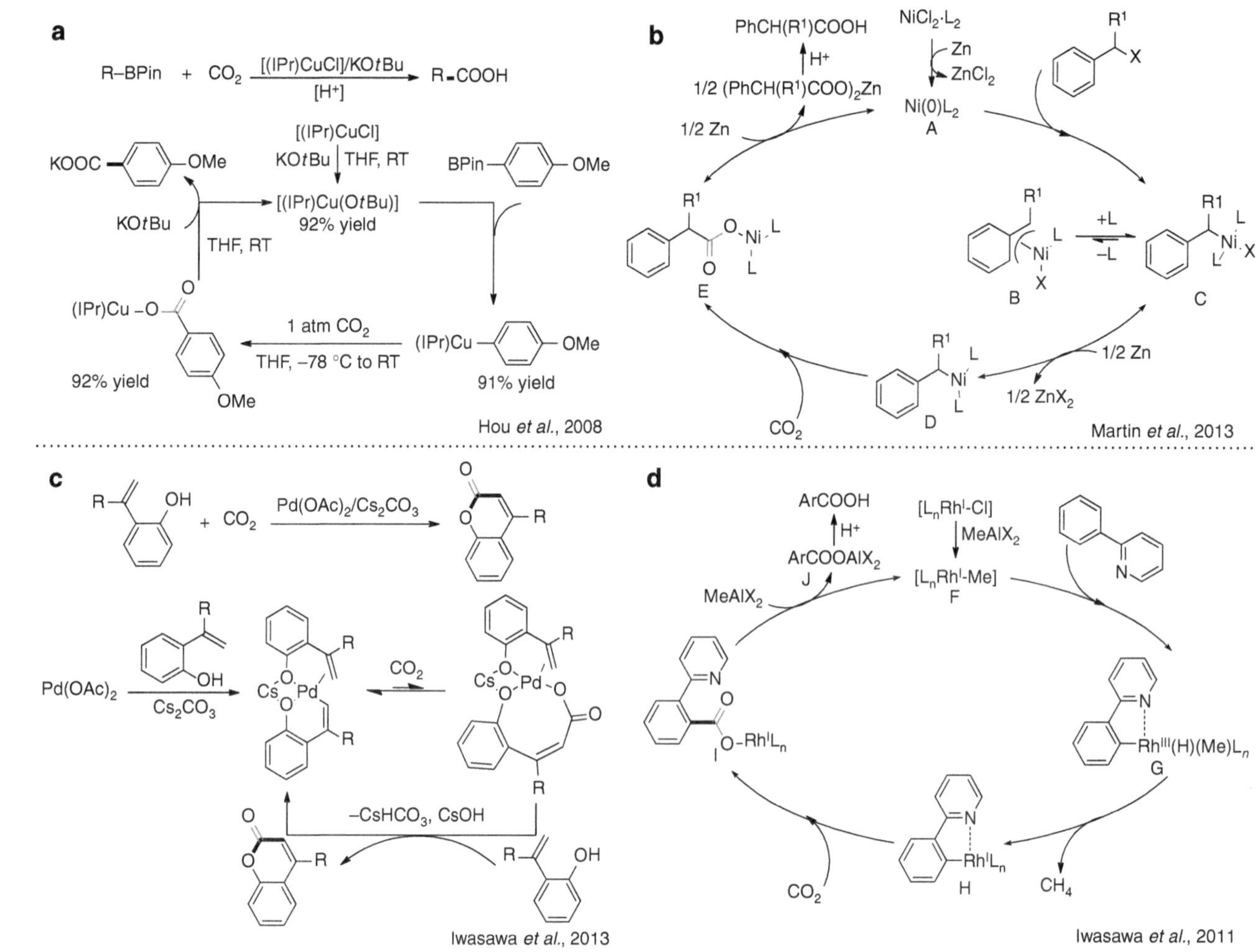


**Figure 5 | Possible reaction mechanisms for typical carboxylation reactions with $CO_2$.** (**a**) Reaction mechanism for Cu-catalysed carboxylation reactions of organoboron compounds. (**b**) Reaction mechanism for Ni-catalysed carboxylation reactions of benzyl halides. (**c,d**) Directing group-assisted carboxylation of C-H bonds.

catalytically active low-valent Pd(0) species. After Martin's work, Tsuji and co-workers found that less-expensive nickel catalysts are active in carboxylation of aryl chlorides and vinyl chlorides. Notably, these reactions proceed under 1 atm $CO_2$ pressure at room temperature (Table 1, entry 16)[70]. The Ni(I) species was involved in the catalytic cycle with the aid of the Mn/$Et_4NI$ system as an efficient reducing agent. Although the carboxylation of aryl bromides and chlorides with $CO_2$ provides a straightforward synthetic route to functionalized benzoic acids, these protocols cannot be utilized to access aliphatic electrophiles, for example, for the synthesis of biologically active phenylacetic acid derivatives. Hence, it was interesting when Martin's group reported in 2013 the first catalytic carboxylation of primary, secondary and tertiary benzyl halides utilizing $CO_2$. Their protocol provides an opportunity to prepare phenylacetic acids from commercially available and cheap starting materials (Table 1, entry 17)[71]. The proposed catalytic cycle starts with an initial reduction of the Ni(II) pre-catalyst by Zn followed by oxidative addition to the corresponding alkyl halides, thus delivering $\eta^3$- and $\eta^1$-bound nickel complexes **B** and **C** in a rapid equilibrium. Generation of the reduced Ni(I) species **D** by Zn and $CO_2$ insertion affords the carboxylate nickel intermediate **E**. Further treatment with Zn regenerates the active Ni(0)$L_2$ **A** species and provides zinc carboxylate, which on hydrolysis delivers the final product (Fig. 5b). Very recently, the same group reported a novel Ni-catalysed carboxylation of aryl esters with less-activated C-O bonds (Table 1, entry 18)[72]. This new protocol deals with the development of a synergistic activation of $CO_2$ and a rather challenging activation of inert $C(sp^2)-O$ and $C(sp^3)-O$ bonds derived from simple and cheap alcohols. In general, all these modern transformations have to be compared with the classic carboxylation of Grignard reagents. The main advantage of these recent methodologies can be the tolerance of functional groups, and thus the possibility to synthesize complex carboxylic acid derivatives. On the other hand, the major drawback of all carboxylation reactions of electrophiles is the necessity to use (over)stoichiometric amounts of metals as reducing agents. Obviously, hydrogen would be a much more favourable reductant for such processes; however, no efficient regeneration of the active low-valent metal complexes have been realized yet.

Significant advancements were made in the catalytic $CO_2$ hydrocarboxylation of unsaturated C-C bonds too. For example, in 2008, Iwasawa and co-workers reported the hydrocarboxylation of allenes and 1,3-dienes with a silyl pincer-type palladium complex. The reactions were carried out at room temperature under 1 atm $CO_2$ and enabled the facile and regioselective synthesis of $\beta,\gamma$-unsaturated carboxylic acids[73]. The reaction is believed to start with generation of a silyl pincer-type palladium hydride complex via transmetalation of $AlEt_3$ to (PSiP)PdOTf followed by β-hydride elimination. Hydrometalation of 1,1-disubstituted allenes preferentially afford the less-congested σ-allylpalladium intermediate, which then undergoes nucleophilic addition to $CO_2$ at the γ-position of the palladium regioselectively to give the carboxylate palladium complex. Finally, a transmetalation/β-hydride elimination sequence regenerates the catalytically active palladium hydride species and releases the carboxylation product as its Al salt (Fig. 6a). Meanwhile, the first direct hydrocarboxylation of alkenes occurred in similar manner. Rovis and co-workers[74] developed a Ni-catalysed hydrocarboxylation of styrenes using $Et_2Zn$ as the hydride donor with $CO_2$ (Fig. 6b). The suggested mechanism involved a nickel hydride complex as active catalyst. Afterwards, a highly regio- and stereoselective three-component Ni-catalysed syn-hydrocarboxylation of alkynes with $Et_2Zn$ and $CO_2$ using $Ni(cod)_2$ as catalyst and CsF as additive was presented by Ma and co-workers[75]. Although these three examples demonstrate interesting carboxylation reactions, strong and extremely air-sensitive reducing agents are indispensable. This problem is partially avoided in the copper-catalysed hydrocarboxylation of alkynes with $CO_2$ using hydrosilanes as milder reducing agent realized by Tsuji and co-workers (Fig. 6c)[76]. The fundamental catalytic steps in their hydrocarboxylation were examined by stoichiometric reactions. Treatment of the catalyst precursor $[Cl_2IPrCuF]$ with silane led to the catalytically active copper hydride species. It reacted with an alkyne substrate smoothly to afford the corresponding copper alkenyl complex, which on $CO_2$ insertion affords the copper carboxylato complex. δ-bond metathesis of this species with hydrosilane regenerated $[Cl_2IPrCuH]$ and provided the corresponding silyl ester. In analogy to hydrocarboxylations, catalytic boracarboxylations of alkynes with diborane and $CO_2$ using an *N*-heterocyclic carbene copper catalyst were recognized by Hou in 2012 (ref. 77). The reaction of [(NHC)Cu(OR)] with $B_2(pin)_2$ instantly gave the borylcopper complex [(NHC)CuB(pin)]. Its subsequent reaction with internal alkyne afforded a β-boryl alkenyl complex, which undergoes the following $CO_2$ insertion and metathesis processes to furnish the final product (Fig. 6d).

**Carbonylation reactions using $CO_2$ as CO surrogates.** Compared with all above strategies to directly use $CO_2$ as C1 building block in chemical synthesis, the *in situ* generation of CO from $CO_2$ is clearly underdeveloped. Nevertheless, this strategy represents a particularly promising tool for more practical applications. We are convinced that such $CO_2$ valorization approach will open up new opportunities to realize industrially important carbonylation reactions. Advantageously, here $CO_2$ is used as a sustainable carbonyl source instead of hazardous CO. For more than 60 years, carbonylation reactions using CO

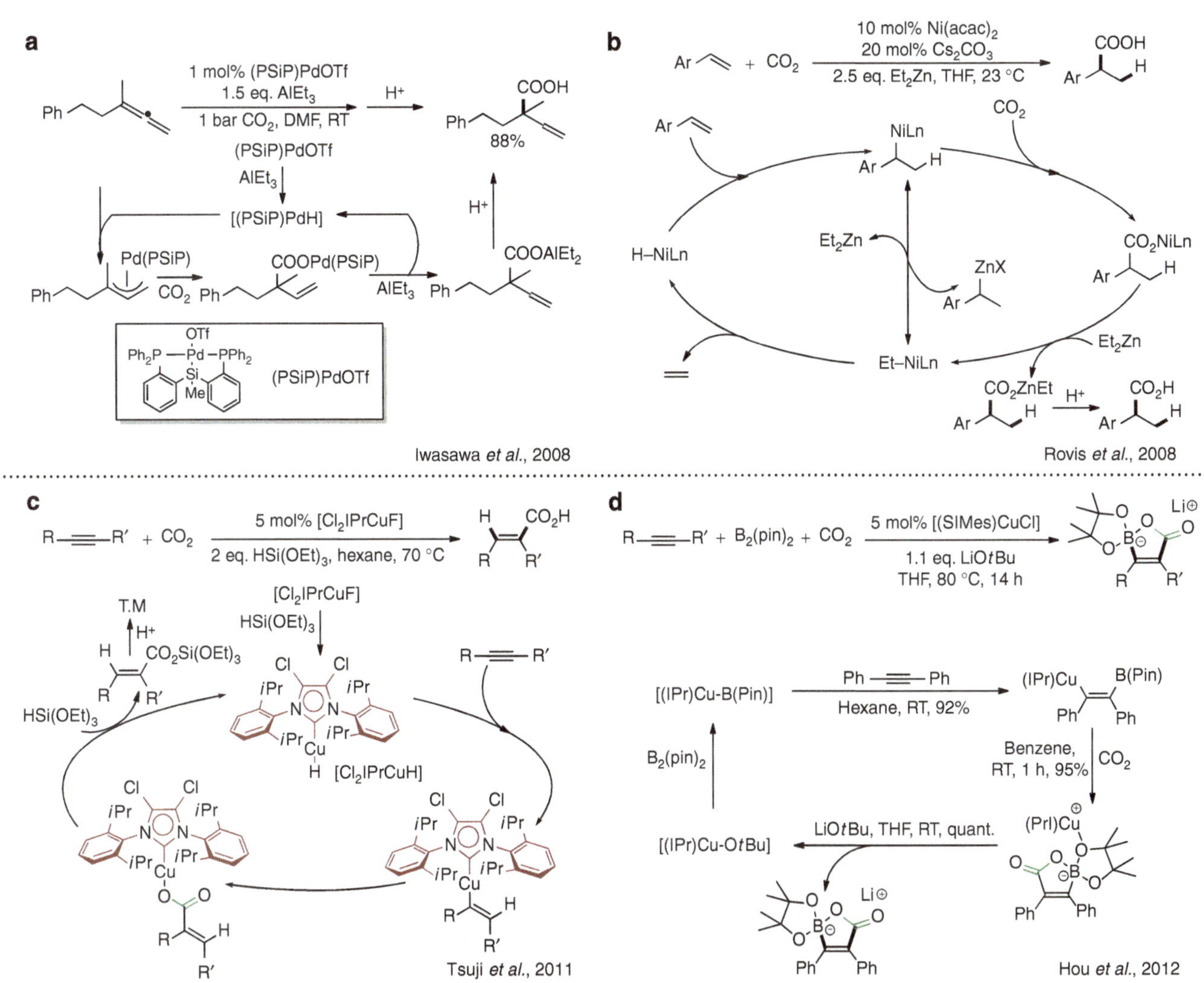


**Figure 6 | Selected examples for hydro(bora)carboxylation reactions of alkenes and alkynes with $CO_2$.** Examples include (**a**) allenes, (**b**) styrenes and (**c,d**) internal alkynes.

represent major technology for the production of value-added bulk and fine chemicals from easily available feedstocks. As an example, over 10 million tons of oxo-products are produced every year from alkenes via hydroformylation and related tandem reactions[78]. Besides, methyl propionate, a key intermediate for methyl methacrylate (annual demand is currently about 3 million tons), is in part produced by palladium-catalysed methoxycarbonylation of ethylene[79]. So far, all these processes make use of CO, which is also difficult to transport on a bulk scale. Notably, the physical properties of carbon monoxide (especially its toxicity) impede its wider use in academia too. In this respect, carbonylations with $CO_2$ offer a more safe and accessible tool for synthetic organic chemists. The general strategy to use carbon dioxide as a carbonyl source is based on the *in situ* reduction of $CO_2$ to CO, which then can be used in subsequent carbonylation reactions. It is long known that combination of carbon dioxide and $H_2$ at high temperature (>300 °C) allows for so-called reverse water gas shift (RWGS) reaction, in which a mixture of carbon monoxide, carbon dioxide, hydrogen and water is formed. On the basis of their previous studies on the RWGS process, Tominaga and Sasaki *et al.*[80] reported in 2000 the first ruthenium-catalysed hydroformylation/reduction of alkenes with $CO_2$ using $Ru_3(CO)_{12}$ as catalyst in the presence of LiCl as additive. It is surprising that this seminal work of Tominaga *et al.* has attracted only minor attention, probably due to the still drastic conditions and the efficiency of the reaction. For example, unwanted alkene hydrogenation side reactions occurred. Recently, our group improved this process by the application of a specific bulky phosphite ligand (Fig. 7a)[81]. Applying the optimized catalytic system, the yields of alcohols, the chemoselectivity of the process and especially the TON of the catalyst was significantly improved compared with the known ligand-free catalysis system. In the above-mentioned carbonylation reactions via RWSG reaction, water is always produced as a side product and is not involved in the following alkene carbonylation reactions. Remarkably, the group of Leitner[82] reported a rhodium-catalysed hydrocarboxylation of alkenes with $CO_2$ and $H_2$ to produce carboxylic acids, in which *in situ*-formed $H_2O$ was utilized as the nucleophile in this transformation (Fig. 7b).

Although $H_2$ can act as an efficient reducing reagent for $CO_2$ reduction, in many cases the following carbonylation reaction does not tolerate $H_2$ due to the hydrogenation of the substrate. As a result, the development of other benign $CO_2$ reductions to generate CO is highly desired for the valorization of $CO_2$. Combining our long-term experience in alcohol dehydrogenation[83] and carbon dioxide reduction reactions[5,84], very recently efficient reforming processes of alcohols and $CO_2$ were developed. This allows for catalytic alkoxycarbonylation reactions using $CO_2$ as the C1 source in the presence of a ruthenium/[Bmim]Cl catalyst system. Key to success was an unprecedented *in situ* formation of CO from $CO_2$ (Fig. 7c)[85]. In contrast to previously reported hydrocarboxylation reactions of alkenes with $CO_2$, similar carboxylated products are obtained, but the presence of stoichiometric amounts of strong and expensive reductants like $Et_2Zn$ and silanes is avoided. The following reaction steps are proposed: initially, $CO_2$ is reduced to CO (path I) or alkyl formates (path II) through transfer hydrogenation with alcohols. Path I is proposed to be the major reaction pathway for this transformation based on the control experiments. Meanwhile, the alcohol undergoes dehydrogenation to form the corresponding carbonyl compound (path III). Interestingly, a minor amount of the CO comes from further decarbonylation reactions of the alcohol as shown by isotope-labelling experiments. It is estimated that this novel methodology is applicable to other carbonylation reactions as well and will contribute substantially to the development of next-generation $CO_2$ utilization. Apart from $CO_2$ reduction with hydrogen and alcohols, Skrydstrup and co-workers shortly afterwards reported the successful identification of an efficient conversion of $CO_2$ to CO performed at room temperature with catalytic caesium fluoride in the presence of a disilane, which can be coupled to Pd-catalysed amino- and alkoxycarbonylations with aryl bromides and iodides[86].

Inspired by nature's efficiency in converting carbon dioxide into carbohydrates and other organic matter through photosynthesis, chemists have also explored opportunities to directly reduce $CO_2$ to CO by photocatalytic means. Since $CO_2$ does not absorb either visible or ultraviolet radiation in the wavelengths of 200–700 nm, this process requires a suitable photocatalyst to absorb ultraviolet–visible radiation and transfer it to $CO_2$. Photosensitizers (PSs) that meet this criterion often include a metal centre, as the *d*-orbital splitting in organometallic species often falls within the energy range of far-ultraviolet and visible light. The reduction process begins with excitation of the PS. This causes the transfer of an electron from the metal centre into the coordinated ligands. This movement is termed as metal-to-ligand charge transfer process. Back-electron transfer from the ligands to the metal after the charge transfer is prevented by including an electron-donating species to quench the exciting state of the photocatalyst, thus forming the one-electron-reduced (OER) intermediate for further $CO_2$ reduction processes[87].

Initial work by Lehn and co-workers in the 1980s led the development of first photocatalytic $CO_2$ reductions. Here, $CoCl_2$ was used as the catalytic site for $CO_2$ reduction with $[Ru(bpy)_3]Cl_2$ as PS (Fig. 7d)[88]. However, $CoCl_2$ has very low selectivity towards the reduction of $CO_2$ to CO, resulting from the reduction of water in the aqueous media to produce $H_2$ that is the main product in this system. The TON of $(CO+H_2)$ is only 32 based on $[Ru(bpy)_3]Cl_2$. The efficiency of these photocatalytic processes is also characterized by its quantum yield $\Phi$, which is defined as the number of times a $CO_2$ molecule is transformed per photon absorbed by the PS. However, this system reached a quantum yield of only 1.2%. Despite the low selectivity and efficiency of this reaction, the simultaneous reduction of $CO_2$ and $H_2O$ produces in fact syngas, which is of interest for large-scale industrial processes. Moreover, it represents an early step in the development of chemical systems capable of artificial photosynthesis and solar energy conversion. Later, the same research group found that rhenium-based complexes $[Re(bipy)(CO)_3X]$ (X = Cl, Br) acting as both PS and catalyst are more efficient and represent more selective catalysts for $CO_2$ reduction[89]. The process has very high $CO/H_2$, selectivity, generating only CO, and affords a quantum yield of up to 14%. It is indicated that the dissociation of the ligand $X^-$ from the unstable 19-electron OER is a key step in this photocatalytic reduction of $CO_2$ (ref. 90). In addition, Fujita and co-workers identified the binuclear [Re-C(O)O-Re] moiety as a key intermediate of the two-electron reduction process from $CO_2$ to CO (ref. 91). In 2008, based on their insightful mechanistic study, the most efficient photocatalytic system was successfully developed by the group of Ishitani using a mixed catalysts' system with *fac*-$[Re(bpy)(CO)_3(CH_3CN)]^+$ and *fac*-$[Re\text{-}4,4'\text{-}(MeO)_2bpy(CO)_3P(OEt)_3]^+$ affording the optimized quantum yield for CO formation was 0.59 (ref. 92).

Despite the improved activity of rhenium polypyridine complexes for $CO_2$ reduction, a major problem with these photocatalysts is the lack of an extended absorption into the visible region. To solve this problem, Ishitani and Bian and co-workers presented the use of covalently linked heteronuclear Ru and Re multimetallic complexes in the photocatalytic reduction of $CO_2$ (refs 93,94). Enhancement of the photocatalytic response to light in the visible region was achieved by specifically designed

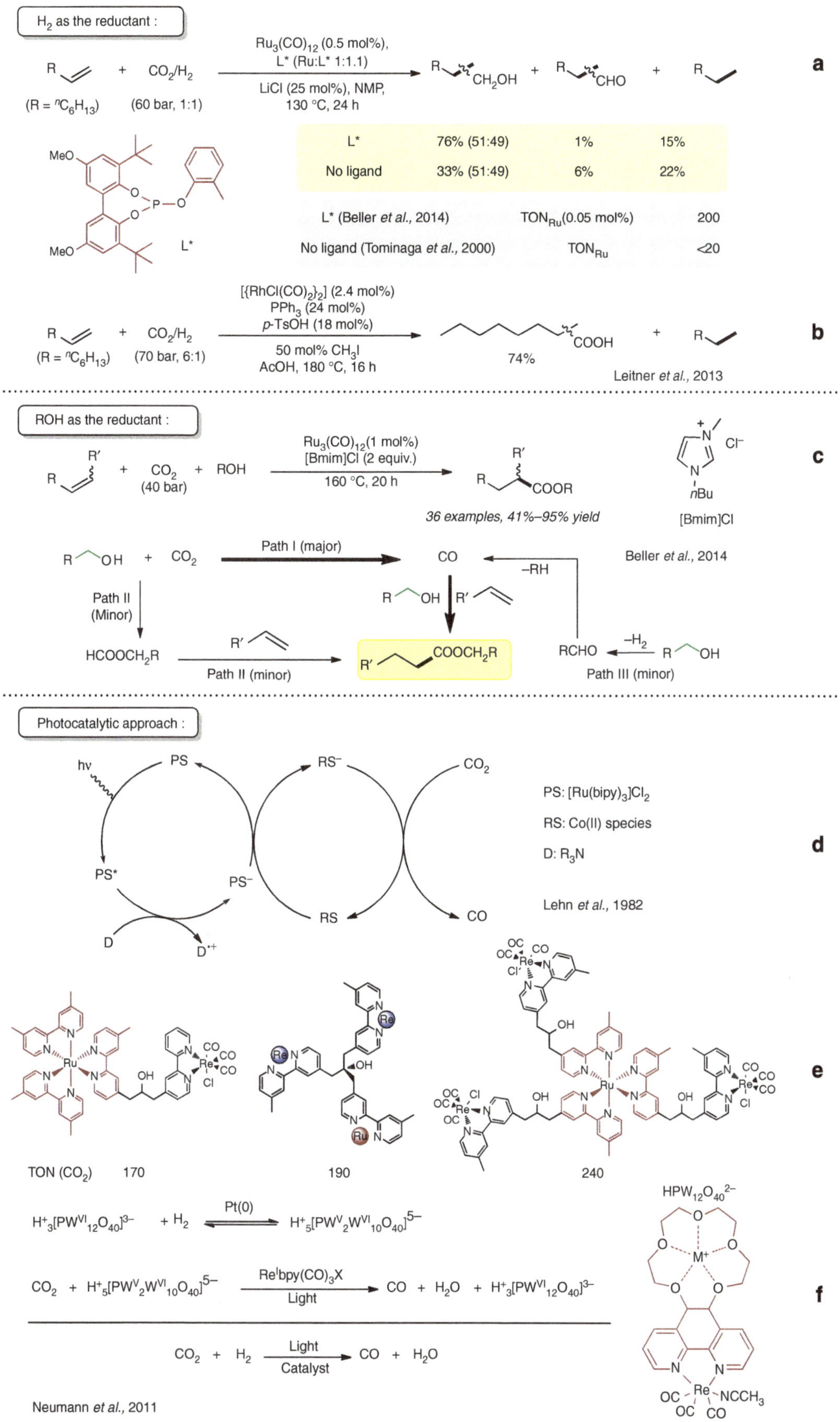


**Figure 7 | Catalytic production of CO from $CO_2$ reduction and its application in carbonylation reactions.** (**a**,**b**) $H_2$ as the reductant. (**c**) Alcohol as the reductant. (**d**–**f**) Photocatalytic approach.

supramolecular complexes, and their high photocatalytic activities were attributed to intramolecular electron transfer from the OER Ru species to the Re moiety. The binuclear complex Ru–Re, trinuclear complex $RuRe_2$ and tetranuclear complex $RuRe_3$ all furnished higher TON of CO (Fig. 7e). Despite recent improvements[95], it should be noted that, in the above-mentioned photochemical $CO_2$ reduction, common sacrificial donors such as triethylamine (TEA), triethanolamine (TEOA) and 1-benzyl-1,4-dihydronicotinamide (BNAH) have to be used. From a practical point of view, it is noteworthy that in 2011 Neumann and co-workers[96] developed a photoreduction of $CO_2$ to CO with $H_2$ catalysed by a rhenium(I) phenanthroline-polyoxometalate hybrid complex. In this study, previously used sacrificial electron donors were replaced by $H_2$. The working catalysis system is summarized in Fig. 7f. On one hand, a polyoxometalate oxidizes $H_2$ to two protons and two electrons in the presence of Pt(0). On the other hand, the reduced form of the polyoxometalate reacts with $CO_2$ with the assistance of a chromophoric rhenium(I) complex and visible light to produce CO and $H_2O$. As a result, the overall reaction is a photocatalytic RWGS reaction, which is of interest for both industry and academia for its further potential application in carbonylation reaction with $CO_2$.

## Future perspectives

Today's chemical production as well as academic organic synthesis relies mainly on the use of fossil-based carbon sources. In contrast, the increased valorization of $CO_2$ as a recyclable and abundant C1 building block would allow for the development of a sustainable chemical industry. In this respect, the development of benign transformations of carbon dioxide is a highly attractive object for synthetic chemistry. Clearly, in the past decades, several effective activation modes for catalytic $CO_2$ fixation and a series of novel transformations have been disclosed.

**Issues need to be resolved**. Despite the impressive progress, the efficiency of most reactions is far from satisfying. Hence, only recent new processes utilizing $CO_2$ as the C1 source have been industrialized. In most methodology-driven work, there remain important issues that need to be addressed in the future, for example, less utilization of very reactive/sensitive reagents, extending substrate scope, milder reaction conditions, improved chemo- and regioselectivity. Due its inherent inertness, not surprisingly, reactions of carbon dioxide under mild conditions need a thermodynamic driving force, for example, energy-rich coupling partners. The copolymerization of epoxides with $CO_2$ to produce polycarbonates is an elegant example for this strategy and permits efficient $CO_2$ valorization. Alternative copolymerizations with easily available chemical feedstocks, such as olefins, are highly desired too. In this respect, the recent report from Nozaki's group of a copolymerization of $CO_2$ and 1,3-butadiene is interesting. This work provides a straightforward approach towards the production of novel polymeric materials from $CO_2$ (ref. 97).

Another promising field might be the use of carbon dioxide as methylation agent. For example, ruthenium-catalysed methylation of amines with $CO_2$ and $H_2$ to afford valuable products has very recently been achieved. However, the current catalyst loading is still too high for large-scale production. Hence, the development of more efficient catalyst systems and non-noble metal-based catalysts for this and related transformations constitutes a demanding goal.

Most of the $CO_2$-fixation reactions, which proceed via oxidative cyclometallation pathway, are restricted to extensive $\pi$ systems (dienes and diynes) as substrates and the use of zinc reagents to cleave the metallacycle intermediates. Consequently, utilization of other simple alkene or alkyne substrates and milder nucleophiles is highly desired to further improve this strategy.

Carboxylation reactions via $CO_2$ insertion into C-M bonds in combination with classical coupling reactions are intensively studied in the past 10 years. Although this strategy furnished a series of interesting transformations, the requirement of stoichiometric amounts of strong bases, sensitive metal reagents or expensive hydrosilanes are evident drawbacks that need to be overcome. The key point to solve such problems in the future will be the invention of new routes to quench metal carboxylate complex (RCOOM) with less-nucleophilic agents to afford the final product and regenerate the original catalytically active species.

Finally, it is worth noting that the catalytic *in situ* generation of CO from $CO_2$ and its recent applications in sequential carbonylation reactions is a more favourable strategy for $CO_2$ valorization. It provides the use of safe and abundant $CO_2$ instead of CO as carbonyl source. Here the improvement of (regio)selectivities and to use milder reaction conditions represent important targets. So far, most catalysts in such reactions are ruthenium-based complexes. Encouraged by the recently developed highly regioselective ruthenium-catalysed hydroformylation reaction[98,99], in the future more efforts should be donated to solve the regioselectivity problem with suitable ligands and/or additives. Last but not the least, similar to other catalysis processes, the discovery of non-noble metal-catalysed reactions, for example, iron catalysts, is an essential issue in this field. Although related photocatalytic approaches proceed under very mild conditions, the TON is too low currently. Probably, cascade reaction sequences that make use of carbonylation processes to consume the *in situ*-generated CO will drive the overall reaction more efficiently.

**Towards artificial photosynthesis**. Photosynthesis basically supplies the carbon source for all organic compounds on Earth. It is an ideal green process for the conversion of light energy into chemical energy, which makes use of $CO_2$ (Fig. 8, equation a). Inspired by the nature's strategy, the development of sustainable and straightforward catalytic C-C bond-forming reactions of $CO_2$ in industrially relevant target molecules will be the 'crown jewel' of organic synthesis and catalysis research in the future. We expect several dream reactions towards artificial photosynthesis ($CO_2$ reduction under very mild conditions) and beyond will become true in the coming years. For example, sodium acrylate is an industrially important acrylate, with a global market volume of about 4 million tons and is commonly used as a monomer in the production of superabsorber polymers. Thus, the direct synthesis of acrylic acid from $CO_2$ and ethylene represents an economically attractive but also most challenging dream reaction. Until now, a TON of only 10 (Ni catalyst and NaO*t*Bu as base) has been realized; nevertheless, the possibility of this reaction was clearly demonstrated (Fig. 8, equation b)[100]. Ethylene glycol is another important raw material in the manufacture of polyester fibres and polyethylene terephthalate resins used in bottling. The direct

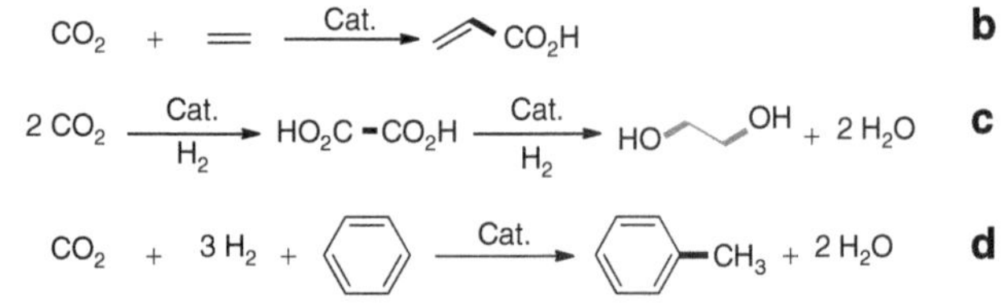


**Figure 8 | Dream reactions towards artificial photosynthesis. (a)** Photosynthesis reaction process. **(b)** Direct synthesis of acrylic acid from ethylene and $CO_2$. **(c)** Ethylene glycol synthesis via reductive coupling of $CO_2$. **(d)** Reductive methylation of benzene using $H_2$ and $CO_2$.

hydrogenation of $CO_2$ to produce this chemical in bulk scale might be a prime example for selective reductive coupling using carbon dioxide and could substitute its traditional synthesis from ethylene oxide (Fig. 8, equation c). As a final example, the reductive methylation of more widely available arenes, for example, benzene (from biogas), to generate toluene and xylenes is also highly interesting (Fig. 8, equation d).

Above all, it is anticipated that more research groups will join this very promising but also challenging field to discover a number of applied organic synthesis approaches using $CO_2$ as the C1 building block, in which the employment of less-active substrates, reduced waste generation and milder reaction conditions should be addressed.

## References


1. Schierbaum, F. in *Carbon Dioxide as Chemical Feedstock* Vol. 62 (ed. Aresta, M.) (Wiley-VCH, 2010).
2. Mikkelsen, M., Jorgensen, M. & Krebs, F. C. The teraton challenge. A review of fixation and transformation of carbon dioxide. *Energy Environ. Sci.* **3,** 43–81 (2010).
3. Sakakura, T., Choi, J.-C. & Yasuda, H. Transformation of carbon dioxide. *Chem. Rev.* **107,** 2365–2387 (2007).
4. Maeda, C., Miyazaki, Y. & Ema, T. Recent progress in catalytic conversions of carbon dioxide. *Catal. Sci. Technol.* **4,** 1482–1497 (2014).
5. Federsel, C., Jackstell, R. & Beller, M. State-of-the-art catalysts for hydrogenation of carbon dioxide. *Angew. Chem. Int. Ed.* **49,** 6254–6257 (2010).
6. Jian, G., Bing, Y. & Liang-Nian, H. in *Production and Purification of Ultraclean Transportation Fuels. ACS Symposium Series* Vol. 1088, Ch. 8, 143–174 (American Chemical Society, 2011).
7. Wang, W., Wang, S., Ma, X. & Gong, J. Recent advances in catalytic hydrogenation of carbon dioxide. *Chem. Soc. Rev.* **40,** 3703–3727 (2011).
8. Coates, G. W. & Moore, D. R. Discrete metal-based catalysts for the copolymerization of $CO_2$ and epoxides: discovery, reactivity, optimization, and mechanism. *Angew. Chem. Int. Ed.* **43,** 6618–6639 (2004).
9. Sakakura, T. & Kohno, K. The synthesis of organic carbonates from carbon dioxide. *Chem. Commun.* **11,** 1312–1330 (2009).
10. Clements, J. H. Reactive applications of cyclic alkylene carbonates. *Ind. Eng. Chem. Res.* **42,** 663–674 (2003).
11. Martin, R. & Kleij, A. W. Myth or reality? Fixation of carbon dioxide into complex organic matter under mild conditions. *ChemSusChem* **4,** 1259–1263 (2011).
12. Aida, T. & Inoue, S. Activation of carbon dioxide with aluminum porphyrin and reaction with epoxide. Studies on (tetraphenylporphinato)aluminum alkoxide having a long oxyalkylene chain as the alkoxide group. *J. Am. Chem. Soc.* **105,** 1304–1309 (1983).
13. Jin, L. *et al.* Metal porphyrin/phenyltrimethylammonium tribromide: high efficient catalysts for coupling reaction of $CO_2$ and epoxides. *J. Mol. Catal. A Chem.* **261,** 262–266 (2007).
14. Ema, T., Miyazaki, Y., Koyama, S., Yano, Y. & Sakai, T. A bifunctional catalyst for carbon dioxide fixation: cooperative double activation of epoxides for the synthesis of cyclic carbonates. *Chem. Commun.* **48,** 4489–4491 (2012).
15. Paddock, R. L. & Nguyen, S. T. Chemical $CO_2$ fixation: Cr(III) salen complexes as highly efficient catalysts for the coupling of $CO_2$ and epoxides. *J. Am. Chem. Soc.* **123,** 11498–11499 (2001).
16. Decortes, A. & Kleij, A. W. Ambient fixation of carbon dioxide using a ZnII salphen catalyst. *ChemCatChem* **3,** 831–834 (2011).
17. North, M. & Pasquale, R. Mechanism of cyclic carbonate synthesis from epoxides and $CO_2$. *Angew. Chem. Int. Ed.* **48,** 2946–2948 (2009).
18. Whiteoak, C. J. *et al.* A powerful aluminum catalyst for the synthesis of highly functional organic carbonates. *J. Am. Chem. Soc.* **135,** 1228–1231 (2013).
19. Li, F., Xiao, L., Xia, C. & Hu, B. Chemical fixation of $CO_2$ with highly efficient $ZnCl_2$/[BMIm]Br catalyst system. *Tetrahedron Lett.* **45,** 8307–8310 (2004).
20. Liang, S. *et al.* Highly efficient synthesis of cyclic carbonates from $CO_2$ and epoxides over cellulose/KI. *Chem. Commun.* **47,** 2131–2133 (2011).
21. Cheng, M., Lobkovsky, E. B. & Coates, G. W. Catalytic reactions involving C1 feedstocks: new high-activity Zn(II)-based catalysts for the alternating copolymerization of carbon dioxide and epoxides. *J. Am. Chem. Soc.* **120,** 11018–11019 (1998).
22. Moore, D. R., Cheng, M., Lobkovsky, E. B. & Coates, G. W. Mechanism of the alternating copolymerization of epoxides and $co_2$ using β-diiminate zinc catalysts: evidence for a bimetallic epoxide enchainment. *J. Am. Chem. Soc.* **125,** 11911–11924 (2003).
23. Kember, M. R., Knight, P. D., Reung, P. T. R. & Williams, C. K. Highly active dizinc catalyst for the copolymerization of carbon dioxide and cyclohexene oxide at one atmosphere pressure. *Angew. Chem. Int. Ed.* **48,** 931–933 (2009).
24. Nakano, K., Kobayashi, K., Ohkawara, T., Imoto, H. & Nozaki, K. Copolymerization of epoxides with carbon dioxide catalyzed by iron–corrole complexes: synthesis of a crystalline copolymer. *J. Am. Chem. Soc.* **135,** 8456–8459 (2013).
25. Nozaki, K., Nakano, K. & Hiyama, T. Optically active polycarbonates: asymmetric alternating copolymerization of cyclohexene oxide and carbon dioxide. *J. Am. Chem. Soc.* **121,** 11008–11009 (1999).
26. Hua, Y.-Z. *et al.* Highly enantioselective catalytic system for asymmetric copolymerization of carbon dioxide and cyclohexene oxide. *Chem. Eur. J.* **20,** 12394–12398 (2014).
27. Martin, C. *et al.* Easily accessible bifunctional Zn(salpyr) catalysts for the formation of organic carbonates. *Catal. Sci. Technol.* **4,** 1615–1621 (2014).
28. Lim, C.-H., Holder, A. M., Hynes, J. T. & Musgrave, C. B. Roles of the Lewis acid and base in the chemical reduction of $CO_2$ catalyzed by frustrated Lewis pairs. *Inorg. Chem.* **52,** 10062–10066 (2013).
29. Tlili, A., Frogneux, X., Blondiaux, E. & Cantat, T. Creating added value with a waste: methylation of amines with $CO_2$ and $H_2$. *Angew. Chem. Int. Ed.* **53,** 2543–2545 (2014).
30. Das Neves Gomes, C. *et al.* A diagonal approach to chemical recycling of carbon dioxide: organocatalytic transformation for the reductive functionalization of $CO_2$. *Angew. Chem. Int. Ed.* **51,** 187–190 (2012).
***A pioneering example of diagonal transformations for $CO_2$ recycling, which combine both reduction of $CO_2$ and formation of C-C, C-N, or C-O bonds with $CO_2$.***
31. Jacquet, O., Das Neves Gomes, C., Ephritikhine, M. & Cantat, T. Recycling of carbon and silicon wastes: room temperature formylation of N–H bonds using carbon dioxide and polymethylhydrosiloxane. *J. Am. Chem. Soc.* **134,** 2934–2937 (2012).
32. Li, Y., Fang, X., Junge, K. & Beller, M. A general catalytic methylation of amines using carbon dioxide. *Angew. Chem. Int. Ed.* **52,** 9568–9571 (2013).
33. Jacquet, O., Frogneux, X., Das Neves Gomes, C. & Cantat, T. $CO_2$ as a C1-building block for the catalytic methylation of amines. *Chem. Sci.* **4,** 2127–2131 (2013).
34. Beydoun, K., vom Stein, T., Klankermayer, J. & Leitner, W. Ruthenium-catalyzed direct methylation of primary and secondary aromatic amines using carbon dioxide and molecular hydrogen. *Angew. Chem. Int. Ed.* **52,** 9554–9557 (2013).
35. Li, Y., Sorribes, I., Yan, T., Junge, K. & Beller, M. Selective methylation of amines with carbon dioxide and $H_2$. *Angew. Chem. Int. Ed.* **52,** 12156–12160 (2013).
36. Cui, X., Dai, X., Zhang, Y., Deng, Y. & Shi, F. Methylation of amines, nitrobenzenes and aromatic nitriles with carbon dioxide and molecular hydrogen. *Chem. Sci.* **5,** 649–655 (2014).
37. Aresta, M., Nobile, C. F., Albano, V. G., Forni, E. & Manassero, M. New nickel-carbon dioxide complex. Synthesis, properties, and crystallographic characterization of (carbon dioxide)bis(tricyclohexylphosphine)nickel. *J. Chem. Soc. Chem. Commun.* 636-637 (1975).
***We consider this paper pivotal in developing the strategy of $CO_2$ acitivation via oxidative cyclometallation; the first metallacycle complex from $CO_2$ was reported.***
38. Hoberg, H., Peres, Y., Krueger, C. & Tsay, Y. H. A 1-oxa-2-nickela-5-cyclopentanone from ethene and carbon dioxide; preparation, structure and reactivity. *Angew. Chem.* **99,** 799–800 (1987).
39. Burkhart, G. & Hoberg, H. Oxanickelacyclopentene derivatives from nickel(0), carbon dioxide, and alkynes. *Angew. Chem.* **94,** 75 (1982).
40. Takimoto, M. & Mori, M. Cross-coupling reaction of oxo-π-allylnickel complex generated from 1,3-diene under an atmosphere of carbon dioxide. *J. Am. Chem. Soc.* **123,** 2895–2896 (2001).
41. Louie, J., Gibby, J. E., Farnworth, M. V. & Tekavec, T. N. Efficient nickel-catalyzed [2 + 2 + 2] cycloaddition of $CO_2$ and diynes. *J. Am. Chem. Soc.* **124,** 15188–15189 (2002).
42. Takimoto, M. & Mori, M. Novel catalytic $CO_2$ incorporation reaction: nickel-catalyzed regio- and stereoselective ring-closing carboxylation of bis-1,3-dienes. *J. Am. Chem. Soc.* **124,** 10008–10009 (2002).
43. Takimoto, M., Nakamura, Y., Kimura, K. & Mori, M. Highly enantioselective catalytic carbon dioxide incorporation reaction: nickel-catalyzed asymmetric carboxylative cyclization of bis-1,3-dienes. *J. Am. Chem. Soc.* **126,** 5956–5957 (2004).
44. Behr, A. & Juszak, K.-D. Palladium-catalyzed reaction of butadiene and carbon dioxide. *J. Organomet. Chem.* **255,** 263–268 (1983).
45. Yeung, C. S. & Dong, V. M. Beyond Aresta's complex: Ni- and Pd-catalyzed organozinc coupling with $CO_2$. *J. Am. Chem. Soc.* **130,** 7826–7827 (2008).
46. Huang, K., Sun, C.-L. & Shi, Z.-J. Transition metal-catalyzed C-C bond formation through the fixation of carbon dioxide. *Chem. Soc. Rev.* **40,** 2435–2452 (2011).
***A seminal introduction to $CO_2$ valorization reactions with C-C bond formation and its application in organic synthesis.***

47. Behr, A. Carbon dioxide as an alternative C1-building block: activation by transition metal complexes. *Angew. Chem.* **100,** 681–698 (1988).
48. de Meijere, A. & Diederich, F. (eds) *Metal-Catalyzed Cross-Coupling Reactions* 2nd, Completely Revised and Enlarged Edition, Vol. 2 (Wiley-VCH, 2004).
49. Correa, A. & Martin, R. Metal-catalyzed carboxylation of organometallic reagents with carbon dioxide. *Angew. Chem. Int. Ed.* **48,** 6201–6204 (2009).
50. Ackermann, L. Transition-metal-catalyzed carboxylation of C-H bonds. *Angew. Chem. Int. Ed.* **50,** 3842–3844 (2011).
51. Dalton, D. M. & Rovis, T. C-H carboxylation takes gold. *Nat. Chem.* **2,** 710–711 (2010).
52. Zhang, Y.-G. & Riduan, S. N. Catalytic hydrocarboxylation of alkenes and alkynes with $CO_2$. *Angew. Chem. Int. Ed.* **50,** 6210–6212 (2011).
53. Ukai, K., Aoki, M., Takaya, J. & Iwasawa, N. Rhodium(I)-catalyzed carboxylation of aryl- and alkenylboronic esters with $CO_2$. *J. Am. Chem. Soc.* **128,** 8706–8707 (2006).
   ***The authors reported the first catalytic carboxylation reaction of weak carbon nucleophiles with*** $CO_2$.
54. Takaya, J., Tadami, S., Ukai, K. & Iwasawa, N. Copper(I)-catalyzed carboxylation of aryl- and alkenylboronic esters. *Org. Lett.* **10,** 2697–2700 (2008).
55. Ohishi, T., Nishiura, M. & Hou, Z. Carboxylation of organoboronic esters catalyzed by N-heterocyclic carbene copper(I) complexes. *Angew. Chem. Int. Ed.* **47,** 5792–5795 (2008).
56. Ohmiya, H., Tanabe, M. & Sawamura, M. Copper-catalyzed carboxylation of alkylboranes with carbon dioxide: formal reductive carboxylation of terminal alkenes. *Org. Lett.* **13,** 1086–1088 (2011).
57. Ohishi, T., Zhang, L., Nishiura, M. & Hou, Z. Carboxylation of alkylboranes by N-heterocyclic carbene copper catalysts: synthesis of carboxylic acids from terminal alkenes and carbon dioxide. *Angew. Chem. Int. Ed.* **50,** 8114–8117 (2011).
58. Zhang, X. *et al.* Silver(I)-catalyzed carboxylation of arylboronic esters with $CO_2$. *Chem. Commun.* **48,** 6292–6294 (2012).
59. Zhang, X., Zhang, W.-Z., Ren, X., Zhang, L.-L. & Lu, X.-B. Ligand-free Ag(I)-catalyzed carboxylation of terminal alkynes with $CO_2$. *Org. Lett.* **13,** 2402–2405 (2011).
60. Yu, D. & Zhang, Y. The direct carboxylation of terminal alkynes with carbon dioxide. *Green Chem.* **13,** 1275–1279 (2011).
61. Yu, D. & Zhang, Y. Copper- and copper-N-heterocyclic carbene-catalyzed C - H activating carboxylation of terminal alkynes with $CO_2$ at ambient conditions. *Proc. Natl Acad. Sci. USA* **107,** 20184–20189 (2010).
62. Goossen, L. J., Rodriguez, N., Manjolinho, F. & Lange, P. P. Synthesis of propiolic acids via copper-catalyzed insertion of carbon dioxide into the C-H bond of terminal alkynes. *Adv. Synth. Catal.* **352,** 2913–2917 (2010).
63. Zhang, L., Cheng, J., Ohishi, T. & Hou, Z. Copper-catalyzed direct carboxylation of C-H bonds with carbon dioxide. *Angew. Chem. Int. Ed.* **49,** 8670–8673 (2010).
64. Vechorkin, O., Hirt, N. & Hu, X. Carbon dioxide as the C1 source for direct C-H functionalization of aromatic heterocycles. *Org. Lett.* **12,** 3567–3569 (2010).
65. Boogaerts, I. I. F. & Nolan, S. P. Carboxylation of C-H bonds using N-heterocyclic carbene gold(I) complexes. *J. Am. Chem. Soc.* **132,** 8858–8859 (2010).
66. Boogaerts, I. I. F., Fortman, G. C., Furst, M. R. L., Cazin, C. S. J. & Nolan, S. P. Carboxylation of N-H/C-H bonds using N-heterocyclic carbene copper(I) complexes. *Angew. Chem. Int. Ed.* **49,** 8674–8677 (2010).
67. Sasano, K., Takaya, J. & Iwasawa, N. Palladium(II)-catalyzed direct carboxylation of alkenyl C-H bonds with $CO_2$. *J. Am. Chem. Soc.* **135,** 10954–10957 (2013).
68. Mizuno, H., Takaya, J. & Iwasawa, N. Rhodium(I)-catalyzed direct carboxylation of arenes with $CO_2$ via chelation-assisted C-H bond activation. *J. Am. Chem. Soc.* **133,** 1251–1253 (2011).
69. Correa, A. & Martin, R. Palladium-catalyzed direct carboxylation of aryl bromides with carbon dioxide. *J. Am. Chem. Soc.* **131,** 15974–15975 (2009).
70. Fujihara, T., Nogi, K., Xu, T., Terao, J. & Tsuji, Y. Nickel-catalyzed carboxylation of aryl and vinyl chlorides employing carbon dioxide. *J. Am. Chem. Soc.* **134,** 9106–9109 (2012).
71. León, T., Correa, A. & Martin, R. Ni-catalyzed direct carboxylation of benzyl halides with $CO_2$. *J. Am. Chem. Soc.* **135,** 1221–1224 (2013).
72. Correa, A., León, T. & Martin, R. Ni-catalyzed carboxylation of C(sp2)– and C(sp3)–O bonds with $CO_2$. *J. Am. Chem. Soc.* **136,** 1062–1069 (2013).
73. Takaya, J. & Iwasawa, N. Hydrocarboxylation of allenes with $CO_2$ catalyzed by silyl pincer-type palladium complex. *J. Am. Chem. Soc.* **130,** 15254–15255 (2008).
74. Williams, C. M., Johnson, J. B. & Rovis, T. Nickel-catalyzed reductive carboxylation of styrenes using $CO_2$. *J. Am. Chem. Soc.* **130,** 14936–14937 (2008).
75. Li, S., Yuan, W. & Ma, S. Highly regio- and stereoselective three-component nickel-catalyzed syn-hydrocarboxylation of alkynes with diethyl zinc and carbon dioxide. *Angew. Chem. Int. Ed.* **50,** 2578–2582 (2011).
76. Fujihara, T., Xu, T., Semba, K., Terao, J. & Tsuji, Y. Copper-catalyzed hydrocarboxylation of alkynes using carbon dioxide and hydrosilanes. *Angew. Chem. Int. Ed.* **50,** 523–527 (2011).
77. Zhang, L., Cheng, J., Carry, B. & Hou, Z. Catalytic boracarboxylation of alkynes with diborane and carbon dioxide by an N-heterocyclic carbene copper catalyst. *J. Am. Chem. Soc.* **134,** 14314–14317 (2012).
78. Franke, R., Selent, D. & Boerner, A. Applied hydroformylation. *Chem. Rev.* **112,** 5675–5732 (2012).
79. Eastham, G. R., Waugh, M., Pringle, P. & Turner, T. P. W. Process for the carbonylation of ethylenically unsaturated compounds, bidentate carbonylation ligands and catalyst systems incorporating such ligands WO2011083305 (2011).
80. Tominaga, K.-I. & Sasaki, Y. Ruthenium complex-catalyzed hydroformylation of alkenes with carbon dioxide. *Catal. Commun.* **1,** 1–3 (2000).
   ***This work creats the benchmark for the directly carbonylation of alkenes with*** $CO_2$***, in which CO is in-situ generated from*** $CO_2$ ***hydrogenation (reverse water gas shift process).***
81. Liu, Q. *et al.* Development of a ruthenium/phosphite catalyst system for domino hydroformylation–reduction of olefins with carbon dioxide. *Chem. Eur. J.* **20,** 6888–6894 (2014).
82. Ostapowicz, T. G., Schmitz, M., Krystof, M., Klankermayer, J. & Leitner, W. Carbon dioxide as a C1 building block for the formation of carboxylic acids by formal catalytic hydrocarboxylation. *Angew. Chem. Int. Ed.* **52,** 12119–12123 (2013).
83. Nielsen, M. *et al.* Low-temperature aqueous-phase methanol dehydrogenation to hydrogen and carbon dioxide. *Nature* **495,** 85–89 (2013).
84. Ziebart, C. *et al.* Well-defined iron catalyst for improved hydrogenation of carbon dioxide and bicarbonate. *J. Am. Chem. Soc.* **134,** 20701–20704 (2012).
85. Wu, L., Liu, Q., Fleischer, I., Jackstell, R. & Beller, M. Ruthenium-catalysed alkoxycarbonylation of alkenes with carbon dioxide. *Nat. Commun.* **5,** 3091 (2014).
   ***This paper presents the first catalytic system for the*** $CO_2$ ***reduction to CO using alcohols as the reductant and also its application in the alkoxycarbonylation of alkenes with*** $CO_2$.
86. Lescot, C. *et al.* Efficient fluoride-catalyzed conversion of $CO_2$ to CO at room temperature. *J. Am. Chem. Soc.* **136,** 6142–6147 (2014).
87. Reithmeier, R., Bruckmeier, C. & Rieger, B. Conversion of $CO_2$ via visible light promoted homogeneous redox catalysis. *Catalysts* **2,** 544–571 (2012).
88. Lehn, J.-M. & Ziessel, R. Photochemical generation of carbon monoxide and hydrogen by reduction of carbon dioxide and water under visible light irradiation. *Proc. Natl Acad. Sci. USA* **79,** 701–704 (1982).
89. Hawecker, J., Lehn, J.-M. & Ziessel, R. Efficient photochemical reduction of $CO_2$ to CO by visible light irradiation of systems containing $Re(bipy)(CO)_3X$ or $Ru(bipy)_3^{2+}$-$Co^{2+}$ combinations as homogeneous catalysts. *J. Chem. Soc. Chem. Commun.* 536–538 (1983).
90. Hawecker, J., Lehn, J.-M. & Ziessel, R. Photochemical and electrochemical reduction of carbon dioxide to carbon monoxide mediated by (2,2′-bipyridine)tricarbonylchlororhenium(I) and related complexes as homogeneous catalysts. *Helv. Chim. Acta* **69,** 1990–2012 (1986).
91. Hayashi, Y., Kita, S., Brunschwig, B. S. & Fujita, E. Involvement of a binuclear species with the Re – C(O)O – Re moiety in $CO_2$ reduction catalyzed by tricarbonyl rhenium(I) complexes with diimine ligands: strikingly slow formation of the Re – Re and Re – C(O)O – Re species from $Re(dmb)(CO)_3S$ (dmb = 4,4′-dimethyl-2,2′-bipyridine, S = solvent). *J. Am. Chem. Soc.* **125,** 11976–11987 (2003).
92. Takeda, H., Koike, K., Inoue, H. & Ishitani, O. Development of an efficient photocatalytic system for $CO_2$ reduction using rhenium(I) complexes based on mechanistic studies. *J. Am. Chem. Soc.* **130,** 2023–2031 (2008).
93. Bian, Z.-Y. *et al.* A novel tripodal ligand, Tris[(4′-methyl-2,2′-bipyridyl-4-yl)methyl]carbinol and its trinuclear RuII/ReI mixed-metal complexes: synthesis, emission properties, and photocatalytic $CO_2$ reduction. *Inorg. Chem.* **47,** 10801–10803 (2008).
94. Gholamkhass, B. *et al.* Architecture of supramolecular metal complexes for photocatalytic $CO_2$ reduction: ruthenium – rhenium Bi- and tetranuclear complexes. *Inorg. Chem.* **44,** 2326–2336 (2005).
95. Sato, S., Morikawa, T., Kajino, T. & Ishitani, O. A highly efficient mononuclear iridium complex photocatalyst for $CO_2$ reduction under visible light. *Angew. Chem. Int. Ed.* **52,** 988–992 (2013).
96. Ettedgui, J., Diskin-Posner, Y., Weiner, L. & Neumann, R. Photoreduction of carbon dioxide to carbon monoxide with hydrogen catalyzed by a rhenium(I) phenanthroline–polyoxometalate hybrid complex. *J. Am. Chem. Soc.* **133,** 188–190 (2011).

97. Nakano, R., Ito, S. & Nozaki, K. Copolymerization of carbon dioxide and butadiene via a lactone intermediate. *Nat. Chem.* **6,** 325–331 (2014).
98. Wu, L., Fleischer, I., Jackstell, R. & Beller, M. Efficient and regioselective ruthenium-catalyzed hydro-aminomethylation of olefins. *J. Am. Chem. Soc.* **135,** 3989–3996 (2013).
99. Wu, L. *et al.* Ruthenium-catalyzed hydroformylation/reduction of olefins to alcohols: extending the scope to internal alkenes. *J. Am. Chem. Soc.* **135,** 14306–14312 (2013).
100. Lejkowski, M. L. *et al.* The first catalytic synthesis of an acrylate from $CO_2$ and an alkene-A rational approach. *Chem. Eur. J.* **18,** 14017–14025 (2012).

## Acknowledgements
We thank the State of Mecklenburg-Vorpommern and the BMBF for financial support. Q.L. also appreciates the grant from Alexander von Humboldt Foundation.

## Additional information

**Competing financial interests:** The authors declare no competing financial interests.

# 7

# Computational ligand design in enantio- and diastereoselective ynamide [5 + 2] cycloisomerization

R.N. Straker[1], Q. Peng[1], A. Mekareeya[1], R.S. Paton[1] & E.A. Anderson[1]

Transition metals can catalyse the stereoselective synthesis of cyclic organic molecules in a highly atom-efficient process called cycloisomerization. Many diastereoselective (substrate stereocontrol), and enantioselective (catalyst stereocontrol) cycloisomerizations have been developed. However, asymmetric cycloisomerizations where a chiral catalyst specifies the stereochemical outcome of the cyclization of a single enantiomer substrate—regardless of its inherent preference—are unknown. Here we show how a combined theoretical and experimental approach enables the design of a highly reactive rhodium catalyst for the stereoselective cycloisomerization of ynamide-vinylcyclopropanes to [5.3.0]-azabicycles. We first establish highly diastereoselective cycloisomerizations using an achiral catalyst, and then explore phosphoramidite-complexed rhodium catalysts in the enantioselective variant, where theoretical investigations uncover an unexpected reaction pathway in which the electronic structure of the phosphoramidite dramatically influences reaction rate and enantioselectivity. A marked enhancement of both is observed using the optimal theory-designed ligand, which enables double stereodifferentiating cycloisomerizations in both matched and mismatched catalyst-substrate settings.

[1]Chemistry Research Laboratory, University of Oxford, 12 Mansfield Road, Oxford OX1 3TA, UK. Correspondence and requests for materials should be addressed to E.A.A. (email: edward.anderson@chem.ox.ac.uk) or to R.S.P. (email: robert.paton@chem.ox.ac.uk) or to Q.P. (email: qian.peng@chem.ox.ac.uk).

Demand for higher efficiency, economy and selectivity in the synthesis of novel molecular scaffolds drives organic chemistry[1,2]. In this context, cycloisomerizations represent ideal methods for the formation of cyclic organic molecules, as they can fulfil all of these criteria. Despite much research into transition metal-catalysed cycloisomerization[3,4], and reports where high enantioselectivity is achieved for the cyclization of prochiral substrates to enantioenriched products[5,6], this important field of synthetic methodology has neglected the development of enantiospecific diastereoselective transformations, where single enantiomer starting materials are subjected to asymmetric cycloisomerization to give specific diastereomer products (that is, double stereodifferentiation, where catalyst and substrate stereocontrol compete)[7–10]. In an age where absolute control of molecular substitution and stereochemistry is paramount for applications, the realization of such processes would represent a major advance in the sustainable synthesis of precision-manufactured target molecules.

Here we describe rationally designed cycloisomerization catalysts that address these challenges. The reaction selected for this study was the [5 + 2] cycloisomerization of ynamide-vinylcyclopropanes to [5.3.0] azabicycles (**1**→**2**, **3**, Fig. 1). Although [5 + 2] processes have a rich history in the field of alkyne-vinylcyclopropanes[11–22], the development of this process using ynamides—or indeed any asymmetric cycloisomerization of ynamide-tethered enynes—has not been explored either experimentally or theoretically[23–27]. More generally, we question whether a catalyst system optimized to achieve an enantioselective cyclization (**1**→**2**, Fig. 1b) can translate to a double stereodifferentiating setting (**1**→**3** or ***epi*-3**, Fig. 1c), particularly should the catalyst be required to overcome powerful substrate stereocontrol. Intrinsic to these studies is a combined theoretical and experimental approach to optimize catalyst design[28–30], which in the event reveals an unexpected mechanistic pathway for rhodium-catalysed [5 + 2] carbocyclizations (Fig. 1a). This work demonstrates the powerful role of density functional level of theory (DFT) computations in understanding asymmetric catalysis, leading to quantitative computational-led design of new, highly selective ligands.

## Results

**Substrate synthesis and reaction optimization.** A selection of ynamide cycloisomerization substrates **1** were readily accessed from allylic esters **4** (Fig. 2) by an Ireland-Claisen/Curtius rearrangement/ynamide formation sequence. Substituents could be introduced at one or both of the carbon atoms on the ene-ynamide backbone, and by incorporating an enzymatic resolution into this synthesis, enantioenriched ynamides could be prepared. Ynamide formation (**6**→**1**) was achieved using copper-catalysed coupling of the sulfonamide with a bromoalkyne[32], or via formation of an intermediate dichloroenamide[33]. Initial screening of ruthenium[14] and rhodium[20,34] catalyst systems (Table 1) revealed that only the latter afforded high yields of azabicycle **7a** from ynamide **1a**; using [Rh(cod)naphthalene]$SbF_6$ (5 mol%) as the catalyst[35,36], the cycloisomerization could be effected within 3 h at room temperature, giving **7a** in 91% yield (Entry 6).

**Substrate scope.** A variety of ynamides **1a–s** were now examined in the cycloisomerization using these optimized conditions (Fig. 3). Aryl-substituted ynamides **1b–d** reacted with high efficiency, and revealed a clear electronic effect on the reaction rate, with electron-deficient ynamides **1b** and **1c** showing reduced reaction times. Alkyl-substituted ynamides **1e–g** also displayed enhanced reactivity compared with phenyl ynamide **1a**, affording the corresponding [5.3.0]-azabicycles **7e–g** in excellent yields within 15 min at ambient temperature. Similar efficient reactivity was observed for the cyclization of the heteroaryl-substituted ynamides **1h** and **1i** to the indolyl- and pyrrolyl-substituted products **7h** and **7i**. The mild conditions of the reaction are emphasized by the cyclization of the aniline-derived ynamide **1j**, which led to tricycle **7j** in 74% yield—notably, this 1,4-diene did not undergo *in situ* isomerization to the indole.

Of clear interest was the level of substrate stereocontrol that might be achieved using ynamides **1l–s**. Excellent levels of substrate stereoinduction were observed, with products **7l–s** afforded in high yield and as single diastereomers for substrates featuring an allylic stereocenter; and, for **7s**, as a single regioisomer as well as diastereoisomer[14]. Cyclization of substrates containing homoallylic stereocenters proved less selective; however, good stereoselectivity ($dr$ = 12:1) could be achieved using [Rh(cod)Cl]$_2$. These collected results are significant in the wider context of [5 + 2] cycloisomerization, where high levels of substrate stereocontrol have previously been generally interpreted to arise from an 'inside alkoxy effect' from oxygen-bearing stereocenters allylic to the vinylcyclopropane[37,38]. In this work, it is apparent that such stereoelectronic effects are not a prerequisite for high stereoselectivity.

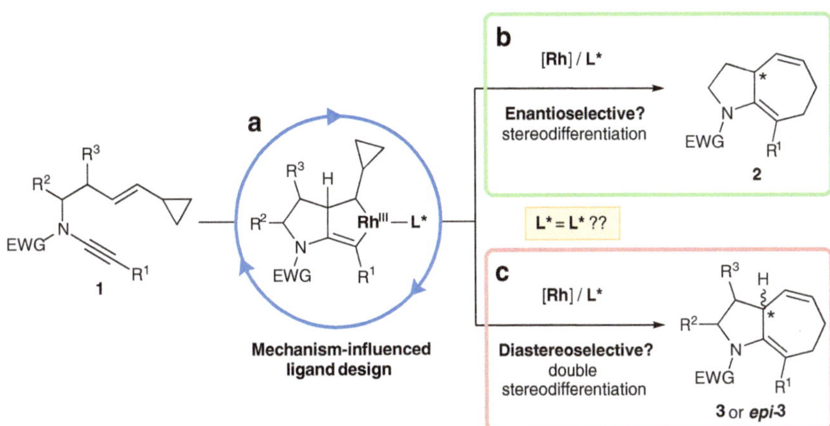


**Figure 1 | A reaction blueprint for enantioselective and/or double stereodifferentiating diastereoselective ynamide-vinylcyclopropane cycloisomerization.** (**a**) We show that mechanistic calculations are crucial in the design of an enantioselective cycloisomerization catalyst. (**b**) Enantioselective higher order ynamide cycloisomerization is realized. (**c**) Despite high levels of substrate stereocontrol, the computationally designed catalyst is able to define product stereochemistry.

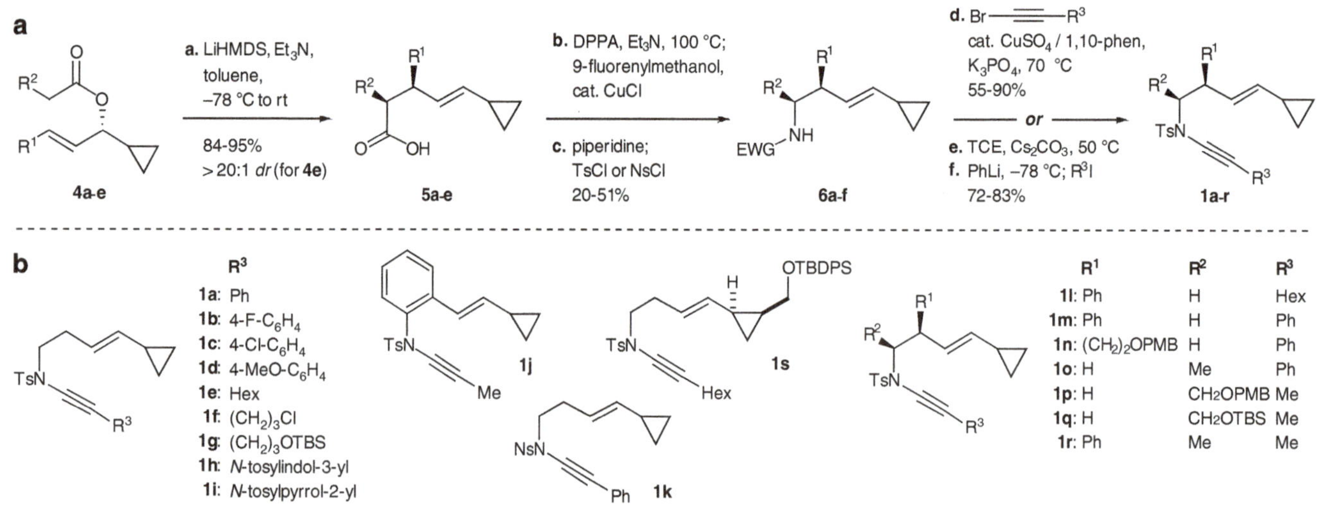


**Figure 2 | Synthesis of ynamide-vinylcyclopropanes 1.** (**a**) The synthetic route employed uses an Ireland-Claisen rearrangement of esters **4a-e** to construct the vinylcyclopropane and install up to two backbone substituents; subsequent Curtius rearrangement / sulfonylation converts the carboxylic acids **5** to sulfonamides **6**; ynamide formation is then achieved using copper-catalysed coupling with a bromoalkyne or, for hindered or aniline-derived ynamides, a two step route via a dichloroenamide. For the preparation of enantioenriched esters **4b**, **4d** and **4e**, and ynamides **1q** and **1s**, see Supplementary Fig. 4. (**b**) Ynamide-vinylcyclopropanes prepared using this strategy.

**Table 1 | Optimization of the [5 + 2] cycloisomerization reaction.**

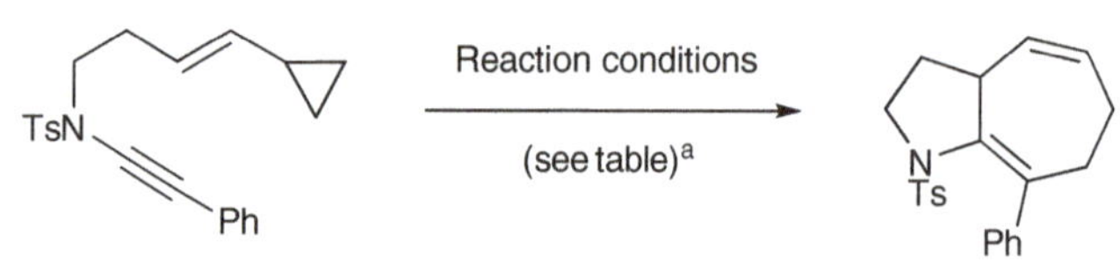


**1a** **7a**

| Entry | Catalyst (mol%) | Solvent | Temp. (°C) | Time (h) | Yield (%) |
|---|---|---|---|---|---|
| 1 | [CpRu(MeCN)$_3$]PF$_6$ (10) | acetone | 50 | 20 | –$^b$ |
| 2 | [RhCl(cod)]$_2$ (10) | toluene | 110 | 20 | –$^b$ |
| 3 | RhCl(PPh$_3$)$_3$ (10) | toluene | 110 | 1 | 72 |
| 4 | [(C$_{10}$H$_8$)Rh(cod)]SbF$_6$ (10) | 1,2-dichloroethane | rt | 1.5 | 92 |
| 5 | [(C$_{10}$H$_8$)Rh(cod)]SbF$_6$ (10) | CH$_2$Cl$_2$ | rt | 1.5 | 94 |
| 6 | [(C$_{10}$H$_8$)Rh(cod)]SbF$_6$ (5) | CH$_2$Cl$_2$ | rt | 3 | 91 |

$^a$Reactions performed at 0.1 M substrate in the solvent stated. $^b$Reaction gave a complex mixture of products.

**Enantioselective cycloisomerization.** In targeting an asymmetric version of the cyclization, we were mindful of the excellent asymmetric [5 + 2] Rh-catalysed cycloisomerization of alkyne-vinylcyclopropanes described by Shintani *et al.*[39,40], which employed the versatile phosphoramidite ligand **L1** (Fig. 4a)[41]. A preliminary screen of a range of phosphoramidite ligands **L1–L4** revealed that **L1** indeed seemed optimal for the enantioselective cycloisomerization of ynamide-vinylcyclopropane **1a**, delivering product (–)-**7a** in excellent yield and enantioselectivity (96%, 98% *ee*) after just 15 min at room temperature (see Supplementary Information for assignment of absolute stereochemistry). However, forays with other substrates suggested that this ligand might not meet our expectations in more challenging settings, and we realized that any further advance would require a combined theoretical and experimental design approach.

To this end, we conducted a computational exploration of the reaction pathway, which began with investigation of the empirical model adopted by Shintani *et al.* to explain the enantioselectivity of vinylcyclopropane-alkyne cycloisomerization. This model (Fig. 4b) is based on an X-ray crystal structure of [Rh(**L1**)norbornadiene]BF$_4$ reported by Mezetti (**8**) (ref. 42), which reveals an $\eta^2$-complexation of one of the $\alpha$-methylbenzylamine arenes to the rhodium cation—the phosphoramidite thus acting as a bidentate ligand. The Shintani–Hayashi model docks the vinylcyclopropane-alkyne onto this ligated rhodium framework as in structure **9**, such that the alkyne coordinates *trans* to the phosphorous atom of the phosphoramidite ligand, and the vinylcyclopropane coordinates *trans* to the $\eta^2$-complexed arene, with an orientation that minimizes steric interactions between the cyclopropane ring and naphthyl group.

**Computational study and ligand optimization.** Computation, particularly at the DFT level, has emerged as a powerful tool for assessing the feasibility of mechanistic steps involved in catalysis[43]. Our theoretical work first explored eight possibilities for the binding of ynamide-vinylcyclopropane **1a** to the [Rh(**L1**)]

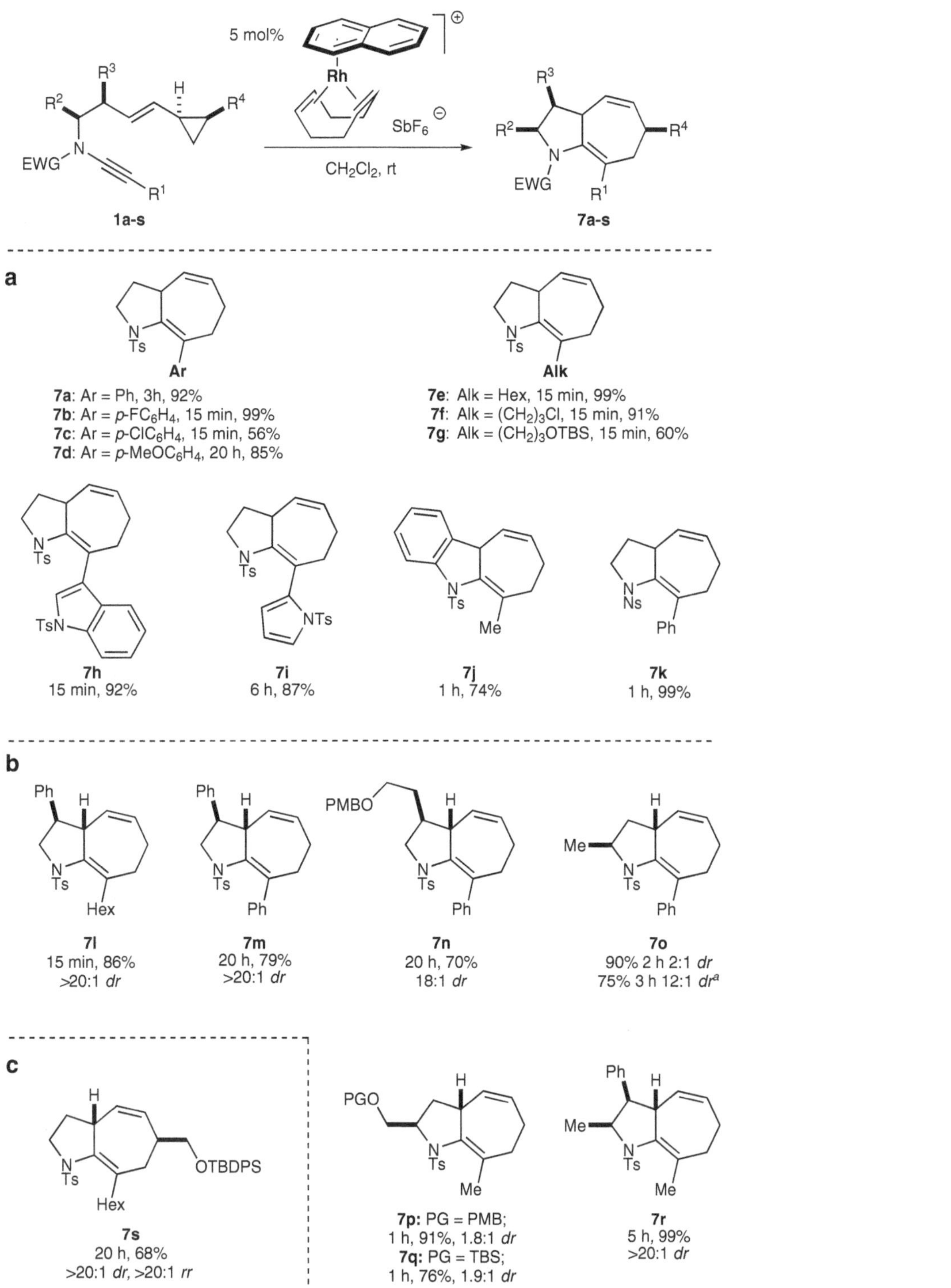


**Figure 3 | Rhodium-catalysed ynamide [5 + 2] cycloisomerization.** (**a**) Ynamide substituent scope. (**b**) Diastereoselective cycloisomerization proceeds with high levels of substrate stereoinduction. (**c**) High regioselectivity and diastereoselectivity is observed with a substituted cyclopropane. [a]Reaction performed using $[Rh(cod)Cl]_2$ (10 mol%), toluene, 110 °C.

cation (Table 2). In contrast to the Shintani–Hayashi model, this suggested that the lowest energy complex of [Rh(**L1**)**1a**] positions the ynamide proximal to the naphthyl ring and *trans-* to the arene ligand, and the vinylcyclopropane *trans-* to the phosphorous atom (**10**, Fig. 5, P-*trans*-ene/Up). The next lowest energy structure maintains this positional selectivity of substrate binding, but inverts its orientation (that is, P-*trans*-ene/Down). Next, calculations were carried out to explore the two widely accepted mechanisms for [5 + 2] cycloisomerization, which initiate either with an oxidative addition into the vinylcyclopropane, followed by alkyne insertion into the resultant σ/π-allyl rhodium(III) complex **11** to give eight-membered rhodacycle **12** (vinylcyclopropane pathway), or via oxidative cycloaddition of Rh(I) with the alkyne and alkene to give rhodacyclopentene **13**, followed by ring expansion into the cyclopropane to give the same intermediate **12** (metallacyclopentene pathway). Where investigations by Yu, Houk and Wender on Rh-catalysed [5 + 2] cycloisomerizations suggest that oxidative addition into the vinylcyclopropane is the first step on the catalytic pathway[17,21], our DFT calculations

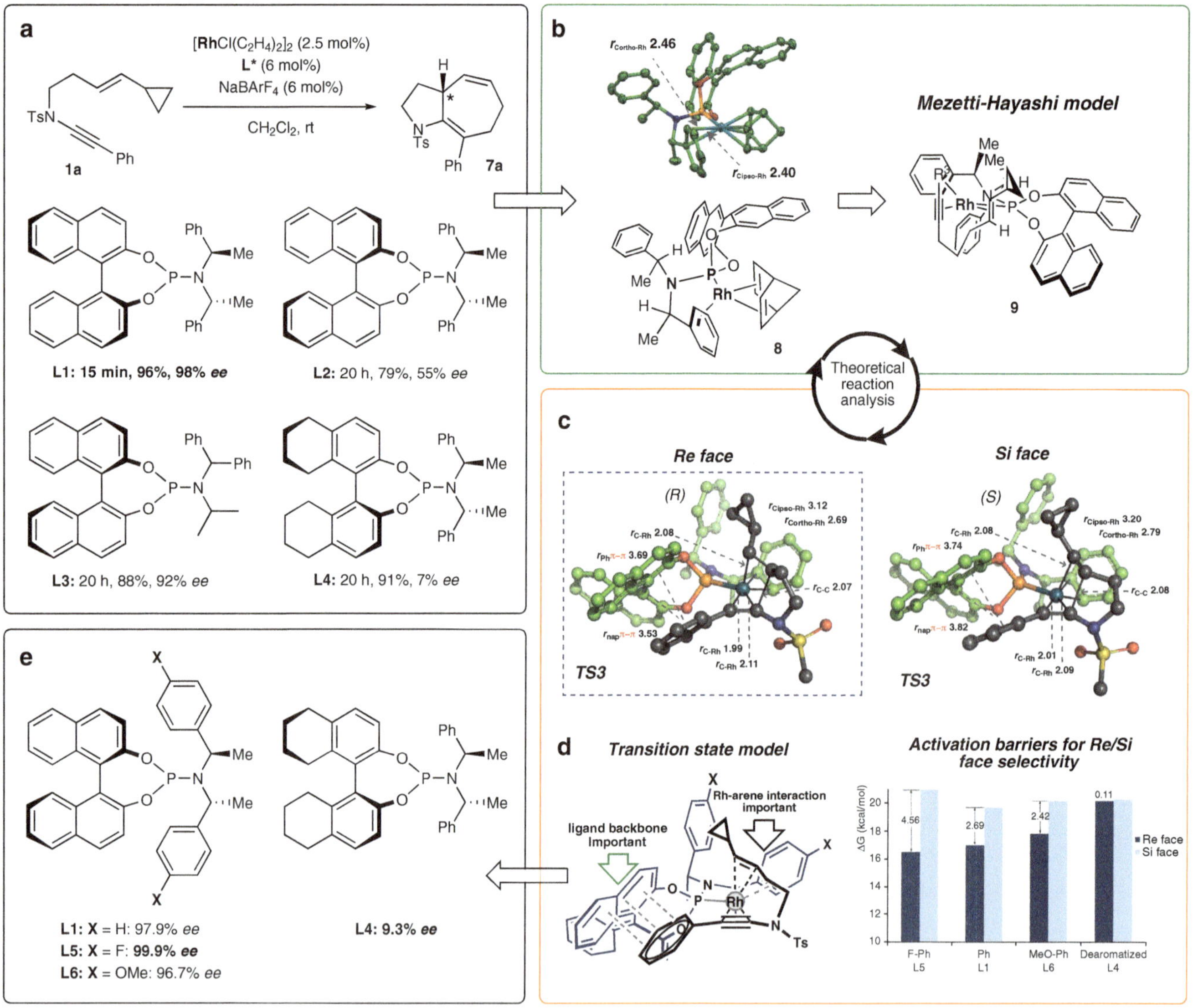

**Figure 4 | Exploration of asymmetric ynamide [5 + 2] cycloisomerization inspires a computation-guided ligand design.** (**a**) A selection of phosphoramidite ligands screened reveals a strong dependence of enantioselectivity and reaction rate on amine and BINOL components' structure and chirality, with **L1** giving superior results. (**b**) The Shintani-Hayashi model for enantioselectivity is based on a crystal structure of [**L1** Rh(norbornadiene)]$^+$. (**c**) Our calculations (Fig. 5) support an alternative mode of binding of the ynamide, and an alternative reaction pathway that initiates with rhodacyclopentene formation. *Re* face binding of the vinylcyclopropane is favoured, with calculated structures, and activation barriers (kcal mol$^{-1}$)/ selectivities depicted (distances in Å). (**d**) A transition state model is shown that identifies key ligand design parameters. (**e**) Variation of the electronic character of the $\eta^2$-bound arene of the phosphoramidite leads to predicted enantioselectivies for **L5** (*p*-F) and **L6** (*p*-OMe), and an excellent correlation of experiment and theory for **L1** and **L4**.

(Fig. 5) suggest that oxidative coupling of the alkene-ynamide to form metallacyclopentene **13** appears to be the preferred reaction pathway for an ynamide-vinylcyclopropane to give **12**, followed by ring expansion of the cyclopropane. Notably, this mechanism has also been calculated to be the preferred pathway in ruthenium-catalysed [5 + 2] cycloisomerization[17]. The preference for this pathway over vinylcyclopropane oxidative addition is in the range of 4–12 kcal mol$^{-1}$, depending on the orientation with which **1a** binds to the [Rh(**L1**)] cation; notably, the transition states for both pathways favour *Re*-face binding of the alkene in a P-*trans*-ene/Down orientation. This sequence of steps is favoured in this intramolecular reaction due to additional stabilization of the forming Rh(III) intermediate in the oxidative coupling step by the electron-rich ynamide. The free energy profile for the catalytic cycle is exergonic by more than 40 kcal mol$^{-1}$, and product inhibition is predicted to be minimal, since the reactant preferentially binds to the catalyst by 6.2 kcal mol$^{-1}$; taken together with a turnover and selectivity determining barrier of 17.0 kcal mol$^{-1}$ for the metallacyclopentene pathway, our computations are consistent with the short reaction times observed at room temperature for the conversion of **1a**–**7a** using **L1**.

The lowest energy transition state for the oxidative coupling of this metallacyclopentene pathway (TS3) is illustrated in Fig. 4c. This transition state leads to a calculated enantioselectivity for cyclization with ligand **L1** of 97.9% *ee* ($\Delta\Delta G^{\ddagger}_{Re/Si}$ = –2.69 kcal mol$^{-1}$), which is in excellent agreement with the experimental value (*R*, 98% *ee*). The calculated enantioselectivity for phosphoramidite **L4** (9.3% *ee*, $\Delta\Delta G^{\ddagger}_{Re/Si}$ = –0.11 kcal mol$^{-1}$) also correlates well with the poor selectivity we had already observed experimentally with this ligand (7% *ee*), supporting the mechanistic model, and implying that the naphthyl ring plays a crucial role—likely related (for aryl ynamides) to a stabilizing dispersive ($\pi - \pi$) interaction between

**Table 2 | Eight possible orientations of enynamide docking onto the L1-Rh cation are explored.**

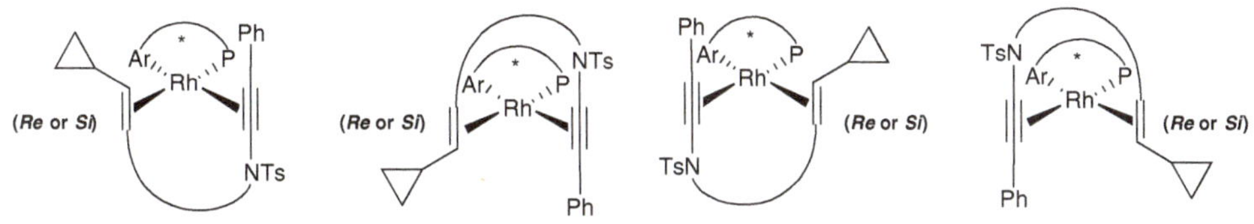


| Orientation 1: | Orientation 2: | Orientation 3: | Orientation 4: |
|---|---|---|---|
| P *trans* to ene | P *trans* to ene | P *trans* to yne | P *trans* to yne |
| Backbone of substrate **Down** to Ar-Rh-P | Backbone of substrate **Up** to Ar-Rh-P | Backbone of substrate **Down** to Ar-Rh-P | Backbone of substrate **Up** to Ar-Rh-P |

| **Orientation** | **Metallacyclopentene Pathway** | | **Vinylcyclopropane pathway** | |
|---|---|---|---|---|
| | ***Re*-selectivity (*R*)** | ***Si*-selectivity (*S*)** | ***Re*-selectivity (*R*)** | ***Si*-selectivity (*S*)** |
| 1 | 18.7 | 31.3 | 23.0 | 22.0 |
| 2 | **17.0** | 19.7 | 21.9 | 24.8 |
| 3 | 23.6 | 23.9 | 29.3 | 24.5 |
| 4 | 23.2 | 30.7 | 35.3 | 41.7 |

Transition state energies (SMD-ωB97XD/6-311+G(d,p)/LanI2TZ//ωB97XD/6-31G(d)/LanI2DZ Gibbs free energies, shown in kcal $mol^{-1}$) associated with each mode of binding are tabulated.

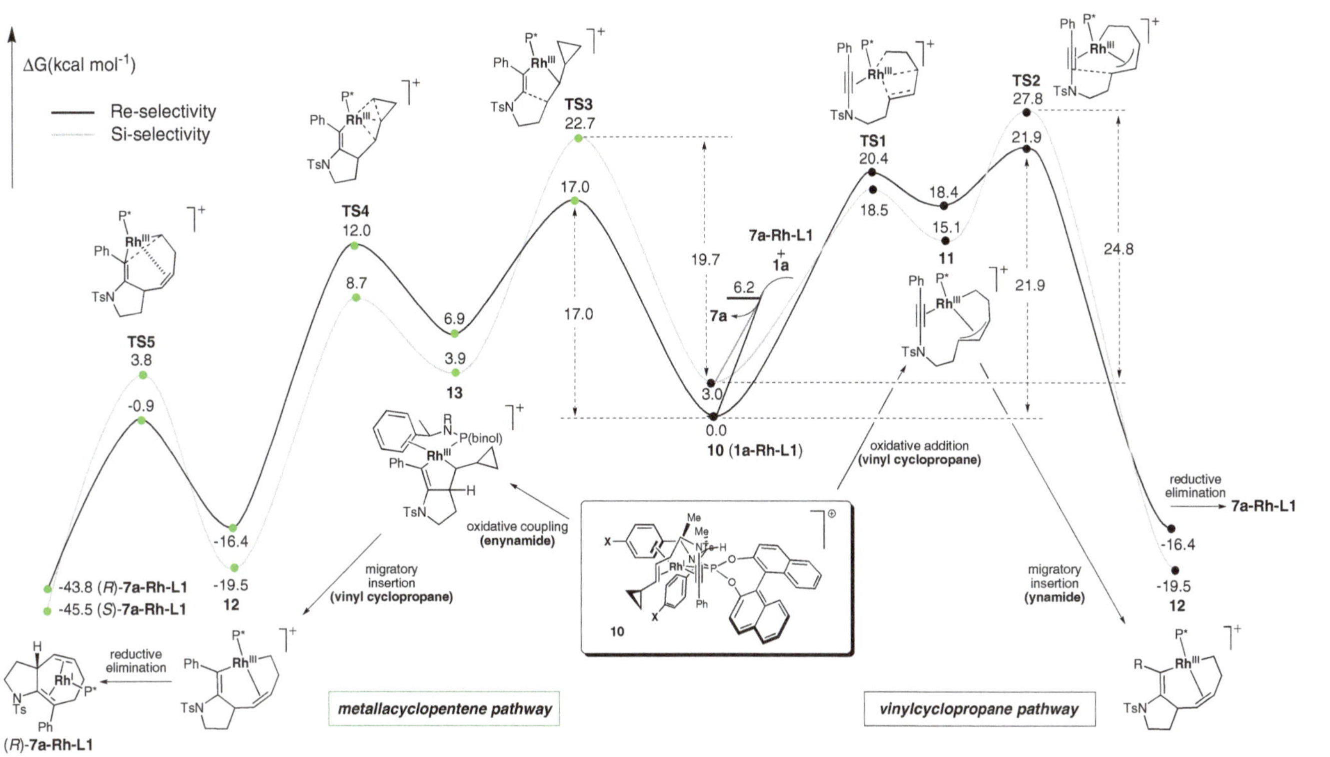


**Figure 5 | Theoretical reaction analysis.** Complex **10** (*Re* face binding) is found to be the lowest energy ground state for ynamide complexation, and transition state for oxidative coupling (metallacyclopentene pathway, see Table 2). The calculated energy profiles of the cycloisomerization for Re-face (black, bold) and Si-face (grey) substrate association via pathways initiating with metallacyclopentene formation (left), or vinylcyclopropane insertion (right) are illustrated; the former is favoured. SMD-ωB97XD/6-311+G(d,p)/LanI2TZ//ωB97XD/6-31G(d)/LanI2DZ Gibbs free energies are shown in kcal $mol^{-1}$.

the ynamide substituent and the naphthyl group in TS3. Although partial saturation of the naphthyl backbone (that is, in **L4**) leads to higher activation barriers, it is notable that the erosion of enantioselectivity for this ligand results from a greater increase (by 2.58 kcal $mol^{-1}$) to the activation barrier for *Re*-face addition compared with *Si*-face addition; this supports the existence, and importance, of stabilizing non-bonding (dispersion) interactions because of the aromatic backbone of **L1** in favouring the major enantiomer. Alkyl substituents are expected to experience similar attractive non-bonding interactions (CH–π); these interactions are further evident from an analysis of the computed non-covalent interaction index (Supplementary Fig. 68).

The oxidative coupling (TS3) is the rate-limiting and enantioselectivity determining step in this rhodacyclopentene mechanism. As observed in the Mezetti norbornene crystal structure (Fig. 4b)[42], one of the ligand benzylamine groups acts as 2π-electron donor (2.6–2.7 Å) in these transition states.

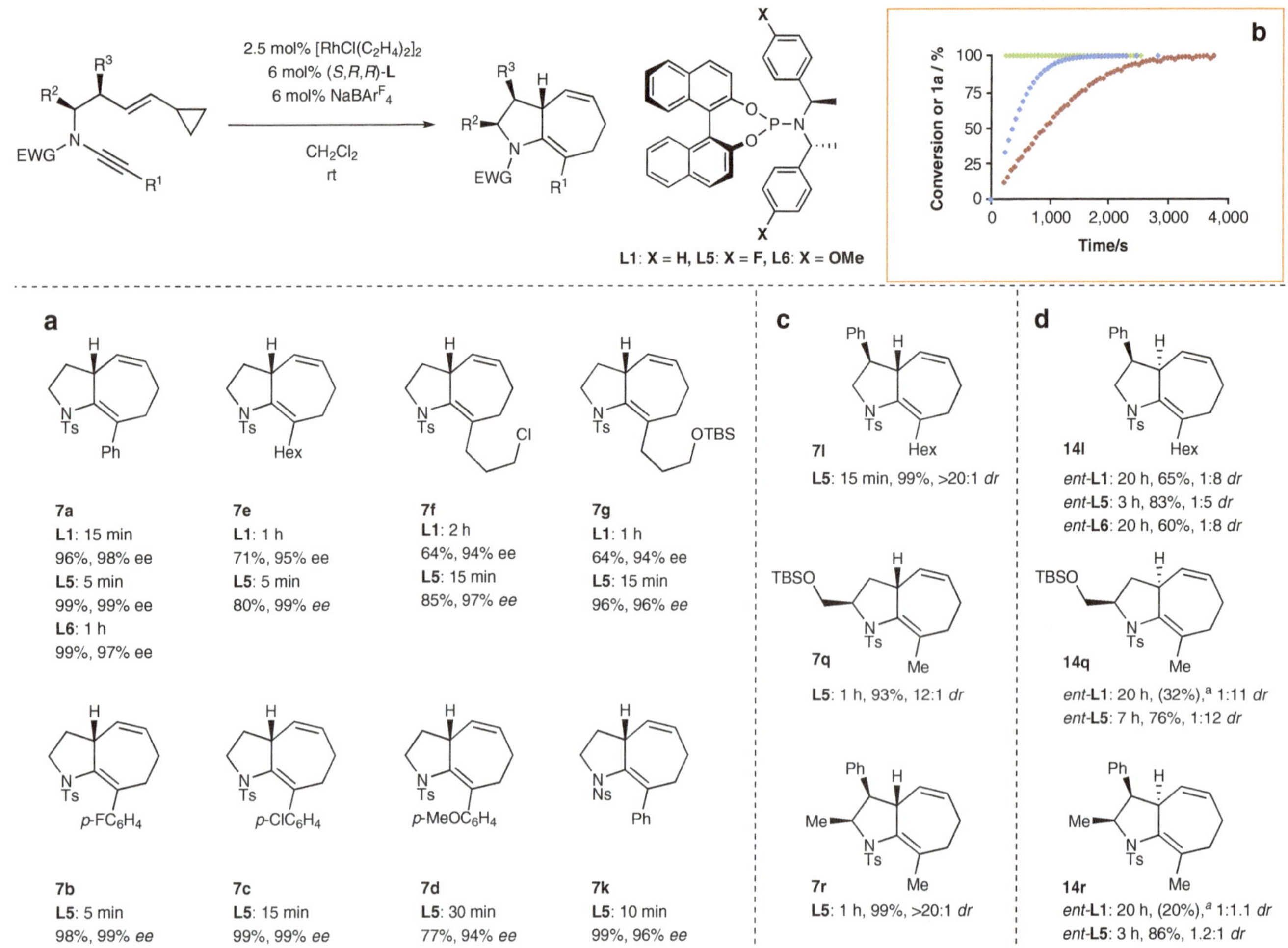


**Figure 6 | Enantioselective and double stereodifferentiating, ynamide [5 + 2] cycloisomerizations.** (**a**) Enantioselective cyclization is tested against a range of ynamide-vinylcyclopropanes, showing an excellent correlation of computation and experiment for the theory-designed ligands **L5** and **L6**. The synthesis of **7e** was performed on 1 mmol scale (1.25 mol% catalyst). (**b**) $^1$H NMR spectroscopic monitoring of reaction progress emphasizes the rate enhancement between ligands **L6**, **L1** and **L5**, also predicted theoretically. (**c**) Matched double stereodifferentiating cycloisomerizations proceed with high selectivity and rate. (**d**) Mismatched double stereodifferentiating cycloisomerizations proceed successfully under catalyst stereocontrol using the enantiomers of ligands **L1**, **L5** and **L6** (that is, (*R,S,S*)-**L** stereochemistry); the major diastereomer is shown. The $^a$ signifies reaction conversion, as judged by $^1$H NMR spectroscopic analysis.

We hypothesized that variation of the electronic character of this $\eta^2$-complexed arene could dramatically influence both the rate, and selectivity, of the reaction. Calculations on the effect of electronically-biasing methoxy and fluoro groups at the *para* position of these arenes suggested that decreasing the electron density on the arene (*p*-F, **L5**, Fig. 4d,e) would lead to an increase in reaction rate ($\Delta\Delta G^{\ddagger}_{L5/L1} = -0.47\,\text{kcal mol}^{-1}$) and enantioselectivity ($\Delta\Delta G^{\ddagger}_{Re/Si} = -4.56\,\text{kcal mol}^{-1}$, 99.9% *ee*), while an electron-donating substituent (*p*-MeO, **L6**) would have the opposite effect ($\Delta\Delta G^{\ddagger}_{L6/L1} = +0.76\,\text{kcal mol}^{-1}$, $\Delta\Delta G^{\ddagger}_{Re/Si} = -2.42\,\text{kcal mol}^{-1}$, 96.7% *ee*). This computed trend in reactivity and enantioselectivity, namely **L5** > **L1** > **L6**, results from a weakened metal-arene interaction in the TS leading to the major enantiomer, which is compensated for through stronger coordination of the alkynyl and alkenyl groups (see Supplementary Information for further details and discussion). These ligand structural modifications also modulate the Lewis basicity at phosphorus, however preferential stabilization of the major enantiomeric pathway confirms that the predominant effect is not inductive in nature, but rather depends on through-space interactions involving the Rh-coordinating aryl group.

**Substrate scope in the asymmetric cycloisomerization.** While phosphoramidite **L6** is known, (ref. 44) **L5** is not, but could be readily synthesized using standard methods[45]. Both were evaluated in the enantioselective cyclization alongside **L1** (Fig. 6a). To our delight, experiment correlated well with the predicted outcomes of these cyclizations; the *p*-fluorobenzyl ligand **L5** indeed showed a dramatic enhancement of rate and selectivity in the cyclization of **1a** compared with the parent ligand **L1**, affording **7a** in under 5 min with 99% *ee* (calc. 99.9% *ee*), while the *p*-methoxybenzyl ligand **L6** exhibited a much reduced rate of reaction, and also lower enantioselectivity (1 h, 97% *ee*, calc. 96.7% *ee*). The cyclization of alkyl-substituted ynamides **1e–g** to give enantioenriched azabicycles **7e–g** also showed a rate and selectivity enhancement between **L1** and **L5**. The potent reactivity of the latter ligand was extended to a variety of aryl-substituted ynamides, where the marked rate difference between electron-poor (**1b**) and electron-rich (**1d**) ynamides supports the hypothesis that the ynamide is intimately involved in the rate-determining step (that is, a metallacyclopentene pathway). In all cases using ligand **L5**, these reactions proceeded in under 30 min with excellent levels of enantioselectivity (94–99% *ee*); furthermore, the synthesis of **7e**

could be achieved on a 1 mmol scale in less than 5 min with 1.25 mol% catalyst loading. Monitoring of reaction conversion by $^1H$ NMR spectroscopy (in $CDCl_3$, Fig. 6b) emphasizes the dramatic rate difference between these three ligands, with the reaction using **L5** complete in under 4 min, compared with the slower cyclizations with **L1** and **L6** (addition of the catalyst solution to the substrate followed by NMR spectroscopic analysis necessitated a four minute delay between reaction initiation and acquisition of the first NMR spectrum. The reaction catalysed by [L5Rh] was complete by this time).

Finally, the performance of these ligands was tested against the challenge of cyclizations that proceed with high levels of substrate stereocontrol, to address the question of how a high enantioselectivity-inducing catalyst would cope with mismatched substrate-catalyst diastereoselective cycloisomerization scenarios. Three substrates were chosen for this challenge: ynamides **1l**, **1q** and **1r**, which gave >20:1, 1.8:1 and >20:1 *dr*, respectively when cyclized using [Rh(cod)naphthalene]$SbF_6$ (Fig. 3). As expected, these single enantiomer substrates cyclized rapidly, and with high efficiency and selectivity, with the matched substrate-catalyst combination (that is, (*S,R,R*)-**L5**, Fig. 6c). To our delight, we found that cycloisomerization of substrate **1l** using the enantiomeric catalyst system ((*R,S,S*)-**L1**) successfully overturned this powerful substrate stereoselectivity, giving **14l** in up to 1:8 *dr*. Interestingly, it was ligand **L1**—and not **L5**—which performed optimally in this challenging situation, albeit requiring an extended reaction time. In the case of **1q**, the catalyst (with either **L1** or **L5**) was able to achieve an equivalent (reversed) level of selectivity to that achieved in the matched sense—in this instance, a modest level of inherent substrate stereocontrol being completely overturned. Finally, the most challenging setting of the reinforcing substituents in **1r** proved a hurdle that could also be partly overcome, again demonstrating significant catalyst influence.

These observations may suggest that tighter substrate-Rh complexation in the case of **L5** (a consequence of a slightly weaker ligand-metal interaction observed in our calculations (see Supplementary Information for details), which improves rate and enantioselectivity), enhances unfavourable (that is, mismatched) steric effects in a double stereodifferentiating setting such that it effectively increases the stereocontrolling influence of the substrate relative to that of the ligand. The most reactive/selective catalysts for enantioselective cyclizations could thus suffer from higher than expected transition state energies in diastereoselective cyclizations, where such steric effects are enhanced compared with 'less enantioselective' catalysts (looser substrate binding); and that different considerations are therefore needed in the development of double stereodifferentiating reactions, with more 'promiscuous' catalysts potentially giving superior selectivity.

## Discussion

In summary, the first example of an enantioselective ynamide-tethered cycloisomerization has been achieved, with a series of highly enantio- and diastereoselective cyclizations giving a range of substituted/enantioenriched [5.3.0] azabicycles. Theoretical reaction analysis crucially influenced ligand design, leading to a catalyst system that displayed enhanced rate and enantioselectivity in the cycloisomerization. The demonstration of the first successful examples of enantiospecific diastereoselective transition metal-catalysed cycloisomerizations in a significantly mismatched substrate-catalyst environment illustrates that cycloisomerization can assemble functionalized ring systems with tuneable selectivity. These studies set the stage for the development of further computationally guided catalyst systems.

## Methods

**Racemic [5 + 2] cycloisomerization.** To an oven-dried vial containing the ynamide vinylcyclopropane (1.0 equiv.) under Ar was added a solution of [($C_{10}H_8$)Rh(cod)]$SbF_6$ (5 mol%) in degassed $CH_2Cl_2$ (10 ml mmol$^{-1}$ of ynamide). The reaction mixture was stirred at room temperature under Ar until consumption of the ynamide was observed by thin layer chromatography (see Fig. 3 for reaction times). The reaction mixture was then concentrated, and the resulting material was purified by flash chromatography ($SiO_2$, petroleum ether/ethyl acetate eluent) (Fig. 3).

**Asymmetric [5 + 2] cycloisomerization.** A solution of $[RhCl(C_2H_4)_2]_2$ (2.5 mol%), $NaBAr^F_4$ (6 mol%) and phosphoramidite ligand (6 mol%) in degassed $CH_2Cl_2$ (10 ml mmol$^{-1}$ of ynamide) was stirred for 20 min under Ar. The solution was filtered (through a PTFE filter-tipped syringe) into an oven-dried vial containing ynamide vinylcyclopropane (1.0 equiv.) under Ar. The reaction mixture was stirred at room temperature under Ar until consumption of the ynamide was observed by thin layer chromatography (see Fig. 6 for reaction times). The reaction mixture was then concentrated, and the resulting material was purified by flash chromatography ($SiO_2$, petroleum ether/ethyl acetate eluent) (Fig. 6).

**Computational methods.** Molecular geometries were fully optimized at the DFT theory level in Gaussian 09 (rev. D.01), using the dispersion-corrected ω-B97XD functional[46] without symmetry constraints. The effective core potentials (ECPs) of Hay and Wadt[47] with a double-ζ basis set (LanL2DZ) were used for Rh, S and P, and the 6-31G(d) basis set was used for H, C, N and O(BS1). The energies were further estimated using a larger basis set (6-311 + G (d, p) basis set for H, C, N, O, S and P) and triple-ζ basis set (LanL2TZ)[48] for Rh (BS2) by single-point calculations, in implicit solvent $CH_2Cl_2$ treated with the SMD universal solvation models[49]. The structures of the ynamide substrate and a series of phosphoramidite ligands were computed in full, while the toluenesulfonamide *p*-tolyl group was modelled as a methyl group in the interests of computational tractability. See Supplementary Methods for further details.

## References


1. Newhouse, T., Baran, P. S. & Hoffmann, R. W. The economies of synthesis. *Chem. Soc. Rev.* **38,** 3010–3021 (2009).
2. Trost, B. M. The atom economy-a search for synthetic efficiency. *Science* **254,** 1471–1477 (1991).
3. Michelet, V., Toullec, P. Y. & Genet, J. P. Cycloisomerization of 1,n-enynes: challenging metal-catalyzed rearrangements and mechanistic insights. *Angew. Chem. Int. Ed.* **47,** 4268–4315 (2008).
4. Inglesby, P. A. & Evans, P. A. Stereoselective transition metal-catalysed higher-order carbocyclisation reactions. *Chem. Soc. Rev.* **39,** 2791–2805 (2010).
5. Marinetti, A., Jullien, H. & Voituriez, A. Enantioselective, transition metal catalyzed cycloisomerizations. *Chem. Soc. Rev.* **41,** 4884–4908 (2012).
6. Watson, I. D. G. & Toste, F. D. Catalytic enantioselective carbon-carbon bond formation using cycloisomerization reactions. *Chem. Sci.* **3,** 2899–2919 (2012).
7. Feducia, J. A., Campbell, A. N., Doherty, M. Q. & Gagné, M. R. Modular catalysts for diene cycloisomerization: Rapid and enantioselective variants for bicyclopropane synthesis. *J. Am. Chem. Soc.* **128,** 13290–13297 (2006).
8. Ham, Y. J. *et al.* Rhodium-catalyzed reductive cyclization of 1,6-enynes and stereoselective synthesis of the putative structure of lucentamycin A and its stereoisomers. *Tetrahedron* **68,** 1918–1925 (2012).
9. Lei, A., He, M. & Zhang, X. Rh-catalyzed kinetic resolution of enynes and highly enantioselective formation of 4-alkenyl-2,3-disubstituted tetrahydrofurans. *J. Am. Chem. Soc.* **125,** 11472–11473 (2003).
10. Trost, B. M., Ryan, M. C., Rao, M. & Markovic, T. Z. Construction of enantioenriched [3.1.0] bicycles via a ruthenium-catalyzed asymmetric redox bicycloisomerization reaction. *J. Am. Chem. Soc.* **136,** 17422–17425 (2014).
11. Wender, P. A., Takahashi, H. & Witulski, B. Transition metal catalyzed [5 + 2] cycloadditions of vinylcyclopropanes and alkynes: a homolog of the Diels-Alder reaction for the synthesis of seven-membered rings. *J. Am. Chem. Soc.* **117,** 4720–4721 (1995).
12. Wender, P. A. *et al.* Transition metal-catalyzed [5 + 2] cycloadditions with substituted cyclopropanes: first studies of regio- and stereoselectivity. *J. Am. Chem. Soc.* **121,** 10442–10443 (1999).
13. Trost, B. M., Toste, F. D. & Shen, H. Ruthenium-catalyzed intramolecular [5 + 2] cycloadditions. *J. Am. Chem. Soc.* **122,** 2379–2380 (2000).
14. Trost, B. M. *et al.* Syntheses of seven-membered rings: ruthenium-catalyzed intramolecular [5 + 2] cycloadditions. *Chem. Eur. J.* **11,** 2577–2590 (2005).
15. Pellissier, H. Recent developments in the [5 + 2] cycloaddition. *Adv. Synth. Catal.* **353,** 189–218 (2011).
16. Ylijoki, K. E. O. & Stryker, J. M. [5 + 2] Cycloaddition reactions in organic and natural product synthesis. *Chem. Rev.* **113,** 2244–2266 (2013).
17. Mustard, T. J. L., Wender, P. A. & Cheong, P. H.-Y. Catalytic efficiency is a function of how rhodium(I) (5 + 2) catalysts accommodate a conserved

substrate transition state geometry: induced fit model for explaining transition metal catalysis. *ACS Catal.* **5,** 1758–1763 (2015).
18. Hong, X., Trost, B. M. & Houk, K. N. Mechanism and origins of selectivity in Ru(II)-catalyzed intramolecular (5 + 2) cycloadditions and ene reactions of vinylcyclopropanes and alkynes from density functional theory. *J. Am. Chem. Soc.* **135,** 6588–6600 (2013).
19. Xu, X. *et al.* Ligand effects on rates and regioselectivities of Rh(I)-catalyzed (5 + 2) cycloadditions: a computational study of cyclooctadiene and dinaphthocyclooctatetraene as ligands. *J. Am. Chem. Soc.* **134,** 11012–11025 (2012).
20. Liu, P. *et al.* Electronic and steric control of regioselectivities in Rh(I)-catalyzed (5 + 2) cycloadditions: experiment and theory. *J. Am. Chem. Soc.* **132,** 10127–10135 (2010; .
21. Liu, P., Cheong, P. H.-Y., Yu, Z.-X., Wender, P. A. & Houk, K. N. Substituent effects, reactant preorganization, and ligand exchange control the reactivity in RhI-catalyzed (5 + 2) cycloadditions between vinylcyclopropanes and alkynes. *Angew. Chem. Int. Ed.* **47,** 3939–3941 (2008).
22. Yu, Z.-X., Wender, P. A. & Houk, K. N. On the Mechanism of $[Rh(CO)_2Cl]_2$-Catalyzed Intermolecular (5 + 2) Reactions between Vinylcyclopropanes and Alkynes. *J. Am. Chem. Soc.* **126,** 9154–9155 (2004).
23. Walker, P. R., Campbell, C. D., Suleman, A., Carr, G. & Anderson, E. A. Palladium- and ruthenium-catalyzed cycloisomerization of enynamides and enynhydrazides: A rapid approach to diverse azacyclic frameworks. *Angew. Chem. Int. Ed.* **52,** 9139–9143 (2013).
24. Couty, S., Meyer, C. & Cossy, J. Diastereoselective gold-catalyzed cycloisomerizations of ene-ynamides. *Angew. Chem. Int. Ed.* **45,** 6726–6730 (2006).
25. Witulski, B. & Alayrac, C. A highly efficient and flexible synthesis of substituted carbazoles by rhodium-catalyzed inter- and intramolecular alkyne cyclotrimerizations. *Angew. Chem. Int. Ed.* **41,** 3281–3284 (2002).
26. Wang, X.-N. *et al.* Ynamides in ring forming transformations. *Acc. Chem. Res.* **47,** 560–578 (2014).
27. Nishimura, T., Takiguchi, Y., Maeda, Y. & Hayashi, T. Rhodium-catalyzed asymmetric cycloisomerization of 1,6-ene-ynamides. *Adv. Synth. Catal.* **355,** 1374–1382 (2013).
28. Tsang, A. S. K., Sanhueza, I. A. & Schoenebeck, F. Combining experimental and computational studies to understand and predict reactivities of relevance to homogeneous catalysis. *Chem. Eur. J.* **20,** 16432–16441 (2014).
29. Nielsen, M. C., Bonney, K. J. & Schoenebeck, F. Computational ligand design for the reductive elimination of $ArCF_3$ from a small bite angle PdII complex: remarkable effect of a perfluoroalkyl phosphine. *Angew. Chem. Int. Ed.* **53,** 5903–5906 (2014).
30. Sperger, T., Sanhueza, I. A., Kalvet, I. & Schoenebeck, F. Computational studies of synthetically relevant homogeneous organometallic catalysis involving Ni, Pd, Ir, and Rh: An overview of commonly employed DFT methods and mechanistic insights. *Chem. Rev* **115,** 9532–9586 (2015).
31. Kocienski, P. J., Christopher, J. A., Bell, R. & Otto, B. Nucleophilic addition of $\alpha$-metallated carbamates to planar chiral cationic $\eta^3$-allylmolybdenum complexes: a stereochemical study. *Synthesis* 75–84 (2005).
32. Zhang, X. *et al.* Copper(II)-catalyzed amidations of alkynyl bromides as a general synthesis of ynamides and Z-enamides. An intramolecular amidation for the synthesis of macrocyclic ynamides. *J. Org. Chem.* **71,** 4170–4177 (2006).
33. Mansfield, S. J., Campbell, C. D., Jones, M. W. & Anderson, E. A. A robust and modular synthesis of ynamides. *Chem. Commun.* **51,** 3316–3319 (2015).
34. Saito, A., Ono, T. & Hanzawa, Y. Cationic Rh(I) catalyst in fluorinated alcohol: mild intramolecular cycloaddition reactions of ester-tethered unsaturated compounds. *J. Org. Chem.* **71,** 6437–6443 (2006).
35. Wender, P. A. & Williams, T. J. $[(arene)Rh(cod)]^+$ Complexes as catalysts for [5 + 2] cycloaddition reactions. *Angew. Chem. Int. Ed.* **41,** 4550–4553 (2002).
36. Wender, P. A. *et al.* Structural complexity through multicomponent cycloaddition cascades enabled by dual-purpose, reactivity regenerating 1,2,3-triene equivalents. *Nat. Chem* **6,** 448–452 (2014).
37. Trost, B. M., Shen, H. C., Schulz, T., Koradin, C. & Schirok, H. On the diastereoselectivity of Ru-catalyzed [5 + 2] cycloadditions. *Org. Lett.* **5,** 4149–4151 (2003).
38. Trost, B. M., Waser, J. & Meyer, A. Total synthesis of (−)-pseudolaric acid B. *J. Am. Chem. Soc.* **130,** 16424–16434 (2008).
39. Shintani, R., Nakatsu, H., Takatsu, K. & Hayashi, T. Rhodium-catalyzed asymmetric [5 + 2] cycloaddition of alkyne-vinylcyclopropanes. *Chem. Eur. J.* **15,** 8692–8694 (2009).
40. Wender, P. A. *et al.* Asymmetric catalysis of the [5 + 2] cycloaddition reaction of vinylcyclopropanes and $\pi$-systems. *J. Am. Chem. Soc.* **128,** 6302–6303 (2006).
41. Teichert, J. F. & Feringa, B. L. Phosphoramidites: Privileged ligands in asymmetric catalysis. *Angew. Chem. Int. Ed.* **49,** 2486–2528 (2010).
42. Mikhel, I. S. *et al.* A chiral phosphoramidite beyond monodentate coordination: Secondary pi-interactions turn a dangling aryl into a two-, four-, or six-electron donor in d(6) and d(8) complexes. *Organometallics* **27,** 2937–2948 (2008).
43. Cheng, G.-J., Zhang, X., Chung, L. W., Xu, L. & Wu, Y.-D. Computational organic chemistry: bridging theory and experiment in establishing the mechanisms of chemical reactions. *J. Am. Chem. Soc.* **137,** 1706–1725 (2015).
44. Polet, D. & Alexakis, A. Kinetic study of various phosphoramidite ligands in the iridium-catalyzed allylic substitution. *Org. Lett.* **7,** 1621–1624 (2005).
45. Roth, P. M. C., Sidera, M., Maksymowicz, R. M. & Fletcher, S. P. Copper-catalyzed asymmetric conjugate addition of alkylzirconium reagents to cyclic enones to form quaternary centers. *Nat. Protoc.* **9,** 104–111 (2014).
46. Chai, J. D. & Head-Gordon, M. Long-range corrected hybrid density functionals with damped atom-atom dispersion corrections. *Phys. Chem. Chem. Phys.* **10,** 6615–6620 (2008).
47. Wadt, W. R. & Hay, P. J. *Ab initio* effective core potentials for molecular calculations. Potentials for main group elements Na to Bi. *J. Chem. Phys.* **82,** 284–298 (1985).
48. Roy, L. E., Hay, P. J. & Martin, R. L. Revised basis sets for the LANL effective core potentials. *J. Chem. Theory Comput.* **4,** 1029–1031 (2008).
49. Marenich, A. V., Cramer, C. J. & Truhlar, D. G. Universal solvation model based on solute electron density and on a continuum model of the solvent defined by the bulk dielectric constant and atomic surface tensions. *.J. Phys. Chem. B* **113,** 6378–6396 (2009).

## Acknowledgements

E.A.A. and R.N.S. thank the EPSRC (EP/K005391/1, EP/H025839/1 and a studentship to R.N.S. (EP/K503113/1)) for support. R.S.P. thanks the Chemical Structure Association Trust, and Q.P. thanks the European Community (FP7-PEOPLE-2012-IIF under grant agreement 330364) for support. We acknowledge the use of the EPSRC UK National Service for Computational Chemistry Software (CHEM773) in carrying out this work. A.M. thanks the Royal Thai Government and the DPST project for a studentship.

## Author contributions

E.A.A. and R.N.S. conceived of the synthetic methodology. R.N.S. carried out the experimental work, with the aid of A.M. Q.P. and R.S.P. carried out the theoretical calculations. R.N.S., E.A.A., Q.P. and R.S.P. analysed the collected results. E.A.A., R.N.S., Q.P. and R.S.P. wrote the manuscript.

## Additional information

**Competing financial interests:** The authors declare no competing financial interests.

# 8

# Direct and selective hydrogenolysis of arenols and aryl methyl ethers

Shuhei Kusumoto[1] & Kyoko Nozaki[1]

For valorization of biomass, the conversion of lignin to deoxygenated bulk aromatic compounds is an emerging subject of interest. Because aromatic rings are susceptible to metal-catalysed hydrogenation, the selective hydrogenolysis of carbon-oxygen bonds still remains a great challenge. Herein we report direct and selective hydrogenolysis of $sp^2$ C-OH bonds in substituted phenols and naphthols catalysed by hydroxycyclopentadienyl iridium complexes. The corresponding arenes were obtained in up to 99% yields, indicating the possible production of arenes from lignin-derived bio-oils. Furthermore, the same catalysts were applied to the unprecedented selective hydrogenolysis of the $sp^3$ C-O bonds in aryl methyl ethers. Thus, the hydrodeoxygenation of vanillylacetone, a lignin model compound, afforded alkylbenzenes as the major products via triple deoxygenation.

[1] Department of Chemistry and Biotechnology, Graduate School of Engineering, The University of Tokyo, 7-3-1 Hongo, Bunkyo-ku, Tokyo 113-8656, Japan. Correspondence and requests for materials should be addressed to K.N. (email: nozaki@chembio.t.u-tokyo.ac.jp).

For more than half a century, the chemical industry has mostly depended on petroleum as chemical feedstock. Because oil is hydrocarbon with a low oxidation state, various selective oxidation processes have been developed for its conversion to value-added chemicals, such as monomers, solvents and intermediates for bulk and fine chemicals[1]. However, oil is a non-renewable resource, and therefore, emerging attention has been focused on the utilization of biomass as a renewable substitute in the last few decades. Unlike petroleum, plant dry matter is mostly composed of highly oxidized carbon compounds such as cellulose and lignin, which are not very amenable for deoxygenation and subsequent processing into chemical compounds on the market. Thus, the development of an efficient technology involving the chemical reduction of biomass has become a great challenge to attain maximum valorization of biomass[2,3]. For the reduction of lignin, a major polymeric component of biomass comprising arenols, alcohols and their ethers, intensive studies have been devoted for the development of efficient catalysts aimed at the upgrading of bio-oils, that is, to break them into a mixture of shorter-fragment liquefied fuels[4–6]. Among the various chemical reduction processes, the use of molecular dihydrogen is the most desirable method because it can be the least expensive and the most atom-economical pathway.

However, more attention should be concurrently paid to the conversion of lignin to aromatic compounds, such as benzene and its derivatives, to effectively utilize lignin as the feedstock for bulk and fine chemicals[3,4]. Currently, heterogeneous catalysts play a major role in the hydrodeoxygenation of lignin-model compounds such as arenols and their ethers; however, the complete hydrogenation of the aromatic rings into saturated hydrocarbons over the catalyst surface was mostly inevitable under these conditions[7]. Thus, the preconversion of arenols to more activated esters was essential for the selective hydrogenolysis of the aryl–oxygen bonds over hydrogenation of the aromatic rings[8]. Recently, a few examples have been reported for the selective reductive cleavage of the less activated aryl–oxygen bonds in the alkyl ethers of arenols using either heterogeneous catalysts[4], homogeneous nickel catalysts with a phosphine or a carbene ligand[9–12], iron catalysts[13] or Lewis bases[14]. It should be noted, however, that none of the reported catalysts were applicable to the selective hydrodeoxygenation of free arenols, which would enable the complete deoxygenation of arenols to simple aromatic chemicals. To solve the problem, we envisioned a strategy utilizing the cooperative activation of arenols using hydroxycyclopentadienyl hydrido metal complexes, as shown in Fig. 1: a cyclopentadienone–metal species, which was reported to cleave the H–H bond of dihydrogen heterolytically[15,16], reacts with dihydrogen, affording a hydroxycyclopentadienyl hydrido metal complex. If the proton on the ligand and the hydride on the metal are delivered to an arenol in a concerted manner, the $sp^2$ C–OH bond in an arenol would be cleaved into an arene and water.

Here in this communication, we report direct and selective hydrogenolysis of $sp^2$ C–OH bonds in arenols by hydroxycyclopentadienyl dihydridoiridium catalysts. Hydrogenolysis of $sp^3$ C–O bonds in aryl methyl ethers were also achieved using the same catalysts. These catalyses were further applied to deoxygenation of a lignin model compound, implying the potential application for mass production of arenes from lignin or its degraded components.

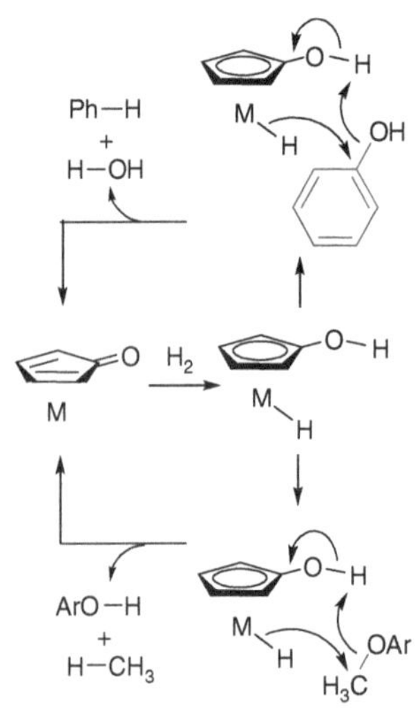


**Figure 1 | Hydrogenolysis of arenols and aryl methyl ethers via metal–ligand cooperative hydrogen transfer.** Dihydrogen is split into a proton on the hydroxycyclopentadienyl ligand and a hydride on the metal, which are transferred to a C–O single bond in a concerted manner.

## Results

**Hydrogenolysis of $sp^2$ C–OH bonds in arenols.** The hydrogenolysis reactions of the $sp^2$ C–OH bond in phenols and naphthols were accomplished using hydroxycyclopentadienyl iridium complexes **A** and **C** as the catalysts under 1 atm of dihydrogen pressure at 200 °C (Table 1; Fig. 2). From *para*-phenylphenol, biphenyl was obtained in 57% yield using the iridium complex **A** (entry 1). The reaction was strongly suppressed with complex **B**, a silyl-protected analogue of complex **A**, indicating that the free hydroxy group plays an important role in this catalysis (entry 2). Most probably, in entry 2, the catalysis was mediated by the species desilylated under the reaction conditions (see Supplementary Fig. 4 and Supplementary Discussion). Triphenylphosphine complex **C** showed a higher activity than the di-*tert*-butylmethylphosphine complex **A**, affording biphenyl in 91% yield (entry 3). In sharp contrast to the results obtained using iridium complexes **A** and **C**, no reaction was observed with Cp* iridium complex **D** without the hydroxy group, or with ruthenium catalyst **E**, a ruthenium analogue of **A**, a combination of $Ni(cod)_2$ and *N,N'*-(2,6-diisopropylphenyl)imidazolinylidene (SIPr) (**F**), which was effective for the hydrogenolysis of $sp^2$ C–OMe bonds, did not show any activity either (Supplementary Table 1)[9]. This reaction was sensitive to steric demand, and the yield of biphenyl was strongly affected by the position of the phenyl substituent. A comparison among the *para*-, *meta*- and *ortho*-substrates indicated that the reactivity decreases in this order (entries 4 and 5). Notably, 1- and 2-naphthols could be successfully converted to naphthalene along with the formation of minimal amounts of 1,2,3,4-tetrahydronaphthalene, a side product resulting from over-reduction. The reaction was never achieved by either heterogeneous or homogeneous catalysts, before; this is because in all the cases, the hydrogenation of the aromatic rings dominated over the C–O bond cleavage[17,18]. Thus, both with catalysts **A** and **C**, the complete consumption of 1- or 2-naphthol was observed (entries 6–9). From 2-naphthol, naphthalene was obtained as the sole product in excellent yields (98% and 99%, entries 6 and 7, respectively), whereas the reactions of 1-naphthol afforded 17–23% of 1,2,3,4-tetrahydronaphthalene as a side product (entries 8 and 9). In entry 7, almost quantitative formation of water (91%) was confirmed by Karl Fisher titration. The reactions of naphthalenediols were either stopped on formation of naphthols (entries 10 and 12) or on conversion to naphthalene (entries 11 and 13) by changing the catalyst. The selectivity for the mono-deoxygenated product by complex **A** may have resulted from the catalyst deactivation during the reaction. (Please see Supplementary Figs 2–7 and Supplementary Table 1 for further experimental results.)

**Table 1 | Hydrogenolysis of hydroxyarenes by iridium complexes*.**

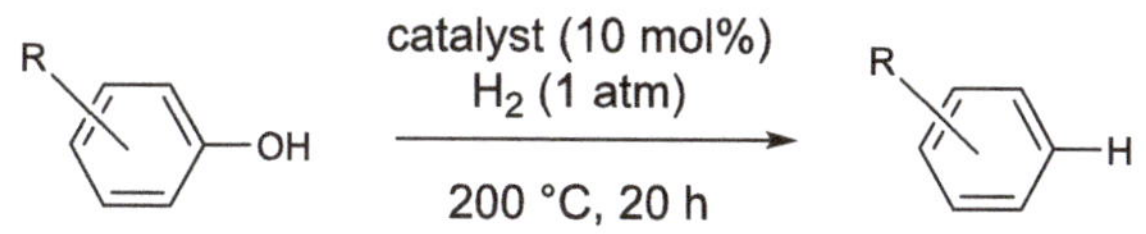


| Entry | Substrate | Catalyst | Conversion† (%) | Products, yield†,‡ (%) |
|---|---|---|---|---|
| | 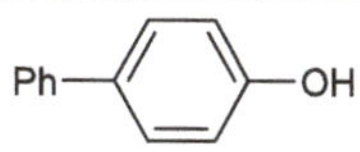 | | | biphenyl (cyclohexylbenzene) |
| 1 | | **A** | 67 | 57 (<1) |
| 2 | | **B** | 11 | 6 (<1) |
| 3 | | **C** | 100 | 91 (3)<br>84§ |
| 4 | 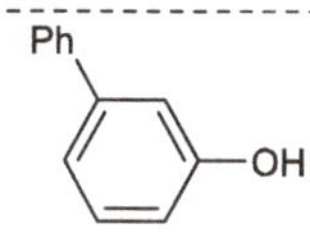 | **C** | 62 | 50 (3) |
| 5 | 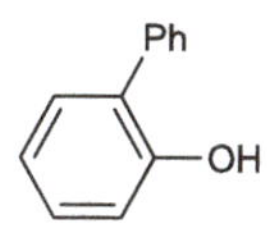 | **C** | 31 | 14 (7) |
| | 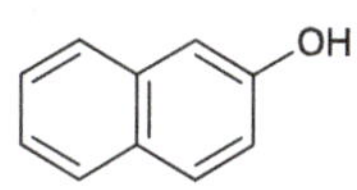 | | | naphthalene |
| 6 | | **A** | 100 | 98 |
| 7 | | **C** | 100 | 99‖ |
| | 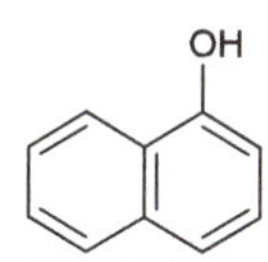 | | | naphthalene (tetrahydronaphthalene) |
| 8 | | **A** | 100 | 83 (17) |
| 9 | | **C** | 100 | 71 (23) |
| | 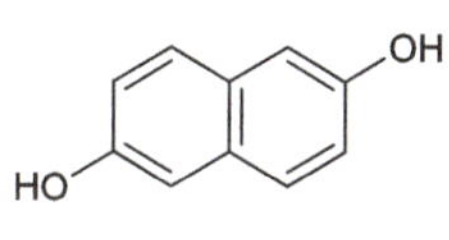 | | | 2-naphthol (5,6,7,8-tetrahydro-2-naphthol) |
| 10 | | **A** | 74 | 49 (9) |
| | | | | naphthalene (tetrahydronaphthalene) |
| 11 | | **C** | 100 | 73 (13) |
| | 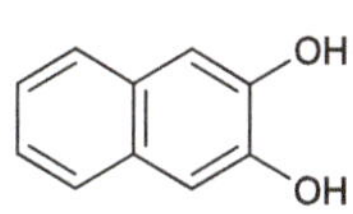 | | | 2-naphthol (naphthalene) |
| 12 | | **A** | 96 | 74 (9) |
| | | | | naphthalene (tetrahydronaphthalene) |
| 13 | | **C** | 100 | 89 (7) |

GC, gas chromatography.
*Reaction conditions: all the reactions were performed without any solvent. A 3-ml glass tube containing substrate (200 μmol) and catalyst (10 mol%) was connected to a 1-l rubber gas bag filled with $H_2$ and stirred at 200 °C.
†Determined by GC with dodecane as the internal standard.
‡Side products and their yields are shown in parenthesis.
§Isolated yield.
‖Formation of water (91%) was observed by Karl Fisher titration.

**Hydrogenolysis of sp³ C–O bonds in aryl methyl ethers.** To our surprise, the selective hydrogenolysis of the sp³ C–O bond, namely, demethylation, proceeded when 4-phenylphenyl methyl ethers and naphthyl methyl ethers were treated with hydroxycyclopentadienyl iridium dihydrido complexes **A** and **C**. For the demethylation of methyl ethers, that is, the deprotection of arenols, strong acids such as $BBr_3$, $AlCl_3$ and trimethylsilyl iodide or basic nucleophiles such as NaSEt are usually used[19].

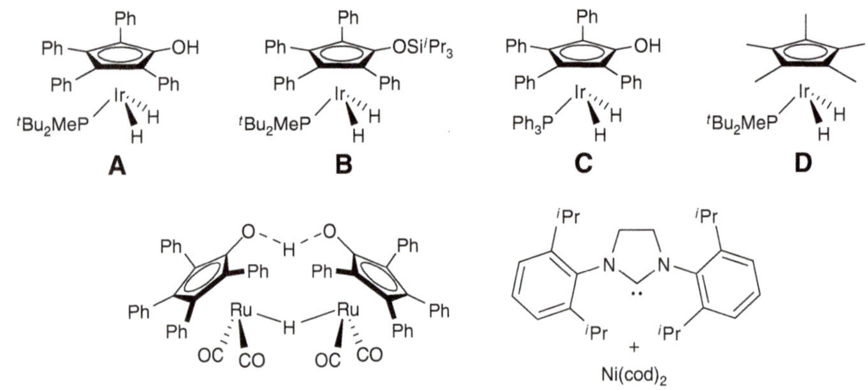


E F

**Figure 2 | Structures of catalysts.** The figure shows molecular structures of the catalysts examined in this study.

**Table 2 | Hydrogenolysis of aryl methyl ethers by iridium complexes A and C*.**

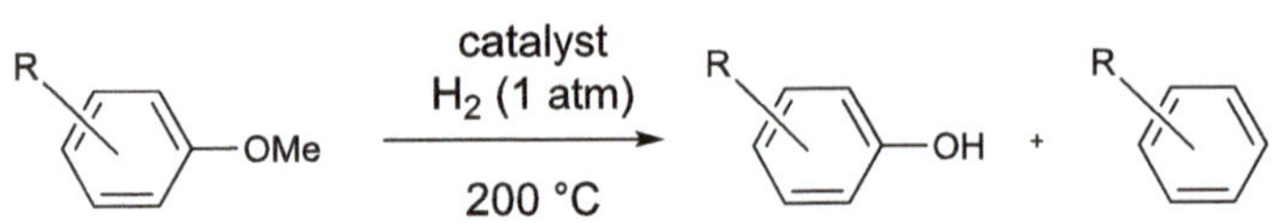


| Entry | Substrate | Catalyst | Time (h) | Conversion† (%) | Products, yield†,‡ (%) |
|---|---|---|---|---|---|
| | 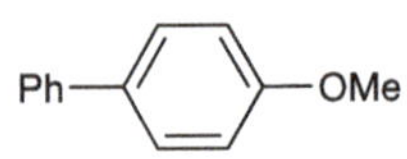 | | | | 4-phenylphenol (biphenyl) |
| 1 | 0.54 mmol | **A** (20 μmol) | 20 | 37 | 33 (4.1) |
| 2 | 0.54 mmol | **C** (20 μmol) | 20 | 50 | 40 (8.7) |
| 3 | 0.54 mmol | **A** (1.0 μmol) | 68 | 55 | 47 (8.5)§ |
| 4 | 0.54 mmol | **C** (1.0 μmol) | 68 | 80 | 57 (20) |
| 5 | 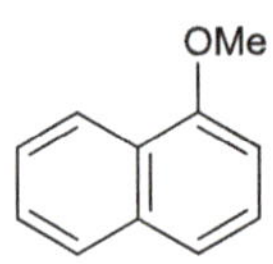 1.38 mmol | **A** (20 μmol) | 20 | 5.1 | 1-naphthol (naphthalene) 5.1 (0) |
| 6 | 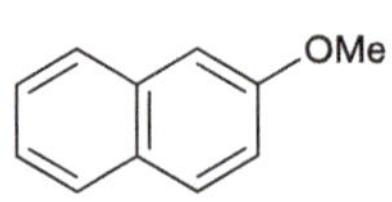 0.63 mmol | **A** (20 μmol) | 20 | 20 | 2-naphthol (naphthalene) 14 (3.8) |
| 7 | 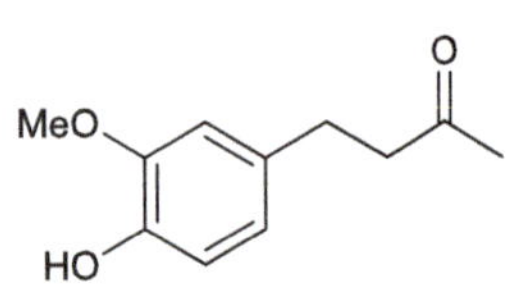 vanillylacetone 0.10 mmol | **C** (20 μmol) | 50 | 100 | 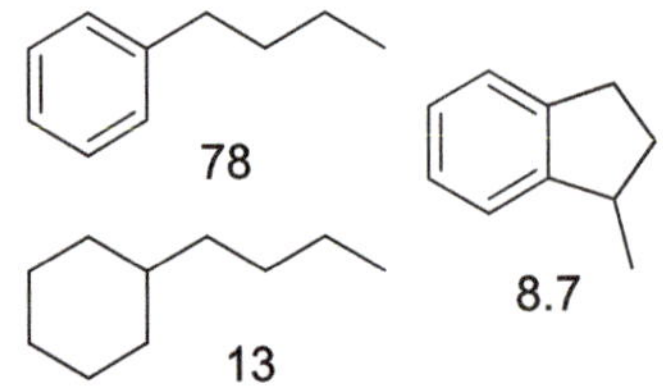 78, 13, 8.7 |

GC, gas chromatography.
*Reaction conditions: all the reactions were performed without any solvent. A 3-ml glass tube containing substrate and catalyst was connected to a 1-l rubber gas bag filled with $H_2$ and stirred at 200 °C.
†Determined by GC with dodecane as the internal standard.
‡Side products and their yields are shown in parenthesis.
§Formation of methane was observed by GC analysis.

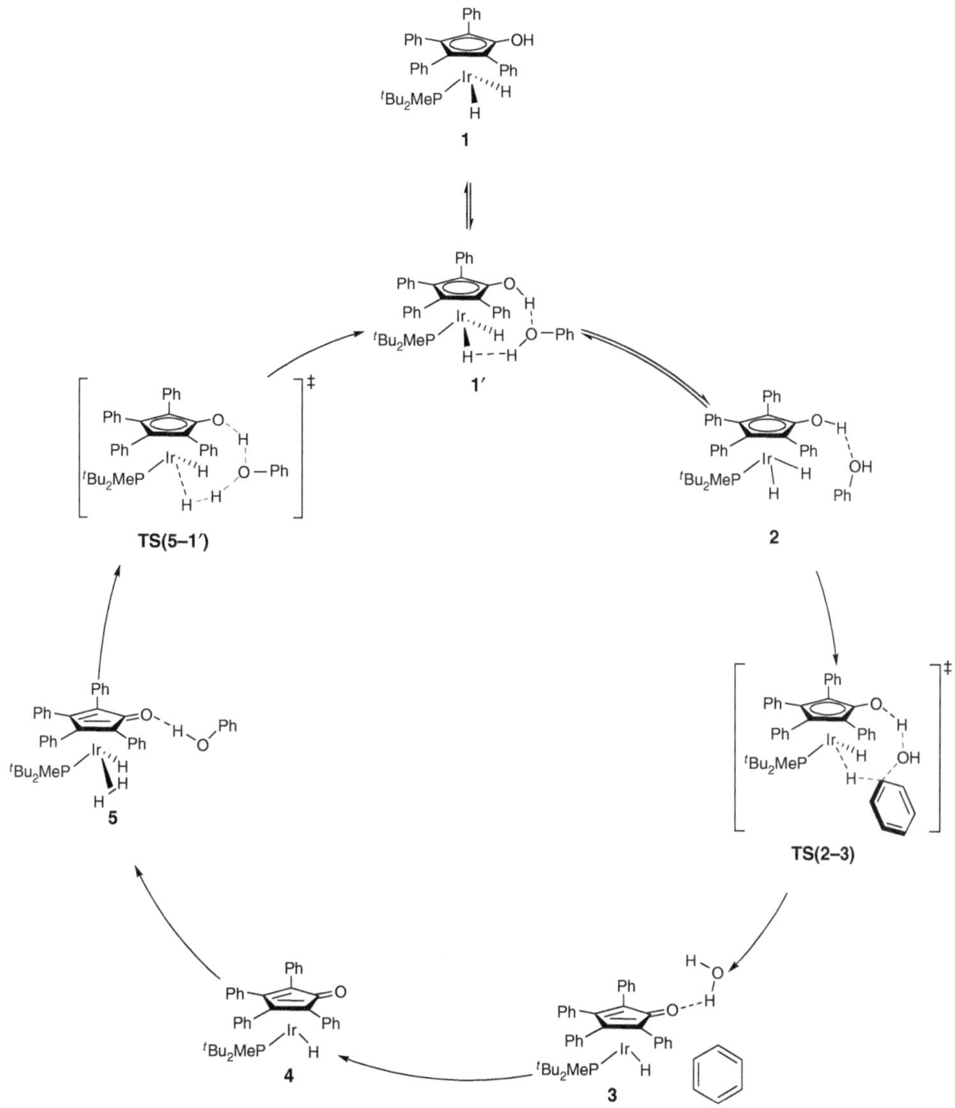

**Figure 3 | Proposed mechanism for the hydrogenolysis of phenol.** The proposed catalytic cycle for hydrogenolysis of phenol is shown. From dihydrido complex, C–O bond in phenol is cleaved via concerted proton and hydride transfer from ligand and iridium (**TS2-3**) to afford monohydrido complex **4**. Heterolytic dihydrogen split mediated by phenol affords the dihydrido complex (**TS(5-1′)**).

Although the metal-catalysed hydrogenolysis of the $sp^3$ methyl–oxygen bond in aryl methyl ethers would be an ideal candidate for a more atom-economical transformation under neutral conditions, all the preceding reduction reactions by transition metal catalysts resulted in the cleavage of $sp^2$ C–O bonds, but not $sp^3$ C–O bonds while only two exceptions were reported in a stoichiometric reaction with nickel(II) and iridium(0)[9–13,20,21]. The reaction also provides a new strategy for the production of arenols from lignin or its shorter fragments. With catalysts **A** and **C**, the hydrogenolysis of 4-methoxybiphenyl proceeded, affording 4-phenylphenol (Table 2, entries 1–4). In entry 3, the formation of methane was observed by gas chromatography (GC) analysis, proving that the demethylation reaction proceeds via the hydrogenolysis reaction. The reaction was significantly decelerated either in the absence of dihydrogen or in the presence of a small amount of water, both of which strongly supported that the reaction is not a simple hydrolysis reaction, but a hydrogenolysis reaction.

We applied the hydrodeoxygenation/demethylation process to a model compound of lignin, vanillylacetone[2] (Table 2, entry 7). Vanillylacetone has three different types of oxygen functionalities, a phenolic hydroxy group, a methyl ether functionality and an aliphatic keto carbonyl group as the representative oxygen component of lignin and its degraded materials. When vanillylacetone was treated with 20 mol% of catalyst **C** under 1 atm of dihydrogen atmosphere for 50 h, completely deoxygenated butylbenzene and cyclized methyl indane were obtained as the two major products in 78% and 9% yields, respectively, while the yield of butylcyclohexane, a side product resulting from over-reduction, was suppressed to as low as 13%. Thus, the hydrodeoxygenation reaction successfully proceeded for phenolic C–OH bonds, aryl methyl ethers and aliphatic ketones in this lignin model compound (please see Supplementary Table 2 and Supplementary Figs 8 and 9 for further experimental results).

## Discussion

The unique reactivity and selectivity in C–O bond hydrogenolysis, here we discovered with complexes **A** and **C**, may be explained by the catalytic cycles shown in Figs 3 and 4. The mechanisms are still speculative at this moment. Shown in Fig. 3 is a proposed mechanism for the hydrogenolysis of $sp^2$ C–O bond in phenols. The starting dihydrido complex **1** is stabilized with two hydrogen bonds with one phenol molecule (**1′**). From

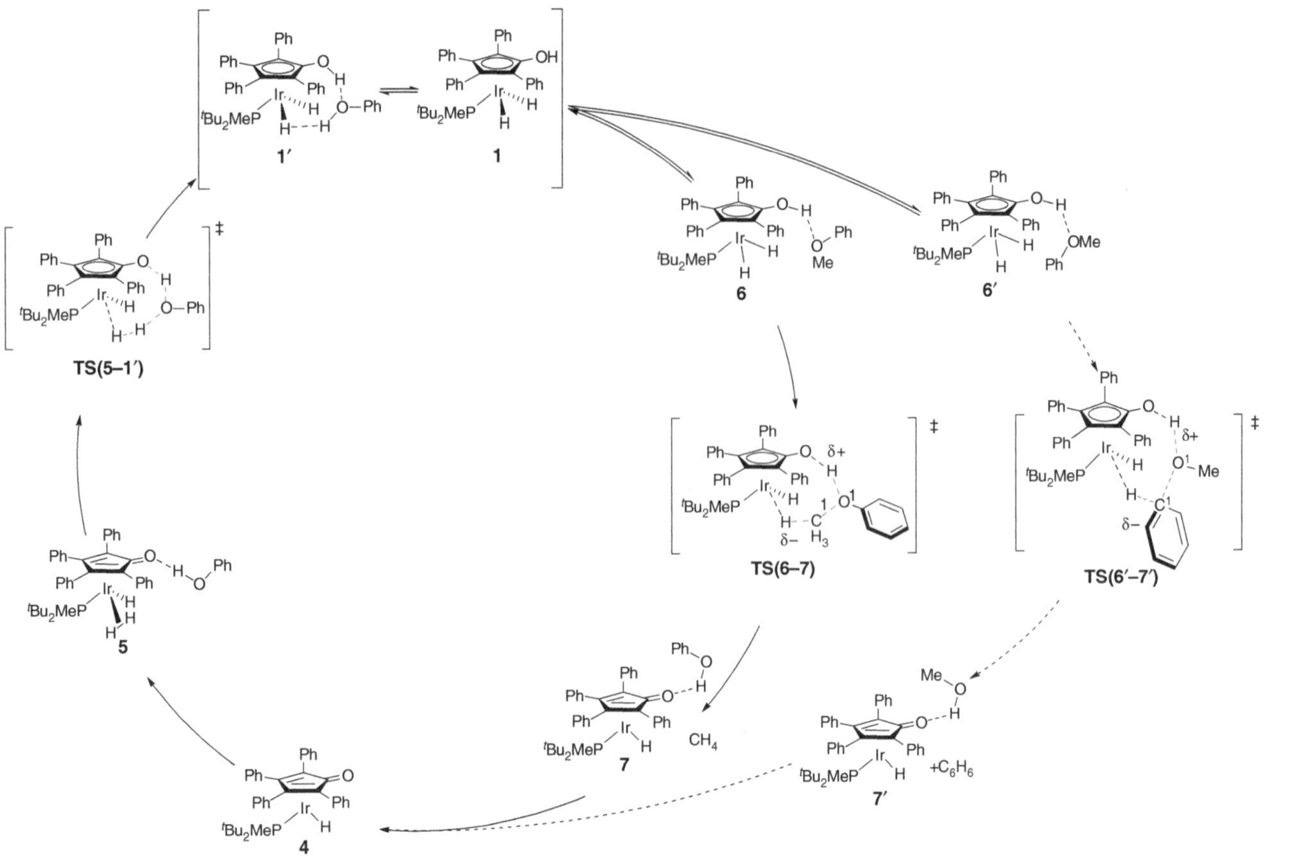

**Figure 4 | Proposed mechanism for the hydrogenolysis of phenyl methyl ether.** Proposed catalytic cycle for hydrogenolysis of $sp^3$ C-O bonds (via **TS(6–7)**) and $sp^2$ C-O bonds (via **TS(6′–7′)**) in anisole are shown. In both pathways, proton and hydride were delivered to C-O bonds in a concerted manner followed by heterolytic cleavage of dihydrogen mediated by phenol molecule (**TS(5–1′)**).

phenol-stabilized complex **1′**, cleavage of a weak H–H bond provides intermediate **2,** which undergoes outer-sphere transfer of a proton on the ligand and a hydride on the metal centre to a C–OH bond in a concerted manner via **TS(2–3)**. The mechanism is similar to the one proposed for the hydrogenation of carbonyl compounds by Shvo's ruthenium catalyst **E**[15]. The successive heterolytic cleavage of dihydrogen assisted by one phenol molecule would afford complex **1′**.

For the hydrogenolysis of phenyl methyl ether, both the $sp^2$ and $sp^3$ C–O bond cleavages should be considered to reveal the reason why hydroxycyclopentadienyl iridium complexes **A** and **C** show the $sp^3$ C–O bond-selective cleavage, which is in contrast to the $sp^2$ C–O bond-selective cleavage reported for low-valent nickel or iron complexes (Fig. 4)[9,10–13]. The outer-sphere dihydrogen transfer would prefer $sp^3$ C–O bond cleavage via **6**→**7** to $sp^2$ C–O bond cleavage via **6′**→**7′**. In the transition state for $sp^3$ C–O bond cleavage (**TS(6–7)**), we are now assuming the hydride on the iridium centre approaches the $\sigma^*$-orbital of the $sp^3$ C–O bond in aryl methyl ether to undergo the $S_N2$ reaction. On the other hand, in the transition state for $sp^2$ C–O bond cleavage (**TS(6′–7′)**), the hydride approaches the $\pi^*$ orbital to make $C^1$ atom have $sp^3$ hybridization, causing the loss of aromaticity of the benzene ring in the transition state.

In conclusion, the direct and selective hydrogenolysis of the $sp^2$ C–OH bonds in phenols and naphthols, and the selective hydrogenolysis of the $sp^3$ C–O bonds in aryl methyl ethers have been achieved using hydroxycyclopentadienyl iridium complexes **A** and **C**. A metal–ligand cooperative outer-sphere mechanism for the C–O bond cleavage was proposed to be responsible for the unique reactivity and selectivity of the reactions. The current chemical reduction technology has great potential for the future utilization of lignin as the chemical feedstock.

## Methods

**General procedure for hydrogenolysis of arenols.** Into a 3-ml glass tube, catalyst (20 μmol) and substrate (200 μmol) were placed. The reaction vessel was connected to a rubber gas bag filled with dihydrogen and heated at 200 °C in a sand bath. After stirring at 200 °C for 20 h, a certain amount of dodecane (ca. 7.4 mg) was added as an internal standard and the products were analysed by GC. The products were also analysed by $^1$H NMR analysis. Please see Supplementary Methods for further information.

**General procedure for hydrogenolysis of aryl methyl ethers.** Into a 3-ml glass tube, catalyst (1.0–20 μmol) and substrate (100 mg for solids, 200 μl for liquids) were placed. The reaction vessel was connected to a rubber gas bag filled with dihydrogen and heated at 200 °C in a sand bath. After stirring at 200 °C for the time indicated, a certain amount of dodecane (ca. 7.4 mg) was added as an internal standard and the products were analysed by GC. The products were also analysed by $^1$H NMR analysis. Please see Supplementary Methods for further information.

## References

1. Weissermel, K. & Arpe, H. J. (Eds *Industrial Organic Chemistry* 4th edn (Wiley VCH, 2003).
2. Tuck, C. O., Pérez, E., Horváth, I. T., Sheldon, R. A. & Poliakoff, M. Valorization of biomass: deriving more value from waste. *Science* **337,** 695–699 (2012).
3. van Haveren, J., Scott, E. L. & Sanders, J. Bulk chemicals from biomass. *Biofuels Bioprod. Bioref.* **2,** 41–57 (2008).
4. Furimsky, E. Catalytic hydrodeoxygenation. *Appl. Catal. A General* **199,** 147–190 (2000).
5. Saidi, M. *et al.* Upgrading of lignin-derived bio-oils by catalytic hydrodeoxygenation. *Energy Environ. Sci.* **7,** 103–129 (2014).
6. Laskar, D. D., Yang, B., Wang, H. & Lee, J. Pathways for biomass-derived lignin to hydrocarbon fuels. *Biofuels Bioprod. Bioref.* **7,** 602–626 (2013).

7. Zhao, C., Kou, Y., Lemonidou, A. A., Li, X. B. & Lercher, J. A. Highly selective catalytic conversion of phenolic bio-oil to alkanes. *Angew. Chem. Int. Ed.* **48,** 3987-3990 (2009).
8. Herrmann, J. M. & König, B. Reductive deoxygenation of alcohols: catalytic methods beyond Barton-McCombie deoxygenation. *Eur. J. Org. Chem.* **2013,** 7017-7027 (2013).
9. Sergeev, A. G. & Hartwig, J. F. Selective, nickel-catalyzed hydrogenolysis of aryl ethers. *Science* **332,** 439-443 (2011).
10. Álvarez-Bercedo, P. & Martin, R. Ni-catalyzed reduction of inert C–O bonds: a new strategy for using aryl ethers as easily removable directing groups. *J. Am. Chem. Soc.* **132,** 17352-17353 (2010).
11. Tobisu, M., Yamakawa, K., Shimasaki, T. & Chatani, N. Nickel-catalyzed reductive cleavage of aryl-oxygen bonds in alkoxy- and pivaloxyarenes using hydrosilanes as a mild reducing agent. *Chem. Commun.* **47,** 2946-2948 (2011).
12. Cornella, J., Gómez-Bengoa, E. & Martin, R. Combined experimental and theoretical study on the reductive cleavage of inert C–O bonds with silanes: ruling out a classical Ni(0)/Ni(II) catalytic couple and evidence for Ni(I) intermediates. *J. Am. Chem. Soc.* **135,** 1997-2009 (2013).
13. Ren, Y., Yan, M., Wang, J., Zhang, Z. C. & Yao, K. Selective reductive cleavage of inert aryl C–O bonds by an iron catalyst. *Angew. Chem. Int. Ed.* **52,** 12674-12678 (2013).
14. Fedorov, A., Toutov, A. A., Swisher, N. A. & Grubbs, R. H. Lewis-base silane activation: from reductive cleavage of aryl ethers to selective ortho-silylation. *Chem. Sci.* **4,** 1640-1645 (2013).
15. Blum, Y., Czarkie, D., Rahamin, Y. & Shvo, Y. (Cyclopentadienone)rutheniuim carbonyl complexes—a new class of homogeneous hydrogenation catalysts. *Organometallics* **4,** 1459-1461 (1985).
16. Kusumoto, S., Akiyama, M. & Nozaki, K. Acceptorless dehydrogenation of C–C single bonds adjacent to functional groups by metal-ligand cooperation. *J. Am. Chem. Soc.* **135,** 18726-18729 (2013).
17. Mine, E., Hiyoshi, N., Sato, O., Chandrashekhar, R. V. & Shirai, M. Selective hydrogenation of naphthols to tetralones over supported palladium catalysts in supercritical carbon dioxide solvent. *Chem. Lett.* **35,** 780-781 (2006).
18. Nishimura, S., Ohbuchi, S., Ikeno, K. & Okada, Y. Hydrogenation and hydrogenolysis. XVII. The selectivities of platinum group metals in catalytic hydrogenation of 2-naphthol and tetrahydro-2-naphthols. *Bull. Chem. Soc. Jpn* **57,** 2557-2564 (1984).
19. Wuts, P. G. M. & Greene, T. W. in *Greene's Protective Groups in Organic Synthesis* 3rd edn, 367-430 (Wiley, 2006).
20. van der Boom, M. E., Liou, S.-Y., David, Y. B., Shimon, L. J. W. & Milstein, D. Alkyl – and aryl – oxygen bond activation in solution by rhodium(I), palladium(II), and nickel(II). transition-metal-based selectivity. *J. Am. Chem. Soc.* **120,** 6531-6541 (1998).
21. Choi, J. *et al.* Cleavage of $sp^3$ C–O bonds via oxidative addition of C–H bonds. *J. Am. Chem. Soc.* **131,** 15627-15629 (2009).

## Acknowledgements

This work was supported by a JSPS Grant-in-Aid for JSPS Fellows. We are grateful to Dr Shingo Ito (The University of Tokyo) for careful proof reading.

## Author contributions

Both the authors designed studies, discussed the results and wrote the paper. S.K. performed all the experimental work.

## Additional information

**Competing financial interests:** The authors declare no competing financial interests.

# 9

# A simple and versatile design concept for fluorophore derivatives with intramolecular photostabilization

Jasper H. M. van der Velde[1], Jens Oelerich[2], Jingyi Huang[3], Jochem H. Smit[1], Atieh Aminian Jazi[1], Silvia Galiani[4], Kirill Kolmakov[5], Giorgos Guoridis[1], Christian Eggeling[4], Andreas Herrmann[3], Gerard Roelfes[2] & Thorben Cordes[1]

Intramolecular photostabilization via triple-state quenching was recently revived as a tool to impart synthetic organic fluorophores with 'self-healing' properties. To date, utilization of such fluorophore derivatives is rare due to their elaborate multi-step synthesis. Here we present a general strategy to covalently link a synthetic organic fluorophore simultaneously to a photostabilizer and biomolecular target via unnatural amino acids. The modular approach uses commercially available starting materials and simple chemical transformations. The resulting photostabilizer-dye conjugates are based on rhodamines, carbopyronines and cyanines with excellent photophysical properties, that is, high photostability and minimal signal fluctuations. Their versatile use is demonstrated by single-step labelling of DNA, antibodies and proteins, as well as applications in single-molecule and super-resolution fluorescence microscopy. We are convinced that the presented scaffolding strategy and the improved characteristics of the conjugates in applications will trigger the broader use of intramolecular photostabilization and help to emerge this approach as a new gold standard.

[1] Molecular Microscopy Research Group, Zernike Institute for Advanced Materials, University of Groningen, Nijenborgh 4, 9747 AG Groningen, The Netherlands. [2] Stratingh Institute for Chemistry, University of Groningen, Nijenborgh 4, 9747 AG Groningen, The Netherlands. [3] Department of Polymer Chemistry, Zernike Institute for Advanced Materials, University of Groningen, Nijenborgh 4, 9747 AG Groningen, The Netherlands. [4] MRC Human Immunology Unit, Weatherall Institute of Molecular Medicine, University of Oxford, Headley Way, Oxford OX3 9DS, UK. [5] Department NanoBiophotonics, Max-Planck-Institute of Molecular Medicine, Am Fassberg 1, 37077 Goettingen, Germany. Correspondence and requests for materials should be addressed to T.C. (email: t.m.cordes@rug.nl).

Organic fluorophores are a major driving force for the recent success of fluorescence-based methods, but they intrinsically suffer from transient excursions to dark states (blinking) and irreversible destruction (photobleaching)[1,2]. Both processes fundamentally limit their applicability and have, for a long time, hampered the development of advanced microscopy techniques with single-molecule sensitivity[2] or optical super-resolution <250 nm (refs 2–4). Lüttke and colleagues[5] introduced covalent binding of triplet-state quenchers and singlet-oxygen scavengers[6] to organic fluorophores as a strategy to reduce the above mentioned effects. Such photostabilizer–dye conjugates with intramolecular triplet-state quenching have 'self-healing'[7] or 'self-protecting'[8] properties, preventing photodamage without the use of solution additives. This non-invasive strategy has clear advantages compared with commonly used approaches, where micro- to millimolar concentrations of organic compounds are added to the buffer system[9–15].

Intramolecular photostabilization was recently revived independently by two groups[16–19], to reduce photobleaching and blinking even in demanding applications such as single-molecule fluorescence microscopy[16–20] or *in vivo* imaging[16]. It was also shown that efficient intramolecular photostabilization can be achieved without additives in the buffer system[18], minimizing potential influences on the biological system of interest. Applications of intramolecular photostabilization in biophysical or microscopy research are, however, still limited to proof-of-principle studies[5,6,16–19]. This is mainly due to the synthetic effort of photostabilizer–dye conjugates, which thus far requires a multi-step synthesis route[16,17]. These synthetic challenges represent a fundamental hurdle for researchers with limited organic chemistry experience to use this concept. Moreover, only a small number of bifunctional cyanine derivatives are currently available to synthesize photostabilizer–dye conjugates on specific biomolecular targets (DNA, RNA, proteins and antibodies)[16,17], strongly restricting the choice of fluorophore type (chemical structure, redox potential, water solubility and so on) and photophysical properties (colour, brightness, fluorescence lifetime and so on). Especially the available cyanine fluorophores suffer from limited brightness and signal-to-noise ratio (SNR) due to *cis/trans* isomerization[21], a fact that emphasizes the urgent need for a synthetic strategy to study and use other classes of organic fluorophores via intramolecular photostabilization[18–22].

Here we introduce a versatile and simple design concept to synthesize photostabilizer–dye conjugates on a specific biomolecular target using unnatural amino acids (UAAs)[23]. UAAs were chosen as a scaffold that links multiple chemical units, that is, the fluorophore and photostabilizer to a specific target. The presented conjugation strategy is based on well-known chemical reactions with (commercially available derivatives of) synthetic organic fluorophores, photostabilizers and UAAs, which can be bound as a single moiety to a biomolecular target. Depending on the UAA scaffold, the chemical nature of the functional groups can be *N*-hydroxysuccinimid esters (NHS), alkynes, azides or other bio-orthogonal reactive functionalities. Two different UAAs were used to bind different rhodamine, carbopyronine and cyanine fluorophores (Alexa555, RhodamineB, KK114, ATTO647N and Cy5) covalently to a photostabilizer on distinct biomolecular targets. The synthesized fluorophore derivatives comprise of either a reducing or oxidizing photostabilizer in the form of the antioxidant Trolox (TX) or a nitrophenyl group[10,12]. We characterized their photophysical properties with single-molecule fluorescence microscopy on the biomolecular target DNA and observed significant increases in photostability for all compounds including suppression of triplet-based blinking. Secondly, photostabilizer–dye conjugates were synthesized, which allow labelling of biomolecules (DNA, antibodies and proteins) in a single step as with commercially available fluorophores. Finally, we demonstrate state-of-the-art applications of photostabilizer–dye conjugates in single-molecule Förster resonance energy transfer (smFRET) and super-resolution stimulated emission depletion (STED) microscopy, with significantly increased sensitivity and photostability of the conjugates compared to their non-stabilized parent fluorophores.

## Results

**Amino acids for intramolecular photostabilization.** Figure 1 shows the central idea to use UAAs as a scaffold that links multiple chemical units, that is, fluorophore and photostabilizer onto a specific biomolecular target. The conjugation strategy is based on established chemical reactions (amide bond formation or click chemistry) using commercially available derivatives of synthetic organic fluorophores (F), photostabilizers (P) and UAAs.

As a proof-of-concept, we synthesized photostabilizer–dye conjugates of fluorophores that could thus far not be tested for intramolecular photostabilization due to limited scaffolding options. We focused on rhodamines and carbopyronines, that is, fluorophores from the ATTO and Alexa series, which are extremely popular for (life science) applications. *(S)*-Nitrophenylalanine, NPA (Fig. 2), was used as a scaffold for the first generation of compounds, which consists of single-stranded DNA (ssDNA) as the biomolecular target, a commercially available organic fluorophore (Alexa555, ATTO647N and Cy5) and the *p*-nitrophenyl group of the known photostabilizer NPA[16,17,19] (Fig. 2, compounds **5**, **6** and **7**). The cyanine fluorophore Cy5 served as a 'positive control' experiment that permits to benchmark the effects of intramolecular photostabilization for different fluorophore classes with respect to published studies[16–20]. The used DNA and its specific base sequence was selected because of the detailed photophysical characterization of various fluorophores on this target[11,18,19,24–27].

The synthesis shown in Fig. 2 was conducted starting from commercially available fluorenylmethyloxycarbonyl (Fmoc)-protected NPA (**1**), which was converted into the corresponding NHS-ester derivative (**2**) to react with a 5′-aminoalkyl functionalized ssDNA, yielding **3**. Subsequent Fmoc deprotection of **3** was followed by a reaction of the resulting primary amine (**4**) with the commercially available NHS-ester derivative of Alexa555, ATTO647N or Cy5, to yield **5**, **6** and **7**, respectively. All compounds comprise an organic fluorophore and a *p*-nitrophenyl group for intramolecular photostabilization linked to the biomolecular ssDNA target (Figs 1 and 2) and are abbreviated 'NPA fluorophore' throughout the text. The non-stabilized control compounds were obtained via a direct reaction of the NHS-ester-activated fluorophores with the respective amino-modified ssDNA-$NH_2$. The DNA–photostabilizer–dye conjugates were isolated by high-performance liquid chromatography (HPLC; see Supplementary Fig. 1) and characterized by ultraviolet–visible absorption spectroscopy and mass

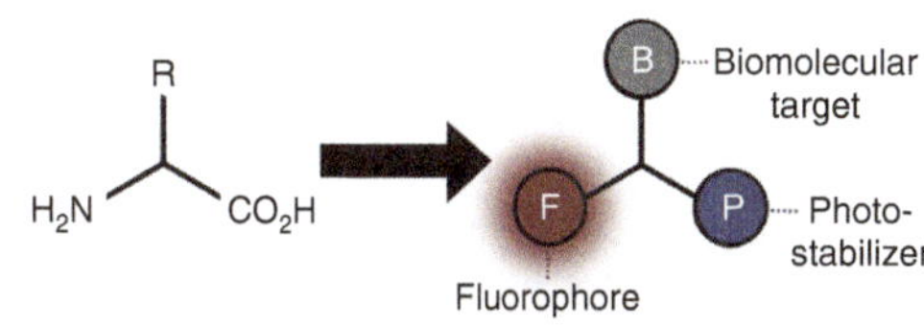


**Figure 1 | Design concept for photostabilizer–dye conjugates.** UAAs are used to combine an organic fluorophore covalently with a photostabilizer on a biomolecular target or linker structure.

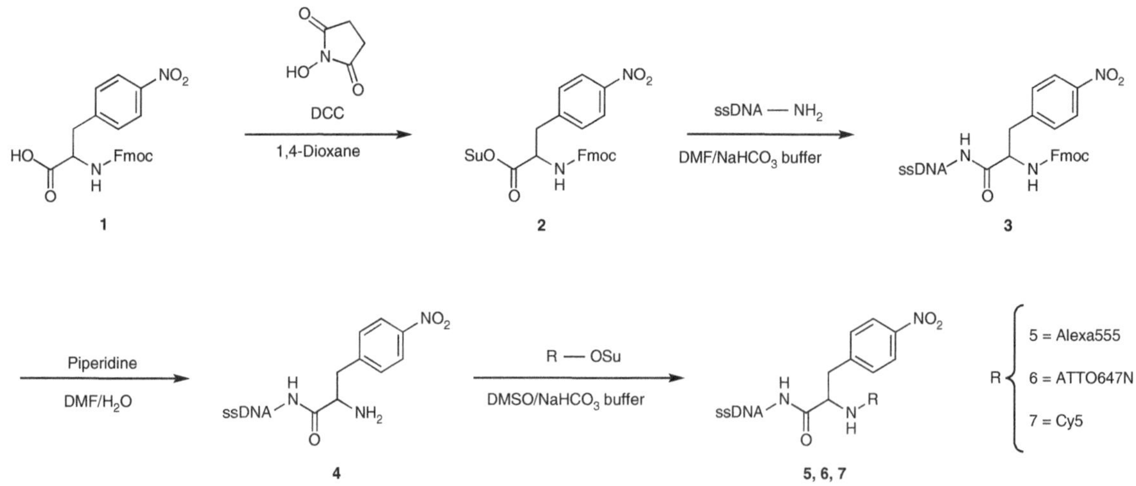

**Figure 2 | Synthesis route towards photostabilizer-dye conjugates on nucleic acids.** NPA-based fluorophores of Alexa555, ATTO647N and Cy5 on ssDNA. Fmoc, fluorenylmethyloxycarbonyl; DCC, *N-N'*-dicyclohexylcarbodiimide; Su, succinimide; DMF, dimethylformamide; $NaHCO_3$ buffer, 200 mM sodium hydrogen carbonate buffer pH 8.35; TBTA, tris[(1-benzyl-1H 1,2,3 triazol-4-yl)methyl]amine; TEAA buffer, 50 mM triethylammonium acetate buffer, pH 7.0.

spectrometry (Supplementary Figs 1 and 2). After purification, all compounds elute as single peaks in the HPLC chromatogram (Supplementary Fig. 1b,c) and display ultraviolet–visible absorption maxima characteristic for DNA (~260 nm) and the respective chromophore (~550/~650 nm) (Supplementary Fig. 2a). In addition, matrix-assisted laser desorption/ionization time-of-flight (MALDI-TOF) mass spectrometry reveals a significant mass increase of the DNA–dye conjugates compared with the non-modified DNA (Supplementary Fig. 2b). See the Methods section and the Supplementary Information for further details on chemical synthesis and characterization of functionalized oligonucleotides and reactive precursor molecules.

Next, single-molecule fluorescence microscopy was used to benchmark the potential of the scaffolding approach with respect to photophysical parameters. Confocal scanning microscopy[11,12,18,24] was used to investigate signal fluctuations and fluorescence lifetime, whereas total-internal-reflection fluorescence (TIRF) microscopy[11,18] was employed to obtain quantitative photophysical values such as bleaching lifetime, fluorescence count rate, total number of photons and SNR ratio. The different fluorophore derivatives were immobilized according to published procedures on a streptavidin-coated microscope coverslip by hybridization to form double-stranded DNA (dsDNA) comprising a 3′-terminal biotin unit (Supplementary Figs 3a–11a)[18,19]. All experiments described in this section were performed in the absence of oxygen, to minimize the convolution of triplet-state quenching by molecular oxygen and intramolecular photostabilization by NPA.

Using confocal microscopy, single immobilized fluorophores on DNA were identified as spots in the images and fluorescence time traces were recorded for each of the spots using single-photon counting detectors. The length of the fluorescence time traces gave an estimate of the photostability (the longer, the more stable), whereas fluctuations in the time traces indicated blinking due to transient population of dark states (such as the triplet state). The calculation of the autocorrelation function of the individual fluorescence time traces revealed the underlying time constants of the blinking kinetics (Fig. 3) and time-correlated single-photon counting was used to determine the fluorescence lifetime. The non-stabilized fluorophores, showed fast photo-bleaching and pronounced blinking on the corresponding dsDNA in deoxygenated PBS buffer (Fig. 3a,c,e and Supplementary Figs 3d, 5d and 7d), whereas the NPA derivatives showed a stable fluorescence signal over extended periods of time (Fig. 3b,d,f and Supplementary Figs 4d, 6d and 7d).

Fluorescence time traces of Alexa555, a rhodamine-based fluorophore, were characterized by short observation times and pronounced blinking on the timescale of $22 \pm 6$ ms in deoxygenated PBS buffer (Fig. 3a and Supplementary Fig. 3). This is a typical behaviour for rhodamines and is caused by population of the triplet state[4]. The fluorescence lifetime of the sample was found to be $1.2 \pm 0.3$ ns (Supplementary Fig. 3). On conjugation of Alexa555 to the NPA scaffold (NPA–Alexa555, compound **5**), the triplet-induced blinking diminished (negligible autocorrelation) and the overall fluorophore photostability and count rate increased to several seconds and 5–10 kHz, respectively, at ~0.3 kW cm$^{-2}$ irradiance (Fig. 3b and Supplementary Fig. 4). Despite the overall increase in count rate (mainly due to abolishment of dark state transitions), the fluorescence lifetime of NPA–Alexa555 decreased to $0.9 \pm 0.1$ ns compared with the parent fluorophore, indicating the presence of singlet quenching by the NPA scaffold (Supplementary Fig. 4).

The photophysical behaviour of ATTO647N in deoxygenated PBS buffer was characterized by a mixture of short and long photobleaching times ranging from seconds to minutes with pronounced blinking on the millisecond timescale (Fig. 3c and Supplementary Fig. 5). The off-state lifetime associated with the blinking events was found to be $29 \pm 5$ ms and is attributed to the triplet state (Fig. 3c and Supplementary Fig. 5e)[4]. This and the observed fluorescence lifetime of $4.2 \pm 0.4$ ns are in agreement with literature[11,28]. Covalent binding of NPA to ATTO647N resulted in a homogeneous, bright, non-blinking and prolonged fluorescence emission (Fig. 3d). The fluorescence lifetime was reduced to $3.4 \pm 0.2$ ns for the majority of traces, again indicating dynamic singlet quenching by the NPA scaffold. Notably, >50% of the observed NPA–ATTO647N molecules did not photobleach within the 2 min observation period at excitation intensities of ≈0.66 kW cm$^{-2}$ (see Fig. 3d and Supplementary Fig. 6). In addition, a small fraction (<30%) of NPA–ATTO647N molecules showed a reduced lifetime of around 2.6 ns in combination with a lower count rate pointing to stronger singlet quenching (Supplementary Fig. 6). Direct switching of a single NPA–ATTO647N molecule between the two states was observed and will be a topic of future research[11].

Results for Cy5 (Fig. 3e,f) were obtained at higher excitation intensities (4 kW cm$^{-2}$), to allow for a direct comparison with

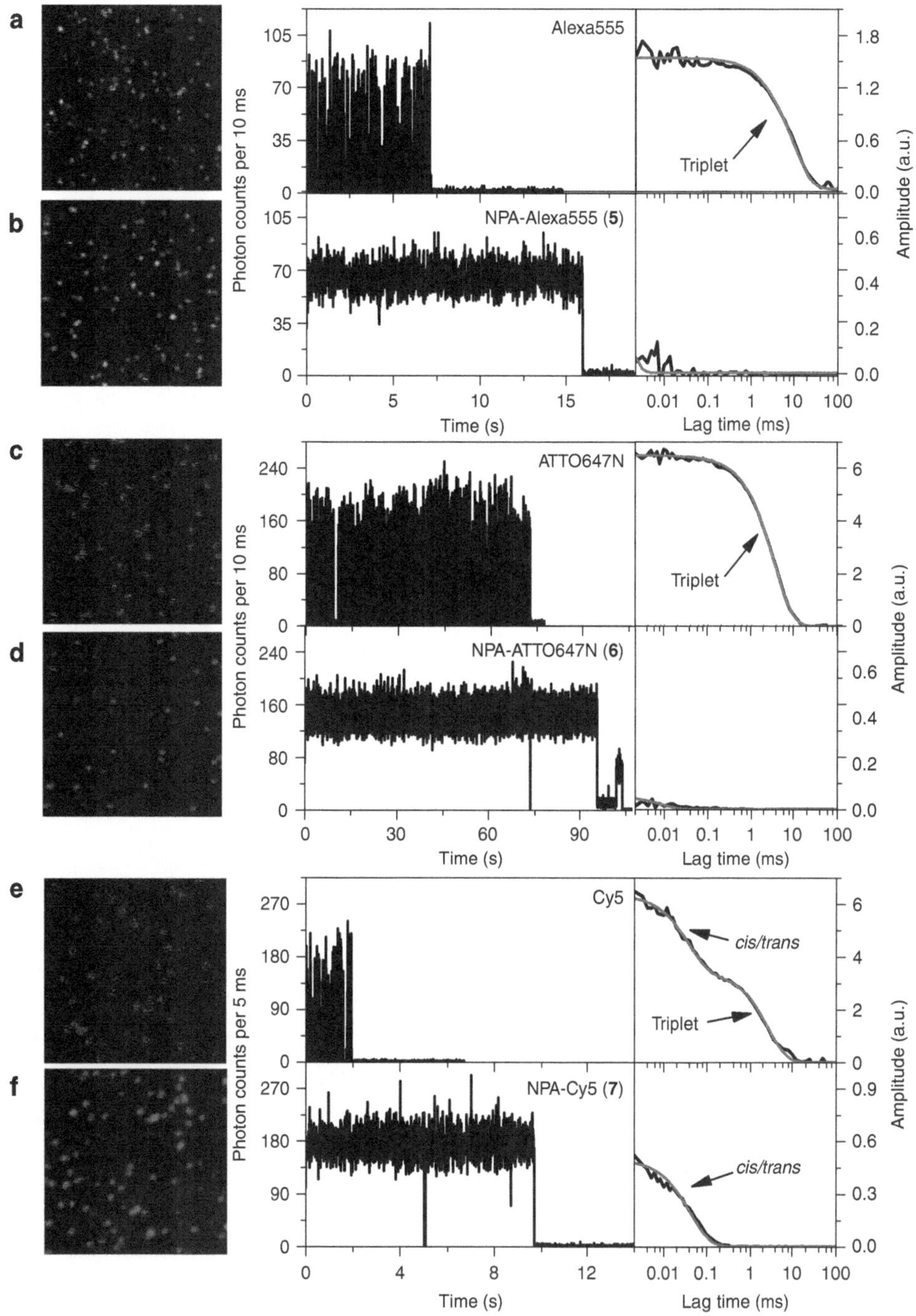


**Figure 3 | Photophysical characterization of photostabilizer-dye conjugates with confocal microscopy.** All data was recorded with a home-built confocal microscope with a sample in aqueous PBS buffer at pH 7.4 in the absence of oxygen. Each panel shows a representative overview image (10 × 10 µm, 50 nm pixel size, 2 ms per pixel) with spots from individual immobilized fluorophores (left), a fluorescence time trace (middle) and the corresponding autocorrelation decay in black, with according fits in grey (right). (**a**,**b**) Alexa555 and NPA-Alexa555 (image intensity scale from 3 to 60 counts, excitation intensity of ≈0.3 kW cm$^{-2}$ at 532 nm). (**c**,**d**) ATTO647N and NPA-ATTO647N (image intensity scale from 5 to 100 counts, excitation intensity of ≈0.66 kW cm$^{-2}$ at 640 nm). (**e**,**f**) Cy5 and NPA-Cy5 image intensity scale from 10 to 300 counts, excitation intensity of 4 kW cm$^{-2}$ at 640 nm. Further experimental details and data for each fluorophore can be found in the Methods section and Supplementary Figs 3-9.

previously published results using different approaches towards intramolecular photostabilization[29]. In agreement with earlier findings, the autocorrelation analysis of Cy5 fluorescent time traces[18] revealed the presence of two different photophysical processes that were attributed to triplet blinking (11 ± 4 ms) and *cis/trans* isomerization (54 ± 12 µs)[4,18,26,27]. The fluorescence lifetime of Cy5 was found to be 1.65 ± 0.15 ns (compare also Supplementary Fig. 7)[18,30]. NPA-Cy5 (7) under identical conditions revealed bright and prolonged fluorescence emission with typical observation times of several seconds and, as revealed by the autocorrelation analysis, with negligible triplet-state blinking for ~75% of the observed emitters (Fig. 3f). The remaining autocorrelation decay revealed the on-off transition due to *cis/trans* isomerization, with a lifetime in the order of ~50–70 µs (see Supplementary Figs 7 and 8), which therefore still restricts the overall achievable count rate for the cyanine fluorophores. In addition, a mono-exponential fit does not fully describe the experimental autocorrelation decay (see the deviations between grey fit and black data in Fig. 3f), suggesting a remaining small triplet population with a lifetime <100 μs. A smaller population of molecules (<25%) showed intensity fluctuations on completely different timescales (Supplementary Fig. 8), which is consistent with earlier reports[18,19], and its origin remains to be explained mechanistically. The fluorescence lifetime of NPA–Cy5 was found to be 1.60 ± 0.15 ns and is similar to the parent fluorophore, indicating that any reduction in the signal brightness of NPA–Cy5 is due to static quenching, that is, transient formation of non-fluorescent complexes between photostabilizer and fluorophore rather than dynamic singlet quenching. These findings are consistent with a subtle blue shift in the Cy5-absorption spectrum when bound to NPA (Supplementary Fig. 2a).

To further quantify the improved performance of the photostabilizer–dye conjugates, we used single-molecule TIRF microscopy to determine various parameters with high statistics. Movies with 100 ms integration time were recorded at laser excitation intensities of ≈50–100 W cm$^{-2}$, that is, a significantly lower excitation intensity than in the confocal scanning microscopy experiments, yielding different count rates and SNR ratios. Figure 4a shows a typical example of a camera frame with single Cy5 fluorophores. The number of fluorescent molecules per video frame was determined and the decay in number of molecules over subsequent image frames was fitted with an exponential decay ($y(t) = C + A \times \exp(-t/\tau_{bleach})$), to obtain the photobleaching lifetime $\tau_{bleach}$ (Fig. 4a)[18]. Background-corrected single-molecule time traces were extracted and used to determine the fluorescence count rate in kHz (Fig. 4a, brightness), the SNR ratio (Fig. 4a) and the total number of detected photons before photobleaching (Fig. 4a, $N_{total}$ = brightness × $\tau_{bleach}$). The mean and s.d. of all values was derived from multiple ($n \geq 3$) independent experiments. We benchmarked the performance of the photostabilizer–dye conjugates against the antioxidant TX as a solution additive in the deoxygenated imaging buffer (2 mM after 20 min ultraviolet treatment), which is a common standard for photostabilization in single-molecule experiments[10,12].

The data of the non-stabilized rhodamine Alexa555 revealed a brightness of 1.5 ± 0.2 kHz, $N_{total} = 5.1 \pm 0.4 \times 10^4$ and an SNR = 3.5 ± 0.3, whereas the photobleaching could be described by a mono-exponential decay with a time constant $\tau_{bleach} = 34 \pm 26$ s (Fig. 4b). An improvement for most of the photophysical parameters was found for both adding 2 mM TX or Alexa555-NPA with up to a 20-fold increase in photostability, 25-fold increase in $N_{total}$ and a 3-fold increase in SNR, whereas the brightness was found to be comparable.

Similar results were obtained for both the carbopyronine ATTO647N and the cyanine dye Cy5 (Fig. 4c and Supplementary Movie 1). The photobleaching time increased from $\tau_{bleach}$ = 138 ± 93 s (ATTO647N) to 212 ± 52 s (NPA–ATTO647N) and 298 ± 45 s (ATTO647N + 2 mM TX), and from $\tau_{bleach} = 7.0 \pm 1.5$ s (Cy5) to 139 ± 55 s (NPA–Cy5) and 384 ± 60 s (Cy5 + 2 mM TX). This results in reduced blinking and an increase in the total number of detected photons ($N_{total}$) and SNR from $N_{total} = 1.9 \pm 0.7 \times 10^5$ and SNR = 1.6 ± 0.2 (ATTO647N) to $N_{total} = 8.6 \pm 0.4 \times 10^5$ and SNR = 11.6 ± 1.7 (NPA–ATTO647N), and $N_{total} = 2.0 \pm 0.7 \times 10^6$ and SNR = 17.4 ± 2.4 (ATTO647N + 2 mM TX), as well as $N_{total} = 3.1 \pm 0.9 \times 10^4$ and SNR = 3.0 ± 0.7 (Cy5) to $N_{total} = 6.4 \pm 1.6 \times 10^5$ and SNR = 11.6 ± 1.7 (NPA–Cy5), and $N_{total} = 2.4 \pm 0.6 \times 10^6$ and SNR = 9.5 ± 1.2 (ATTO647N + 2 mM TX), that is, 20- to 50-fold increases.

The investigations in Figs 3 and 4 regarding the photophysical properties of rhodamines, carbopyronines and cyanines with intramolecular photostabilization revealed the following: (i) photobleaching and blinking could be efficiently removed using UAA scaffolding and (ii) all photophysical parameters were improved substantially compared with the parent fluorophore. (iii) Despite showing no decrease in brightness, rhodamines and carbopyronines showed singlet quenching, whereas cyanines show no reduction of the excited state lifetime. (iv) Solution-based healing using 2 mM TX remained more efficient than intramolecular photostabilization.

Despite the slightly lower photostability of compounds with intramolecular photostabilization compared with solution additives such as TX, this approach has unique advantages that compensates for these shortcomings. The UAA approach is thus far the only possible method for photostabilization when organic fluorophores are used under live-cell conditions, in experiments where the addition of a diffusion-based photostabilizer is not tolerated due to its toxicity or when the photostabilizer has an unwanted influence on properties of the system of interest[31]. It is also the only viable option when diffusion-based photostabilization remains ineffective which could be caused by lack of collisions between the photostabilizer and the fluorophore[32,33].

**UAAs as a general scaffold for photostabilizer–dye conjugates.** The scaffolding strategy presented in Fig. 2 is restricted by the availability of commercial UAAs with different (photostabilizing) residues and does not allow to use custom-made photostabilizers[19]. We hence set out to generalize the approach and to allow linkage of three arbitrary moieties. The amino acid propagylglycine (PG (**8**); Fig. 5) represents a more versatile scaffold that can link three chemical groups via NHS and click chemistry (here, a [3 + 2]-Huysgens cycloaddition between alkyne and azide derivatives). Hence, PG (**8**) provides the flexibility needed to combine any photostabilizer and fluorophore on a (bio)molecular target, assuming their availability as NHS- or click-reactive derivatives.

In the first synthetic step, racemic **8** was reacted with the NHS-ester derivative of TX (**9**). The resulting adduct **10** was converted into the corresponding NHS-ester derivative **11** and reacted with the 5′-aminoalkyl functionalized oligonucleotide to yield **12**. Finally, the fluorophore Cy5 was bound using a copper-catalysed click reaction[34] yielding **13**, called TX–PG–Cy5 throughout the manuscript.

TX–PG–Cy5 (**13**) showed a very similar photophysical behaviour compared with NPA–Cy5, that is, removal of blinking and an increased photobleaching lifetime (Fig. 4d) with significant improvement factors (up to 11-fold). However, a higher heterogeneity was observed with ~60% of the molecules showing fluorescent traces with little intensity fluctuations (shorter bleaching lifetime; Supplementary Fig. 9). The remaining ~40% exhibited longer observation times with an increased amount of blinking events (Supplementary Fig. 9). A unifying fluorescence lifetime of 1.65 ± 0.15 ns was observed, which is equal to the parent Cy5. Several TX–PG–Cy5 molecules showed a typical behaviour for fluorophores with intramolecular photostabilization, where 'bleaching' or irreversible destruction of the stabilizer occurred before photobleaching of the fluorophore. Here the fluorophore drastically changed its emission pattern from stable to a blinking emission pattern that closely resembles the behaviour in the absence of photostabilizer (Supplementary Fig. 9d, bottom row, left panel: stable: 0–8.5 s; blinking: 8.5–10 s).

The PG-based fluorophore system showed to be a versatile scaffold where the photostabilizer and fluorophore can be varied independently of each other. This has the advantage that the fluorophore and photostabilizer can be matched, to obtain maximum photostabilization as was shown above for the fluorophore Cy5 and the photostabilizer TX.

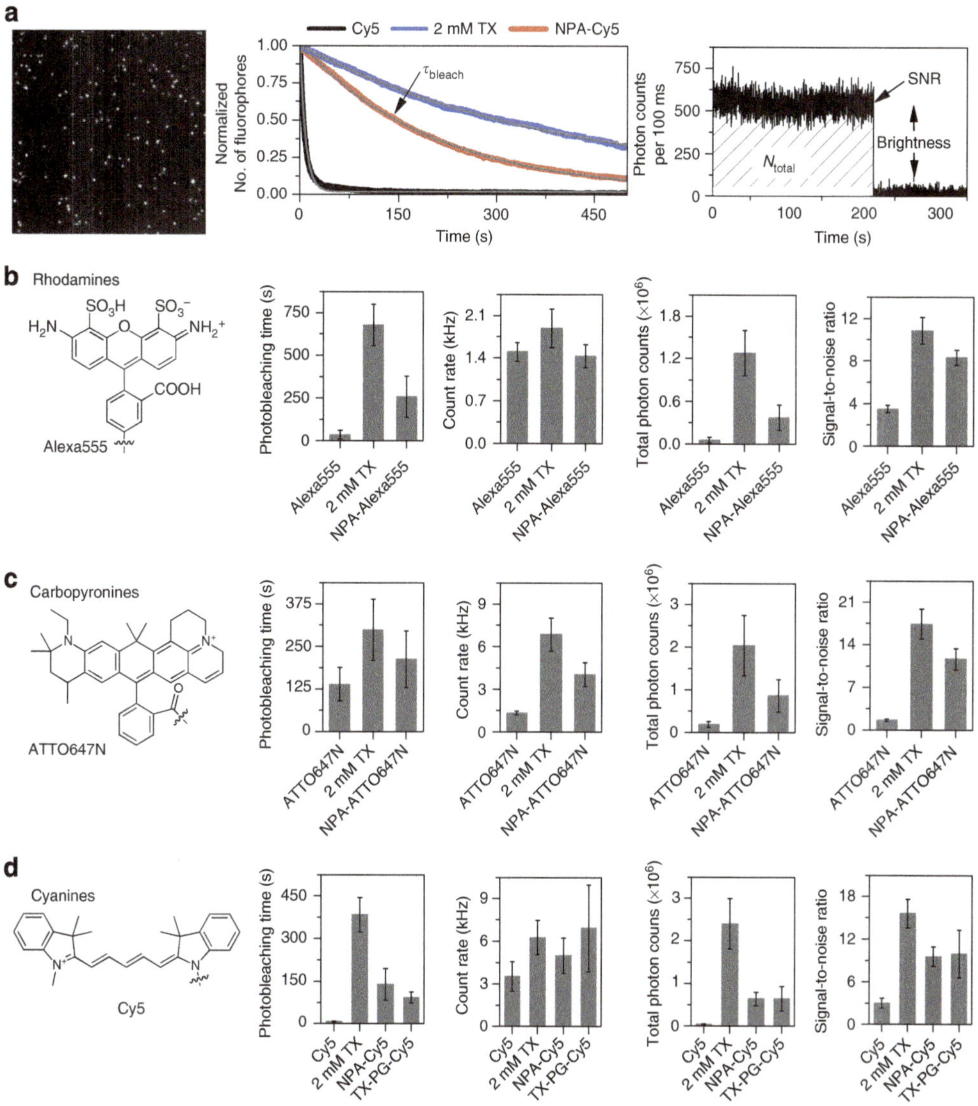


**Figure 4 | Photophysical characterization of photostabilizer-dye conjugates and their parent fluorophores with single-molecule TIRF microscopy.** Data was recorded in aqueous PBS buffer at pH 7.4 in the absence of oxygen under continuous 640 nm excitation with ≈50 W cm$^{-2}$ or 532 nm excitation with ≈20 W cm$^{-2}$. (**a**) Representative image frame (left panel) showing single fluorescent molecules (10 × 10 μm, exemplarily for Cy5). Subsequent images recorded over a period of 500 s showed an exponential decrease in the number of fluorescing molecules with a photobleaching lifetime $\tau_{bleach}$, as shown for Cy5 (black), Cy5 with 2 mM TX (blue) and NPA-Cy5 (red) (middle panel). The curves shown were obtained by averaging over >5 TIRF movies. (**b**,**c**,**d**) Right panel: chemical structures (left panels) and respective photophysical parameters obtained from background-corrected fluorescence traces for Alexa555 (**b**, rhodamines), ATTO647N (**c**, carbopyronines) and Cy5 (**d**, cyanines). Values and error bars (s.d.) in bar graphs obtained from $N>500$ molecules. For further details of the experimental techniques, data acquisition and analysis, see the Methods section.

**Direct labelling of biomolecules.** The synthesis and photophysical characterization of the photostabilizer–dye conjugates in the previous sections demonstrated the principle improvements in fluorescence emission properties by the UAAs scaffolding approach. Yet, our overall goal is to show the potential of the new fluorophores in biological research. For this, the most important step is labelling biomolecules such as proteins, nucleic acids or antibodies. As shown in ‘UAAs as a general scaffold for photostabilizer–dye conjugates’, this usually requires a complex (multi-step) synthesis process, demanding larger amounts of substances, that is, both fluorophore and biomolecular target. However, the amount of biomolecules is often limited. We hence reduced the chemical steps needed for labelling to a minimum. For this we altered the synthesis strategy of the photostabilizer–dye conjugates, introducing a functionalized group such as a NHS ester or a maleimide, which allow using the fluorophore derivative straightforward for labelling of a biomolecular target via primary amines or cysteine residues (as is done conventionally). NPA was again used as a scaffold for this second generation of photostabilizer–dye conjugates.

Owing to its ease and cost-effective accessibility, we first used the NHS-ester derivative of the dye RhodamineB (**14**), which was subjected to NPA (**15**) under basic conditions to yield **16** (Fig. 6). Subsequent activation with NHS/*N*-*N*′-dicyclohexylcarbodiimide in dimethylformamide (DMF) gave the NHS-ester derivative of the NPA–RhodamineB conjugate (**17**). In a first step, we (as for the previous fluorophores) targeted this conjugate to ssDNA (**18**), which was done straightforward from **17** without further purification. The photophysical characteristics of the resulting compound were characterized when immobilized on glass in deoxygenated buffer.

Similar to the fluorophores studied before (see Fig. 3), RhodamineB showed strong blinking in deoxygenated PBS buffer with typical observation times until photobleaching on the

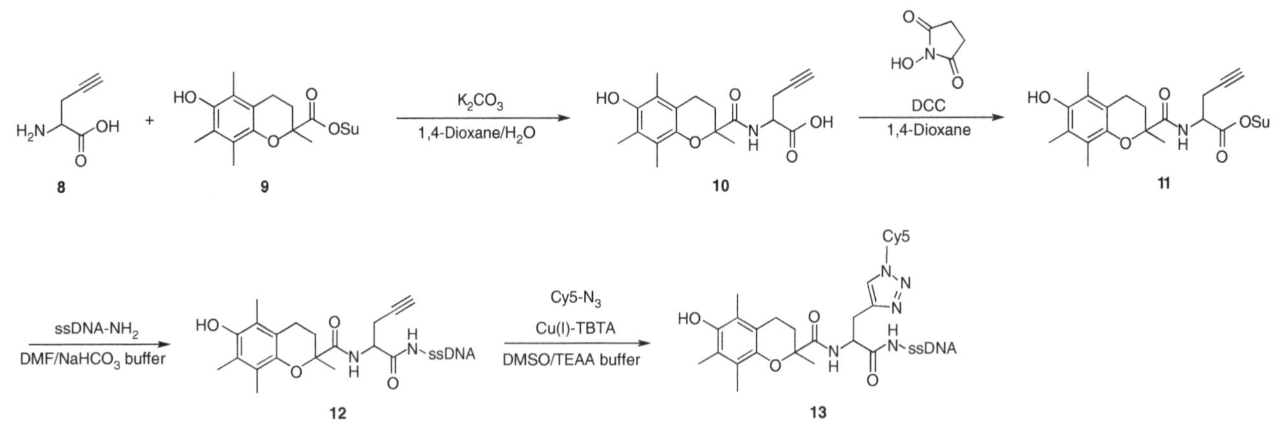

**Figure 5 | Synthesis route to obtain TX-PG-Cy5 (13) on ssDNA.** DCC, *N-N'*-dicyclohexylcarbodiimide; Su, succinimide; DMF, dimethylformamide; $NaHCO_3$ buffer, 200 mM sodium hydrogen carbonate buffer pH 8.35; TBTA, tris[(1-benzyl-1H-1,2,3-triazol-4-yl)methyl]amine; TEAA buffer, 50 mM triethylammonium acetate buffer, pH 7.0.

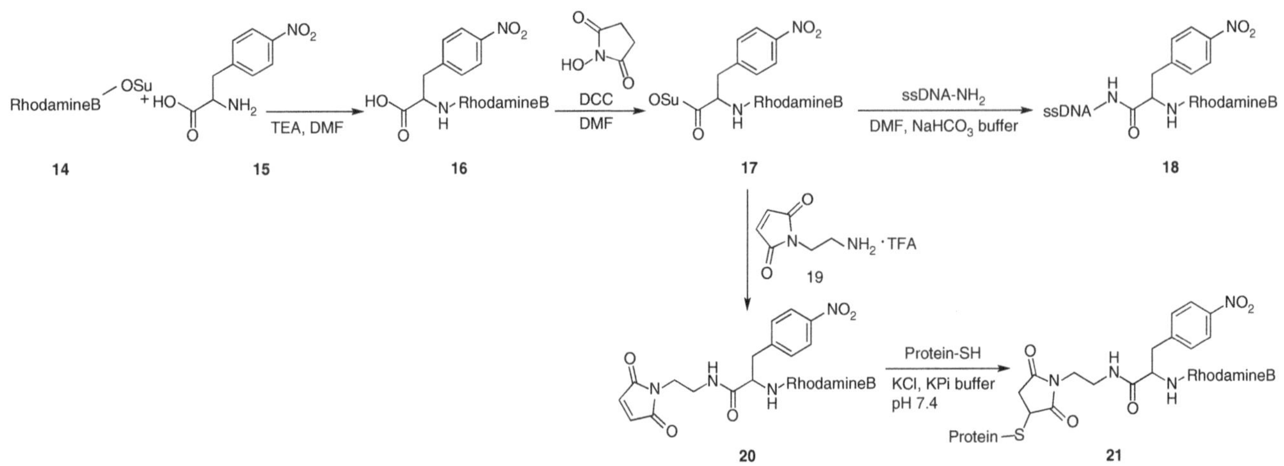

**Figure 6 | Synthesis of reactive photostabilizer-dye conjugates of RhodamineB for direct labelling of primary amines and thiol residues.** The strategy can be extended to other biochemical targets by a varying the linker of molecule 19.

timescale of 10–20 s (Supplementary Fig. 10). In contrast, RhodamineB displayed strongly heterogeneous blinking characteristics with very short or longer off times within one trace (Supplementary Fig. 10). Consequently, the autocorrelation function of this fluctuating signal could only be described by a bi-exponential decay with average off times peaking at $7 \pm 4$ ms and a significant fraction of values $> 20$ ms, deviating from a normal distribution (Supplementary Fig. 10). This indicates the presence of multiple dark states or heterogeneous dye environments. Still the blinking off times were in the same range as determined before for the triplet state of the structurally related rhodamine fluorophores, for example, Alexa555 ($22 \pm 6$ ms, Supplementary Fig. 3) or ATTO565 ($6 \pm 2$ ms)[4] and are hence considered to be triplet related. Yet, the fluorescence lifetime of RhodamineB was found to be rather homogeneously distributed with an average of $3.1 \pm 0.4$ ns (Supplementary Fig. 10). Strikingly, NPA–RhodamineB on ssDNA (**18**) showed strongly reduced blinking, an increased photostability resulting in a stable and non-blinking emission pattern with observation times of up to minutes (Supplementary Fig. 11) and an increased brightness (Supplementary Fig. 12).

Next, we intended to use our NPA-based photostabilizer–dye conjugates for direct labelling of proteins. For this purpose, two different synthesis strategies were developed, accounting for the available quantity of the fluorophore. In the most straightforward case, that is, larger amounts of amine-reactive photostabilizer–dye conjugate are available (for example, $> 50$ mg of compound **17** was available), the NHS ester of the fluorophore (**17**) can be coupled directly with 2-maleimidoethylamine (**19**), to yield a maleimide derivative of, for example, NPA–RhodamineB (**20**) in Fig. 6. The second strategy is also feasible for small quantities of reactive fluorophore species ($< 10$ mg) that could be due to high prices of commercially available precursors or complicated synthesis. ATTO647N is a good example for a fluorophore that is often used in demanding fluorescence applications but is not readily available in large amounts. For these cases we optimized the synthesis route as shown in Fig. 7, to yield a thiol-reactive derivative of ATTO647N containing the photostabilizer NPA (Fig. 7, compound **25**). As shown in 'Biomolecular FRET study with photostabilizer–dye conjugates', both maleimide derivatives can covalently bind to recombinant proteins via solvent-exposed cysteine residues (Fig. 6 compound **21** and Fig. 7 compound **26**).

**Biomolecular FRET study with photostabilizer–dye conjugates.** To show the benefits of intramolecular photostabilization in fluorescence applications with proteins, we studied the substrate-binding domain 2 (SBD2) of the Lactococcus lactis ABC transporter GlnPQ[35]. Using smFRET and alternating laser excitation (ALEX) spectroscopy, the conformational states of the protein were monitored (Fig. 8a, open unliganded and closed liganded). The structural rearrangements of SBD2 on ligand binding causes a change of $\sim 0.9$ nm regarding the distance between two selected amino acids in the protein[35]. As the FRET

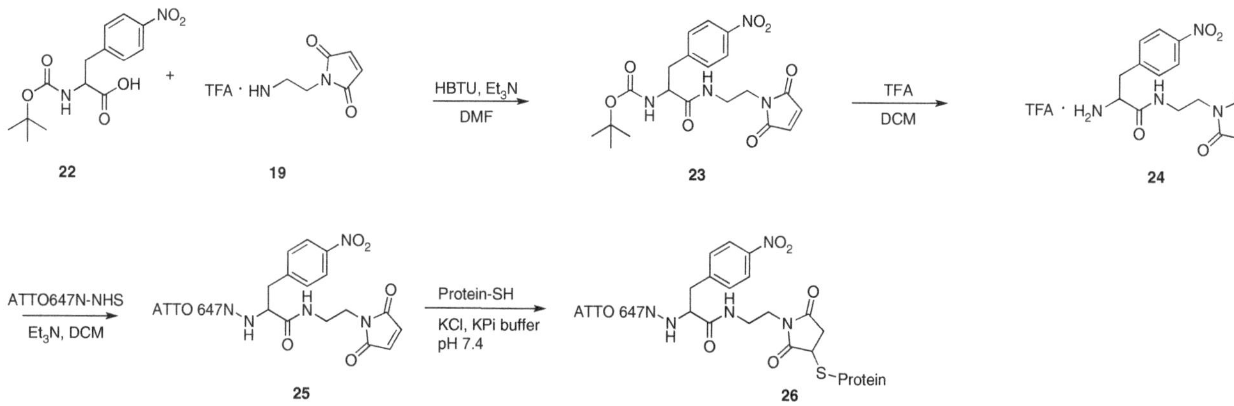

**Figure 7 | Simplified synthesis of reactive photostabilizer-dye conjugates where only small quantities of fluorophore are available.** The resulting NPA-ATTO647N conjugate can be used for direct labelling of thiol residues, for example, in proteins (compound **25**). The strategy can be extended to other biochemical targets by a variation of the linker molecule 19.

donor and acceptor are attached via maleimide chemistry at these positions in the protein (mutant of SBD2: T369C/S451C), the transfer efficiency $E^*$ reports on the conformational state of the protein. As described previously[35], SBD2 was labelled stochastically using appropriate mixtures of donor and acceptor fluorophores (details see Methods). To understand the effects of intramolecular photostabilization in FRET-based assays, we used different fluorophore combinations: Cy3B or RhodamineB as donor fluorophores and ATTO647N as the acceptor. In experiments described below, either the donor (RhodamineB) or the acceptor (ATTO647N) was stabilized via covalent linkage to NPA (synthesis see Figs 6 and 7).

We used smFRET and ALEX spectroscopy, as described in refs 33,35,36, to investigate the photophysical properties of the fluorophores when bound covalently to the protein and the biomolecular function of SBD2. ALEX is a valuable tool for both purposes, as it allows to distinguish the desired protein molecules containing both fluorophores, thus monitoring protein conformation (donor–acceptor species; Fig. 8b, $0.9<S<0.4$), from those labelled with only one species not providing distance information (donor only, $S>0.9$; acceptor only, $S<0.4$). Although $S$ relates to the relative fluorophore brightness and labelling stoichiometry, $E^*$ indicates the FRET efficiency and thus distance between the donor and acceptor (with larger values indicating small donor–acceptor distances), which is the final read out of the protein conformation. In addition, ALEX reveals photophysical artefacts such as blinking of donor or acceptor in the form of bridges between the three different subpopulations[37]. For such measurements, the fluorescence emission of the donor $F(DD)$ under green excitation, that of the acceptor $F(DA)$ when excited via FRET from the donor and that of the acceptor via direct red excitation light $F(AA)$ was determined (see Methods and refs 33,35,36). In our experiments, individual biomolecules were studied for short time periods of a few milliseconds, while diffusing through the excitation volume of a confocal microscope. The challenge of such an experiment is to acquire intense fluorescent bursts during the short observation time under the high excitation intensities of $>10\,\text{kW cm}^{-2}$.

The combination of Cy3B and ATTO647N is known for excellent photophysical performance resulting in fast data acquisition and superior histogram quality in smFRET with little bleaching artefacts and narrow distributions (Fig. 8c). For apo-SBD2 we find mean $E^*$ and $S$ of $0.48\pm0.08$ and $0.59\pm0.07$, respectively, with these labels[35]. These values are in good agreement with our published work and the mean $E^*$ correlates with the expected interprobe distance of $\sim4.9\,\text{nm}$ derived from the crystal structure.

Results of such quality are, however, only available when using 2 mM TX as a photostabilizer in solution, seen from comparison of Fig. 8c,d. Here we show data of apo-SBD2 (labelled with Cy3B/ATTO647N) in the presence and absence of TX. In agreement with Kong *et al.*[37], the high excitation intensities used in our experiments promote acceptor signal fluctuations, that is, blinking and/or bleaching. Cy3B and ATTO647N can hence be seen as a FRET couple where the acceptor photostability is limiting. This appears as a prominent bridge between the donor-only and donor-acceptor population (Fig. 8d), altering both $E^*/S$-values substantially. Closer inspection of the histogram reveals that a significant portion of the molecules show these unwanted photophysical effects. Under these conditions, neither mean $E^*$ nor correction factors for accurate FRET determination are directly accessible. Besides the complete loss of information, the overall acquisition time has also increased in the absence of photostabilizer to obtain sufficient statistics from the relevant donor–acceptor species. It should be noted that such photophysical artefacts of the acceptor (Fig. 8d) are extremely problematic for data interpretation, as they suggest the existence of (non-biological) species in between the donor only and the actual FRET species (see $1D-E^*$ in Fig. 8d that can only be fitted by the sum of two Gaussians with $E^*=0.19\pm0.07$ and $0.43\pm0.1$). Furthermore, it remains challenging to identify those when performing smFRET with green excitation in continuous-wave mode.

Strikingly, the bridge population can be removed by sole photostabilization of the acceptor fluorophore via scaffolding of ATTO647N to NPA. The ALEX data of apo-SBD2 with Cy3B/NPA–ATTO647N is shown in Fig. 8e. Here the bridge between the donor–acceptor and donor-only population is fully removed without the addition of stabilizer to the solution (Fig. 8d versus Fig. 8e). The resulting mean $E^*/S$ values are changed compared to apo-SBD2 with Cy3B/ATTO647N (2 mM TX in solution, Fig. 8b) accounting for decreased acceptor brightness; mean $E^*$ is now found at $0.34\pm0.09$ and mean $S$ at $0.67\pm0.08$ (Fig. 8e). These differences are, however, expected considering the results from ATTO647N on DNA, where a decrease in the overall brightness is observed on conjugation to NPA (Fig. 4c). It should be mentioned that differences in the donor–acceptor population relative to donor and acceptor only comparing the samples Cy3B/ATTO647N and Cy3B/NPA–ATTO647N are not solely due to photophysics but also due to different labelling ratios of the protein.

Next, we used Cy3B/NPA–ATTO647N to study the biomolecular function of SBD2 (Fig. 8b). On addition of the ligand glutamine, the mean $E^*$ changes from $0.36\pm0.1$ (open, interprobe

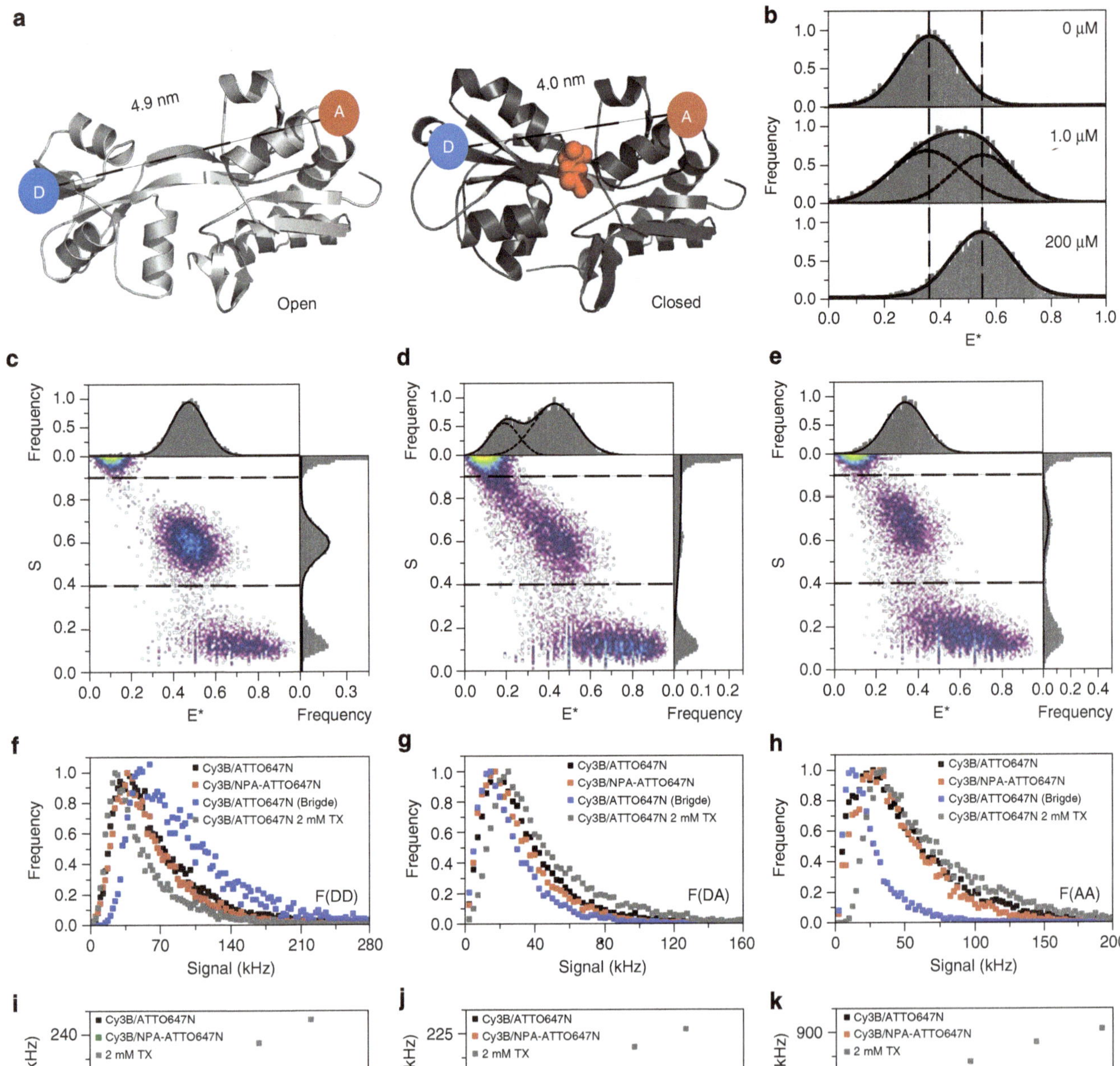


**Figure 8 | Improving smFRET-ALEX measurements by using NPA-based fluorophore as FRET acceptor on the protein GlnPQ-SBD2.** **(a)** Crystal structures of the SBD2 (T369C/S451C) open (left panel, PDB: 4KR5) and closed state (right panel, PDB:4KQP, after binding of the ligand glutamine shown in red) with label positions of donor (D) and acceptor (A). **(b)** Corresponding one-dimensional histograms of $E^*$-values for increasing amounts of ligand glutamine where an increased $E^* = 0.55 \pm 0.1$ (dashed line, closed conformation) becomes visible (as opposed to the open conformation, $E^* = 0.36 \pm 0.1$, dashed line). Excitation intensities of 30 kW cm$^{-2}$ at 532 nm and 20 kW cm$^{-2}$ at 640 nm; data evaluated with dual colour burst search and displayed with additional per-bin thresholds of $F(DD) + F(DA) + F(AA) > 75$ and minimal number of counts per bin in ALEX histogram of 3. **(c,d,e)** 2D histograms of joint pair values of $S$ (labelling stoichiometry) and $E^*$ (FRET efficiency, that is, interprobe distance) of Cy3B/ATTO647N in the presence **(c)** and absence **(d)** of 2 mM TX and Cy3B/NPA-ATTO647N **(e)** without ligand glutamine. Excitation intensities of 30 kW cm$^{-2}$ at 532 nm and 20 kW cm$^{-2}$ at 640 nm; data evaluated with all photon burst search and displayed with additional per-bin thresholds of $F(DD) + F(DA) + F(AA) > 100$ and minimal number of counts per bin in ALEX histogram of 3. **(f,g,h)** Histogram of fluorophore brightness values as determined from photon-counting histograms (PCHs) on single-molecule transits of labelled SBD2 diffusing through the observation volume, comparison of donor brightness $F(DD)$ **(f)**, acceptor brightness when excited via FRET, $F(DA)$ **(g)** and acceptor brightness via direct red excitation $F(AA)$ **(h)**. Excitation intensities of 30 kW cm$^{-2}$ at 532 nm and 20 kW cm$^{-2}$ at 640 nm. **(i,j,k)** Dependence of the mean count rate of $F(DD)$, $F(DA)$ and $F(AA)$ of the different samples for increasing excitation intensity, respectively.

distance of ~4.9 nm) to 0.55 ± 0.1 (closed, with a decreased interprobe distance of ~4.0 nm). A concentration of 200 µM saturates ligand binding and therefore results in a 100% population of the closed state (Fig. 8b). A ligand concentration of 1 µM, which is close to the $K_d$-value of the protein[38], consequently results in a mix of open and closed states (Fig. 8b).

As fluorophore brightness and the resulting photon budget ultimately determine the quality of the final histograms including the necessary measurement time, we quantitatively analysed Cy3B/ATTO647N and Cy3B/NPA–ATTO647N by means of photon-counting histograms. Figure 8f,g,h shows the three relevant photon streams used to determine $E^*$ and $S$ of Cy3B/ATTO647N (donor–acceptor: $0.9 > S > 0.4$; bridge: $0.9 > S > 0.7$) and Cy3B/NPA–ATTO647N (donor–acceptor: $0.9 > S > 0.4$). $F$(DD) shows that the strongest donor quenching and hence the most efficient FRET is found for the addition of TX to Cy3B/ATTO647N, whereas molecules in the bridge (Cy3B/ATTO647N, no TX) show inefficient donor quenching due to a non-functional acceptor (Fig. 8f). Cy3B/NPA–ATTO647N and healthy molecules from the Cy3B/ATTO647N population with no TX show a similar performance of the acceptor-based donor quenching. $F$(DA) correlates directly with quenching in $F$(DD) as seen in Fig. 8g. Again, the best performance is observed from Cy3B/ATTO647N in the presence of TX, while molecules in the bridge show the lowest counts. A striking difference between the bridge and all other conditions is seen in $F$(AA), that is, direct excitation of the acceptor (Fig. 8h). The overall comparison suggest that NPA-based acceptor stabilization is sufficient to remove photophysical artefacts and hence make the smFRET useful for biomolecular studies. Further optimization of the data quality could, however, be obtained by additional donor stabilization of, for example, Cy3B in this case. This interpretation is further supported by excitation intensity-dependent count rates of Cy3B/ATTO647N and Cy3B/NPA–ATTO647N (Fig. 8i,j,k). NPA-based acceptor stabilization improves the saturation characteristics in all three channels but the addition of TX to the solution (resulting in stabilization of both donor and acceptor at the same time) remains superior in terms of achievable count rates.

To study the donor dependence in more detail, we repeated the above described experiments using RhodamineB/ATTO647N and NPA–RhodamineB/ATTO647N. Here, the photostability of the donor is the limiting factor for smFRET data quality. For RhodamineB/ATTO647N we find prominent donor blinking in the absence of photostabilizer (Supplementary Fig. 13). The bridge between the donor–acceptor and acceptor-only population can be removed by addition of TX in solution (Supplementary Fig. 13) or conjugation of RhodamineB to NPA (Supplementary Fig. 13). The overall magnitude of the observed effects is lower than for the case of acceptor bleaching/blinking in Fig. 8. Photon-counting histograms and the intensity dependence show a similar trend as before, that is, correlation between $F$(DD) and $F$(DA) with the wish to increase donor photostability as much as possible. Our data makes clear that NPA–RhodamineB can be used at significantly higher excitation intensities than RhodamineB, as triplet-state population was minimized and would allow for faster and better data acquisition (Supplementary Fig. 13).

Our results show that the simple addition of the NPA unit to the acceptor fluorophore is sufficient to gather reliable results from solution-based smFRET (Fig. 8). Although the addition of a stabilizer on the donor fluorophore can increase the overall available photon budget (which should always be maximized), the acceptor strictly requires a solution-based or covalently linked photostabilizer. We consequently suggest to stabilize the acceptor as a minimum and (if possible) also to stabilize the donor to maximize available count rates and hence data quality. We finally note that all smFRET results of SBD2 with the different pairs of fluorophores shown here are in good agreement with biochemical data from isothermal calorimetry and with our previous results, supporting our hypothesis of an induced-fit-type mechanism in GlnPQ[35]. In addition, it confirms successful labelling of the protein with the custom-made NPA derivative, and that the results from the established FRET assay are indeed independent of the fluorophore pair that is used to monitor the protein conformation[35]. It is noteworthy that spectrally uncorrected apparent FRET is reported here accounting for absolute differences in mean $E^*$-values from varying Förster radius $R_0$ of the fluorophore pairs used or variations in the setup alignment[35].

**Photostabilizer–dye conjugates in super-resolution microscopy.** A common tool in cellular far-field fluorescence microscopy is the use of immunofluorescence, where the majority of cells are fixed and individual structures or proteins are tagged using specific primary antibodies and fluorophore-labelled secondary antibodies. In recent years, pioneering developments in far-field fluorescence microscopy, so-called super-resolution microscopes, have revolutionized cellular imaging, allowing the visualization of biological structures with nanometre resolution, that is, beyond the physical diffraction limit that thus far prevented to resolve structures with a precision better than ~250 nm[2]. Given the importance of photostability and brightness in such techniques, we investigated the potential of fluorophores with intramolecular photostabilization for immunofluorescence and specifically super-resolution imaging. For this, we have used the NHS ester of the dye KK114 (ref. 39) and its derivative NPA–KK114 (Supplementary Fig. 14) to directly tag secondary antibodies using standard procedures. These antibodies were used to specifically immunolabel nuclear pore complexes (NPCs) in fixed mammalian PtK2 cells. Figure 9a and Supplementary Fig. 15 depict that we could successfully apply NPA–KK114 to visualize the spatial distribution of these NPCs as shown in the representative confocal scanning images. However, it also becomes obvious that these complexes appear as rather large (>250 nm), blurred spots due to the diffraction limit. To increase resolution, we employed super-resolution STED microscopy[3,40], which in its most common application adds a STED laser to the confocal microscope that features a focal intensity distribution with a central zero, to allow imaging with subdiffraction spatial resolution. Figure 9b and Supplementary Fig. 15 clearly show the succesfull implementation of NPA–KK114 in STED microscopy. The NPCs were much better resolved and appeared as much smaller spots (85–90 nm, which is a reasonable value considering the use of primary and secondary antibodies). When compared with conventional KK114, the NPA–KK114 showed an increased photostability under STED conditions. Repeated scanning of the same area of the cell revealed that fading of fluorescence was reduced for NPA–KK114, while brightness was only subtly increased (Fig. 9d–f, Supplementary Fig. 15 and Supplementary Movie 2). The improvement in photobleaching resistance under STED conditions via an intramolecular mechanism highlights future directions towards improved dynamic STED imaging without the need of adding (potential toxic) chemical compounds[41].

## Discussion

Our results show that effective intramolecular photostabilization and removal of (triplet induced) dark states can be achieved for NPA- and PG-based fluorophore derivatives with similar or better photostabilization effects as previously introduced photostabilizer–dye conjugates. This shows that the UAA scaffold

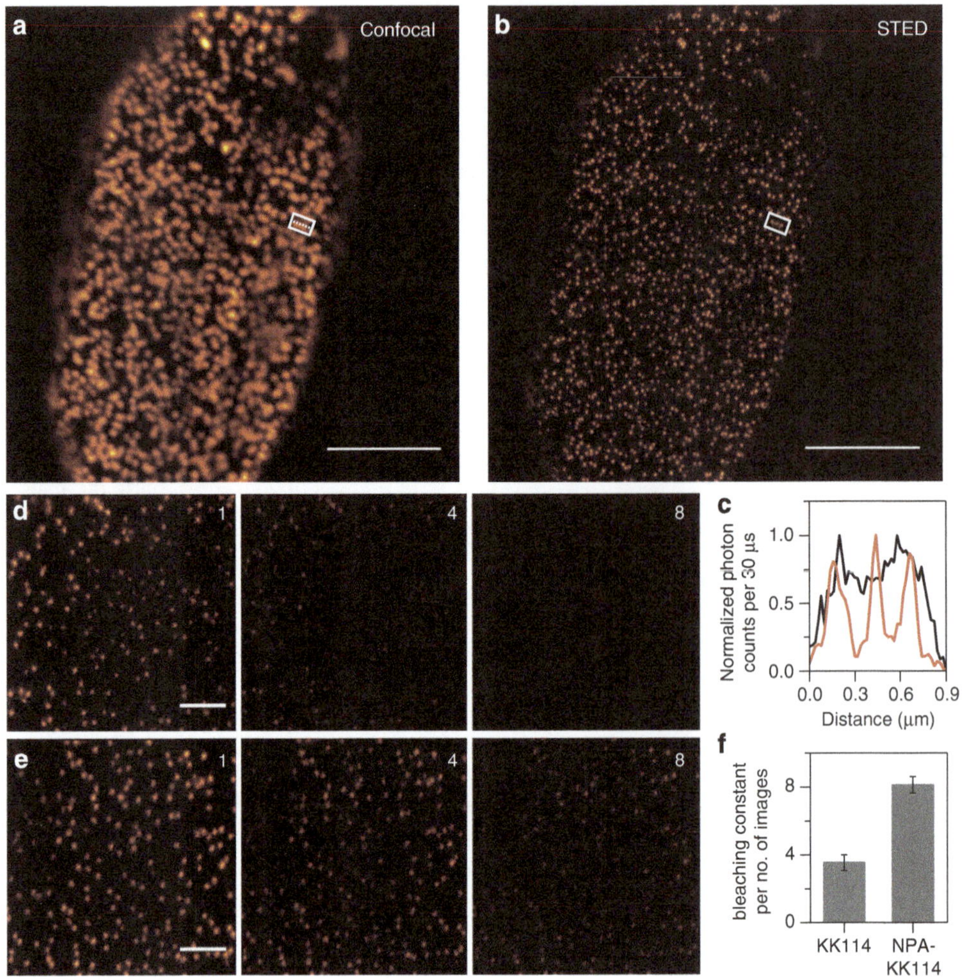


**Figure 9 | Use of photostabilizer-dye conjugates in confocal and super-resolution STED microscopy of immunolabelled cells.** Data from NPCs in fixed mammalian PtK2 cells immunolabelled with KK114 and NPA-KK114 Confocal (**a**) and STED (**b**) images of a representative cell stained with NPA-KK114 (scale bars, 5 µm). (**c**) Normalized intensity profile along the dashed line in the white boxes marked in the images of **a**,**b**, exemplifying how neighbouring NPCs can be much better resolved in the STED (red) compared with the confocal (black) recordings. (**d**-**f**) Repeated scanning of the same area of the cells in the STED mode indicates reduced photobleaching in the case of NPA-KK114 compared with KK114: images 1, 4 and 8 for KK114 (**d**) and NPA-KK114 (**e**) (scale bars, 1 µm), and (**f**) bleaching constant from an exponential fit in terms of number of images for KK114 and NPA-KK114 (total number of subsequent images recorded = 30, $n = 4$).

is a simple and useful tool to covalently link fluorophore, photostabilizer and biomolecular target in a modular manner. The presented approach has many advantages over existing strategies, as it is cost effective, requires low amounts of material, uses only well-known chemical transformations such as amide bond formation or click chemistry, requires standard purification procedures and has the potential to address various important biomolecular targets.

Moreover, our conjugation strategy opens the possibility to apply intramolecular photostabilization to fluorophore classes that could not be studied before. This possibility broadens our understanding of intramolecular photostabilization and might provide a basis for further improvement of the photostabilizer–dye conjugates by studying the effects of fluorophore properties such as charge, redox potential or absorption/emission wavelength. For certain types of fluorophores, for example, oxazines and perylenes, conjugation to NPA is expected to show effects different from those described here considering their interaction with antioxidants[4,24,25], DNA bases and amino acids such as tryptophane[42–44]. Intramolecular quenching of Alexa fluorophores observed with natural amino acids[45] even suggests that labelling of proteins at strategic locations, for example, in the vicinity of aromatic amino acids, could be used for fluorophore stabilization without the use of UAAs, provided that a suitable combination of fluorophore and amino acids can be found.

Intramolecular photostabilization remains, however, less effective compared to the use of diffusion-based photostabilization (see Fig. 4) as reported before[18]. This shortcoming of intramolecular photostabilization is compensated by the fact that it is often the only viable option for applications such as live-cell imaging, for assays where the addition of photostabilizer remains ineffective or is not tolerated by the biological system. We are convinced that future work—also facilitated by the presented synthesis strategy—will allow solving this problem. We hypothesize that the properties of photostabilizer–dye conjugates can be optimized by parameters such as the linking geometry between fluorophores and photostabilizer. This is supported by published data of Cy5 derivatives, where we linked the aromatic nitro group to the fluorophore in a different way[19], compared with the compounds shown in this study (compound 7). These architectures show a significantly higher photostability using basically the same photostabilizer moiety. We are in the process of varying the linker length of NPA-based fluorophore derivatives, to optimize the effects of intramolecular photostabilization and to become fully competitive with solution additives as shown in ref. 19.

We finally suggest that UAA scaffolding could be extended beyond photostabilization, to provide a general framework for the manipulation of fluorophore properties. Potentially interesting UAAs feature antioxidants, triplet sensitizers, photoswitchable molecules to induce blinking for localization-based super-resolution microscopy[2–4,46,47], alter water solubility[48] or the affinity to membranes and residues of natural amino acids such as thiols (see Lui *et al.*[23] for an overview of UAA residues). UAAs with 'clickable' functionalities could be used to link the fluorophore to a biomolecular target and an affinity tag (strep-/his-tag) to simplify purification of labelled protein species.

In summary, we introduced a versatile and simple design concept to synthesize photostabilizer–dye conjugates on a specific target using UAAs. The strategy is based on a straightforward and modular synthesis route using amide bond formation and click chemistry with commercially available starting materials. The UAAs NPA and PG were used as a scaffold to link rhodamine, carbopyronine and cyanine fluorophores covalently to a photostabilizer on a biomolecular target, such as DNA, antibodies and proteins in our case study. We are convinced, however, that other targets (RNA, affinity tags and so on) can also be labelled via similar means, while maintaining the positive effects of photostabilization on these targets. All studied compounds show a considerable increase in photostability and a suppression of triplet-based blinking compared to the non-stabilized parent fluorophore. With this, we are the first to test intramolecular photostabilization for various classes of organic fluorophores and show that intramolecular triplet-state quenching is a generally applicable method. The approach allows labelling of biomolecules in a single step with the same effort as for commercially available reactive fluorophore derivatives. Finally, photostabilizer–dye conjugates were used in two state-of-the-art applications, that is, the study of conformational changes in proteins via smFRET and super-resolution imaging using STED microscopy. We are convinced that the presented strategy will stimulate broader use of intramolecular photostabilization and help to emerge this strategy to the new gold standard for photostabilization.

## Methods

**Synthesis of photostabilizer-dye conjugates and precursors.** All reagents were purchased from commercial suppliers and used without further purification, unless stated otherwise. Cy5 fluorophores were obtained from Lumiprobe (Germany), *Alexa555-NHS* was obtained from Life Technologies (USA) and ATTO647N-NHS from ATTOTEC (Germany). Synthetic oligomers ($NH_2$-C6-5′-TAA TAT TCG ATT CCT TAC ACT TAT ATT GCA TAG CTA TAC G-3′) were received in HPLC-purified quality from IBA or Eurofins (Germany). TX-NHS **9** was synthesized following a published procedure[49]. A Varian 400 (400 and 100 MHz) and Varian 500 MHz were used to record $^1$H-NMR and $^{13}$C-NMR spectra. Chemical shifts ($\delta$) are denoted in p.p.m. using residual solvent peaks as internal standard ($\delta_H = 7.26$ and $\delta_C = 77.0$ for $CDCl_3$; $\delta_H = 3.31$, 4.78 and $\delta_C = 49.15$ for $CD_3OD$). High-resolution mass spectra were recorded on an Orbitrap XL (Thermo Fisher Scientific; ESI pos. or neg. mode). Liquid chromatography–mass spectra were recorded on Waters Acquity H-class UPLC equipped with TUV detector and a Xevo G2 TOF mass detector from Waters Chromatography BV. Flash chromatography was performed using a Grace Reveleris Flash System (40 µm silica column).

*Purification of functionalized oligonucleotides*. The functionalized oligonucleotides were purified on an Illustra NAP 5 column loaded with Sephadex G-25 DNA Grade material obtained from GE Healthcare. Illustra NAP 5 columns were equilibrated with 3 × 5 ml of eluent before use (50 mM triethylammonium acetate (TEAA) buffer, pH 7.0, or water for Fmoc-protected oligonucleotide **3** and **12**). After the reaction, the oligonucleotide sample was diluted with eluent to a final volume of 0.5 ml and added to the column. After the sample was loaded, 1 mL of eluent was added and the purified oligonucleotide was collected in one portion. The oligonucleotide solution was lyophilized (Christ Alpha 2–4 LD plus freeze dyer) directly after collection.

*Isolation and characterization of functionalized oligonucleotides*. Reversed-phase HPLC (rp-HPLC) analysis and preparative purifications (isolation) was performed on a Shimadzu LC-10AD VP machine equipped with Waters Xterra MS C18 column (3.0 × 150 mm, particle size 3.5 µm) and Waters Xterra MS C18 prep column (7.8 × 150 mm, particle size 10 µm) using a gradient of acetonitrile/TEAA buffer (50 mM, pH 7.0). Gradient 1: 05/95, 0–10 min, to 65/35 at 60 min, to 75/25 at 65 min, to 05/95 at 75 min for 15 min. Gradient 2: 05/95, 0–10 min, to 35/65 at 60 min, to 65/35 at 65 min, to 05/95 at 70 min for 20 min. Flow 0.5 ml min$^{-1}$ analytical run or 1.0 ml min$^{-1}$ preparative run. The DNA was isolated by collecting the major peak of interest (see Supplementary Fig. 1). The resulting compounds were characterized by ultraviolet-visible absorption spectroscopy (Supplementary Fig. 2a) and MALDI-TOF mass spectrometry (Supplementary Fig. 2b). Spectra were recorded on an ABI Voyager DE-PRO MALDI-TOF (delayed extraction reflector) Biospectrometry Workstation mass spectrometer.

*Synthesis of ssDNA-NPA-Alexa555/ATTO647N/Cy5*. Step 1: the lyophilized ssDNA $NH_2$ was resuspended in MilliQ water and adjusted to 80 µM in 0.2 M $NaHCO_3$ buffer, pH 8.35. To 100 µl of this solution, the same volume of a 20 mg ml$^{-1}$ solution of **2** in DMF was added and the mixture was vortexed thoroughly. If necessary, additional DMF was added in 10 µl portions, to obtain a clear solution. After the reaction at room temperature overnight, **3** was purified on Illustra NAP 5 column (*vide supra*, MilliQ water was used as eluent for purification on Illustra NAP 5 columns, to prevent partial deprotection of the Fmoc group), and isolated by preparative rp-HPLC (Supplementary Fig. 2, gradient 1, *vide supra*), to yield ssDNA NPA-Fmoc **3**. Step 2: Fmoc deprotection of **3** was performed as follows. The HPLC-purified and lyophilized oligonucleotide was resuspended in 50 µl of 50 mM TEAA buffer, pH 7.0. Deprotection was achieved by addition of 40 µl DMF and 10 µl piperidine. The mixture was vortexed and allowed to react for 2 h at room temperature. The deprotected oligonucleotide was purified on Illustra NAP 5 column (*vide supra*) and lyophilized to yield **4**. Step 3: coupling with fluorophore NHS was achieved as follows. Cy5-NHS as obtained from Lumiprobe, *Alexa555-NHS* was obtained from Life Technologies and ATTO647N-NHS from ATTOTEC; all samples were evenly distributed into ~300 nmol portions inside a glove box. Each portion was sealed with Parafilm and kept in the dark at −18 °C. Lyophilized **4** (~2 nmol) was resuspended into 50 µl of 0.2 M $NaHCO_3$ buffer, pH 8.35. To this solution, one portion of fluorophore NHS dissolved in 10 µl of dimethyl sulfoxide (DMSO) was added and the mixture was vortexed thoroughly. After incubation overnight the oligonucleotide was purified on Illustra NAP 5 column (*vide supra*) and isolated by preparative rp-HPLC (gradient 1, *vide supra*) to yield **5**, **6** and **7** (see Fig. 2 and Supplementary Fig. 1).

*Synthesis of ssDNA TX–PG–Cy5 (13)*. Step 1: the lyophilized ssDNA was resuspended in MilliQ water and adjusted to 80 µM in 0.2 M $NaHCO_3$ buffer, pH 8.35. To 100 µl of this solution, the same volume of a 20 mg ml$^{-1}$ solution of **11** in DMF was added and the mixture was vortexed thoroughly. If necessary, additional DMF was added in 10 µl portions, to obtain a clear solution. After the reaction at room temperature overnight, the oligonucleotide was purified on Illustra NAP 5 column and isolated by preparative rp-HPLC (Supplementary Fig. 1, gradient 2, *vide supra*) to yield **12**. Step 2: coupling with Cy5–$N_3$ was achieved following a modified manufacturer protocol (Lumiprobe). After every addition step the mixture was vortexed briefly. The HPLC-purified and -lyophilized oligonucleotide **12** (~2 nmol) was resuspended in 30 µl of 2 M TEAA buffer, pH 7.0 and added into a 0.5 ml tube. To this solution, 8 µl MilliQ water and 7 µl of DMSO were added. A 15-nmol portion of Cy5–$N_3$ was dissolved in 40 µl of DMSO and added. To this mixture, 10 µl of 5 mM stock solution of ascorbic acid in MilliQ water was added. The solution was degassed by a stream of $N_2$ for 30 s and finally 5 µl of a 10-mM stock solution of Cu(II)-tris[(1-benzyl-1H-1,2,3-triazol-4-yl)methyl]amine in 55% DMSO/MilliQ water was added. The vial was flushed with $N_2$, closed and sealed with Parafilm. After incubation overnight the oligonucleotide was purified on Illustra NAP 5 column (gradient 2, *vide supra*) and isolated by preparative rp-HPLC (*vide supra*) to yield **13** (Supplementary Fig. 1).

*Synthesis of ssDNA Alexa555/ATTO647N*. Step 1: *Alexa555-NHS* was obtained from Life Technologies and ATTO647N-NHS from ATTOTEC; all were evenly distributed into ~300 nmol portions inside a glove box. Each portion was sealed with Parafilm and kept in the dark at −18 °C. ssDNA $NH_2$ (~2 nmol) was resuspended into 50 µl of 0.2 M $NaHCO_3$ buffer, pH 8.35. To this solution, one portion of *Alexa555-NHS/ATTO647N-NHS* dissolved in 10 µl of DMSO was added and the mixture was vortexed thoroughly. In case of RhodamineB, an NHS crude mixture in 100 µl of DMF was added to ssDNA $NH_2$ and the mixture was vortexed thoroughly. After incubation overnight the oligonucleotide was purified with an Illustra NAP 5 column (*vide supra*) and isolated by preparative rp-HPLC (gradient 1, *vide supra*).

*Synthesis of ssDNA NPA–RhodamineB*. Step 1: the lyophilized ssDNA–$NH_2$ was resuspended in MilliQ water and adjusted to 20 µM in 0.2 M $NaHCO_3$ buffer, pH 8.35. To this solution, RhoB–NPA–NHS (**20**) in 100 µl of DMF was added and the mixture was vortexed thoroughly. After incubation overnight the oligonucleotide was purified on Illustra NAP 5 column (*vide supra*) and isolated by preparative rp-HPLC (gradient 1, *vide supra*) to yield **18** (see Fig. 6). The yield was found to be 25% for coupling NPA–RhodamineB–NHS to ssDNA.

*Synthesis of ssDNA RhodamineB*. Step 1: the lyophilized ssDNA–$NH_2$ was resuspended in MilliQ water and adjusted to 20 µM in 0.2 M $NaHCO_3$ buffer, pH 8.35. To this solution, RhodamineB–NHS in 100 µl of DMF was added and the mixture was vortexed thoroughly. After incubation overnight the oligonucleotide was purified on Illustra NAP 5 column (*vide supra*) and isolated by preparative rp-HPLC (gradient 1, *vide supra*) (see Supplementary Fig. 1).

*Labelling of GlnPQ-SBD2*. SBD2 cysteine-containing derivative was obtained as described in ref. 35 and stored at −20 °C in 100-µl aliquots of 20–40 mg ml$^{-1}$ in 50 mM KPi, pH 7.4, 50 mM KCl and 50% glycerol plus 1 mM dithiothreitol (DTT).

Stochastic labelling with maleimide derivatives of donor and acceptor fluorophores was carried out on ~5 nmol of protein; $SBD_2$ derivatives were labelled with RhodamineB–/NPA–RhodamineB–maleimde (donor) and KK114–maleimide (acceptor) in a ratio of protein:donor:acceptor of 1:6:8. Briefly, purified proteins were treated with DTT (10 mM, 30 min), to fully reduce oxidized cysteines. After dilution of the protein sample to a DTT concentration of 1 mM, the reduced protein was bound to a $Ni^{2+}$-Sepharose resin (GE Healthcare) and washed with ten column volumes of 50 mM KPi, pH 7.4, 150 mM KCl and 5% glycerol (buffer B). Simultaneously, the applied fluorophore stocks (50 nmol (s)) dissolved in 5 µl of anhydrous DMSO were added at appropriate amounts to buffer B and immediately applied to the protein bound to the $Ni^{2+}$-Sepharose resin (keeping the final DMSO concentration below 1%). The resin was incubated overnight and kept at 4 °C (under mild agitation). After labelling, unreacted dye was removed by sequential washing with ten column volumes of buffer B, and this was followed by 100 column volumes of 50 mM KPi, pH 7.4, 1 M KCl and 50% glycerol. The protein was eluted in 1 ml of 50 mM KPi, pH 7.4, 150 mM KCl, 5% glycerol and 500 mM imidazole, and was analyzed on a Superdex-200 column (GE Healthcare) equilibrated with 50 mM KPi, pH 7.4, and 200 mM KCl.

**Microscopy and sample preparation.** *Sample preparation and surface immobilization of oligonucleotides.* Immobilization and investigation of single fluorophores was achieved using a dsDNA scaffold comprising two 40-mer oligonucleotides, that is, ssDNA fluorophore and ssDNA biotin. Sequences of both oligomers were adapted from refs 12,24,25. As a non-stabilized control we used ssDNA fluorophore (Cy5-C6-5′-TAA TAT TCG ATT CCT TAC ACT TAT ATT GCA TAG CTA TAC G-3′; as received from Eurofins and IBA). Single immobilized fluorophore molecules were studied in Lab-Tek eight-well-chambered cover slides (Nunc/VWR, The Netherlands) with a volume of 750 µl as described in ref. 11. After cleaning with 0.1% hydrofluoric acid (HF) and washing with PBS buffer (one PBS tablet was dissolved in deionized water yielding a 10 mM phosphate buffer with 2.7 mM potassium chloride and 137 mM sodium chloride at pH 7.4; Sigma Aldrich, The Netherlands), each chamber was incubated with a mixture of 5 mg per 800 ml BSA and 1 mg per 800 ml BSA/biotin (Sigma Aldrich) at 4 °C in PBS buffer overnight. After rinsing with PBS buffer, a 0.2-mg $ml^{-1}$ solution of streptavidin was incubated for 10 min and subsequently rinsed with PBS buffer.

The immobilization of dsDNA was achieved via a biotin–streptavidin interaction using pre-annealed dsDNA with the aim to observe single emitters for prolonged time periods and to guaranty free rotation of fluorophores. For this, 5–50 µl of a 1 µM solution of ssDNA fluorophore, ssDNA-NPA-fluorophore or ssDNA-TX-PG-Cy5 was mixed with the complementary ssDNA-biotin at the same concentration (Biotin-5′-CGT ATA GCT ATG CAA TAT AAG TGT AAG GAA TCG AAT ATT A-3′, used as received from IBA). The respective mixtures of two oligomers were heated to 98 °C for 4 min and cooled down to 4 °C with a rate of 1 °C $min^{-1}$ in annealing buffer (500 mM sodium chloride, 20 mM TRIS-HCL and 1 mM EDTA at pH 8). The treated LabTek coverslides were incubated with a 50–100 pM solution of pre-annealed dsDNA for 1–2 min, leading to a typical surface coverage of fluorophore-labelled dsDNA as shown in Figs 3 and 4 and Supplementary Figs 3–11.

All single-molecule experiments were carried out at room temperature (22 ± 1 °C). Oxygen was removed from the buffer system using an oxygen-scavenging system (PBS, pH 7.4, containing 10% (wt/vol) glucose and 10% (vol/vol) glycerine, 50 µg $ml^{-1}$ glucose oxidase, 100–200 µg $ml^{-1}$ catalase and 0.1 mM Tris(2-carboxyethyl)phosphine hydrochloride (TCEP)). As shown before, such low concentrations of the reducer TCEP has no noticeable effect on the photophysics of organic fluorophores[25,50] and hence do not convolute with effects from intramolecular stabilization. Glucose oxidase catalase[11,29] was used instead of a combination of protocatechuic acid and protocatechuate-3,4-dioxygenase[29], to avoid convolution of inter- and intramolecular photostabilization with protocatechuic acid[51]. When adding TX to the imaging buffer (PBS with 2 mM TX), the latter was irradiated with ultraviolet light for 30 min before adding 10% (wt/vol) glucose and 10% (vol/vol) glycerine, 50 µg $mL^{-1}$ glucose oxidase, 100–200 µg $mL^{-1}$ catalase and 0.1 mM TCEP.

*Confocal scanning microscopy and data analysis.* A custom-built confocal microscope, described in ref. 18, was used to study fluorescence properties of organic fluorophores on the level of single molecules. Briefly, excitation was achieved with a spectrally filtered laser beam of a pulsed supercontinuum source (SuperK Extreme, NKT Photonics, Denmark) with an acousto-optical tunable filter (AOTFnc-VIS, EQ Photonics, Germany), leading to ≈2 nm broad excitation pulses centred at 640 nm or 532 nm. The spatially filtered beam was coupled into an oil-immersion objective ( × 60, numerical aperture (NA) 1.35, UPLSAPO 60XO mounted on an IX71 microscope body, both from Olympus, Germany) by a dichroic beam splitter (zt532/642rpc, AHF Analysentechnik, Tuebingen, Germany). Surface scanning was performed using a XYZ-piezo stage with 100 × 100 × 20 µm range (P-517-3CD with E-725.3CDA, Physik Instrumente, Germany). Fluorescence was collected by the same objective, focused onto a 50-µm pinhole and detected by two avalanche photodiodes (τ-spad, <50 dark counts per second, Picoquant, Germany) with appropriate spectral filtering (green: HC582/75, red: ET700/75 AHF, both from Analysentechnik). The detector signal was registered using a Hydra Harp 400 ps event timer and a module for time-correlated single-photon counting (both from Picoquant). The data were evaluated using custom-made LabVIEW software[24,25]. Blinking kinetics were extracted from fluorescent time traces in the form of ON and OFF times according to established procedures[24]. Fluorescence lifetimes were determined using time-correlated single-photon counting as described before[18].

*TIRF microscopy including data analysis.* Widefield fluorescence imaging was conducted on an inverted microscope (Olympus IX-71 with UPlanSApo × 100, NA 1.49, Olympus, Germany) in objective type TIRF) configuration as described before[18]. Images were collected with a back-illuminated electron multiplying charge-coupled device camera (512 × 512 pixel, C9100-13 (Hammamatsu, Japan), in combination with either ET585/50 or ET700/75 (AHF Analysentechnik)). Excitation from a diode laser (Sapphire and Cube (Coherent, Germany), filtered either by ET535/70 or ZET640/10 (Chroma, USA)) was at 532 nm and 640 nm with ≈50 W $cm^{-2}$ at the sample location. To quantitatively characterize photostability, we imaged areas with the size of ≈25 × 35 µm containing >100 molecules. A movie was typically recorded for 300–600 s with an integration time of 100 ms. Fluorescent time traces were extracted by identifying pixels, which showed at least 2–3 s.d. above background noise (s.d. of all pixels over all frames of the movie) and summing the intensity in a 3 × 3 pixel area. Neighbouring peaks closer than 5 pixels were not taken into account (see typical examples of fluorophore density and fluorescent time traces in Supplementary Figs 3–11). The number of fluorescent spots in each frame image was determined using an absolute threshold criterion. The number of fluorescent emitters per image were than plotted over time and fitted to a mono-exponential decay $y(t) = C + A \times e^{-c \times t}$ (with $c = 1/\tau_{bleach}$ and $\tau_{bleach}$ being the characteristic bleaching time constant). For some samples with a more complicated behaviour, a double exponential decay of similar form was used and $\tau_{bleach}$ was calculated according to $\tau_{bleach} = A_1 \times \tau_1 + A_2 \times \tau_2$ with amplitude normalization to 1. Bleaching times and associated s.d. were derived from multiple repeats of the same experiment on different days, where each compound was tested in ≥5 movies. The SNR ratio was determined using fluorescent time traces, by dividing the s.d. of the signal before photobleaching with the average fluorescence intensity during that period. The total number of detected photons before photobleaching was calculated by multiplying the obtained count rate and photobleaching lifetime.

*Single-molecule FRET and ALEX spectroscopy.* For data acquisition we used the same home-built confocal microscope as described above. Fluorescence photons arriving at the two detection channels (donor detection channel D or acceptor detection channel A) were assigned to either donor- or acceptor-based excitation, based on their photon arrival time as described previously[33,35,36]. Three relevant photon streams were extracted from the data corresponding to donor-based donor emission *F*(DD), donor-based acceptor emission *F*(DA) and acceptor-based acceptor emission *F*(AA). During diffusion, stoichiometry *S* and apparent FRET efficiencies *E** were calculated for each fluorescent burst above a certain threshold yielding a two-dimensional (2D) histogram. Uncorrected FRET efficiency *E** monitors the proximity between the two fluorophores and is calculated according to $E^* = F(DA)/(F(DD) + F(DA))$. Stoichiometry *S* is defined as the ratio between the overall fluorescence intensity during the green excitation period over the total fluorescence intensity during both green and red periods, and describes the ratio of donor-to-acceptor fluorophores in the sample: $S = (F(DA) + F(DD))/(F(DD) + F(DA) + F(AA))$. Using published procedures to identify bursts corresponding to single molecules, we obtained bursts characterized by three parameters (*M*, *T* and *L*). A fluorescent signal is considered a burst provided it meets the following criteria: a total of *L* photons are collected, having *M* neighbouring photons within a time interval of *T* microseconds. We applied different burst search algorithms on the data sets: the sum of all three detection channels (all photons) using parameters $M = 15$, $T = 500$ µs and $L = 50$ or a dual-colour burst search with $M = 12$, $T = 500$ µs and $L = 30$; additional per-bin thresholding removed spurious changes in fluorescence intensity and selected for intense single-molecule bursts; precise treatment of each data set is given in each figure caption. Binning of the detected bursts into a 2D *E**/*S* histogram yielded subpopulations that can be separated according to their *S*-values. Photon-counting histograms were obtained using similar thresholds as for the 2D *E**/*S* histograms.

ALEX experiments were carried out at room temperature using 25–50 pM of double-labelled protein in imaging buffer: 50 mM KPi, pH 7.4, 150 mM KCl, pH 7.4, containing 10% (wt/vol) glucose and 10% (vol/vol) glycerine, 50 µg $ml^{-1}$ glucose oxidase, 100–200 µg $ml^{-1}$ catalase and 0.1 mM TCEP. For experiments with TX added to the imaging buffer, 2 mM TX, 50 mM KPi, pH 7.4, and 150 mM KCl were irradiated with ultraviolet light for 30 min before adding 10% (wt/vol) glucose and 10% (vol/vol) glycerine, 50 µg $ml^{-1}$ glucose oxidase, 100–200 µg $ml^{-1}$ catalase and 0.1 mM TCEP. Titration experiments were done via adding increasing concentrations of ligand to the imaging buffer (0, 1.0 and 200 µM of glutamine).

*Count rate versus excitation intensity.* The count rate per molecule dependence on the excitation intensity was recorded using 20–100 nM of labelled SBD2 protein in imaging buffer (10 mM KPi, 2.7 mM KCl, 137 mM NaCl, pH 7.4, 10% w/v glucose, 50 µg $ml^{-1}$ glucose oxidase, 100–200 µg $ml^{-1}$ catalase, and 0.1 mM TCEP, both in the presence and the absence of 2 mM aged TX. The count rate per molecule was acquired via excitation from the 543 and 633 nm laser on a LSM 710 confocal laser scanning microscope (Carl Zeiss, Jena, Germany) through a C-apochromat × 401.20 w Korr M27 water-immersion objective (NA = 1.2). Excitation light from both lasers (at different intensities) was coupled into the objective by a dichroic beam splitter (MBS 458/543/633). Fluorescence was collected through the same objective with appropriate spectral filtering (NFT 635 VIS and LP 580).

*STED and confocal microscopy.* STED microscopy and the corresponding confocal microscopy was performed on an Abberior Instruments Resolft microscope (Abberior Instruments, Goettingen, Germany) to which a Ti:Sa STED laser was added (MaiTai, Newport-Spectra Physics; as outlined in detail in ref. 52). Excitation was performed at 640 nm (LDH-D-C-640P laser diode, Picoquant) and STED at 780 nm, at 80 MHz. Pulsing of the excitation laser was triggered by the Ti:Sa laser using a photodiode (APS-100-01), an amplifier (CON-TTL, both from Becker & Hickl, Berlin, Germany) along with a ps-delay unit (either Abberior Instruments or LAS-015617, MPD/Laser2000 UK). A donut-shaped intensity distribution of the STED laser focus was realized by the incorporation of a vortex phase plate (VPP-1a, RPC Photonics, Rochester, NY) into the collimated STED beam. Laser focusing and fluorescence collection was done by an oil-immersion microscope objective (UPlanSapo × 100/1.4 oil, Olympus, Japan). The collected fluorescence was detected by an avalanche photodiode (SPCM-AQRH-13, Excelitas Technologies) with a 650–690 nm bandpass filter (670/40, AHF Analysentechnik).

*Immunolabelling.* Gm5756T human fibroblast were fixed with 3% paraformaldehyde for 10 min, rinsed several times in PBS, permeabilized in 0.1% Triton X-100 for 10 min, rinsed again several times in PBS and blocked in 2% BSA/5% FCS in PBS for 1 h. Samples were incubated with primary mouse antibodies, to stain nuclear pore diluted 1:500 in blocking buffer for 1 h at room temperature. Coverslips were washed five times in 1% BSA in PBS and then incubated for 30 min with goat anti-mouse antibodies (Abberior GmbH, Gottingen, Germany) diluted 1:250 in blocking buffer. Samples were washed five times in 1% BSA in PBS and mounted in the buffer system using an oxygen-scavenging system as used before in the single-molecule experiments.

## References


1. Weiss, S. Fluorescence spectroscopy of single biomolecules. *Science* **283,** 1676–1683 (1999).
2. Ha, T. & Tinnefeld, P. Photophysics of fluorescent probes for single-molecule biophysics and super-resolution imaging. *Annu. Rev. Phys. Chem.* **63,** 595–617 (2012).
3. Hell, S. W. Far-field optical nanoscopy. *Science* **316,** 1153–1158 (2007).
4. Vogelsang, J. *et al.* Make them blink: probes for super-resolution microscopy. *Chemphyschem.* **11,** 2475–2490 (2010).
5. Liphardt, B., Liphardt, B. & Luttke, W. Laser dyes with intramolecular triplet quenching. *Opt. Commun.* **38,** 207–210 (1981).
6. Liphardt, B., Liphardt, B. & Luttke, W. Laser dyes III: concepts to increase the photostability of laser dyes. *Opt. Commun.* **48,** 129–133 (1983).
7. Tinnefeld, P. & Cordes, T. 'Self-healing' dyes: intramolecular stabilization of organic fluorophores. *Nat. Methods* **9,** 426–427 (2012).
8. Blanchard, S. C. Reply to "Self-healing' dyes: intramolecular stabilization of organic fluorophores'. *Nat. Methods* **9,** 427–428 (2012).
9. Dave, R., Terry, D. S., Munro, J. B. & Blanchard, S. C. Mitigating unwanted photophysical processes for improved single-molecule fluorescence imaging. *Biophys. J.* **96,** 2371–2381 (2009).
10. Rasnik, I., McKinney, S. A. & Ha, T. Nonblinking and long-lasting single-molecule fluorescence imaging. *Nat. Methods* **3,** 891–893 (2006).
11. Vogelsang, J. *et al.* A reducing and oxidizing system minimizes photobleaching and blinking of fluorescent dyes. *Angew. Chem. Int. Ed. Engl.* **47,** 5465–5469 (2008).
12. Cordes, T., Vogelsang, J. & Tinnefeld, P. On the mechanism of Trolox as antiblinking and antibleaching reagent. *J. Am. Chem. Soc.* **131,** 5018–5019 (2009).
13. Widengren, J., Chmyrov, A., Eggeling, C., Löfdahl, P.-A. & Seidel, C. a. M. Strategies to improve photostabilities in ultrasensitive fluorescence spectroscopy. *J. Phys. Chem. A* **111,** 429–440 (2007).
14. Lemke, E. A. *et al.* Microfluidic device for single-molecule experiments with enhanced photostability. *J. Am. Chem. Soc.* **131,** 13610–13612 (2009).
15. Cordes, T., Maiser, A., Steinhauer, C., Schermelleh, L. & Tinnefeld, P. Mechanisms and advancement of antifading agents for fluorescence microscopy and single-molecule spectroscopy. *Phys. Chem. Chem. Phys.* **13,** 6699–6709 (2011).
16. Altman, R. B. *et al.* Cyanine fluorophore derivatives with enhanced photostability. *Nat. Methods* **9,** 68–71 (2012).
17. Altman, R. B. *et al.* Enhanced photostability of cyanine fluorophores across the visible spectrum. *Nat. Methods* **9,** 428–429 (2012).
18. Van der Velde, J. H. M. *et al.* Mechanism of intramolecular photostabilization in self-healing cyanine fluorophores. *Chemphyschem.* **14,** 4084–4093 (2013).
19. Van der Velde, J. H. M. *et al.* The power of two: covalent coupling of photostabilizers for fluorescence applications. *J. Phys. Chem. Lett.* **5,** 3792–3798 (2014).
20. Zheng, Q. *et al.* Ultra-stable organic fluorophores for single-molecule research. *Chem. Soc. Rev.* **43,** 1044–1056 (2014).
21. Levitus, M. & Ranjit, S. Cyanine dyes in biophysical research: the photophysics of polymethine fluorescent dyes in biomolecular environments. *Q. Rev. Biophys.* **44,** 123–151 (2011).
22. Juette, M. F. *et al.* The bright future of single-molecule fluorescence imaging. *Curr. Opin. Chem. Biol.* **20,** 103–111 (2014).
23. Liu, C. C. & Schultz, P. G. Adding new chemistries to the genetic code. *Annu. Rev. Biochem.* **79,** 413–444 (2010).
24. Vogelsang, J., Cordes, T., Forthmann, C., Steinhauer, C. & Tinnefeld, P. Controlling the fluorescence of ordinary oxazine dyes for single-molecule switching and superresolution microscopy. *Proc. Natl Acad. Sci. USA* **106,** 8107–8112 (2009).
25. Vogelsang, J., Cordes, T. & Tinnefeld, P. Single-molecule photophysics of oxazines on DNA and its application in a FRET switch. *Photochem. Photobiol. Sci.* **8,** 486–496 (2009).
26. Stein, I. H. *et al.* Linking single-molecule blinking to chromophore structure and redox potentials. *Chemphyschem.* **13,** 931–937 (2012).
27. Steinhauer, C., Forthmann, C., Vogelsang, J. & Tinnefeld, P. Superresolution microscopy on the basis of engineered dark states. *J. Am. Chem. Soc.* **130,** 16840–16841 (2008).
28. Kasper, R., Heilemann, M., Tinnefeld, P. & Sauer, M. in *Biophotonics 2007 Optics in Life Science* 6633_71 (OSA Publishing, 2007).
29. Aitken, C. E., Marshall, R. A. & Puglisi, J. D. An oxygen scavenging system for improvement of dye stability in single-molecule fluorescence experiments. *Biophys. J.* **94,** 1826–1835 (2008).
30. Cordes, T. *et al.* Single-molecule redox blinking of perylene diimide derivatives in water. *J. Am. Chem. Soc.* **132,** 2404–2409 (2010).
31. Alejo, J. L., Blanchard, S. C. & Andersen, O. S. Small-molecule photostabilizing agents are modifiers of lipid bilayer properties. *Biophys. J.* **104,** 2410–2418 (2013).
32. Mickler, M., Hessling, M., Ratzke, C., Buchner, J. & Hugel, T. The large conformational changes of Hsp90 are only weakly coupled to ATP hydrolysis. *Nat. Struct. Mol. Biol.* **16,** 281–286 (2009).
33. Cordes, T. *et al.* Sensing DNA opening in transcription using quenchable Förster resonance energy transfer. *Biochemistry* **49,** 9171–9180 (2010).
34. Hein, J. E. & Fokin, V. V. Copper-catalyzed azide-alkyne cycloaddition (CuAAC) and beyond: new reactivity of copper(I) acetylides. *Chem. Soc. Rev.* **39,** 1302–1315 (2010).
35. Gouridis, G. *et al.* Conformational dynamics in substrate-binding domains influences transport in the ABC importer GlnPQ. *Nat. Struct. Mol. Biol.* **22,** 57–64 (2014).
36. Kapanidis, A. N. *et al.* Fluorescence-aided molecule sorting: analysis of structure and interactions by alternating-laser excitation of single molecules. *Proc. Natl Acad. Sci. USA* **101,** 8936–8941 (2004).
37. Kong, X., Nir, E., Hamadani, K. & Weiss, S. Photobleaching pathways in single-molecule FRET experiments. *J. Am. Chem. Soc.* **129,** 4643–4654 (2007).
38. Fulyani, F. *et al.* Functional diversity of tandem substrate-binding domains in ABC transporters from pathogenic bacteria. *Structure* **21,** 1879–1888 (2013).
39. Kolmakov, K. *et al.* Red-emitting rhodamine dyes for fluorescence microscopy and nanoscopy. *Chemistry* **16,** 158–166 (2010).
40. Hell, S. W. & Wichmann, J. Breaking the diffraction resolution limit by stimulated emission: stimulated-emission-depletion fluorescence microscopy. *Opt. Lett.* **19,** 780 (1994).
41. Kasper, R. *et al.* Single-molecule STED microscopy with photostable organic fluorophores. *Small* **6,** 1379–1384 (2010).
42. Marmé, N., Knemeyer, J.-P., Sauer, M. & Wolfrum, J. Inter- and intramolecular fluorescence quenching of organic dyes by tryptophan. *Bioconjug. Chem.* **14,** 1133–1139 (2003).
43. Doose, S., Neuweiler, H. & Sauer, M. A close look at fluorescence quenching of organic dyes by tryptophan. *Chemphyschem.* **6,** 2277–2285 (2005).
44. Doose, S., Neuweiler, H. & Sauer, M. Fluorescence quenching by photoinduced electron transfer: a reporter for conformational dynamics of macromolecules. *Chemphyschem.* **10,** 1389–1398 (2009).
45. Chen, H., Ahsan, S. S., Santiago-Berrios, M. B., Abruña, H. D. & Webb, W. W. Mechanisms of quenching of Alexa fluorophores by natural amino acids. *J. Am. Chem. Soc.* **132,** 7244–7245 (2010).
46. Schwering, M. *et al.* Far-field nanoscopy with reversible chemical reactions. *Angew. Chem. Int. Ed. Engl.* **50,** 2940–2945 (2011).
47. Nanguneri, S., Flottmann, B., Herrmannsdörfer, F., Thomas, K. & Heilemann, M. Single-molecule super-resolution imaging by tryptophan-quenching-induced photoswitching of phalloidin-fluorophore conjugates. *Microsc. Res. Tech.* **00,** 1–7 (2014).
48. Yang, S. K., Shi, X., Park, S., Ha, T. & Zimmerman, S. C. A dendritic single-molecule fluorescent probe that is monovalent, photostable and minimally blinking. *Nat. Chem.* **5,** 692–697 doi:10.1038/nchem.1706 (2013)).
49. Koufaki, M. *et al.* Synthesis of chroman analogues of lipoic acid and evaluation of their activity against reperfusion arrhythmias. *Bioorg. Med. Chem.* **12,** 4835–4841 (2004).
50. Vaughan, J. C., Dempsey, G. T., Sun, E. & Zhuang, X. Phosphine quenching of cyanine dyes as a versatile tool for fluorescence microscopy. *J. Am. Chem. Soc.* **135,** 1197–1200 (2013).

51. Le Gall, A. *et al.* Improved photon yield from a green dye with a reducing and oxidizing system. *Chemphyschem.* **12,** 1657–1660 (2011).
52. Clausen, M. P. *et al.* Pathways to optical STED microscopy. *NanoBioImaging* **1,** 1–12 (2013).

## Acknowledgements

This work was financed by the Zernike Institute for Advanced Materials, the Centre for Synthetic Biology (University of Groningen, start-up grant to T.C.). J.H.M.v.d.V. acknowledges Ubbo-Emmius funding (University of Groningen). J.O. and G.R. thank the NRSC-C and the European Commission (ERC Starting Grant Number 280010) for financial support. A.H. thanks the European Commission (ERC Starting Grant and STREP project MICREAGENTS) and the Netherlands Organization for Scientific Research (NWO-Vici) for funding. C.E. and S.G. acknowledge funding by the MRC, BBSRC and ESPRC. G.G. acknowledges support by an EMBO (long-term fellowship ALF 47-2012) and is currently supported by a NWO-VENI grant (grant number 722.012.012). We thank L.J. Ugen, E.M. Warszawik and E. Ploetz for discussions. We thank S.W. Hell and V. Belov for the generous gift of KK114 carboxylic acid and C. Wurm for help with labelling antibodies. Finally, we are grateful to A.M. van Oijen for enduring and generous support, friendly advice and thoughtful comments on the manuscript.

## Author contributions

T.C. conceived the study. J.H.M.v.d.V., J.O., A.H., G.R. and T.C. designed experiments and synthesis routes. J.H.M.v.d.V., J.H. and A.A.J. performed single-molecule imaging and ALEX experiments. J.O., J.H., J.H.S., K.K., A.H. and G.R. provided new reagents including chemical characterization. S.G. and C.E. performed STED microscopy. G.G. provided proteins for FRET studies. J.H.M.v.d.V. and T.C. analysed data and all authors contributed to writing the paper.

## Additional information

**Competing financial interests:** The authors declare no competing financial interests.

# 10

# Design concept for α-hydrogen-substituted nitroxides

Michal Amar[1,*], Sukanta Bar[1,*], Mark A. Iron[2], Hila Toledo[1], Boris Tumanskii[1], Linda J.W. Shimon[2], Mark Botoshansky[1], Natalia Fridman[1] & Alex M. Szpilman[1]

Stable nitroxides (nitroxyl radicals) have many essential and unique applications in chemistry, biology and medicine. However, the factors influencing their stability are still under investigation, and this hinders the design and development of new nitroxides. Nitroxides with tertiary alkyl groups are generally stable but obviously highly encumbered. In contrast, α-hydrogen-substituted nitroxides are generally inherently unstable and rapidly decompose. Herein, a novel, concept for the design of stable cyclic α-hydrogen nitroxides is described, and a proof-of-concept in the form of the facile synthesis and characterization of two diverse series of stable α-hydrogen nitroxides is presented. The stability of these unique α-hydrogen nitroxides is attributed to a combination of steric and stereoelectronic effects by which disproportionation is kinetically precluded. These stabilizing effects are achieved by the use of a nitroxide co-planar substituent in the γ-position of the backbone of the nitroxide. This premise is supported by a computational study, which provides insight into the disproportionation pathways of α-hydrogen nitroxides.

[1] Schulich Faculty of Chemistry, Technion-Israel Institute of Technology, 32000 Haifa, Israel. [2] Department of Chemical Research Support, Weizmann Institute of Science, 76100 Rehovot, Israel. * These authors contributed equally to this work. Correspondence and requests for materials should be addressed to A.M.S. (email: szpilman@tx.technion.ac.il).

Persistent nitroxides have intriguing structural and chemical properties[1–3] and possess a multitude of highly important applications, such as oxidation catalysts[4–7], reagents in living polymerization reactions[8], spin probes in biology[9], radical traps in mechanistic investigations, spin trapping agents, medicinal agents[10,11] and in materials chemistry[12]. Designing and preparing stable nitroxides, as well as predicting their properties, remains an important challenge that must be met through a combination of computational methods and creativity[12]. Most research has been conducted on the archetypical nitroxide TEMPO ((2,2,6,6-tetramethyl piperidin-1-yl)oxyl, Fig. 1) and structurally related compounds. Given these factors, the development of new classes of nitroxides is an enticing and a superbly challenging goal[13].

TEMPO (**3**) is a shelf-stable compound and is the catalyst of choice in industry[14] as well as in academia for the oxidation of primary alcohols to aldehydes[1–7]. It has many applications in evaluating radical formation in biological tissues[15,16] and is also the trapping agent of choice in mechanistic investigations of chemical processes when a radical intermediate is suspected[17,18]. Conjugates of TEMPO and other nitroxides are also being evaluated for various medicinal uses[10,11]. Despite its popularity, it suffers from several shortcomings. For example, it often fails to catalyse the oxidation of secondary or hindered alcohols to carbonyl compounds and is a poor catalyst for polymerization of non-styrenic monomers. This is, at least in part, due to the four methyl groups adjacent to the reactive locus: the nitroxyl moiety. Since most known cyclic nitroxides conserve the four methyl groups flanking the radical function, the preparation of cyclic nitroxides with other substitution patterns would be highly desirable. In addition, chiral nitroxides would be desirable both for catalytic as well as biological applications.

One avenue of investigation for developing more active and tuneable systems would be to reduce the steric hindrance by replacing one or more of the methyl groups with a hydrogen atom. However, α-hydrogen-substituted nitroxyl compounds are notoriously and inherently unstable, and undergo fast disproportionation in which two radicals react to give one molecule of nitrone and one of hydroxylamine (Fig. 2)[19]. The magnitude of this challenge was succinctly described in a 2009 review: '…nitroxides are prepared by the oxidation of secondary amines that contain no hydrogen atoms on the α-carbons… If the amines carry α-hydrogens, the oxidation products are nitrones, not free-radical nitroxides…'[1].

Given their instability, which imposes severe structural design limitations, it is not surprising that examples of α-hydrogen nitroxides are extremely rare. This is especially evident when compared to the development of, for example, amine and phosphine ligands for transition metal catalysis of which many thousands have been developed[20]. In fact, we are aware of only two successful designs of α-hydrogen nitroxides that have been reported before this work.

More than 40 years ago, Rassat and Dupeyere[21] described the first example of bridged bicyclic α-hydrogen nitroxyls, namely **4** and AZADO (**5**; Fig. 1). These compounds proved to be exceptionally stable. Their stability is believed to be derived from the inability to form a nitrone C = N double bond to the bridgehead—an elegant application of Bredt's rule[22]. AZADO (**5**) has recently been shown to be a powerful oxidation catalyst and is now commercially available because of the efforts of Iwabuchi and co-workers[23–25], who also reported a number of other nitroxides inspired by the same design concept. Unfortunately, the bridged bicyclic design is not universally applicable. For example, compound **6** reported by Rychnovsky and co-workers[26] decomposes rapidly, possibly via C–N bond scission.

The only other reported structural family are a group of acyclic α-hydrogen nitroxides first described in 1996 (refs 27,28) typified here by the polymerization catalysts TIPNO (**7**) and SG1 (**8**)[29,30]. Many examples conforming to this design have been prepared and successfully applied in living polymerization reactions[8]; however, we are unaware of any examples of them being used in oxidation chemistry.

These examples, bicyclic and acylic, constitute compelling examples of chemical knowledge-based design overcoming the natural tendency of α-hydrogen nitroxides to decompose. It would be desirable to develop a design that would allow the preparation of monocyclic α-hydrogen nitroxides, both from the

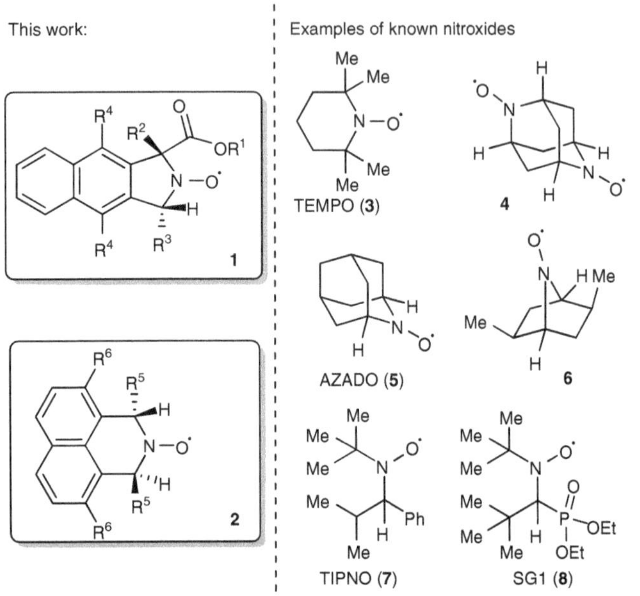


**Figure 1 | Structures of the α-hydrogen nitroxides.** Novel nitroxide families **1** and **2** introduced in this work, TEMPO (**3**), examples of bridged bicyclic α-hydrogen nitroxides (**4**), AZADO (**5**) and (**6**), and acyclic nitroxides TIPNO (**7**) and SG1 (**8**).

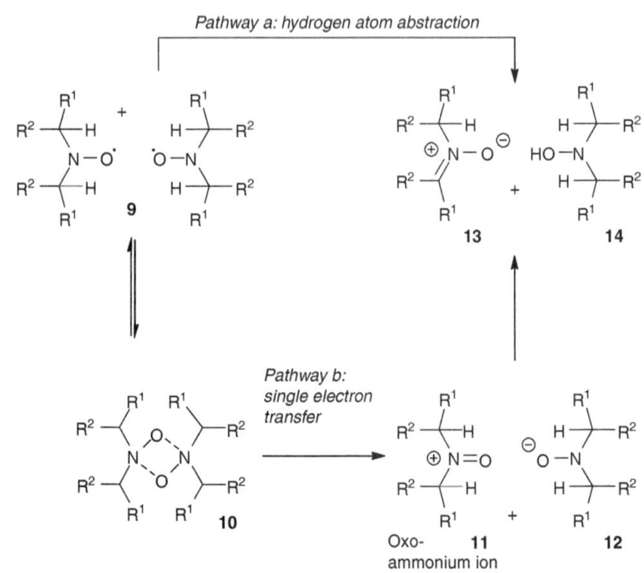


**Figure 2 | Disproportionation mechanism for α-hydrogen nitroxides.** Proposed disproportionation mechanisms for nitroxides. (*Pathway a*) direct hydrogen atom abstraction[31]. (*Pathway b*) SET via the head-to-tail dimer **10** to give ion pair **11/12** followed by proton transfer[19].

basic scientific as well as a practical point of view. One stumbling block to designing new α-hydrogen nitroxides is that there are a few proposed decomposition pathways and there is no consensus as to which one is operative. Ingold and co-workers[31] first proposed that α-hydrogen nitroxyls decompose by hydrogen abstraction (that is, disproportionation) to give the isolable nitrones **13** and hydroxylamine **14** (Fig. 2, pathway a). An alternative proposal by Braslau and co-workers[19] is that an initial single electron transfer (SET) step takes place to afford an oxoammonium ion **11** and a hydroxylamine anion **12** (Fig. 2, pathway b). The SET may involve the initial formation of a head-to-tail nitroxide dimer **10**. Subsequent proton transfer, from **11** to **12**, would lead to the observed products (Fig. 2)[19]. A third possibility[31] involving a rearrangement of **10** was ruled out by Braslau and co-workers[19] based, *inter alia*, on thermodynamic and stereochemical considerations. It is not clear which of these mechanisms is operative, and the decomposition mechanism may depend on the structure of the nitroxide in question (see detailed discussion below).

It should be noted that, while oxoammonium cations such as **11** are important species inferred in many catalytic oxidative processes, only a few oxoammonium salts, mainly of TEMPO derivatives, have been isolated as discrete stable materials[1–3]. For fully alkyl-substituted radicals such as **3**, it has been shown that decomposition of the oxoammonium salt occurs predominantly via β-deprotonation by a base[1,32]. For **3**, the equilibrium between two nitroxides and the oxoammonium ion and hydroxylamine can also be catalysed by acid[1,32]. Other fully α-substituted nitroxides with highly sterically demanding flanking groups may also undergo C–N bond scission[19].

Herein, a novel design concept based on a combination of stereoelectronic and steric guidelines is presented. A proof-of-concept is provided; it is shown that the design is applicable to two different systems (**1** and **2**, Fig. 1), and two different examples for each class are given. To the best of our knowledge, this paper constitutes the first example of a stable α-hydrogen piperidin-1-yl-oxy and pyrrolidin-1-yl-oxy nitroxides that are not bicyclic. A computational study that underpins the design principles outlined herein, which may be of broad general interest for the understanding of nitroxides, is also reported.

## Results

**Design concept**. A nitroxyl design along the lines of **15** (Fig. 3) was envisioned that would achieve stability by incorporating a stabilizing substituent $R^5$ on a planar rigid (aromatic) radical backbone. This would lead to an increased activation barrier for disproportionation through a combination of three synergistic effects: (i) The $R^5$ substituent would push the $R^4$ substituent out of the backbone/N–O plane. This would place the $R^4$ substituent in **15** right in the path of approach of a second molecule of **15**, sterically inhibiting hydrogen atom abstraction. (ii) Forcing the $R^4$ substituent out of the backbone/N–O plane would lead to compression of the H–C–N–O angle. Angle compression below the optimal 90° angle for hydrogen atom abstraction would increase the energy barrier for hydrogen atom abstraction and nitrone formation. This is due to poor overlap between the C–H σ-bond being broken and the singly-occupied molecular orbital localized on the N–O bond that will be part of the nitrone double bond being formed[33]; similar considerations have been put forth to explain the stability of the acyclic radicals of the TIPNO (**7**)/SG1 (**8**) family based on conformational stabilities[27,34,35]. (iii) During nitrone formation the $R^4$ substituent would move into the plane of the $R^5$ substituent. This would lead to incipient allylic strain, which would further increase the kinetic barrier to disproportionation irrespective of the mechanism. This conjecture was inspired by the observation that naphthalenes substituted in the 1 and 8 positions with sterically hindered alkyl groups are deformed out of plane, with concomitant loss of aromaticity, due to allylic strain[36–39].

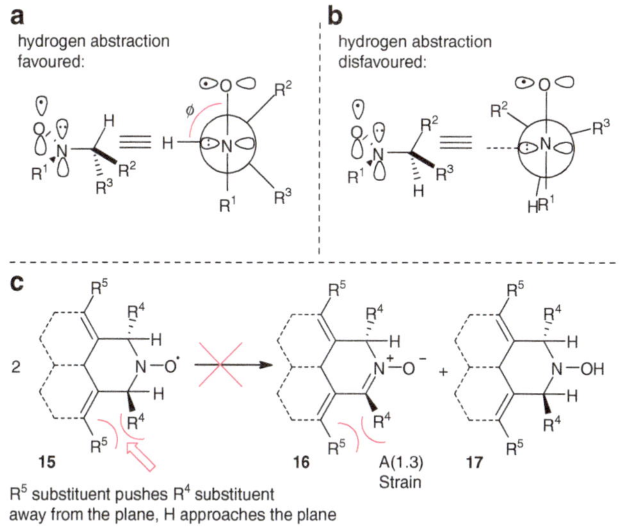


**Figure 3 | The design principles used in the design of α-hydrogen nitroxides 1 and 2.** (**a**,**b**) Avoiding the optimal 90° dihedral angle for hydrogen atom abstraction (as in **b** versus **a**). (**c**) Impedement of the disproportionation of **15** to nitrone **16** and hydroxylamine **17** through predominance of a conformer in which the $R^4$ of two opposing radicals would sterically repulse each other and in which incipient A(1,3) strain would kinetically disfavour nitrone formation. These three kinetically stabilizing effects are proposed to be enhanced by placing a sterically hindered $R^5$ substituent on the backbone of structure **15** that will dispose the $R^4$ group out of the naphthalene/N-O plane, thus forcing the α-hydrogen atom closer to the plane of the N-O. See text for discussion.

By combining these three diverse and mutually enforcing elements into our design, it was hoped that sufficient kinetic stability of the resulting nitroxides would be achieved to make them stable at ambient temperature. These considerations led to the proposed isoindole structure **1** with a single α-hydrogen and *iso*-azaphenalene **2** with two α,α′-hydrogen atoms (Fig. 1). It is important to note that simple α-hydrogen nitroxides that do not conform to this design are inherently unstable; for instance, oxidation of piperidine N-oxides with single substituents on C(2) or C(4) results in nitrones[40–43].

In order for any nitroxide design to be attractive from a practical point of view, it is essential that the nitroxide be prepared in high yield and in few synthetic steps. Flexible syntheses of both target structures **1** and **2** have been developed. These syntheses potentially afford access to a wide range of compounds of structures with different substituents (Figs 4 and 5). The syntheses described herein are amenable to gram scale assembly of the nitroxide scaffolds and are highly stereoselective.

**Preparation of stable α-hydrogen nitroxyl radicals**. The challenge of synthesizing isoindole structure **1** bearing a single α-hydrogen atom was addressed first (Fig. 4)[44]. The entire scaffold was assembled in a single synthetic step through the [3 + 2] cycloaddition of commercially available naphthoquinone **18** with imines **19** (R = mesityl (mes) or naphthyl (naph))[45,46]. It was found that the initial bisnaphthol product **21a,b** is rapidly oxidized to the quinone **22a,b** when exposed to air during workup. Quinones **22a** and **22b** were isolated in 28% and 63%

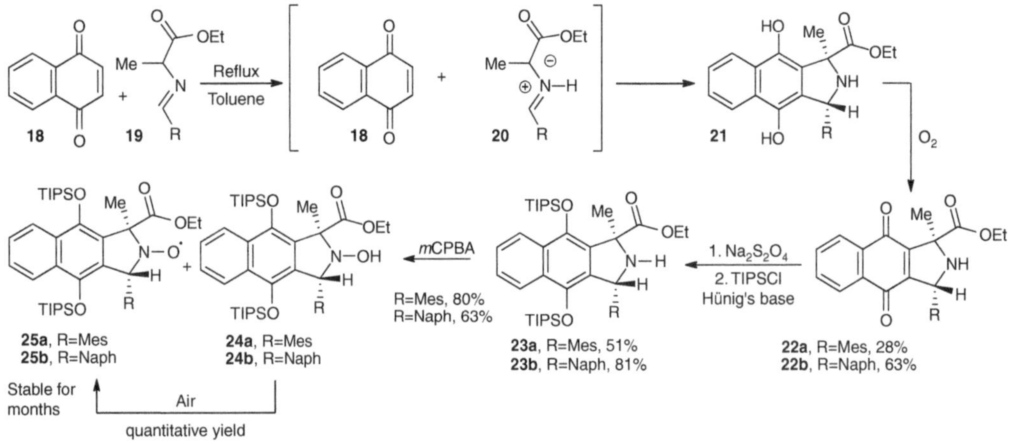


**Figure 4 | Synthesis of isoindole nitroxides 25a and 25b.**

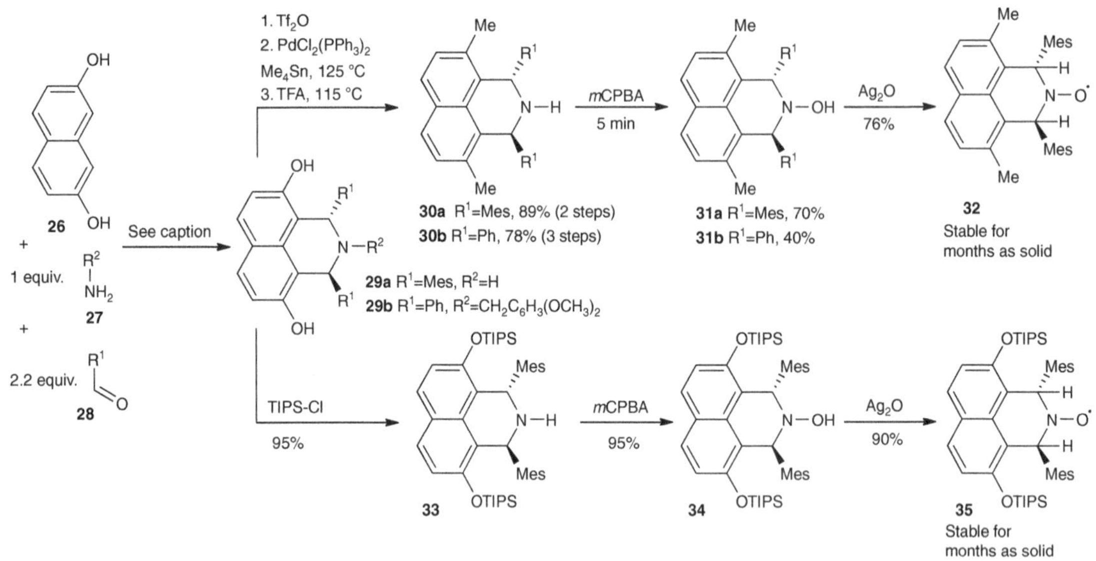


**Figure 5 | Synthesis of *iso*-azaphenylene nitroxides 32 and 35.** Conditions for preparing **29a**, $R^1 = Mes$, $R^2 = H$: mesityl aldehyde (2.5 equiv.), $H_2N\text{-}R^2 = NH_4OAc$ (1.2 equiv.), EtOH, reflux 2 days, 70% yield, >17:1 d.r. Conditions for preparing **29b**, $R^1 = Ph$, $R^2 =$ 2,4-dimethoxybenzylamine: 1.1 equiv. 2,4-dimethoxybenzylamine (1.5 equiv.), 6 equiv. benzaldehyde, 2 days, and then toluene sulphonic acid (1 equiv.), dichloromethane, 1 day, 95% yield, >17:1 d.r.

yield, respectively, on a gram scale. It was hypothesized that appending TIPS (tri-*iso*-propylsilyl) groups to the hydroquinone oxygen atoms, affording OTIPS group, would provide sufficient bulk for stability. To this end, reduction of the quinones with $Na_2S_2O_4$ followed by treatment of the unpurified hydroquinones with TIPSCl and Hünig's base led to the stable silylated products **23a** and **23b** in 51% and 81% yields, respectively. Oxidation using *meta*-chloroperbenzoic acid (*m*CPBA) afforded the hydroxylamines **24a** and **24b** as mixtures with the corresponding nitroxides **25a** and **25b** respectively. The relative configuration of **24a** was determined by X-ray crystallography (see Supplementary Fig. 1). Further oxidation of the hydroxylamine/nitroxide mixtures to the pure radicals using silver oxide or other oxidizing reagents led to decomposition only. However, simply exposing the mixtures to air led to quantitative formation of the desired nitroxides **25a** and **25b**. These radicals were characterized by electron spin resonance (ESR; Supplementary Fig. 2a,b), infrared and high-resolution mass spectrometry (HR-MS). Both radicals are highly stable, remaining unchanged in benzene solution for 2 months without any reduction in ESR signal amplitudes, nor any sign of decomposition products. The radicals can be isolated as solids by simple evaporation of the solvent and kept as solids for at least 2 months. *This constitutes a proof-of-principle of our design concept.* The ESR spectra have the expected splitting of the N–O system ($a_N = 15.2$ G for **25a** and $a_N = 13.9$ G for **25b**, Supplementary Fig. 2a,b) with additional splitting by the adjacent C–H group ($a_H = 21.8$ G for **25a** and $a_H = 16.3$ G for **25b**), consistent with simulated spectra. In contrast to the marked stability of **25a** and **25b**, oxidation of **22a** with *m*CPBA leads to the formation of a nitroxide that decomposes within minutes (see Supplementary Fig. 17 for the ESR of this radical). Thus, the electronic nature of the naphthyl and the steric requirements of the naphthyl $R^4$ substituent in **1** (Fig. 1) are important for stability.

In order to test the generality of the design concept, the study was extended to a second design based on the same principles, but imposing an even greater challenge by having two α-hydrogen atoms—one on either side of the nitroxyl function. This led to the *iso*-azaphenalene family **2** (Fig. 1)[47]. This attractive structural backbone can be efficiently assembled in a single synthetic step (Fig. 5). Betti-type reaction of 2,7-naphthalene diol (**26**) with ammonium acetate or 2,4-dimethoxybenzylamine and variety of aryl aldehydes **28** afforded condensates **29** in excellent yields and with high stereoselectivities (>17:1)[48]. The aldehyde may be benzaldehyde or derivatives bearing electron-withdrawing or -donating substituents (see Supplementary Methods). Subsequent triflation of **29a** ($R^1$ = mesityl) proceeded in 94% yield. Stille coupling of the triflate with tetramethyltin gave **30a** in 95% yield. Phenyl-substituted **29b** ($R^1$ = phenyl) was converted into the triflate in 95% yield and methylated in 95% yield. Removal of the 2,4-dimethoxybenzyl protecting group using trifluoroacetic acid (TFA) proceeded in 86% yield to give the free amine **30b**.

Oxidation of amines **30ab** to hydroxylamines **31ab** was carried out using *m*CPBA. In both cases, the reaction was complete within minutes, yet the stability of the products was remarkably dissimilar. The less sterically congested hydroxylamine **31b** was isolated in 40% yield and decomposed after purification to give the nitrone. In contrast, the mesityl derivative **31a** was isolated in 70% yield after flash chromatography and proved to be completely stable for more than 9 months as a solid. Both hydroxylamines contained minor amounts of the corresponding nitroxide, observed by ESR. The stable mesityl derivative **31a** was characterized by X-ray crystallography (Supplementary Fig. 1) revealing the *trans* relationship of the two $R^1$ substituents.

Treatment of the hydroxylamine **31b** with silver oxide did not result in the isolation of the corresponding radical. Presumably, the transient radical immediately disproportionated. In stark contrast, oxidation of **31a** led to the formation of stable nitroxide **32**. This radical was isolated in 76% yield after purification by standard flash chromatography with no special precautions and was characterized by HR-MS and ESR. The ESR spectrum (Supplementary Fig. 2c) corroborates the structure with splitting of the N–O system ($a_N$ = 14.5 G) with additional splitting by the two adjacent C–H groups ($a_{2H}$ = 17.5 G), consistent with a simulated spectrum. In a benzene solution exposed to air, it has a half-life of 3 days. However, ~2% of the radical still remained in solution (as detected by ESR) after 2 months. In the solid state, the yellow radical is stable for at least 9 months at −15 °C exposed to air without any detectable degradation. The compound possesses slight solvatochromism, appearing orange in some solutions, similar to TEMPO (**3**).

A TIPS substituted *iso*-azaphenalene radical **35** was also prepared in this study. Silylation of **29a** proceeded smoothly to afford **33** (Fig. 5). Oxidation with *m*CPBA led to the stable hydroxylamine **34** in 95% yield. Oxidation of **34** with silver oxide gave the free nitroxide **35**. The nitroxide was characterized by ESR (Supplementary Fig. 2d), infrared and HR-MS. X-ray diffraction studies of single crystals of **34** showed that the unit cell contained two different structures differing only by the conformation of the TIPS groups (Supplementary Fig. 1c). Interestingly, this is also reflected in the ESR spectrum of **35** (Supplementary Fig. 2d) that clearly shows two radicals, indicating that interconversion is slow on the ESR timescale. The overlapping spectra were deconvoluted using simulated ESR spectra (Supplementary Fig. 16). Partial coalescence of the ESR spectra could be observed by heating the radical to 80 °C in toluene. On cooling to room temperature, the signals again separated, and resolved further by cooling to −20 °C (Supplementary Fig. 16). Nitroxide **35** is stable for at least 9 months as a solid.

The applicability and generality of the design concept have thus been demonstrated by the preparation of two different families of nitroxides (**1** and **2**) and two stable examples of each family (**25a**, **25b**, **32** and **35**). We have shown that OTIPS and methyl groups on the radical backbone stabilize the isoindoline and *iso*-azaphenalene nitroxides. In contrast, the nitroxide derived by oxidation of quinone **22b** decomposes rapidly as do other piperidine hydroxylamines upon oxidation[40–43]. The design principles outlined in the introduction were then evaluated, taking advantage of the power of computational studies of the nitroxides already demonstrated to be stable[12,49,50].

**Computational study into the stability of the nitroxides.** The principal decomposition pathway of α-hydrogen nitroxides is disproportionation to give one molecule of nitrone **13** and one of hydroxylamine **14** as shown in Fig. 2. Three different mechanisms of disproportionation have been proposed, namely: (*i*) direct hydrogen atom abstraction by one radical from the other to give the products in a single step (Fig. 2, pathway a)[31], (*ii*) head-to-tail dimer formation (**10**), SET to give one molecule of oxoammonium cation **11** and one of hydroxylamine anion **12** and subsequent proton transfer to give the products (Fig. 2, pathway b)[19] and (*iii*) formation of a head-to-head (O–O) dimer **10** followed by a concerted rearrangement in a second step to give **13** and **14** (ref. 31). The third mechanism has been rejected based on thermodynamical and stereochemical arguments[19]. However, the other two mechanisms are still under consideration. Indeed, it is possible that different nitroxides may decompose by different mechanisms depending, *inter alia*, on substituent effects, much like the mechanism of nucleophilic substitution changes from $S_N2$ to $S_N1$ on going from methyl iodide to *tert*-butyl iodide.

All density functional theory (DFT) calculations were performed at the SMD(acetonitrile)-RIJCOSX-DSD-PBEB95/def2-TZVP//DF-$PBE_{D3BJ}$/def2-SVP level of theory (Supplementary Methods, Computational Methods section). On performing the calculations, certain assumptions were made in order to evaluate computationally pliable systems at as high a level of theory as practically possible. The TIPS groups, while hydrolytically stable, result in a large number of degrees of freedom, as noted in the X-ray crystallography of **34** and ESR spectrum of **35** (*vide supra*), making calculations very demanding. Consequently, it was assumed that calculations on models in which the TIPS groups in radicals **25a** and **35** are replaced by the simpler trimethylsilyl (TMS) groups, as in **36** and **37** (Fig. 6), would only have a minor impact on calculated energies and properties but would significantly reduce computational time. The validity of this assumption can be seen in that the overall driving force for the disproportionation of **35** and **37** are within 1 kcal $mol^{-1}$ of each other (Table 1, column 2). For comparison, the disproportionation of the hypothetical *iso*-azaphenalene **38** (Fig. 6) of family **2** was also studied. Compound **38** should be unstable based on the experimental observation that the phenyl-substituted **31b** spontaneously decomposes. Compound **38** would have a much smaller tendency for incipient allylic strain when transitioning into the corresponding nitrone **13/16** (Fig. 3), less steric hindrance and higher conformational flexibility, thus being able to achieve the orbital overlap required for hydrogen abstraction (Fig. 2, pathway a and Fig. 3a). In contrast, it was not expected to have a dramatically different ability to undergo SET (Fig. 2, pathway b) compared with the other radicals studied.

Importantly, in all cases, the disproportionation is energetically favourable (Table 1, column 2). This shows clearly that the stability of the nitroxides is due to kinetic factors. In contrast, the SET from the radicals to give the oxoammonium **11** and hydroxylamine anion **12** (Fig. 2, pathway b) are all strongly endergonic (Table 1, column 3). Therefore, it is reasonable to rule out the SET pathway, leaving the hydrogen-atom-transfer mechanism (Fig. 2, pathway a) as the most likely mechanism. Further support may be obtained from the plots of the Mulliken

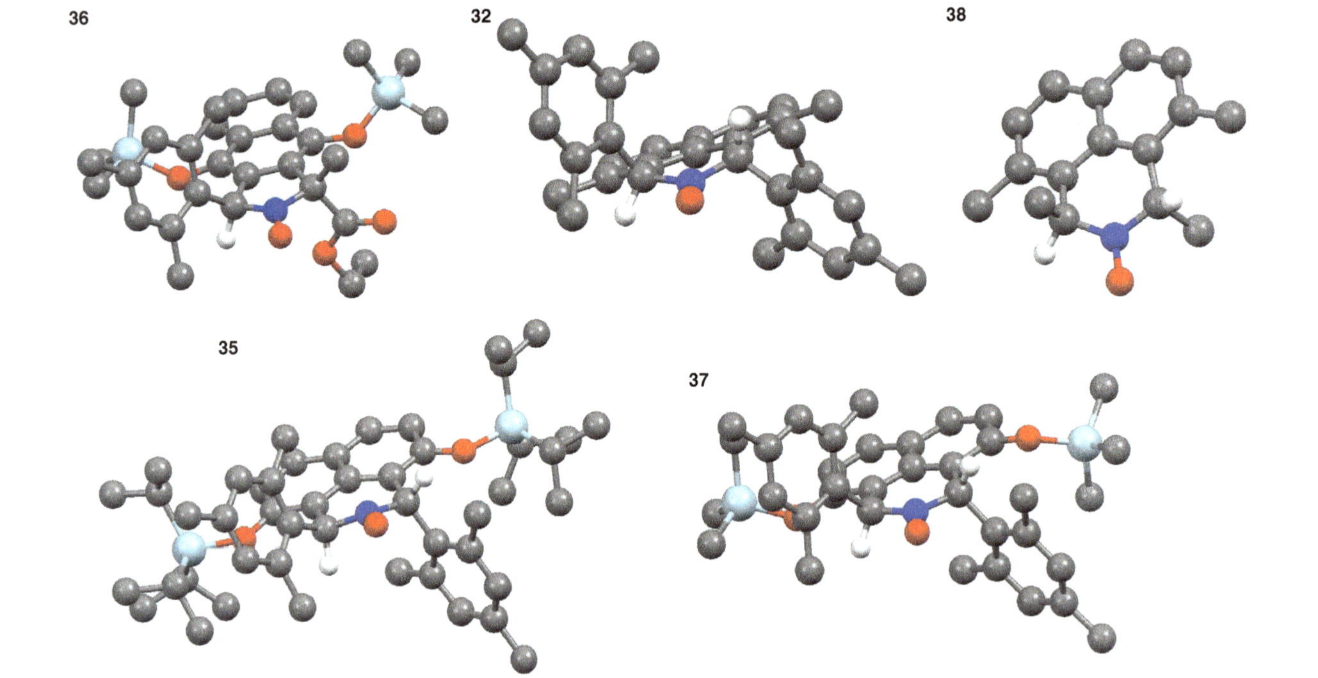


**Figure 6 | DFT optimized structures of the nitroxides 32 and 35–38.** Hydrogen atoms, other than the α-hydrogens, have been omitted for clarity. (Colour code: C—grey, H—white, N—blue, O—red, Si—light blue).

**Table 1 | Calculated energies for radicals 32, 35–38*.**

| Radical | Disproportionation | SET to give 11/12 | Radical pair precomplex | $\Delta G^{\ddagger}_{298}$ |
|---|---|---|---|---|
| **36** | −20.8 | 76.6 | 27.2 | 38.4 |
| **32** | −25.4 | 72.3 | 13.0 | 44.5 |
| **38** | −12.5 | 69.6 | −17.2† | 6.6 |
| **35** | −24.3 | 73.3 | —‡ | —‡ |
| **37** | −25.0 | 71.4 | —‡ | —‡ |

SET, single electron transfer.
*Reaction free energies ($\Delta G_{298,sol}$, kcal mol$^{-1}$) for the disproportionation of two nitroxides to one nitrone and one hydroxylamine, SET to give **11** and **12**, formation of the radical pair precomplex and the height of the transition state barrier for hydrogen atom abstraction via pathway a ($\Delta G^{\ddagger}_{298}$).
†Attempts at optimization of the radical dimer led directly to the nitrone/hydroxylamine hydrogen-bonded dimer instead.
‡Not calculated.

charges (Supplementary Figs 3–5) on the fragments during the hydrogen-transfer processes (*vide infra*). The charge on the hydrogen atom being transferred remains fairly constant at ~0.2e$^{-}$, while there is not any significant charge accumulation on either fragment.

The direct hydrogen abstraction mechanism (Fig. 2, pathway a) is associated with a large reaction barrier for the stable radicals (Table 1, column 5). The structures of the **32** and **36** radical pair precomplexes, where the oxygen atom of one radical is pointing towards the α-hydrogen of the second, were first optimized (see Table 1, column 4 for energies of formation). The hydrogen-transfer transition states could not be found for these nitroxides, likely because of the flatness of the potential energy surface at the level of theory used for geometry optimizations and because of the strong exergonicity of the disproportionation reaction. Therefore, relaxed scans along the hydrogen abstraction reaction coordinate were performed, fitting the points around the maximum to a third-order polynomial, giving barriers, approximating the transition states, of $\Delta G^{\ddagger}_{298} \approx 44.5$ and 38.4 kcal mol$^{-1}$, respectively (Fig. 7 and Table 1, Column 5); these values account for the endergonic formation of the initial radical dimers (Table 1, column 4). Note that the value of 38.4 kcal mol$^{-1}$ is for the truncated radical **36** where the TIPS of **25a** (Fig. 4) is replaced by the smaller TMS (Fig. 6). One would expect that inclusion of TIPS instead of TMS would substantially increase the barrier. In contrast, the related relaxed scan for hypothetical radical **38** revealed a barrier of only $\Delta G^{\ddagger}_{298} \approx 6.6$ kcal mol$^{-1}$, indicating that this reaction should proceed very rapidly; in fact, optimization of the radical pair precomplex did not lead to the analogous radical pair but rather hydrogen-bonded nitrone and hydroxylamine products.

In order to verify the electronic state of the radical dimer pair, NEVPT2-CASSCF(8,8) calculations were performed for the radical pair precomplex of **38** for the structure from the relaxed scan where the C–H bond is closest to a typical C–H bond. It is reasonable to assume that the other radicals will have similar behaviour. Unfortunately, the other nitroxides are far too large to be amenable to a CASSCF treatment. These calculations show that the electronic ground state has a closed-shell singlet and the dominant configuration has all electrons paired with a weight of 84% and only small contributions from other configurations. The triplet state is 3.538 eV (3.293 eV for CASSCF) higher in energy. Thus, it is reasonable to treat the radical dimers as closed-shell singlets. If one examines the electronic structure of the precomplex, one notes that there is a hydrogen bond between the oxygen of one nitroxide and the hydrogen being abstracted on the other; the Mayer bond index between these two atoms is 0.33. In forming the hydrogen bond, 0.58 e$^{-}$ of charge (Mulliken partial atomic charges) is transferred from one nitroxide to the other. Nevertheless, this would be consistent with the formation of the hydrogen bond as an integral first step in hydrogen abstraction rather than SET.

A major contribution to the large barriers is unfavourable steric interactions between the two radicals in the radical pairs. Supplementary Table 1 lists the steric interaction energies ($\Delta E_{steric}$) of the various radical dimers. While **38** shows a very small $\Delta E_{steric}$, the radical pair of **32** and the disproportionation TS are relatively high in energy. This will have a significant impact on the reaction, making the disproportionation of **32**

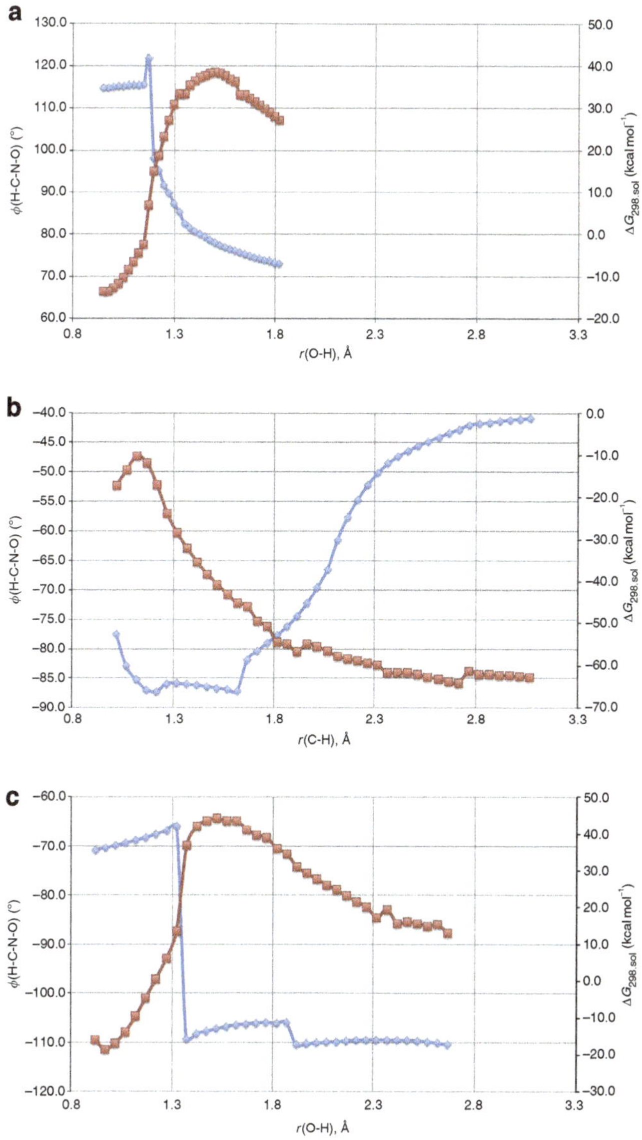


**Figure 7 | Comparison of dihedral angle changes during disproportionation.** Relaxed scans of the reaction energy for the hydrogen atom abstraction for (**a**) **36**, (**b**) **38**, (**c**) **32** as a function of the O-H (**36** and **32**) or C-H (**38**) distance (right axis, red). Also shown are plots of the $\phi$(H-C-N-O) dihedral angle (left axis, blue) during the reaction.

unfavourable. The same tendency is observed for **36**. This is also apparent from the NCI plots (Supplementary Fig. 7). In the NCI plot for **32**, there are significantly more weak interactions between the two fragments than in the NCI plot for **38**. In addition, from the structures presented, one notes that the radical pair of **32** involves two distinct radicals, whereas the equivalent structure for **38** is an O–H···O hydrogen-bonded hydroxylamine–nitrone pair—the disproportionation products (Fig. 2). The equivalent radical pair could not be located for **38**, likely because of the low barrier and high exergonicity for disproportionation. In the disproportionation transition state for stable radical **32** (Supplementary Fig. 8), the structure is clearly an early transition state with the C–H bond maintained and with a strongly attractive (blue) O···H interaction.

Figure 7 shows changes in the angles between the C–H bond being broken in the disproportionation step and the N–O bond (that is, the $\phi$(H–C–N–O) dihedral angle). As noted above, the ideal situation for hydrogen atom abstraction would be when there is maximal overlap between the C–H bond and the N–O singly-occupied molecular orbital (Fig. 2), since one of the electrons from the C-H bond forms part of the C=N double bond in the nitrone product, the other electron coming from the N–O radical. Thus, this process would be facilitated by a ±90° dihedral angle between these two bonds. This angle ($\phi$) is plotted against the reaction coordinate for each radical in Fig. 7. For the radical monomer of **32**, the $\phi$ angles are −51° and −69°. As the dimer is formed, the N–O is pushed out of the plane of the naphthalene ring, reaching −110° when the hydrogen is abstracted (Fig. 7c). Similar tendencies are observed for the other radicals. For **38** (Fig. 7b), the $\phi$ angles are −42– and −53°. As the dimer is formed, it moves towards −90° as the hydrogen atom is abstracted. This conformationally flexible compound has the lowest barrier of the studied systems. For all the radical dimers, $\phi$ is 15–20° from the ideal energy conformer for hydrogen abstraction when the abstraction takes place (Compare Fig. 7 with Figs 2b and 5). For *iso*-azaphenalene **32**, it appears that it is not possible for the C–H bond to move into periplanarity with the N–O orbital and thus when abstraction takes place the barrier is higher than it would be in the ideal case as found in **38** (Fig. 7c). For radical **36**, the situation is less clear as $\phi$ continues to approach 90° during the hydrogen abstraction (Fig. 7a). This particular effect is likely less important for the stability of this specific radical.

In combination, these results indicate that hydrogen atom abstraction is the active disproportionation mechanism for systems **1** and **2**. Ingold and co-workers[31,51] in a classical study showed that the decomposition of diethylnitroxide takes place via second-order or pseudo-second-order dependence on the radical concentration, that diamagnetic dimers (presumably of type **10**) are formed, and that the rate of radical decay is inversely proportional to solvent polarity (being roughly 30 times faster in *iso*-pentane than water). This solvent effect would appear to be inconsistent with a mechanism involving the formation of ion pair **11**/**12** (Fig. 2, pathway b) and could be in line with a hydrogen atom abstraction mechanism (Fig. 2, pathway a). It has been proposed that head-to-tail dimers occur on the path to product formation, specifically as a step preceding ion pair formation. Nonetheless, neither the observation of radical dimers in solution nor the low pre-exponential factors found in Ingold's Arrhenius analysis imply that dimer **10** is an intermediate on the direct reaction pathway to the nitrone and hydroxylamine (Fig. 2, pathway b) rather than simply a competing equilibrium. Our calculations did not proceed through dimers such as **10**, rather at the highest point along the reaction coordinates (Fig. 7 and Table 1), radical···radical dimers were observed. Obviously, more work is required to make any definite conclusions, however, based on the present calculations and the earlier observations by Ingold, it is reasonable that the hydrogen atom abstraction mechanism may be the predominant disproportionation pathway in other systems as well.

**Nitroxide properties and applications.** In order to evaluate the potential of the new radicals in applications such as catalysis, the bond dissociation energies (BDE) and redox potentials[52] of the radicals were calculated. The calculated redox potentials of **32**, **35–38** obtained are similar to that calculated for TEMPO (Table 2). The ability of the radicals to serve as a catalyst in oxidation reaction might therefore be envisioned.

Indeed, nitroxides **25a**, **25b**, **32** and **35** are able to reversibly cycle between their reduced (hydroxylamine) and nitroxide states, as shown by their use as a co-catalyst in the aerobic copper-catalysed oxidation of benzyl alcohol using the conditions developed by Hoover and Stahl[53]. These conditions have been shown to involve the nitroxides alternating between these two

**Table 2 | Calculated homo- and heterolytic BDE (kcal mol$^{-1}$) of the radicals in the gas phase and in acetonitrile, and the redox potentials (eV) in acetonitrile.**

| Radical | 3 | 36 | 32 | 35 | 37 | 38 |
|---|---|---|---|---|---|---|
| *Solution BDE* | | | | | | |
| Heterolytic | 57.8 | 57.9 | 56.4 | 59.6 | 63.9 | 48.2 |
| Homolytic | 57.6 | 55.3 | 59.0 | 60.5 | 58.5 | 51.8 |
| *Gas phase BDE* | | | | | | |
| Heterolytic | 365.1 | 350.8 | 355.4 | 355.9 | 359.9 | 355.3 |
| Homolytic | 57.7 | 52.3 | 60.1 | 61.6 | 59.4 | 50.7 |
| *Redox potential* | | | | | | |
| Oxidation | − 0.35 | − 0.89 | − 0.60 | − 0.45 | − 0.66 | − 0.61 |
| Reduction | − 2.18 | − 2.29 | − 2.06 | − 2.14 | − 2.41 | − 2.02 |

**Table 3 | Catalytic aerobic oxidation of benzyl alcohol.**

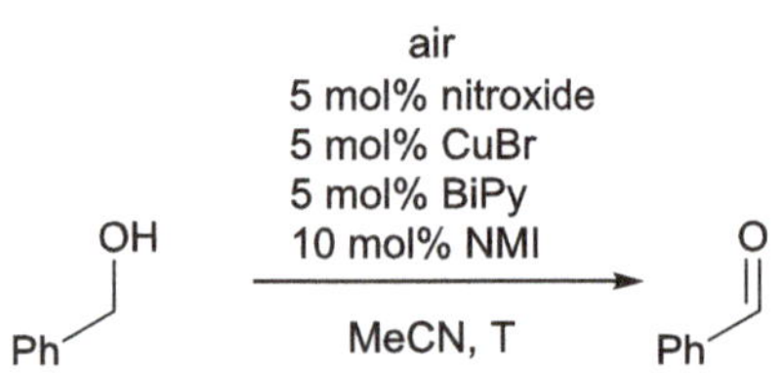


| Nitroxide | Reaction time (h) | *T* (°C) | Yield (%) |
|---|---|---|---|
| **3** | 3 | 25 | 95 |
| **25a** | 6 | 55 | 82 |
| **25b** | 1.5 | 25 | 92 |
| **32** | 2 | 25 | 96 |
| **35** | 1 | 25 | 96 |

BiPy, Bipyridine; NMI,N-methylimidazole.

oxidation states, and between a copper bound and a free state[54]. Oxidation of benzyl alcohol using nitroxides **25b**, **32** and **35** is complete within 1-2 h at room temperature. For comparison, oxidation using TEMPO, performed in parallel, requires 3 h for complete conversion under the same conditions.

In contrast, nitroxide radical **25a** requires heating to 55 °C and 6 h reaction time. This is because of the precipitation of a nitroxide–copper complex at room temperature. Partial dissolution of the precipitate could be observed along with a colour change from green (copper(II)) back to red-brown (copper(I)) at 55 °C. Thus, the reaction time and the temperature required for **25b** results from solubility limitations and not reactivity. Less than 10% conversion takes place with CuBr/bipyridine/NMI at 55 °C for 6 h in the absence of a nitroxide.

This study demonstrates the ability of all four radicals to cycle between hydroxylamine and radical oxidation states at least 20 times (that is, with a TON of at least 20). The preliminary results presented here are useful for establishing the potential feasibility and future work on this catalytic system. The nitroxides in this study were studied with stability in mind rather than catalytic activity; however, the data shown in Table 3 indicates that high catalytic efficiencies are possible. Since the syntheses of radicals **1** and **2** (Figs 4 and 5) are highly modular, this allows for the easy preparation of a diverse family of nitroxides in terms of the $R^1$–$R^5$ substituents (Fig. 1). It should therefore be possible to design nitroxides of type **1** and **2** with high catalytic efficiencies. Research in this direction is already underway.

## Conclusions

A new concept for the design of α-hydrogen-substituted nitroxides of two distinct structural families—**1** and **2**—has been developed (Fig. 1). This design is based on the ability of a substituent in the γ position of the radical ($R^4$ in **1** and $R^6$ in **2**) to interact with the α substituent, thereby inhibiting disproportionation through a combination of steric and stereoelectronic effects. The compounds prepared are stable for months in the solid form.

The principles outlined and verified experimentally are supported by a thorough computational study. To the best of our knowledge, this study is the first to investigate *in silico* the two prevalent mechanisms for nitroxide disproportionation. The hydrogen atom abstraction mechanism proposed by Ingold and co-workers[31] was identified as being the most likely mechanism for the decomposition of nitroxides of types **1** and **2**. These results may have significance for other α-hydrogen nitroxides. The nitroxides of types **1** and **2** are shown to possess BDE and redox potentials as expected for nitroxides, which indicates that they should be useful in many applications for which nitroxides are suitable. Catalytic proficiency of these compounds is demonstrated in a preliminary study.

This report constitutes a design concept for cyclic α-hydrogen nitroxides, that is, nitroxides that are not bicyclic like AZADO (**5**) or acyclic like TIPNO (**7**). The design principles outlined in this paper are likely not restricted to nitroxides **1** and **2**, and may be applied to the preparation of other nitroxides. The novel nitroxides of types **1** and **2** can be prepared in a highly efficient manner and in high yields. The flexibility of the synthesis allows for simple modification of the parent structures for tailored applications. In the present paper, the catalytic activity of the novel nitroxides was demonstrated, proving their ability to cycle between radical and hydroxylamine oxidation states. Since these compounds are chiral, it is especially attractive to consider potential applications in asymmetric catalysis[25,26,55] and as probes for biological studies. Indeed, multiple applications in catalysis, biology and medicine may be foreseen; however, this is beyond the scope of the present communication and will form the basis of further reports.

## Methods

For general, synthetic and computational methods and characterization data see Supplementary Methods section.

For NMR spectra of the compounds in this article, see Supplementary Figs 18–67.

For ESR spectra of the compounds in this article, see Supplementary Figs 2 and 9–17.

## References

1. Bobbitt, J. M., Bruckner, C. & Merbouh, N. Oxoammonium- and nitroxide-catalyzed oxidations of alcohols. *Org. Reactions* **74**, 103–424 (2009).
2. Tansakul, C. & Braslau, R. *Nitroxides in Synthetic Radical Chemistry* Vol. 2 1095–1129 (John Wiley & Sons Ltd., 2012).
3. Tebben, L. & Studer, A. Nitroxides: applications in synthesis and in polymer chemistry. *Angew. Chem. Int. Ed. Engl.* **50**, 5034–5068 (2011).
4. Lucio Anelli, P., Biffi, C., Montanari, F. & Quici, S. Fast and selective oxidation of primary alcohols to aldehydes or to carboxylic acids and of secondary alcohols to ketones mediated by oxoammonium salts under two-phase conditions. *J. Org. Chem.* **52**, 2559–2562 (1987).
5. Wertz, S. & Studer, A. Nitroxide-catalyzed transition-metal-free aerobic oxidation processes. *Green Chem.* **15**, 3116–3134 (2013).
6. Toledo, H., Pisarevsky, E., Abramovich, A. & Szpilman, A. M. Organocatalytic oxidation of aldehydes to mixed anhydrides. *Chem. Commun.* **49**, 4367–4369 (2013).
7. Abramovich, A., Toledo, H., Pisarevsky, E. & Szpilman, A. M. Organocatalytic oxidative dimerization of alcohols to esters. *Synlett* **23**, 2261–2265 (2012).
8. Gigmes, D. & Marque, S. R. A. *Encyclopedia of Radicals in Chemistry, Biology and Materials* Vol. 4, 1813–1850 (John Wiley & Sons Ltd., 2012).
9. Yan, G. P., Peng, L., Jian, S. Q., Li, L. & Bottle, S. E. Spin probes for electron paramagnetic resonance imaging. *Chin. Sci. Bull.* **53**, 3777–3789 (2008).

10. Bi, W. *et al.* Novel TEMPO-PEG-RGDs conjugates remediate tissue damage induced by acute limb ischemia/reperfusion. *J. Med. Chem.* **55,** 4501–4505 (2012).
11. Castagna, R. *et al.* Nitroxide radical TEMPO reduces ozone-induced chemokine IL-8 production in lung epithelial cells. *Toxicol. In Vitro* **23,** 365–370 (2009).
12. Gryn'ova, G., Barakat, J. M., Blinco, J. P., Bottle, S. E. & Coote, M. L. Computational design of cyclic nitroxides as efficient redox mediators for dye-sensitized solar cells. *Chem. Eur. J.* **18,** 7582–7593 (2012).
13. Hamada, S., Furuta, T., Wada, Y. & Kawabata, T. Chemoselective oxidation by electronically tuned nitroxyl radical catalysts. *Angew. Chem. Int. Ed. Engl.* **52,** 8093–8097 (2013).
14. Ciriminna, R. & Pagliaro, M. Industrial oxidations with organocatalyst TEMPO and its derivatives. *Org. Process Res. Dev.* **14,** 245–251 (2010).
15. Ishii, K., Kubo, K., Sakurada, T., Komori, K. & Sakai, Y. Phthalocyanine-based fluorescence probes for detecting ascorbic acid: phthalocyaninatosilicon covalently linked to TEMPO radicals. *Chem. Commun.* **47,** 4932–4934 (2011).
16. Pattison, D. I., Lam, M., Shinde, S. S., Anderson, R. F. & Davies, M. J. The nitroxide TEMPO is an efficient scavenger of protein radicals: cellular and kinetic studies. *Free Radic. Biol. Med.* **53,** 1664–1674 (2012).
17. Paul, N. D. *et al.* Metalloradical approach to 2H-chromenes. *J. Am. Chem. Soc.* **136,** 1090–1096 (2014).
18. Amos, R. I. J., Gourlay, B. S., Schiesser, C. H., Smith, J. A. & Yates, B. F. A mechanistic study on the oxidation of hydrazides: application to the tuberculosis drug isoniazid. *Chem. Commun.* 1695–1697 (2008).
19. Nilsen, A. & Braslau, R. Nitroxide decomposition: implications toward nitroxide design for applications in living free-radical polymerization. *J. Polym. Sci., A Polym. Chem* **44,** 697–717 (2005).
20. Kamer, P. C. J. & Leeuwen, P.W.N.M.v. *Phosphorus(III) Ligands in Homogeneous Catalysis: Design and Synthesis* (John Wiley & Sons, Ltd, 2012).
21. Dupeyre, R. M., Rassat, A. & Ronzaud, J. Nitroxides. LII. Synthesis and electron spin resonance studies of N,N'-dioxy-2,6-diazaadamantane, a symmetrical ground state triplet. *J. Am. Chem. Soc.* **96,** 6559–6568 (1974).
22. Bredt, J. Über sterische Hinderung in Brückenringen (Bredtsche Regel) und über die meso-trans-Stellung in kondensierten Ringsystemen des Hexamethylens. *Justus Liebigs Ann. Chem.* **437,** 1–13 (1924).
23. Shibuya, M., Tomizawa, M., Suzuki, I. & Iwabuchi, Y. 2-Azaadamantane N-Oxyl (AZADO) and 1-Me-AZADO: highly efficient organocatalysts for oxidation of alcohols. *J. Am. Chem. Soc.* **128,** 8412–8413 (2006).
24. Shibuya, M., Tomizawa, M., Sasano, Y. & Iwabuchi, Y. An expeditious entry to 9-Azabicyclo[3.3.1]nonane N-Oxyl (ABNO): another highly active organocatalyst for oxidation of alcohols. *J. Org. Chem.* **74,** 4619–4622 (2009).
25. Tomizawa, M., Shibuya, M. & Iwabuchi, Y. Highly enantioselective organocatalytic oxidative kinetic resolution of secondary alcohols using chirally modified AZADOs. *Org. Lett.* **11,** 1829–1831 (2009).
26. Graetz, B., Rychnovsky, S., Leu, W.-H., Farmer, P. & Lin, R. C2-Symmetric nitroxides and their potential as enantioselective oxidants. *Tetrahedron* **16,** 3584–3598 (2005).
27. Reznikov, V. A., Gutorov, I. A., Gatilov, Y. V., Rybalova, T. V. & Volodarsky, L. B. Stable nitroxyl radicals with a hydrogen atom at alpha-carbon atom of nitroxyl group. *Russ. Chem. Bull.* **45,** 384–392 (1996).
28. Grimaldi, S. *et al.*Polymerisation in the presence of a beta-substituted nitroxide radical WO 96124620 (1996).
29. Benoit, D., Chaplinski, V., Braslau, R. & Hawker, C. J. Development of a universal alkoxyamine for "living" free radical polymerizations. *J. Am. Chem. Soc.* **121,** 3904–3920 (1999).
30. Benoit, D. *et al.* Controlled free-radical polymerization in the presence of a novel asymmetric nitroxyl radical. *Polym. Prepr. (Am. Chem. Soc. Div. Polym. Chem.)* **38,** 729–730 (1997).
31. Adamic, K., Bowman, D. F. & Ingold, K. U. Self-reaction of diethylnitroxide radicals. *J. Am. Chem. Soc.* **92,** 1093–1094 (1970).
32. Tikhonov, I. V. *et al.* Effect of the structure of nitroxyl radicals on the kinetics of their acid-catalyzed disproportionation. *J. Phys. Org. Chem.* **27,** 114–120 (2014).
33. Breuer, E., Aurich, H. G. & Nielsen, A. *Nitrones, Nitronates, and Nitroxides* (New York,Wiley, 1989).
34. Studer, A., Harms, K., Knoop, C., Mueller, C. & Schulte, T. New sterically hindered nitroxides for the living free radical polymerization: X-ray structure of an α-H-bearing nitroxide. *Macromolecules* **37,** 27–34 (2004).
35. Grimaldi, S., Siri, D., Finet, J. P. & Tordo, P. N-tert-butyl-N-(1-dibenzylphosphono-2,2-dimethylpropyl)nitroxide.. *Acta Crystallogr.Sect. C Cryst. Struct. Commun.* **54,** 1712–1714 (1998).
36. Franck, R. W. & Leser, E. G. 1,8-Di-tert-butylnaphthalenes. *J. Am. Chem. Soc.* **91,** 1577–1578 (1969).
37. Franck, R. W. & Leser, E. G. Synthesis of 1,8-di-tert-butylnaphthalenes. *J. Org. Chem.* **35,** 3932–3939 (1970).
38. Anderson, J. E., Franck, R. W. & Mandella, W. L. Peri interactions in some 1,8-di-tert-butylnaphthalene compounds. Rotation and flipping of the tert-butyl groups. *J. Am. Chem. Soc.* **94,** 4608–4614 (1972).
39. Aikawa, H., Takahira, Y. & Yamaguchi, M. Synthesis of 1,8-di(1-adamantyl)naphthalenes as single enantiomers stable at ambient temperatures. *Chem. Commun.* **47,** 1479–1481 (2011).
40. Carruthers, W., Coggins, P. & Weston, J. B. Nitrone cycloadditions: synthesis of (±)-andrachamine. *J. Chem. Soc. Perkin. Trans.* **1,** 2323–2327 (1990).
41. Carruthers, W., Coggins, P. & Weston, J. B. Nitrone cycloaddition: an approach to the cyclophane alkaloid (±)-lythranidine. *J. Chem. Soc. Perkin. Trans.* **1,** 611–616 (1991).
42. Brandi, A., Cicchi, S., Paschetta, V., Gomez Pardo, D. & Cossy, J. A new nitrone from C2 symmetric piperidine for the synthesis of hydroxylated indolizidinone. *Tetrahedron Lett.* **43,** 9357–9359 (2002).
43. Moosa, B. A. & Ali, S. A. Regioselective transformation of 6/5-fused bicyclic isoxazolidines to second-generation cyclic aldonitrones. *ARKIVOC* 132–148 (2010).
44. Sysoeva, N. A., Sholle, V. D., Rozantsev, E. G. & Buchachenko, A. L. NMR spectra of nitroxyl radicals of the isoindoline series. *Izv. Akad. Nauk SSSR Ser. Khim.* **20,** 1716–1717 (1971).
45. Amornraksa, K., Grigg, R., Gunaratne, H. Q. N., Kemp, J. & Sridharan, V. X = Y-ZH systems as potential 1,3-dipoles. 8. Pyrrolidines and delta-5-Pyrrolines (3,7-Diazabicyclo 3.3.0 Octenes) from the reaction of imines of alpha-amino-acids and their esters with cyclic diploarophiles - mechanism of racemization of alpha-amino-acids and their esters in the presence of aldehydes. *J. Chem. Soc. Perkin. Trans.* **1,** 2285–2296 (1987).
46. Szpilman, A. M. & Carreira, E. M. Cycloaddition reactions, in *Silver in Organic Chemistry* (ed. Harmata, M.) 43–82 (John Wiley & Sons Inc., 2010).
47. Blinco, J. P., McMurtrie, J. C. & Bottle, S. E. The first example of an azaphenalene pro-fluorescent nitroxide. *Eur. J. Org. Chem.* **2007,** 4638–4641 (2007).
48. Foroughifar, N., Mobinikhaledi, A. & Moghanian, H. Simple and efficient procedure for the synthesis of novel 1,3-diphenyl-2-azaphenalene derivatives via one-pot multicomponent reactions. *Synth. Commun.* **40,** 1812–1821 (2010).
49. Gryn'ova, G., Ingold, K. U. & Coote, M. L. New insights into the mechanism of amine/nitroxide cycling during the hindered amine light stabilizer inhibited oxidative degradation of polymers. *J. Am. Chem. Soc.* **134,** 12979–12988 (2012).
50. Gryn'ova, G., Lin, C. Y. & Coote, M. L. Which side-reactions compromise nitroxide mediated polymerization? *Polym. Chem.* **4,** 3744–3754 (2013).
51. Ingold, K. U., Adamic, K., Bowman, D. F. & Gillan, T. Kinetic applications of electron paramagnetic resonance spectroscopy. I. Self-reactions of diethyl nitroxide radicals. *J. Am. Chem. Soc.* **93,** 902–908 (1971).
52. Blinco, J. P. *et al.* Experimental and theoretical studies of the redox potentials of cyclic nitroxides. *J. Org. Chem.* **73,** 6763–6771 (2008).
53. Hoover, J. M. & Stahl, S. S. Highly practical copper(I)/TEMPO catalyst system for chemoselective aerobic oxidation of primary alcohols. *J. Am. Chem. Soc.* **133,** 16901–16910 (2011).
54. Ryland, B. L., McCann, S. D., Brunold, T. C. & Stahl, S. S. Mechanism of alcohol oxidation mediated by copper(II) and nitroxyl radicals. *J. Am. Chem. Soc.* **136,** 12166–12173 (2014).
55. Rychnovsky, S. D., McLernon, T. L. & Rajapakse, H. Enantioselective oxidation of secondary alcohols using a chiral nitroxyl (N-oxoammonium salt) catalyst. *J. Org. Chem.* **61,** 1194–1195 (1996).

## Acknowledgements

This research was supported by a German Israel Foundation Young Investigator Grant (grant no. I-2234-2067/2009). A.M.S. is the incumbent of the Chaya Career Development Chair.

## Author contributions

M.A. and S.B. designed and carried out the experiments. M.A.I. did the DFT calculations and helped to write the manuscript. H.T. participated in designing and carrying out the experiments. B.T. performed the ESR measurements. L.J.W.S., M.B. and N.F. performed the X-ray crystallography. A.M.S. invented the concepts presented here, designed the experiments, supervised the work and wrote the manuscript.

## Additional information

**Competing financial interests:** The authors declare no competing financial interests.

# 11

# Total synthesis of periploside A, a unique pregnane hexasaccharide with potent immunosuppressive effects

Xiaheng Zhang[1], Yu Zhou[2], Jianping Zuo[2] & Biao Yu[1]

Periploside A is a pregnane hexasaccharide identified from the Chinese medicinal plant *Periploca sepium*, which features a unique seven-membered formyl acetal bridged orthoester (FABO) motif and potent immunosuppressive activities. Here, we show the synthesis of this molecule in a total of 76 steps with the longest linear sequence of 29 steps and 9.2% overall yield. The FABO motif is constructed via a combination of Sinaÿ's and Crich's protocol for the formation of orthoester and acetal glycosides, respectively. The 2-deoxy-β-glycosidic linkages are assembled stereoselectively with judicious choice of the glycosylation methods. The epimer at the spiro-quaternary carbon in the FABO motif has also been elaborated in a stereo-controlled manner. This epimer, as well as the synthetic analogues bearing the FABO motif, retain largely the inhibitory activities of periploside A against the proliferation of T-lymphocyte, indicating the importance of the chemical connection of the FABO motif to their immunosuppressive activity.

[1]State Key Laboratory of Bioorganic and Natural Products Chemistry, Shanghai Institute of Organic Chemistry, Chinese Academy of Sciences, 345 Lingling Road, Shanghai 200032, China. [2] State Key Laboratory of New Drug Research, Shanghai Institute of Materia Medica, Chinese Academy of Sciences, Shanghai 201203, China. Correspondence and requests for materials should be addressed to B.Y. (email: byu@mail.sioc.ac.cn).

Periploside A (or Periplocoside E) is the prototypical member of a group of pregnane glycosides isolated from *Periploca sepium* and *P. forrestii* (Asclepiadaceae), which features a unique seven-membered formyl acetal bridged orthoester (FABO) linkage between two sugar units[1–7]. This natural product shows immunosuppressive activities, that is, it inhibits the ConA-induced T cell proliferation as potently as rapamycin and cyclosporin A while showing reasonably low toxicity ($IC_{50} = 0.64\,\mu M$ and $CC_{50} = 10.1\,\mu M$)[5,6]. The immunosuppressive effects of periploside A has also been found significant in mice models[8–10], that validates the folkloric reputation of the plant as a traditional Chinese medicine for rheumatoid arthritis[11]. Such autoimmune diseases as rheumatoid arthritis, multiple sclerosis, systemic lupus erythematosus and Crohn's disease, are notoriously malignant and refractory, thus efficacious new immunosuppressant drugs have long been a quest[12].

The FABO motif in periplosides, which has been found critical to their activities[6], remains ambiguous in both the chemical connection and stereochemistry for over two decades. In 1987, Hikino *et al.* reported the first structural assignment of periploside A, in that the connection of the orthoester was proposed on the basis of extensive chemical degradation and spectroscopic analysis[1]. However, the stereochemistry of the quaternary anomeric carbon was overlooked but drawn in a configuration comprising an unusual α-(1→4)-glycosidic linkage (**1a**). Shortly afterwards, Itokawa *et al.*[2–4] suggested a novel peroxy linkage (**1b**) for periplosides (and named as periplocosides), as misled by a conventional colour reaction of peroxides, although all the analytical data of periplocoside E were found identical to those of the previous periploside A (ref. 3). This peroxide structure was then accepted till 2011 (refs 5,6), Zhao and coworkers acquired an X-ray diffraction of a single crystal of a periploside congener, confirming the originally assigned FABO connection[6]. Unfortunately, the stereochemistry of the spiro-quaternary carbon remained incorrect until we examined it carefully during our journey toward the synthesis of periploside A (ref. 7).

Herein, we report the first total synthesis of periploside A (**1**), employing total of 76 steps of transformations from glucal, methyl α-D-glucopyranoside and dehydroepiandrosterone, with the longest linear sequence of 29 steps and in 9.2% overall yield. The stereoselective construction of the FABO motif with either the natural or unnatural configuration at the anomeric spiro-quaternary center is achieved via alternative combination of Crich's protocol for acetal glycoside synthesis and Sinaÿ's protocol for glycosyl orthoester formation. The 2-deoxy-β-glycosidic linkages are synthesized by glycosylation with trifluoroacetimidate donors (**6**/**27**) equipped with 2-iodide, with digitoxosyl *ortho*-alkynylbenzoate donor (**8**) installed with bulky TBDPS group on the remote 3,4-OH, and with cymarosyl *ortho*-alkynylbenzoate donor (**2**) under the promotion of $Ph_3PAuOTf$/TTBP. In addition, a preliminary structure–activity relationship of the synthetic periplosides against the proliferation of T-lymphocyte is provided, indicating the importance of the chemical connection of the FABO motif to the activities.

## Results

**Retrosynthetic analysis.** Given the big size and linear structure of the target molecule, a convergent synthesis was desired[13]. Thus, periploside A (**1**) was disconnected into two fragments of similar complexity, that is, tetrasaccharide donor **2** and pregnane FABO disaccharide **3** (Fig. 1). The (1→4)-cymarosyl linkage between

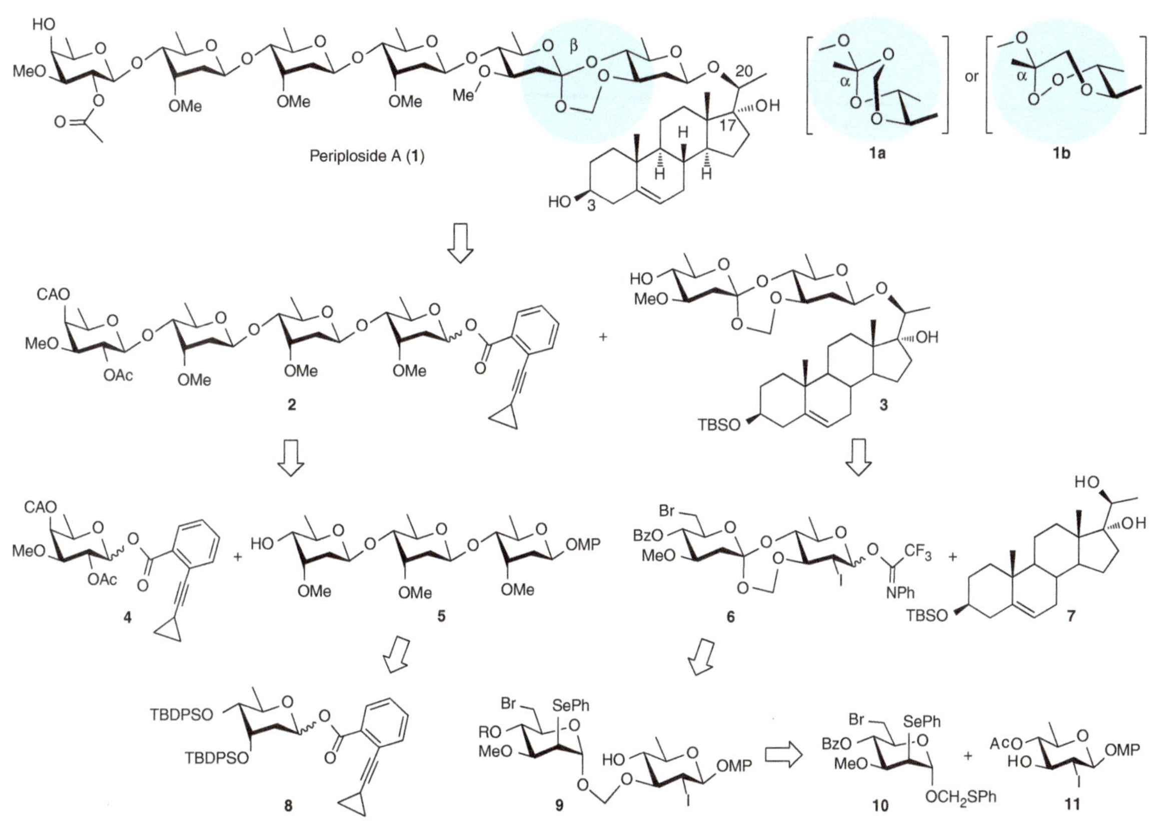


**Figure 1 | Periploside A 1 and the retrosynthetic analysis.** Shadowed in pale blue are the FABO motifs that have been previously assigned; highlighted in red are leaving groups in donors and in blue are hydroxyl groups in acceptors. Ac, acetyl; CA, chloroacetyl; MP, 4-methoxyphenyl; TBDPS, *tert*-butyldiphenylsilyl; TBS, *tert*-butyldimethylsilyl.

the tetrasaccharide unit and 3 would be difficult to construct in favour of formation of the thermodynamically unfavored β-anomer[14–17]; the glycosylation protocol with *ortho*-alkynylbenzoates as leaving groups (as in **2**) under the catalysis of a gold(I) complex could address this problem, in that substitution via an α-glycosyloxypyrylium or α-triflate intermediate might get invoked[18]. The β-(1→4)-cymarosyl linkages in tetrasaccharide **2**, however, could be built stereoselectively under the influence of a substituent at the axial 3-OH of a digitoxosyl donor, which could subsequently be converted into the required methyl group. In this respect, we had screened carefully during the synthesis of digitoxin and gordonoside F and found gold(I)-catalysed glycosylation with digitoxosyl *ortho*-cyclopropylethynylbenzoate equipped with two bulky TBDPS groups at 3,4-OH (that is, **8**) to be an optimal choice[15,17]. After furnishing the cymarose trisaccharide **5**, its coupling with the terminal digitalose unit could be realized with donor **4**, in that the formation of the β-linkage should be secured by participation of the 2-*O*-acetyl group; and the 4-*O*-chloroacetyl group was selectively removable afterwards. The *p*-methoxyphenyl (MP) group was employed throughout (as in **5**, **9** and **11**) as the anomeric protecting group, which could remain intact before cleavage and subsequently converting the saccharides into donors (for example, **2** and **6**).

Uncertainty lies in the synthesis of FABO disaccharide **3**. Condensation of 3β-silyloxy-pregnene-17α,20α-diol **7** with disaccharide trifluoroacetimidate **6**, one of the most reliable type of glycosylation donors[19], which was equipped with an equatorial iodide at C2 (ref. 20), would lead to the desired C20-*O*-β-glycoside **3** in a regio- and stereo- selective manner. However, few clues occur in the literatures, which could lead to the construction of a disaccharide precursor consisting of the FABO linkage. The most promising approach turned out to be a combination of Crich's protocol to form the formyl acetal glycosidic linkage (from a thiomethyl glycoside donor such as **10**; ref. 21) and Sinaÿ's protocol to synthesize the anomeric orthoester via a ketene acetal intermediate (derived from a 2-selenoglycoside such as **9**; refs 22–24). However, the tolerance of the formyl acetal (in **9**) towards the strong conditions required for Sinaÿ's oxidation–elimination–cyclization sequence and the stereochemistry in the formation of the seven-membered spiro-orthoester were beyond estimation.

**Synthesis of tetrasaccharide donor 2.** The desired digitalosyl and digtoxosyl *ortho*-cyclopropylethynylbenzoates **4** and **8** were prepared in a robust manner from D-fucose (13 steps and 37% overall yield) (see Supplementary Methods) and methyl α-D-glucopyranoside (10 steps and 47% overall yield)[15], respectively. Subjection of **8** to glycosidation with *p*-methoxyphenol under optimized conditions (0.1 equivalent (equiv) $Ph_3PAuNTf_2$, toluene, 4 Å molecular sieves (MS), −40 °C) led to β-glycoside **12** as the sole anomer in an excellent 95% yield (Fig. 2). Removal of the 3,4-di-*O*-TBDPS group followed by selective acetylation of the axial 3-OH via orthoester formation/hydrolysis provided glycoside **13** bearing a free 4-OH (92%, two steps)[25]. Similar transformations were then applied to extend monosaccharide **13** to trisaccharide **17**, in that the glycosidation of **8** with monosaccharide acceptor **13** and disaccharide acceptor **15** led to the corresponding coupled 2-deoxy-β-saccharides nearly quantitatively. Trisaccharide 3″,4″-diol **17** was then subjected to a tin-mediated selective benzylation on the equatorial 4″-OH (ref. 26); cleavage and scramble of the acetyl groups on the axial hydroxyl groups were detected, nevertheless, exposure of the resultant mixture to LiOH afforded **18** in excellent yield (87%, three steps). The three axial hydroxyl groups in **18** were then methylated (99%). Unexpectedly, hydrogenolysis of the 4″-*O*-benzyl group in trisaccharide **19** under conventional conditions

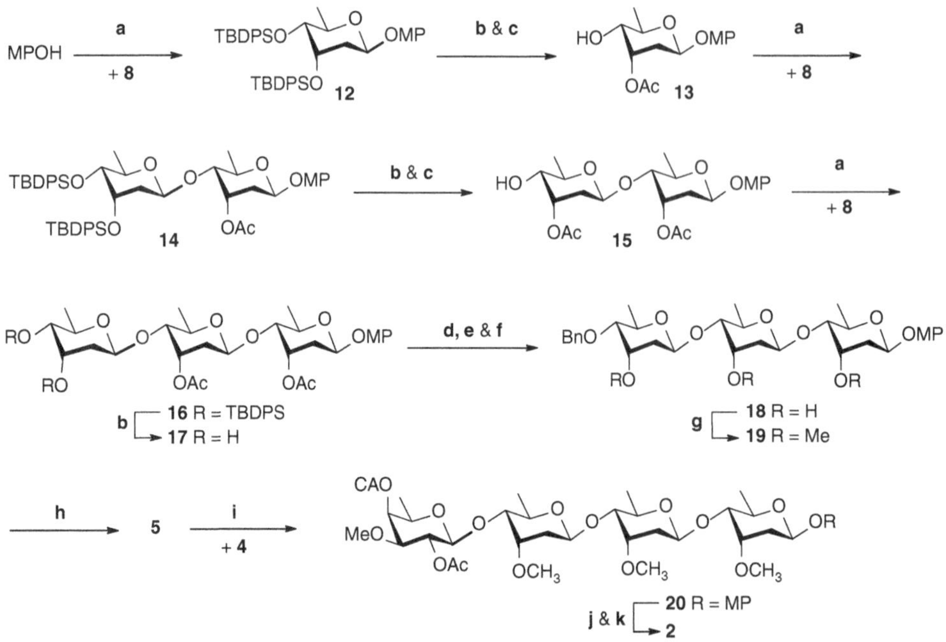


**Figure 2 | Synthesis of tetrasaccharide donor 2.** Highlighted in red are the nascent glycosidic bonds. (**a**) $Ph_3PAuNTf_2$ (0.1 equiv), toluene, 4 Å MS, −40 °C; 95% and β only (for **12**); 99% and β only (for **14** and **16**); (**b**) TBAF, THF, 0 °C to RT; 99% (from **12**); 94% (from **14**); 96% (from **16**); (**c**) $CH_3(OMe)_3$, *p*-TsOH, RT, 93% for (**13**); 91% (for **15**); (**d**) $^nBu_2SnO$, MeOH, reflux; (**e**) BnBr, DMF, CsF, RT; (**f**) LiOH, THF, $H_2O$, RT; 87% (for three steps); (**g**) MeI, NaH, DMF, 0 °C to RT, 99%; (**h**) $Pd(OH)_2/C$, $H_2$ (1 atm), $Et_3N$, EtOAc, MeOH, 50 °C, 93%; (**i**) $Ph_3PAuNTf_2$ (0.2 equiv), toluene, 5 Å MS, −50 °C to RT, 93%, β/α = 4/1; (**j**) $Ag(DPAH)_2$, $CH_3CN$, $H_2O$, 0 °C to RT, 95%; (**k**) *o*-cyclopropylethynylbenzoic acid, EDCI, DMAP, 4 Å MS, $CH_2Cl_2$, RT, 99%. DMAP, 4,4-dimethylaminopyridine; DPAH, bis(hydrogen dipicolinate); EDCI, *N*-ethyl-*N*′-(3-dimethylaminopropyl)carbodiimide hydrochloride; MS, molecular sieves; TBAF, tetrabutylammonium fluoride.

($Pd(OH)_2/C$, $H_2$, room temperature (RT)) led to cleavage of the glycosidic linkages, testifying the vulnerability of the cymarosyl-β-(1→4)-linkage. Thus, $Et_3N$ was added to sequester any nascent proton; the hydrogenolysis proceeded sluggishly, nevertheless, leading to the desired trisaccharide **5** in a satisfactory 93% yield at 50 °C for 2 days.

Given the acid lability of the cymarosyl-β-(1→4)-linkage, it was not fully surprising to find that glycosylation of trisaccharide acceptor **5** with digitalosyl donor **4** in the presence of $Ph_3PAuOTf$ led to considerable cleavage of the trisaccharide. $Ph_3PAuNTf_2$ was found to be more stable towards moisture (thus less acidic)[27], therefore, it catalysed the condensation of **5** and **4** smoothly. However, in spite of the presence of 2-*O*-acetyl group in donor **4**, the glycosylation led to anomeric mixtures of the coupled tetrasaccharide. Under optimized conditions (0.2 equiv $Ph_3PAuNTf_2$, toluene, 5Å MS, − 50 °C to RT), tetrasaccharide **20** was obtained in 93% yield with β/α ratio of 4:1, which was separable on silica gel column chromatography. Removal of the anomeric MP group in tetrasaccharide **20** with CAN was not successful under a variety of conditions, resulting unavoidably degradation of the cymarosyl-β-(1→4)-linkages. Fortunately, a mild oxidizing agent, $Ag(DPAH)_2$, could achieve this task ($CH_3CN/H_2O$, 0 °C to RT) to afford smoothly the corresponding hemiacetal (95%) (ref. 28), which was then subjected to the formation of *ortho*-cyclopropylethynylbenzoate **2** (99%) (see Supplementary Methods).

**Attempts at construction of the FABO motif.** We have tried a number of approaches to the construction of the FABO disaccharide motif, those include condensation of sugar lactones with vicinal diols in the presence of formaldehyde or its surrogates[29,30], transacetalation of sugar lactones with sugar 4-hydroxyl-3-methyoxymethyl ether derivatives[31], and intramolecular hydrogen atom transfer reactions of formyl acetal linked disaccharides[21,32,33]. However, no coupled FABO products have ever been detected, that prompted us to focus our attention on the planned Crich–Sinaÿ approach (Fig. 3). Thus, starting from D-glucal triacetate, 3-*O*-methyl-6-bromo-glucal **21** and *p*-methoxyphenyl 2,6-dideoxy-2-iodo-glucoside **11** were prepared in seven steps (66% overall yield) and 13 steps (18% overall yield), respectively (see Supplementary Methods). Glucal **21** was converted into the desired phenylthiomethyl 2-phenylseleno-α-D-mannopyranoside **10** in three steps (45%), that is, addition with PhSeCl to provide 2-phenylseleno-D-mannopyranose **22**, conversion of the lactol into fluoride with DAST, and subsequent glycosidation with $PhSCH_2OH$ in the presence of $SnCl_2$ (refs 23,34). Condensation of **10** with sugar acceptor **11** under Crich's conditions (NIS, TfOH, $CH_2Cl_2$, − 30 °C) did lead to the desired acetal glycoside **24**, albeit in only a moderate yield of ~30%; the major product turned out to be the α-glycoside **23** (~65%). A wide variety of the promoters for glycosidation of thioglycosides[35] were screened in the present coupling (see Supplementary Table 1), including NBS/TfOH[36], $Tf_2O$/BSP/TTBP[37], DMTST/TTBP[38], MeOTf[39] and NIS/AgOTf[40]. However, the yield of acetal glycoside **24** was not improved. These results indicate that the transient glycosyloxymethyl cation (generated from **10**) decomposes favourably to the glycosyl oxocarbenium species (and a molecule of formaldehyde); capture of the oxocarbenium intermediate by the incoming hydroxyl group (in **11**) leads to disaccharide **23**. Adjustment of the substituting groups in donor **10**, that is, removal of the 6-bromide and/or replacement of the 4-*O*-benzoyl group with benzyl group also failed to improve the yields of the corresponding acetal glycosides in condensation with **11**. Gratifyingly, when we applied the inverse procedure[41] (addition of donor **10** to a mixture of the acceptor **11** and promoter NIS/TfOH at − 30 °C) and increased the equivalent of acceptor **11** (to 5.0 equiv), acetal glycoside **24** was obtained in a satisfactory 75% yield. The 4-*O*-acetyl and 4′-*O*-benzoyl group in **24** were removed simultaneously in the presence of MeONa in MeOH, providing diol **9** (93%), which was ready for Sinaÿ's orthoester formation.

Thus, 2′-seleno-disaccharide **9** was subjected to oxidation ($NaIO_4$, $NaHCO_3$, $MeOH/CH_2Cl_2/H_2O$, RT) to provide the selenoxide cleanly, subsequent *syn*-elimination and intramolecular cyclization of the resultant ketene acetal (**A**) took place sluggishly under forced conditions (vinyl acetate/toluene/DIPA, 140 °C), leading to an orthoester product (**25**) in a moderate 20% yield for 12 h. By carefully screening the reaction conditions, we finally managed to obtain **25** in an excellent 85% yield under microwave irradiation (145 °C, 20 min). The addition of the 4-hydroxyl group onto C1′ of the ketene acetal in **A** could proceed from both the top and the bottom faces, thus two diastereoisomers should be provided. Surprisingly, only one isomer (**25**) was isolated under various conditions. We had expected that the diastereoselectivity bestowed by the two sugar units could lead to formation of the native configuration in the natural product (opposite to that drawn in **25**). However, in the previous synthesis of the Sinaÿ-type five- and six-membered orthoesters, cyclization of the incipient ketene acetals was found to favour formation of the axial α-linkage, so as to maximize the anomeric effect[22–24] or to follow the trajectory of addition onto Woerpel's low-energy conformer of the dioxocarbenium intermediate[42,43]. Determination of the configuration of the quaternary C1′ in **25** was not possible by spectroscopic methods, especially without comparison to its epimer, and attempts at acquisition of a single crystal of **25** or its derivatives for X-ray diffraction analysis were unsuccessful. Therefore, this problem was not solved until completion of the total synthesis.

**Synthesis of C1″-*epi*-periploside A (31).** Temporary protection of the 4′-OH in **25** with benzoyl group led to **26** (99%), which was subjected to selective cleavage of the anomeric MP group ($Ag(DPAH)_2$, $CH_3CN$, $H_2O$, 90%) and subsequent formation of trifluoroacetimidate (95%) to afford the desired donor **27** (Fig. 3). Glycosylation of pregnane diol **7**, which was prepared from hydroxyandrost-5-en-17-one in three steps (see Supplementary Methods and Supplementary Fig. 59), with disaccharide imidate **27** proceeded smoothly under the catalysis of TBSOTf (0.1 equiv) in the presence of 5 Å MS at − 78 °C in $CH_2Cl_2$, affording pregnane 20-*O*-disaccharide **28** in 85% (96%, b.r.s.m.) yield and β/α ratio of 3:1. The erosion of 1,2-*trans*-selectivity in the glycosylation with a donor (that is, **27**) equipped with an iodide at C2 could be contributed to the presence of the FABO motif, which might cause conformational restraint for the neighbouring participation. Notably, the present reaction must be quenched at − 78 °C (with $Et_3N$), so as to avoid the decomposition of the acid-labile FABO motif. Reductive removal of the bromide and iodide in **28** under the conventional radical conditions (AIBN/$Bu_3SnH$, toluene, 80 °C) led to decomposition of the substrate. $Et_3B$ could initiate the radical reaction at RT[44], which thus enabled the reduction of **28** with $Bu_3SnH$ to proceed smoothly to afford the corresponding deoxydisaccharide (99%). Subsequent removal of the 4′-*O*-benzoyl group provided the desired pregnane 20-*O*-disaccharide **29** (99%).

The expected challenge in stereoselective construction of the 2-deoxy-β-cymarosyl-(1→4)-linkage in the union of tetrasaccharide **2** and pregnane disaccharide **29** was further manifested by the fact that both the 2,6-dideoxyglycosidic linkages and the FABO motif were shown to be extremely labile toward acid. Thus, we screened carefully the coupling conditions, including the gold(I)

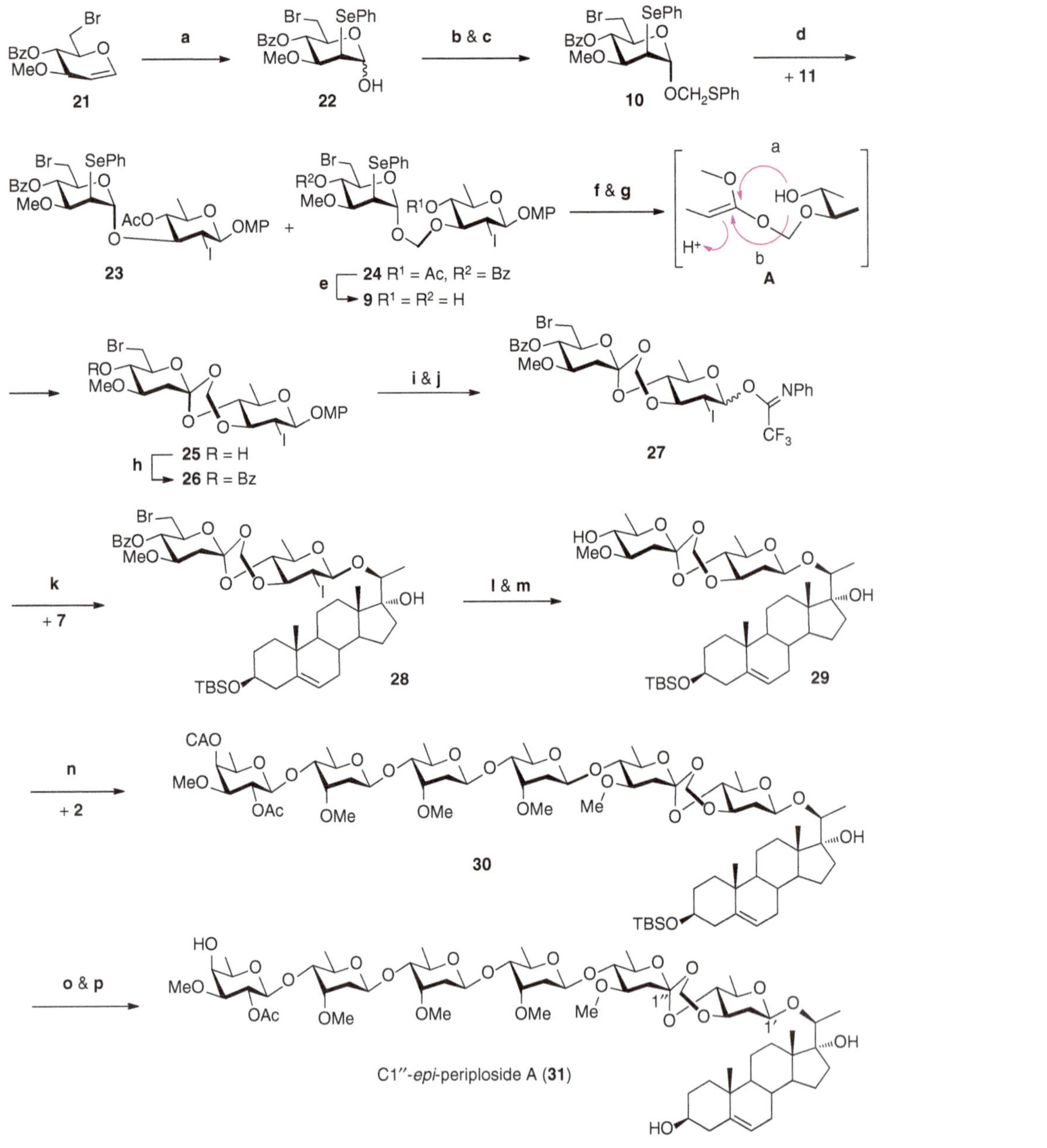

**Figure 3 | Construction of the FABO motif and synthesis of C1″-*epi*-periploside A (31).** Highlighted in red are the nascent glycosidic bonds. (**a**) PhSeCl, $CH_3CN$, − 40 °C, 78%; (**b**) DAST, THF, − 30 °C to RT; (**c**) $PhSCH_2OH$, $SnCl_2$, $CH_2Cl_2$, 4 Å MS, − 40 °C to RT; 58% (for two steps); (**d**) NIS, TfOH, $CH_2Cl_2$, − 30 °C, 75% (for **24**), 20% (for **23**); (**e**) MeONa, MeOH, $CH_2Cl_2$, RT, 93%; (**f**) $NaIO_4$, $NaHCO_3$, MeOH, $CH_2Cl_2$, $H_2O$, RT, 99%; (**g**) vinyl acetate, toluene, DIPA, Mw, 145 °C, 20 min, 85%; (**h**) BzCl, $Et_3N$, DMAP, $CH_2Cl_2$, 0 °C to RT, 99%; (**i**) $Ag(DPAH)_2$, $CH_3CN$, $H_2O$, 0 °C to RT, 90%; (**j**) *N*-phenyl-2,2,2-trifluoroacetimidoyl chloride, $Cs_2CO_3$, $CH_2Cl_2$, RT, 95%; (**k**) TBSOTf, $CH_2Cl_2$, 5 Å MS, − 78 °C, 85%, $\beta/\alpha = 3/1$; (**l**) $Et_3B$, $Bu_3SnH$, toluene, RT, 97%; (**m**) MeONa, MeOH, $CH_2Cl_2$, RT, 99%; (**n**) $Ph_3PAuOTf$ (0.8 equiv), TTBP, $CH_2Cl_2$, 4 Å MS, − 20 °C to − 10 °C, 64% (87% b.r.s.m.), $\beta/\alpha = 2.1/1$; (**o**) thiourea, pyridine, EtOH, 80 °C; (**p**) HF · py, pyridine; THF, 0 °C to RT; 91% (for two steps). DAST, (diethylamino)sulfur trifluoride; DIPA, diisopropylamine; TBSOTf, *tert*-butyldimethylsilyl trifluoromethane sulfonate; Mw, microwave; TTBP, 2,4,6-tri-*tert*-butylpyrimidine.

catalyst, solvent and temperature. When the coupling of **2** and **29** was conducted in the presence of $Ph_3PAuNTf_2$ (0.2 equiv), the coupled hexasaccharide was obtained in good yield (50%) but in favour of the thermodynamically more stable α-anomer ($\beta/\alpha \sim 1:3$). With $Ph_3PAuOTf$ as the catalyst, however, the reaction of **2** and **29** provided a messy mixture. Careful isolation led to identification of a pentasaccharide lactone, which was derived from cleavage of the FABO linkage in the coupled hexasaccharide (**30**). Encouragingly, the originally formed cymarosyl-(1→4)-linkage in the resultant pentasaccharide lactone was found in the required β configuration, implying that the glycosylation reaction took place via a glycosyl α-triflate intermediate[18]. Nevertheless, the HOTf generated after glycosylation, before being quenched by the isochromen-4-yl gold(I) complex[45], degraded the FABO linkage. On the basis of this rational, we introduced a hindered base, that is, 2,4,6-tri-*tert*-butylpyrimidine (TTBP), into the present $Ph_3PAuOTf$-catalysed glycosylation to rest the incipient HOTf, so as to avoid the cleavage of the FABO linkage. The sequester of the HOTf also retarded the gold(I) catalytic cycle, therefore, requiring almost equivalent of the gold complex to drive the reaction to completion[45,46]. In fact, the condensation of **2** and **29** proceeded smoothly in the presence of $Ph_3PAuOTf$ (0.8 equiv) and TTBP (1.5 equiv) in $CH_2Cl_2$ at −10 °C, leading to the coupled hexasaccharide **30** in a satisfactory yield of 64% (87%, b.r.s.m.) and β/α ratio of 2.1:1. Finally, removal of the terminal CA group and TBS group was achieved with thiourea[47] and pyridine-buffered HF · pyridine, respectively, furnishing the target pregnane hexasaccharide **31** cleanly (91%, two steps).

The $^1H$ and $^{13}C$ NMR spectra of **31** are similar to those of the authentic periploside A (**1**) (see Supplementary Figs 60 and 61). However, discrepancies occur for the signals from the formyl acetyl $CH_2$ residue (88.4 p.p.m. for the $^{13}C$ signal in **31** versus 86.4 p.p.m. in **1**; 4.93 and 4.82 p.p.m. for the $^1H$ signals in **31** versus 5.13 and 4.74 p.p.m. in **1**) (see Supplementary Table 3). In addition, the adjacent C4′, C2″ and C3″ are also in disparate chemical shifts (C4′: 81.5 versus 79.2 p.p.m.; C2″: 37.9 versus 36.7 p.p.m.; C3″: 77.4 versus 78.3 p.p.m., in **31** and **1**, respectively). These discrepancies indicate that the synthetic **31** is indeed the epimer of periploside A (**1**) with the opposite configuration at the quaternary C1″. The problem of diastereoselectivity in the previous formation of the FABO motif is thus addressed.

**Construction of the FABO motif of the natural configuration.** The addition of the hydroxyl group onto the ketene acetal in **A** proceeded exclusively from the α face, leading to the FABO disaccharide with the unnatural configuration (**9**→**25**). The presence of a nucleophile capable of interception of the oxo-carbenium species developed from the ketene acetal might lead to a thermodynamically favored α-intermediate; substitution of the intermediate *in situ* by the 4-hydroxyl group would then proceed from the β face to provide the FABO disaccharide with the natural configuration. Therefore, we tried the elimination/cyclization (from **9**) in various nucleophilic solvents, including $CH_3CN$, 1,4-dioxane, THF, DME, DMF and MeOH, under varied temperatures; however, the only FABO disaccharide identified was **25**. Addition of nucleophilic additives, such as 4-dimethylaminopyridine (DMAP), LiBr, NaBr and NaI, was also found futile. The presence of an electrophilic reagent, such as NIS, NBS and $I_2$, might convert the ketene acetal into a 1,2-halonium intermediate, which could then be attacked by the 4-hydroxyl group to give the FABO disaccharide. These attempts were again unsuccessful.

The failure in formation of the seven-membered orthoester in the desired stereochemistry forced us to construct the orthoester before cyclization of the formyl acetal (Fig. 4). Thus, 2-phenylseleno-D-mannopyranoside **22** was converted into the fluoride (with DAST), which was then coupled with sugar acceptor **32** (see Supplementary Methods) under the action of $SnCl_2$ to give α-disaccharide **33** (85%). Gratifyingly, after intensive attempts, we managed to obtain the ketene acetal **34** in excellent yield (91%) from disaccharide **33** via selenoxide formation and *syn*-elimination under the modified Sinaÿ conditions ($NaIO_4$, $NaHCO_3$, $MeOH/CH_2Cl_2/H_2O$, RT; then vinyl acetate/toluene/DIPA, microwave 140 °C, 40 min). Although purification of **34** on silica gel required the addition of 1% $Et_3N$ in the eluent to prevent hydrolysis, the purified ketene acetal **34** remained stable at −20 °C for several days. To the best of our knowledge, this is the first time a ketene acetal linked disaccharide (that is, **34**) being isolated. Noteworthy is the decisive role played by the 3-*O*-Lev group (in **33**–**35**), which facilitated the procurement of the ketene acetal disaccharide as well as the selective removal afterwards; in contrast, analogues of **33** bearing an electron-donating group (for example, TBS) at the 3-OH led to decomposed monosaccharide derivatives under identical conditions.

Addition of $PhSCH_2OH$ onto ketene acetal **34** was first attempted in $CDCl_3$ at 50 °C (ref. 48), the starting ketene acetal decomposed largely in 12 h, with the desired orthoester being isolated in low yield (20%) as a single α-isomer. By raising the

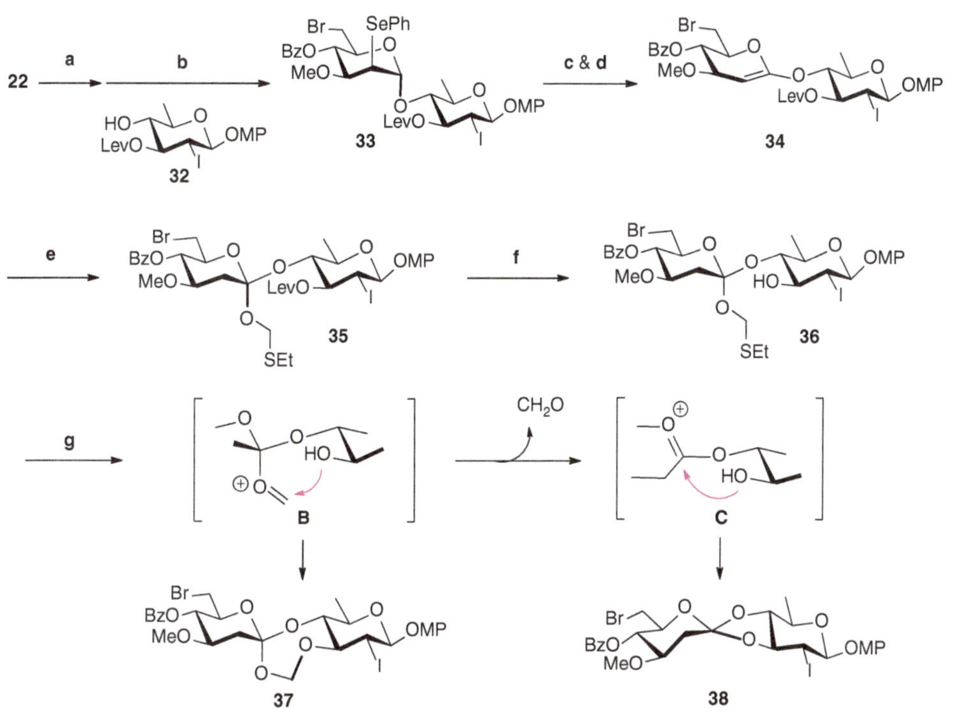


**Figure 4 | Construction of the FABO motif with the natural configuration.** Highlighted in red are the nascent glycosidic bonds. (**a**) DAST, THF, −30 °C to RT; (**b**) $SnCl_2$, $Et_2O$, 4 Å MS, 0 °C to RT; 85% (for two steps); (**c**) $NaIO_4$, $NaHCO_3$, MeOH, $CH_2Cl_2$, $H_2O$, RT, 99%; (**d**) vinyl acetate, toluene, DIPA, Mw, 140 °C, 40 min, 92%; (**e**) $EtSCH_2OH$, $DCCl_3$, Mw, 10 min, 110 °C, 81%; (**f**) $H_2NNH_2 \cdot H_2O$, pyridine, HOAc, 0 °C to RT, 92%; (**g**) BSP, $Tf_2O$, DTBP, 5 Å MS, $Et_2O$, −114 °C, 64% (for **37**), 18% (for **38**). BSP, 1-benzenesulfinyl piperidine; DTBP, 2,6-di-*tert*-butylpyridine; Lev, levuloyl; Mw, microwave.

temperature and shortening the reaction duration ($CDCl_3$, microwave 100 °C, 1 h), the addition of **34** and $PhSCH_2OH$ led to the desired orthoester in a good 70% yield but in a moderate diastereoselectivity of 1.4/1. Modification of the reaction conditions by the addition of acidic promoters (*p*-TsOH or PPTS) or variation of solvent (toluene, $CH_2Cl_2$ or $CHCl_3$) failed to improve the yield or the diastereoselectivity. Fortunately, replacement of $PhSCH_2OH$ with the more nucleophilic $EtSCH_2OH$, the addition with ketene acetal **34** under the optimal conditions ($CDCl_3$, microwave 110 °C, 10 min) furnished the desired orthoester **35** in 81% yield as a single α-isomer. It should be noted that the replacement of $CDCl_3$ with $CHCl_3$ or toluene led to lower yields of **35** (<65%). Selective removal of the 3-*O*-Lev group in **35** was achieved with $H_2NNH_2 \cdot H_2O$ to provide orthoester **36** (92%).

Treatment of ethylthiomethyl glycoside **36** with the mild promoters for glycosidation of thioglycosides (see Supplementary Table 2), such as MeOTf/TTBP, DMTST/TTBP, and $CuBr_2$/$Bu_4NBr$[49], failed to afford the seven-membered FABO derivative **37**; instead, the five-membered orthoester **38** was isolated in high yield (>80%) as a mixture of the two diastereoisomers. These results indicate again (*cf.*, **10→23/24**) that the glycosyloxymethyl cation (**B**) generated via activation of the thioacetal could readily undergo decomposition to the glycosyl oxocarbenium (**C**) and formaldehyde; intramolecular addition of the proximal hydroxyl group to **C** led to the five-membered orthoester. Thus, a strong promoter which can activate the thioacetal under milder conditions might be able to allow the transient glycosyloxymethyl cation **B** to be captured by the hydroxyl group before decomposition. Using Crich's conditions ($Tf_2O$/BSP/TTBP, −60 °C, $CH_2Cl_2$)[37], however, the five-membered orthoester **38** was again obtained nearly quantitatively as a mixture of the diastereoisomers. Encouragingly, when lowering the reaction temperature to −78 °C, a trace amount of the desired seven-membered FABO product **37** was detected. Optimizing along this direction, we finally managed to obtain the desired **37** in a satisfactory 64% yield in the presence of $Tf_2O$/BSP/DTBP in $Et_2O$ at −114 °C, whereas the five-membered orthoester **38** was isolated in 18% yield as a single diastereoisomer. Comparison of the NMR data of the present FABO disaccharide **37** with those of the previous **26** supported that we had fixed the desired configuration by the present approach.

**Completion of the synthesis of periploside A (1).** The transformations previously developed for the synthesis of C1″-*epi*-periploside A (**26→31**, eight steps, 28% overall yield) were applied to the synthesis of periploside A (Fig. 5). Thus, FABO disaccharide **37** was subjected to selective cleavage of the anomeric MP group (with $Ag(DPAH)_2$, 91%) and conversion into the trifluoroacetimidate donor **6** (90%). Coupling of pregnane diol **7** with donor **6** under the similar conditions as that for **7+27→28** (85%, β/α = 3:1) led to the corresponding disaccharide **39** in a similar yield (87%) but much higher β selectivity (β/α = 6.1:1). This result supports the previous assumption that the conformational restraint provided by the FABO motif affects the glycosidation transition state of the disaccharide. The bromide and iodide in **39** were removed cleanly with $Et_3B$/$Bu_3SnH$ (95%); subsequent cleavage of the 4′-*O*-benzoyl group (NaOMe, HOMe, 96%) provided the desired disaccharide acceptor **3**. Condensation of **3** with tetrasaccharide donor **2** under the optimized conditions for **29+2→30** (64%, β/α = 2.1:1) afforded hexasaccharide **40** in a higher 80% (92%, b.r.s.m.) yield and similar β/α ratio of 2:1. Finally, the terminal CA and TBS groups were removed successfully with thiourea and pyridine-buffered HF-pyridine, respectively, furnishing the target periploside A (**1**) (93%, two steps). Gratifyingly, all the analytical data of **1** are in full agreement with those obtained for the natural product[1] (see Supplementary Table 3).

**The immunosuppressive activities of the synthetic compounds.** The synthetic periploside A (**1**) showed similar toxicity and

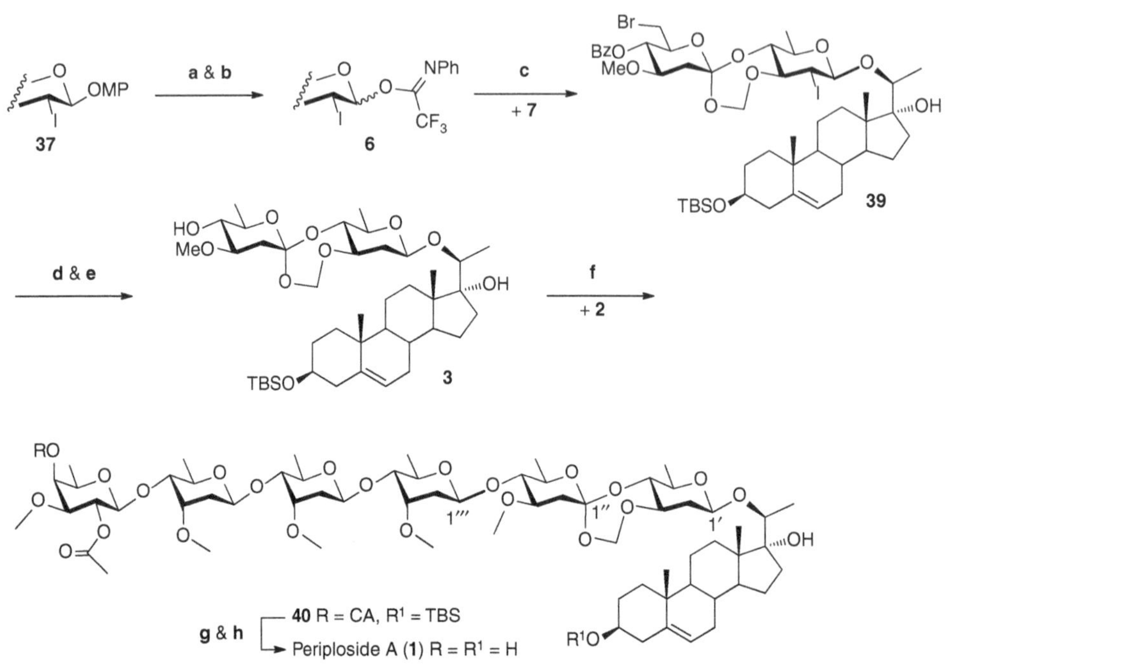


**Figure 5 | Completion of the total synthesis of periploside A (1).** Highlighted in red are the nascent glycosidic bonds. (**a**) $Ag(DPAH)_2$, $CH_3CN$, $H_2O$, 0 °C, 91%; (**b**) *N*-phenyl-trifluoroacetimidoyl chloride, $Cs_2CO_3$, $CH_2Cl_2$, RT, 90%; (**c**) TBSOTf, $CH_2Cl_2$, 5 Å MS, −78 °C, 87%, β/α = 6.1/1; (**d**) $Et_3B$, $Bu_3SnH$, toluene, RT, 95%; (**e**) MeONa, MeOH, $CH_2Cl_2$, RT, 96%; (**f**) $Ph_3PAuOTf$ (0.8 equiv), TTBP, $CH_2Cl_2$, 4 Å MS, −10 °C, 80%, β/α = 2/1; (**g**) thiourea, pyridine, EtOH, 80 °C; (**h**) HF·py, pyridine, THF, 0 °C to RT, 93% (for two steps).

**Table 1 | Inhibitory activity of periploside A and analogues against T-lymphocyte proliferation*.**

| Compound | $CC_{50}$ (μM) | $IC_{50}$ (μM) | SI |
|---|---|---|---|
| Periploside A (**1**) | 2.10 ± 0.32 | 0.17 ± 0.04 | 12 |
| C1″-*epi*-periploside A (**31**) | 1.58 ± 0.19 | 0.41 ± 0.15 | 3.9 |
| C1‴-*epi*-periploside A (**41**) | 11.40 ± 3.66 | 1.96 ± 0.43 | 5.8 |
| Disaccharide **42** | 12.80 ± 0.52 | 6.20 ± 3.20 | 2.1 |
| Disaccharide **43** | 27.70 ± 6.61 | 3.70 ± 2.25 | 7.5 |

*See Supplementary Methods and Supplementary Figs 62 and 63 for details.

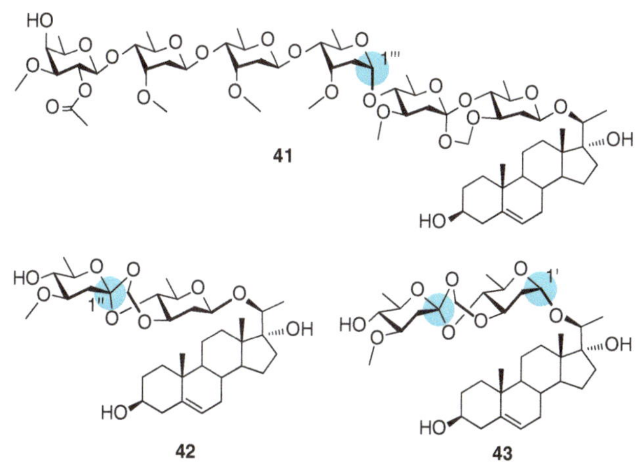


**Figure 6 | Synthetic analogues of periplosides.** Shadowed in pale blue are the carbon centres with opposite configurations to those in the nature product.

activity as the natural product against T-lymphocyte and their proliferation induced by ConA[5,6]. The C1″-epimer (**31**) which has the opposite configuration at the FABO motif as in periploside A was also found active ($IC_{50}$ = 0.41 μM), although being ~2.4-fold less potent than the natural product (Table 1). The C1‴-epimer of periploside A (**41**) was also active ($IC_{50}$ = 1.96 μM) (Fig. 6). Interestingly, disaccharide **42**, with the tetrasaccharide in **31** being truncated, still had certain activity ($IC_{50}$ = 6.20 μM); and its α-anomer **43** even showed a higher activity ($IC_{50}$ = 3.70 μM). Previous assays have demonstrated that reductive cleavage of the formyl acetal in the FABO motif in periploside A abolished the activity completely[6]. Taken together, it demonstrates that the chemical connection of the FABO motif, rather than the stereochemistry, has a critical role in the immunosuppressive activities of periplosides. The capability of releasing a formylaldehyde might be taken into account in the future researches into the mechanism of action of this unique type of compounds.

## Methods

**General methods.** All the reactions were carried out under nitrogen or argon with anhydrous solvents in flame-dried glassware, unless otherwise noted. All the glycosylation reactions were performed in the presence of 4 or 5 Å MS, which were flame-dried immediately before use in the reaction under high vacuum. Glycosylation solvents were dried using a solvent purification system and used directly without further drying. The chemicals used were reagent grade as supplied, except where noted. For details of the synthetic procedures and the characterization data of compounds, see Supplementary Methods. For $^{1}H$ and $^{13}C$ NMR spectra of the compounds prepared in this study, see Supplementary Figs 1–58.

**Synthesis of acetal glycoside 24.** To a solution of sugar alcohol **11** (620 mg, 1.47 mmol) in $CH_2Cl_2$ (5 ml) was added acid-washed 3 Å MS at RT. After stirring for 30 min at RT, NIS (142 mg, 0.63 mmol) and TfOH (2.6 μl, 0.029 mmol) were added to the mixture at −30 °C, followed by the addition of a solution of phenylthiomethyl glycoside **10** (183 mg, 0.29 mmol) in $CH_2Cl_2$ (3 ml) via a syringe pump. The reaction mixture was stirred at −30 °C for 1 h. When TLC showed the donor **10** had been consumed, saturated aqueous $NaHCO_3$ was added at 0 °C. The mixture was filtered. The filtrate was washed with a solution of $NaS_2O_3$ and brine, respectively, and was then dried over $Na_2SO_4$ and concentrated. The residue was purified by flash chromatography (petroleum ether/$CH_2Cl_2$/EtOAc = 8:1:1) to afford **24** (206 mg, 75%) and glycoside **23** (53 mg, 20%) as syrups.

**Synthesis of FABO disaccharide 25.** To a solution of acetal glycoside **9** (187 mg, 0.24 mmol) in MeOH/$CH_2Cl_2$/$H_2O$ (3 ml/2 ml/1 ml) were added $NaIO_4$ (507 mg, 2.37 mmol) and $NaHCO_3$ (159 mg, 1.90 mmol) at RT. After stirring for 3 h, the mixture was diluted with $CH_2Cl_2$, washed with saturated $NH_4Cl$ solution and brine, respectively, and was then dried over $Na_2SO_4$ and concentrated. The resultant selenoxide was azeotroped with toluene (3 × 5 ml) and dried under high vacuum for 2 h to afford a white solid (190 mg, 99%). The above selenoxide (125 mg, 0.155 mmol) was dissolved in toluene (6 ml). Diisopropylamine (3 ml) and vinyl acetate (6 ml) were added, and the reaction was conducted under microwave at 145 °C for 20 min. The mixture was cooled to RT and concentrated. The residue was purified by flash chromatography on silica gel (petroleum ether/EtOAc = 1.5:1) to afford **25** (83 mg, 85%) as a colourless syrup.

**Synthesis of α-disaccharide 33.** To a solution of lactol **22** (603 mg, 1.21 mmol) in THF (10 ml) was added DAST (0.44 ml, 3.63 mmol) at −30 °C. After stirring for 1.5 h while warming to RT, a saturated $NaHCO_3$ solution was added slowly to the mixture. The resulting mixture was extracted with $CH_2Cl_2$. The combined organic layer was washed with brine, dried over $Na_2SO_4$ and concentrated. The crude glycosyl fluoride was azeotroped with toluene (3 × 5 ml). After drying under high vacuum for 2 h, the above product was dissolved in $Et_2O$ (10 ml), 4 Å MS (1.2 g) and sugar alcohol **32** (330 mg, 0.69 mmol) were added and the reaction mixture was stirred at 0 °C for 30 min. $SnCl_2$ (235 mg, 1.24 mmol) was added in one portion and the reaction mixture was allowed to warm to RT and stirred for 4 h. The mixture was quenched with $Et_3N$ (1 ml) and filtered. The solution was diluted with EtOAc and washed with water. The water layer was extracted with EtOAc twice. The combined organic layer was washed with saturated $NaHCO_3$ solution and brine, respectively, and was then dried over $Na_2SO_4$ and concentrated. The residue was purified by flash chromatography (petroleum ether/EtOAc = 4:1) to afford **33** (560 mg, 85%) as a colourless syrup.

**Synthesis of ketene acetal 34.** To a solution of **33** (248 mg, 0.26 mmol) in MeOH/$CH_2Cl_2$/$H_2O$ (3 ml/2 ml/1 ml) were added $NaIO_4$ (552 mg, 2.58 mmol) and $NaHCO_3$ (173 mg, 2.06 mmol) at RT. After stirring for 12 h at RT, the mixture was diluted with $CH_2Cl_2$, and washed with saturated $NH_4Cl$ solution and brine, respectively. The organic layer was dried over $Na_2SO_4$ and concentrated. The crude selenoxide was azeotroped with toluene (3 × 5 ml) and dried under high vacuum for 2 h to afford a colourless syrup (251 mg, 99%). The above selenoxide (104 mg, 0.11 mmol) was dissolved in toluene (2 ml). Diisopropylamine (1 ml) and vinyl acetate (2 ml) were added. The reaction was conducted under microwave at 140 °C for 40 min. The mixture was cooled to RT and concentrated. The residue was purified by flash chromatography (petroleum ether/EtOAc = 5:1, containing 1% $Et_3N$) to afford **34** (85 mg, 92%) as a colourless syrup.

**Synthesis of orthoester 35.** To a solution of **34** (75 mg, 0.093 mmol) in $CDCl_3$ (3 ml) was added $EtSCH_2OH$ (0.05 ml) at RT. The reaction was conducted under microwave at 110 °C for 10 min. The mixture was cooled to RT and concentrated. The residue was purified by flash chromatography (petroleum ether/EtOAc = 5:1, containing 1% $Et_3N$) to afford **35** (67 mg, 81%) as a colourless syrup.

**Synthesis of orthoester 36.** To a solution of **35** (110 mg, 0.120 mmol) in pyridine/HOAc (3 ml/2 ml) was added $H_2NNH_2$ $H_2O$ (0.10 ml, 1.60 mmol) at 0 °C. After stirring at RT for 5 h, the mixture was diluted with $CH_2Cl_2$, and washed with ice water and then with a saturated $NaHCO_3$ solution and brine, respectively. The organic layer was dried over $Na_2SO_4$ and concentrated. The residue was purified by flash chromatography (petroleum ether/EtOAc = 5:1, containing 1% $Et_3N$) to afford **36** (90 mg, 92%) as a colourless syrup.

**Synthesis of FABO disaccharide 37.** To a solution of **36** (19.1 mg, 0.024 mmol) in $Et_2O$ (3 ml) were added BSP (8.2 mg, 0.036 mmol), 2,6-di-*tert*-butylpyridine (16.0 μl, 0.072 mmol) and 5 Å MS at RT. After stirring at −114 °C (liq. $N_2$-EtOH) for 20 min, $Tf_2O$ (6.0 μl, 0.036 mmol) was added to the mixture. The mixture was stirred at −114 °C for 1 h, and was then warmed to RT and filtered. The filtrate was washed with a saturated aqueous $NaHCO_3$ solution and brine, respectively, and was then dried over $Na_2SO_4$ and concentrated. The residue was purified by flash chromatography on silica gel (petroleum ether/EtOAc = 5:1) to afford **37** (11.3 mg, 64%) and the five-membered orthoester **38** (3.0 mg, 18%) as colourless syrups.

**Synthesis of FABO disaccharide trifluoroacetimidate 6.** To a solution of *p*-methoxyphenyl glycoside **37** (36 mg, 0.049 mmol) in $CH_3CN/H_2O$ (3 ml/3 ml) was added $Ag(DPAH)_2$ (79 mg, 0.17 mmol) at 0 °C. After stirring for 30 min at this temperature, the mixture was filtered. The filtrate was diluted with $CH_2Cl_2$, washed with saturated $NaHCO_3$ solution and brine, respectively, and was then dried over $Na_2SO_4$ and concentrated. The residue was purified by flash chromatography (petroleum ether/EtOAc = 3:1) to yield the corresponding hemiacetal (28 mg, 91%) as a colourless syrup. To a solution of the hemiacetal (29 mg, 0.046 mmol) in $CH_2Cl_2$ (3 ml) were added $Cs_2CO_3$ (75 mg, 0.23 mmol) and *N*-phenyl-2,2,2-trifluoroacetimidoyl chloride (14 µl, 0.14 mmol) at RT. After stirring for 3 h, the mixture was filtered. The filtrate was evaporated in vacuo to give a residue, which was subjected to chromatography on Davisil silica (pH = 7.0, petroleum ether/EtOAc, 5:1) to give **6** (33 mg, 90%) as a colourless syrup. This compound was used directly without further characterization.

**Synthesis of pregnane β-disaccharide 39.** To a solution of imidate **6** (33.0 mg, 0.041 mmol) and pregnane diol **7** (15.6 mg, 0.035 mmol) in $CH_2Cl_2$ (3 ml) was added 5 Å MS at RT. After stirring at − 78 °C for 30 min, TBSOTf (1.2 µl, 0.0052 mmol) was added to the mixture. After stirring for 6 h at this temperature, $Et_3N$ was added to quench the reaction. The resulting mixture was filtered. The filtrate was evaporated in vacuo to give a residue, which was purified by flash chromatography (petroleum ether/$CH_2Cl_2$/EtOAc = 10:5:1) to afford **39** (27.6 mg, 75%) and its α-anomer (4.5 mg, 12%) as white solids.

**Synthesis of pregnane disaccharide 3.** To a solution of **39** (21.0 mg, 0.020 mmol) in toluene (2 ml) were added $Bu_3SnH$ (32 µl, 0.12 mmol) and $Et_3B$ (12 µl, 0.012 mmol) at 0 °C. After stirring for 1 h at RT, the mixture was concentrated in vacuo. The residue was purified by flash chromatography (petroleum ether/EtOAc = 5:1) to afford the corresponding 2′,6″-deoxy derivative (16.5 mg, 97%) as a white solid (see Supplementary Methods). To a solution of this compound (16.5 mg, 0.019 mmol) in $CH_2Cl_2$/MeOH (1.5 ml/1.5 ml) was added NaOMe (20 mg, 0.37 mmol) at RT. After stirring for 40 h, the mixture was filtered through silica gel. The filtrate was evaporated in vacuo to give a residue, which was purified by flash chromatography (petroleum ether/EtOAc = 2:1) to afford **3** (14.0 mg, 97%) as a white solid.

**Synthesis of pregnane hexasaccharide 40.** To a solution of tetrasaccharide *ortho*-cyclopropylethynylbenzoate **2** (22.3 mg, 0.025 mmol), pregnane disaccharide **3** (8.0 mg, 0.011 mmol) and 2,4,6-tri-*tert*-butylpyrimidine (TTBP) (4.0 mg, 0.016 mmol) in $CH_2Cl_2$ (2 ml) was added 4 Å MS at RT. After stirring for 30 min at − 20 °C, a solution of $PPh_3AuOTf$ in $CH_2Cl_2$ (0.05 ml, 0.1 M) was added to the mixture. The mixture was stirred for 2 h while warming to − 10 °C, then another portion of $PPh_3AuOTf$ in $CH_2Cl_2$ (0.05 ml, 0.1 M) was added to the mixture. After stirring for 4 h at − 10 °C, $Et_3N$ was added to quench the reaction. The resulting mixture was filtered and concentrated. The residue was purified by flash chromatography (petroleum ether/$CH_2Cl_2$/EtOAc = 1:1:1) to afford **40** (8.3 mg, 53%), its α-anomer (4.2 mg, 27%) as white foams, and recovered **3** (1.0 mg, 13%).

**Synthesis of Periploside A 1.** To a solution of **40** (4.7 mg, 3.2 µmol) in pyridine/EtOH (1.0 ml/1.0 ml) was added thiourea (10 mg, 0.13 mmol) at RT. After stirring for 2 h at 80 °C, the mixture was concentrated in vacuo to give a residue, which was purified by flash chromatography ($CHCl_3$/MeOH = 30:1) to afford a colourless syrup. The syrup was dissolved in THF/pyridine (1.5 ml/0.75 ml). HF-py (70% HF in pyridine, 0.10 ml) was added dropwise at 0 °C. After stirring for 40 h at RT, a saturated $NaHCO_3$ solution was added slowly to the mixture at RT. The resulting mixture was diluted with $CH_2Cl_2$, washed with saturated $NaHCO_3$ solution and was then extracted with $CH_2Cl_2$ twice. The combined organic layer was washed with brine, dried over $Na_2SO_4$ and concentrated. The residue was purified by flash chromatography ($CHCl_3$/MeOH = 30:1) to afford periploside A (**1**) (3.8 mg, 93%) as a white foam.

## References


1. Oshima, Y., Hirota, T. & Hikino, H. Periplosides A, B and C, steroidal glycosides of *Periploca sepium* root barks. *Heterocycles* **26,** 2093–2098 (1987).
2. Itokawa, H. *et al.* Studies on chemical constituents of antitumor fraction from *Periploca sepium*. II. Structures of new pregnane glycosides, periplocosides A, B and C. *Chem. Pharm. Bull.* **36,** 982–987 (1988).
3. Itokawa, H. *et al.* Studies on chemical constituents of antitumor fraction from *Periploca sepium*. IV. Structures of new pregnane glycosides, periplocosides D, E, L and M. *Chem. Pharm. Bull.* **36,** 2084–2089 (1988).
4. Itokawa, H. *et al.* Studies on chemical constituents of antitumor fraction from *Periploca sepium*. V. Structures of new pregnane glycosides, periplocosides J, K, F and O. *Chem. Pharm. Bull.* **36,** 4441–4446 (1988).
5. Feng, J. Q. *et al.* Immunosuppressive pregnane glycosides from *Periploca sepium* and *Periploca forrestii*. *Phytochemistry* **69,** 2716–2723 (2008).
6. Wang, L.-Y. *et al.* Structural revision of periplocosides and periperoxides, natural immunosuppressive agents from the genus *Periploca*. *Phytochemistry* **72,** 2230–2236 (2011).
7. Wang, L.-Y. *et al.* Corrigendum to "Structural revision of periplocosides and periperoxides, natural immunosuppressive agents from the genus *Periploca*". *Phytochemistry* **95,** 445 (2013).
8. Zhu, Y.-N. *et al.* Periplocoside E, an effective compound from *Periploca sepium* Bge, inhibited T cell proliferation *in vitro* and *in vivo*. *J. Pharmacol. Exp. Ther.* **316,** 662–669 (2006).
9. Zhu, Y.-N. *et al.* Periplocoside E inhibits experimental allergic encephalomyelitis by suppressing IL-12-dependent CCR5 expression and IFN-γ-dependent CXCR3 expression in T lymphocytes. *J. Pharmacol. Exp. Ther.* **318,** 1153–1162 (2006).
10. Zhao, W. *et al.* Use of periplocoside A and periplocoside E for preparing immunosuppressants used for preventing and treating autoimmune diseases such as rheumatoid arthritis, autoimmune hepatitis and insulin-dependent diabetes. WO2007036113-A1; CN1939324-A; CN1939324-B.
11. Jiangsu New Medical College. *Dictionary of Chinese Herb Medicines* 1683 and 2388A (Shanghai Scientific and Technologic Press, 1998).
12. Mann, J. Natural products as immunosuppressive agents. *Nat. Prod. Rep.* **18,** 417–430 (2001).
13. Yu, B., Sun, J. & Yang, X. Assembly of naturally occurring glycosides, evolved tactics, and glycosylation methods. *Acc. Chem. Res.* **45,** 1227–1236 (2012).
14. Borovika, A. & Nagorny, P. Recent advances in the synthesis of natural 2-deoxy-β-glycosides. *J. Carbohydr. Chem.* **31,** 255–283 (2012).
15. Ma, Y., Li, Z., Shi, H., Zhang, J. & Yu, B. Assembly of digitoxin by gold(I)-catalyzed glycosidation of glycosyl *o*-alkynylbenzoates. *J. Org. Chem.* **76,** 9748–9756 (2011).
16. Zhang, J., Shi, H., Ma, Y. & Yu, B. Expeditious synthesis of saponin P57, an appetite suppressant from *Hoodia* plants. *Chem. Commun.* **48,** 8679–8681 (2012).
17. Zhang, S., Ma, Y., Ma, J., Yu, B. & Xie, X. Molecular matchmaking between the popular 'weight loss' herb *Hoodia gordonii* and GPR119, a potential drug target for metabolic disorders. *Proc. Natl Acad. Sci. USA* **111,** 14571–14576 (2014).
18. Tang, Y., Li, J., Zhu, Y., Li, Y. & Yu, B. Mechanistic insights into the gold(I)-catalyzed activation of glycosyl *ortho*-alkynylbenzoates for glycosidation. *J. Am. Chem. Soc.* **135,** 18396–18405 (2013).
19. Yu, B. & Sun, J. Glycosylation with glycosyl *N*-phenyltrifluoroacetimidates (PTFAI) and a perspective of the future development of new glycosylation methods. *Chem. Commun.* **46,** 4668–4679 (2010).
20. Roush, W. R. & Bennett, C. E. A highly stereoselective synthesis of 2-deoxy-β-glycosides using 2-deoxy-2-iodo-glucopyranosyl acetate donors. *J. Am. Chem. Soc.* **121,** 3541–3542 (1999).
21. Crich, D. & Yang, F. Phenylthiomethyl glycosides: convenient synthons for the formation of azidomethyl and glycosylmethyl glycosides and their derivatives. *Angew. Chem. Int. Ed.* **48,** 8896–8899 (2009).
22. Jaurand, G., Beau, J.-M. & Sinaÿ, P. Organoselenium chemistry: stereoselective conversion of glycals into anomeric spiro-orthoesters using a glycosyloxyselenation-oxidation elimination sequence. *J. Chem. Soc. Chem. Commun.* 701–703 (1982).
23. Nicolaou, K. C. *et al.* Total synthesis of everninomicin 13,384-1-part 2: synthesis of the $FGHA_2$ fragment. *Angew. Chem. Int. Ed.* **38,** 3340–3345 (1999).
24. Nicolaou, K. C. *et al.* 1,2-Seleno migrations in carbohydrate chemistry: solution and solid-phase synthesis of 2-deoxy glycosides, orthoesters, and allyl orthoesters. *Angew. Chem. Int. Ed.* **39,** 1089–1093 (2000).
25. Zhou, M. & O'Doherty, G. A. A stereoselective synthesis of digitoxin and digitoxigen mono- and bisdigitoxoside from digitoxigenin via a palladium-catalyzed glycosylation. *Org. Lett.* **8,** 4339–4342 (2006).
26. Kadota, I. & Yamamoto, Y. Stereocontrolled total synthesis of hemibrevetoxin B. *J. Org. Chem.* **63,** 6597–6606 (1998).
27. Tang, Y. & Yu, B. Identification of (phosphine)gold(I) hydrates and their equilibria in wet solutions. *RSC Adv.* **2,** 12686–12689 (2012).
28. Noshita, T., Sugiyama, T., Kitazumi, Y. & Oritani, T. Phenolic Ferrier reaction and its application to the natural product synthesis. *Tetrahedron Lett.* **35,** 8259–8262 (1994).
29. Horito, S., Asano, K., Umemura, K., Hashimoto, H. & Yoshimura, J. A new synthesis of 2,3,4,6-tetra-*O*-benzyl-D-glucopyranosylidene acetals, using trimethylsilyl trifluoromethanesulfonate as the catalyst. *Carbohydr. Res.* **121,** 175–185 (1983).
30. Ohtake, H., Ichiba, N. & Ikegami, S. Synthesis and structure of 4-*O*,6-*O*-glycosylidene glycosides. *J. Org. Chem.* **65,** 8164–8170 (2000).
31. Durham, T. B., Blanchard, N., Savall, B. M., Powell, N. A. & Roush, W. R. Total synthesis of formamicin. *J. Am. Chem. Soc.* **126,** 9307–9317 (2004).
32. Praly, J.-P. & Descotes, G. Photocyclisation d'hydroxyalkyl glycosides en orthoesters glycosidiques. *Tetrahedron Lett.* **23,** 849–852 (1982).
33. Martín, A., Pérez-Martín, I., Quintanal, L. M. & Suárez, E. Intramolecular 1,8-versus 1,6-hydrogen atom transfer between pyranose units in α (1→4)-

disaccharide model promoted by alkoxyl radicals. Conformational and stereochemical requirements. *Org. Lett.* **9,** 1785–1788 (2007).
34. Nicolaou, K. C. *et al.* Total synthesis of apoptolidin: construction of enantiomerically pure fragments. *J. Am. Chem. Soc.* **125,** 15433–15442 (2003).
35. Lian, G., Zhang, X. & Yu, B. Thioglycosides in carbohydrate research. *Carbohydr. Res.* doi:10.1016/j.carres.2014.06.009 (in the press).
36. Sasaki, M. & Tachibana, K. An efficient and stereocontrolled synthesis of the nephritogenoside core structure. *Tetrahedron Lett.* **32,** 6873–6876 (1991).
37. Crich, D. & Smith, M. 1-Benzenesulfinyl piperidine/trifluoromethanesulfonic anhydride: a potent combination of shelf-stable reagents for the low-temperature conversion of thioglycosides to glycosyl triflates and for the formation of diverse glycosidic linkages. *J. Am. Chem. Soc.* **123,** 9015–9020 (2001).
38. Fügedi, P. & Garegg, P. J. A novel promoter for the efficient construction of 1,2-trans linkages in glycoside synthesis, using thioglycosides as glycosyl donors. *Carbohydr. Res.* **149,** C9–C12 (1986).
39. Ishiwata, A., Sakurai, A., Nishimiya, Y., Tsuda, S. & Ito, Y. Synthetic study and structural analysis of the antifreeze agent xylomannan from *Upis ceramboides. J. Am. Chem. Soc.* **133,** 19524–19535 (2011).
40. Konradsson, P., Udodong, U. E. & Fraser-Reid, B. Iodonium promoted reactions of disarmed thioglycosides. *Tetrahedron Lett.* **31,** 4313–4316 (1990).
41. Schmidt, R. R. & Toepfer, A. Glycosylation with highly reactive glycosyl donors: efficiency of the inverse procedure. *Tetrahedron Lett.* **32,** 3353–3356 (1991).
42. Yang, M. T. & Woerpel, K. A. The effect of electrostatic interactions on conformational equilibria of multiply substituted tetrahydropyran oxocarbenium ions. *J. Org. Chem.* **74,** 545–553 (2009).
43. Chamberland, S., Ziller, J. W. & Woerpel, K. A. Structural evidence that alkoxy substituents adopt electronically preferred pseudoaxial orientations in six-membered ring dioxocarbenium ions. *J. Am. Chem. Soc.* **127,** 5322–5323 (2005).
44. Miura, K. *et al.* Triethylborane-induced hydrodehalogenation of organic halides by tin hydrides. *Bull. Chem. Soc. Jpn* **62,** 143–147 (1989).
45. Zhu, Y. & Yu, B. Characterization of the isochromen-4-yl-gold(I) intermediate in the gold(I)-catalyzed glycosidation of glycosyl *ortho*-alkynylbenzoates and enhancement of the catalytic efficiency thereof. *Angew. Chem. Int. Ed.* **50,** 8329–8332 (2011).
46. Nie, S., Li, W. & Yu, B. Total synthesis of nucleoside antibiotic A201A. *J. Am. Chem. Soc.* **136,** 4157–4160 (2014).
47. Cook, A. F. & Maichuk, D. T. Use of chloroacetic anhydride for the protection of nucleoside hydroxyl groups. *J. Org. Chem.* **35,** 1940–1943 (1970).
48. Adam, W. & Wang, X. *E/Z* Isomerization, solvolysis, addition, and cycloaddition reactions of *(E)-tert*-butylketene methyl *tert*-butyldimethylsilyl acetal. *J. Org. Chem.* **56,** 7244–7250 (1991).
49. Sawada, D. & Ito, Y. A new method for formacetal linkage formation: protection of alcohols, phenols and carboxylic acids. *Tetrahedron Lett.* **42,** 2501–2504 (2001).

## Acknowledgements

This work is financially supported by the National Natural Science Foundation of China (21432012 and 91213301), the Ministry of Sciences and Technology of China (2012ZX09502-002) and the E-Institute of Shanghai Municipal Education Commission (E09013). We are grateful to Professor Wei-Min Zhao for providing an authentic sample of the natural product periploside A and to Professor Tian-Xiang Li for providing a photo of the plant.

## Author contributions

X.Z. and B.Y. conceived the synthetic route. X.Z. conducted the synthetic work. Y.Z. and J.Z. conducted bioassay. B.Y. and X.Z. wrote the manuscript.

## Additional information

**Competing financial interests:** The authors declare no competing financial interests.

# 12

# Nucleophilic arylation with tetraarylphosphonium salts

Zuyong Deng[1], Jin-Hong Lin[1] & Ji-Chang Xiao[1]

Organic phosphonium salts have served as important intermediates in synthetic chemistry. But the use of a substituent on the positive phosphorus as a nucleophile to construct C–C bond remains a significant challenge. Here we report an efficient transition-metal-free protocol for the direct nucleophilic arylation of carbonyls and imines with tetraarylphosphonium salts in the presence of caesium carbonate. The aryl nucleophile generated from phosphonium salt shows low basicity and good nucleophilicity, as evidenced by the successful conversion of enolizable aldehydes and ketones. The reaction is not particularly sensitive to water, shows wide substrate scope, and is compatible with a variety of functional groups including cyano and ester groups. Compared with the arylmetallic reagents that are usually moisture sensitive, the phosphonium salts are shelf-stable and can be easily handled.

[1] Key Laboratory of Organofluorine Chemistry, Shanghai Institute of Organic Chemistry, Chinese Academy of Sciences, 345 Lingling Road, Shanghai 200032, China. Correspondence and requests for materials should be addressed to J.-C.X. (email: jchxiao@sioc.ac.cn).

Significant advances in C-arylation have emerged in the past few decades because this reaction can effectively form C–C bonds to provide valuable molecules that are the core structures of natural products or pharmaceuticals[1–5]. Recently, intensive studies have been devoted to the exploration of efficient methods for arylation of carbonyl and imino compounds, allowing access to α-aryl-alcohols and α-aryl-amines, which are abundant in biologically active molecules[1,2]. Traditional arylmetallic reagents, which are usually generated *in situ* under Barbier-type conditions[6–8] or prepared in advance like arylmagnesium reagents[9–11], have long been used for nucleophilic addition to carbonyl or imino groups. However, this approach often suffers from narrow substrate scope, or the high moisture sensitivity and the strong basicity of the reagents. Transition-metal-catalysed arylation of carbonyl or imino groups with aryl halides[12,13], arylboronic acid derivatives[14–20] or other reactive reagents[21–25] showed good functional group tolerance under mild reaction conditions. However, the high cost of noble transition metal complexes and possible contamination of the end products by trace amounts of heavy metals probably limit the wide application of this approach in industrial and pharmaceutical processes. Therefore, it is highly desirable to explore broadly applicable transition-metal-free methods for the arylation of carbonyl and imino compounds.

Our research interest in the chemistry of organic phosphonium salts[26–29] prompted us to investigate their application in arylation reactions. Organic phosphonium salts have served as important intermediates in synthetic chemistry[30–35]. Although phosphonium salts are electrophilic species and the positive phosphorus could enhance the electrophilicity[36] or acidity[34,35,37] of its substituents because of the inductive effect, we speculated that the substituent on the phosphorus might be converted to a nucleophile initiated by another appropriate nucleophile, which may firstly attack the phosphorus and then result in the cleavage of a substituent with nucleophilic ability (Fig. 1).

Herein we report a nucleophilic arylation of carbonyl and imino compounds with tetraarylphosphonium salts in the presence of caesium carbonate (Fig. 1). The generated aryl nucleophiles exhibit low basicity and good nucleophilicity. The arylation protocol shows wide substrate scope and a high level of functional group tolerance. Mechanistic investigation reveals that the reaction is predominantly initiated by caesium carbonate.

## Results

**Optimization of reaction conditions.** Considering the high affinity between phosphorus and oxygen, an O-nucleophile was used to initiate the arylation of 4-phenylbenzaldehyde (**1a**) with tetraphenylphosphonium iodide (**2a**) (Table 1, entries 1–4). Neither $Na_2CO_3$ nor $K_2CO_3$ could initiate the transformation (entries 1 and 2). Interestingly, $Cs_2CO_3$ was found to be a good initiator (entry 3), which is likely because of the better solubility of the caesium salt. Because the caesium salt was effective, another O-initiator with caesium as the cation (CsOAc; entry 4) was examined. To our surprise, no desired product was detected, which might be caused by the lower nucleophilicity of the acetate anion. Further screening of non-O-initiators (entries 5 and 6) showed that $Cs_2CO_3$ was the best initiator (entry 3), indicating that the high P–O affinity may have an important role in this reaction. The yield was increased to 65% with the elongation of reaction time (entry 7 versus entry 3). The effects of other solvents were also investigated (entries 8–10). The reaction in 1,4-dioxane gave the product in comparable yield (entry 8 versus entry 7), while acetonitrile was not as effective, probably because its acidic α-proton can quench the leaving phenyl group from salt **2a** (entry 10). Increasing the loading of salt **2a** and caesium carbonate (entries 11 and 12) increased the yield significantly (entry 12), while lowering the temperature led to significantly lower yield (entry 13). The reaction was not particularly sensitive to water, as evidenced by the moderate yield obtained with the addition of water into the reaction system (entry 14). Both of the chloride ($Ph_4P^+$ $Cl^-$) and bromide ($Ph_4P^+$ $Br^-$) instead of the iodide **2a** gave lower yields, which should be due to the lower solubility of the chloride and bromide (entries 15–16).

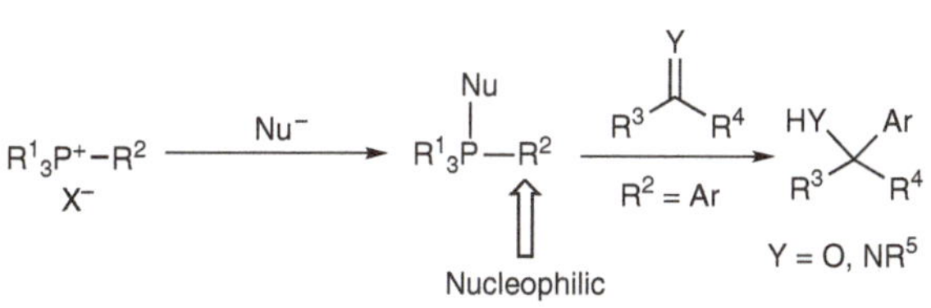


**Figure 1 | Design of arylation reaction with phosphonium salt.** An aryl substituent on phosphorus can act as a nucleophile to realize the nucleophilic arylation of carbonyls and imines.

**Substrate scope for phenylation of aldehydes.** With the optimized reaction conditions in hand (Table 1, entry 12), we then explored the substrate scope for the phenylation of aldehydes with phosphonium salt **2a** (Table 2). The transformation can be applied to a variety of aldehydes and gave the desired products in moderate to excellent yields. Increasing the scale of the reaction by 20-fold still afforded the desired product **3aa** in good yields, which is highly important in the synthetic application, albeit the longer reaction time or higher loading of $Cs_2CO_3$ was required. Investigation of the electronic effects showed that the electron-donating or -withdrawing groups

**Table 1 | Screening reaction conditions*.**

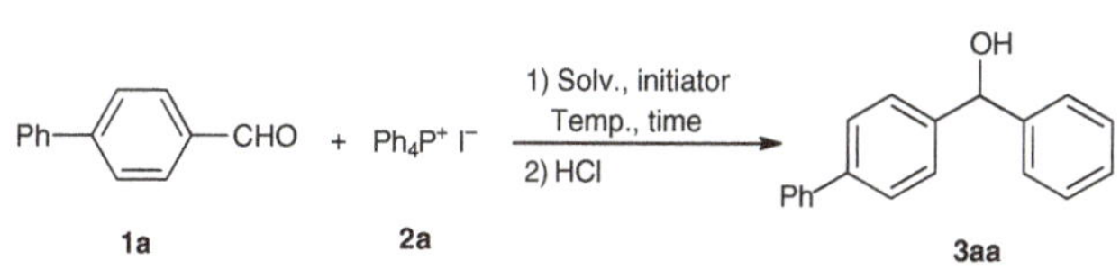


| Entry | Initiator | Ratio† | Solvent | Temperature (°C) | Time (h) | Yield (%)‡ |
|---|---|---|---|---|---|---|
| 1 | $Na_2CO_3$ | 1:1.5:2.0 | THF | 65 | 6 | 0 |
| 2 | $K_2CO_3$ | 1:1.5:2.0 | THF | 65 | 6 | 0 |
| 3 | $Cs_2CO_3$ | 1:1.5:2.0 | THF | 65 | 6 | 56 |
| 4 | CsOAc | 1:1.5:2.0 | THF | 65 | 6 | 0 |
| 5 | CsF | 1:1.5:2.0 | THF | 65 | 6 | 13§ |
| 6 | $Et_3N$ | 1:1.5:2.0 | THF | 65 | 6 | 0 |
| 7 | $Cs_2CO_3$ | 1:1.5:2.0 | THF | 65 | 12 | 65 |
| 8 | $Cs_2CO_3$ | 1:1.5:2.0 | 1,4-dioxane | 65 | 12 | 64 |
| 9 | $Cs_2CO_3$ | 1:1.5:2.0 | DMF | 65 | 12 | 52 |
| 10 | $Cs_2CO_3$ | 1:1.5:2.0 | $CH_3CN$ | 65 | 12 | 0 |
| 11 | $Cs_2CO_3$ | 1:2.0:2.5 | THF | 65 | 12 | 80 |
| 12 | $Cs_2CO_3$ | 1:2.5:3.0 | THF | 65 | 12 | 90 |
| 13 | $Cs_2CO_3$ | 1:2.5:3.0 | THF | 40 | 12 | 68 |
| 14‖ | $Cs_2CO_3$ | 1:2.5:3.0 | THF | 65 | 12 | 49 |
| 15¶ | $Cs_2CO_3$ | 1:2.5:3.0 | THF | 65 | 12 | 29 |
| 16# | $Cs_2CO_3$ | 1:2.5:3.0 | THF | 65 | 12 | 44 |

*Conditions: **1a** (0.5 mmol), **2a** and $Cs_2CO_3$ in solvent (4 ml).
†Molar ratio of **1a**:**2a**:initiator.
‡Isolated yield.
§Yield determined by $^1H$ NMR.
‖2 equivalents of $H_2O$ added.
¶Chloride salt ($Ph_4P^+$ $Cl^-$) was used instead of **2a**.
#Bromide salt ($Ph_4P^+$ $Br^-$) was used instead of **2a**.

**Table 2 | Substrate scope of phenylation of aldehydes*.**

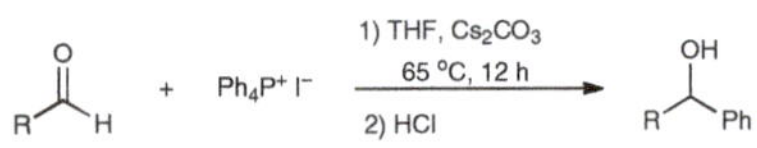


1 2a 3

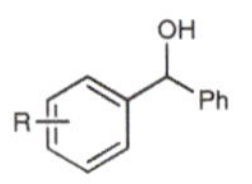

3aa–3oa

**3aa** R = *p*-Ph, 90 % (84 %†, 75 %‡)
**3ba** R = *p*-NMe$_2$, 69 %
**3ca** R = *p*-OMe, 89 %
**3da** R = *o*-OMe, 80 %
**3ea** R = 2,4,6-Me$_3$, 86 %
**3fa** R = *p*-F, 94 %
**3ga** R = *p*-Br, 69 %
**3ha** R = *m*-Br, 62 %
**3ia** R = *p*-CF$_3$, 86 %
**3ja** R = *m*-CF$_3$, 84 %
**3ka** R = *o*-CF$_3$, 88 %
**3la** R = *m*-CN, 73 %
**3ma** R = *p*-CN, 68 %
**3na** R = *p*-CO$_2$CH$_3$, 75 %
**3oa** R = *o*-Ph, 90 %

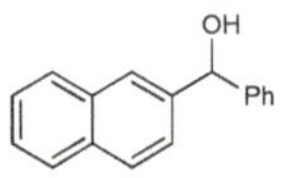
**3pa**, 88%

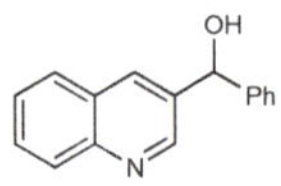
**3qa**, 91%

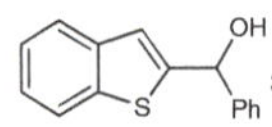
**3ra**, 93%

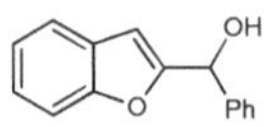
**3sa**, 95%

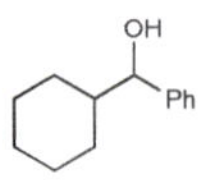
**3ta**, 85%

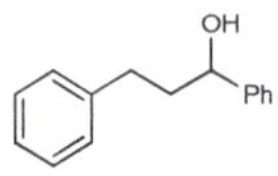
**3ua**, 60%

*Conditions: **1** (0.5 mmol), **2a** (2.5 equivalents) and $Cs_2CO_3$ (3.0 equivalents) in THF (4 ml). Yields are isolated yields;
†increasing the scale of the reaction by 20-fold (**1a**, 10 mmol, 1.8 g), 5 days of reaction time was needed to obtain **3aa** in 84% isolated yield with the use of 3 equivalents of $Cs_2CO_3$;
‡increasing the scale of the reaction by 20-fold (**1a**, 10 mmol, 1.8 g), 12 h of reaction time can give **3aa** in 75% isolated yield with the use of 5 equivalents of $Cs_2CO_3$.

on the aryl ring did not significantly influence the reaction (**3aa–3ra**). A broad functionality tolerance was demonstrated for aryl aldehydes bearing different substituents, including cyano (**3la–3ma**) and ester groups (**3na**). The conversion was not quite sensitive to steric effects, given that high yields were obtained for products **3ea**, **3ka** and **3oa**. It is noteworthy that the enolizable aldehydes were also converted into the expected products in good yields (**3ta** and **3ua**) without self-aldol condensation.

**Substrate scope for phenylation of ketones**. Although ketone is less reactive than aldehyde, the phenylation of ketones with salt **2a** still proceeded smoothly to give the desired products (Table 3). However, increased loading of salt **2a** and caesium carbonate to 4 and 4.5 equivalents, respectively, was necessary to achieve good yields. Both aromatic (**5a–5l**) and aliphatic ketones (**5m** and **5n**) were reactive under the optimal conditions. The successful conversion of hindered ketones indicates that the reaction is moderately tolerant to steric effects (**5f–5h, 5j**). In addition, no reaction at the carbonyl-activated ester group of **5k** suggests that the transformation shows high chemoselectivity.

Compared with the traditional arylation of carbonyl groups with arylmetallic reagents, which are strong nucleophiles and can lead to the arylation of cyano[38] and ester[12] groups, this mild reaction clearly exhibits a higher level of functional group tolerance. More importantly, the reaction is applicable to enolizable aldehydes and ketones, indicating that the *in situ*-generated phenyl-anion equivalent exhibits high nucleophilicity and low basicity.

**Substrate scope for phenylation of imines**. The successful conversion of aldehydes and ketones prompted us to examine the phenylation of imines with salt **2a** (Table 4). Imines containing electron-donating groups on the phenyl ring were effectively converted into the desired products, but for imines with a strong electron-withdrawing substituent such as cyano, the reaction failed to give the desired product. It is likely that under these reaction conditions, the imine is prone to hydrolysis to give the aldehyde, which would then react with salt **2a** to afford the alcohol. In contrast to tosyl-protected imines (*N*-Ts),

**Table 3 | Substrate scope of phenylation of ketones*.**

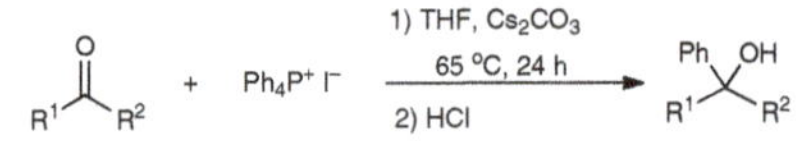


4 2a 5

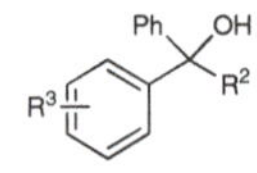

5a–5k

**5a** R$^3$ = *p*-Ph, R$^2$ = Me, 75%
**5b** R$^3$ = *p*-F, R$^2$ = Me, 74%
**5c** R$^3$ = *p*-CN, R$^2$ = Me, 65%
**5d** R$^3$ = *p*-CF$_3$, R$^2$ = Me, 86%
**5e** R$^3$ = *p*-CF$_3$, R$^2$ = Et, 87%
**5f** R$^3$ = H, R$^2$ = $^i$Pr, 65%
**5g** R$^3$ = H, R$^2$ = cyclohexyl, 60%
**5h** R$^3$ = H, R$^2$ = $^t$Bu, 50%
**5i** R$^3$ = H, R$^2$ = CH$_2$Ph, 60%
**5j** R$^3$ = H, R$^2$ = (*p*-CF$_3$)C$_4$H$_4$, 89%
**5k** R$^3$ = H, R$^2$ = COOEt, 63%

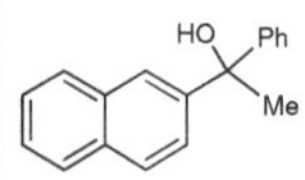
**5l**, 66%

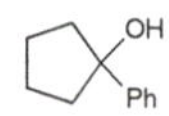
**5m**, 60%

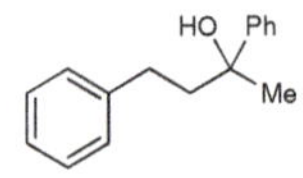
**5n**, 56%

*Conditions: **4** (0.4 mmol), **2a** (4.0 equivalents) and $Cs_2CO_3$ (4.5 equivalents) in THF (4 ml). Yields are isolated yields.

*p*-methoxyphenyl-protected imine (*N*-PMP) is inert under the reaction conditions.

**The investigation of the reactivity of phosphonium salts**. The reactivities of different tetraarylphosphonium salts were also investigated. Compared with tetraphenylphosphonium iodide (**3aa**, Table 2), tetra-*p*-methoxyphenylphosphonium iodide (Table 5, entry 1) and tetra-*p*-methylphenylphosphonium iodide (entry 2) showed lower reactivity towards arylation and gave the corresponding arylation products in lower yields. This is probably because the *in situ*-generated aryl-anion equivalent containing an electron-donating group is less stable and would readily abstract a proton from the trace amount of water present.

To this point of our study, the four aryl groups on the phosphorus were identical. It will be interested to know the different leaving ability when one aryl group was different from the others. To address this issue, a variety of aryl triphenylphosphonium salts were prepared and tested for the arylation of aldehyde **1a** (Table 5, entries 3–9). If the aryl moiety contained a weak electron-withdrawing group, the corresponding arylation product was obtained in low yield (entry 3), with simultaneous formation of more phenylation product **3aa** (43% yield). This might be because the leaving ability between the phenyl and aryl group is similar owing to their comparable electron-withdrawing effect, and three phenyl groups compete with one aryl group, thus leading to more phenylation product **3aa**. However, if the aryl moiety was substituted with a strong electron-withdrawing group, the phenylation product **3aa** would be greatly suppressed and the arylation became the main reaction, indicating that electron-withdrawing ability has an important role in the selectivity (entries 4–7). In the case of **3ai**, the low yield was not because of the phenylation reaction to produce **3aa**, but because of a side reaction to give 4′-phenylacetophenone (27% yield based on aldehyde; entry 8). Pyridyl triphenylphosphonium iodide was also suitable for the desired pyridylation (entry 9).

In the case of the salts containing strong electron-withdrawing groups (**2e′–2i′**), it would be better to change the salt anion from iodide to bis(trifluoromethanesulfonyl)imide anion ($Tf_2N^-$) and the procedure for arylation should be modified accordingly. For example, the *in situ*-generated aryl-anion equivalent would readily attack the phosphonium salt to give the diaryl by-product (33% based on aldehyde) because of the increased electrophilicity of the salt. To avoid this problem, the phosphonium salts were added slowly to the reaction to keep the substrate aldehyde in constant excess. And this required the salts to be soluble in tetrahydrofuran (THF) so as to realize the addition of the salts via

**Table 4 | Substrate scope of phenylation of imines*.**

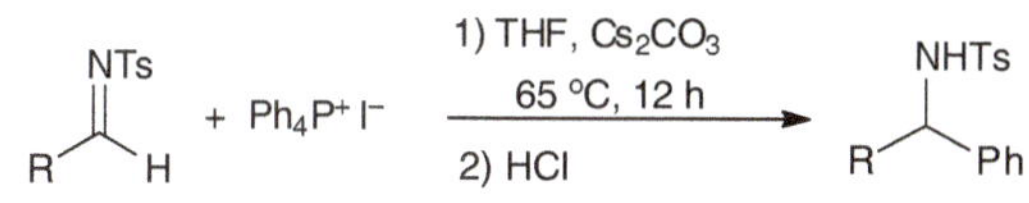


6 2a 7

| Entry | R | 7, Yield (%)† |
|---|---|---|
| 1 | Ph | **7a**, 72 |
| 2 | 4-$PhC_6H_4$ | **7b**, 75 |
| 3 | 4-$MeC_6H_4$ | **7c**, 74 |
| 4 | 2,4,6-$Me_3C_6H_2$ | **7d**, 86 |
| 5 | 4-$MeOC_6H_4$ | **7e**, 71 |
| 6 | 2-$MeOC_6H_4$ | **7f**, 74 |
| 7 | 4-$CF_3OC_6H_4$ | **7g**, 43 |
| 8 | 4-$BnOC_6H_4$ | **7h**, 85 |
| 9 | 4-$FC_6H_4$ | **7i**, 70 |
| 10 | 2-naphthyl | **7j**, 75 |

*Conditions: **6** (0.4 mmol), **2a** (2.5 equivalents), and $Cs_2CO_3$ (3 equivalents) in THF (4 ml).
†Isolated yields.

**Table 5 | Substrate scope of arylation of 4-phenylbenzaldehyde*.**

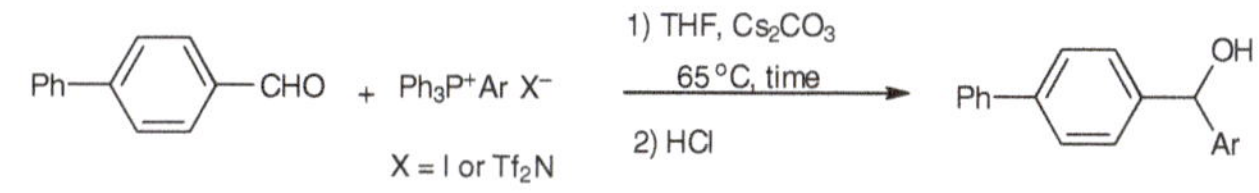


1a 2 3

| Entry | 2, X | Ar | 3, Yield (%)† |
|---|---|---|---|
| 1‡ | **2b**, I | 4-$MeOC_6H_4$ | **3ab**, 55 |
| 2§ | **2c**, I | 4-$MeC_6H_4$ | **3ac**, 49 |
| 3‖ | **2d**, I | 4-$PhC_6H_4$ | **3ad**, 35 |
| 4 | **2e′**, $Tf_2N$ | 4-$CNC_6H_4$ | **3ae**, 73 |
| 5 | **2f′**, $Tf_2N$ | 3-$CNC_6H_4$ | **3af**, 75 |
| 6 | **2g′**, $Tf_2N$ | 4-$CF_3C_6H_4$ | **3ag**, 78 |
| 7 | **2h′**, $Tf_2N$ | 4-$EtO_2CC_6H_4$ | **3ah**, 66 |
| 8 | **2i′**, $Tf_2N$ | 4-$AcC_6H_4$ | **3ai**, 35 |
| 9 | **2j**, I | 2-Pyridyl | **3aj**, 73 |

*Conditions: **1a** (0.5 mmol), **2** (2.5 equivalents), and $Cs_2CO_3$ (3 equivalents) in THF (4 ml).
†Isolated yields.
‡$[(4\text{-}MeOC_6H_4)_4P]^+\ I^-$ (**2b**) was used.
§$[(4\text{-}MeC_6H_4)_4P]^+\ I^-$ (**2c**) was used.
‖The reaction also gave **3aa** in 43% isolated yield.

syringe. The phosphonium salts containing $Tf_2N^-$ were soluble in THF and the corresponding arylation proceeded smoothly by the addition of the phosphonium salt solution into the reaction (Table 5, entries 4–8).

**Mechanistic investigation**. As for the reaction mechanism, the analysis of trace transition metal in phosphonium salts and caesium carbonate by ICP spectrometry (Supplementary Table 1) revealed that this arylation is a transition-metal-free reaction. Although salt **2a** is quite stable, an equilibrium might be established between this salt and $Ph_3P$/PhI under the phenylation conditions. But the use of $Ph_3P$/PhI system instead of salt **2a** for the phenylation of aldehyde **1a** failed to give the desired product at all (equation 1 Fig. 2), meaning that the equilibrium cannot account for the arylation. Neither radical scavenger TEMPO [(2,2,6,6-tetramethylpiperidin-1-yl)oxyl] nor single-electron-transfer inhibitor *p*-dinitrobenzene can dramatically suppress the phenylation of aldehyde **1a** with salt **2a**, indicating that the arylation may not proceed via a single-electron-transfer mechanism (equation 2, Fig. 2). Nothing happened while refluxing aldehyde **1a**/salt **2a** system or aldehyde **1a**/$Cs_2CO_3$ system in THF. But the full conversion of salt **2a** into $Ph_3PO$ was observed by refluxing salt **2a** and $Cs_2CO_3$ in THF, suggesting that the interaction between salt **2a** and $Cs_2CO_3$ initiates this arylation reaction.

The alkaline hydrolysis of phosphonium salts has been well-studied for the past decades[30,39–47]. As a third-order reaction (first order with respect to phosphonium salt, second order to hydroxide)[40,41], the alkaline hydrolysis of tetraphenylphosphonium salt would lead to the formation of triphenylphosphine oxide ($Ph_3PO$) (refs 46,48). Since $Ph_3PO$ was detected by gas chromatography mass spectrometry (GC–MS) in every phenylation reaction system using **2a** ($Ph_4P^+\ I^-$) and $Ph_3PO$ was isolated in high yield for the phenylation of substrate **1a** (equation 1, Fig. 3), it is reasonable to conceive that the arylation might proceed via this hydrolysis process. Although the solvent THF purchased from commercial source was extra dry ('Extra Dry over Molecular Sieve', <0.005% water content) and the arylation was performed under $N_2$ atmosphere, trace amount of water may still be unavoidable in the reaction system, thus leading to the hydrolysis of caesium carbonate (equation 2). The resulting hydroxide anion would decompose the phosphonium salt to generate $Ph_3PO$, suggesting that the oxygen in $Ph_3PO$ could partially come from water. Indeed, the reaction of phosphonium salt with caesium carbonate in the presence of $H_2^{18}O$ gave both $Ph_3PO$ and $Ph_3P^{18}O$ (equation 3). The molar ratio of $Ph_3PO$: $Ph_3P^{18}O$ determined by $^{31}P$ NMR was 1:1.26. This ratio was consistent with elemental analysis of the mixture (see Supplementary Methods for the formation of $Ph_3P^{18}O$). It seems that this hydrolysis process might be involved in the arylation, which was further supported by the observation that the direct use of caesium hydroxide instead of caesium carbonate also gave the expected product in 63% yield (equation 4). Compared with the 90% yield while using $Cs_2CO_3$, lower yield might result from the presence of hydrated water in CsOH (note: anhydrous CsOH is not commercially available). Although water can promote the hydrolysis of $Cs_2CO_3$ to produce CsOH, excessive water would suppress the subsequent arylation through the capture of the *in situ*-generated phenyl nucleophile. In addition, this possible hydrolysis process of $Cs_2CO_3$ could also explain why moderate yield (49%) of phenylation product could be obtained with the deliberate addition of water (entry 14 of Table 1).

Nevertheless, all reagents including highly dry $Cs_2CO_3$ (99.994% purity, metals basis) and the solvent THF (<0.005% water content) were stored in a glove box, and the arylation were conducted under $N_2$ atmosphere, suggesting that the water content in the reaction system should be extremely low. This trace amount of water cannot cause the complete hydrolysis of $Cs_2CO_3$, indicating that another reaction process should exist. As shown in Table 2, for the large-scale phenylation to give **3aa**, significantly less reaction time (12 h versus 120 h) was needed to achieve comparable yield (75% versus 84%) with the use of more $Cs_2CO_3$ (5 equivalents versus 3 equivalents). The water contents (trace) in both reaction systems should be the same since these reactions were conducted under the same reaction conditions. The faster arylation with the use of more $Cs_2CO_3$ means that $Cs_2CO_3$ may be able to promote the arylation without the involvement of water. If this is the case, more $Cs_2CO_3$ would result in a faster conversion of $Ph_4P^+\ I^-$ into $Ph_3PO$. Indeed, for the reaction of $Ph_4P^+\ I^-$ with $Cs_2CO_3$ in THF for 4 h, the more $Cs_2CO_3$ was used, the more $Ph_3PO$ was produced (note: $Ph_4P^+\ I^-$ was not completely consumed in every reaction; Fig. 4). These results indicate that $Cs_2CO_3$, for the most part, directly promoted this arylation.

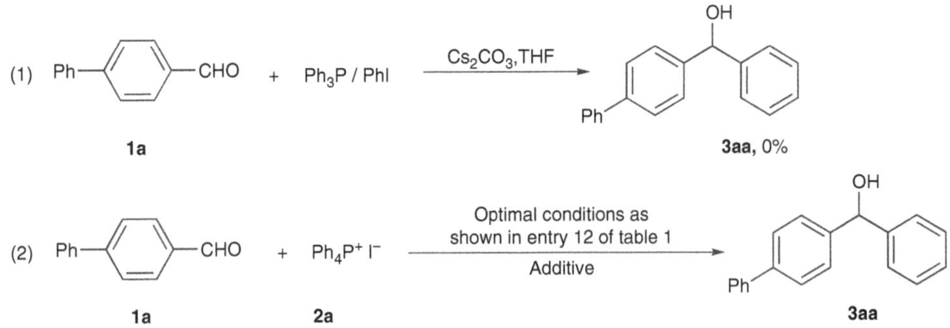


| Additive | Yield[a] |
|---|---|
| None | 90% |
| TEMPO (2 equiv.) | 84% |
| *p*-Dinitrobenzene (2 equiv.) | 70% |

**Figure 2 | Mechanistic experiments.** (1) The use of $Ph_3P$/PhI system instead of salt **2a** failed to convert aldehyde **1a** into product **3aa**. (2) Neither radical scavenger TEMPO nor single-electron-transfer inhibitor *p*-dinitrobenzene can obviously suppress the phenylation reaction. [a]Isolated yields.

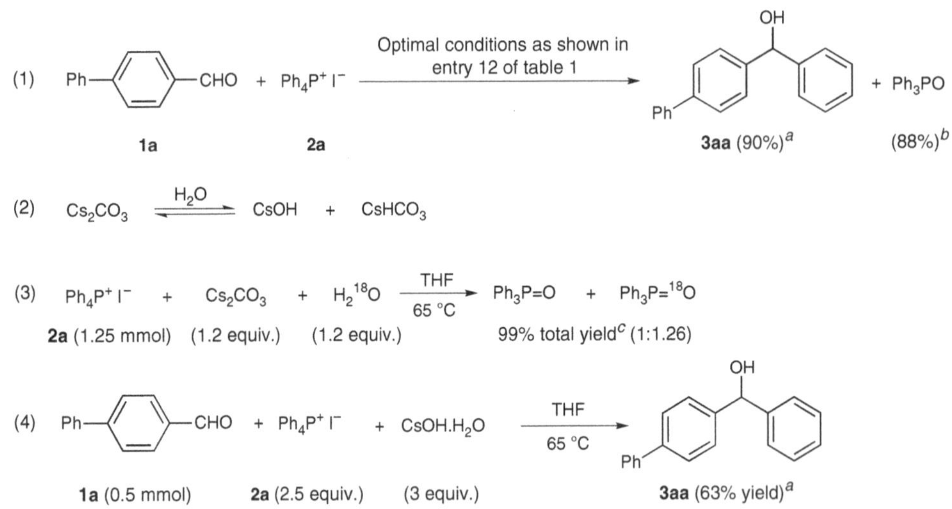


**Figure 3 | Evidence to support the path via alkaline hydrolysis.** (1) $Ph_3PO$ was isolated as a by-product for phenylation reaction. [a]Isolated yield based on **1a**; [b]Isolated yield based on **2a**; (2) An equilibrium between $Cs_2CO_3$ and CsOH may be established under the reaction conditions. (3) The presence of $H_2^{18}O$ led to the formation of $Ph_3P^{18}O$. [c]Isolated yield. (4) CsOH can also initiate the phenylation reaction.

Apparently, nucleophilic phenyl species was formed in the process of the arylation. Salt **2a**, also as an electrophilic species, might be able to trap this nucleophilic phenyl group to give $Ph_5P$ (ref. 49). To ascertain whether $Ph_5P$ is generated or not, more evidences were collected. $^{31}P$ NMR measurement of the reaction of **1a** with salt **2a** only detected the formation of $Ph_3PO$. Furthermore, instead of the phosphonium salt **2a**, $Ph_5P$ was directly used in its reaction with aldehyde (See Supplementary Methods for the procedure for arylation of **1a** with $Ph_5P$)[50,51]. However, no desired product **3aa** was detected without the presence of $Cs_2CO_3$. And even in the presence of $Cs_2CO_3$, the yield of **3aa** was very low (11%), indicating that $Ph_5P$ may not be involved in the above arylation.

On the basis of the above results, we proposed a plausible mechanism shown in Fig. 5. The direct attack of $Cs_2CO_3$ at the phosphonium cation is the predominant path (Path I). Nucleophilic addition to the positive phosphorus would always be along the axial co-ordinate and pseudorotation would occur afterwards to place the electronegative substituent in the other axial position[30,39,47,52], generating trigonal bipyramidal tetraaryloxyphosphorane **A**. The simultaneous decarboxylation and nucleophilic attack of Ar in intermediates **A** to carbonyl or imino groups afford $Ph_3PO$ and the adduct **B**, which is then protonated by acid to furnish the final product. However, owing to the unavoidable presence of trace amount of water, an equilibrium between $Cs_2CO_3$ and $CsHCO_3$/CsOH may be established (path II). $CsHCO_3$ might gradually decompose into $Cs_2CO_3$, water and carbon dioxide at the reaction temperature. The generated CsOH would promote the subsequent arylation via an alkaline-hydrolysis path[30,39–47]. Nucleophilic addition of hydroxide to phosphorus and the subsequent pseudorotation generate intermediate **C**, deprotonation of which by CsOH gives an oxyanionic phosphorane **D**. The expulsion of the aryl anion from intermediate **D** and the nucleophilic addition of this Ar group to substrate also afford the adduct **B** and $Ph_3PO$. Therefore, the two paths might coexist. It should be noted that none of the intermediate **A**, **C** or **D** was detected by $^{31}P$ NMR, although *P*-hydroxytetraorganophosphorane has recently been observed and characterized by low-temperature NMR[53]. Considering that cyano and ester groups are stable under the optimized reaction conditions and the enolizable carbonyl compounds were successfully converted, the naked aryl anion is not believed to be formed in the arylation process. Carbon dioxide cannot be detected by GC–MS in the gas phase of the reaction system because the gas phase was full of the solvent THF. Fortunately, when DMF was used instead of THF in the reaction of aldehyde **1a** with salt **2a**, GC–MS analysis of the gas phase of the reaction mixture successfully detected the generation of $CO_2$ (see 'determination of $CO_2$ by GC–MS spectroscopy' in Supplementary Methods) albeit that the desired product was isolated in only 60% yield, further supporting the proposed reaction mechanism.

## Discussion

Although phosphonium salts are electrophilic species, we found that a substituent on the positive phosphorus can act as a nucleophile for the construction of C–C bond while initiated by another appropriate nucleophile. The successful application of tetraarylphosphonium salts to the nucleophilic transition-metal-free arylation of carbonyl and imino compounds in the presence of caesium carbonate is described herein. The high P–O affinity has an important role in this arylation strategy. This practical protocol is attractive because the reaction is not particularly sensitive to water, shows wide substrate scope under mild conditions, and is compatible with a variety of functional groups. It provides a highly valuable method for C-arylation and easy access to α-aryl-alcohols and α-aryl-amines. In addition, the success of the strategy means that phosphonium salts may find applications in other areas of chemistry.

$$\underset{1\ \text{mmol}}{Ph_4P^+\ I^-} + \underset{X\ \text{equiv.}}{Cs_2CO_3} \xrightarrow[65\ °C,\ 4\ h]{THF\ (1.5\ ml)} Ph_3PO$$

| X (equiv.) | Yield[a] |
|---|---|
| 1.0 | 34% |
| 1.5 | 44% |
| 2.0 | 50% |
| 3.0 | 66% |

**Figure 4 | Evidence to support the $Cs_2CO_3$ promoted arylation without water.** The more $Cs_2CO_3$ was used for the reaction of $Ph_4P^+\ I^-$ with $Cs_2CO_3$ for the same period of time, the more $Ph_3PO$ was produced, [a]yields were determined by $^{31}P$ NMR.

## Methods

**General procedure for phenylation of aldehydes.** Under $N_2$ atmosphere, the mixture of aldehyde (0.50 mmol), tetraphenylphosphonium iodide (583.0 mg, 1.25 mmol) and $Cs_2CO_3$ (488.7 mg, 1.50 mmol) in THF (4 ml) and stirred at 65 °C for 12 h. The reaction was quenched by 3 N HCl (0.5 ml). The resulting mixture was extracted with dichloromethane (DCM; 3 × 30 ml). The combined organic phase was dried over $Na_2SO_4$. After filtration, the solvent was removed by concentration, and the residue was subjected to column chromatography to afford the pure product.

**General procedure for phenylation of ketones.** Under $N_2$ atmosphere, the mixture of ketone (0.40 mmol), tetraphenylphosphonium iodide (745.8 mg, 1.60 mmol) and $Cs_2CO_3$ (586.4 mg, 1.80 mmol) in THF (4 ml) was stirred at 65 °C for 24 h. The reaction was quenched by 4.5 N HCl (1.5 ml). The resulting mixture was extracted with DCM (3 × 30 ml). The combined organic phase was dried over $Na_2SO_4$. After filtration, the solvent was removed by concentration, and the residue was subjected to column chromatography to afford the pure product.

**General procedure for phenylation of imines.** Under $N_2$ atmosphere, the mixture of imine (0.50 mmol), tetraphenylphosphonium iodide (583.0 mg, 1.25 mmol) and $Cs_2CO_3$ (488.7 mg, 1.50 mmol) in THF (4 ml) was stirred at 65 °C for 12 h. The reaction was quenched by 3 N HCl (0.5 ml). The resulting mixture was extracted with DCM (3 × 30 ml). The combined organic phase was dried over $Na_2SO_4$. After filtration, the solvent was removed by concentration. The residue was subjected to column chromatography to afford the pure product.

**Arylation of 1a with various teraarylphosphonium salts.** *Method A.* Under $N_2$ atmosphere, the mixture of [1,1′-biphenyl]-4-carbaldehyde (**1a**; 91.2 mg, 0.50 mmol), tetraarylphosphonium iodide (1.25 mmol) and $Cs_2CO_3$ (488.7 mg, 1.50 mmol) in THF (4 ml) was stirred at 65 °C until phosphonium salt was completely consumed as monitored by $^{31}P$ NMR. The reaction was quenched by 3 N HCl (0.5 ml). The resulting mixture was extracted with DCM (3 × 30 ml). The combined organic phase was dried over $Na_2SO_4$, and the residue was subjected to column chromatography to afford the pure product.

*Method B.* Under $N_2$ atmosphere, into the mixture of [1,1′-biphenyl]-4-carbaldehyde (**1a**; 91.2 mg, 0.50 mmol) and $Cs_2CO_3$ (488.7 mg, 1.50 mmol) in THF (2 ml) at 65 °C was added the solution of phosphonium bis(trifluoromethanesulfonyl)amide (1.25 mmol) in THF (2 ml) slowly for 6 h. On completion of addition, the reaction was stirred for another 10 min. The mixture was quenched by 3 N HCl (0.5 ml) and extracted with DCM (3 × 30 ml). The combined organic phase was dried over $Na_2SO_4$. After filtration, the solvent was removed by concentration, and the residue was subjected to column chromatography to afford the pure product.

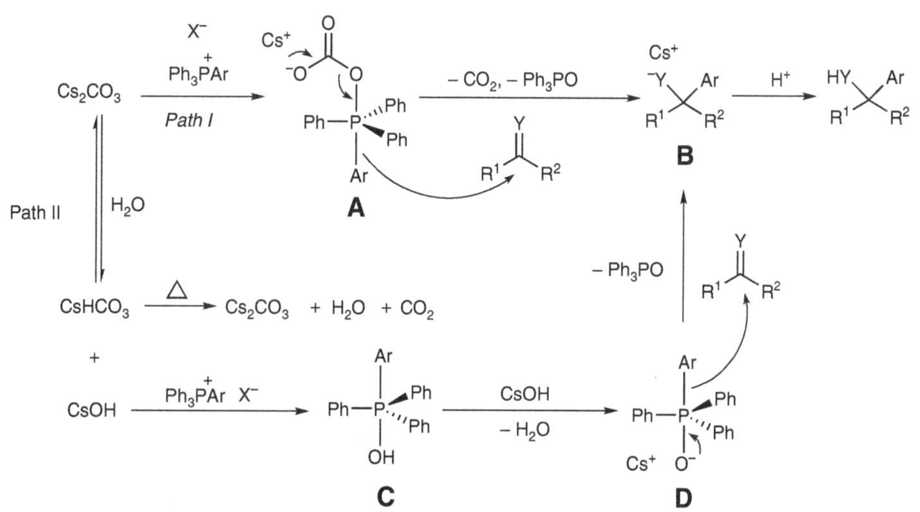

**Figure 5 | Proposed nucleophilic arylation mechanism.** The reaction should be predominantly initiated by $Cs_2CO_3$ (path I), and may also be promoted by CsOH due to the trace amount of water present in the reaction system (path II).

## References


1. *Modern Arylation Methods* (ed. Ackermann, Lutz) (Wiley-VCH, Verlag GmbH & Co. KGaA, 2009).
2. Bruke, A. J. & Marques, C. S. *Catalytic Arylation Methods - from the Academic Lab to Industrial Processes* (Wiley-VCH Verlag GmbH & Co. KGaA, 2015).
3. Cuthbertson, J. D. & MacMillan, D. W. C. The direct arylation of allylic $sp^3$ C-H bonds via organic and photoredox catalysis. *Nature* **519,** 74–77 (2015).
4. He, J. *et al.* Ligand-controlled C($sp^3$)-H arylation and olefination in synthesis of unnatural chiral α-amino acids. *Science* **343,** 1216–1220 (2014).
5. Pirnot, M. T., Rankic, D. A., Martin, D. B. C. & MacMillan, D. W. C. Photoredox activation for the direct β-arylation of ketones and aldehydes. *Science* **339,** 1593–1596 (2013).
6. Kunishima, M., Hioki, K., Kono, K., Sakuma, T. & Tani, S. Babier-type reactions of aryl halides with ketones by samarium diiodide. *Chem. Pharm. Bull.* **42,** 2190–2192 (1994).
7. Sugimoto, O., Yamada, S. & Tanji, K. Preparation of nitrogen-containing π-deficient heteroaromatic grignard reagents: oxidative magnesiation of nitrogen-containing π-deficient halogenoheteroaromatics using active magnesium. *J. Org. Chem.* **68,** 2054–2057 (2003).
8. Therkelsen, F. D., Rottländer, M., Thorup, N. & Pedersen, E. B. 4-metalated condensed pyrimidines: their preparation and reaction with aldehydes under barbier-type conditions. *Org. Lett.* **6,** 1991–1994 (2004).
9. Hatano, M., Ito, O., Suzuki, S. & Ishihara, K. Zinc(II)-catalyzed addition of grignard reagents to ketones. *J. Org. Chem.* **75,** 5008–5016 (2010).
10. Hatano, M., Ito, O., Suzuki, S. & Ishihara, K. Zinc(II)-catalyzed Grignard additions to ketones with RMgBr and RMgI. *Chem. Commun.* **46,** 2674–2676 (2010).
11. Zong, H., Huang, H., Liu, J., Bian, G. & Song, L. Added-metal-free catalytic nucleophilic addition of grignard reagents to ketones. *J. Org. Chem.* **77,** 4645–4652 (2012).
12. Gao, F. *et al.* A simple and efficient copper oxide-catalyzed Barbier–Grignard reaction of unactivated aryl or alkyl bromides with ester. *Tetrahedron Lett.* **55,** 880–883 (2014).
13. Zhou, F. & Li, C.-J. The Barbier-Grignard-type arylation of aldehydes using unactivated aryl iodides in water. *Nat. Commun.* **5,** 4254–4260 (2014).
14. Sakai, M., Ueda, M. & Miyaura, N. Rhodium-catalyzed addition of organoboronic acids to aldehydes. *Angew. Chem. Int. Ed.* **37,** 3279–3281 (1998).
15. Tetsuya, Y., Tetsuo, O. & Yoshihiko, I. Palladium-catalyzed addition of arylboronic acids to aldehydes. *Org. Lett.* **7,** 4153–4155 (2005).
16. Liao, Y.-X., Xing, C.-H. & Hu, Q.-S. Rhodium(I)/diene-catalyzed addition reactions of arylborons with ketones. *Org. Lett.* **14,** 1544–1547 (2012).
17. Huang, R. & Shaughnessy, K. H. Alkynes as activators in the nickel-catalysed addition of organoboronates to aldehydes. *Chem. Commun.* 1459–1461 (2005).
18. Liao, Y.-X., Xing, C.-H., He, P. & Hu, Q.-S. Orthoplatinated Triarylphosphite as a highly efficient catalyst for addition reactions of arylboronic acids with aldehydes: low catalyst loading catalysis and a new tandem reaction sequence. *Org. Lett.* **10,** 2509–2512 (2008).
19. Tomita, D., Kanai, M. & Shibasaki, M. Nucleophilic activation of alkenyl and aryl boronates by a chiral $Cu^{I}F$ complex: catalytic enantioselective alkenylation and arylation of aldehydes. *Chem. Asian J.* **1,** 161–166 (2006).
20. Zou, T., Pi, S.-S. & Li, J.-H. $FeCl_3$-catalyzed 1,2-addition reactions of aryl aldehydes with arylboronic acids. *Org. Lett.* **11,** 453–456 (2009).
21. Oi, S., Moro, M. & Inoue, Y. Rhodium-catalysed arylation of aldehydes with arylstannanes. *Chem. Commun.* 1621–1622 (1997).
22. Li, C.-J. & Meng, Y. Grignard-type carbonyl phenylation in water and under an air atmosphere. *J. Am. Chem. Soc.* **122,** 9538–9539 (2000).
23. Oi, S., Moro, M. & Inoue, Y. Rhodium-catalyzed addition of phenylmethyldifluorosilane to aldehydes. *Organometallics* **20,** 1036–1037 (2001).
24. Fujii, T., Koike, T., Mori, A. & Osakada, K. Rhodium-catalyzed addition of aryl- and alkenylsilanediols to aldehydes. *Synlett* **2,** 298–300 (2002).
25. Yang, L., Correia, C. A. & Li, C.-J. Grignard-type arylation of aldehydes via a rhodium-catalyzed C-H activation under mild conditions. *Adv. Synth. Catal.* **353,** 1269–1273 (2011).
26. Zheng, J., Cai, J., Lin, J.-H., Guo, Y. & Xiao, J.-C. Synthesis and decarboxylative Wittig reaction of difluoromethylene phosphobetaine. *Chem. Commun.* **49,** 7513–7515 (2013).
27. Zheng, J., Lin, J.-H., Cai, J. & Xiao, J.-C. Conversion between difluorocarbene and difluoromethylene ylide. *Chem. Eur. J.* **19,** 15261–15266 (2013).
28. Zheng, J., Lin, J.-H., Deng, X.-Y. & Xiao, J.-C. 1,8-diazabicyclo[5.4.0]undec-7-ene (DBU)-promoted decomposition of difluorocarbene and the subsequent trifluoromethylation. *Org. Lett.* **17,** 532–535 (2015).
29. Deng, X.-Y., Lin, J.-H., Zheng, J. & Xiao, J.-C. Difluoromethylation and gem-difluorocyclopropenation with difluorocarbene generated by decarboxylation. *Chem. Commun.* **51,** 8805–8808 (2015).
30. Cristau, H. J. & Plenat, F. in *The Chemistry of Organophosphorus Compounds: Phosphonium Salts, Ylides and Phosphoranes* **Vol. 3,** 45–184 (ed. Hartley, Frank R.) (John Wiley & Sons Ltd (1994).
31. Holmes, R. R. in *A Guide to Organophosphorus Chemistry* (ed. Quin, Louis D.) (John Wiley & Sons, Inc., 2000).
32. Marsden, S. P. Organic synthesis: the Wittig reaction cleans up. *Nat. Chem.* **1,** 685–687 (2009).
33. Zaragoza, F. One-step conversion of alcohols into nitriles with simultaneous two-carbon chain elongation. (Cyanomethyl)trimethylphosphonium iodide as a reagent with a dual mode of action. *J. Org. Chem.* **67,** 4963–4964 (2002).
34. Ye, L.-W., Sun, X.-L., Wang, Q.-G. & Tang, Y. Phosphine-catalyzed intramolecular formal [3 + 2] cycloaddition for highly diastereoselective synthesis of bicyclo[n.3.0] compounds. *Angew. Chem. Int. Ed.* **46,** 5951–5954 (2007).
35. Wang, P., Liao, S., Zhu, J.-B. & Tang, Y. Double [gamma]-alkylation of allylic phosphorus ylides: a unique access to oxa-bicyclic[3.3.0] diene skeletons. *Chem. Commun.* **50,** 808–810 (2014).
36. Zhang, Y., Aguirre, S. L. & Klumpp, D. A. Reactive, dicationic electrophiles: electrophilic activation involving the phosphonium group. *Tetrahedron Lett.* **43,** 6837–6840 (2002).
37. Blakemore, P. R. in *Comprehensive Organic Synthesis II* 2nd edn. 516–608 (ed. Knochel, Paul) (Elsevier Ltd., 2014).
38. Zhang, L., Ang, G. Y. & Chiba, S. Copper-catalyzed benzylic C – H oxygenation under an oxygen atmosphere via N-H imines as an intramolecular directing group. *Org. Lett.* **13,** 1622–1625 (2011).
39. Fenton, G. W. & Ingold, C. K. CCCVII.-influence of poles and polar linkings on the course pursued by elimination reactions. Part V. The mechanism of thermal decomposition of quaternary phosphonium hydroxides. *J. Chem. Soc.* 2342–2357 (1929).
40. Zanger, M., Vander Werf, C. A. & McEwen, W. E. Kinetic study of the decomposition of quaternary phosphonium hydroxides. *J. Am. Chem. Soc.* **81,** 3806–3807 (1959).
41. McEwen, W. E., Kumli, K. F., Blade-Font, A., Zanger, M. & VanderWerf, C. A. Mechanisms of substitution reactions at phosphorus. X. The Wittig reaction and the decomposition of quaternary phosphonium hydroxides. *J. Am. Chem. Soc.* **86,** 2378–2384 (1964).
42. McEwen, W. E., Axelrad, G., Zanger, M. & VanderWerf, C. A. Mechanisms of substitution reactions at phosphorus. XII. A kinetic study of the decomposition of quaternary phosphonium hydroxides. *J. Am. Chem. Soc.* **87,** 3948–3952 (1965).
43. Marsi, K. L. & Oberlander, J. E. Alkaline cleavage reactions of tetraalkylphosphonium salts. *J. Am. Chem. Soc.* **95,** 200–204 (1973).
44. Allen, D. W. & Millar, I. T. The alkaline hydrolysis of some cyclic phosphonium salts: ring-opening and ring-expansion reactions. *J. Chem. Soc. C Org.* 252–258 (1969).
45. Hawes, W. & Trippett, S. Alkaline hydrolysis of phosphonium salt with retention of configuration at phosphorus. *Chem. Commun.* 295–296 (1968).
46. Dawber, J. G., Skerratt, R. G., Tebby, J. C. & Waite, A. A. C. Kinetics of alkaline hydrolysis of quateranry phosphonium salt. The influence of protic and aprotic solents on the hydrolysis of alkyl phenylaphosphonium salt. *Phosphorus, Sulfur Silicon Relat. Elem.* **187,** 1261–1268 (2012).
47. Allen, D. W. The alkaline hydrolysis of some tri-(2-thienyl)phosphonium salts. Inductive effects on the rate of nucleophilic attack at phosphorus. *J. Chem. Soc. B. Phy. Org.* 1490–1493 (1970).
48. Songstad, G. A. A. J. Kinetic study of the reaction between phosphonium compounds and hydroxyl respectively alkoxides ions. *Acta Chem. Scand.* **16,** 1426–1432 (1962).
49. Wittig, G. & Rieber, M. Darstellung und eigenschaften des pentaphenyl-phosphors. *Justus Liebigs Ann. Chem.* **562,** 187–192 (1949).
50. Sharutin, V. V., Egorova, I. V., Ivanenko, T. K. & Ettenko, E. N. New synthesis of tetraphenylphosphonium halides. *Russ. J. Org. Chem.* **37,** 1794–1794 (2001).
51. Sharutin, V. V. *et al.* Synthesis and structure of tetraphenylstibonium and tetraphenylphosphonium hydrogen sulfates. *Russ. J. Org. Chem.* **73,** 536–540 (2003).
52. Albright, T. A., Burdett, J. K. & Whangbo, M.-H. *Orbital Interactions In Chemistry* (John Wiley & Sons, Inc., 2013 359–400.
53. Byrne, P. A., Ortin, Y. & G. Gilheany, D. First ever observation of the intermediate of phosphonium salt and ylide hydrolysis: P-hydroxytetraorganophosphorane. *Chem. Commun.* **51,** 1147–1150 (2015).


## Acknowledgements

This work was financially supported by the National Natural Science Foundation (21172240, 21421002, 21472222, 21502214), National Basic Research Program of China

(2015CB931900, 2012CBA01200) and the Chinese Academy of Sciences (XDA02020105, XDA02020106).

## Author contributions

D.Z. performed the experiments. L.J.-H. analysed the data and wrote the manuscript. X.J.-C. designed the experiments and wrote the manuscript.

## Additional information

**Competing financial interests:** The authors declare no competing financial interests.

# 13

# Total synthesis and antiviral activity of indolosesquiterpenoids from the xiamycin and oridamycin families

Zhanchao Meng[1], Haixin Yu[1], Li Li[2], Wanyin Tao[2], Hao Chen[1], Ming Wan[1], Peng Yang[1], David J. Edmonds[3], Jin Zhong[2] & Ang Li[1,4]

Indolosesquiterpenoids are a growing class of natural products that exhibit a wide range of biological activities. Here, we report the total syntheses of xiamycin A and oridamycins A and B, indolosesquiterpenoids isolated from *Streptomyces*. Two parallel strategies were exploited to forge the carbazole core: 6π-electrocyclization/aromatization and indole C2-H bond activation/Heck annulation. The construction of their *trans*-decalin motifs relied on two diastereochemically complementary radical cyclization reactions mediated by Ti(III) and Mn(III), respectively. The C23 hydroxyl of oridamycin B was introduced by an $sp^3$ C-H bond oxidation at a late stage. On the basis of the chemistry developed, the dimeric congener dixiamycin C has been synthesized for the first time. Evaluation of the antiviral activity of these compounds revealed that xiamycin A is a potent agent against herpes simplex virus-1 (HSV-1) *in vitro*.

[1] State Key Laboratory of Bioorganic and Natural Products Chemistry, Shanghai Institute of Organic Chemistry, Chinese Academy of Sciences, 345 Lingling Road, Shanghai 200032, China. [2] Unit of Viral Hepatitis, Key Laboratory of Molecular Virology and Immunology, Institut Pasteur of Shanghai, Shanghai Institutes for Biological Sciences, Chinese Academy of Sciences, Shanghai 200032, China. [3] Pfizer Worldwide Research and Development, Cambridge, Massachusetts 02139, USA. [4] Collaborative Innovation Center of Chemistry for Life Sciences, 22 Hankou Road, Nanjing, Jiangsu 210093, China. Correspondence and requests for materials should be addressed to A.L. (email: ali@sioc.ac.cn).

Indolosesquiterpenoids have attracted increasing attention from chemical and biosynthetic perspectives[1–7]; however, their biological activities are yet to be systematically explored through chemical synthesis[8–14]. Xiamycin A (**1**, Fig. 1), isolated independently from *Streptomyces* species by Hertweck[15–18] and Zhang[19–21], displays selective anti-HIV activity. The naturally occurring antibiotics oridamycins A and B (**2** and **3**) share a similar scaffold to that of **1**, though with the opposite stereochemistry at quaternary C16 (ref. 22). Several congeners of **1** were also identified from the same species, including the C6–N1′ dimer dixiamycin C (**4**), the C11–C12 bond oxidized and ring-expanded oxiamycin (**5**) and a putative biosynthetic precursor, indosespene (**6**)[17–21]. Considerable efforts have been made towards revealing the biogenesis of this class of natural products, including the elucidation of the relevant gene clusters and enzymes[17–20]. The chemical synthesis of these compounds[13,14] may help further the understanding of the biosynthetic relationships within this indolosesquiterpenoid family, as well as the discovery of their unknown biological activities. Very recently, Baran *et al.*[13] disclosed the first total synthesis of xiamycin A (**1**) and dixiamycin B.

In this paper, we report the total syntheses of xiamycin A, oridamycins A and B, dixiamycin C (**1–4**) and indosespene (**6**). The syntheses of **2–4** were accomplished for the first time. A 6π-electrocyclization strategy and an oxidative Heck annulation strategy were developed for assembling the carbazole scaffold. The antiviral properties of the synthetic natural products were evaluated *in vitro*, and **1** exhibits potent inhibitory activity against herpes simplex virus-1 (HSV-1).

## Results

**Retrosynthetic analysis**. The synthetic challenge of the xiamycin and oridamycin families may be separated into two distinct problems, namely the stereocontrolled assembly of the *trans*-decalin motif bearing the C16 quaternary centre and the construction of the carbazole core. Inspired by the structural correlation between xiamycin A (**1**) and indosespene (**6**), our retrosynthetic analysis of **1** (Fig. 2) began with a disconnection at the C2–C21 bond. A 6π-electrocyclization[23] (path a) and a Heck-type annulation[24–29] (path b) were envisioned as two alternatives to forge the carbazole motif. The former draws on our recent syntheses of tubingensin A and daphenylline[30,31], while the latter relies on activation of the indole C2–H bond by palladium. Correspondingly, triene (**7**) and indosespene ester (**8**) are considered as the key intermediates for the two strategies. These functionalized decalin systems are expected to arise from a cascade cyclization of dienylepoxide precursors, such as **9** and **10**. This type of radical cyclization may be initiated by reduction of the epoxide by a Ti(III) species[32–47], although, to our knowledge, α,β-epoxy esters/ketones have not been fully explored as the substrates for such cyclization[14,48]. Notably, Cuerva and coworkers[49] disclosed that enol radicals derived from α,β-unsaturated ketones inhibit the cyclization reaction. The stereochemical outcome of the cyclization cascade (including the stereochemistry at C16) would be determined by the chirality and geometry of the precursor, which may allow a convenient access to oridamycins A and B (**2** and **3**) from an alternative precursor. Successful execution of the route outlined above would also be anticipated to provide access to indosespene (**6**) and dixiamycin C (**4**) through minor modifications.

**The 6π-electrocyclization/aromatization approach**. Fig. 3 illustrates the 6π-electrocyclization/aromatization approach for the synthesis of xiamycin A (**1**). Starting from optically active epoxide **11** (96% ee)[50], a sequence of 2-azaadamantane *N*-oxyl (AZADO) oxidation[51,52] and esterification led to compound **12** in 76% overall yield. On the basis of seminal work on the Ti(III)-catalysed bioinspired epoxypolyene cyclization by Cuerva and Oltra[38–41], we investigated the cyclization of the α,β-epoxy ester **12**. Under the optimized conditions [$Cp_2TiCl_2$ (20 mol%), Mn, $iPr_2NEt$, TMSCl], the desired *trans*-decalin **13** was obtained as a single diastereomer. TBS protection followed by acetyl deprotection gave **14** (44% yield for the three steps), which was subjected to a sequence of DMP oxidation, Grignard addition (with reagent **15**)[30], and dehydration (MsCl/$iPr_2NEt$) to furnish triene **16** in 75% overall yield. On microwave irradiation (120 °C, air), **16** underwent a thermal 6π-electrocyclization and a subsequent aromatization to generate carbazole **17** (80% yield). Similarly, when using Grignard reagent **18** with a bromine substituent, we readily obtained compound **19** via the corresponding triene **20**. This bromocarbazole served as an

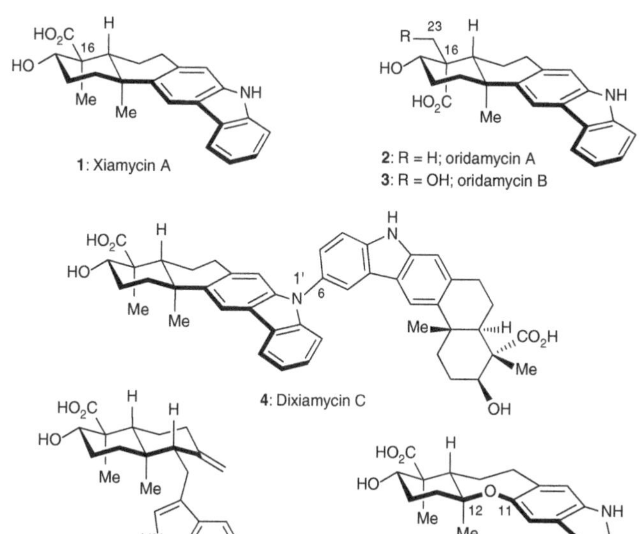


**Figure 1 | The structures of representative natural products from the xiamycin and oridamycin families.** The two families share a similar scaffold but differ in the C16 stereochemistry. Indosespene (**6**) is a putative biosynthetic precursor of xiamycin A (**1**).

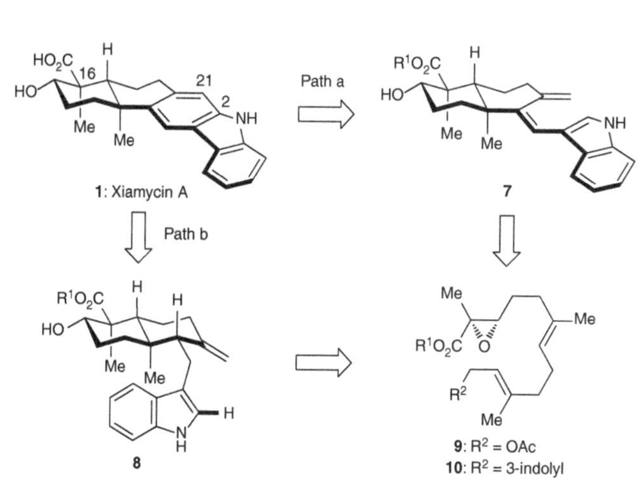


**Figure 2 | Two retrosynthetic analyses of xiamycin A.** One (path a) features a 6π-electrocyclization/aromatization strategy, and the other (path b) involves an indole C2-H bond activation/Heck annulation strategy. The devised strategies are anticipated to be applicable to the synthesis of other members from the xiamycin and oridamycin families.

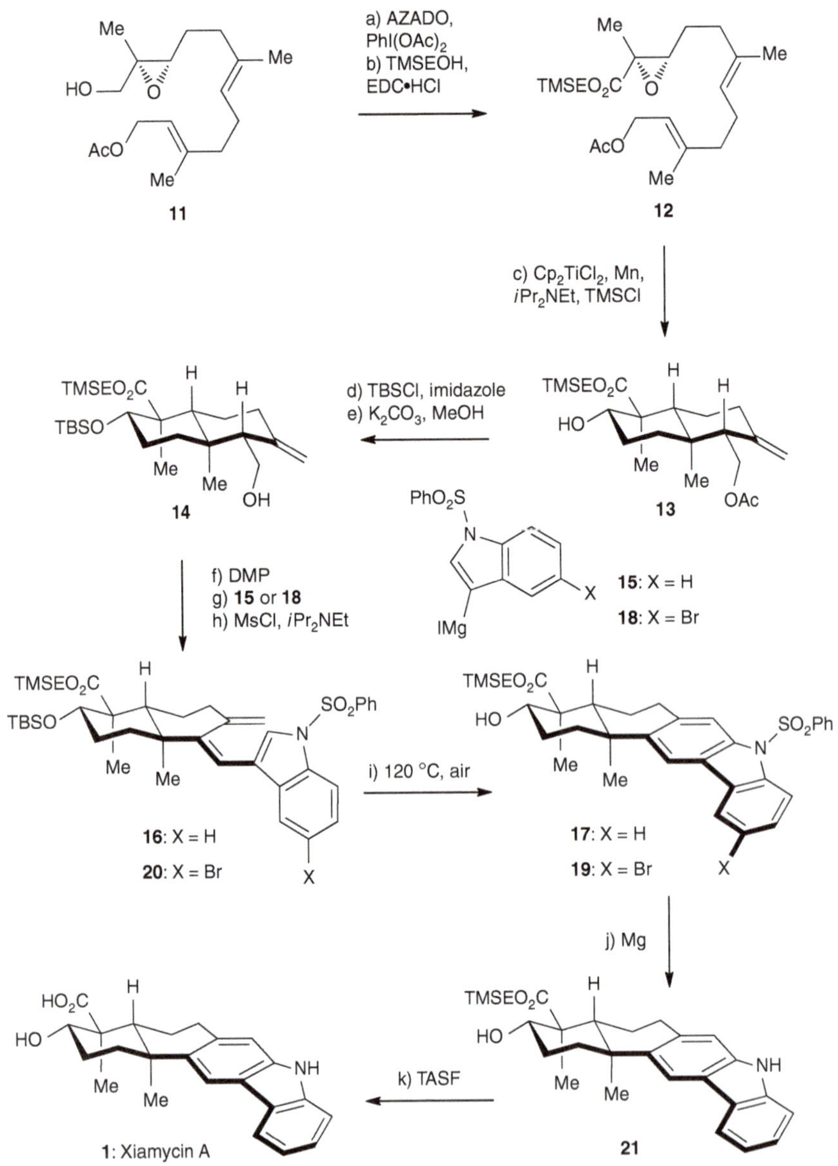


**Figure 3 | The 6π-electrocyclization/aromatization approach towards the total synthesis of xiamycin A.** Reagents and conditions: (a) AZADO (5 mol%), PhI(OAc)$_2$ (3.0 eq), CH$_2$Cl$_2$, 22 °C, 6 h; (b) TMSEOH (1.2 eq), EDC•HCl (1.1 eq), 4-DMAP (1.0 eq), CH$_2$Cl$_2$, 22 °C, 5 h, 76% (two steps); (c) Cp$_2$TiCl$_2$ (20 mol%), Mn (8.0 eq), *i*Pr$_2$NEt (6.0 eq), TMSCl (5.0 eq), THF, 22 °C, 4 h; (d) TBSCl (1.2 eq), imidazole (1.5 eq), DMF, 22 °C, 10 h; (e) K$_2$CO$_3$ (1.0 eq), MeOH, 22 °C, 3 h, 44% (3 steps); (f) DMP (1.5 eq), CH$_2$Cl$_2$, 22 °C, 30 min; (g) **15** or **18** (3.0 eq), THF, 22 °C, 40 min; (h) MsCl (1.2 eq), *i*Pr$_2$NEt (1.5 eq), CH$_2$Cl$_2$, 0 °C, 1 h, 75% for **16** (three steps), 71% for **20** (three steps); (**i**) air, DMSO, 120 °C, 2 h, 80% for **17**, 76% for **19**; (**j**) Mg (2.0 eq), MeOH, 22 °C, 1 h, 98%; (k) TASF (2.0 eq), DMF, 50 °C, 95%.

appropriate substrate for the crucial C–N bond formation in our dixiamycin C synthesis (*vide infra*). Desulfonylation of **17** followed by TMSE deprotection of the resulting compound **21** afforded xiamycin A (**1**) in a good overall yield, the structure of which was confirmed by X-ray crystallographic analysis (Fig. 4).

The mechanism of the cascade cyclization is postulated in Fig. 5. The epoxide C16–O bond is reductively cleaved by the Ti(III) species to generate **22**, which triggers two sequential cyclization reactions, and the resulting radical **23** could undergo a mixed disproportionation[53] to give the intermediate **24** with an exocyclic C=C bond. In fact, two side reactions occurred: 1, over reduction of **22** to the corresponding enolate[48] followed by β-elimination to form an α,β-unsaturated ester; 2, hydrogen transfer to **23** to form the saturated derivative of **24**. Organic bases had subtle influence on these undesired reactions[54], and *i*Pr$_2$NEt was found to suppress the latter effectively in our case. Considering the challenges for the acid-promoted epoxy polyene

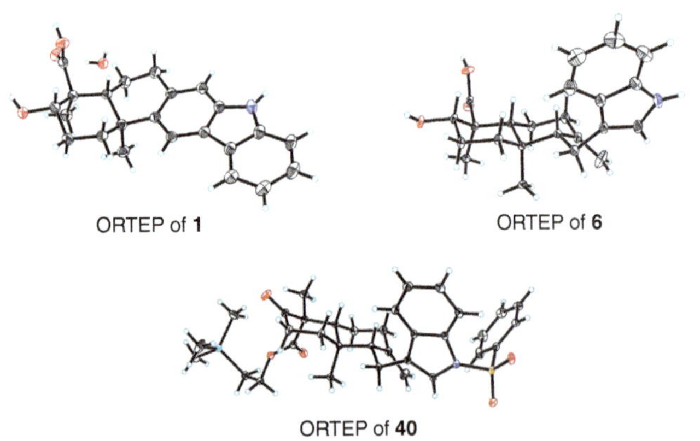


**Figure 4 | ORTEP drawings of compounds 1, 6 and 40.** Non-hydrogen atoms are shown as 30% ellipsoids. The stereochemical outcomes of the radical cyclization cascades are unambiguously confirmed.

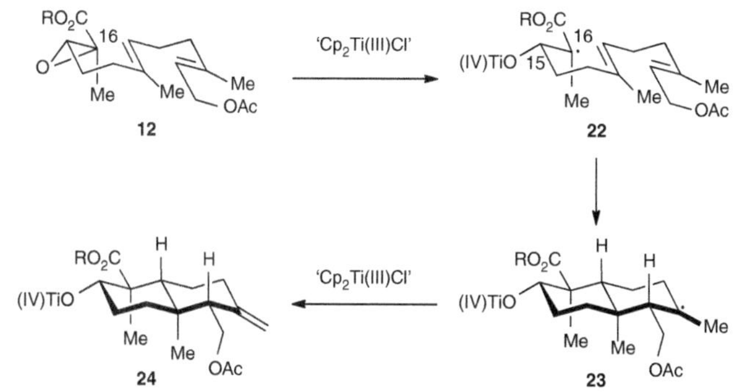


**Figure 5 | A postulated mechanism of the Ti(III)-mediated radical cyclization.** The epoxy ester serves as a suitable substrate for this reaction. The *trans*-decalin system with an equatorial alkoxycarbonyl and an exocyclic methylene groups is obtained.

cyclization, such as the preference to the formation of products with tri-/tetrasubstituted olefins and difficulty generating carbocations with electron-withdrawing substituents, this radical chemistry offers a useful and complementary route for preparing functionalized sesquiterpenoid synthons.

We then focused on the synthesis of oridamycin A bearing an axial carboxylic substituent on C16 (Fig. 6). The *cis*-epoxy analogue of **12** was first prepared and subjected to the Ti(III) conditions. However, the compound **13** with an equatorial carboxylic substituent was detected as the major product, presumably due to the fast rotation of the C15–C16 bond to adopt the sterically favoured conformation (Fig. 5) for the cyclization. Thus, we made recourse to a mechanistically different oxidative radical cyclization mediated by Mn(III)[55–59] using β-ketoester **25** (ref. 56) as a substrate. Treatment with $Mn(OAc)_3$/$Cu(OAc)_2$ in DMSO effected the radical cyclization with an acceptable yield of decalin **26** (48%, a single diastereomer

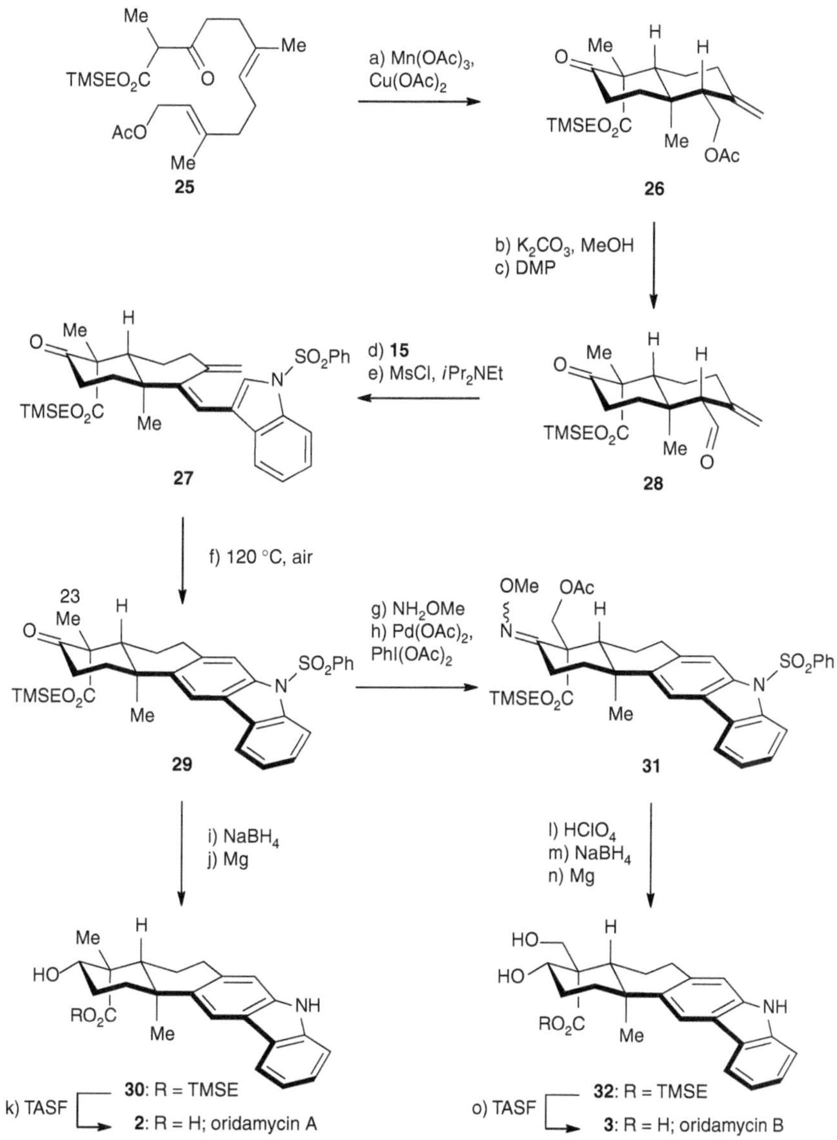


**Figure 6 | The 6π-electrocyclization/aromatization approach towards the total syntheses of oridamycins A and B.** Reagents and conditions: (a) $Mn(OAc)_3•2H_2O$ (2.0 eq), $Cu(OAc)_2•H_2O$ (1.0 eq), DMSO, 22 °C, 12 h, 48%; (b) $K_2CO_3$ (1.5 eq), MeOH, 22 °C, 3 h; (c) DMP (2.0 eq), $CH_2Cl_2$, 22 °C, 30 min, 92% (two steps); (d) **15** (3.0 eq), THF, 22 °C, 40 min; (e) MsCl (1.2 eq), $iPr_2NEt$ (1.5 eq), $CH_2Cl_2$, 0 °C, 1 h, 74% (2 steps); (f) air, DMSO, 120 °C, 2 h, 83%; (g) $NH_2OMe•HCl$ (2.0 eq), pyridine, 22 °C, 2 h, 94%; (h) $Pd(OAc)_2$ (20 mol%), $PhI(OAc)_2$ (1.0 eq), $Ac_2O$/AcOH (1:1), 110 °C, 1 h, 82%; (i) $NaBH_4$ (3.0 eq), MeOH, 0 °C, 30 min; (j) Mg (8.0 eq), MeOH, 22 °C, 1 h, 83% (two steps); (k) TASF (2.0 eq), DMF, 50 °C, 92%; (l) aq. $HClO_4$ (6.0 M)/acetone (1:3), 22 °C, 12 h; (m) $NaBH_4$ (3.0 eq), MeOH, 0 °C, 30 min, 62% (2 steps); (n) Mg (8.0 eq), MeOH, 22 °C, 1 h, 90%; (o) TASF (2.0 eq), DMF, 50 °C, 91%.

detected). This compound was subjected to a sequence similar to that described above to afford triene **27** with good overall efficiency, via the intermediacy of aldehyde **28**. Under thermal conditions (120 °C) and in presence of air, **27** was converted to carbazole **29** (83% yield) in one pot. Sequential ketone reduction ($NaBH_4$) and desulfonylation (Mg/MeOH) provided compound **30** in 83% overall yield. Final TMSE deprotection led to the completion of the first synthesis of (±)-oridamycin A (**2**).

We accomplished the synthesis of oridamycin B for the first time by adjusting the C23 oxidation state of **29** (Fig. 6), taking advantage of the neighbouring ketone carbonyl group. *O*-Methyloxime formation afforded the substrate suitable for a directed C–H bond oxidation, and Sanford's conditions [$Pd(OAc)_2$ (20 mol%), $PhI(OAc)_2$][60,61] were effective to give the desired acetate **31** with good overall efficiency. It should be noted that stoichiometric Pd-mediated $Csp^3$–H bond oxidation reactions have been exploited in terpenoid synthesis[41,62,63]. Deprotection of the oxime and acetate, followed by reduction and desulfonylation, afforded diol **32** with acceptable overall efficiency. Removal of TMSE with TASF provided (±)-oridamycin B (**3**) in 91% yield. The spectroscopic data of the synthetic samples of oridamycins A and B matched those reported for the natural products.

**The C–H bond activation/Heck annulation approach.** The second approach to xiamycin A (**1**) fulfilled the strategy of indole C2–H activation/Heck-type annulation (Fig. 7). Stille coupling[64] of **12** with *N*-Boc-3-(tributylstannyl)indole **33**[65] followed by deprotection of Boc[66] provided a fully functionalized precursor **34** (66% overall yield), which underwent a Ti(III)-catalysed radical cyclization under the same conditions as above to give decalin **35** as a single diastereomer in 60% yield[39]. In the presence of $Pd(OAc)_2$ and *p*-benzoquinone at 50 °C (refs 28,29), **35** was converted to carbazole **21** in 81% yield, presumably through the devised oxidative Heck/aromatization sequence. TMSE deprotection of **35** and **21** with TASF then afforded indosespene (**6**) and xiamycin A (**1**), respectively. The structure of **6** was verified by X-ray crystallographic analysis (Fig. 4).

The above strategy was further applied to alternative syntheses of oridamycins A and B (Fig. 8). In this case, an indole-containing β-ketoester **36** was employed as the substrate for the Mn(III) mediated cyclization. *N*-Sulfonylation of 3-geranylindole **37** (ref. 67; NaOH, $PhSO_2Cl$) and subsequent allylic oxidation ($SeO_2$, *t*BuOOH) gave alcohol **38** in 78% overall yield, which underwent bromination ($Et_3N$, MsCl; LiBr) followed by

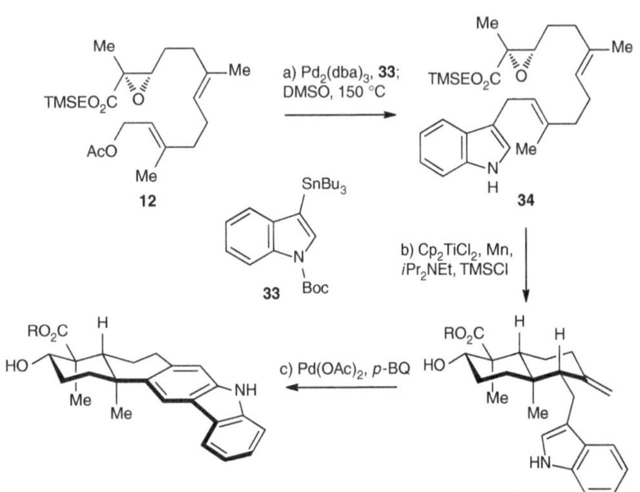


**Figure 7 | The oxidative Heck annulation approach towards the total synthesis of xiamycin A.** Reagents and conditions: (a) $Pd_2(dba)_3$ (5 mol%), LiCl (3.0 eq), **33** (2.0 eq), DMF, 80 °C, 2 h, then DMSO, 150 °C, 30 min, 66%; (b) $Cp_2TiCl_2$ (20 mol%), Mn (8.0 eq), $iPr_2NEt$ (6.0 eq), TMSCl (5.0 eq), THF, 22 °C, 4 h, 60%; (c) $Pd(OAc)_2$ (20 mol%), *p*-benzoquinone (2.0 eq), AcOH/toluene (1:4), 50 °C, 2 h, 81%; (d) TASF (2.0 eq), DMF, 50 °C, 95% for **1**, 91% for **6**.

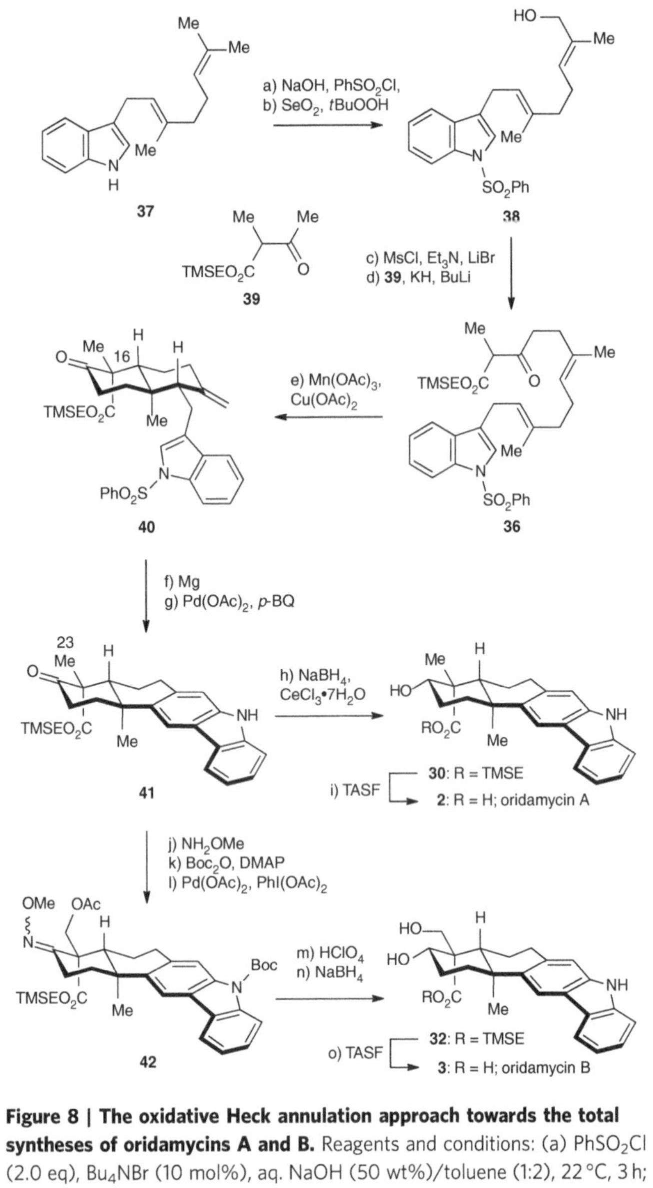


**Figure 8 | The oxidative Heck annulation approach towards the total syntheses of oridamycins A and B.** Reagents and conditions: (a) $PhSO_2Cl$ (2.0 eq), $Bu_4NBr$ (10 mol%), aq. NaOH (50 wt%)/toluene (1:2), 22 °C, 3 h; (b) $SeO_2$ (20 mol%), *t*BuOOH (2.0 eq), $CH_2Cl_2$, 0 °C, 10 h, 78% (two steps); (c) $Et_3N$ (1.5 eq), MsCl (1.2 eq), LiBr (5.0 eq), THF, 0 °C, 30 min; (d) **39**, (1.0 eq), KH (1.2 eq), BuLi (1.2 eq), THF/HMPA (5:1), 0 °C, 3 h, 80% (two steps); (e) $Mn(OAc)_3$•$2H_2O$ (2.0 eq), $Cu(OAc)_2$•$H_2O$ (1.0 eq), DMSO, 22 °C, 12 h, 52%; (f) Mg (3.0 eq), $NH_4Cl$ (1.2 eq), MeOH, 22 °C, 1 h, 96%; (g) $Pd(OAc)_2$ (20 mol%), *p*-benzoquinone (2.5 eq), AcOH/toluene (1:4), 50 °C, 2 h, 65%; (h) $NaBH_4$ (3.0 eq), $CeCl_3$•$7H_2O$ (3.0 eq), MeOH, 0 °C, 30 min; (i) TASF (2.0 eq), DMF, 50 °C, 2 h, 85% (two steps); (j) $NH_2OMe$•HCl (2.0 eq), pyridine, 22 °C, 2 h, 93%; (k) $Boc_2O$ (2.0 eq), 4-DMAP (2.0 eq), THF, 22 °C, 1 h, 89%; (l) $Pd(OAc)_2$ (20 mol%), $PhI(OAc)_2$ (1.1 eq), $Ac_2O$/AcOH (1:1), 110 °C, 1 h, 81%; (m) aq. $HClO_4$ (6.0 M)/acetone (1:3), 22 °C, 8 h; (n) $NaBH_4$ (3.0 eq), MeOH, 0 °C, 30 min, 54% (two steps); (o) TASF (2.0 eq), DMF, 50 °C, 2 h, 91%.

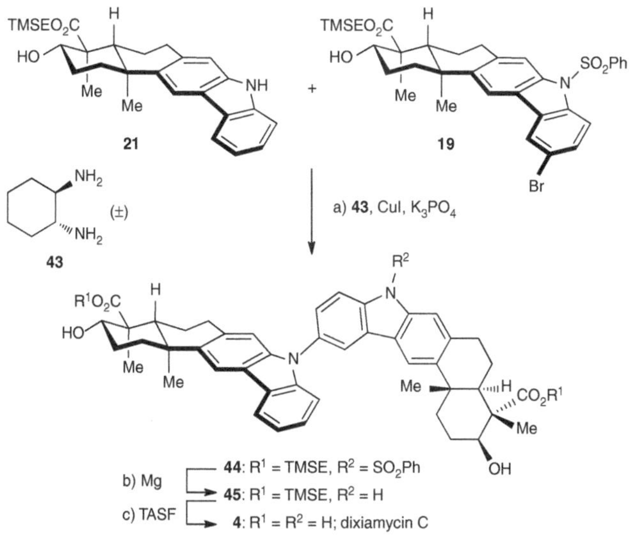


**Figure 9 | Total synthesis of dixiamycin C.** Reagents and conditions: (a) CuI (1.5 eq), **21** (1.0 eq), **19** (1.5 eq), (±)-**43** (3.0 eq), $K_3PO_4$ (7.5 eq), 1,4-dioxane, 110 °C, 12 h; (b) Mg (6.0 eq), MeOH, 22 °C, 1 h, 51% for the two steps; (c) TASF (4.0 eq), DMF, 50 °C, 5 h, 96%.

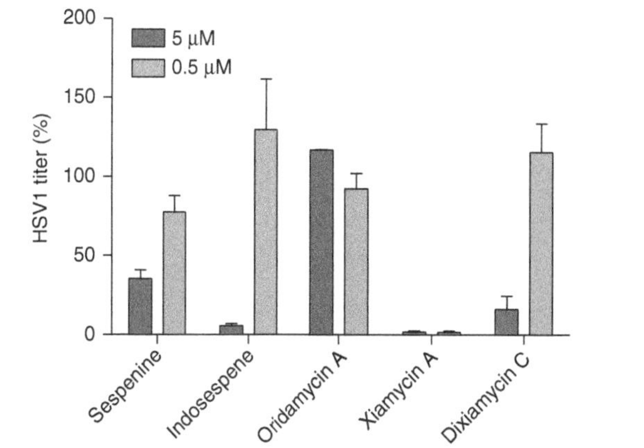


**Figure 10 | Effects of the synthetic compounds on HSV-1 propagation.** Vero cells were infected with HSV1 (strain KOS-ATCC VR1493) for 3 h and then treated with different concentrations of compounds for 20 h. HSV-1 in the culture supernatants were then quantified by using $TCID_{50}$ assay. Virus titres were presented as the percentages of the mock DMSO treatment. Data represent means ± s.e.m. of six individual infections.

alkylation with the dianion generated from β-ketoester **39** (KH, then BuLi) to furnish **36** (80% yield for the two steps). Under the similar conditions used for the cyclization of compound **25** (Fig. 6), decalin **40** was obtained in 52% yield as a single detectable diastereomer. Its C16 stereochemistry was secured by X-ray crystallographic analysis (Fig. 4). Desulfonylation (Mg/MeOH, 96% yield) and subsequent oxidative Heck annulation (*vide supra*) afforded carbazole **41** (65% yield), which was subjected to a two-step sequence of ketone reduction and TMSE deprotection to give (±)-oridamycin A (**2**). Sanford $Csp^3$–H bond oxidation was exploited again for the synthesis of oridamycin B from **41**; deactivation of the carbazole motif with an electron-withdrawing *N*-protecting group was required. Thus, **41** was subjected to $NH_2OMe$ condensation and Boc protection to afford the oxidation precursor. The same conditions [$Pd(OAc)_2$ (20 mol%), $PhI(OAc)_2$] as used before delivered acetate **42** smoothly (67% yield for the three steps). Global deprotection of the oxime, acetyl and Boc groups under acidic conditions followed by reduction of the exposed carbonyl group furnished the late intermediate **32** in 54% overall yield, which was converted to (±)-oridamycin B (**3**) through the removal of TMSE group.

**The synthesis of dixiamycin C.** Having synthesized the monomeric natural products, we directed our attention to the synthesis of dixiamycin C (Fig. 9). Starting from coupling partners **19** and **21**, we investigated a variety of conditions for the transition metal promoted C–N bond formation. Buchwald's protocol[68] [CuI, (±)-**43**, $K_3PO_4$, 110 °C] proved to be optimal to provide dimer **44**. Deprotection of the benzenesulfonyl and TMSE groups furnished dixiamycin C (**4**) in a good overall yield, via the intermediacy of **45**. The spectra and physical properties of the synthetic dixiamycin C were consistent with those reported for the natural products.

**The antiviral activity.** The activities of synthetic xiamycin A (**1**), oridamycins A (**2**), dixiamycin C (**4**), indosespene (**6**) and sespenine[14,16] against HSV-1 and hepatitis C virus (HCV) were evaluated. Notably, xiamycin A was reported to possess moderate anti-HIV activity but no significant cytotoxicity against a number of human tumour cell lines[16], and a structurally relevant carbazole natural product, named tubingensin A, was found to exhibit anti-HSV-1 activity (without details disclosed)[69]. We treated HSV-1 (KOS strain)-infected vero cells with 5.0 and 0.50 μM solutions of the compounds for 20 h and then quantified the infectious HSV-1 titres in the supernatants by using $TCID_{50}$ assay[70]. As shown in Fig. 10, **1**, **4**, **6** and sespenine can suppress HSV-1 propagation in a dose-dependent manner. Notably, xiamycin A (**1**) is the most potent compound that reduces >95% of virus propagation at a concentration of 0.50 μM. Indosespene (**6**) significantly inhibits HSV-1 propagation at the concentration of 5.0 μM and displays essentially no antiviral activity at the concentration of 0.50 μM. Dixiamycin C (**4**) and sespenine show moderate inhibitory activity against HSV-1 at 5.0 μM but no activity at 0.50 μM. For the anti-HCV test (using a replicon assay), only dixiamycin C (**6**) exhibited inhibitory activity at a high concentration (50 μM, see Supplementary Fig. 81). It should be mentioned that these compounds did not exhibit cytotoxicity against the cells used for both anti-HSV-1 and HCV tests (see Supplementary Fig. 82).

## Discussion

We have accomplished the total synthesis of indolosesquiterpenoids xiamycin A, oridamycins A and B, dixiamycin C and indosespene. Two unified strategies for the framework construction and a series of transition-metal-mediated reactions formed the basis of the synthesis. Antiviral tests of the synthetic natural products revealed that xiamycin A exhibits a potent inhibitory activity against HSV-1.

## Methods

**General.** All reactions were carried out under an argon atmosphere with dry solvents under anhydrous conditions, unless otherwise noted. Tetrahydrofuran (THF), toluene and 1,4-dioxane were distilled immediately before use from sodium-benzophenone ketyl. Methylene chloride ($CH_2Cl_2$), *N,N*-dimethylformamide (DMF), dimethyl sulfoxide (DMSO), hexamethylphosphoramide (HMPA), triethylamine ($Et_3N$), *N,N*-diisopropylethylamine ($iPr_2NEt$) and pyridine were distilled from calcium hydride and stored under an argon atmosphere. Methanol (MeOH) was distilled from magnesium and stored under an argon atmosphere. Acetone was dried over drierite and distilled before use. Reagents were purchased at the highest commercial quality and used without further purification, unless otherwise stated. Solvents for chromatography were used as supplied by Titan Chemical. Reactions were monitored by thin layer chromatography (TLC) carried out on S-2 0.25 mm E. Merck silica gel plates (60F-254) using UV light as a visualizing agent and aqueous ammonium cerium nitrate/ammonium molybdate or basic aqueous potassium permanganate as a developing agent. E. Merck silica gel

(60, particle size 0.040–0.063 mm) was used for flash column chromatography. Preparative thin layer chromatography separations were carried out on 0.25 or 0.50 mm E. Merck silica gel plates (60F-254). NMR spectra were recorded on Bruker AV-400, DRX-600 or Agilent 500/54/ASP instrument and calibrated by using residual undeuterated chloroform ($\delta_H = 7.26$ p.p.m.) and $CDCl_3$ ($\delta_C = 77.16$ p.p.m.), or undeuterated methanol ($\delta_H = 3.31$ p.p.m.) and methanol-$d_4$ ($\delta_C = 49.00$ p.p.m.), as internal references. IR spectra were recorded on a Thermo Scientific Nicolet 380 FT-IR spectrometer. Melting points (m.p.) are uncorrected and were recorded on a SGW X-4 apparatus. High-resolution mass spectra (HRMS) were recorded on a Bruker APEXIII 7.0 Tesla ESI-FT mass spectrometer.

For $^1H$ and $^{13}C$ NMR spectra of compounds, see Supplementary Figs 1–70. For the comparisons of $^1H$ and $^{13}C$ NMR spectra of the natural and synthetic natural products, see Supplementary Figs 71–80. For the comparisons of $^1H$ and $^{13}C$ NMR spectroscopic data of the natural and synthetic natural products, see Supplementary Tables 1–10. For the experimental procedures, and spectroscopic and physical data of compounds, the antiviral assay and cytotoxicity assays, and the crystallographic data of compounds **1**, **6** and **40**, see Supplementary Methods.

## References


1. O'Connor, S. E. & McCoy, E. *Terpene Indole Alkaloid Biosynthesis in Recent Advances in Phytochemistry* Vol. 40 (ed. Romeo, J. T.) 1–22 (Elsevier, 2006).
2. Marcos, I. S., Moro, R. F., Costales, I., Basabe, P. & Díez, D. Sesquiterpenyl indoles. *Nat. Prod. Rep.* **30,** 1509–1526 (2013).
3. Lacey, E., Power, M., Wu, Z. & Rickards, R. W. Terpenylated Diketopiperazines, (Drimentines). WO 1998/009968, March 12, 1998.
4. Uchida, R., Kim, Y. P., Namatame, I, Tomoda, H. & Ōmura, S. Sespendole, a new inhibitor of lipid droplet synthesis in macrophages, produced by Pseudobotrytis terrestris FKA-25. *J. Antibiot.* **59,** 93–97 (2006).
5. Roll, D. M. *et al.* The lecanindoles, nonsteroidal progestins from the terrestrial fungus Verticillium lecanii 6144. *J. Nat. Prod.* **72,** 1944–1948 (2009).
6. Che, Q. *et al.* Hybrid isoprenoids from a reeds rhizosphere soil derived actinomycete Streptomyces sp. CHQ-64. *Org. Lett.* **14,** 3438–3441 (2012).
7. Che, Q. *et al.* Polycyclic hybrid isoprenoids from a reed rhizosphere soil derived Streptomyces sp. CHQ-64. *J. Nat. Prod.* **76,** 759–763 (2013).
8. Velthuisen, E. J. & Danishefsky, S. J. Total synthesis of ( + )-suaveolindole: establishment of its absolute configuration. *J. Am. Chem. Soc.* **129,** 10640–10641 (2007).
9. Marcos, I. S. *et al.* Synthesis of 12-epi-ent-polyalthenol an antitumour indole sesquiterpene alkaloid. *Tetrahedron* **68,** 7932–7940 (2012).
10. Marcos, I. S. *et al.* Biomimetic synthesis of an antitumour indole sesquiterpene alkaloid, 12-epi-ent-pentacyclindole. *Tetrahedron* **69,** 7285–7289 (2013).
11. Asanuma, A., Enomoto, M., Nagasawa, T. & Kuwahara, S. Synthesis of ( ± )-lecanindole D. *Tetrahedron Lett.* **54,** 4561–4563 (2013).
12. Sun, Y., Li, R., Zhang, W. & Li, A. Total synthesis of indotertine A and drimentines A, F, and G. *Angew. Chem. Int. Ed.* **52,** 9201–9204 (2013).
13. Rosen, B. R., Werner, E. W., O'Brien, A. G. & Baran, P. S. Total synthesis of dixiamycin B by electrochemical oxidation. *J. Am. Chem. Soc.* **136,** 5571–5574 (2014).
14. Sun, Y. *et al.* Bioinspired total synthesis of sespenine. *Angew. Chem. Int. Ed.* **53,** 9012–9016 (2014).
15. Ding, L. *et al.* Xiamycin, a pentacyclic indolosesquiterpene with selective anti-HIV activity from a bacterial mangrove endophyte. *Bioorg. Med. Chem. Lett.* **20,** 6685–6687 (2010).
16. Ding, L., Maier, A., Fiebig, H. H., Lin, W. H. & Hertweck, C. A family of multicyclic indolosesquiterpenes from a bacterial endophyte. *Org. Biomol. Chem.* **9,** 4029–4031 (2011).
17. Xu, Z., Baunach, M., Ding, L. & Hertweck, C. Bacterial synthesis of diverse indole terpene alkaloids by an unparalleled cyclization sequence. *Angew. Chem. Int. Ed.* **51,** 10293–10297 (2012).
18. Baunach, M., Ding, L., Bruhn, T., Bringmann, G. & Hertweck, C. Regiodivergent N–C and N–N aryl coupling reactions of indoloterpenes and cycloether formation mediated by a single bacterial flavoenzyme. *Angew. Chem. Int. Ed.* **52,** 9040–9043 (2013).
19. Li, H. *et al.* Identification and characterization of xiamycin A and oxiamycin gene cluster reveals an oxidative cyclization strategy tailoring indolosesquiterpene biosynthesis. *J. Am. Chem. Soc.* **134,** 8996–9005 (2012).
20. Zhang, Q. *et al.* Carboxyl formation from methyl via triple hydroxylations by xiaM in xiamycin A biosynthesis. *Org. Lett.* **14,** 6142–6145 (2012).
21. Zhang, Q. *et al.* N–N-Coupled indolo-sesquiterpene atropo-diastereomers from a marine-derived actinomycete. *Eur. J. Org. Chem.* **27,** 5256–5262 (2012).
22. Takada, K., Kajiwara, H. & Imamura, N. Oridamycins A and B, anti-saprolegnia parasitica indolosesquiterpenes isolated from Streptomyces sp. KS84. *J. Nat. Prod.* **73,** 698–701 (2010).
23. Beaudry, C. M., Malerich, J. P. & Trauner, D. Biosynthetic and biomimetic electrocyclizations. *Chem. Rev.* **105,** 4757–4778 (2005).
24. Ferreira, E. M., Zhang, H. & Stoltz, B. M. *Oxidative Heck-Type Reactions (Fujiwara–Moritani Reactions), in The Mizoroki-Heck Reaction* (ed. Oestreich, M.) 345–382 (John Wiley & Sons, 2009).
25. Baran, P. S. & Corey, E. J. A short synthetic route to ( + )-austamide, ( + )-deoxyisoaustamide, and ( + )-hydratoaustamide from a common precursor by a novel palladium-mediated indole→dihydroindoloazocine cyclization. *J. Am. Chem. Soc.* **124,** 7904–7905 (2002).
26. Ferreira, E. M. & Stoltz, B. M. Catalytic C–H bond functionalization with palladium(II): aerobic oxidative annulations of indoles. *J. Am. Chem. Soc.* **125,** 9578–9579 (2003).
27. Grimster, N. P., Gauntlett, C., Godfrey, C. R. A. & Gaunt, M. J. Palladium-catalyzed intermolecular alkenylation of indoles by solvent-controlled regioselective C–H functionalization. *Angew. Chem. Int. Ed.* **44,** 3125–3129 (2005).
28. Kong, A., Han, X. & Lu, X. Highly efficient construction of benzene ring in carbazoles by palladium-catalyzed endo-mode oxidative cyclization of 3-(3′-alkenyl)indoles. *Org. Lett.* **8,** 1339–1342 (2006).
29. Han, X. & Lu, X. Novel Palladium-catalyzed acyloxylation/cyclization of 2-(3′-alkenyl)indoles. *Org. Lett.* **11,** 2381–2384 (2009).
30. Bian, M. *et al.* Total syntheses of anominine and tubingensin A. *J. Am. Chem. Soc* **134,** 8078–8081 (2012).
31. Lu, Z., Li, Y., Deng, J. & Li, A. Total synthesis of the Daphniphyllum alkaloid daphenylline. *Nat. Chem.* **5,** 679–684 (2013).
32. Nugent, W. A. & RajanBabu, T. V. Transition-metal-centered radicals in organic synthesis. Titanium(III)-induced cyclization of epoxy olefins. *J. Am. Chem. Soc.* **110,** 8561–8562 (1988).
33. RajanBabu, T. V. & Nugent, W. A. Selective generation of free radicals from epoxides using a transition-metal radical. A powerful new tool for organic synthesis. *J. Am. Chem. Soc.* **116,** 986–997 (1994).
34. Gansäuer, A., Pierobon, M. & Bluhm, H. Catalytic, Highly regio- and chemoselective generation of radicals from epoxides: titanocene dichloride as an electron transfer catalyst in transition metal catalyzed radical reactions. *Angew. Chem. Int. Ed.* **37,** 101–103 (1998).
35. Gansäuer, A., Bluhm, H. & Pierobon, M. Emergence of a novel catalytic radical reaction: Titanocene-catalyzed reductive opening of epoxides. *J. Am. Chem. Soc.* **120,** 12849–12859 (1998).
36. Fernández-Mateos, A., Martin de la Nava, E., Pascual Coca, G., Ramos Silva, A. & Rubio González, R. Radicals from epoxides. Intramolecular addition to aldehyde and ketone carbonyls. *Org. Lett.* **1,** 607–610 (1999).
37. Barrero, A. F., Cuerva, J. M., Herrador, M. M. & Valdivia, M. V. A new strategy for the synthesis of cyclic terpenoids based on the radical opening of acyclic epoxypolyenes. *J. Org. Chem.* **66,** 4074–4078 (2001).
38. Justicia, J. *et al.* Titanocene-catalyzed cascade cyclization of epoxypolyprenes: straightforward synthesis of terpenoids by free-radical chemistry. *Chem. Eur. J.* **10,** 1778–1788 (2004).
39. Justicia, J., Oltra, J. E. & Cuerva, J. M. General approach to polycyclic meroterpenoids based on Stille couplings and titanocene catalysis. *J. Org. Chem.* **69,** 5803–5806 (2004).
40. Justicia, J. *et al.* 7-endo Radical cyclizations catalyzed by Titanocene(III). Straightforward synthesis of terpenoids with seven-membered carbocycles. *J. Am. Chem. Soc.* **127,** 14911–14921 (2005).
41. Justicia, J., Oltra, J. E. & Cuerva, J. M. Exploiting $Pd^{II}$ and $Ti^{III}$ chemistry to obtain γ-dioxygenated terpenoids: synthesis of rostratone and novel approaches to aphidicolin and pyripyropene A. *J. Org. Chem.* **70,** 8265–8272 (2005).
42. Gansäuer, A. & Narayan, S. Titanocene-catalysed electron transfer-mediated opening of epoxides. *Adv. Synth. Catal.* **344,** 465–475 (2002).
43. Barrero, A. F., Quílez del Moral, J. F., Sánchez, E. M. & Arteaga, J. F. Titanocene-mediated radical cyclization: an emergent method towards the synthesis of natural products. *Eur. J. Org. Chem.* **7,** 1627–1641 (2006).
44. Cuerva, J. M., Justicia, J., Oller-López, J. L. & Oltra, J. E. $Cp_2TiCl$ in natural product synthesis. *Top. Curr. Chem.* **264,** 63–91 (2006).
45. Gansäuer, A., Justicia, J., Fan, C.-A., Worgull, D. & Piestert, F. Reductive C–C bond formation after epoxide opening via electron transfer. *Top. Curr. Chem.* **279,** 25–52 (2007).
46. Justicia, J. *et al.* Bioinspired terpene synthesis: a radical approach. *Chem. Soc. Rev.* **40,** 3525–3537 (2011).
47. Morcillo, S. P. *et al.* Recent applications of $Cp_2TiCl$ in natural product synthesis. *Org. Chem. Front.* **1,** 15–33 (2014).
48. Hardouin, C., Chevallier, F., Rousseau, B. & Doris, E. $Cp_2TiCl$-mediated selective reduction of α,β-epoxy ketones. *J. Org. Chem.* **66,** 1046–1048 (2001).
49. Morcillo, S. P. *et al.* Ti(III)-catalyzed cyclizations of ketoepoxypolyprenes: control over the number of rings and unexpected stereoselectivities. *J. Am. Chem. Soc.* **136,** 6943–6951 (2014).
50. Lan, J. *et al.* First total synthesis and absolute configuration of marine cembrane diterpenoid ( + )-11,12-epoxysarcophytol A. *Tetrahedron Lett.* **41,** 2181–2184 (2000).
51. Shibuya, M., Tomizawa, M., Suzuki, I. & Iwabuchi, Y. 2-Azaadamantane N-oxyl (AZADO) and 1-Me-AZADO: highly efficient organocatalysts for oxidation of alcohols. *J. Am. Chem. Soc.* **128,** 8412–8413 (2006).
52. Kuranaga, T. *et al.* Total synthesis and complete structural assignment of yaku'amide A. *J. Am. Chem. Soc.* **135,** 5467–5474 (2013).

53. Justicia, J., Jiménez, T., Morcillo, S. P., Cuerva, J. M. & Oltra, J. E. Mixed disproportionation versus radical trapping in titanocene(III)-promoted epoxide openings. *Tetrahedron* **65,** 10837–10841 (2009).
54. Paradas, M. *et al.* Understanding the exceptional hydrogen-atom donor characteristics of water in $Ti^{III}$-mediated free-radical chemistry. *J. Am. Chem. Soc.* **132,** 12748–12756 (2010).
55. Snider, B. B. Manganese(III)-based oxidative free-radical cyclizations. *Chem. Rev.* **96,** 339–364 (1996).
56. Snider, B. B., Mohan, R. & Kates, S. A. Manganese (III) based oxidative free-radical cyclizations. 2. Polycyclization reactions proceeding through tertiary cations. *Tetrahedron Lett.* **28,** 841–844 (1987).
57. Zoretic, P. A., Weng, X., Biggers, M. S., Caspar, M. L. & Davis, D. G. Manganese(III)-promoted oxidative free-radical cyclizations of β-keto imides: asymmetric induction with oppolzer's chiral sultam. *Tetrahedron Lett.* **33,** 2637 (1992).
58. Zoretic, P. A., Fang, H. & Ribeiro, A. A. Synthesis of d,l-norlabdane oxide and related odorants: an intramolecular radical approach. *J. Org. Chem.* **63,** 4779–4785 (1998).
59. Barrero, A. F., Herrador, M. M., Quílez del Moral, J. F. & Valdivia, M. V. A convenient synthesis of A-ring-functionalized podolactones. Revision of the structure of wentilactone B. *Org. Lett.* **4,** 1379–1382 (2002).
60. Desai, L. V., Hull, K. L. & Sanford, M. S. Palladium-catalyzed oxygenation of unactivated sp3C–H bonds. *J. Am. Chem. Soc.* **126,** 9542–9543 (2004).
61. Neufeldt, S. R. & Sanford, M. S. O-Acetyl oximes as transformable directing groups for Pd-catalyzed C–H bond functionalization. *Org. Lett.* **12,** 532–535 (2010).
62. Baldwin, J. E., Jones, R. H., Najera, C. & Yus, M. Functionalization of unactivated methyl groups through cyclopalladation reactions. *Tetrahedron* **41,** 699–711 (1985).
63. Zhu, C., Tang, P. & Yu, B. Total synthesis of lobatoside E, a potent antitumor cyclic triterpene saponin. *J. Am. Chem. Soc.* **130,** 5872–5873 (2008).
64. Del Valle, L., Stille, J. K. & Hegedus, L. S. Palladium-catalyzed coupling of allylic acetates with aryl- and vinylstannanes. *J. Org. Chem.* **55,** 3019–3023 (1990).
65. McNeil, D. W. *et al.* Novel non-nucleoside inhibitors of human immunodeficiency virus type 1 reverse transcriptase. 6. 2-indol-3-yl- and 2-azaindol-3-yl- dipyridodiazepinones. *J. Med. Chem.* **40,** 2430–2433 (1997).
66. Im, G. J. *et al.* Indolyne experimental and computational studies: Synthetic applications and origins of selectivities of nucleophilic additions. *J. Am. Chem. Soc.* **132,** 17933–17944 (2010).
67. Westermaier, M. & Mayr, H. Electrophilic allylations and benzylations of indoles in neutral aqueous or alcoholic solutions. *Org. Lett.* **8,** 4791–4794 (2006).
68. Klapars, A., Antilla, J. C., Huang, X. & Buchwald, S. L. A general and efficient copper catalyst for the amidation of aryl halides and the N-arylation of nitrogen heterocycles. *J. Am. Chem. Soc.* **123,** 7727–7729 (2001).
69. TePaske, M. R., Gloer, J. B., Wicklow, D. T. & Dowd, P. F. Tubingensin A: an antiviral carbazole alkaloid from the sclerotia of Aspergillus tubingensis. *J. Org. Chem.* **54,** 4743–4746 (1989).
70. Reed, L. J. & Muench, H. A simple method of estimating fifty percent endpoints. *Am. J. Hyg.* **27,** 493–497 (1938).

## Acknowledgements

We thank Professors N. Imamura and C. Zhang for providing the $^{13}C$ NMR spectrum of oridamycin A and an authentic sample of dixiamycin C, respectively. We are grateful to Professor Z.-J. Yao for his assistance with the cryoprobe NMR facility. Financial support was provided by Ministry of Science & Technology (2013CB836900), National Natural Science Foundation of China (21290180, 21172235 and 21222202), Pujiang Program (12PJ1410800) and Chinese Academy of Sciences.

## Author contributions

A.L. directed the project. A.L., Z.M., H.Y. and D.J.E. conceived the synthetic route. Z.M., H.Y., H.C. and M.W. conducted the synthetic work. L.L. and W.T. conducted the biological work. A.L., Z.M., H.Y., H.C., P.Y., D.J.E., L.L., W.T. and J.Z. analysed the results. A.L., D.J.E., L.L. and J.Z. wrote the manuscript.

## Additional information

**Competing financial interests:** The authors declare no competing financial interest.

# 14

# Late-stage C–H functionalization of complex alkaloids and drug molecules via intermolecular rhodium-carbenoid insertion

Jing He[1], Lawrence G. Hamann[1], Huw M.L. Davies[2] & Rohan E.J. Beckwith[1]

Alkaloids constitute a large family of natural products possessing diverse biological properties. Their unique and complex structures have inspired numerous innovations in synthetic chemistry. In the realm of late-stage C–H functionalization, alkaloids remain a significant challenge due to the presence of the basic amine and a variety of other functional groups. Herein we report the first examples of dirhodium(II)-catalysed intermolecular C–H insertion into complex natural products containing nucleophilic tertiary amines to generate a C–C bond. The application to a diverse range of alkaloids and drug molecules demonstrates remarkable chemoselectivity and predictable regioselectivity. The capacity for late-stage diversification is highlighted in the catalyst-controlled selective functionalizations of the alkaloid brucine. The remarkable selectivity observed, particularly for site-specific C–H insertion at *N*-methyl functionalities, offers utility in a range of applications where efficient installation of synthetic handles on complex alkaloids is desired.

[1]Department of Global Discovery Chemistry, Novartis Institutes for BioMedical Research, 250 Massachusetts Avenue, Cambridge, Massachusetts 02139, USA. [2]Department of Chemistry, Emory University, 1515 Dickey Drive, Atlanta, Georgia 30322, USA. Correspondence and requests for materials should be addressed to H.M.L.D. (email: hmdavie@emory.edu) or to R.E.J.B. (email: rohan.beckwith@novartis.com).

The decline in the number of new drug approvals in recent years despite high levels of R&D investment has spurred a number of alternative approaches towards drug discovery within the pharmaceutical industry[1]. This includes a resurgence in phenotypic screens, and along with that a renewed interest in exploiting the unique structural diversity of natural products to modulate disease-relevant processes[2–5]. Drug-discovery efforts around natural product optimization have been underexplored, owing in part to the synthetic challenge of effecting useful modifications in the face of structural complexity and diverse functionality. Advances in synthetic methodology are therefore essential to enable progression of natural product hits of compelling phenotype into suitable probes for elucidating mechanism of action and towards the development of new therapeutics[6,7].

C–H functionalization is one such approach that has the potential to revolutionize how complex organic molecules, such as natural products, are made through the ability to selectively and efficiently transform C–H bonds in a predictable manner under mild conditions. Over the last two decades, the field of C–H functionalization has experienced explosive growth, and a number of highly effective transformations have been developed. Many compelling examples have been reported on the utilization of C–H functionalization as key strategic reactions in total synthesis, illustrating novel retrosynthetic analysis and step-economical synthesis[8–13]. In recent years, considerable efforts have been expended on the late-stage C–H functionalization of biologically compelling natural products and drug-like molecules[14–22]. This has been challenging, because although the pace of advances in C–H functionalization has been impressive, in many instances the new methodology is limited to a narrow range of substrates and functional groups. Many of the most effective C–H functionalization processes rely on the use of directing groups, which need to be introduced and then removed[23]. Consequently, the late-stage C–H functionalization examples demonstrated to date often use rather specialized substrates and are of limited generality.

Our aim is to select structurally complex molecules of compelling biological activity and then attempt to devise approaches to execute site-selective modification of C–H bonds in such compounds bearing multiple functional groups. The overarching goal is to establish a toolkit of reactions and reagents amenable to effecting functionalizations on a variety of complex molecules in a relatively predictive and selective manner for broad application. Alkaloids represent an extremely challenging class of natural products for direct C–H functionalization, because they typically contain a basic amine and a variety of other reactive functionality. In particular, a basic amine can impede many C–H-activation methodologies, as the nitrogen functionality may coordinate to and poison the catalyst[24–26] or generate undesirable side reactions[27,28]. Only a few examples of selective oxidation[29,30] or C–H amination of alkaloid derivatives have been reported, although in the case of the latter aza-ylide products are also formed and in many cases are the exclusive product[27]. One of the most effective methods for site-selective C–H functionalization, without resorting to the use of directing groups, has been the C–H insertion chemistry of rhodium-bound donor/acceptor carbenes[31]. Herein we demonstrate for the first time that such highly reactive intermediates are indeed capable of undergoing site-selective C–H functionalization of a range of complex alkaloids and drug molecules possessing a tertiary amine. We show that our effective strategy enables C–H functionalization at methyl, methylene or methine sites proximal to the basic amine, with little to no aza-ylide formation. This study constitutes a proof of concept that we envision can be readily extended to a wide range of applications, where efficient installation of synthetic handles on complex alkaloids may find utility. For instance, such a paradigm would be expected to greatly facilitate the ability to derivatize payloads for attachment to monoclonal antibodies for antibody–drug conjugate systems[32], and in probe design for target identification or mechanism of action studies in a chemical genetics context[27,33].

## Results

**Rhodium-carbenoid-mediated C–H functionalization of brucine.** Brucine (**1**) is a readily available complex natural product, which contains 22 C–H bonds in different chemical environments. The presence of two doubly activated allylic sites, a complex multi-ring architecture, a lactam functionality, an electron-rich phenyl ring, methyl ethers, as well as a basic tertiary amine offered a challenging target on which to commence these studies. A metal-free carbene approach to derivatize brucine was recently reported, which yielded a ring-expanded product through formation of an aza-ylide species followed by a [1,2]-Stevens rearrangement[34]. Indeed, the use of metallocarbenoids to access an aza-ylide intermediate followed by ring expansion has been widely described[35–37]. Interested in effecting C–H insertions via a carbenoid approach, we were intrigued as to whether the presence of a suitable rhodium catalyst and appropriate temperature might influence the nature of the reaction with brucine, allowing for C–H insertion over aza-ylide formation or catalyst poisoning, despite the presence of the nucleophilic amine. To the best of our knowledge, the ability to effect such chemistry on substrates possessing a tertiary amine, especially in structurally complex alkaloids, has not been previously reported[31,38].

As illustrated in Fig. 1a, we first compared the influence of a dirhodium catalyst in contrast to a metal-free system. Brucine was treated with excess methyl-*p*-bromophenyl diazoacetate (**2**) in the presence or absence of the well-established carbenoid C–H insertion catalyst $Rh_2(S\text{-DOSP})_4$ (ref. 31) in trifluorotoluene at 83 °C (oil bath temperature). We observed that the $Rh_2(S\text{-DOSP})_4$-catalysed reaction provided C–H insertion product **3** in 20% yield as a single diastereomer, in addition to the expected Stevens rearrangement product **4** in 58% yield with 1.2:1 diastereomeric ratio (dr). When the same reaction was carried out in the absence of $Rh_2(S\text{-DOSP})_4$, formation of **3** and **4** was not observed, strongly suggesting that effective functionalization of brucine was mediated via the dirhodium catalyst. Attempting the $Rh_2(S\text{-DOSP})_4$-catalysed reaction at ambient temperature in either dichloromethane or trifluorotoluene failed to afford compound **3** or **4**. The differential outcome observed when conducting the reaction at room temperature as opposed to 83 °C led us to rationalize that a higher temperature enables a more rapid kinetic dissociation of the amine from the dirhodium catalyst. No longer incapacitated, the catalyst is free to effect diazo decomposition and carbenoid formation.

**Control experiment.** Having isolated aza-ylide **5**, we postulated whether formation of C–H insertion product **3** actually proceeded through a non-rhodium-bound ylide intermediate (as proposed in the formation of **4**)[34] aided by thermal activation. Accordingly, a control experiment was conducted as presented in Fig. 1b. Aza-ylide **5** was alternatively prepared *in situ* through deprotonation of the corresponding trifluoroacetate salt **6** at 0 °C. The reaction was heated to 83 °C in a pressure vessel for 30 min, yielding exclusively [1,2]-Stevens rearrangement product **4** in 1:1 dr. Since C–H insertion product **3** was not observed, it suggests that ylide formation and carbenoid C–H insertion proceed through different reaction pathways. Indeed, we speculate that product **3** is formed via direct C–H insertion

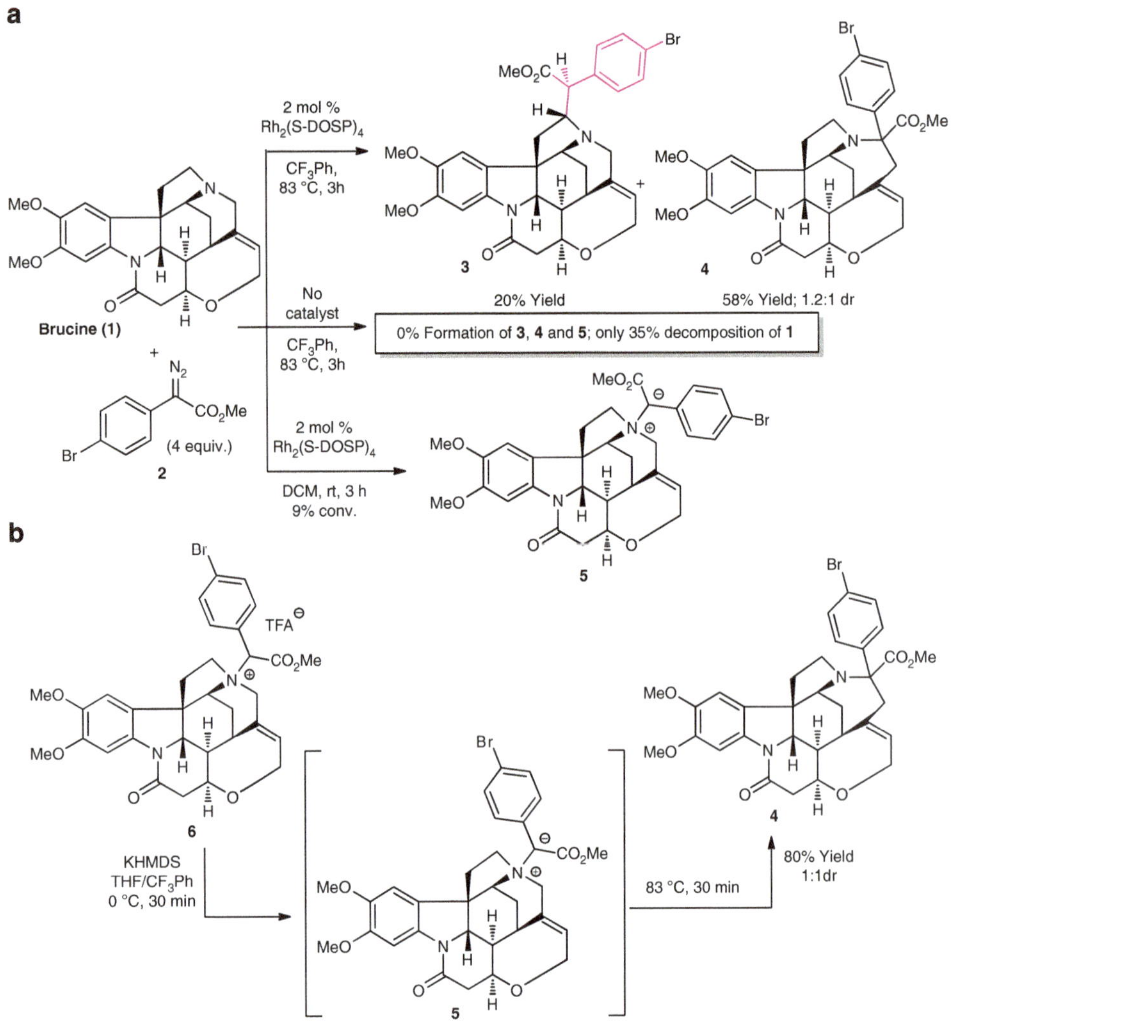


**Figure 1 | Donor/acceptor rhodium-carbenoid-mediated functionalization of brucine (1).** (**a**) Investigations into the role of the catalyst and temperature. (**b**) Control experiment provided mechanistic insights for the formation of C-H insertion product **3**. Conversion, yield and dr were determined by $^1$H NMR with addition of 4-dimethylaminopyridine (DMAP) after work-up as an internal standard. rt, room temperature.

induced by the metal carbene[39,40]. This being the case, it is remarkable that **3** is the sole C–H insertion product observed from a molecule that has numerous potential sites for C–H functionalization. The subtle balance between electronic, steric and stereoelectronic effects can often result in exquisite regiocontrol in the C–H functionalization reactions of donor/acceptor rhodium carbenes[31], but such control has not been previously demonstrated with such a complex substrate.

**Influence of dirhodium catalysts.** The electrophilicity and steric environment of the metallocarbenoids can be affected by the nature of the catalyst[31]. Hence, we were eager to explore what influence various dirhodium catalysts would have on the site selectivity of the reaction (Table 1). The bulky dirhodium catalyst $Rh_2(TPA)_4$ promoted formation of the C–H insertion product, giving **3** in 50% yield as a single diastereomer. Subsequent investigations into other achiral rhodium catalysts (Table 1, entries 2–5) revealed that $Rh_2(TPA)_4$ is the most effective catalyst for carbenoid-mediated C–H insertion. In addition, the formation of Stevens rearrangement product **4** is a competing pathway for which the chemoselectivity between the formation of compounds **3** and **4** is catalyst dependent. Furthermore, the two enantiomers of chiral catalysts afforded different results. For example, only the use of the *S*-enantiomer of the DOSP ligand allowed for C–H insertion (Table 1, entries 7 and 8). A most unexpected result was obtained with the very bulky chiral catalyst $Rh_2(BTPCP)_4$ refs 41,42. Under the standard reaction conditions, the $Rh_2(BTPCP)_4$-catalysed reactions suffered very poor conversion, irrespective of whether the *R*- or *S*-enantiomer of the catalyst was used (Table 1, entries 11 and 12). Careful $^1$H nuclear magnetic resonance (NMR) analysis of the crude reaction mixture revealed an alternative C–H insertion product (**7**) had been generated in low yield with $Rh_2(S\text{-}BTPCP)_4$. Increasing the catalyst loading of $Rh_2(S\text{-}BTPCP)_4$ to 20 mol% afforded compound **7** in 39% yield as a single diastereomer, albeit with moderate conversion (Fig. 2a). Extensive one- and two-dimensional NMR experiments confirmed that this new compound **7** was generated through C–H insertion into the tertiary C–H bond adjacent to the amine. The ability for $Rh_2(S\text{-}BTPCP)_4$ to enable selective C–H insertion at the methine position is unexpected, because a bulky catalyst should generally favour functionalization of a less-sterically encumbered C–H bond[42]. The catalyst screen reveals that one can manipulate the favoured site for functionalization in a catalyst-controlled manner, allowing one to rapidly probe three alternative sites on a complex molecule (Fig. 2a).

**Table 1 | Influence of dirhodium catalyst on C-H carbene insertion of brucine (1).**

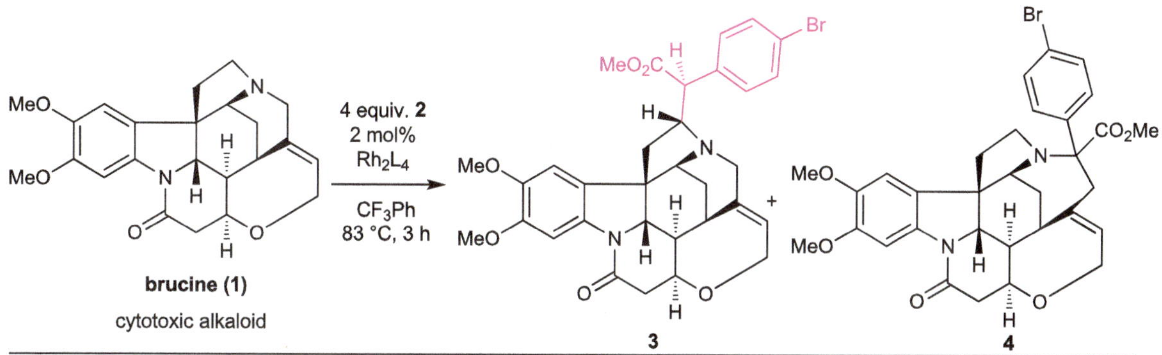


| Entry | $Rh_2L_4$ | Conversion | Yield 3 | Yield 4 | Entry | $Rh_2L_4$ | Conversion | Yield 3 | Yield 4 |
|---|---|---|---|---|---|---|---|---|---|
| 1 | $Rh_2(TPA)_4$ | 100% | 50% | 20% (1:1 dr) | 7* | $Rh_2(S\text{-}DOSP)_4$ | 95% | 16% | 49% (1.2:1 dr) |
| 2 | $Rh_2(OAc)_4$ | 57% | 0% | 38% (1:1 dr) | 8† | $Rh_2(R\text{-}DOSP)_4$ | 86% | 0% | 46% (1.1:1 dr) |
| 3 | $Rh_2(TFA)_4$ | 100% | 4% | 26% (1:1 dr) | 9 | $Rh_2(S\text{-}PTAD)_4$ | 36% | 7% | 20% (1:1 dr) |
| 4 | $Rh_2(Oct)_4$ | 100% | 0% | 74%[b] (1.2:1 dr) | 10 | $Rh_2(R\text{-}PTAD)_4$ | 100% | 13% | 65% (1.2:1 dr) |
| 5 | $Rh_2(esp)_2$ | 100% | 0% | 24% (1.2:1 dr) | 11 | $Rh_2(S\text{-}BTPCP)_4$ | 28% | 0% | 8% (1:1 dr) |
| 6 | $Rh_2(S\text{-}DOSP)_4$ | 100% | 20% | 58% (1.2:1 dr) | 12 | $Rh_2(R\text{-}BTPCP)_4$ | 35% | 0% | 11% (1.2:1 dr) |

Conversion, yield and dr were determined by $^1$H NMR analysis with addition of DMAP during work-up as an internal standard, unless otherwise specified.
*1,2-Dichloroethane was used as solvent.
†Isolated yield.

**Diazo scope.** Studies with a small set of aryldiazoacetates of distinct electronic properties indicate that the nature of the carbenoid also plays a role in the chemoselectivity of the rhodium-carbenoid insertion (Table 2, entries 1–3). Donor/acceptor carbenoids bearing electron-deficient aryl substituents favoured C–H insertion product **9** over aza-ylide formation (Table 2, entries 1 and 3), whereas an electron-donating substituent provides a less-electrophilic rhodium-carbenoid for which exclusive formation of ring-expanded product **10** was observed (Table 2, entry 2). Reaction with methyl diazomalonate (**8c**) or ethyl diazoacetate (**8d**) failed to generate any C–H insertion or Stevens rearrangement product (Table 2, entries 4 and 5), suggesting the importance of the donor group in the efficiency of intermolecular C–H insertion reactions under these conditions.

**C-H functionalization of securinine.** To explore reaction generality, we applied the $Rh_2(TPA)_4$-mediated approach to other structurally complex alkaloids and tertiary amine-containing drug molecules (Figs 3 and 4). The $GABA_A$ antagonist securinine (**12**) is a tricyclic alkaloid possessing two olefins in conjugation with a lactone functionality, in addition to the tertiary amine (Fig. 3a). Treatment of **12** with methyl-*p*-bromophenyl diazoacetate (**2**) and $Rh_2(TPA)_4$ in trifluorotoluene at 83 °C selectively gave C–H insertion product **13** (44% yield, 2.2:1 dr), with no competing cyclopropanation at either olefin[43]. Furthermore, although securinine contains four C–H bonds adjacent to the amine (two methine and two diastereotopic methylene in nature), only a single methylene C–H bond undergoes carbene insertion. Unlike the case of brucine in which substrate control afforded a highly diastereoselective C–H insertion product, with securinine the diastereoselectivity was rather moderate, and was influenced by the dirhodium catalyst itself (Fig. 3a).

**C-H functionalization of apovincamine.** Further application of this method to apovincamine (**14a**) led to bis-cyclopropanation of the electron-rich indole ring **16,** with only minimal formation of the C–H insertion product (Fig. 3b)[44]. Attempts with other dirhodium catalysts provided similar reaction outcomes. It is well established that aromatic rings are sterically protected from reaction with rhodium donor/acceptor carbenoids when they are at least 1,4-disubstituted[45]. Accordingly, the iodinated apovincamine analogue **14b** was prepared and evaluated in the reaction, and in this instance the C–H insertion product **15b** was obtained in improved yield (47% yield) as a single diastereomer. Although **14b** contains one methine and four methylene C–H bonds adjacent to the amine, the C–H insertion occurred selectively at the benzylic methylene site rather than alpha to the amine. Previously, it has been reported that benzylic C–H bonds are also favourable sites for C–H insertion, due to the capacity of $\pi$-systems to stabilize the neighbouring buildup of positive charge in the transition state[46].

**Selective *N*-methyl C-H insertion.** C–H functionalization induced by dirhodium-bound donor/acceptor carbenoids has been shown to be initiated by a hydride transfer event, and sites that are capable of stabilizing positive charge buildup at carbon are electronically favoured[40]. However, the general order of reactivity of C–H bonds in competition reactions is typically methine ~ methylene > methyl, because although tertiary sites are

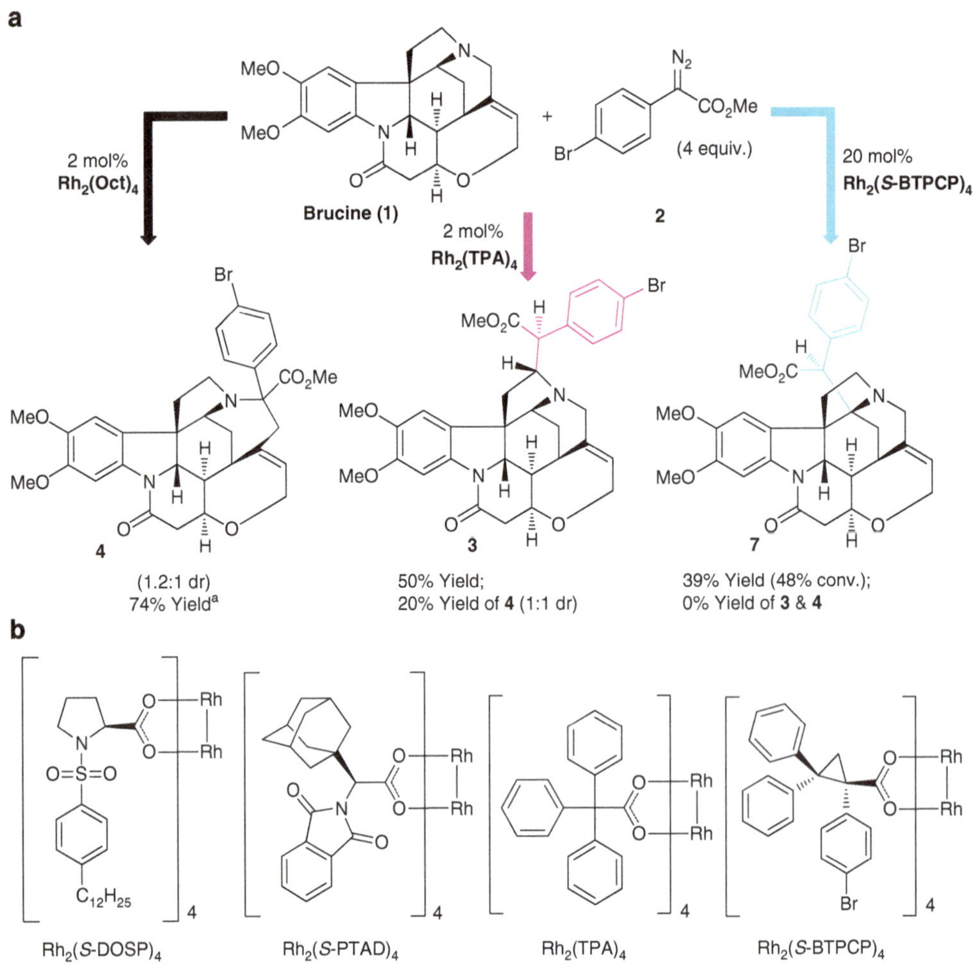


**Figure 2 | Optimization of rhodium-catalysed C-H carbene insertion of brucine (1).** (**a**) Catalyst influence on site-selective C-H functionalization of brucine (**1**). (**b**) Structures of dirhodium catalysts. Conversion, yield and dr were determined by $^1$H NMR analysis with addition of 4-dimethylaminopyridine after work-up as an internal standard, unless otherwise specified. [a]Isolated yield.

electronically most activated, donor/acceptor carbenoids are sterically very demanding[31]. In the alkaloids studied so far the reactions have been remarkably site selective, suggesting that the majority of electronically activated methine and methylene sites in these alkaloids are sterically inaccessible. Therefore, we chose to explore the reactions of alkaloids containing *N*-methyl groups. Functionalization of electronically activated methyl C–H bonds has been observed, but only a few examples of C–H insertion into methyl C–H bonds in the presence of activated methylene and/or methine C–H bonds have been reported[42,47–49], and none of those feature a basic amine. We found for alkaloids possessing an *N*-Me functionality, the most favoured product in each case arose from C–H insertion into the said *N*-methyl group (Fig. 4). With dextromethorphan (**17**), for example, carbene insertion at the *N*-methyl C–H bond was the most favoured product despite methylene sites adjacent to the nitrogen atom and the phenyl ring. It is worth noting that no insertion was observed at the accessible methyl ether site. Indeed, C–H insertion at the *N*-methyl site was such a favoured pathway that we were able to effect the reaction at room temperature affording 87% yield of C–H insertion product (1:1 dr), with complete conversion of **17**. The structurally related yet more elaborate thebaine (**18**) smoothly underwent C–H insertion at the *N*-methyl site (52%, 1.4:1 dr), without any competing reactivity despite the electron-rich aryl group and the 1,3-diene functionality. It appears that regardless of conformation or neighbouring functionalities, in all systems explored the donor/acceptor carbenoid derived from **2** selectively inserts into the primary C–H bond adjacent to nitrogen. In addition, the apparent ease of insertion into the *N*-methyl group enables some reactions to be conducted at room temperature, with diazo **2** used as the limiting reagent, as was the case with noscapine (**19**), although the structurally related bicuculline (**20**) required more forcing conditions. Generally, the diastereoselectivity of the $Rh_2(TPA)_4$-mediated *N*-methyl insertion reactions afforded diastereomeric ratios in the 1:1-2:1 range, and the application of chiral dirhodium catalysts failed to improve on this (Table 3). Sercloremine (**21**) represents an interesting substrate in that its relatively simple structure suggests little potential for steric differentiation between the methylene sites and the terminal methyl site adjacent to the amine, however, the *N*-methyl C–H insertion product is again exclusively formed in 62% yield. It should be noted that Stevens-type rearrangement products were not observed in any of the *N*-methyl-containing systems explored in this study and presumably is no longer a competing pathway. It is likely that the ease of accessibility the primary C–H bond offers has a significant impact on the site-selectivity of this reaction.

We have successfully devised and implemented an effective strategy for non-directed C–H functionalization of nucleophilic tertiary amine-containing complex natural products and drug molecules. We demonstrate that these traditionally challenging substrates are capable of undergoing rhodium-

**Table 2 | Influence of diazo reagents on rhodium-catalysed carbene insertion of brucine (1).**

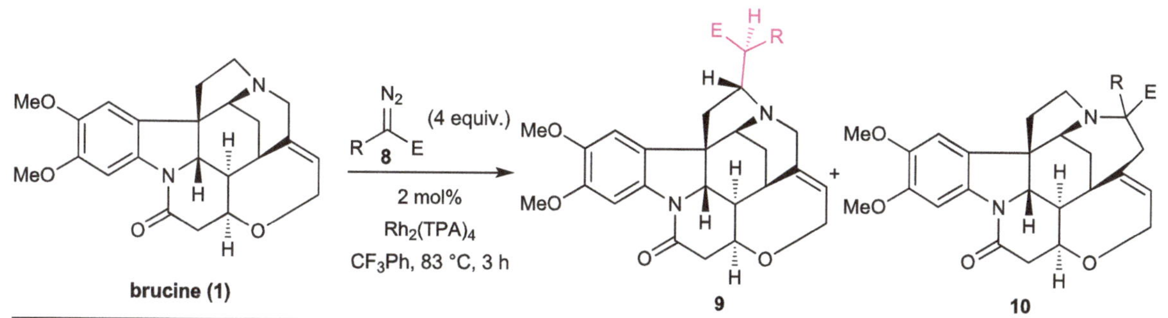


| Entry | Diazo reagent | Conv.* | Yield* 9 | Yield* 10 | Entry | Diazo reagent | Conv.* | Yield* 9 | Yield* 10 |
|---|---|---|---|---|---|---|---|---|---|
| 1 | 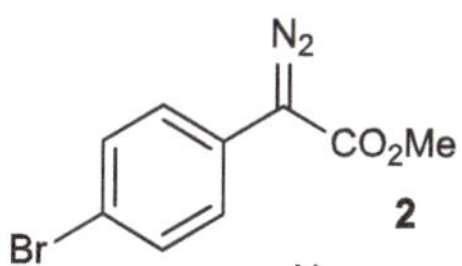 **2** | 100% | 50% (**3**) | 20% (**4**) (1:1 dr) | 4 | 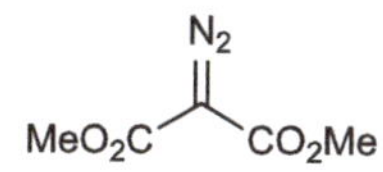 **8c** | <5% | 0% | 0% |
| 2 | 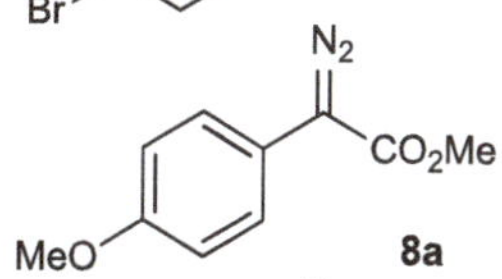 **8a** | 100% | ND† | 48% (**10a**) (1.8:1 dr) | 5‡ | 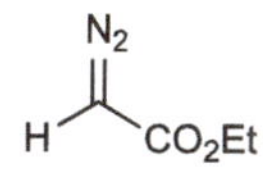 **8d** | 13% | 0% | 0% |
| 3 | 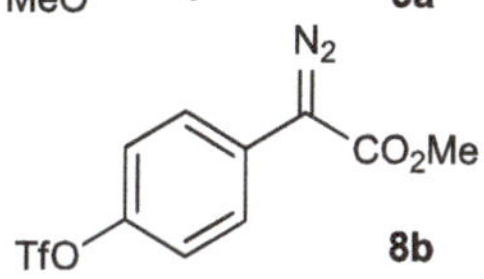 **8b** | 100% | 34% (**9b**) | ND† | | | | | |

DMAP, 4-dimethylaminopyridine; NMR, nuclear magnetic resonance.
*Conversion, yield and dr were determined by $^1H$ NMR analysis with addition of DMAP during work-up as an internal standard, and where formed, the corresponding product is given in bold font in parentheses next to the yield.
†Not determined, but cannot be excluded owing to the presence of a small amount of material isolated as an intractable mixture of products.
‡The corresponding aza-ylide **11** was obtained as trifluoroacetate salt in 4% isolated yield.

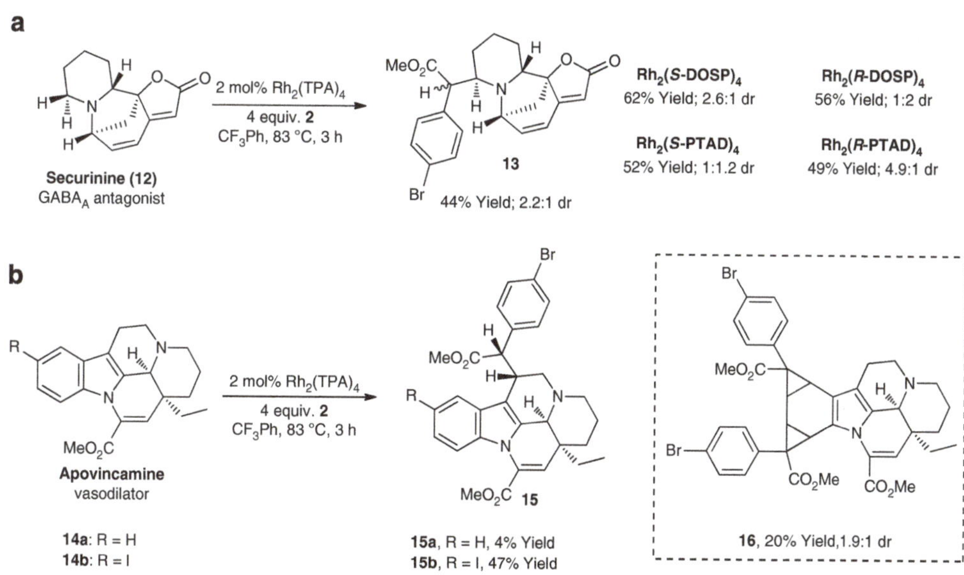


**Figure 3 | Rhodium-catalysed C-H insertion of alkaloids securinine and apovincamine.** (**a**) Diastereoselective C-H carbene insertion is achieved by using chiral rhodium catalysts. (**b**) Introduction of steric hindrance could protect electron-rich aromatic rings from cyclopropanation and enable desired C-H insertion. Yield and dr were determined by $^1H$ NMR analysis with addition of 4-dimethylaminopyridine after work-up as an internal standard, unless otherwise specified.

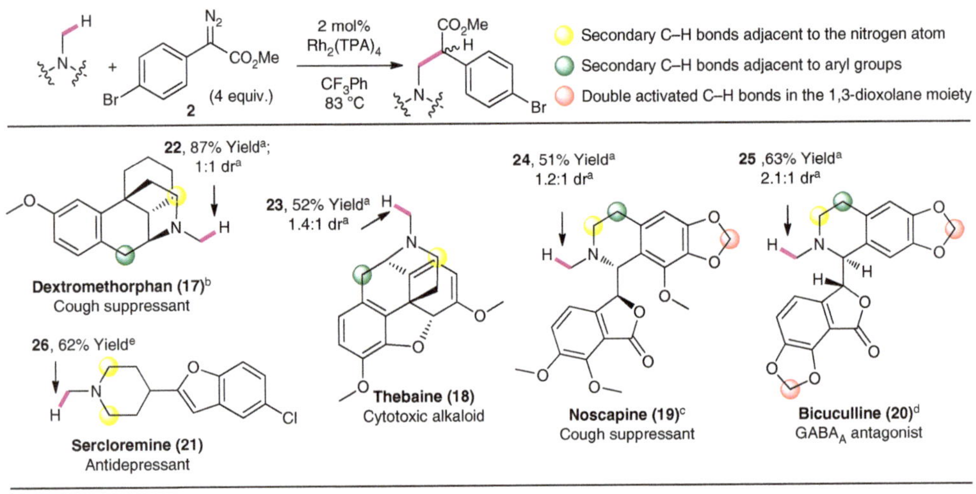


**Figure 4 | C-H insertion in *N*-methyl-containing natural products and drug molecules.** [a]Yield and dr were determined by $^1$H NMR with 4-dimethylaminopyridine as an internal standard. [b]Reaction was carried out in dichloromethane under room temperature. [c]Reaction was carried out with 0.5 equiv. of **2** in dichloromethane at room temperature (54% recovered SM). [d]Reaction was carried out with 0.5 equiv. of **2** (52% recovered SM). [e]Isolated yield.

**Table 3 | Chiral dirhodium catalysts screen using bicuculline (20) as a substrate.**

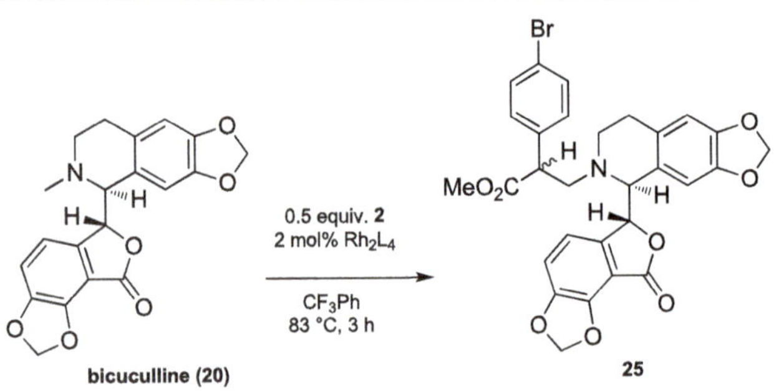


| Entry | $Rh_2L_4$ | Yield 25 |
|---|---|---|
| 1 | $Rh_2(TPA)_4$ | 63% (2.2:1 dr) |
| 2 | $Rh_2(S\text{-}DOSP)_4$ | 45% (2:1 dr) |
| 3 | $Rh_2(R\text{-}DOSP)_4$ | 57% (1:1.4 dr) |
| 4 | $Rh_2(S\text{-}PTAD)_4$ | 62% (1:1.1 dr) |
| 5 | $Rh_2(R\text{-}PTAD)_4$ | 66% (2.3:1 dr) |

DMAP, 4-dimethylaminopyridine; NMR, nuclear magnetic resonance.
Yield and dr were determined by $^1$H NMR analysis with addition of DMAP during work-up as an internal standard.

catalysed C–H insertion reactions to selectively install a new C–C bond. The methods we describe are effectively implemented on a diverse set of alkaloids possessing a wide variety of architectures, functionalities and potentially reactive sites. The use of donor/acceptor carbenoids is the key to our ability to achieve such remarkable chemoselectivity and predictable regioselectivity. In addition, our approach obviates the challenging issue of undesired aza-ylide formation commonly observed with carbenoids in the presence of basic nitrogen-containing substrates, and in doing so the work significantly expands and enables the scope of this field[27]. The highly efficient generation of three distinct brucine derivatives in a catalyst-controlled fashion (Fig. 2a), showcases the potential of this approach to offer late-stage diversification of complex molecules. It is worth noting that the site of C–H insertion is routinely in proximity to the amine moiety. We look upon this consistency, in particular the remarkable selectivity for C–H insertion into *N*-methyl-containing alkaloids, as a robust and highly predictable approach for site-specific functionalization of complex molecules, which could be applied late on in a synthetic route for direct derivatization adjacent to an amine, or for the introduction of a synthetic handle for use in bioconjugation strategies for chemical biology studies.

## Methods

**Materials.** Materials were obtained from commercial suppliers and used as received, or prepared according to standard procedures, unless otherwise noted. Sercloremine (**21**) and apovincamine (**14a**) were obtained from the Novartis compound archive. Methyl-*p*-bromophenyl diazoacetate (**2**) as well as its analogues **8a** and **8b** were synthesized according to a previously reported procedure[43]. All reactions were conducted under an inert atmosphere of dry nitrogen. Analytical thin-layer chromatography was performed on Kieselgel 60 F254 (250 μm silica gel) glass plates and compounds were visualized with ultraviolet light 254 nm). Flash column chromatography was performed using Kieselgel 60 (230–400 mesh) silica gel with ethyl acetate/hexanes as eluent, unless indicated otherwise.

**General spectroscopic methods.** $^1$H NMR spectra were measured at 400 MHz on a Bruker Avance instrument and reported in parts per million (*δ*, p.p.m.). Coupling constants (*J*-values) were reported in Hertz (Hz), with multiplicity reported following usual convention: s = singlet, d = doublet, t = triplet, q = quartet, dd = doublet of doublets, m = multiplet and br = broad. The proton signal of the residual, non-deuterated solvent (*δ* 7.26 for $CHCl_3$) was used as an internal reference for $^1$H NMR spectra. $^{13}$C NMR spectra were completely hetero-decoupled and measured at 100 MHz. Residual chloroform (*δ* 77.23) was used as an internal reference. Preparative high-performance liquid chromatography was performed using Waters Autopurification system with a photodiode array detector. Preparative supercritical fluid chromatography was performed using a Thar (Waters) SFC 80 preparative system with a Waters 2489 UV/visible detector. All tested compounds were found to be >95% pure (unless stated otherwise) as determined by liquid chromatography-UV (LC-UV)/electrospray ionization (ESI)-mass spectrometry (MS), recorded using an Acquity G2 Xevo OTof mass spectrometer (accuracy <5 p.p.m.) with an electrospray ionization source and Acquity ultra performance liquid chromatograph (conditions: Acquity UPLC BEH C18 1.7 μm 2.1 × 50 mm column, solvent A: water + 0.1% formic acid, solvent B: acetonitrile + 0.1% formic acid, gradient: from 2 to 98% B in 4.4 min, 1.0 ml min$^{-1}$ flow rate and 50 °C). Ammonium salts, which were generated in preparative high-performance liquid chromatography with trifluoroacetic acid or formic acid as a modifier, were converted to the corresponding free amine with 4 equiv. MP-carbonate (Biotage) in 0.1 M dichloromethane for 1 h at room temperature. $^1$H NMR, $^{13}$C NMR and high-resolution mass spectra (HRMS) are provided for all compounds. For NMR spectra and detailed analysis of NMR assignments, see

Supplementary Figs 1–28; Supplementary Notes 1–5, respectively. For supercritical fluid chromatography analysis and ion chromatography analysis of **6** see Supplementary Notes 6,7. See Supplementary Methods for the characterization data of compounds not listed in this section.

**Synthesis of 3.** An oven-dried 10 ml two-necked round-bottom flask fitted with a condenser was charged with brucine (**1**) (133.2 mg, 0.338 mmol) and $Rh_2(TPA)_4$ (9.7 mg, 6.8 μmol) in degassed trifluorotoluene (3.0 ml) under a nitrogen atmosphere. A solution of methyl-*p*-bromophenyl diazoacetate (**2**) (345 mg, 1.35 mmol) in degassed trifluorotoluene (3.8 ml) was slowly added via a syringe pump over 1 h under nitrogen atmosphere at 83 °C (oil bath temperature). The mixture was stirred at this temperature for 2 h and then concentrated *in vacuo*. $^1$H NMR analysis with addition of 5.0 mg 4-dimethylaminopyridine as an internal standard indicated **3** was obtained in 50% yield as a single diastereomer and **4** in 20% yield (1:1 dr). The resulting crude was then purified through flash column chromatography (0–50% EtOAc/heptane) twice to afford **3** as a white solid (92.9 mg, 44% isolated yield, Rf = 0.04 in 50% EtOAc/heptane). To aid analysis, **3** was isolated as the corresponding TFA salt. $^1$H NMR (400 MHz, chloroform-*d*) δ 7.76 (s, 1H), 7.59 (d, *J* = 8.0 Hz, 2H), 7.44 (d, *J* = 8.0 Hz, 2H), 6.91 (s, 1H), 6.01 (t, *J* = 4.0 Hz, 1H), 4.77 (br, 1H), 4.36 (d, *J* = 8.0 Hz, 1H), 4.35 (m, 1H), 4.24 (dd, *J* = 16.0, 8.0 Hz, 1H), 4.06 (d, *J* = 12.0 Hz, 1H), 3.96–4.06 (m, 2H), 3.91 (s, 3H), 3.90 (s, 3H), 3.86 (d, *J* = 16.0 Hz, 1H), 3.63 (s, 3H), 3.26 (br, 1H), 3.20 (dd, *J* = 16.0, 8.0 Hz, 1H), 2.73 (dd, *J* = 16.0, 4.0 Hz, 1H), 2.42–2.51 (m, 2H), 2.29 (t, *J* = 16.0 Hz, 1H), 2.17 (d, *J* = 16.0 Hz, 1H), 1.67 (d, *J* = 16.0 Hz, 1H), 1.39 (dt, *J* = 8.0, 4.0 Hz, 1H). $^{13}$C NMR (100 MHz, chloroform-*d*) δ 170.9, 169.2, 161.7 ($CF_3$), 150.7, 147.5, 135.9, 135.5, 133.2, 132.9, 132.7, 130.3, 124.1, 118.8, 105.1, 101.3, 76.9, 66.7, 64.2, 63.5, 59.7, 56.5, 53.2, 52.5, 52.0, 51.4, 47.1, 46.1, 42.1, 30.5, 25.6. HRMS (ESI-time of flight (TOF)) $[M + H]^+$ calculated for $C_{32}H_{34}BrN_2O_6$: 621.1600; found: 621.1575. LC-UV/ESI-MS retention time: 1.70 min. Purification of the crude reaction mixture also afforded **4** contaminated with small amount of impurity. A subsequent purification through flash column chromatography (0–50% EtOAc/heptane) provided the two diastereomers of **4** as separate compounds (24.2 mg, 11% isolated yield, Rf = 0.30 in 50% EtOAc/heptane; 17.2 mg, 8% isolated yield, Rf = 0.25 in 50% EtOAc/heptane). **4a**, diastereomer 1: $^1$H NMR (400 MHz, chloroform-*d*) δ 7.82 (s, 1H), 7.54 (d, *J* = 8.6 Hz, 2H), 7.40 (d, *J* = 8.6 Hz, 2H), 6.72 (s, 1H), 5.86 (q, *J* = 3.8 Hz, 1H), 4.38 (dd, *J* = 16.4, 5.0 Hz, 1H), 4.27 (m 1H), 4.10 (dm, *J* = 16.4 Hz, 1H), 3.98 (d, *J* = 11.1 Hz, 1H), 3.94 (s, 1H), 3.90 (s, 3H), 3.89–3.85 (m, 4H), 3.63 (s, 3H), 3.33–3.22 (m, 1H), 3.03 (d, *J* = 16.0 Hz, 1H), 2.97 (t, *J* = 7.5 Hz, 1H), 2.93 (m, 2H), 2.83 (s, 1H), 2.00 (dt, *J* = 16.0, 4.0 Hz, 1H), 1.96–1.81 (m, 2H), 1.60 (dt, *J* = 8.0, 4.0 Hz, 1H), 1.41 (d, *J* = 15.5 Hz, 1H). $^{13}$C NMR (100 MHz, chloroform-*d*) δ 173.3, 170.0, 149.4, 146.3, 142.8, 138.3, 135.9, 131.7, 128.8, 128.4, 123.8, 121.5, 105.9, 101.1, 75.8, 68.7, 65.2, 62.9, 59.6, 56.8, 56.4, 53.2, 53.1, 49.4, 48.3, 47.0, 42.4, 39.5, 35.3, 31.7. HRMS (ESI-TOF) $[M + H]^+$ calculated for $C_{32}H_{34}BrN_2O_6$: 621.1600; found: 621.1594. LC-UV/ESI-MS retention time: 2.88 min. **4b**, diastereomer 2: $^1$H NMR (400 MHz, chloroform-*d*) δ 7.81 (s, 1H), 7.48 (d, *J* = 8.2 Hz, 2H), 7.26 (br, 2H), 6.67 (s, 1H), 5.52 (s, 1H), 4.28–4.13 (m, 2H), 4.09 (d, *J* = 10.4 Hz, 1H), 4.00 (dm, *J* = 14.8 Hz, 1H), 3.94 (s, 3H), 3.90 (s, 3H), 3.89 (s, 3H), 3.68 (d, *J* = 15.2 Hz, 1H), 3.46 (d, *J* = 4.2 Hz, 1H), 3.10 (dd, *J* = 16.2, 8.3 Hz, 1H), 2.96 (s, 1H), 2.88–2.59 (m, 3H), 2.50–2.29 (m, 2H), 2.07–1.95 (m, 1H), 1.75–1.61 (m, 2H), 1.43 (dt, *J* = 10.5, 4.5 Hz, 1H). $^{13}$C NMR (100 MHz, chloroform-*d*) δ 173.9, 170.5, 149.1, 146.5, 142.9, 142.1, 136.6, 131.5, 128.3, 128.1 (br), 124.1, 121.6, 105.7, 101.2, 79.3, 68.5, 67.0, 64.5, 60.4, 56.8, 56.4, 53.0, 52.1, 49.1, 47.9, 45.8, 42.2, 40.8, 35.2, 29.3. HRMS (ESI-TOF) $[M + H]^+$ calculated for $C_{32}H_{34}BrN_2O_6$: 621.1600; found: 621.1600. LC-UV/ESI-MS retention time: 3.03 min.

## References


1. Bunnage, M. E. Getting pharmaceutical R&D back on target. *Nat. Chem. Biol.* **7**, 335–339 (2011).
2. Romo, R. & Robles, O. Chemo- and site-selective derivatizations of natural products enabling biological studies. *Nat. Prod. Rep.* **31**, 318–334 (2014).
3. Carlson, E. E. Natural products as chemical probes. *ACS Chem. Biol.* **5**, 639–653 (2010).
4. Metz, J. T. & Hajduk, P. J. Rational approaches to targeted polypharmacology: creating and navigating protein–ligand interaction networks. *Curr. Opin. Chem. Biol.* **14**, 498–504 (2010).
5. Fishman, M. C. & Porter, J. A. Pharmaceuticals: a new grammar for drug discovery. *Nature* **437**, 491–493 (2005).
6. Kesavan, S. & Marcaurelle, L. A. Translational synthetic chemistry. *Nat. Chem. Biol.* **9**, 210–213 (2013).
7. Bauer, A. & Brönstrup, M. Industrial natural product chemistry for drug discovery and development. *Nat. Prod. Rep.* **31**, 35–60 (2014).
8. Godula, K. & Sames, D. C–H bond functionalization in complex organic synthesis. *Science* **312**, 67–72 (2006).
9. Yamaguchi, J., Yamaguchi, A. D. & Itami, K. C–H bond functionalization: emerging synthetic tools for natural products and pharmaceuticals. *Angew. Chem. Int. Ed.* **51**, 8960–9009 (2012).
10. McMurray, L., O'Hara, F. & Gaunt, M. J. Recent developments in natural product synthesis using metal-catalyzed C–H bond functionalisation. *Chem. Soc. Rev.* **40**, 1885–1898 (2011).
11. Wencel-Delord, J. & Glorius, F. C–H bond activation enables the rapid construction and late-stage diversification of functional molecules. *Nat. Chem.* **5**, 369–375 (2013).
12. Davies, H. M. L. & Denton, J. R. Application of donor/acceptor-carbenoids to the synthesis of natural products. *Chem. Soc. Rev.* **38**, 3061–3071 (2009).
13. Gutekunst, W. R. & Baran, P. S. C–H functionalization logic in total synthesis. *Chem. Soc. Rev.* **40**, 1976–1991 (2011).
14. Wender, P. A., Hilinski, M. K. & Mayweg, A. V. W. Late-Stage intermolecular CH activation for lead diversification: a highly chemoselective oxyfunctionalization of the C-9 position of potent bryostatin analogues. *Org. Lett.* **7**, 79–82 (2005).
15. Lapointe, D., Markiewicz, T., Whipp, C. J., Toderian, A. & Fagnou, K. Predictable and site-selective functionalization of poly(hetero)arene compounds by palladium catalysis. *J. Org. Chem.* **76**, 749–759 (2011).
16. Dai, H.-X., Stepan, A. F., Plummer, M. S., Zhang, Y.-H. & Yu, J.-Q. Divergent C–H functionalizations directed by sulfonamide pharmacophores: late-stage diversification as a tool for drug discovery. *J. Am. Chem. Soc.* **133**, 7222–7228 (2011).
17. O'Hara, F., Blackmond, D. G. & Baran, P. S. Radical-based regioselective C–H functionalization of electron-deficient heteroarenes: scope, tunability, and predictability. *J. Am. Chem. Soc.* **135**, 12122–12134 (2013).
18. Sawayama, A. M. *et al.* A panel of cytochrome P450 BM3 variants to produce drug metabolites and diversify lead compounds. *Chem.Eur. J.* **15**, 11723–11729 (2009).
19. Gormisky, P. E. & White, M. C. Catalyst-controlled aliphatic C–H oxidations with a predictive model for site-selectivity. *J. Am. Chem. Soc.* **135**, 14052–14055 (2013).
20. Zhang, K., Shafer, B. M., Demars, II M. D., Stern, H. A. & Fasan, R. Controlled oxidation of remote $sp^3$ C–H bonds in artemisinin *via* P450 catalysts with fine-tuned regio- and stereoselectivity. *J. Am. Chem. Soc.* **134**, 18695–18704 (2012).
21. Lee, J. S., Cao, H. & Fuchs, P. L. Ruthenium-catalyzed mild C–H oxyfunctionalization of cyclic steroidal ethers. *J. Org. Chem.* **72**, 5820–5823 (2007).
22. Larsen, M. A. & Hartwig, J. F. Iridium-catalyzed C–H borylation of heteroarenes: scope, regioselectivity, application to late-stage functionalization, and mechanism. *J. Am. Chem. Soc.* **136**, 4287–4299 (2014).
23. Rousseau, G. & Breit, B. Removable directing groups in organic synthesis and catalysis. *Angew. Chem. Int. Ed.* **50**, 2450–2494 (2011).
24. Malik, H. A. *et al.* Non directed allylic C–H acetoxylation in the presence of Lewis basic heterocycles. *Chem. Sci.* **5**, 2352–2361 (2014).
25. Davies, H. M. L. & Townsend, R. J. Catalytic asymmetric cyclopropanation of heteroaryldiazoacetates. *J. Org. Chem.* **66**, 6595–6603 (2001).
26. Aller, E. *et al.* N-H insertion reactions of rhodium carbenoids. Part 1. Preparation of α-amino acid and α-aminophosphonic acid derivatives. *J. Chem. Soc. Perkin Trans.* **1**, 2879–2884 (1996).
27. Li, J. *et al.* Simultaneous structure–activity studies and arming of natural products by C–H amination reveal cellular targets of eupalmerin acetate. *Nat. Chem.* **5**, 510–517 (2013).
28. Padwa, A. & Hornbuckle, S. F. Ylide formation from the reaction of carbenes and carbenoids with heteroatoms lone pairs. *Chem. Rev.* **91**, 263–309 (1991).
29. Kim, J., Ashenhurst, J. A. & Movassaghi, M. Total synthesis of (+)-11,11′-dideoxyverticillin A. *Science* **324**, 238–241 (2009).
30. Su, S., Seiple, I. B., Young, I. S. & Baran, P. S. Total syntheses of (±)-massadine and massadine chloride. *J. Am. Chem. Soc.* **130**, 16490–16491 (2008).
31. Davies, H. M. L. & Beckwith, R. E. J. Catalytic enantioselective C–H activation by means of metal-carbenoid-induced C–H insertion. *Chem. Rev.* **103**, 2861–2903 (2003).
32. Zhou, Q. *et al.* Bioconjugation by native chemical tagging of C-H bonds. *J. Am. Chem. Soc.* **135**, 12994–12997 (2013).
33. Ziegler, S., Pries, V., Hedberg, C. & Waldmann, H. Target identification for small bioactive molecules: finding the needle in the haystack. *Angew. Chem. Int. Ed.* **52**, 2744–2792 (2013).
34. Hansen, S. R., Spangler, J. E., Hansen, J. H. & Davies, H. M. L. Metal-free N-H insertions of donor/acceptor carbenes. *Org. Lett.* **14**, 4626–4629 (2012).
35. Vanecko, J. A., Wan, H. & West, F. G. Recent advances in the Stevens rearrangement of ammonium ylides. Application to the synthesis of alkaloid natural products. *Tetrahedron* **62**, 1043–1062 (2006).
36. Sharma, A., Guénée, L., Naubron, J.-V. & Lacour, J. One-step catalytic asymmetric synthesis of configurationally stable Tröger bases. *Angew. Chem. Int. Ed.* **50**, 3677–3680 (2011).
37. Osipov, S. N., Sewald, N., Kolomiets, A. F., Fokin, A. V. & Burger, K. Synthesis of α-trifluoromethyl substituted α-amino acid derivatives from methyl 3,3,3-trifluoro-2-diazopropionate. *Tetrahedron Lett.* **37**, 615–618 (1996).
38. Davies, H. M. L. & Morton, D. Guiding principles for site selective and stereoselective intermolecular C–H functionalization by donor/acceptor rhodium carbenes. *Chem. Soc. Rev.* **40**, 1857–1869 (2011).

39. Nakamura, E., Yoshikai, N. & Yamanaka, M. Mechanism of C–H bond activation/C-C bond formation reaction between diazo compound and alkane catalyzed by dirhodium tetracarboxylate. *J. Am. Chem. Soc.* **124,** 7181–7192 (2002).
40. Hansen, J., Autschbach, J. & Davies, H. M. L. Computational study on the selectivity of donor/acceptor-substituted rhodium carbenoids. *J. Org. Chem.* **74,** 6555–6563 (2009).
41. Qin, C. *et al.* $D_2$-Symmetric dirhodium catalyst derived from a 1,2,2-triarylcyclopropanecarboxylate ligand: design, synthesis and application. *J. Am. Chem. Soc.* **133,** 19198–19204 (2011).
42. Qin, C. & Davies, H. M. L. Role of sterically demanding chiral dirhodium catalysts in site-selective C–H functionalization of activated primary C–H bonds. *J. Am. Chem. Soc.* **136,** 9792–9796 (2014).
43. Wang, H., Guptill, D. M., Varela-Alvarez, A., Musaev, D. G. & Davies, H. M. L. Rhodium-catalyzed enantioselective cyclopropanation of electron-deficient alkenes. *Chem. Sci.* **4,** 2844–2850 (2013).
44. Hedley, S. J., Ventura, D. L., Dominiak, P. M., Nygren, C. L. & Davies, H. M. L. Investigation into factors influencing stereoselectivity in the reactions of heterocycles with donor-acceptor-substituted rhodium carbenoids. *J. Org. Chem.* **71,** 5349–5356 (2006).
45. Davies, H. M. L., Jin, Q., Ren, P. & Kovalevsky, A. Y. Catalytic asymmetric benzylic C–H activation by means of carbenoid-induced C–H insertions. *J. Org. Chem.* **67,** 4165–4169 (2002).
46. Davies, H. M. L., Hansen, J. H. & Churchill, M. R. Catalytic asymmetric C–H activation of alkanes and tetrahydrofuran. *J. Am. Chem. Soc.* **122,** 3063–3070 (2000).
47. Davies, H. M. L. & Venkataramani, C. Catalytic enantioselective synthesis of $\beta^2$-amino acids. *Angew. Chem. Int. Ed.* **41,** 2197–2199 (2002).
48. Davies, H. M. L. & Yang, J. Influence of a β-alkoxy substituent on the C–H activation chemistry of alkyl ethers. *Adv. Synth. Catal.* **345,** 1133–1138 (2003).
49. Davies, H. M. L. & Jin, Q. Double C–H activation strategy for the asymmetric synthesis of $C_2$-symmetric anilines. *Org. Lett.* **6,** 1769–1772 (2004).

## Acknowledgements

We acknowledge support from Novartis Institutes for BioMedical Research and the NSF under the CCI Center for Selective C–H Functionalization, CHE-1205646. We thank Professor Djamaladdin G. Musaev (Emory University), Dr Andrew Patterson and Dr Hasnain Malik for insightful discussion. J.H. gratefully acknowledges the Education Office of the Novartis Institutes for BioMedical Research Inc. for receipt of a Presidential Postdoctoral Fellowship. We thank Changming Qin (Emory University) for providing $Rh_2(BTPCP)_4$ catalyst. We thank Jinhai Gao and Melissa Grondine for providing NMR and preparative SFC separation support, respectively.

## Author contributions

R.E.J.B. and H.M.L.D. conceived and supervised this study. J.H., L.G.H., H.M.L.D. and R.E.J.B. were involved with experimental design and results discussion. J.H. conducted the experiments and analysed the data. J.H., H.M.L.D. and R.E.J.B. wrote the manuscript. All authors edited the manuscript.

## Additional information

**Competing financial interests**: The authors declare no competing financial interests.

# 15

# Dynamic control of chirality in phosphine ligands for enantioselective catalysis

Depeng Zhao[1], Thomas M. Neubauer[1] & Ben L. Feringa[1]

Chirality plays a fundamental role in biology and chemistry and the precise control of chirality in a catalytic conversion is a key to modern synthesis most prominently seen in the production of pharmaceuticals. In enantioselective metal-based catalysis, access to each product enantiomer is commonly achieved through ligand design with chiral bisphosphines being widely applied as privileged ligands. Switchable phosphine ligands, in which chirality is modulated through an external trigger signal, might offer attractive possibilities to change enantioselectivity in a catalytic process in a non-invasive manner avoiding renewed ligand synthesis. Here we demonstrate that a photoswitchable chiral bisphosphine based on a unidirectional light-driven molecular motor, can be used to invert the stereoselectivity of a palladium-catalysed asymmetric transformation. It is shown that light-induced changes in geometry and helicity of the switchable ligand enable excellent selectivity towards the racemic or individual enantiomers of the product in a Pd-catalysed desymmetrization reaction.

[1]Centre for System Chemistry, Stratingh Institute for Chemistry, Faculty of Mathematics and Natural Sciences, University of Groningen, Nijenborgh 4, 9747AG Groningen, The Netherlands. Correspondence and requests for materials should be addressed to B.L.F. (email: b.l.feringa@rug.nl).

Responsive catalytic systems in which reactivity or selectivity can be modulated via external non-invasive signals are highly promising to enable tuning of catalytic function, allow spatio-temporal control in chemical transformations and ultimately arrive at multitasking catalysts. In recent years, the development of stimuli- responsive catalysts has attractive considerable attention[1–3] and important efforts are related to on–off switching of catalytic activity[4,5]. Remarkable reversal of enantioselectivity in asymmetric catalysis has been achieved using solvent responsive helical polymers[6], light-triggered organocatalysts[7,8] and redox sensitive metal complexes[9]. These findings offer intriguing opportunities towards the design of broadly applicable responsive catalysts realizing that chiral transition metal catalysis centres among the most powerful and widely used methodologies to access enantiomerically pure compounds[10]. The asymmetric transformations rely on the well-defined chirality provided by the metal-bound ligands with a prominent role for chiral bisphosphines that encompass huge structural diversity[11]. A major challenge in asymmetric catalysis is to use a single enantiomer of a ligand or catalyst to obtain each enantiomer of a chiral product on demand[12,13]. Reported methodologies in which switching of enantioselectivity is observed are usually serendipitous or restricted to specific modifications, which include changing metals, solvents, temperature or other reaction parameters[14–20]. To achieve dual stereocontrol, we envision a rational and more general approach by designing chiral phosphine ligands that can exist in two (pseudo-)enantiomeric forms, which can be interconverted by light.

Although it is difficult to photochemically switch the chirality of conventional chiral ligands, artificial light-driven molecular motors provide a unique platform to achieve this goal[21–23]. Unidirectional rotary molecular motors based on overcrowded alkenes are intrinsically multistage chiral switches as we have recently shown in the design of three-stage organocatalysts[7,8].

Herein, we demonstrate that a chiral bisphosphine with an intrinsic molecular motor core structure can be used as a multistage chiral switchable ligand for Pd catalysis. The stereoselectivity in a catalytic desymmetrization reaction involving an allylic substitution can be readily modulated by switching the helicity of the bisphosphine ligand on irradiation. This allows access to both enantiomers as well as the racemate of a functionalized cyclopentene using a single chiral ligand and these findings bring responsive chiral catalysts into the vast domain of phosphine-based transition metal catalysis[10,24].

## Results

**Design and synthesis.** As a starting point for ligand design, we noted the advantages of $C_2$-symmetric structures in many privileged bisphosphine ligands (Fig. 1a) and reasoned that a first generation molecular motor with two pending phosphine moieties using an appropriate linker might provide an optimal choice to achieve dynamic control over geometry and helicity of such

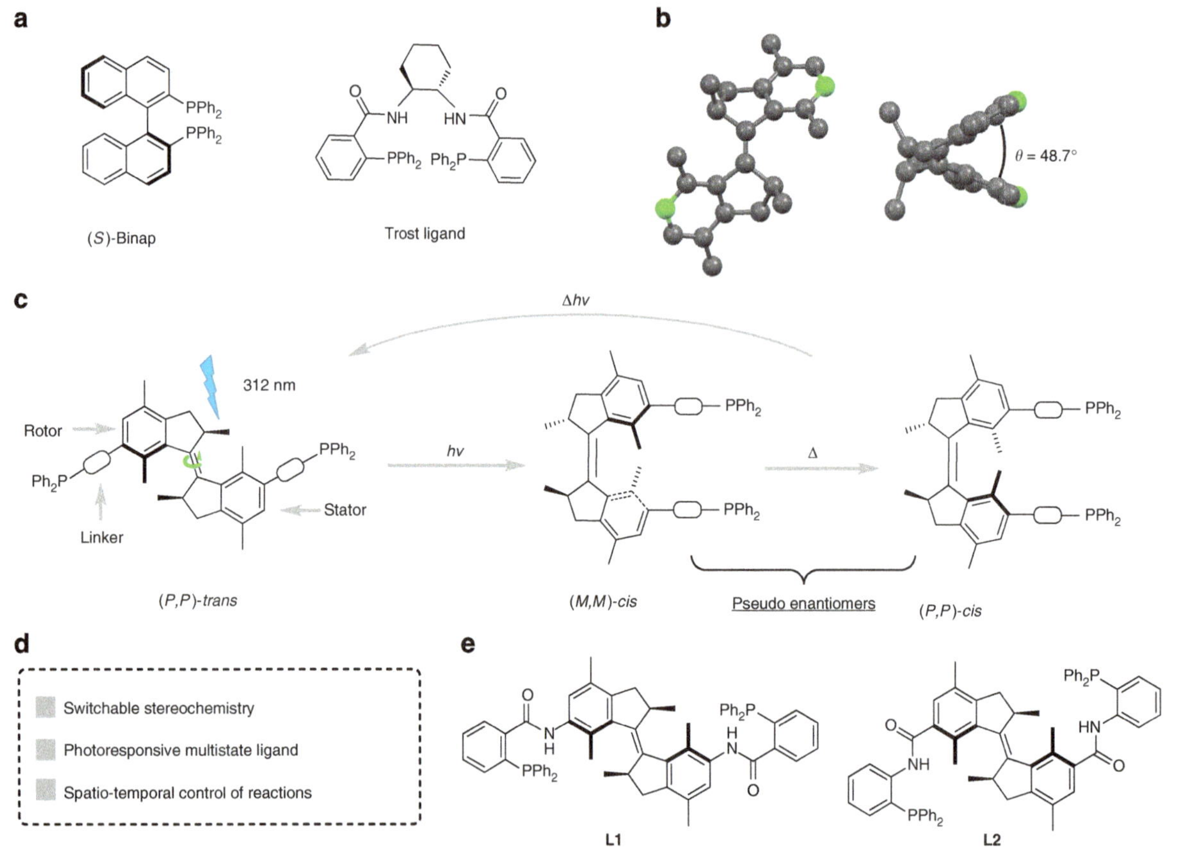


**Figure 1 | Chiral photoresponsive phosphine ligands.** (**a**) Traditional $C_2$-symmetric privileged ligands: (*S*)-Binap and Trost ligand. (**b**) Crystal structure of the parent molecular motor[25]; left: (*S,S*)-(*M,M*)-*trans* isomer, right: (*R,R*)-(*P,P*)-*cis* isomer. (**c**) Novel three-state photoresponsive phosphine ligand based on unidirectional molecular motor. The (*P,P*)-*trans* ligand based system might actually represent several catalytic complexes in the multi-state system, that is, oligomeric/polymeric P-Pd-ligand-Pd-ligand-Pd species due to phosphine ligand coordination in a monodentate manner to the metal. (**d**) Key features of novel chiral photoresponsive phosphine ligands. (**e**) Two potential photoresponsive phosphine ligands **L1** and **L2** incorporating amide linkers. *R* and *S* are used to define the chirality of the stereogenic centres; *P* (plus) and *M* (minus) are used to define the helicity of right and left-handed helix, respectively.

bidentate chiral ligands (Fig. 1b,c). The structural design for the $C_2$-symmetric bisphosphine ligands **L1** and **L2** with unique dynamic properties is shown in Fig. 1e.

The light-driven molecular motor unit is expected to undergo a unidirectional four-stage rotation in a 360° rotary cycle reaching (*P,P*)-*trans*, (*M,M*)-*cis*, (*P,P*)-*cis* and (*M,M*)-*trans* states, inducing both large geometrical and helical changes[21,22]. This allows both the spatial distance of the phosphine moieties and the chirality to be precisely controlled in each step of the cycle. We anticipated that in the *trans*-isomers the two phosphine groups are far apart from each other and unable to achieve intramolecular metal coordination, whereas in the *cis*-isomers they can effectively cooperate to form a chiral bisphosphine–metal complex (Fig. 1c). As the (*P,P*)-*cis* ligand–metal complex and (*M,M*)-*cis* ligand–metal complex are pseudo enantiomers, it can be expected that chiral products with opposite absolute configuration are obtained when these isomers are used in a catalytic asymmetric transformation. A crucial design feature is the nature of the linker moiety between motor and phosphine: (i) The linker should not contain additional chiral moieties to avoid match-mismatch effects for the *cis*-isomers and be relatively short so that the chiral orientation of the phosphines reflects the chirality of the motor; (ii) Small dihedral angle of the two phenyl rings in the *cis*-states based on the previous published crystal structure of a related first generation motor (Fig. 1b, 48.7° for (*P,P*)-*cis* isomer)[25]. The dihedral angle is closely related to the bite angle of the ligand–metal complex (P–metal–P angle), which is known to influence the catalytic behaviour and stereoselectivity in many asymmetric transformations[26,27]. In view of these considerations, amide linkers were chosen taking inspiration from Trost-ligands[28], as these chiral ligands feature bisphosphines, amide linkers and in several cases show good tolerance for the chiral core units and are effective in numerous asymmetric transformations[29]. **L1** incorporates the amide linker in the same bond order as for the Trost ligand as shown in Fig. 1a, whereas **L2** adopts the reversed amide connection (Fig. 1e).

The synthesis of molecular motor **L1** involved coupling of diamine **1** (ref. 8) with 2-(diphenylphosphino) benzoic acid **2** in the presence of BOP reagent ((Benzotriazol-1-yloxy)tris(dimethylamino)phosphonium hexafluorophosphate) and triethylamine (TEA) to afford bisphosphine **L1** in 65% yield (Fig. 2a). The asymmetric synthesis of bisphosphine ligand (*R,R*)-(*P,P*)-*trans*-**L2** started from chiral ketone **3**, which was prepared according to our recently reported gold-catalysed enantioselective protonation of the corresponding silyl enol ether in 97% *ee* (refs 30,31). McMurry coupling using titanium trichloride and zinc provided the chiral dibromo motor **4** as a *trans*/*cis* mixture in 75% yield (*trans*/*cis* = 3/1; 99% *ee* for both isomers). Subsequent palladium-catalysed carbonylation of dibromo motor **4** afforded diester **5** in 79% yield (*trans*/*cis* = 3/1; isolated yield for the separated isomers: 59% for (*P,P*)-*trans*-**5** and 20% for (*P,P*)-*cis*-**5**). Hydrolysis of (*P,P*)-*trans* diester **5** with aq. NaOH gave diacid **6** in nearly quantitative yield as a single enantiomer. Treatment of the diacid **6** with oxalyl chloride in a mixture of THF and $CH_2Cl_2$ with catalytic amount of DMF gave the acid chloride intermediate, which was directly used in the next condensation step with (2-aminophenyl)diphenylphosphane **7** to furnish the final product **L2** in 44% yield over two steps. Molecular motor-based bisphosphines **L1** and **L2** were characterized by $^1H$, $^{13}C$ and $^{31}P$ NMR spectroscopy and high-resolution mass spectrometry (see Supplementary Figs 3–16).

**Photochemical and thermal isomerization.** With the switchable bisphosphines in hand, their photochemical and thermal isomerization properties were investigated. Surprisingly, photoisomerization of **L1** at $\lambda_{max}$ = 312 nm (– 15 °C, THF) showed that only 7% of (*P,P*)-*trans*-**L1** was converted to (*M,M*)-*cis*-**L1** after reaching the photostationary state (PSS), based on $^1H$ NMR analysis. On the contrary, ligand **L2** with a reversed amide bond linker, was functioning as an effective molecular motor, which underwent the expected four-stage rotary cycle uncompromised (Fig. 3). The 360° unidirectional rotation cycle of **L2** includes two

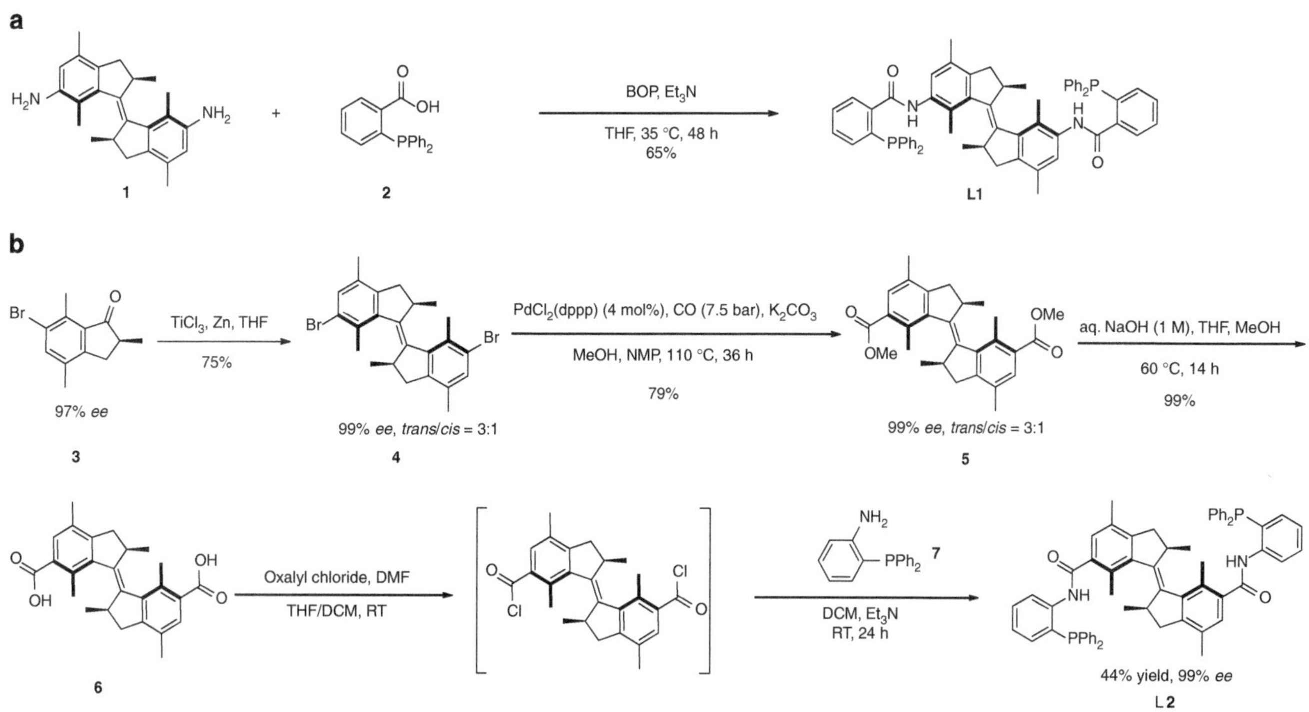


**Figure 2 | Procedure for the synthesis of ligands L1 and L2.** (**a**) Procedure for synthesis of bisphosphine ligand (*P,P*)-*trans*-**L1**. (**b**) Asymmetric synthesis of bisphosphine ligand (*R,R*)-(*P,P*)-*trans*-**L2**. (For details, see Supplementary Methods).

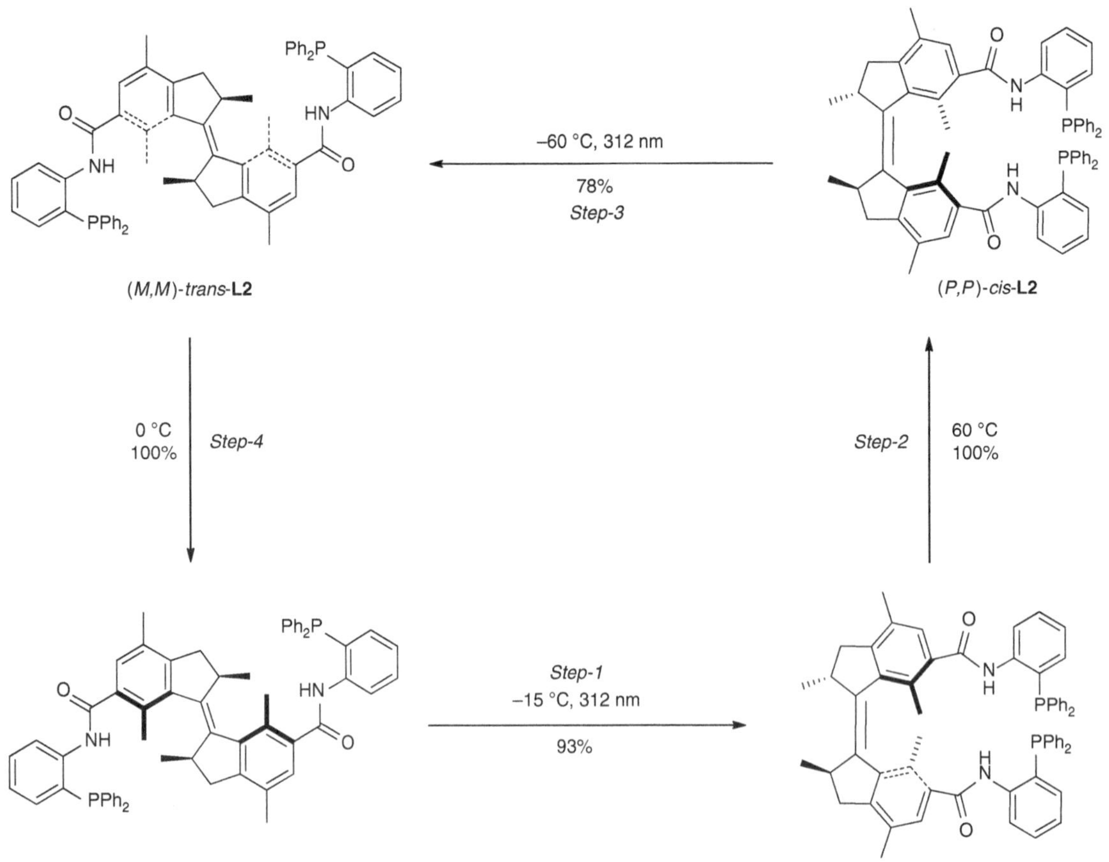


**Figure 3 | Schematic representation of unidirectional four-step rotary cycle of molecular motor L2.** Step-1: photoisomerization of (*P,P*)-*trans*-**L2** resulted in (*M,M*)-*cis*-**L2** in which the two phosphine groups are brought into proximity to one another. Step-2: thermal helix inversion of (*M,M*)-*cis*-**L2** provides (*P,P*)-*cis*-**L2** in which the helicity changed from *M* to *P* while the two phosphine groups remain spatially in proximity. Step-3 and step-4: Photoisomerization of (*P,P*)-*cis*-**L2** and subsequent thermal helix inversion of (*M,M*)-*trans*-**L2** completes the full cycle.

photoisomerization steps and two thermal isomerization steps as characterized by ultraviolet–visible, circular dichroism (CD), $^1$H NMR and $^{31}$P NMR spectroscopy. The first photoisomerization step was performed by irradiation of enantiomerically pure (*P,P*)-*trans*-**L2** with ultraviolet light ($\lambda_{max}$ = 312 nm, − 15 °C) in THF, which resulted in a significant decrease in the intensity of the absorption band at 280 nm and the appearance of a new absorption band at 350 nm (Fig. 4a). This red shift is typical for the formation of the (*M,M*)-*cis*-**L2** isomer[7,8,21,32]. $^1$H NMR and $^{31}$P NMR studies in $CD_2Cl_2$ also confirmed this structure as is evident from the downfield shift of all the aliphatic ring protons in $^1$H NMR and the shift of the phosphorus absorption in $^{31}$P NMR from − 18.88 to − 19.58 p.p.m. (Fig. 5). After reaching the PSS, a ratio of 93% (*M,M*)-*cis*-**L2** and 7% (*P,P*)-*trans*-**L2** was established by $^1$H NMR and $^{31}$P NMR. The observed changes in CD spectrum also confirm the *P,P* to *M,M* helix inversion during this step (new band at 350 nm, Fig. 4e). The excellent (*M,M*)-*cis*-**L2** to (*P,P*)-*trans*-**L2** ratio in the PSS state is highly beneficial in applications of catalytic asymmetric reactions. On heating of (*M,M*)-*cis*-**L2** in THF at 65 °C for 1 h, the thermal helix inversion from (*M,M*)-*cis*-**L2** to (*P,P*)-*cis*-**L2** resulted in a blue shift in the ultraviolet–visible spectrum (Fig. 4b). The thermal helix inversion was also confirmed by the upfield shifts of all the signals of the aliphatic ring protons in the $^1$H NMR spectrum, the $^{31}$P shift from − 19.58 to − 18.86 p.p.m. and the CD spectral changes (Figs 4f and 5). The helix inversion step-2 resulted in a quantitative transformation from (*M,M*)-*cis*-**L2** to (*P,P*)-*cis*-**L2** based on $^1$H NMR and $^{31}$P NMR measurements. A kinetic study for the thermal isomerization step-2 provided the standard Gibbs energy of activation ($\Delta^{\ddagger}G^{\circ}$ = 100.2 kJ mol$^{-1}$) and half-life of (*M,M*)-*cis*-**L2** ($t_{½}$ = 362 h at 0 °C). The long half-life is essential for further application of this isomer in enantioselective catalysis studies. The subsequent photoisomerization step-3, which was carried out at − 60 °C on irradiation at 312 nm, and the following thermal isomerization step-4 at 0 °C regenerated the initial (*P,P*)-*trans*-**L2** and completed the full four-step rotary cycle. These two final steps were also monitored with ultraviolet-vis, CD and $^1$H NMR spectroscopy (Fig. 4c,d,g,h). (M,M)-*trans*-**L2** is highly unstable ($\Delta^{\ddagger}G^{\circ}$ = 83.4 kJ mol$^{-1}$, $t_{½}$ = 11 min at 0 °C) under ambient conditions and as a consequence not suitable for further catalytic studies (see Supplementary Figs 1 and 2).

**Switchable asymmetric catalysis.** Having confirmed the four-step unidirectional rotary cycle of bisphosphine **L2**, we investigated its performance as a chiral switchable ligand in asymmetric catalysis. Based on the stability of three out of four isomers only, (*P,P*)-*trans*-**L2**, (*M,M*)-*cis*-**L2** and (*P,P*)-*cis*-**L2** were used in the palladium-catalysed desymmetrization of *meso*-cyclopent-2-en-1,4-diol bis(carbamate) **8,** a well-established model reaction to determine the enantiodiscrimination abilities of chiral ligands[28,33]. Initially, the desymmetrization reaction of biscarbamate **8** was carried out with 2.5 mol% of $Pd_2(dba)_3$ and 7.5 mol% of (*R,R*)-(*P,P*)-*cis*-L2 in THF. To our delight, the reaction proceeds with high selectivity affording chiral

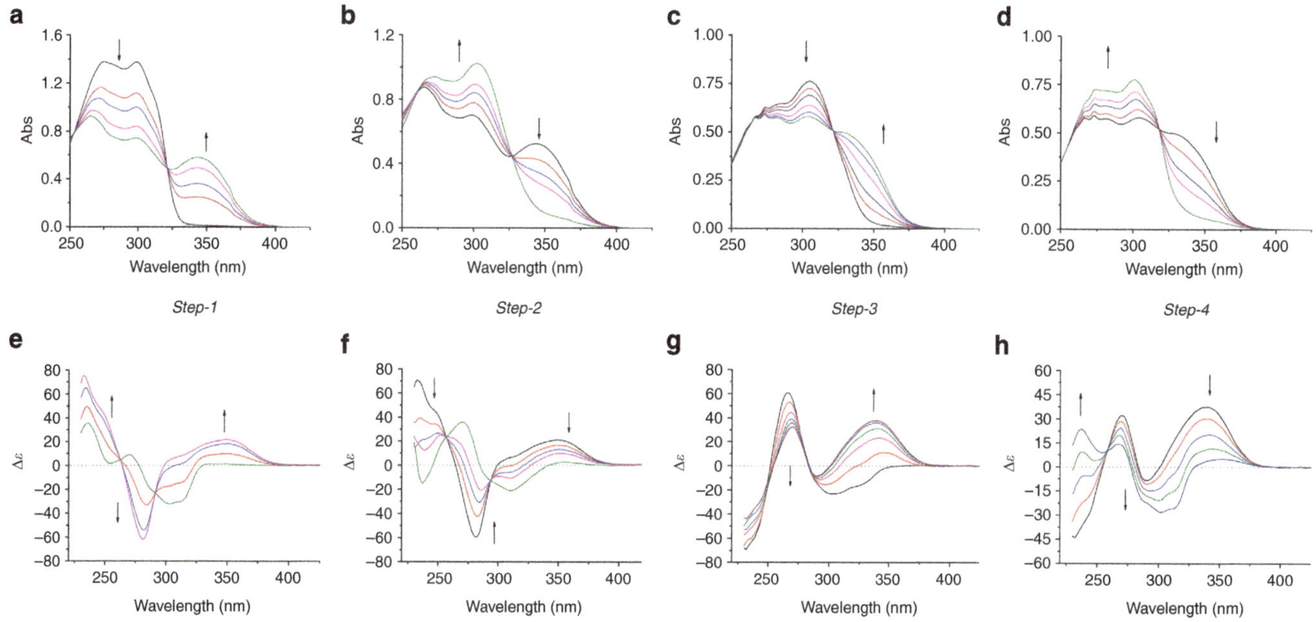


**Figure 4 | Ultraviolet-visible and CD spectral changes of L2 in each isomerization step.** (**a**) Ultraviolet-visible spectral changes during step-1 on irradiation at 312 nm in THF at −15 °C. (**b**) Ultraviolet-visible spectral changes during step-2 on heating in THF at 60 °C. (**c**) Ultraviolet-visible spectral changes during step-3 on irradiation at 312 nm in THF at −60 °C. (**d**) Ultraviolet-visible spectral changes during step-4 ($t_{½}$ = 11 min at 0 °C) in THF at 0 °C. Isosbestic points of step-1 to step-4: 321, 327, 322 and 318 nm. (**e-h**) The corresponding CD spectral changes during each of the four isomerization steps, respectively.

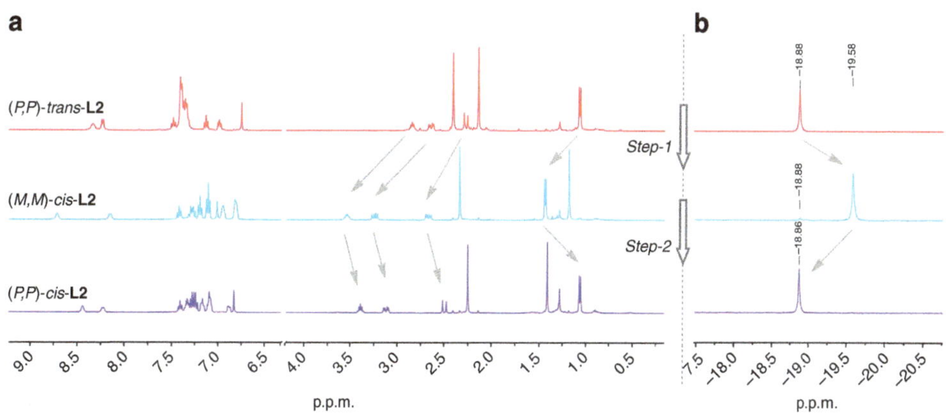


**Figure 5 | NMR spectra of switching step-1 and step-2.** (**a**) Partial $^1$H NMR ($CD_2Cl_2$) spectra of compounds (*P,P*)-*trans*-**L2**, (*M,M*)-*cis*-**L2** and (*P,P*)-*cis*-**L2** on irradiation of (*P,P*)-*trans*-**L2** at 312 nm in $CH_2Cl_2$ and after thermal helix inversion ((*R,R*)-(*M,M*)-*cis*-**L2**→(*R,R*)-(*P,P*)-*cis*-**L2**) on heating at 45 °C for 14 h. (**b**) The corresponding $^{31}$P NMR ($CD_2Cl_2$) spectra of compounds (*P,P*)-*trans*-**L2**, (*M,M*)-*cis*-**L2** and (*P,P*)-*cis*-**L2**, respectively ($H_3PO_4$ was used as an external standard).

oxazolidinone **9** in 89% yield with an enantiomeric ratio (e.r.) of 11/89. Encouraged by this initial result, several parameters were examined including temperature and base to further enhance the selectivity (see the Supplementary Table 1)[34]. The reaction in the presence of 2.0 equiv. TEA (pKa = 10.8) as the base led to a slight increase in the enantioselectivity. Since the base involves deprotonation of nonionized urethane (pKa ≈ 3.7; ref. 35) in the cyclization reaction, screening of a number of other bases revealed that *N,N*-diisopropylethylamine (pKa = 10.8) was the best choice increasing the e.r. significantly to 6/94.

Having established the optimized conditions, the catalytic performance of this tunable chiral ligand was investigated with different isomers (Fig. 6). It should be noted that the *in situ* switching experiment in the presence of Pd is complicated so far due to the stability of the catalysts generated *in situ* under ultraviolet irradiation and heating, leading to a significant decrease in selectivity. When the (*R,R*)-(*P,P*)-*trans*-**L2** isomer was used in the Pd-catalysed asymmetric desymmetrization, as expected, nearly racemic product **9** was obtained in 65% yield and e.r. of 53/47 (*3R,4S*/*3S,4R*). Most probably each phosphine in *trans*-**L2** acts as a monodentate ligand as is supported by precipitation of substantial amounts of oligomeric palladium complexes during the reaction. Much to our delight, after photoisomerization to (*R,R*)-(*M,M*)-*cis*-**L2**, the Pd-complex of **L2** was able to catalyse this reaction with excellent stereocontrol providing (*3R,4S*)-product **9** (e.r. = 93/7, *3R,4S*/*3S,4R*). In contrast, after isomerization to (*R,R*)-(*P,P*)-*cis*-**L2**, the opposite enantiomer (*3S,4R*)-**9** is obtained again with excellent enantioselectivities (e.r. = 6/94, *3R,4S*/*3S,4R*). From these results, it is clear that ligand **L2** shows excellent performance as a chiral responsive bidentate phosphine ligand. The combination of photochemical isomerization between *trans* and *cis*-isomers and the thermal

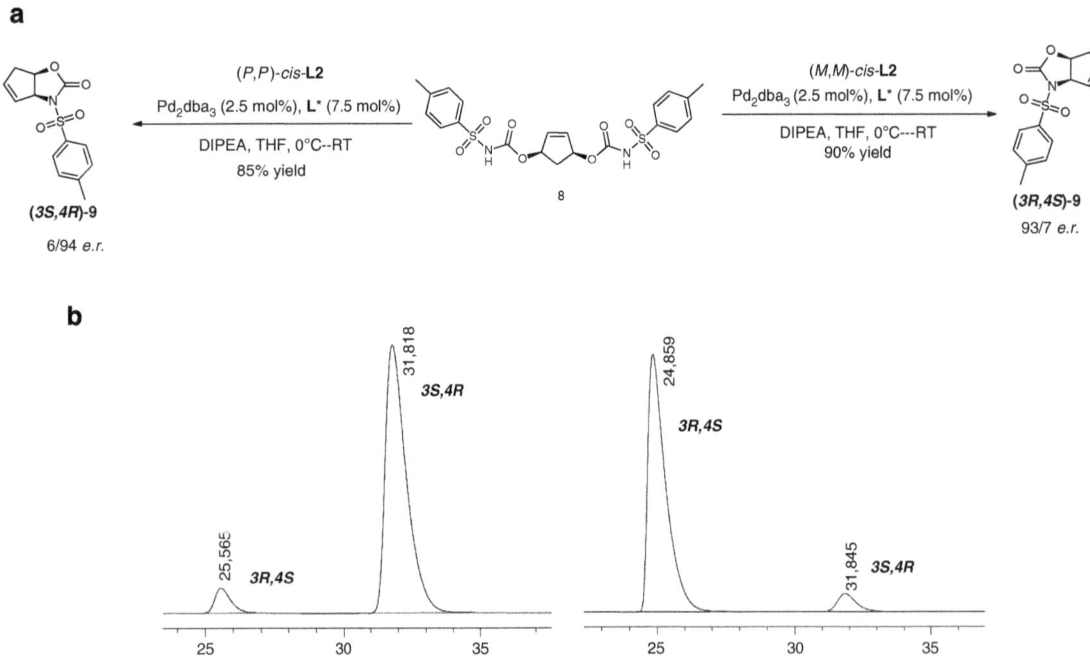


**Figure 6 | Switchable asymmetric catalysis.** (**a**) Stereodivergent synthesis of (3*S*,4*R*)-and (3*R*,4*S*)-oxazolidinone **9** by switching the chirality of ligand *cis*-**L2**. (**b**) HPLC trace for **9** from (*R*,*R*)-(*P*,*P*)-*cis*-**L2** and (*R*,*R*)-(*M*,*M*)-*cis*-**L2**. Enantiomeric excess was determined by HPLC using a Chiracel OD-H column. The absolute configuration of the product **9** was confirmed by comparing the sign of optical rotation with literature reported data[34].

helix inversion of (*M*,*M*)-*cis* to (*P*,*P*)-*cis*-isomers enables a single chiral ligand to control the formation of nearly racemic, and each enantiomer of the product of a representative Pd-catalysed transformation. In addition, the order in which the different chiral ligands can be formed (in this case *PP*—*MM*—*PP*) is dictated by the stereogenic centres of the chiral motor core unit.

## Discussion

We have successfully designed a photoresponsive chiral bisphosphine ligand that enables the formation of each enantiomer of a product of a Pd-catalysed reaction with a single chiral ligand. The unique combination of a light-driven molecular motor bridging two phosphine moieties allows switching between multiple stereochemical forms with distinct ligand properties. The photochemical and thermal isomerization during unidirectional rotation around the motor central double bond enable to achieve stepwise control over the helicity of the bisphosphine ligand and spatial distance between these two phosphine groups. The stereoselectivity of the Pd-catalysed desymmetrization reaction can be modulated from nearly racemic for (*P*,*P*)-*trans*-state to up to 93/7 e.r. for (*M*,*M*)-*cis*-state and 6/94 e.r. for (*P*,*P*)-*cis*-state, with enantioselectivities comparable to those of related conventional chiral ligands. These results highlight the proof-of-principle of multistage dynamically tunable and responsive chiral ligands at the molecular level for transition-metal-catalysed asymmetric synthesis. These responsive ligand systems have considerable potential for a wide range of enantioselective transformations based on transition metal bisphosphine catalysts. In addition, these systems hold promise to modulate catalysts activity and switch stereoselectivity with high spatio-temporal control ultimately arriving at a catalyst that can perform multiple functions controlled by light.

## Methods

**Procedure for Pd-catalysed intramolecular cyclization of 8.** To a Schlenk tube containing $Pd_2dba_3$ (2.3 mg, 2.5 mol%) and the specific isomer of **L2** (7.0 mg, 7.5 mol%), anhydrous THF (0.4 ml) was added, and the mixture was stirred at 0 °C for 30 min under a nitrogen atmosphere. Then, *N*,*N*-diisopropylethylamine (35 μl, 0.20 mmol) and biscarbamate **8** (49.5 mg, 0.10 mmol) were added under nitrogen. The resulting solution was stirred at 0 °C and allowed to slowly warm to room temperature overnight. The mixture was directly submitted to flash chromatography on silica gel with pentane–EtOAc (3/1) as eluent affording product **9** (65% yield for (*P*,*P*)-*trans*-**L2**; 90% yield for (*M*,*M*)-*cis*-**L2**; 85% yield for (*P*,*P*)-*cis*-**L2**).

**3-Tosyl-3,3a,6,6a-tetrahydro-2H-cyclopenta[*d*]oxazol-2-one (9).** $^1$H NMR (400 MHz, $CDCl_3$) δ = 7.94 (d, *J* = 8.3 Hz, 2H), 7.35 (d, *J* = 8.1 Hz, 2H), 6.07–5.96 (m, 2H), 5.28 (d, *J* = 7.3 Hz, 1H), 5.10 (t, *J* = 6.8 Hz, 1H), 2.81 (dd, *J* = 18.7, 6.3 Hz, 1H), 2.69 (d, *J* = 18.7 Hz, 1H), 2.44 (s, 3H). $^{13}$C NMR (101 MHz, $CDCl_3$) δ = 151.3, 145.5, 135.0, 133.8, 129.7, 128.3, 128.0, 77.3, 77.0, 76.8, 76.7, 66.3, 39.0, 21.7. HRMS (ESI +, *m/z*) calculated for $C_{13}H_{13}NO_4SNa$ $[M + Na]^+$ 302.0458; found 302.0460. Enantiomeric excess was determined by HPLC (Chiracel OD-H), *n*-heptane/*i*-propanol = 85/15, 40 °C, 254 nm, 0.5 ml min$^{-1}$, retention times: $t_{R(3R,4S)} = 25$ min, $t_{R(3S,4R)}$ 32 min.

## References

1. Kumagai, N. & Shibasaki, M. Catalytic chemical transformations with conformationally dynamic catalytic systems. *Catal. Sci. Technol.* **3,** 41–57 (2013).
2. Stoll, R. S. & Hecht, S. Artificial light-gated catalyst systems. *Angew. Chem. Int. Ed.* **49,** 5054–5075 (2010).
3. Neilson, B. M. & Bielawski, C. W. Illuminating photoswitchable catalysis. *ACS Catal.* **3,** 1874–1885 (2013).
4. Peters, M. V., Stoll, R. S., Kühn, A. & Hecht, S. Photoswitching of basicity. *Angew. Chem. Int. Ed.* **47,** 5968–5972 (2008).
5. Nojiri, A., Kumagai, N. & Shibasaki, M. *In situ* manipulation of catalyst performance via the photocontrolled aggregation/dissociation state of the catalyst. *Chem. Commun.* **49,** 4628–4630 (2013).
6. Yamamoto, T., Yamada, T., Nagata, Y. & Suginome, M. High-molecular-weight polyquinoxaline-based helically chiral phosphine (PQXphos) as chirality-switchable, reusable, and highly enantioselective monodentate ligand in catalytic asymmetric hydrosilylation of styrenes. *J. Am. Chem. Soc.* **132,** 7899–7901 (2010).
7. Wang, J. & Feringa, B. L. Dynamic control of chiral space in a catalytic asymmetric reaction using a molecular motor. *Science* **331,** 1429–1432 (2011).
8. Vlatković, M., Bernardi, L., Otten, E. & Feringa, B. L. Dual stereocontrol over the Henry reaction using a light- and heat-triggered organocatalyst. *Chem. Commun.* **50,** 7773–7775 (2014).
9. Mortezaei, S., Catarineu, N. R. & Canary, J. W. A redox-reconfigurable, ambidextrous asymmetric catalyst. *J. Am. Chem. Soc.* **134,** 8054–8057 (2012).
10. Jacobsen, E. N., Pfaltz, A. & Yamamoto, H. (eds) *Comprehensive Asymmetric Catalysis, and Supplements 1 and 2* (Springer, 2004).
11. Börne, A. *Phosphorus Ligands in Asymmetric Catalysis* (Wiley-VCH, 2008).

12. Bartók, M. Unexpected inversions in asymmetric reactions: reactions with chiral metal complexes, chiral organocatalysts, and heterogeneous chiral catalysts. *Chem. Rev.* **110,** 1663–1705 (2010).
13. Escorihuela, J., Burguete, M. I. & Luis, S. V. New advances in dual stereocontrol for asymmetric reactions. *Chem. Soc. Rev.* **42,** 5595–5617 (2013).
14. Yabu, K. *et al.* Switching enantiofacial selectivities using one chiral source: catalytic enantioselective synthesis of the key intermediate for (20*S*)-camptothecin family by (*S*)-selective cyanosilylation of ketones. *J. Am. Chem. Soc.* **123,** 9908–9909 (2001).
15. Inagaki, T., Ito, A., Ito, J. & Nishiyama, H. Asymmetric iron-catalysed hydrosilane reduction of ketones: effect of zinc metal upon the absolute configuration. *Angew. Chem. Int. Ed.* **49,** 9384–9387 (2010).
16. Kim, H., Li, J., Kim, S. & Oh, K. Stereodivergency in catalytic asymmetric conjugate addition reactions of glycine (ket)imines. *J. Am. Chem. Soc.* **133,** 20750–20753 (2011).
17. Zhou, J. & Tang, Y. Enantioselective Friedel–Crafts reaction of indoles with arylidene malonates catalysed by *i*Pr-bisoxazoline–Cu(OTf)$_2$. *Chem. Commun.* 432–433 (2004).
18. Inoue, Y., Yokoyama, T., Yamasaki, N. & Tai, A. An optical yield that increases with temperature in a photochemically induced enantiomeric isomerization. *Nature* **341,** 225–226 (1989).
19. Casey, C. P., Martins, S. C. & Fagan, M. A. Reversal of enantioselectivity in the hydroformylation of styrene with [2*S*,4*S*-BDPP]Pt(SnCl$_3$)Cl at high temperature arises from a change in the enantioselective-determining step. *J. Am. Chem. Soc.* **126,** 5585–5592 (2004).
20. Chan, V. S., Chiu, M., Bergman, R. G. & Toste, F. D. Development of ruthenium catalysts for the enantioselective synthesis of P-stereogenic phosphines via nucleophilic phosphido intermediates. *J. Am. Chem. Soc.* **131,** 6021–6032 (2009).
21. Koumura, N., Zijlstra, R. W. J., van Delden, R. A., Harada, N. & Feringa, B. L. Light-driven monodirectional molecular rotor. *Nature* **401,** 152–155 (1999).
22. Feringa, B. L. The art of building small: from molecular switches to molecular motors. *J. Org. Chem.* **72,** 6635–6652 (2007).
23. Schliwa, M. (ed.) *Molecular Motors* (VCH-Wiley, 2003).
24. Walsh, P. J. & Kozlowski, M. C. *Fundamentals of Asymmetric Catalysis* (Univ. Science Books, 2009).
25. Pollard, M. M., Meetsma, A. & Feringa, B. L. A redesign of light-driven rotary molecular motors. *Org. Biomol. Chem.* **6,** 507–512 (2008).
26. Tang, W. J. & Zhang, X. M. New chiral phosphorus ligands for enantioselective hydrogenation. *Chem. Rev.* **103,** 3029–3069 (2003).
27. Kean, Z. S. *et al.* Photomechanical actuation of ligand geometry in enantioselective catalysis. *Angew. Chem. Int. Ed.* **53,** 14508–14511 (2014).
28. Trost, B. M., Van Vranken, D. L. & Bingel, C. A modular approach for ligand design for asymmetric allylic alkylations via enantioselective palladium-catalysed ionizations. *J. Am. Chem. Soc.* **114,** 9327–9343 (1992).
29. Trost, B. M. & Crawley, M. L. Asymmetric transition-metal-catalysed allylic alkylations: applications in total synthesis. *Chem. Rev.* **103,** 2921–2944 (2003).
30. Neubauer, T. M. *et al.* Asymmetric synthesis of first generation molecular motors. *Org. Lett.* **16,** 4220–4223 (2014).
31. Cheon, C. H., Kanno, O. & Toste, F. D. Chiral Brønsted acid from a cationic gold(I) complex: catalytic enantioselective protonation of silyl enol ethers of ketones. *J. Am. Chem. Soc.* **133,** 13248–13251 (2011).
32. Chen, J., Kistemaker, J. C. M., Robertus, J. & Feringa, B. L. Molecular stirrers in action. *J. Am. Chem. Soc.* **136,** 14924–14932 (2014).
33. Butts, C. P. *et al.* Structure-based rationale for selectivity in the asymmetric allylic alkylation of cycloalkenyl esters employing the Trost 'Standard Ligand' (TSL): isolation, analysis and alkylation of the monomeric form of the cationic $\eta^3$-cyclohexenyl complex $[(\eta^3\text{-c-}C_6H_9)Pd(TSL)]^+$. *J. Am. Chem. Soc.* **131,** 9945–9957 (2009).
34. Trost, B. M. & Patterson, D. E. Enhanced enantioselectivity in the desymmetrization of *meso*-biscarbamates. *J. Org. Chem* **63,** 1339–1341 (1998).
35. Taylor, L. D., Pluhar, M. & Rubin, L. E. Reaction of sulfonyl isocyanates with polymeric alcohols to produce a polymeric acid. *J. Polym. Sci. B Polym. Lett.* **5,** 77–78 (1967).


## Acknowledgements

This work was supported financially by the European Research Council (Advanced Investigator Grant, No. 227897 to B.L.F.); The Netherlands Organization for Scientific Research (NWO-CW); funding from the Ministry of Education, Culture and Science (Gravitation program 024.001.035); The Royal Netherlands Academy of Arts and Sciences (KNAW); and NRSC-Catalysis are gratefully acknowledged. T.M.N. acknowledges the DAAD for funding.

## Author contributions

D.Z. and B.L.F. conceived the project. D.Z. and T.M.N. carried out the experimental work. All the authors contributed to the design of the experiments, the analysis of the data and the writing of the paper.

## Additional information

**Competing financial interests:** The authors declare no competing financial interests.

# 16

# The evolution of cyclopropenium ions into functional polyelectrolytes

Yivan Jiang[1,*], Jessica L. Freyer[1,*], Pepa Cotanda[2,3], Spencer D. Brucks[1], Kato L. Killops[4], Jeffrey S. Bandar[1], Christopher Torsitano[1], Nitash P. Balsara[3,5], Tristan H. Lambert[1] & Luis M. Campos[1]

Versatile polyelectrolytes with tunable physical properties have the potential to be transformative in applications such as energy storage, fuel cells and various electronic devices. Among the types of materials available for these applications, nanostructured cationic block copolyelectrolytes offer mechanical integrity and well-defined conducting paths for ionic transport. To date, most cationic polyelectrolytes bear charge formally localized on heteroatoms and lack broad modularity to tune their physical properties. To overcome these challenges, we describe herein the development of a new class of functional polyelectrolytes based on the aromatic cyclopropenium ion. We demonstrate the facile synthesis of a series of polymers and nanoparticles based on monomeric cyclopropenium building blocks incorporating various functional groups that affect physical properties. The materials exhibit high ionic conductivity and thermal stability due to the nature of the cationic moieties, thus rendering this class of new materials as an attractive alternative to develop ion-conducting membranes.

[1]Department of Chemistry, Columbia University, New York, New York 10027, USA. [2]Joint Center for Artificial Photosynthesis, Lawrence Berkeley National Laboratory, Berkeley, California 94720, USA. [3]Department of Chemical and Biomolecular Engineering, University of California, Berkeley, California 94720, USA. [4]Edgewood Chemical Biological Center, Aberdeen Proving Ground, Aberdeen, Maryland 21010, USA. [5]Materials Sciences Division, Environmental Energy Technologies Division and Joint Center for Energy Storage, Lawrence Berkeley National Laboratory, Berkeley, California 94720, USA. * These authors contributed equally to this work. Correspondence and requests for materials should be addressed to L.M.C. (email: lcampos@columbia.edu).

Modularly designed polymeric materials can be engineered to suit a broad range of applications representing an attractive platform for technological advancement[1]. Materials that possess both inherent compositional versatility and ready accessibility via robust and scalable synthetic pathways are of particular import to the field[2,3]. In this regard, cationic polyelectrolytes have emerged as a versatile class of materials that have been exploited in a broad array of applications[4–6], ranging from gene delivery[7,8] to ion-conducting membranes[9–11], and water purification[12,13]. Development in the area of cationic polyelectrolytes has thus far focused on a limited menu of monomeric functionalities, including ammonium, phosphonium, imidazolium, pyridinium and guanidinium ions[14–16]. These heteroatomic systems, while valuable, are application specific and are limited in terms of the ability to finely tune their physical properties. Thus, the identification of new modular cationic polyelectrolytes, with superior characteristics for processing, controllable self-assembly and function, represents an important goal for this field[17,18].

In developing a new family of polyelectrolytes, certain criteria must be met[11,13,17] including: (1) thermodynamic stability; (2) ease and scalability of polymerisations by controlled methods; (3) incorporation of accessible chemical handles to allow for diversity and intimate control of physical properties and (4) tunable Coulombic interactions. As an outgrowth of ongoing research efforts, we postulated that polyelectrolytes based on the cyclopropenium ion could satisfy these design criteria, while offering a highly distinct structural architecture and electronic properties. We further recognized that such cyclopropenium-based systems possess unique characteristics that distinguish them from existing cationic polyelectrolytes, namely: enhanced dispersion of the positive charge (compared with ammonium, phosphonium and guanidinium systems) and weaker H-bond donor capacity (compared with imidazolium and pyridinium ions)[19].

As the smallest of the Hückel aromatics[20], the cyclopropenium (CP) ion possesses significant stability despite its carbocationic nature (Fig. 1a)[21,22]. This remarkable degree of stability may be further enhanced through the incorporation of amino substituents onto the CP ring[23]. Indeed, with $pK_{R+}$ values estimated at $>13$, aminocyclopropenium ions are stable even in strongly alkaline aqueous solutions[24,25]. Moreover, thermal decomposition ($T_{dec}$) of the tris(dialkylamino)CP chloride salts has been measured at $>300\,°C$[19], significantly exceeding that of dialkylimidazolium chloride salts ($T_{dec}\sim250\,°C$)[26]. These unique structural features have already inspired the development of aminocyclopropenium ions for a range of applications, including as metal ligands[27], organocatalysts[28–30] and ionic liquids[19]; however, the incorporation of these cations into a polymeric backbone has only led to polymers with unstable CP ions as intermediates[31]. Given the tunable functionality and robust, efficient and orthogonal chemistry characterizing CP ions, it is desirable to exploit them in polymeric materials.

Here, we describe the synthesis and evaluation of a new family of cationic polyelectrolytes. As outlined in Fig. 1b, our vision for the design of the parent ionic monomer includes a polymerisable unit, a spectator group (which could also serve as a functional handle) and four additional modular groups that provide the means to tune the physical properties of the resulting macromolecules. In these initial studies, we elected to focus on styrenic CP monomers (termed, **CPR**) bearing a series of dialkylamino ($NR_2$) substituents. Styrene-based monomers can be subjected to various reversible-deactivation radical polymerization strategies[32–34]. We used reversible-addition fragmentation chain transfer (RAFT) polymerization[34] to assemble homopolymers, statistical copolymers and diblock copolymers of different

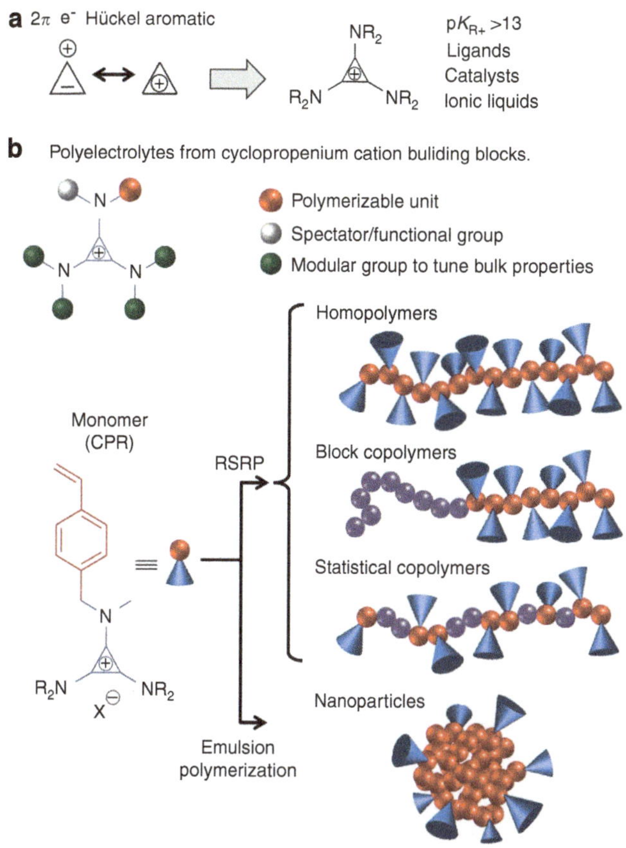


**Figure 1 | CP ion building blocks.** (**a**) Structure of the CP ion, including the dialkylamino groups that can be used to stabilize and vary the application of this diverse building block. (**b**) Types of polyelectrolytes that can be synthesized from CP monomers by reversible-deactivation radical polymerization (RDRP) strategies and emulsion polymerization.

compositions ranging from 20 to 50 mol % of CP functionality. As will be shown, we demonstrate that macromolecular assemblies of these materials can be used as ion-conducting membranes, and that the physical properties of these assemblies may be tuned through variation of the dialkylamino handles. In addition, we demonstrate that **CPR** monomers undergo a surfactant-free emulsion polymerization with styrene, yielding well defined, sub-100 nm nanoparticles with charged surfaces.

## Results

**Monomer synthesis.** Our exploration of the CP functional group in the context of cationic polyelectrolytes was originally inspired by its ionic liquid properties[19] and the straightforward elaboration of the CP ion with various functional groups. Thus far, however, there are no reports on the incorporation of this thermodynamically stable carbocation into macromolecules; CP derivatives have only appeared in polymers as transient species[31,35]. Derivatives of the CP ion are made from inexpensive reagents and can be easily prepared on a multi-gram scale under ambient conditions[19]. As robust chemistry is requisite for large-scale production of materials, we devised a viable synthetic strategy *en route* to the polymerisable CP ion monomers. The general approach to synthesise CP ion-containing monomers is based on the facile preparation of asymmetric amino-substituted CP ions (for example, **CPR**, Fig. 1)[36]. This procedure allows us to intimately change functionality while maintaining cationic properties and thermal stability. Notably, synthetic routes to aminocyclopropenium derivatives are modular and highly scalable[22], with efficiency levels approaching those attained via click chemistry[37].

The preparation of the **CPR** monomers begins with pentachlorocyclopropane (**1**), which is commercially available or inexpensively synthesized in hundred-gram quantities[38]. Reaction of **1** with a secondary amine leads to near-quantitative yields of a corresponding CP cation (**2** or **3**). Thus, amines with high steric hindrance (dicyclohexylamine, Cy and diisopropylamine, iP) add twice to **1**, preventing addition of a third bulky amine and leading directly to **2**. Conversely, less sterically hindered amines, such as morpholine (Mo), add thrice to **1**, resulting in a *tris*-amino CP (**3**). The latter is readily hydrolyzed to its corresponding cyclopropenone in hot, aqueous base, which is subsequently chlorinated to obtain **2**. To underscore the accessibility of these materials, we note that the monomers are obtained by simple purification techniques (see Supplementary Methods for details). Using this process, we prepared multigram quantities of **2** incorporating three different secondary amines, as depicted in Fig. 2 (bottom). Importantly, the chemistry in Fig. 2 is highly amenable to a wide range of nucleophilic secondary amines incorporating a variety of functional groups, including elements of asymmetry. We specifically chose to examine dicyclohexylamine, diisopropylamine and morpholine, given that they differ significantly in their degrees of hydrophilicity and steric hindrance.

The synthesis of the **CPR** monomers from the precursor (**2**) was readily achieved in 10–20 g quantities. The chlorinated 1-position of **2** is highly susceptible to addition of a secondary amine bearing a polymerisable unit, such as compound **5**. A styrene-based polymerisable group was chosen as it is a well-behaved building block in polymer chemistry, and its hydrophobicity relative to the CP ion moiety could stabilize an emulsion of the type used in nanoparticle synthesis. We postulate that other polymerisable moieties would yield functional monomers as well.

**Polymer synthesis and characterization**. The three chosen monomers (**CPCy**, **CPiP** and **CPMo**, Fig. 2) were polymerized by RAFT in multigram quantities yielding linear polymers (**PCPCy**, **PCPiP** and **PCPMo**, respectively, Fig. 3). **PCPCy** was purified through precipitation into 1,4-dioxane from $CH_2Cl_2$ with 88% recovered yield. **PCPiP** was precipitated from acetone or $CH_2Cl_2$ into cold ethyl acetate with 70% recovered yield. Due to their solubility in water, both **PCPiP** and **PCPMo** can be purified by dialysis. Purification of **PCPMo** resulted in a 51% recovered yield. Each of these polymers was isolated as a powder, and **PCPiP** and **PCPMo** were observed to be extremely hygroscopic. Through end-group analysis of the $^1H$ nuclear magnetic resonance (NMR) spectra, we calculated the degree of polymerization (DP) and molecular mass of each of the homopolymers (Supplementary Table 1). Due to the cationic nature of the CP groups, polymers (and copolymers) cannot be characterized using size exclusion chromatography eluted with organic solvents, as the polymers adhere to the column. We attempted to characterise the dispersity (*Đ*) of the hydrophilic homopolymers (**PCPiP** and **PCPMo**) on an acetate-buffered aqueous size exclusion chromatography; however, only **PCPMo** successfully eluted owing to its greater hydrophilicity (Supplementary Fig. 1, Supplementary Table 2). The *Đ* of **PCPMo** was determined to be 1.3, but we note that this value may not accurately reflect the DP control, given that the polyelectrolyte may still be interacting with the column as it is eluted. We note that the synthetic accessibility of these various CP-based polymers is straightforward and highly efficient, rivalling that of ammonium, phosphonium and imidazolium polymers[39–42].

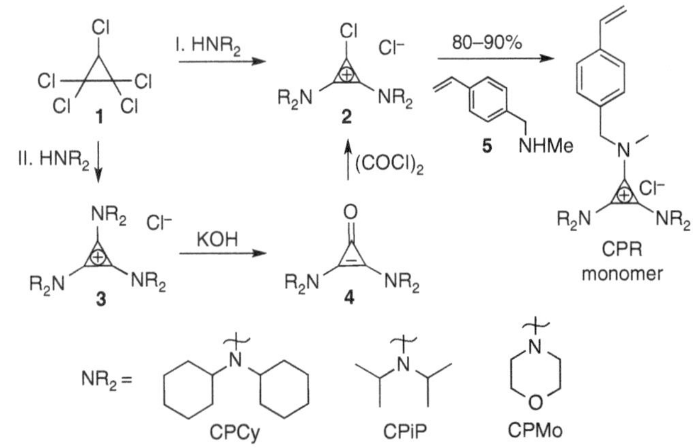


**Figure 2 | Synthesis of CP monomers for RAFT polymerisations.** The monomers **CPCy** and **CPiP** were synthesized by addition of dicyclohexylamine (92%) and diisopropylamine (85%) to **1**, followed by substitution of styrenic-type amine **5** under basic conditions (86% and 88%, respectively). **CPMo** was similarly synthesized. After addition of morpholine to **1**, subsequent hydrolysis and treatment with oxalyl chloride (42%, 3 steps), **5** was substituted to yield **CPMo** (59%).

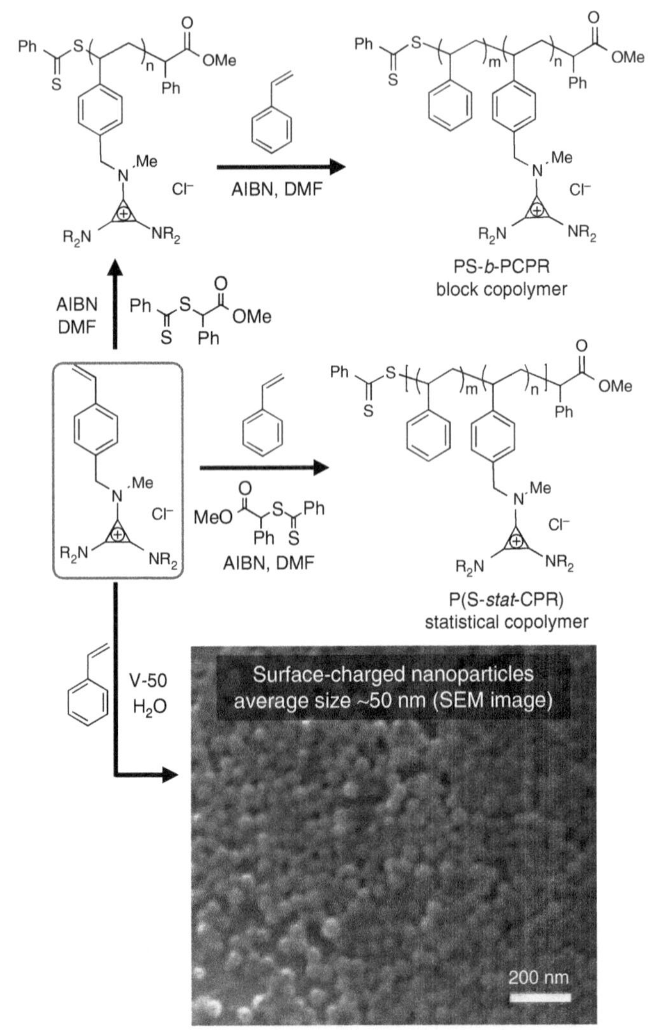


**Figure 3 | Synthesis of CP-containing polymers. CPR** is polymerized by RAFT yielding both homopolymers, **PCPR**, and statistical copolymers, **P(S-*stat*-CPR)**. **PCPR** is reacted further to form block copolymers **PS-*b*-PCPR** of varying styrene content. Nanoparticles are synthesized by surfactant-free emulsion polymerization with styrene using the water-soluble thermal initiator V-50.

As expected, a significant relationship was observed between the nature of the amino substituent and the physical properties of the resultant homopolymer. We observed that the decomposition temperature ($T_{dec}$), glass transition temperature ($T_g$) and solubility of the homopolymers varied as a function of substituent (Supplementary Table 1). Through characterization by thermogravimetric analysis we found that the $T_{dec}$ of the homopolymers increased as the amino substituents became less sterically hindered. Of particular note, **PCPMo** decomposed at 310 °C, which is comparable to the $T_{dec}$ of imidazolium-based polymers[43]. Differential scanning calorimetry was performed to identify the $T_g$ for the homopolymers, as ion-conducting membranes are frequently melt-processed. Both **PCPCy** and **PCPiP** have no observable $T_g$ before decomposition, but **PCPMo** exhibited a $T_g$ of 160 °C. Previous reports have revealed a connection between the nature of the counterion and the accessible temperature window; replacement of the chloride with an alternative, typically bulkier counterion is expected to increase the $T_{dec}$ while decreasing the $T_g$[19,43,44]. Importantly, however, our preliminary data clearly demonstrate a similar relationship between alkyl chain identity and observed $T_{dec}$ and $T_g$; thus, by modifying the CP substituents, we are able to significantly broaden the temperature window in which these materials are processable, without the need to adjust the counterion. In addition, solubility of **PCPR** homopolymers is highly dependent on amino substituents, again reflecting the influence of building block composition on macromolecular properties. Characterization of the homopolymers, including thermal data and solubility information, is summarized in Supplementary Table 1.

Statistical copolymers were readily synthesized by RAFT, using styrene and **CPR** monomers, **P(S-*stat*-CPR)**. When styrene was copolymerized with each monomer in a 1:1 mole ratio, we observed some disparities in the percent incorporation of functional CP monomers in the resulting copolymer. For **CPCy**, **CPiP** and **CPMo**, the degree of incorporation was ~50%, 48%, and 45%, respectively. Further detailed studies with other monomers are underway to uncover insights into reactivity ratios as a function of R-substituents.

Considering the ability to copolymerize styrene and **CPR** monomers, we sought to synthesise cationic nanoparticles based on the **CPiP** monomer via surfactant-free emulsion polymerization. Many traditional strategies rely on the use of surfactants or additional solvents[45] to obtain sub-100 nm cationic particles. By simply mixing styrene and **CPiP** at various weight percent values (1, 2.5, 5, 10 and 20% of **CPiP**) and using a thermally activated radical initiator (V-50), we were able to obtain particles ranging from 30 to 90 nm (as characterized by dynamic light scattering Supplementary Fig. 2). Higher loadings of **CPiP** compared with styrene resulted in smaller, albeit more disperse, particles. Figure 3 shows the scanning electron microscope image of nanoparticles made from 5% **CPiP**/95% styrene. The average diameter obtained by dynamic light scattering was found to be 50 nm. Furthermore, the particles from stable dipersions, since the zeta potential of the 5% **CPiP** nanoparticles was found to remain above 30 mV over the range of >10 pH units (Supplementary Fig. 2). As a control, particles synthesized with styrene only (without any surfactants or **CPR** monomers) were much larger and exhibited a bimodal size distribution (Supplementary Fig. 2). These data demonstrate that the **CPiP** monomer effectively stabilises oil-in-water droplets, and that the charge is present on the particle surface. A more detailed study of this behaviour will follow, including the incorporation of other **CPR** monomers into cationic nanoparticles. In general, the ability to make charged nanoparticles in a surfactant-free, large-scale process could have far-reaching potential towards interfacial additives and biological applications[46,47].

While the **CPR** monomers were successfully copolymerized to make statistical polymers and nanoparticles, we sought to augment our library of **CPR**-based macromolecules with block copolyelectrolytes (BCPEs). We synthesized block copolymers **PS-*b*-PCPR(CP mol%)** by growing styrene onto **PCPR** macro-chain transfer agents (macro-CTA). By varying the DP of the polystyrene block, we effectively controlled the different mole fractions of the CP functional block. We note that block copolymers **PS-*b*-PCPR** could also be obtained by the reverse process of growing the functional monomer **CPR** onto polystyrene macro-CTAs.

**Morphology of BCPEs.** After synthesizing block copolymers of various compositions, we characterized the morphology of bulk films comprising various **CPR**-building blocks. Recent studies suggest that nanostructured BCPEs have broad implications in materials chemistry, specifically for fuel cells and batteries, if they undergo microphase segregation. For example, Winey and Elabd recently reported that BCPEs with a lamellar morphology conduct ions more effectively than cationic homopolymers, as water and ions confined within nanochannels may accelerate transport[48]. Computational studies from Olvera de la Cruz and coworkers suggest this effect may be enhanced if the conducting path is a continuous, percolating structure[17]; a microphase segregated morphology of charged and neutral blocks observed in ion-containing block copolymers[49]. With this in mind, we characterized block copolymer samples by small-angle X-ray scattering (SAXS) and TEM to understand microphase segregation in **PCPR**-containing BCPEs.

In Fig. 4, we show SAXS[50] profiles of three representative diblock copolymers. The primary scattering peaks seen in each sample (indicated by filled triangles) is attributed to microphase separation. The scattering profile of PS-*b*-PCPMo(35) contains a higher-order peak at $q = 3q^*$. This suggests the presence of a

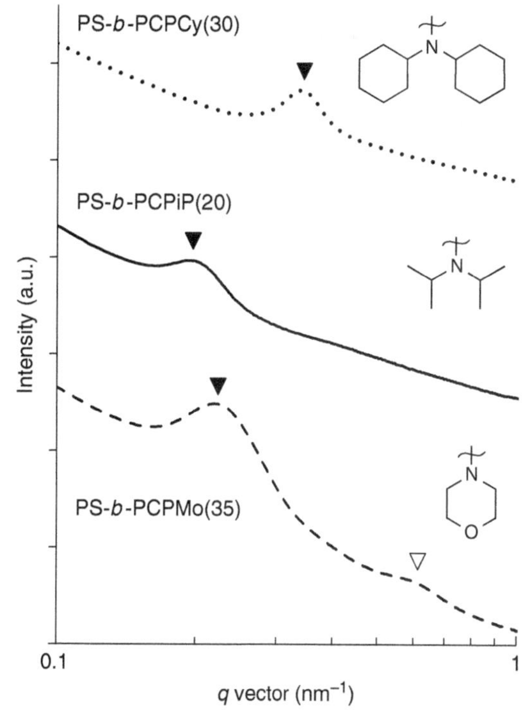


**Figure 4 | SAXS profiles of microphase separated diblock copolymers collected at 25 °C.** Scattering intensity is plotted as a function of the magnitude of the scattering vector, ***q***. Filled triangles represent the primary scattering peaks and the open triangles represent the higher-order scattering peaks.

symmetric lamellar phase. The scattering profiles of the other polymers contain only one peak, which indicates a lack of long-range order. The domain spacing, *d*, of the microphase separated diblock copolymers is calculated by the equation $d = 2\pi/q^*$. The domain spacing values corresponding to each diblock copolymer are given in Table 1. As expected, *d*-spacing increases with molecular mass and molar fraction of styrene. We next sought to probe the potential application of CP-based polyelectrolytes in electrochemical devices[11,17].

Ion conductivity experiments were performed on **PS-*b*-PCPiP(20)** using electrochemical impedance spectroscopy. As conductivity is closely related to morphology, we complemented our SAXS experiments with transmission electron microscopy (TEM). A **PS-*b*-PCPiP(20)** sample (drop cast, no annealing) was microtomed and imaged by TEM (Fig. 5a,b). Even without staining, we clearly observe microphase segregation (cylindrical morphology) (Supplementary Fig. 4). Staining with $RuO_4$ vapour for 2 min preferentially stains the cationic block and helps to visualize the internal structure. The electron micrographs obtained (Fig. 5a,b) show hexagonally packed cylinders in different orientations. The domain spacing by TEM was 29 nm, which is consistent with the domain spacing determined by SAXS (31 nm; see Table 1). The lighter colour of the cylinders with respect to the matrix in Fig. 5a,b indicates that PS cylinders are embedded in a **PCPiP** matrix. The stained **PCPiP** block scatters more electrons, and therefore appears darker by TEM (see Supplementary Fig. 4). The non-functionalised PS cylinders in Fig. 5a,b occupy a very large fraction of the image because the mole fraction of the functional block in this polymer is only 20%. The continuous nature of the conducting phase matrix observed in charged diblock copolymers is expected to facilitate ion transport[49], and is consistent with the ideal percolating structure found by computational modelling[17].

**Table 1 | Characterization of PS-*b*-PCPR (CP mol%) block copolymers.**

| Sample Name | MM/kg mol$^{-1}$ | CP% by DP | CP% by MM | SAXS domain spacing (nm) |
|---|---|---|---|---|
| **PS-*b*-PCPCy(45)** | 40 | 45% | 80% | 15 |
| **PS-*b*-PCPCy(30)** | 50 | 30% | 70% | 18 |
| **PS-*b*-PCPiP(50)** | 20 | 50% | 80% | 18 |
| **PS-*b*-PCPiP(30)** | 27 | 30% | 60% | 24 |
| **PS-*b*-PCPiP(20)** | 30 | 20% | 50% | 31 |
| **PS-*b*-PCPMo(50)** | 38 | 50% | 80% | 26 |
| **PS-*b*-PCPMo(35)** | 42 | 35% | 70% | 28 |

CP, cyclopropenium; DP, degree of polymerization; NMR, nuclear magnetic resonance; SAXS, small-angle X-ray scattering
Block copolymers were synthesized by the addition of styrene to the three homopolymers, and domain spacing was calculated by SAXS (Supplementary Fig. 3). Molecular mass (MM) was determined by $^1H$ NMR spectroscopy.

## Discussion

The in-plane conductivity, $\sigma$, of **PS-*b*-PCPiP(20)** equilibrated in humid air with 90% relative humidity (RH) was measured as a function of increasing temperature from 25 to 65 °C (Fig. 5c). To ensure equilibration, samples were initially annealed for 1 week at 90% RH at 25 °C and for 48 h at each subsequent temperature of interest. The straight line in Fig. 5c is the least-squares fit through the equilibrated conductivity data at each temperature value. In principle, the change in conductivity with temperature, shown in Fig. 5c, could either be due to changes in the mobility of chloride ions or to a change in ion concentration in the membrane. The fact that ion concentration in the membranes is constant indicates that the slope of the line provides an estimate

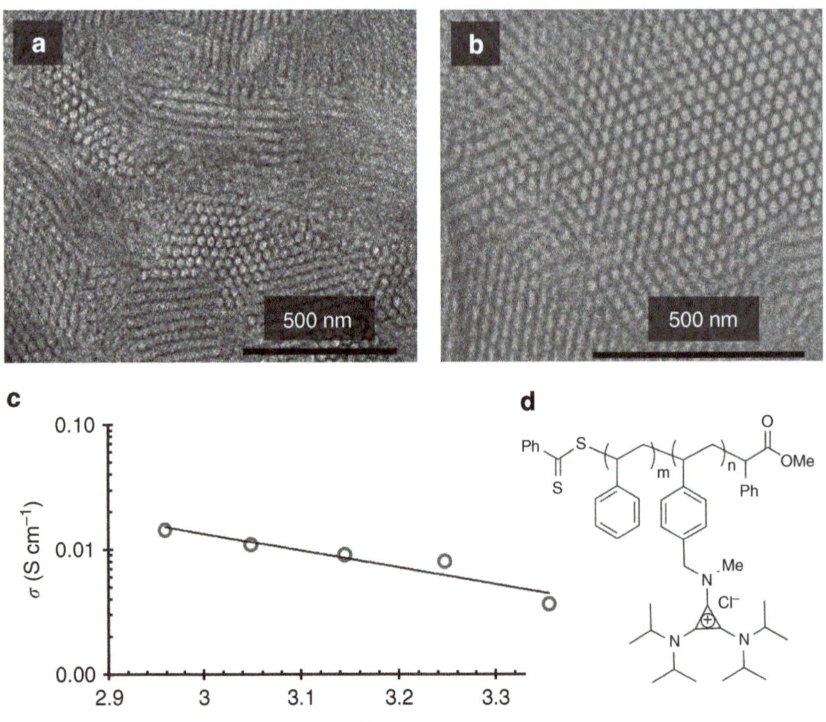


**Figure 5 | Morphology and ionic conductivity of bulk PS-*b*-PCPiP films.** (**a** and **b**) Two representative TEM images of **PS-*b*-PCPiP(20)** reveal a morphology of hexagonally packed cylinders (*d*-spacing = 29 nm; the light colour corresponds to PS). (**c**) Ionic conductivity as a function of inverse temperature, from 25 to 65 °C, for (**d**) **PS-*b*-PCPiP(20)** (ion-exchange capacity, IEC = 1.3, at 90% RH).

of the activation energy for transport of chloride ions through the membrane (Arrhenius law). The estimated activation energy for this system is 25 kJ mol$^{-1}$. This value is comparable to that reported previously for the imidazolium-containing diblock copolymer analogue in water, poly(styrene-*b*-4-vinylbenzyltrimethylimidazolium chloride) (PS-*b*-PIm(35)), 27 kJ mol$^{-1}$ (ref. 43). However, at room temperature, the conductivity of the PS-*b*-PCPiP(20) polymer (ion-exchange capacity, IEC = 1.3 meq g$^{-1}$) is rather high, 0.004 S cm$^{-1}$, considering the low water uptake, $\lambda_w = 7$, of this membrane ($\lambda_w$ is the number of water molecules per chloride ion in the membrane). This value of $\lambda_w$ is four times lower than the value obtained for PS-*b*-PIm(35) immersed in water, for the same conductivity ($\lambda_w = 30$, $\sigma = 0.004$ S cm$^{-1}$), and higher IEC (2.1 meq g$^{-1}$)[43]. These results indicate that the CP-based polyelectrolytes conduct ions more effectively than the optimized membranes from imidazolium-containing polymers, with a minimum amount of water present. Further tuning of the functional groups, backbone structure and morphology is expected to result in polyelectrolytes with exceptionally high ion conductivities[51–54].

In conclusion, we have introduced a new family of electron-rich cationic polyelectrolytes based on the CP ion building block. The robust, efficient and orthogonal chemistry to synthesise the monomers provides facile access to a variety of polymers by RAFT and to well-defined cationic nanoparticles by surfactant-free emulsion polymerization. The nanoparticles exhibit high charge density on the surface and stability over a wide range of pH values. The family of linear polymers is characterized by widely variable physical properties, which are highly dependent on the amino substituents flanking the aromatic cation. Through TEM and SAXS measurements, we observed microphase segregation in bulk samples of the BCPE. In diblock copolymers, the domain spacing increased with increasing styrene content (the length of the functional block was fixed). Compared with imidazolium analogues, CP-based BCPEss show superior properties as ion conductive materials, and we anticipate that further optimization will lead to improved performance. Future studies will be aimed at probing the structure–property relationships of these materials by expanding our **PCPR** library through adjustment of the polymer backbone, modular functional groups, block copolymer composition and CP counterion. Moreover, other **CPR** monomers will be incorporated into nanoparticles via the one-pot emulsion polymerization to assess their efficacy in various biomedical applications and membrane technologies. With such modularity, this new class of CP-based polyelectrolytes offers a wealth of functionality that translates to significant potential across a broad array of applications.

## Methods

**Synthesis of *N*-methyl-1-(2,3-bis(dicyclohexylamino)cyclopropenium)-4-vinylbenzylamine chloride (CPCy).** To a dry round bottom flask of 2,3-bis(dicyclohexylamino)-1-chlorocyclopropenium chloride (22.1 g, 47.3 mmol, 1.0 equiv) under argon was added $CH_2Cl_2$ (150 ml) and triethylamine (6.54 ml, 47.3 mmol, 1.0 equiv). *N*-Methyl-4-vinylbenzylamine (7.74 g, 47.3 mmol and 1.0 equiv) was then slowly added to the solution and the reaction was stirred overnight. $CH_2Cl_2$ (700 ml) was added and the mixture was washed with 1 M HCl (3 × 200 ml) and brine (1 × 200 ml), dried with anhydrous sodium sulfate and concentrated *in vacuo* to yield a crude solid. The crude product was purified with silica gel chromatography (EtOAc then 5% iPrOH in $CH_2Cl_2$) to yield an off-white solid (23.5 g, 40.6 mmol, 86% yield). Monomers were characterized by $^1$H and $^{13}$C NMR (see Supplementary Figs 5–10), and all new compounds were characterized by NMR and mass spectrometry (see Supplementary Notes and Methods).

**Sample synthesis of PCPR: synthesis of PCPCy.** To a dry 20 ml (see Supplementary Notes and Methods for details) scintillation vial, **CPCy** (6.0 g, 10.4 mmol, 60.0 equiv), methyl 2-(phenylcarbonothioylthio)-2-phenylacetate (52.3 mg, 1.73 mmol, 1.0 equiv), AIBN (4.3 mg, 0.26 mmol, 0.15 equiv) and DMF (6.0 ml) were added and vortexed to form a homogenous solution. This solution was transferred to a flame-dried ampule. After 4 freeze-pump-thaw cycles, the ampule was sealed under vacuum. The polymerization was run for 12 h at 80 °C with vigorous stirring. The reaction mixture was precipitated from $CH_2Cl_2$ into 1,4-dioxane three times to remove monomer. The polymer was then precipitated an additional three times into hexanes to remove residual 1,4-dioxane. Drying *in vacuo* yielded the pure polymer as a pink powder (5.3 g, 88% yield). Homopolymers were characterized by $^1$H NMR (see Supplementary Figs 11–13).

**Small-angle X-ray scattering.** Thick polymer samples (1 mm) were prepared by pressing the powder into a teflon washer. Synchrotron SAXS measurements were performed using the 7.3.3 beamline at the Advanced Light Source (ALS, Lawrence Berkeley National Laboratory). The wavelength $\lambda$ of the incident X-ray beam was 0.124 nm ($\Delta\lambda/\lambda = 10^{-4}$) and a sample-to-detector distance of 4 m. The resulting two-dimensional scattering data were averaged azimuthally to obtain intensity versus magnitude of the scattering wave vector $\boldsymbol{q}$ ($\boldsymbol{q} = 4\pi \sin(\theta/2)/\lambda$, where $\theta$ is the scattering angle). All of the scattering profiles were azimuthally symmetric. The scattering data were corrected for the detector dark current and the scattering from air and Kapton windows.

**Electrochemical impedance spectroscopy.** In-plane chloride conductivity of hydrated membranes with dimensions 2 cm × 1 cm × 450 μm was measured by AC impedance spectroscopy using platinum electrodes in the standard four probe configuration using a BekkTech sample clamp. Polymer films of PS-b-PCPiP(20) were prepared by drop casting a 100 mg ml$^{-1}$ solution of polymer onto a clean Teflon substrate. In-plane chloride conductivity of a hydrated membrane composed of **PS-*b*-PCPiP(20)** (calculated molecular mass = 31 KDa, DP = 174) with dimensions 2 cm × 1 cm × 450 μm was measured by AC impedance spectroscopy using platinum electrodes in the standard four probe configuration using a BekkTech sample clamp. Conductivities were collected under humidified conditions, and temperature and RH were controlled by an environmental chamber (Qualitest). Data were collected using 10 mV amplitude over a frequency range of 1 Hz–10 MHz. Separate experiments were conducted to ensure that the response of the sample was linear in this window. Samples were annealed at the temperature of interest for 24–48 h until the measured impedance did not change. Conductivity, $\sigma$, is given by equation (1):

$$\sigma = \frac{w}{rS} \quad (1)$$

where $S$ is the cross-sectional area of sample film, $r$ is the intercept of the Nyquist semi-circle on the real axis (Ω) and $w$ is the distance between the inner platinum electrodes.

**Transmission electron microscopy.** Films of **PS-*b*-PCPiP(20)** (calculated molecular mass = 33.4 KDa, DP = 200) were prepared by drop casting a 100 mg ml$^{-1}$ solution of polymer onto a clean Teflon substrate. After allowing to dry for 24 h, the film was sectioned with Leica UltraCut 6 ultramicrotome at −40 °C, nominal thickness 70 nm using a Diatome Cryo 35° diamond knife. Sections were placed on 300 mesh copper grids with homemade lacey carbon film on top. The sections were stained with RuO4 vapour for 2 min, which preferentially stained the **PCPiP** block. Sections were imaged with FEI Tecnai F20 TEM operated at 200 kV. Images were analysed using ImageJ 1.48v software.

## References


1. Hawker, C. J. & Wooley, K. L. The convergence of synthetic organic and polymer chemistries. *Science* **309**, 1200–1205 (2005).
2. Hunt, J. N. *et al.* Tunable, High modulus hydrogels driven by ionic coacervation. *Adv. Mater.* **23**, 2327–2331 (2011).
3. Leibfarth, F. A. *et al.* A facile route to ketene-functionalized polymers for general materials applications. *Nat. Chem.* **2**, 207–212 (2010).
4. Lodge, T. P. A unique platform for materials design. *Science* **321**, 50–51 (2008).
5. Gao, R., Wang, D., Heflin, J. R. & Long, T. E. Imidazolium sulfonate-containing pentablock copolymer-ionic liquid membranes for electroactive actuators. *J. Mater. Chem.* **22**, 13473–13476 (2012).
6. Hallinan, D. T. & Balsara, N. P. Polymer electrolytes. *Annu. Rev. Mater. Res.* **43**, 503–525 (2013).
7. De Smedt, S., Demeester, J. & Hennink, W. Cationic polymer based gene delivery systems. *Pharm. Res.* **17**, 113–126 (2000).
8. Samal, S. K. *et al.* Cationic polymers and their therapeutic potential. *Chem. Soc. Rev.* **41**, 7147–7194 (2012).
9. Pan, J., Chen, C., Zhuang, L. & Lu, J. Designing advanced alkaline polymer electrolytes for fuel cell applications. *Acc. Chem. Res.* **45**, 473–481 (2011).
10. Hickner, M. A., Herring, A. M. & Coughlin, E. B. Anion exchange membranes: current status and moving forward. *J. Polym. Sci. B Polym. Phys.* **51**, 1727–1735 (2013).
11. Chen, Y. *et al.* Enhancement of anhydrous proton transport by supramolecular nanochannels in comb polymers. *Nat. Chem.* **2**, 503–508 (2010).

12. Elimelech, M. & Phillip, W. A. The future of seawater desalination: energy, technology, and the environment. *Science* **333,** 712–717 (2011).
13. Gin, D. L. & Noble, R. D. Designing the next generation of chemical separation membranes. *Science* **332,** 674–676 (2011).
14. Jangu, C. & Long, T. E. Phosphonium cation-containing polymers: from ionic liquids to polyelectrolytes. *Polymer* **55,** 3298–3304 (2014).
15. Qiu, B., Lin, B., Qiu, L. & Yan, F. Alkaline imidazolium- and quaternary ammonium-functionalized anion exchange membranes for alkaline fuel cell applications. *J. Mater. Chem.* **22,** 1040–1045 (2012).
16. Yuan, J., Mecerreyes, D. & Antonietti, M. Poly(ionic liquid)s: an update. *Prog. Polym. Sci.* **38,** 1009–1036 (2013).
17. Sing, C. E., Zwanikken, J. W. & Olvera de la Cruz, M. Electrostatic control of block copolymer morphology. *Nat. Mater.* **13,** 694–698 (2014).
18. Steele, B. C. H. & Heinzel, A. Materials for fuel-cell technologies. *Nature* **414,** 345–352 (2001).
19. Curnow, O. J., MacFarlane, D. R. & Walst, K. J. Triaminocyclopropenium salts as ionic liquids. *Chem. Commun.* **47,** 10248–10250 (2011).
20. Hückel, E. Grundzüge der Theorie ungesättiger und aromatischer Verbindungen. *Z. Physik* **70,** 77–85 (1938).
21. Breslow, R. Synthesis of the s-triphenylcyclopropenyl cation. *J. Am. Chem. Soc.* **79,** 5318–5318 (1957).
22. Bandar, J. S. & Lambert, T. H. Aminocyclopropenium ions: synthesis, properties, and applications. *Synthesis* **45,** 2485–2498 (2013).
23. Yoshida, Z. & Tawara, Y. Aminocyclopropenium ion. *J. Am. Chem. Soc.* **93,** 2573–2574 (1971).
24. Yoshida, Z.-I., Tawara, Y., Hirota, S. & Ogoshi, H. Vibrational spectrum of trisdimethylaminocyclopropenium perchlorate. *Bull. Chem. Soc. Jpn* **47,** 797–800 (1974).
25. Kerber, R. C. & Hsu, C.-M. Substituent effects on cyclopropenium ions. *J. Am. Chem. Soc.* **95,** 3239–3245 (1973).
26. Huddleston, J. G. *et al.* Characterization and comparison of hydrophilic and hydrophobic room temperature ionic liquids incorporating the imidazolium cation. *Green Chem.* **3,** 156–164 (2001).
27. Bruns, H. *et al.* Synthesis and coordination properties of nitrogen(i)-based ligands. *Angew. Chem. Int. Ed.* **49,** 3680–3683 (2010).
28. Bandar, J. S. & Lambert, T. H. Enantioselective brønsted base catalysis with chiral cyclopropenimines. *J. Am. Chem. Soc.* **134,** 5552–5555 (2012).
29. Bandar, J. S. & Lambert, T. H. Cyclopropenimine-catalyzed enantioselective mannich reactions of tert-butyl glycinates with n-boc-imines. *J. Am. Chem. Soc.* **135,** 11799–11802 (2013).
30. Wilde, M. M. D. & Gravel, M. Bis(amino)cyclopropenylidenes as organocatalysts for acyl anion and extended umpolung reactions. *Angew. Chem. Int. Ed.* **52,** 12651–12654 (2013).
31. Weidner, C. H. & Long, T. E. Synthesis and characterization of 3-aryl-2-(polystyryl)cyclopropenones via cyclopropenium ion substitution on polystyrene. *J. Polym. Chem. A Polym. Chem.* **33,** 1–6 (1995).
32. Hawker, C. J., Bosman, A. W. & Harth, E. New polymer synthesis by nitroxide mediated living radical polymerizations. *Chem. Rev.* **101,** 3661–3688 (2001).
33. Matyjaszewski, K. & Xia, J. Atom transfer radical polymerization. *Chem. Rev.* **101,** 2921–2990 (2001).
34. Chiefari, J. *et al.* Living free-radical polymerization by reversible addition-fragmentation chain transfer: the RAFT process. *Macromolecules* **31,** 5559–5562 (1998).
35. Peart, P. A. & Tovar, J. D. Poly(cyclopropenone)s: formal inclusion of the smallest Hückel aromatic into π-conjugated polymers. *J. Org. Chem.* **75,** 5689–5696 (2010).
36. Curnow, O. J., Holmes, M. T., Ratten, L. C., Walst, K. J. & Yunis, R. A facile route to functionalised, protic and chiral ionic liquids based on the triaminocyclopropenium cation. *RSC Adv.* **2,** 10794–10797 (2012).
37. Kolb, H. C., Finn, M. G. & Sharpless, K. B. Click chemistry: diverse chemical function from a few good reactions. *Angew. Chem., Int. Ed.* **40,** 2004–2021 (2001).
38. Tobey, S. W. & West, R. Pentachlorocyclopropane. *J. Am. Chem. Soc.* **88,** 2478–2481 (1966).
39. Hemp, S. T. *et al.* Comparing ammonium and phosphonium polymerized ionic liquids: thermal analysis, conductivity, and morphology. *Macromol. Chem. Phys.* **214,** 2099–2107 (2013).
40. Yuan, J. & Antonietti, M. Poly(ionic liquid)s: polymers expanding classical property profiles. *Polymer* **52,** 1469–1482 (2011).
41. Wang, R. & Lowe, A. B. RAFT polymerization of styrenic-based phosphonium monomers and a new family of well-defined statistical and block polyampholytes. *J. Polym. Sci. A Polym. Chem.* **45,** 2468–2483 (2007).
42. Texter, J. Anion responsive imidazolium-based polymers. *Macromol. Rapid Commun.* **33,** 1996–2014 (2012).
43. Sudre, G., Inceoglu, S., Cotanda, P. & Balsara, N. P. Influence of bound ion on the morphology and conductivity of anion-conducting block copolymers. *Macromolecules* **46,** 1519–1527 (2013).
44. Weber, R. L. *et al.* Thermal and ion transport properties of hydrophilic and hydrophobic polymerized styrenic imidazolium ionic liquids. *J. Polym. Sci. B Polym. Phys* **49,** 1287–1296 (2011).
45. Ramos, J., Forcada, J. & Hidalgo-Alvarez, R. Cationic polymer nanoparticles and nanogels: from synthesis to biotechnological applications. *Chem. Rev.* **114,** 367–428 (2013).
46. Rothberg, J. M. *et al.* An integrated semiconductor device enabling non-optical genome sequencing. *Nature* **475,** 348–352 (2011).
47. Nederberg, F. *et al.* Biodegradable nanostructures with selective lysis of microbial membranes. *Nat. Chem.* **3,** 409–414 (2011).
48. Ye, Y., Sharick, S., Davis, E. M., Winey, K. I. & Elabd, Y. A. High hydroxide conductivity in polymerized ionic liquid block copolymers. *ACS Macro Lett.* **2,** 575–580 (2013).
49. Park, M. J. *et al.* Increased water retention in polymer electrolyte membranes at elevated temperatures assisted by capillary condensation. *Nano Lett.* **7,** 3547–3552 (2007).
50. Alexander, H. *et al.* A SAXS/WAXS/GISAXS beamline with multilayer monochromator. *J. Phys. Conf. Ser* **247,** 012007 (2010).
51. Schmidt-Rohr, K. & Chen, Q. Parallel cylindrical water nanochannels in Nafion fuel-cell membranes. *Nat. Mater.* **7,** 75–83 (2008).
52. Choi, J.-H., Ye, Y., Elabd, Y. A. & Winey, K. I. Network structure and strong microphase separation for high ion conductivity in polymerized ionic liquid block copolymers. *Macromolecules* **46,** 5290–5300 (2013).
53. Hoarfrost, M. L., Tyagi, M. S., Segalman, R. A. & Reimer, J. A. Effect of confinement on proton transport mechanisms in block copolymer/ionic liquid membranes. *Macromolecules* **45,** 3112–3120 (2012).
54. Ye, Y., Choi, J.-H., Winey, K. I. & Elabd, Y. A. Polymerized ionic liquid block and random copolymers: effect of weak microphase separation on ion transport. *Macromolecules* **45,** 7027–7035 (2012).


## Acknowledgements

This work was supported in part by the National Science Foundation (CAREER, DMR-1351293 to L.M.C. and CHE-0953259 to T.H.L.), ACS Petroleum Research Fund (54471-DNI7 to L.M.C.), and 3M Corporation (to L.M.C.). Y.J. thanks the Columbia Amgen Scholars Program and the Columbia Science Research Fellows. J.S.B is grateful for NDSEG and NSF fellowships. K.L.K. thanks the Army Basic Research Program and ECBC for financial support. Transmission electron micrographs were collected at NYSBC, and were made possible by a grant from NYSTAR. The investigation was conducted in a facility constructed with support from Research Facilities Improvement Program Grant number C06 RR017528-01-CEM from the National Center for Research Resources, National Institutes of Health. This material is in part based upon work performed by the Joint Center for Artificial Photosynthesis, a DOE Energy Innovation Hub, supported through the Office of Science of the U.S. Department of Energy under Award Number DE-SC0004993. SAXS experiments were performed at Lawrence Berkeley National Laboratory's Advance Light Source, Beamline 7.3.3. Beamline 7.3.3 of the Advanced Light Source is supported by the Director of the Office of Science, Office of Basic Energy Sciences, of the U.S. Department of Energy under Contract No. DE-AC02-05CH11231. We thank Dr Jos Paulusse for performing size exclusion chromatography on the homopolymers. We also thank Qi Jiang for performing thermogravimetric analysis on the homopolymers.

## Author contributions

All authors contributed to the conception and execution of experiments. Materials were synthesized by J.L.F., Y.J., S.D.B. and K.L.K. Ionic conductivity measurements and X-ray scattering were performed by P.C. All authors contributed to data analysis. The paper was written by J.L.F., Y.J. and L.C., with input from all authors.

## Additional information

**Competing financial interests:** The authors declare no competing financial interests.

# 17

# Pd(II)-catalysed *meta*-C–H functionalizations of benzoic acid derivatives

Shangda Li[1,*], Lei Cai[1,*], Huafang Ji[1], Long Yang[1] & Gang Li[1,2]

Benzoic acids are highly important structural motifs in drug molecules and natural products. Selective C–H bond functionalization of benzoic acids will provide synthetically useful tools for step-economical organic synthesis. Although direct *ortho*-C–H functionalizations of benzoic acids or their derivatives have been intensely studied, the ability to activate *meta*-C–H bond of benzoic acids or their derivatives in a general manner via transition-metal catalysis has been largely unsuccessful. Although chelation-assisted *meta*-C–H functionalization of electron-rich arenes was reported, chelation-assisted *meta*-C–H activation of electron-poor arenes such as benzoic acid derivatives remains a formidable challenge. Herein, we report a general protocol for *meta*-C–H olefination of benzoic acid derivatives using a nitrile-based sulfonamide template. A broad range of benzoic acid derivatives are *meta*-selectively olefinated using molecular oxygen as the terminal oxidant. The *meta*-C–H acetoxylation, product of which is further transformed at the *meta*-position, is also reported.

[1] State Key Laboratory of Structural Chemistry, Fujian Institute of Research on the Structure of Matter, Chinese Academy of Sciences, Fujian 350002, China.
[2] Key Laboratory of Coal to Ethylene Glycol and Its Related Technology, CAS, Fujian 350002, China. * These authors contributed equally to this work. Correspondence and requests for materials should be addressed to G.L. (email: gangli@fjirsm.ac.cn).

Benzoic acids are highly valuable structural motifs and precursors for synthesizing other organic substances. Direct and selective transformation of the C–H bonds of this class of compounds would be very attractive for developing step-economical organic syntheses (Fig. 1)[1–14]. To date, direct *ortho*-C–H functionalizations of benzoic acids or their derivatives have been intensely studied with transition-metal catalysts[1–12] or by directed *ortho* metalation[13]. Traditionally, the *meta*-position of benzoic acids were functionalized by electrophilic aromatic substitution, which often requires harsh conditions as benzoic acids are generally deactivated towards this reaction[14]. Transition-metal catalysts have been used to functionalize *meta*-C–H bonds of limited benzoic acid derivatives. Hartwig, Yu, Sanford and others have achieved modest to high selectivity because of substrate steric or electronic control[15–25]. In prior examples, an excess of the benzoic acid derivative, which was often used as the solvent, was generally required with a Pd catalyst (Fig. 1a)[17–18,21]. Despite these highly important pioneering studies, the ability to activate the *meta*-C–H bond of benzoic acids or their derivatives in a general manner via transition-metal catalysis has been largely unsuccessful[25]. Therefore, a general approach to *meta*-C–H functionalization of benzoic acids or their derivatives regardless of the substitution patterns is highly desirable to provide synthetic short-cuts.

Controlling the site-selectivity of C–H activation reactions is an outstanding challenge in the development syntheticially useful C–H functinalization methodology[26–27]. So far only a limited number of approaches are available for addressing *meta*-C–H functionalizations of arenes[28–63]. These approaches include inherent substrate control via steric and/or electronic factors[28–35], chelating group-assisted Cu(II)-catalysed arylation[36–37], ruthenium(II) complex facilitated *meta*-C–H functionalizations[38–43], the use of transient norbornene mediator[44–45] and formal *meta*-C–H functionalizations utilizing traceless directing groups[46–51]. Another unique method is the use of nitrile-based templates for *meta*-C–H functionalizations of electron-rich arenes, such as hydrocinnamic acids and phenylacetic acids, which was pioneered by the group of Yu[54–63]. However, chelation-assisted *meta*-C–H activation of electron-poor arenes such as benzoic acid derivatives remains a formidable challenge, possibly due to the low reactivity of the electron-poor arenes towards palladation in this type of C–H activation.

Herein, we disclose our discovery of a recyclable nitrile-based sulfonamide template that promotes the olefination and acetoxylation of *meta*-C–H bonds of a broad range of benzoic acid derivatives (Fig. 1b). Notably, a protocol is developed that enables the use of environmentally benign molecular oxygen as the terminal oxidant for chelation-assisted *meta*-C–H olefination, which previously required the use of costly silver salt oxidants.

## Results

**Development of *meta*-C–H olefination reaction conditions.** As benzoic acids are both electronically and structurally distinct from other electron-rich arenes that undergo template-assisted *meta*-C–H functinalizations, the elaboration of a compatible template-directing group is required to accommodate their unique properties. After investigating several newly designed templates (see Supplementary Table 1 for details), we found amide **1a** bearing a highly electron-withdrawing nosyl group was the most promising substrate. After **1a** was subjected to the similar reaction conditions that were developed by us for *meta*-C–H olefination[64–67] of phenylethylamines[63], much to our delight, excellent yields of desired products were obtained with

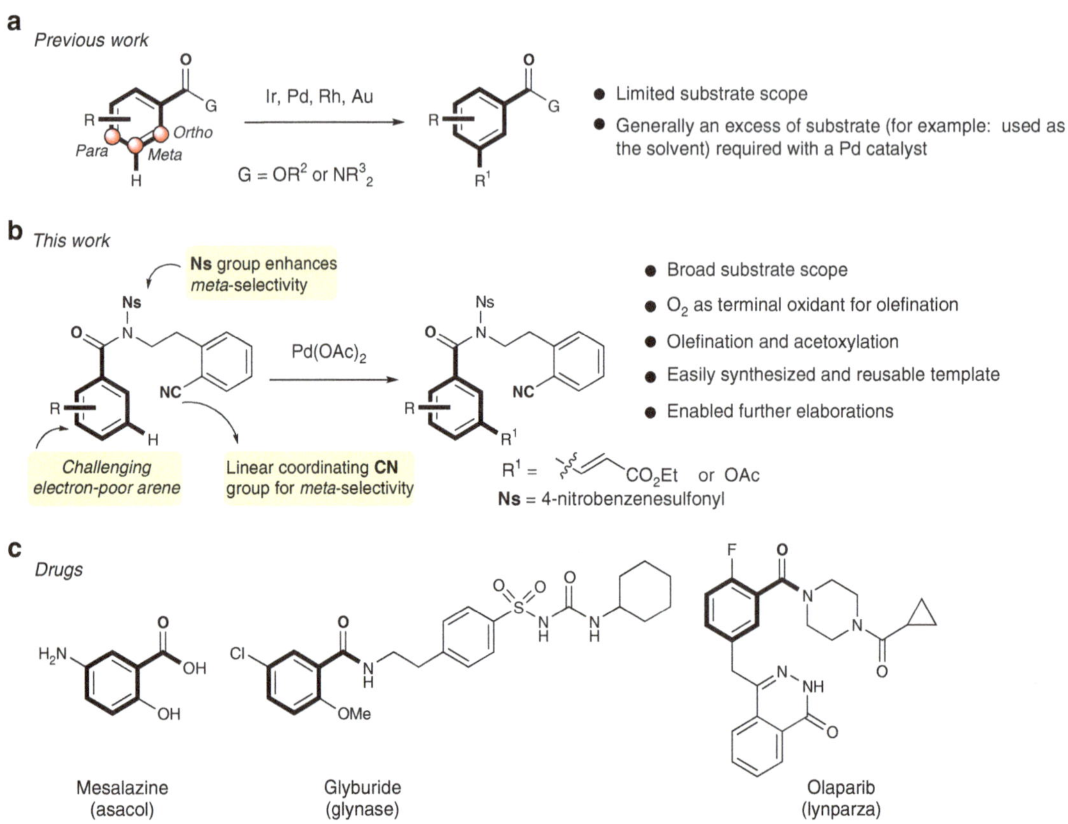


**Figure 1 | Transition-metal-catalysed *meta*-C–H functionalizations of benzoic acid derivatives.** (**a**) Previous reports on *meta*-C–H functionalizations of limited benzoic acid derivatives. (**b**) Our design of template for *meta*-C–H functionalizations of electron-poor benzoic acid derivatives and the highlights of our work. (**c**) Representative drugs of benzoic acid derivatives with *meta*-substituents.

**Table 1 | Optimization of reaction conditions.**

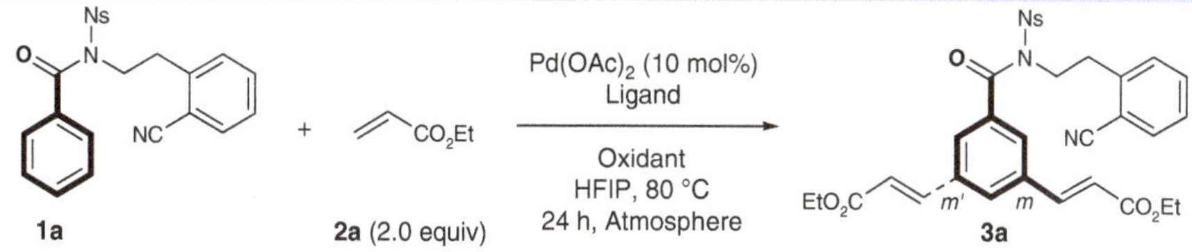


| Entry | Ligand (equiv) | Oxidant (equiv) | Atmosphere (1 atm) | Yield (%; mono/di) |
|---|---|---|---|---|
| 1* | Ac-Gly-OH (0.2) | AgOAc (3) | Air | 90 (1.5/1)† |
| 2 | Ac-Gly-OH (0.2) | $Cu(OAc)_2$ (2) | Air | 61 (5.8/1) |
| 3 | Ac-Gly-OH (0.2) | $CuBr_2$ (2) | Air | —‡ |
| 4 | Ac-Gly-OH (0.2) | $MnO_2$ (3) | Air | 17 (1/—) |
| 5 | Ac-Gly-OH (0.2) | BQ (2) | Air | —‡ |
| 6 | Ac-Gly-OH (0.2) | $Cu(OAc)_2$ (2) | $O_2$ | 72 (5/1) |
| 7 | Ac-Gly-OH (0.4) | $Cu(OAc)_2$ (2) | $O_2$ | 83 (2.3/1) |
| 8 | Ac-Gly-OH (0.6) | $Cu(OAc)_2$ (2) | $O_2$ | 96 (1/1.1) |
| 9 | Ac-Gly-OH (0.6) | $Cu(OAc)_2$ (1) | $O_2$ | 94 (1.1/1) |
| 10 | Ac-Gly-OH (0.6) | $Cu(OAc)_2$ (0.5) | $O_2$ | 83 (1/3.6) |
| 11*,§ | Ac-Gly-OH (0.6) | $Cu(OAc)_2$ (0.2) | $O_2$ | 68 (4.7/1) |
| 12 | Ac-Gly-OH (0.6) | — | $O_2$ | 12 (1/—) |
| **13** *,§,‖ | **Ac-Gly-OH (0.6)** | **$Cu(OAc)_2$ (0.5)** | **$O_2$** | **92 (1.1/1)**† |
| 14¶ | Formyl-Gly-OH (0.6) | $Cu(OAc)_2$ (0.2) | $O_2$ | 72 (5/1)† |
| **15**# | **Formyl-Gly-OH (0.6)** | **$Cu(OAc)_2$ (0.2)** | **$O_2$** | **78 (3.9/1)**† |
| 16 | Formyl-Gly-OH (0.6) | $Cu(OAc)_2$ (0.2) | $O_2$ | 78 (1/3.3) |

HFIP, hexafluoro-2-propanol; Ns (nosyl group), 4-nitrobenzenesulfonyl; Gly, glycine; BQ, 1,4-Benzoquinone.
Reaction conditions: **1a** (0.1 mmol), **2a** (0.2 mmol), $Pd(OAc)_2$ (10 mol%), Ac-Gly-OH (20–60 mol%), oxidant (0.2–3.0 equiv), HFIP (1 ml), 24 h, 80 °C, air or $O_2$. Yield was determined by $^1H$ NMR with $CH_2Br_2$ as internal standard. (mono/di) denotes the ratio of mono-*meta*-olefininated and di-(*meta,meta'*)-olefinated products. The bold 'entry 13' represents the optimal conditions for obtaining highest overall yield. (Note: (a) Although the mono versus di-olefination selectivity is not good, the isolated overall yield is best. (b) Although entry 8 led to higher overall yield, it does not use catalytic amount of $Cu(OAc)_2$). The bold 'entry 15' represents the conditions that result in highest overall yield with relatively good mono versus di-olefination selectivity.
****1a** (0.2 mmol scale).
†Isolated yield.
‡No reaction.
§The reaction was run for 48 h.
‖The reaction was run at 90 °C.
¶$K_2HPO_4$ (0.5 equiv) was added.
#$KH_2PO_4$ (0.5 equiv) was added.

tiny traces of other olefinated isomers (Table 1, entry 1). To avoid using the costly silver acetate as the oxidant, we continued to optimize the reaction conditions to search for a less costly oxidant. After screening a few inorganic as well as organic oxidants (entries 2–5), we were very surprised to find $Cu(OAc)_2$, which might compete with $Pd(OAc)_2$ in coordination with the weakly coordinating nitrile group, could be used as an effective oxidant (entry 2). The yield was increased to 72% when the reaction was run under oxygen atmosphere (entry 6)[10,68,69]. Notably, better results were obtained by increasing the loading of Ac-Gly-OH ligand (entries 7–8), leading to almost full conversion of the substrate with 60 mol% of the ligand (entry 8). By further tuning the loading of $Cu(OAc)_2$, the reaction time and temperature (entries 9–13), the best result was achieved in 48 h at 90 °C using oxygen as the terminal oxidant with catalytic amount of $Cu(OAc)_2$ (entry 13). It was found that the reaction was almost shut down without adding $Cu(OAc)_2$ as co-oxidant (entry 12). The use of oxygen is notable as all the previous chelation-assisted *meta*-C–H olefination reactions required the use of silver salt as the oxidant[54–63]. Finally, the reaction conditions were carefully tuned to improve the mono versus di-olefination selectivity (see also Supplementary Table 1 for details), although the two products could be easily separated. Pleasingly, the use of Formyl-Gly-OH ligand with an inorganic base would result in good mono versus di-olefination selectivities (entries 14–15)[59], albeit in lower overall yield. Interestingly, this selectivity was switched when no inorganic base was added (entry 16).

**Substrate scope of *meta*-C–H olefination.** With the optimized conditions in hand, we carried out olefinations on a variety of benzoic acid derivatives, which were easily prepared in one step from benzoic acids using routine conditions (see Supplementary Methods for details). It was found that both electron-donating and electron-withdrawing *ortho*-substituents were well tolerated (Fig. 2a, **3b**–**3e**), and a good yield of mono-olefinated product was obtained with substrate **1b**. Di-olefination occurred predominantly with less hindered amides **1c**–**1e**. Notably, our method provided a direct access to an aspirin derivative (**3e**). The reaction also proceeded smoothly with *meta*-substituted substrates (**3f**–**3h**). *Para*-substituted benzamides with electron-donating methoxy (**3i**) and methyl (**3j**) groups as well as electron-withdrawing fluoro (**3k**), chloro (**3l**) and bromo (**3m**) groups were all suitable substrates, affording mono-olefinated products selectively. Interesting, the bromo group (**3m**) was tolerated in our protocol, which is synthetically useful for further elaborations of the product. It should be noted di-olefination products with high overall yields could also be produced when $KH_2PO_4$ was used as the base for substrates **1j**–**1l** (see Supplementary Methods for details). Importantly, the method was compatible with a range of substrates carrying two substituents (**3n**–**3w**), generally producing high yields of desired products. It is surprising that tri-substituted substrates were also able to afford desired products in moderate to high yields (**3x**–**3z**). Such highly substituted patterns were not observed in all previous transition-metal catalysed *meta*-C–H functionalizations. Desired products were also generated with other electron-deficient olefin-coupling partners (**3fa**–**3fe**), although production of **3fc** and **3fd** required silver acetate as the oxidant since the standard conditions only afforded low yields of these two products. The *meta*-selectivity was generally excellent, although

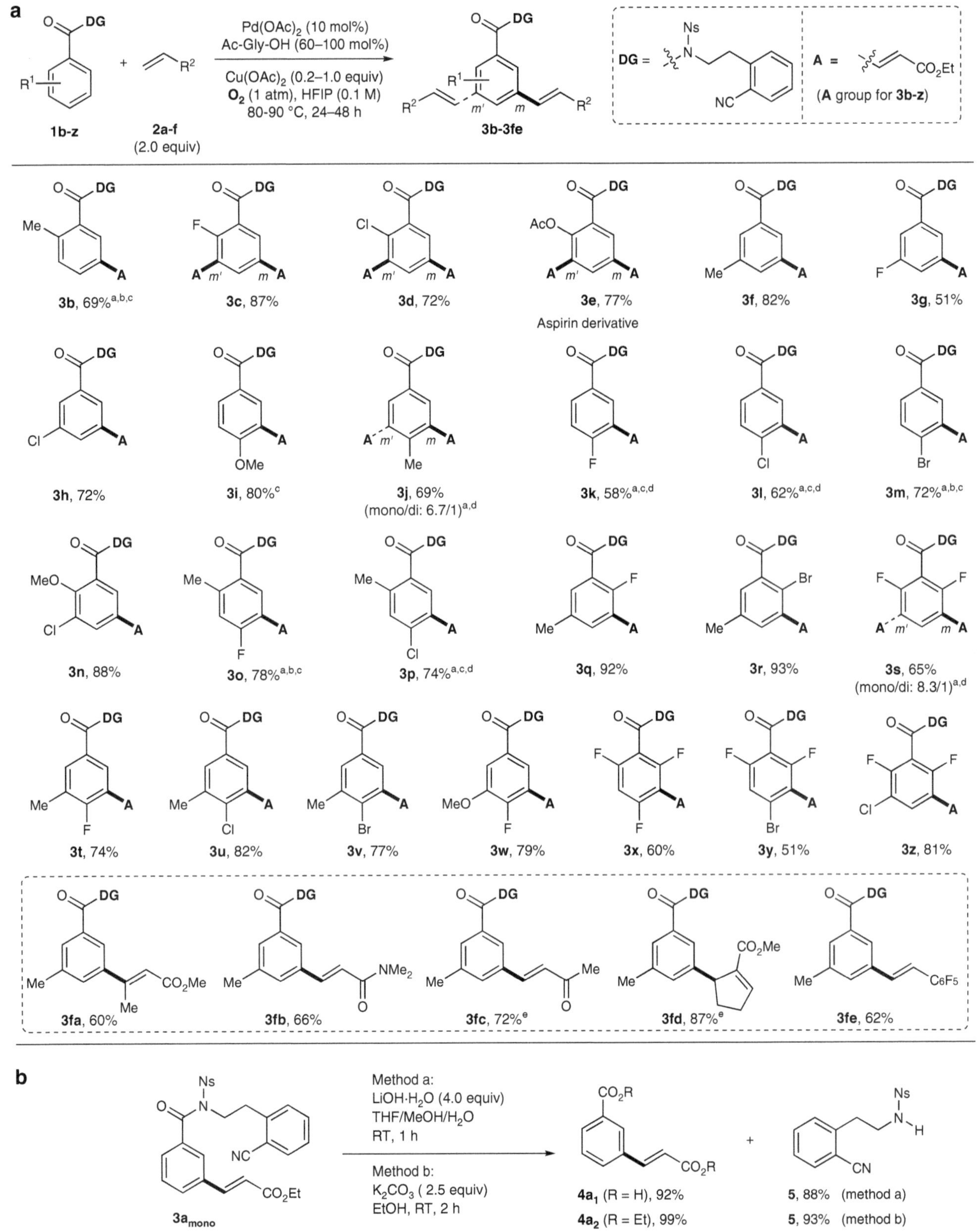


**Figure 2 | Substrate scope and removal and recycle of the template.** (**a**) Substrate scope of *meta*-C-H olefination. Reaction conditions: **1** (0.1-0.2 mmol), **2** (2.0 equiv), $Pd(OAc)_2$ (10 mol%), Ac-Gly-OH (60-100 mol%), $Cu(OAc)_2$ (0.2-1.0 equiv), $O_2$ (1 atm), HFIP (1-2 ml), 80-90 °C, 24-48 h, see the Supplementary Methods for details. Isolated yields are reported. [a]Formyl-Gly-OH (60 mol%) was used instead of Ac-Gly-OH. [b]$KH_2PO_4$ (0.5 equiv) was added. [c]A trace of (*m,m'*)-diolefinated product was observed. [d]$K_2HPO_4$ (0.5 equiv) was added. [e]AgOAc (3.0 equiv) was used instead of $O_2/Cu(OAc)_2$. (**b**) Two mild methods available for regenerating auxiliary **5**.

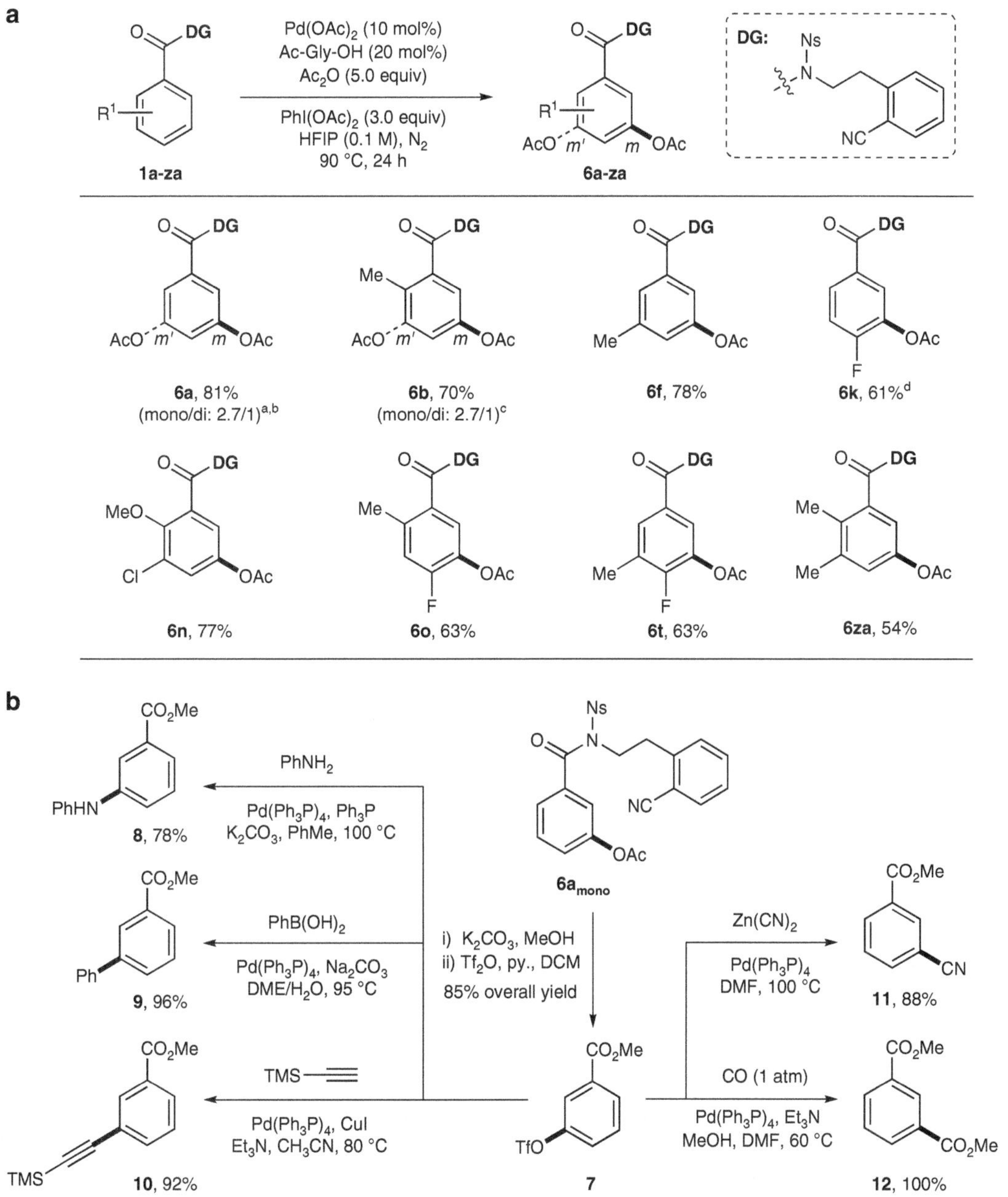


**Figure 3 | *Meta*-C-H acetoxylation and synthetic elaboration.** (**a**) *Meta*-C-H acetoxylation. Reaction conditions: **1** (0.1 mmol), $Pd(OAc)_2$ (10 mol%), Ac-Gly-OH (20 mol%), $Ac_2O$ (5.0 equiv), $PhI(OAc)_2$ (3.0 equiv), HFIP (1 ml), $N_2$, 90 °C, 24 h. Isolated yields are reported; see the Supplementary Methods for details. [a]A trace of regioisomer was observed. [b]The yield was 60% (mono/di: 4.8:1) in 1.3 mmol scale. [c]6 h. [d]A trace of (*m,m'*)-di-acetoxylated product was observed. (**b**) Application potential of *meta*-acetoxylation demonstrated with further elaborations.

traces of isomers were observed for some substrates. However, it was hard to assign the peaks of isomers that were only traces from the crude $^1H$ NMR and thus we did not attempt to calculate the exact ratios of isomers. Finally, the template-directing group could be removed and recycled readily with LiOH or $K_2CO_3$ in high yields as shown in Fig. 2b. Moreover, the auxiliary **5** could be synthesized in multi-gram scale from inexpensive chemicals (see Supplementary Methods for details).

***Meta*-C–H acetoxylation.** The versatility of our sulfonamide template with different catalytic cycles was investigated briefly with *meta*-acetoxylation of benzoic acid derivatives using the previously established oxidation conditions (Fig. 3a)[70]. Although the acetoxylation was generally less efficient than olefination, several substrates with different substitution patterns underwent *meta*-acetoxylation successfully to give desired products in moderate to good yields (**6a–6za**). Moreover, 60% combined yield of products was obtained when the reaction was performed in 1.3 mmol scale (**6a**).

**Synthetic elaboration.** To demonstrate the utility of our *meta*-C–H functionalizations, we attempted further elaboration of the acetoxylated product (Fig. 3b). Thus, triflate **7** was prepared in high yield from **6a**$_{\mathbf{mono}}$ by hydrolysis of the template and the acetoxy group in one step, which was followed by triflation of the resulting hydroxyl group. Triflate **7** was then transformed to a range of synthetically useful substances with well-established coupling reactions, namely, amination (**8**), arylation (**9**), alkynylation (**10**), cyanation (**11**) and carbonylation (**12**), greatly expanding the application potential of our method.

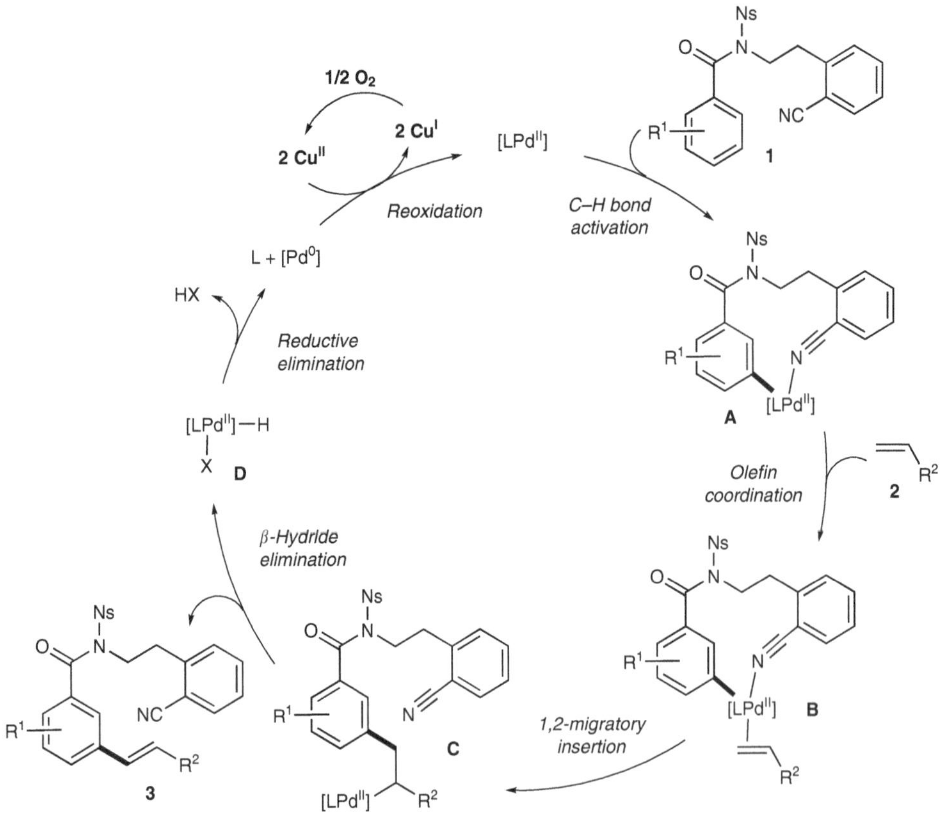


**Figure 4 | Proposed catalytic cycle of *meta*-C–H olefination.** The plausible mechanism involves regeneration of a Pd(II) catalyst with only catalytic amount of Cu(II) species.

**Proposed catalytic cycle.** On the basis of the above results, a plausible catalytic cycle is proposed for *meta*-C–H olefination (Fig. 4). The complex **A** is generated by coordination of substrate **1** to a Pd(II) species followed by template-directed insertion of Pd(II) into the *meta*-C–H bond of **1**. Following coordination of **A** with olefin **2**, the resulting complex **B** undergoes 1,2-migratory insertion to give intermediate **C**. Product **3** is then afforded by *β*-hydride elimination of **C**. Reductive elimination of hydride **D** produces Pd(0), which is reoxidized to Pd(II) by two equivalents of Cu(II). The resulting Cu(I) is then oxidized by molecular oxygen to re-enter the catalytic cycle.

## Discussion

In summary, a general protocol for *meta*-C–H bond olefination of a broad range of benzoic acid derivatives assisted by a nitrile-based sulfonamide template has been developed. Desired *meta*-functionalized products were obtained regardless of the substitution patterns and steric biases of the substrates. Notably, the challenging tri-substituted substrates were tolerated, which was not observed in previous transition-metal catalysed *meta*-C–H functionalizations[15–63]. Importantly, the new protocol was compatible with molecular oxygen as the terminal oxidant. Moreover, the sulfonamide template auxiliary could be efficiently removed and recycled under mild conditions. Finally, the versatility of our template was demonstrated with *meta*-C–H acetoxylation, which enabled the access to five synthetically useful major classes of substitents at the *meta*-position of benzoic acid derivatives. It is expected that the protocol disclosed herein will soon inspire the development of more synthetically useful *meta*-C–H transformations for benzoic acid derivatives.

## Methods

**General methods.** For $^1$H and $^{13}$C NMR spectra of compounds in this manuscript and details of the synthetic procedures, see Supplementary Figs 1–102 and Supplementary Methods.

**General procedure for *meta*-C–H olefination.** To a 50-ml Schlenk sealed tube (with a Teflon cap) equipped with a magnetic stir bar was charged with amide **1** (0.10 mmol, 1.0 equiv), $Pd(OAc)_2$ (2.3 mg, 0.010 mmol, 10 mol%), Ac-Gly-OH (20–100 mol%) and $Cu(OAc)_2$ (0.2–1.0 equiv) sequentially. Hexafluoro-2-propanol (HFIP; 1.0 ml) was added to the mixture along the inside wall of the tube, followed by the corresponding alkene **2** (2.0 equiv). The reaction tube was capped, then evacuated briefly under vacuum and charged with $O_2$ (1 atm, balloon, × 3). The tube was then submerged into a preheated 80 or 90 °C oil bath. The reaction was stirred for 24–48 h and cooled to room temperature. The crude reaction mixture was diluted with EtOAc (5 ml) and filtered through a short pad of Celite. The sealed tube and Celite pad were washed with an additional 20 ml of EtOAc. The filtrate was concentrated *in vacuo*, and the resulting residue was purified by flash silica gel chromatography or preparative thin layer chromatography using petroleum ether/EtOAc as the eluent. The site selectivity was assigned by NMR analysis of the product or the hydrolysed product. Full experimental details and characterization of new compounds can be found in the Supplementary Methods.

**General procedure for *meta*-C–H acetoxylation.** To a 50-ml Schlenk sealed tube (with a Teflon cap) equipped with a magnetic stir bar was charged with amide **1** (0.10 mmol, 1.0 equiv), $Pd(OAc)_2$ (2.3 mg, 0.010 mmol, 10 mol%), Ac-Gly-OH (2.4 mg, 0.020 mmol, 20 mol%) and $PhI(OAc)_2$ (96.6 mg, 0.30 mmol, 3 equiv). HFIP (1.0 ml) was added to the mixture along the inside wall of the tube, followed by $Ac_2O$ (47 µl, 5 equiv). The reaction tube was capped, then evacuated briefly under vacuum and charged with $N_2$ (1 atm, balloon, × 3). The tube was then submerged into a preheated 90 °C oil bath. The reaction was stirred for 24 h and cooled to room temperature. The crude reaction mixture was diluted with EtOAc (5 ml) and filtered through a short pad of Celite. The sealed tube and Celite pad were washed with an additional 20 ml of EtOAc. The filtrate was concentrated *in vacuo*, and the resulting residue was purified by flash silica gel chromatography using petroleum ether/EtOAc as the eluent. Full experimental details and characterization of new compounds can be found in the Supplementary Methods.

## References


1. Lyons, T. W. & Sanford, M. S. Palladium-catalyzed ligand-directed C–H functionalization reactions. *Chem. Rev.* **110,** 1147–1169 (2010).
2. Hartwig, J. F. Borylation and silylation of C–H bonds: a platform for diverse C–H bond functionalizations. *Acc. Chem. Res.* **45,** 864–873 (2011).
3. Engle, K. M., Mei, T.-S., Wasa, M. & Yu, J.-Q. Weak coordination as a powerful means for developing broadly useful C–H functionalization reactions. *Acc. Chem. Res.* **45,** 788–802 (2012).
4. Patureau, F. W., Wencel-Delord, J. & Glorius, F. Cp*Rh-catalyzed C–H activations: Versatile dehydrogenative cross-couplings of $C^2_{SP}$ C–H positions with olefins, alkynes, and arenes. *Aldrichimica Acta* **45,** 31–41 (2012).
5. Rouquet, G. & Chatani, N. Catalytic functionalization of C($sp^2$)–H and C($sp^3$)–H bonds by using bidentate directing groups. *Angew. Chem. Int. Ed.* **52,** 11726–11743 (2013).
6. Ackermann, L. Carboxylate-assisted ruthenium-catalyzed alkyne annulations by C–H/Het–H bond functionalizations. *Acc. Chem. Res.* **47,** 281–295 (2014).
7. Zhang, X.-S., Chen, K. & Shi, Z.-J. Transition metal-catalyzed direct nucleophilic addition of C–H bonds to carbon–heteroatom double bonds. *Chem. Sci* **5,** 2146–2159 (2014).
8. Daugulis, O., Roane, J. & Tran, L. D. Bidentate, monoanionic auxiliary-directed functionalization of carbon–hydrogen bonds. *Acc. Chem. Res.* **48,** 1053–1064 (2015).
9. Satoh, T. & Miura, M. Transition-metal-catalyzed regioselective arylation and vinylation of carboxylic acids. *Synthesis (Mass)* **2010,** 3395–3409 (2010).
10. Miura, M., Tsuda, T., Satoh, T., Pivsa-Art, S. & Nomura, M. Oxidative cross-coupling of N-(2'-Phenylphenyl)benzene- sulfonamides or benzoic and naphthoic acids with alkenes using a palladium – copper catalyst system under air. *J. Org. Chem.* **63,** 5211–5215 (1998).
11. Giri, R. *et al.* Palladium-catalyzed methylation and arylation of $sp^2$ and $sp^3$ C – H bonds in simple carboxylic acids. *J. Am. Chem. Soc.* **129,** 3510–3511 (2007).
12. Chiong, H. A., Pham, Q.-N. & Daugulis, O. Two methods for direct *ortho*-arylation of benzoic acids. *J. Am. Chem. Soc.* **129,** 9879–9884 (2007).
13. Hartung, C. G. & Snieckus, V. in *Modern Arene Chemistry*. (ed. Astruc, D.) 330–367 (Wiley-VCH, 2002).
14. Taylor, R. *Electrophilic Aromatic Substitution* (Wiley, 1990).
15. Cho, J.-Y., Tse, M. K., Holmes, D., Maleczka, R. E. & Smith, M. R. Remarkably selective iridium catalysts for the elaboration of aromatic C–H bonds. *Science* **295,** 305–308 (2002).
16. Ishiyama, T., Takagi, J., Hartwig, J. F. & Miyaura, N. A stoichiometric aromatic C–H borylation catalyzed by iridium(I)/2,2'-bipyridine complexes at room temperature. *Angew. Chem. Int. Ed.* **41,** 3056–3058 (2002).
17. Zhang, Y.-H., Shi, B.-F. & Yu, J.-Q. Pd(II)-catalyzed olefination of electron-deficient arenes using 2,6-dialkylpyridine ligands. *J. Am. Chem. Soc.* **131,** 5072–5074 (2009).
18. Emmert, M. H., Cook, A. K., Xie, Y. J. & Sanford, M. S. Remarkably high reactivity of $Pd(OAc)_2$/pyridine catalysts: nondirected C–H oxygenation of arenes. *Angew. Chem. Int. Ed.* **50,** 9409–9412 (2011).
19. Ball, L. T., Lloyd-Jones, G. C. & Russell, C. A. Gold-catalyzed direct arylation. *Science* **337,** 1644–1648 (2012).
20. Wencel-Delord, J., Nimphius, C., Wang, H. & Glorius, F. Rhodium(III) and hexabromobenzene – A catalyst system for the cross-dehydrogenative coupling of simple arenes and heterocycles with arenes bearing directing groups. *Angew. Chem. Int. Ed.* **51,** 13001–13005 (2012).
21. Shrestha, R., Mukherjee, P., Tan, Y., Litman, Z. C. & Hartwig, J. F. Sterically controlled, palladium-catalyzed intermolecular amination of arenes. *J. Am. Chem. Soc.* **135,** 8480–8483 (2013).
22. Boursalian, G. B., Ngai, M.-Y., Hojczyk, K. N. & Ritter, T. Pd-catalyzed aryl C–H imidation with arene as the limiting reagent. *J. Am. Chem. Soc.* **135,** 13278–13281 (2013).
23. Cong, X., Tang, H., Wu, C. & Zeng, X. Role of mono-N-protected amino acid ligands in palladium(II)-catalyzed dehydrogenative Heck reactions of electron-deficient (hetero)arenes: Experimental and computational studies. *Organometallics* **32,** 6565–6575 (2013).
24. Cheng, C. & Hartwig, J. F. Rhodium-catalyzed intermolecular C–H silylation of arenes with high steric regiocontrol. *Science* **343,** 853–857 (2014).
25. Kuninobu, Y., Ida, H., Nishi, M. & Kanai, M. A *meta*-selective C–H borylation directed by a secondary interaction between ligand and substrate. *Nat. Chem* **7,** 712–717 (2015).
26. Yang, J. Transition metal catalyzed *meta*-C-H functionalization of aromatic compounds. *Org. Biomol. Chem.* **13,** 1930–1941 (2015).
27. Li, J., Sarkar, S. D. & Ackermann, L. *meta*- and *para*-Selective C–H Functionalization by C–H Activation. *Top. Organomet. Chem.* **55,** 217–257 (2016).
28. Ishiyama, T. *et al.* Mild iridium-catalyzed borylation of arenes. High turnover numbers, room temperature reactions, and isolation of a potential intermediate. *J. Am. Chem. Soc.* **124,** 390–391 (2002).
29. Hull, K. L. & Sanford, M. S. Catalytic and highly regioselective cross-coupling of aromatic C – H substrates. *J. Am. Chem. Soc.* **129,** 11904–11905 (2007).
30. Lyons, T. W., Hull, K. L. & Sanford, M. S. Controlling site selectivity in Pd-catalyzed oxidative cross-coupling reactions. *J. Am. Chem. Soc.* **133,** 4455–4464 (2011).
31. Patureau, F. W., Nimphius, C. & Glorius, F. Rh catalyzed C–H activation and oxidative olefination without chelate assistance: on the reactivity of bromoarenes. *Org. Lett.* **13,** 6346–6349 (2011).
32. Wencel-Delord, J., Nimphius, C., Patureau, F. W. & Glorius, F. [$Rh^{III}Cp^*$]-catalyzed dehydrogenative aryl–aryl bond formation. *Angew. Chem. Int. Ed.* **51,** 2247–2251 (2012).
33. Ye, M., Gao, G.-L. & Yu, J.-Q. Ligand-promoted C-3 selective C–H olefination of pyridines with Pd catalysts. *J. Am. Chem. Soc.* **133,** 6964–6967 (2011).
34. Li, B.-J. & Shi, Z.-J. Ir-catalyzed highly selective addition of pyridyl C–H bonds to aldehydes promoted by triethylsilane. *Chem. Sci* **2,** 488–493 (2011).
35. Zhou, L. & Lu, W. Palladium(II)-catalyzed coupling of electron-deficient arenes via C–H activation. *Organometallics* **31,** 2124–2127 (2012).
36. Phipps, R. J. & Gaunt, M. J. A *meta*-selective copper-catalyzed C–H bond arylation. *Science* **323,** 1593–1597 (2009).
37. Duong, H. A., Gilligan, R. E., Cooke, M. L., Phipps, R. J. & Gaunt, M. J. Copper(II)-catalyzed *meta*-selective direct arylation of α-aryl carbonyl compounds. *Angew. Chem. Int. Ed.* **50,** 463–466 (2011).
38. Ackermann, L., Hofmann, N. & Vicente, R. Carboxylate-assisted ruthenium-catalyzed direct alkylations of ketimines. *Org. Lett.* **13,** 1875–1877 (2011).
39. Saidi, O. *et al.* Ruthenium-catalyzed *meta* sulfonation of 2-phenylpyridines. *J. Am. Chem. Soc.* **133,** 19298–19301 (2011).
40. Hofmann, N. & Ackermann, L. *meta*-Selective C–H bond alkylation with secondary alkyl halides. *J. Am. Chem. Soc.* **135,** 5877–5884 (2013).
41. Paterson, St A. J., John-Campbell, S., Mahon, M. F., Press, N. J. & Frost, C. G. Catalytic *meta*-selective C–H functionalisation to construct quaternary carbon centres. *Chem. Commun.* **51,** 12807–12810 (2015).
42. Teskey, C. J., Lui, A. Y. W. & Greaney, M. F. Ruthenium-catalyzed *meta*-selective C–H bromination. *Angew. Chem. Int. Ed.* **54,** 11677–11680 (2015).
43. Li, J. *et al.* N-acyl amino acid ligands for ruthenium(II)-catalyzed *meta*-C–H tert-alkylation with removable auxiliaries. *J. Am. Chem. Soc.* **137,** 13894–13901 (2015).
44. Wang, X.-C. *et al.* Ligand-enabled *meta*-C–H activation using a transient mediator. *Nature* **519,** 334–338 (2015).
45. Dong, Z., Wang, J. & Dong, G. Simple amine-directed *meta*-selective C–H arylation via Pd/norbornene catalysis. *J. Am. Chem. Soc.* **137,** 5887–5890 (2015).
46. Mochida, S., Hirano, K., Satoh, T. & Miura, M. Synthesis of stilbene and distyrylbenzene derivatives through rhodium-catalyzed *ortho*-olefination and decarboxylation of benzoic acids. *Org. Lett.* **12,** 5776–5779 (2010).
47. Cornella, J., Righi, M. & Larrosa, I. Carboxylic acids as traceless directing groups for formal *meta*-selective direct arylation. *Angew. Chem. Int. Ed.* **50,** 9429–9432 (2011).
48. Luo, J., Preciado, S. & Larrosa, I. Overriding *ortho–para* selectivity via a traceless directing group relay strategy: the *meta*-selective arylation of phenols. *J. Am. Chem. Soc.* **136,** 4109–4112 (2014).
49. Zhang, Y., Zhao, H., Zhang, M. & Su, W. Carboxylic acids as traceless directing groups for the rhodium(III)-catalyzed decarboxylative C–H arylation of thiophenes. *Angew. Chem. Int. Ed.* **54,** 3817–3821 (2015).
50. Lee, D. & Chang, S. Direct C–H amidation of benzoic acids to introduce *meta*- and *para*-amino groups by tandem decarboxylation. *Chem. Eur. J* **21,** 5364–5368 (2015).
51. Shi, X.-Y. *et al.* A convenient synthesis of N-aryl benzamides by rhodium-catalyzed *ortho*-amidation and decarboxylation of benzoic acids. *Chem. Eur. J* **21,** 1900–1903 (2015).
52. Zhang, J. *et al.* Palladium(II)-catalyzed *meta*-selective direct arylation of *O*-β-naphthyl carbamate. *Chem. Commun* **51,** 1297–1300 (2015).
53. Martinez-Martinez, A. J., Kennedy, A. R., Mulvey, R. E. & O'Hara, C. T. Directed *ortho-meta'*- and *meta-meta'*-dimetalations: A template base approach to deprotonation. *Science* **346,** 834–837 (2014).
54. Leow, D., Li, G., Mei, T.-S. & Yu, J.-Q. Activation of remote *meta*-C–H bonds assisted by an end-on template. *Nature* **486,** 518–522 (2012).
55. Dai, H.-X., Li, G., Zhang, X.-G., Stepan, A. F. & Yu, J.-Q. Pd(II)-catalyzed *ortho*- or *meta*-C–H olefination of phenol derivatives. *J. Am. Chem. Soc.* **135,** 7567–7571 (2013).
56. Wan, L., Dastbaravardeh, N., Li, G. & Yu, J.-Q. Cross-coupling of remote *meta*-C–H bonds directed by a U-shaped template. *J. Am. Chem. Soc.* **135,** 18056–18059 (2013).
57. Tang, R.-Y., Li, G. & Yu, J.-Q. Conformation-induced remote *meta*-C–H activation of amines. *Nature* **507,** 215–220 (2014).
58. Yang, G. *et al.* Pd(II)-catalyzed *meta*-C-H olefination, arylation, and acetoxylation of indolines using a U-shaped template. *J. Am. Chem. Soc.* **136,** 10807–10813 (2014).
59. Deng, Y. & Yu, J.-Q. Remote *meta*-C–H olefination of phenylacetic acids directed by a versatile U-shaped template. *Angew. Chem. Int. Ed.* **54,** 888–891 (2015).

60. Lee, S., Lee, H. & Tan, K. L. *meta*-Selective C–H functionalization using a nitrile-based directing group and cleavable Si-tether. *J. Am. Chem. Soc.* **135,** 18778–18781 (2013).
61. Bera, M., Modak, A., Patra, T., Maji, A. & Maiti, D. *meta*-Selective arene C–H bond olefination of arylacetic acid using a nitrile-based directing group. *Org. Lett.* **16,** 5760–5763 (2014).
62. Bera, M., Maji, A., Sahoo, S. K. & Maiti, D. Palladium(II)-catalyzed *meta*-C–H olefination: Constructing multisubstituted arenes through homo-diolefination and sequential hetero-diolefination. *Angew. Chem. Int. Ed.* **54,** 8515–8519 (2015).
63. Li, S., Ji, H., Cai, L. & Li, G. Pd(II)-catalyzed remote regiodivergent *ortho*- and *meta*-C–H functionalisations of phenylethylamines. *Chem. Sci* **6,** 5595–5600 (2015).
64. Jia, C., Kitamura, T. & Fujiwara, Y. Catalytic functionalization of arenes and alkanes via C – H bond activation. *Acc. Chem. Res.* **34,** 633–639 (2001).
65. Satoh, T. & Miura, M. Oxidative coupling of aromatic substrates with alkynes and alkenes under rhodium catalysis. *Chem. Eur. J* **16,** 11212–11222 (2010).
66. Yeung, C. S. & Dong, V. M. Catalytic dehydrogenative cross-coupling: forming carbon – carbon bonds by oxidizing two carbon – hydrogen bonds. *Chem. Rev.* **111,** 1215–1292 (2011).
67. Le Bras, J. & Muzart, J. Intermolecular dehydrogenative Heck reactions. *Chem. Rev.* **111,** 1170–1214 (2011).
68. Wang, D.-H., Engle, K. M., Shi, B.-F. & Yu, J.-Q. Ligand-enabled reactivity and selectivity in a synthetically versatile aryl C–H olefination. *Science* **327,** 315–319 (2010).
69. Liu, B., Jiang, H.-Z. & Shi, B.-F. Palladium-catalyzed oxidative olefination of phenols bearing removable directing groups under molecular oxygen. *J. Org. Chem.* **79,** 1521–1526 (2014).
70. Dick, A. R., Hull, K. L. & Sanford, M. S. A highly selective catalytic method for the oxidative functionalization of C–H bonds. *J. Am. Chem. Soc.* **126,** 2300–2301 (2004).

## Acknowledgements

We gratefully acknowledge NSFC (21402198), 'The 1000 Youth Talents Program', and FJIRSM for financial support.

## Author contributions

S.L., L.C. and H.J. performed the experiments and developed the reactions. L.Y. helped collecting some experimental data. G.L. designed and directed the project and wrote the manuscript with the feedback of S.L., L.C and H.J.

## Additional information

**Competing financial interests:** The authors declare no competing financial interests.

# 18

# Enantioselective acyl transfer catalysis by a combination of common catalytic motifs and electrostatic interactions

Hiroki Mandai[1], Kazuki Fujii[1], Hiroshi Yasuhara[1], Kenko Abe[1], Koichi Mitsudo[1], Toshinobu Korenaga[2] & Seiji Suga[1,3,4]

Catalysts that can promote acyl transfer processes are important to enantioselective synthesis and their development has received significant attention in recent years. Despite noteworthy advances, discovery of small-molecule catalysts that are robust, efficient, recyclable and promote reactions with high enantioselectivity can be easily and cost-effectively prepared in significant quantities (that is, >10 g) has remained elusive. Here, we demonstrate that by attaching a binaphthyl moiety, appropriately modified to establish H-bonding interactions within the key intermediates in the catalytic cycle, and a 4-aminopyridyl unit, exceptionally efficient organic molecules can be prepared that facilitate enantioselective acyl transfer reactions. As little as 0.5 mol% of a member of the new catalyst class is sufficient to generate acyl-substituted all-carbon quaternary stereogenic centres in quantitative yield and in up to 98:2 enantiomeric ratio (er) in 5 h. Kinetic resolution or desymmetrization of 1,2-diol can be performed with high efficiency and enantioselectivity as well.

[1] Division of Applied Chemistry, Graduate School of Natural Science and Technology, Okayama University, 3-1-1 Tsushima-naka, Kita-ku, Okayama 700-8530, Japan. [2] Faculty of Engineering, Department of Chemistry and Bioengineering, Iwate University, Morioka, Iwate 020-8551, Japan. [3] Research Center of New Functional Materials for Energy Production, Storage and Transport Okayama University, 3-1-1 Tsushima-naka, Kita-ku, Okayama 700-8530, Japan. [4] Japan Science and Technology Agency, ACT-C, 4-1-8 Honcho, Kawaguchi, Saitama 332-0012, Japan. Correspondence and requests for materials should be addressed to H.M. (email: mandai@cc.okayama-u.ac.jp) or to S.S. (email: suga@cc.okayama-u.ac.jp).

N,*N*-4-dimethylaminopyridine (DMAP) has long been recognized as a uniquely effective and broadly applicable nucleophilic catalyst for a number of important reactions in organic chemistry[1]. One set of transformations for which DMAP is commonly utilized is the acylation of hydroxyl group in the presence of acid anhydride. The commonly accepted mechanism for this class of reactions, which supported by experimental as well as computational findings, is presented in Fig. 1a (refs 2,3). The nucleophilic is believed to react first with the acylating agent **i** to generate *N*-acylpyridinium salt intermediate **ii**, which is likely subjected to nucleophilic attack by the alcohol substrate ($R^2OH$) via intermediate **iii**. A critical feature of this general class of catalytic processes is that the acetate anion that resides within complex **ii** serves as a Brønsted base to enhance the reactivity of the otherwise relatively mildly nucleophilic hydroxyl unit. Ester product **iv** and pyridinium salt **v** are thus formed, and the latter is then neutralized by the stoichiometric base (B; typically $Et_3N$) to re-generate the nucleophilic catalyst. The efficiency with which *N*-acylpyridinium ion **ii** is generated[4], the Lewis basicity of the counteranion unit as well as the degree to which it remains associated with the positively charged acylating agent (versus loose ion pair) are critical to the facility of the overall transformation[5]. A number of strategies have been adopted to modulate the effectiveness of acyl transfer agents for which DMAP serves as the parent compound. Among these is the utilization of electronic factors to extend the lifetime of the *N*-acylpyridinium salt, or by manipulation of conformational effects to enhance electron donation by the amino substituent; there are also instances where a combination of the aforementioned approaches has been adopted (for example, 4-pyrrolidinopyridine (PPY) or 9-azajulolidine)[6]. A variety of chiral variants, employed for kinetic resolution of alcohol or enantioselective acyl transfer processes, have also been introduced[7,8]. Such investigations, which have led to the development of various enantiomerically pure promoter molecules that are based on DMAP or PPY scaffolds, as originally put forth by the notable advances reported by Vedejs[9] and Fu[10], may be classified in four major categories (Fig. 1b–e). One strategy entailed the use of *N*-acylpyridinium salt with a chiral substituent at the pyridyl ring's C2 position; however, this structural alteration proved to be detrimental to reaction efficiency, requiring the use of stoichiometric amounts of the catalyst (Fig. 1b)[9]. A similar approach but involving the C3 site of the heterocyclic ring, has been extensively examined[11–19]. Nonetheless, reactivity levels were again generally reduced, probably as a result of hampering of proper electron donation by the amino substituent, which raises the energy of the critical pyridinium ion intermediates (cf. **ii**, Fig. 1a)[20,21]. The issue of diminished catalyst efficiency applies to DMAP or PPY derivatives that carry a ring that connects the C2 and C3 carbons of the pyridyl ring (Fig. 1d)[22]. In most cases, diminished stability of the key ion-pair intermediate because of steric repulsion between substituents and 4-amino moiety and/or *N*-acetyl group was a complication[21]. An exception was the ferrocene-based catalysts developed by Fu *et al.*[23,24]; comparatively high catalyst activity and enantioselectivity was observed in a number of different applications. Obtaining this set of chiral catalysts in the enantiomerically pure form, however, requires costly resolution procedures[25]. The efficiency and considerable longevity of *N*-acylpyridinium ion has been attributed to the exceptional electron-donating ability of transition metal framework; what's more, the cyclopentadienyl moiety attached to pyridyl ring is sufficiently small to prevent unfavourable interaction with the *N*-acetyl and/or the *N*-dimethyl- and pyrrolidino groups[21]. Chiral amino derivatives have been positioned at the pyridyl group's C4 site as well (Fig. 1e)[26–29], but these distally positioned moieties did not generate high degrees of stereochemical differentiation for a wide variety of substrates and, for electronic reasons already mentioned (cf. Fig. 1c), this came at the cost of significant diminution in efficiency. Thus, relatively large substituents are required to construct chiral environment to achieve high enantioselectivity. Another notable concept (dual catalysis/anion-binding approach), which is not classified into aforementioned categories (Fig. 1b–e), was also very effective for enantioselective acyl transfer reactions[30–38].

## Results

**Catalyst design**. According to our continuous efforts for the development of chiral nucleophilic catalysts[17–19] to design an efficient and highly enantioselective acyl transfer catalyst that can be prepared in significant quantities without the need for expensive and/or specialized techniques, we envisioned chiral DMAP derivatives that might contain a binaphthyl unit at C4 position of a pyridine ring (Fig. 2). We reasoned that this blueprint would have several noteworthy advantages. First, either enantiomeric form of 1,1′-bi-2-naphthol (BINOL) is inexpensively available, thus obviating resolution procedures. Second, the catalyst platform allows synthesis and screening of $C_1$- and $C_2$-symmetric variants from a common intermediate. Third, the proposed catalyst construct easily lends to steric and/or electronic modification through the use of well-established protocols[39,40]. Finally, the absence of substituents at the C2 or C3 sites of the pyridyl ring (cf. Fig. 1b,c) would ensure that efficiency levels remain high. The suggested line of attack poses several significant challenges, however. One is that because the source of stereogenicity is somewhat distal from the pyridine ring, an attribute can result in minimal enantioselectivity. Thus, appropriate structural modification would be needed if the stereochemical bias inherent in the binaphthyl moiety is to exert a meaningful role in the acyl bond forming event. Towards this end, preliminary examination of molecular models implied that the substituents (functional group) at C3 and C3′ positions of the binaphthyl moiety are located sufficiently proximal to the reaction site, offering an attractive option for influencing enantioselectivity through catalyst structure alteration. Thus, we surmised that attractive interactions (for example, $\pi$–$\pi$ stacking, cation-$\pi$ affinity[27,41] or hydrogen-bonding[26]) between an appropriate unit within the catalyst structure and counteranion of *N*-acylpyridinium salt (design option 1, Fig. 2), or the nucleophile (design option 2, Fig. 2) might give rise to enhancement of enantioselectivity and perhaps reaction efficiency. Below, we outline the successful realization of the above plan.

**Catalyst synthesis**. We began by preparing a series of enantiomerically pure DMAP derivatives (**1a** – **p**) that contain a 1,1′-binaphthyl unit with different substituents pattern at their C3 and C3′ sites with reference to chiral quaternary ammonium salt syntheses from BINOL (refs 39,40; Fig. 3a, Supplementary Figs 1–26 and Supplementary Methods). The representative route for synthesis of catalyst **1j** is presented in Fig. 3b. Ortho-lithiation of MOM-protected compound **2** derived from (*S*)-BINOL with *n*-BuLi, followed by quenching with ethyl chloroformate, and deprotection of MOM group under acidic conditions gave the desired BINOL with 3,3′-diesters **3** in 98% yield over 2 steps (78.9 mmol scale). Then, two hydroxy groups of **3** were converted to the corresponding ditriflate, and subjected to Migita–Kosugi–Stille coupling reaction in the presence of tetra *n*-butyl ammonium chloride and lithium chloride as additives to afford **5** in 83% yield. Our studies revealed that Migita–Kosugi–Stille coupling

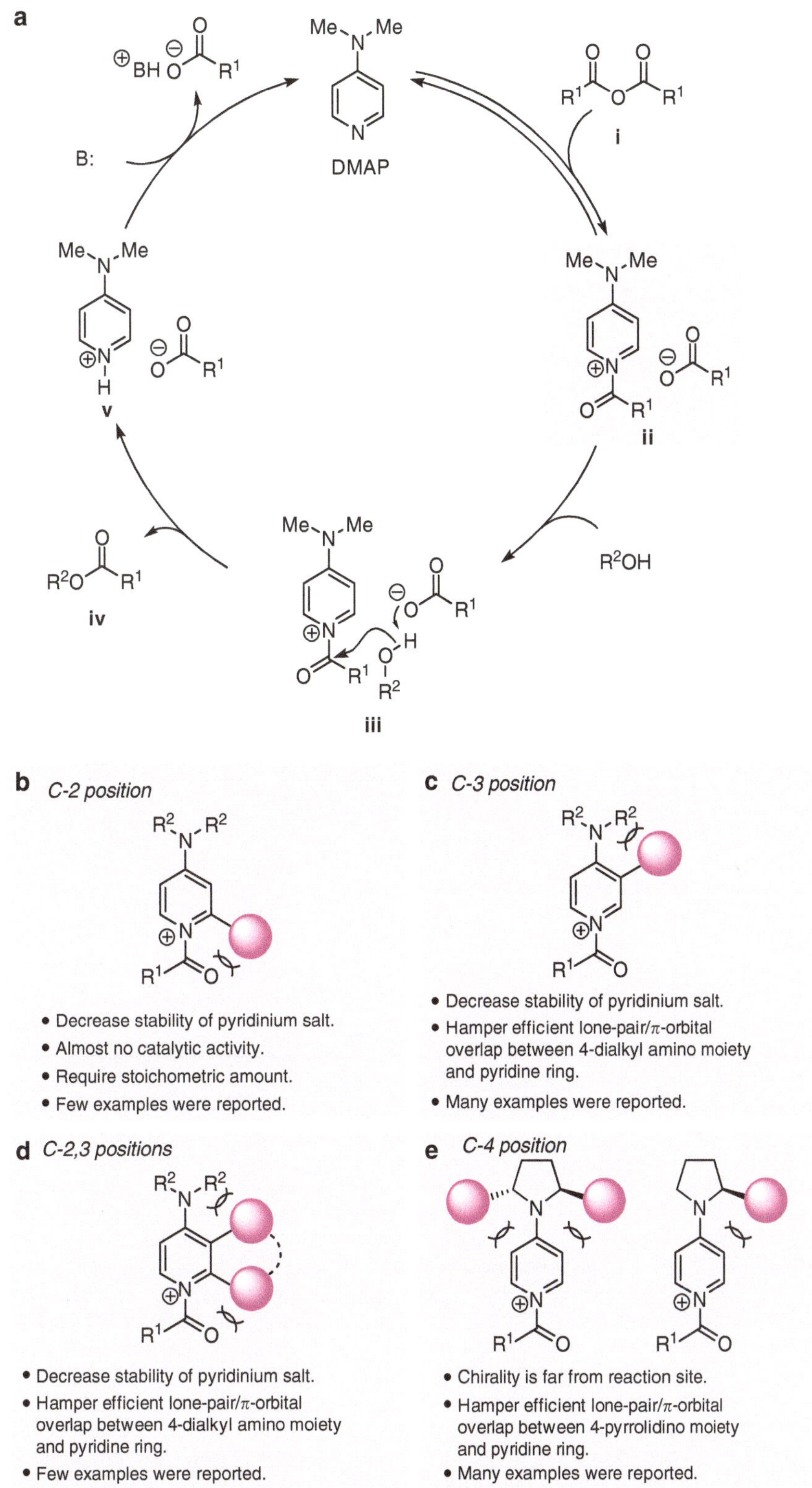


**Figure 1 | Nucleophilic catalyst in acylation reaction.** (**a**) The proposed and generally accepted catalytic cycle for acyl transfer reactions in the presence of acid anhydride promoted by DMAP; this general scheme was used as the framework for the catalyst development studies described in this report. The anionic component within the loose ion-pair intermediate (**ii**) is believed to serve as a general base, playing a critical role in determining the rate of the overall processes. (**b** – **e**) General strategies, and their specific attributes, of the previously employed strategies involving structural alteration of DMAP-containing molecules to generate enantioselective acyl transfer catalysts. B, base; $R^1$, $R^2$, FG, various functional groups.

reaction using a Pd nanoparticle generated *in situ* was found to be superior to Negishi coupling reaction[40] in large scale reaction because $Me_2Zn$ is costly and pyrophoric reagent. The benzylic positions of **5** were brominated using *N*-bromosuccinimide in the presence of 2,2′-azodiisobutyronitrile to give the desired dibromide **6** in 97% yield. Seven-membered ring formation from dibromide **6** using allylamine gave rise to the formation of amine **7** in 82% yield, and deprotection of ally group afforded

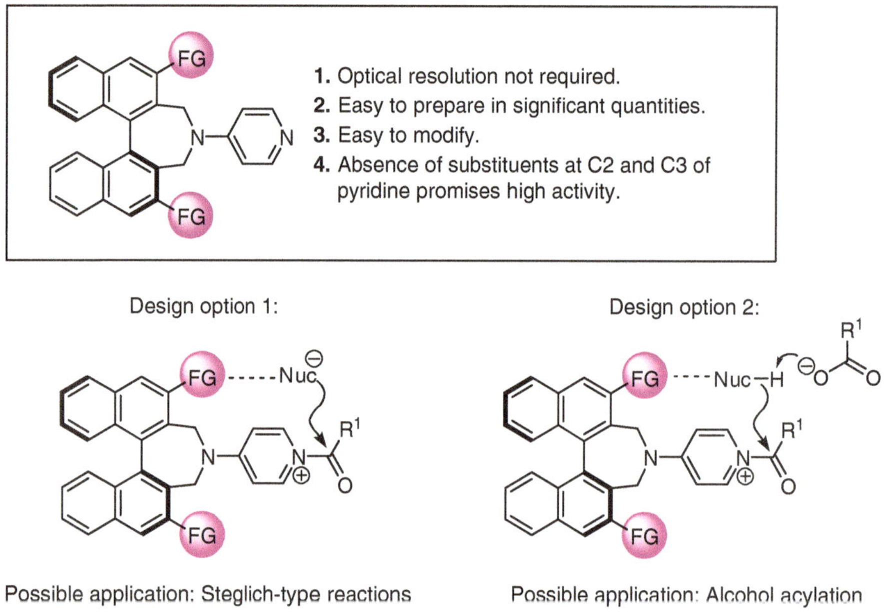


**Figure 2 | Initial consideration and options in catalyst design.** The basic design of the new acyl transfer catalysts, their projected advantages and the possible means by which electrostatic interactions can facilitate transformation. $R^1$, FG, various functional groups.

amine **8** in 95% yield. Installation of pyridine ring was accomplished by using Buchwald–Hartwig amination of 4-bromopyridine hydrochoride with amine **8** to provide the key intermediate **1e** in 70% yield. The formation of bis-tertiary alcohols **1j** was readily accomplished by the addition of ArLi to bis-ester **1e** in 91% yield. All transformations can be easily carried out in large scale (>10 g scale for **1e** from BINOL throughout the whole process, and >1 g scale for **1j** from **1e**), and overall yield of **1j** from (*S*)-BINOL was 38% (10 steps, >90% average yield for each step) with only 4 times of silica gel column chromatography purifications. The present synthetic route sufficiently secures facile accessibility to the key catalyst. The structures of **1e** and **1g** were identified by X-ray single-crystal analysis (Supplementary Figs 27 and 28).

**Enantioselective steglich rearrangement.** With a collection of possible catalysts in hand, we chose Steglich rearrangement[1,15,19,34,42,43] of oxindole derivative **9a** (Design option 1, Fig. 2) as the model transformation for identifying the optimal promoter molecule. Reactions were performed using 5 mol % of **1a – p** in tetrahydrofuran (THF) (0.1 M) at 0 °C for 12 h (Fig. 4a). Reaction with 4-amino pyridine **1a** was efficient but generated nearly racemic product (99% conv., 43:57 enantiomeric ratio (er)). Structurally modified versions **1b – f**, which contain methoxy, phenyl, 2-naphthyl, carboxyl ester or amide groups at their C3 and C3′ positions were either equally or less efficient than **1a**, and were all ineffective in generating **10a** with significant enantioselectivity (no more than 72:28 er). The surprising breakthrough came when we evaluated compounds that contain tertiary alcohols unit within their chiral binaphthyl moiety (**1g – m**); under otherwise identical conditions, use of these catalysts led to the formation of the all-carbon quaternary stereogenic centre not only with exceptional efficiency (>98% conv.) but also in up to 98:2 er. In sharp contrast, $C_1$-symmetric systems, represented by **1n**, **1o** and **1p** proved to be substantially less effective, and afforded nearly racemic mixture of product. Apparently, $C_2$-symmetric scaffold are essential for the establishment of proper enantiofacial discrimination. Subsequent optimization studies (Supplementary Figs 29–38 and Supplementary Tables 1–5), revealed that THF is indeed the optimal solvent and when the transformation is performed at -20 °C with no more than 0.5 mol % **1j**, the reaction proceeds to complete conversion within five hours to afford **10a** in 98:2 er. Finally, we find that use of 15 g of substrate **9a** under the latter conditions (193 mg of **1j**) delivers the desired product in quantitative yield (purification by standard silica gel chromatography) and 98:2 er (Fig. 4b, Supplementary Figs 39 and 40); furthermore, catalyst **1j** could be recovered in 95% yield, and used in same reaction (0.1 mmol scale) to afford the desired product **10a** in >98% yield with 98:2 er without loss of any catalytic activity (Supplementary Fig. 41 and Supplementary Method).

A variety of different substrates can be used in the enantioselective Steglich rearrangement process (Fig. 5 and Supplementary Figs 42–66). Thus, 3-alkyl-, 3-allyl- or 3-propargyl-substituted oxindole products **10b – f** were isolated in quantitative yields and in 96:4–99:1 er. Higher catalyst loading (3.0 versus 0.5 mol%) and temperature (25 versus -20 °C) was required in the case of **10c**, probably as a result of the presence of the more sterically demanding *iso*-propyl group. The stereochemical identity of **10c** was established through an X-ray structure (Supplementary Fig. 67). Transformation with substrates that contain an alkyl chain that bears a functionalizable polar group, such as an amide (**10g**), a silyl ether (**10h**) or a cyano unit (**10i**) proved to be exceptionally efficient (>98% yield) and highly enantioselective (91:9–97:3 er). Rearrangements involving 3-phenyl- (**10j**) or 3-thienyl-substituted oxindoles (**10k**) were more sluggish (3.0 mol % **1j** at − 20 °C was needed) likely due to their sizeable nature, but were also less enantioselective than the previous examples (79:21–84:16 er). Efficient and enantioselective preparation of bromo-substituted **10l** (>98% yield, 92:8 er) and *N*-Me oxindole **10m** (>98% yield, 87:13 er) further illustrates the scope of the catalytic process.

**Mechanistic studies.** To investigate the mechanism of the catalytic enantioselective rearrangement process, we performed a number of key experiments. First, cross-over studies (Fig. 6) between two marked substrates **9a** and **9n** clearly indicated that reactions proceed via an ion-pair complex, identifying the recovered starting material **9a** and **9n** (starting material); **9a′** and **9n′** (scrambled starting material) along with corresponding products **10a**, **10n**, **10a′** and **10n′** (Supplementary Fig. 68)[44].

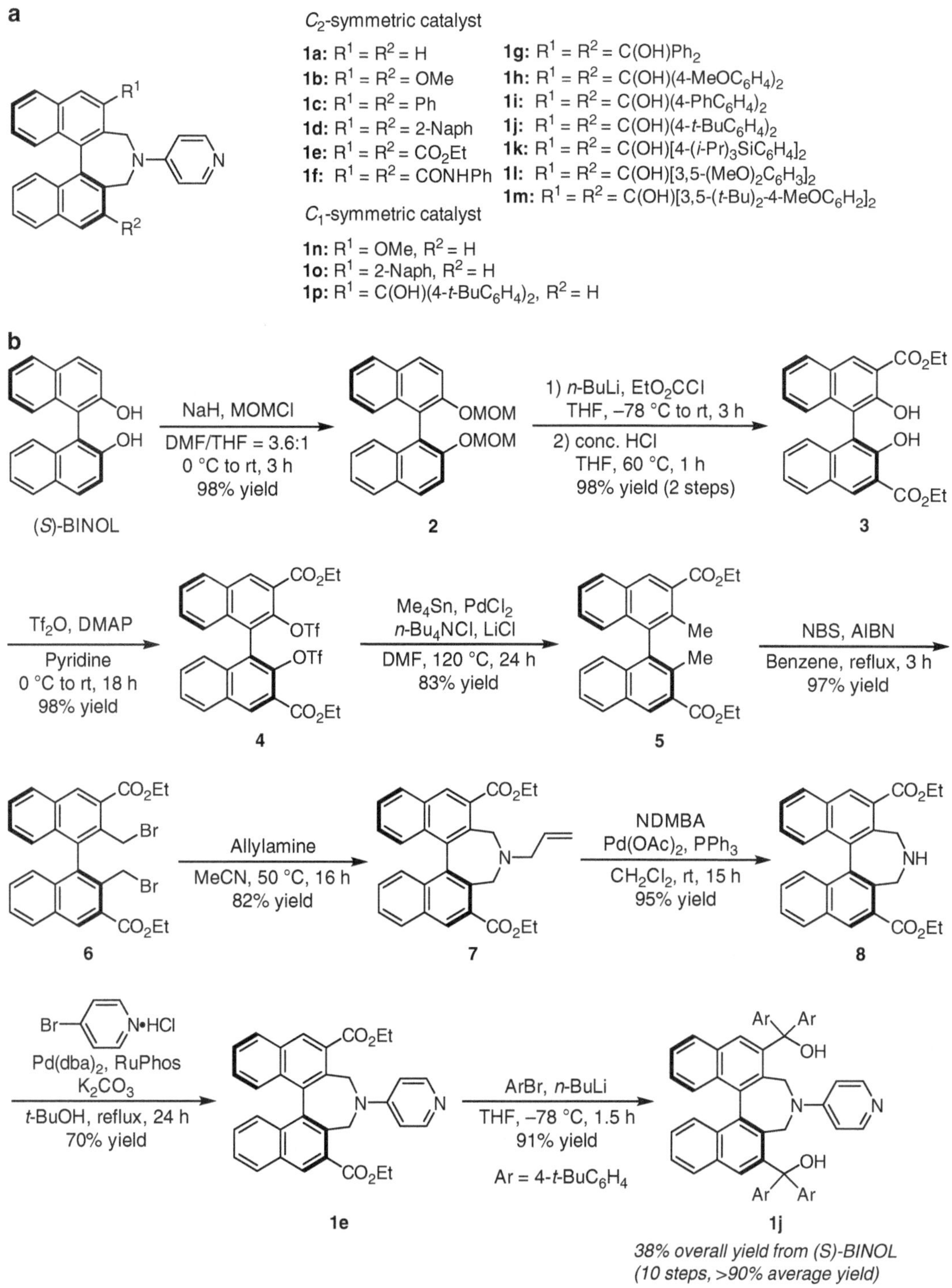


**Figure 3 | Synthesis of binaphthyl-based chiral nucleophilic catalyst candidates.** (**a**) Catalyst libraries containing $C_2$- and $C_1$-symmetic catalyst with different substitution pattern were used in the Steglich rearrangement of *O*-acylated oxindole derivatives. (**b**) $C_2$-symmetric catalyst with polar functional group at 3,3′-positions of binaphthyl moiety can be prepared by a synthesis scheme that is high yielding and is readily amenable to scale-up with minimal column chromatography purification, as the representative example clearly illustrates. AIBN, 2,3′-azodiisobutyronitrile; NBS, *N*-bromosuccinimide; NDMBA, N,N′-dimethylbarbituric acid; RuPhos, 2-dicyclohexylphosphino-2′,6′-di-i-propoxy-1,1′-biphenyl.

Second, we carried out kinetic measurements (Fig. 7) with transformations promoted by catalyst **1j**, the corresponding bis-methyl ether **1j**′, and DMAP; oxindole **9a** served as the substrate under the optimal conditions (0.5 mol% catalyst, THF (0.4 M), −20 °C, 5 h). In all cases, the reactions were found to be first-order with respect to the substrate (Supplementary Table 6). Moreover, substantially higher activity was observed with catalyst **1j** ($k_{\mathbf{1j}} = 1.22\,h^{-1}$) versus its derived bis-methyl ether **1j**′ ($k_{\mathbf{1j'}} = 2.62 \times 10^{-2}\,h^{-1}$), or DMAP ($k_{\mathbf{DMAP}} = 7.47 \times 10^{-2}\,h^{-1}$). Thus, acyl rearrangement was nearly 50 times faster with **1j** compared with **1j**′ and 16 times faster than DMAP. Additionally, whereas the desired product was obtained in 98:2 er when **1j** was use, there was hardly any enantiofacial selectivity when the hydroxyl units were protected (66:34 er with **1j**′). The presence of tertiary alcohols clearly has a significant impact on the rate as well as enantioselectivity of the catalytic process.

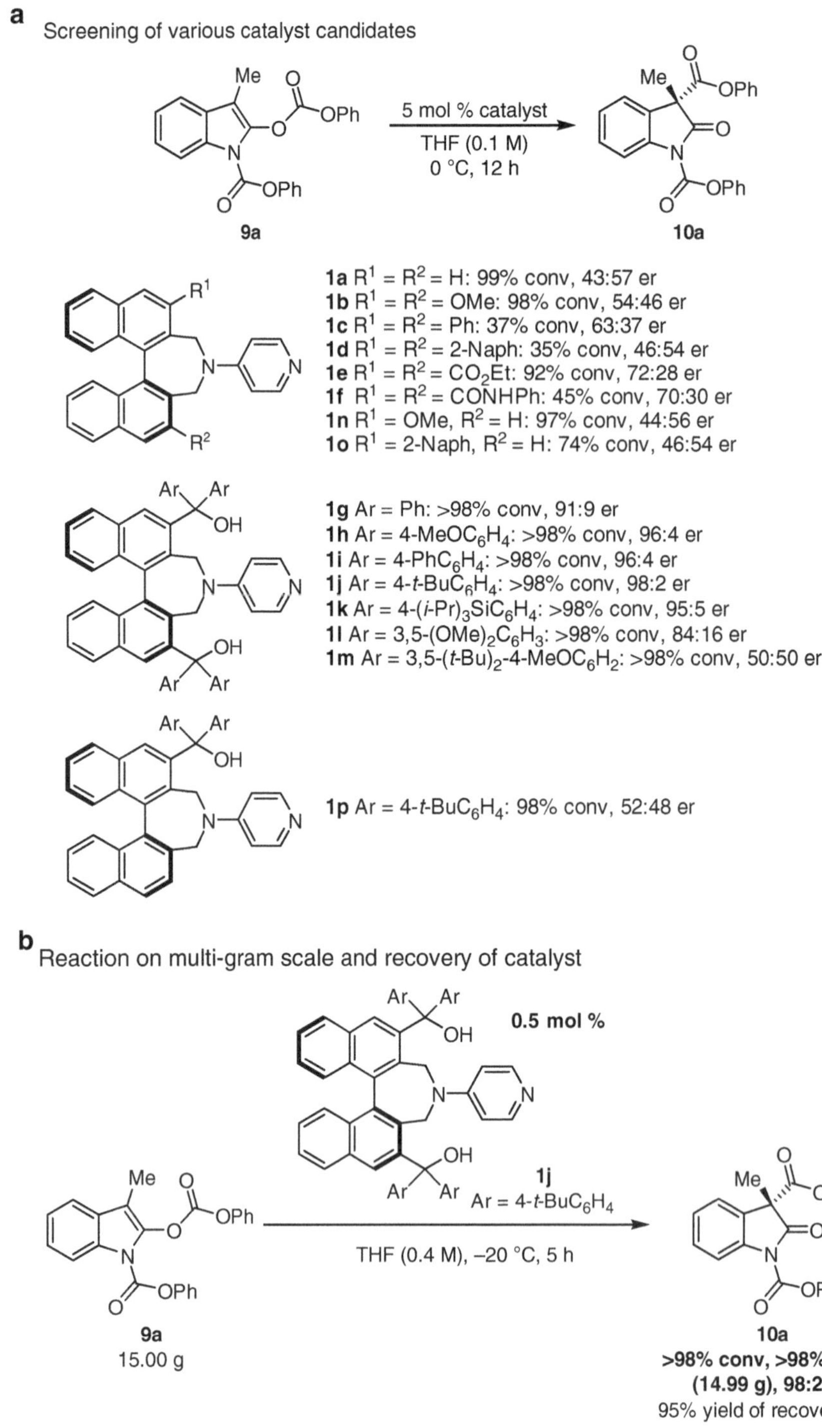


**Figure 4 | Enantioselective Steglich-type reactions promoted by catalyst 1j.** (**a**) Initial catalysts screening for the Steglich rearrangement of O-acylated oxindole derivative with $C_2$- or $C_1$-symmetric binaphthyl-based chiral nucleophilic catalyst. The results clearly indicated that tertiary alcohol unit at C3 and C3′ of binaphthyl are essential for achieving high yield and enantioselectivity. (**b**) The enantioselective reaction can be easily performed in multigram scale without any adverse effect on the efficiency or enantioselectivity of the process. Moreover, the same silica gel chromatography procedure delivers the recovered chiral catalyst in 95% yield.

To gain further insight regarding the role of the chiral catalyst's tertiary hydroxyl groups, density functional theory (DFT) calculations were performed at the B3LYP/6-31G(d) level, with **9a** as the substrate and compound **1g** as the catalyst, which represent simplified catalyst of **1j**. These investigations point to two critical attractive interactions, which directly involve the tertiary alcohol moiety, between the chiral catalyst and tight ion-pair intermediate. In the lowest energy transition state TS-I (Fig. 8 and Supplementary Table 7), which consistent with the experimental observations leads to the observed major enantiomer, there appears to be an H-bonding between the enolate oxygen of the ion pair and the catalyst's hydroxy unit; it is likely that the conformational rigidity imposed by the presence of the two aryl unit on the same carbon, forces the hydroxyl unit to be properly positioned to associate with the negatively charged oxygen. Additionally, there appears to be an additional electrostatic attraction[45,46] between the enolate and an ortho hydrogen of one of the aforementioned aryl units at the C3 site of the chiral

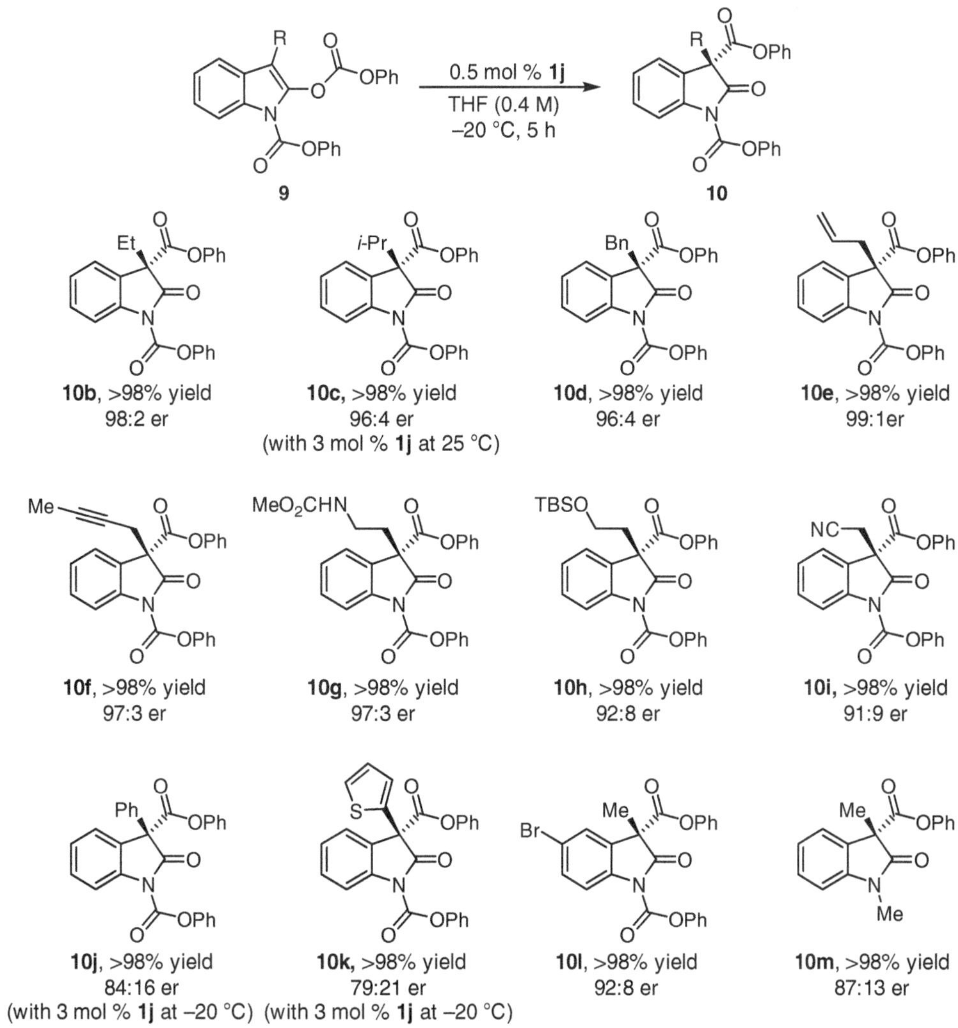


**Figure 5 | Enantioselective Steglich-type reactions of various substrates promoted by catalyst 1j.** A wide range of substrates may be subjected to enantioselective acyl rearrangement processes that afford quaternary carbon stereogenic centres in quantitative yield and in up to 98:2 er. The level of enantioselectivity can depend on the nature of the substituents within the oxindole ring. Reactions were performed on a 0.1 or 0.2 mmol (**10b**, **10d** and **10l**) scale in THF (0.4 M) under an argon atmosphere. Yields are of isolated and purified products after silica gel chromatography (±2%). Er values (±1%) were determined by high-performance liquid chromatography (HPLC) analysis.

catalyst. The second lowest mode of addition, as indicated by DFT calculations, is transition state **II** (Fig. 8 and Supplementary Table 8). While there appears to be similar electrostatic attractive forces operative here as well, **II** seems to suffer from a significant steric repulsion between the substrates phenoxy unit and the backbone of the catalyst's biaryl moiety. Nevertheless, DFT calculations indicate that there is in all likelihood steric repulsion between phenyl unit and the tertiary hydroxyl fragment that is involved in attractive interactions with the enolate group (see transition state (TS)-**I**, Fig. 8); this is supported by the findings that a catalyst with meta-(3,5)-substituted aryl groups give lower enantioselectivities in the reaction of **9a** (cf. **1l** or **1m** in Fig. 4). Further, as indicated by the transformation that afforded **10m** (Fig. 5) involving a more diminutive *N*-Me group, while less enantioselective than when a benzoate group is present (**9a**), there is still appreciable enantiofacial differentiation observed (98:2 er versus 87:13 er for **10a** and **10m**, respectively). This suggests that other, non-steric, factors are at play here as well. Towards this end, second order perturbation theory analysis in NBO at the M062X/6-31g** level indicates that C–H/enolate attraction might be notably stronger in the favoured TS-**I** (Supplementary Table 8).

**The scope of the reaction**. The utility of the present class of catalysts is not confined to Steglich rearrangements. For examples, as the representative data in Fig. 9 indicate, we find that binaphthyl-based DMAP derivatives with *tert*-alcohols at 3,3′-positions of a 1,1′-binaphthyl unit is not only effective in kinetic resolution of secondary carbinol and *d,l*-1,2-diol but in enantioselective desymmetrization of meso-1,2-diol as well. Initially, kinetic resolution of secondary carbinol rac-**11** with an array of binaphthyl-based DMAP derivatives was performed. The reaction in the presence of 5 mol % **1j** proceeded smoothly with moderate *s*-factor ($s = 11$, Fig. 9a, and Supplementary Figs 69–72). We then turned our attention to 1,2-diol which might be more suitable than simple carbinols, expecting positive interactions between a substrate and a catalyst through the hydrogen bonding.

Its merits note that, although a number of organocatalysts for kinetic resolution and desymmetrization of 1,2-diol have been disclosed, the significant majority involve the use of cyclic substrates because enantioselectivity was usually high compared with acyclic substrate[8]. The successful organocatalytic approaches with high enantioselectivity for kinetic resolution[47] and desymmetrization[48,49] of acyclic 1,2-diols were only reported by

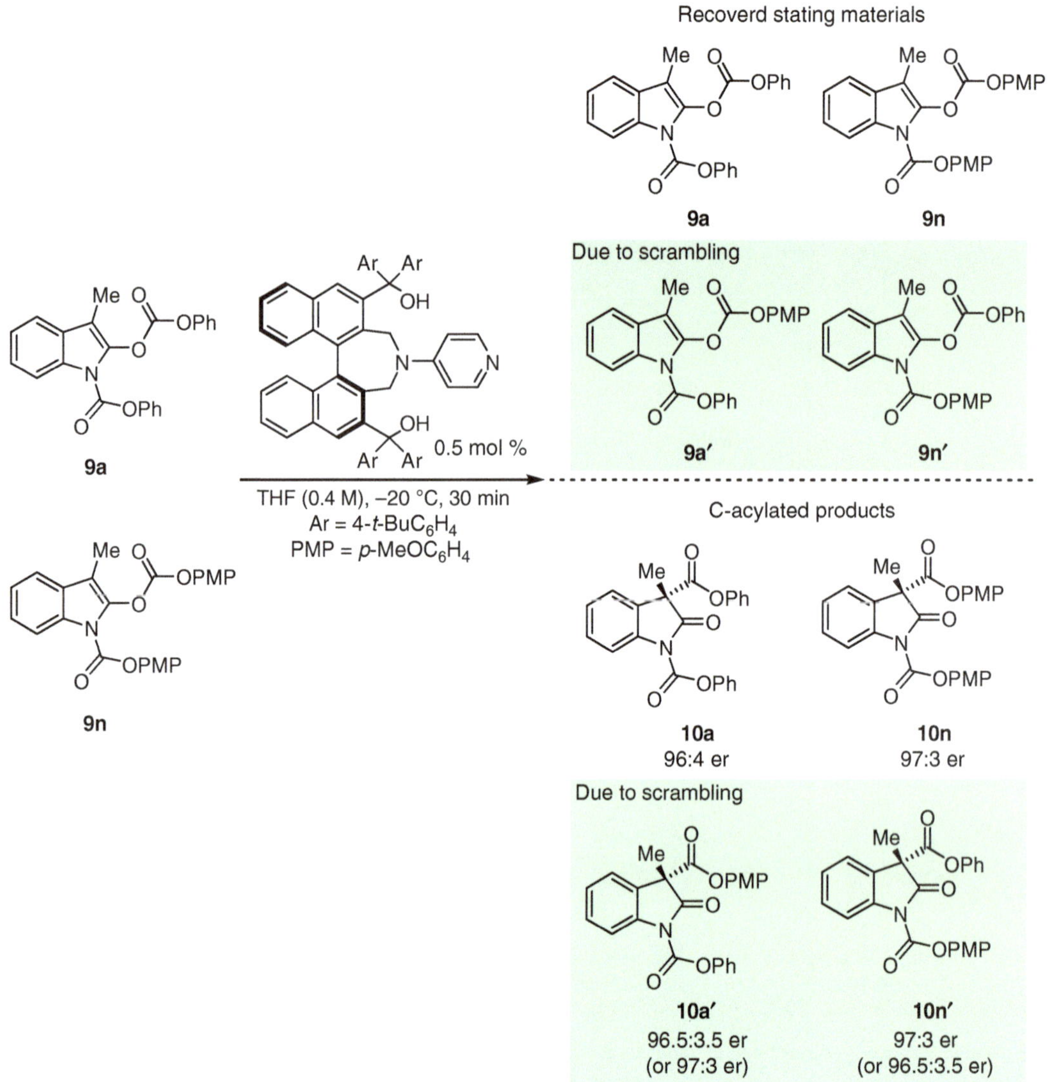


**Figure 6 | Cross-over studies of Steglich rearrangement of oxindole derivatives 9a and 9n.** Cross-over studies between two marked substrates **9a** and **9n** with optimal catalyst **1j** under optimal conditions for 30 min. Four starting materials **9a**, **9n**, **9a′** and **9n′**, and four products **10a**, **10n**, **10a′** and **10n′** were observed, indicating that formation of an ion-pair is involved in the reaction proceed.

Fujimoto utilizing cinchona alkaloids-based catalyst. However, higher catalyst loading (30 mol %) were required for these transformation. Other examples for desymmetrization of acyclic meso-1,2-diols resulted in moderate enantioselectivity of mono-acylated product along with significant amount of undesired bis-acylated product[14,50–52]. Thus, an efficient resolution ($s = 125$) of *d,l*-1,2-diol **13** was observed in an acylation reaction promoted by 0.5 mol % **1g** performed in THF at −78 °C for 9 h (Fig. 9b and Supplementary Figs 73–76). None of the di-acylation product was observed. Furthermore, desymmetrization of acyclic meso-1, 2-diol **15** required only 0.1 mol% **1g** and proceeded to furnish the desired mono-ester **16** in 79% yield with 97:3 er along with small amount of di-acylated product **17** and recovered meso-**15** in 7% and 12% yield, respectively (Fig. 9c and Supplementary Figs 77 and 78). The substantially faster rate of the first acylation reaction points to a high degree of catalyst-substrate molecular recognition that might be largely due to precise arrangements that are required for the type of H-bonding interactions depicted for 'design option 2' in Fig. 2. These preliminary results regarding enantioselective acylation processes, along with the aforementioned Steglich rearrangement reactions, suggest that the present set of catalysts may be applicable to a broad range of acyl transfer-type process.

## Discussion

The studies described above illustrate that the combination of a binaphthyl moiety, a widely utilized chiral motif in enantio-selective catalysis, a 4-aminopyridine, a most effective nucleophilic catalyst, can be effectively used in the development of exceptionally effective small organic molecules that can serve as chiral catalysts for acyl transfer processes. Another crucial design element relates to the presence of suitably positioned hydroxy groups that can facilitate reaction and promote enantioselectivity by establishing electrostatic association with an anionic Lewis basic fragment. The need for incorporation of additional substituents within the pyridyl ring, which can cause significant lowering of catalyst activity, may therefore be obviated. With the present blueprint in hand, development of highly efficient and selective catalysts that belong to the same class and may be used for related transformations such as intermolecular phosphorylation[53], silylation[54–58] or tosylation[59] of polyhydroxylated organic molecules may be envisioned. Investigations along these lines are now underway.

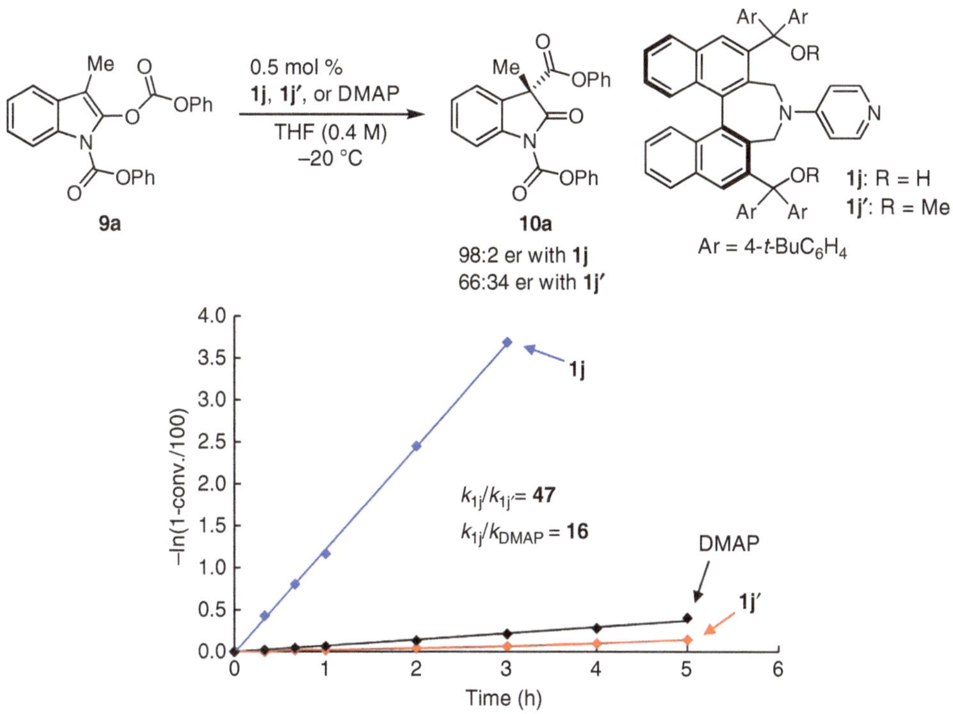


**Figure 7 | Kinetic studies in enantioselective Steglich rearrangement with catalyst 1j, 1j′ and DMAP.** Kinetic profiles with catalyst **1j**, **1j′** and DMAP in Steglich rearrangement of **9a** under the optimal conditions. When **1j** was used as catalyst, higher catalytic activity ($k_{1j} = 1.22\ h^{-1}$) and enantioselectivity (98:2 er) were observed compared with bis-methyl ether catalyst **1j′** ($k_{1j'} = 2.62 \times 10^{-2}\ h^{-1}$ and 66:34 er).

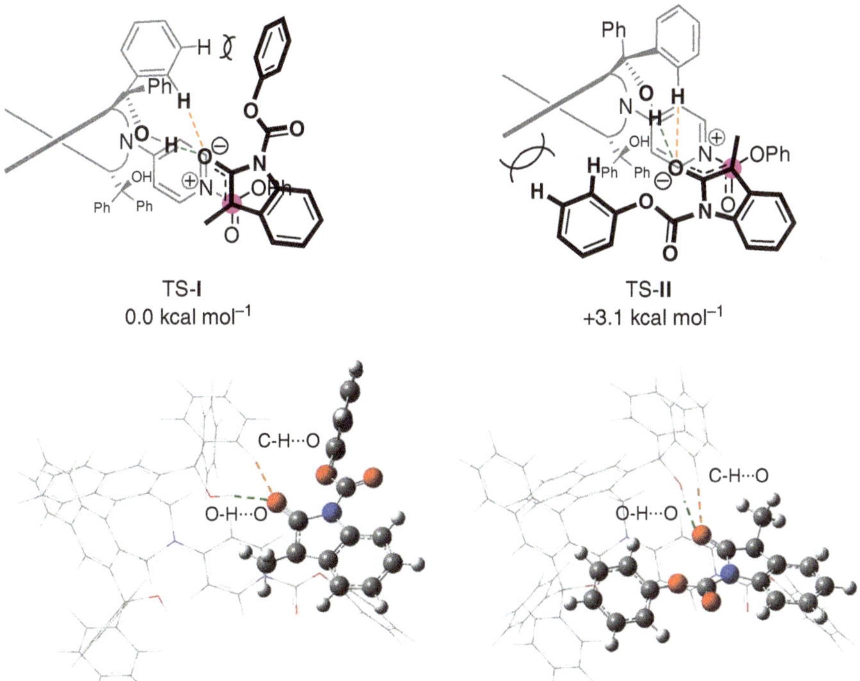


**Figure 8 | DFT calculations of two transition states TS-I and II in enantioselective Steglich rearrangement of 9a with 1g.** Geometries of transition states TS-**I** and **II** from **9a** with **1g** were fully optimized by calculation using the B3LYP (Becke's three-parameter hybrid method using the Lee-Yang-Parr correlation functional) DFT with the 6-31G(d) basis set. Harmonic vibrational frequencies were computed for all stationary points to characterize them as saddle points.

## Method

**General procedure fur enantioselective Steglich rearrangement under optimal conditions.** A solution of the catalyst in THF (50.0 mM) was prepared in advance using the following reaction.

To a solution of substrate **9** (0.2 mmol) in THF (0.5 ml, 0.4 M) was added a solution of the catalyst (20 µl, 50 mM in THF) at − 20 °C. The reaction mixture was stirred for 5 h and then 1 M aqueous HCl was added. The resulting mixture was extracted with EtOAc, dried over $MgSO_4$ and concentrated *in vacuo*. The purification

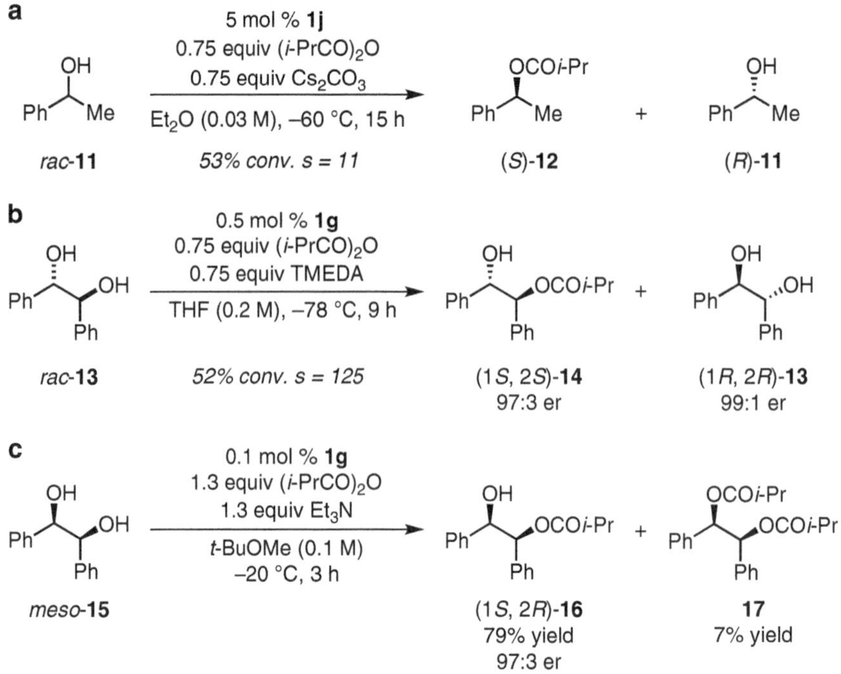


**Figure 9 | Application of the catalyst with *tert*-alcohols at 3,3′-positions of a 1,1′-binaphthyl unit to intermolecular acylations.** (**a**) Kinetic resolution of secondary carbinol with 5 mol% **1j**. The reaction proceeded smoothly with moderate selectivity factor. Conversion values (±1%) and er values (±1%) were determined by high-performance liquid chromatography (HPLC) analysis. (**b**) Kinetic resolution of acyclic *d,l*-1,2-diol **13** with 0.5 mol % **1g**. The reaction proceeded smoothly to afford (1*S*, 2*S*)-**14** and (1*R*, 2*R*)-**13** with high enantioselectivity. None of the di-acylated product was observed. Conversion values (±2%) were determined by $^{1}H$ nuclear magnetic resonance (NMR) analysis of the unpurified reaction mixtures. Er values (±1%) were determined by HPLC analysis. (**c**) Desymmetrization of acyclic meso-1,2-diol **15** with 0.1 mol% **1g**. Monoprotected diol (1*S*, 2*R*)-**16** was obtained in 79% yield with high enantioselectivity (97:3 er). Small amount of di-acylated product **17** and meso-**15** was obtained in 7 and 12% yield, respectively. Er values (±1%) were determined by HPLC analysis.

of the crude product by flash column chromatography on a short pad of silica gel (eluent:hexane/$Et_2O$ = 1/1, v/v) gave the corresponding product **10** (>98% yield). For nuclear magnetic resonance and high-performance liquid chromatography analyses of the products, see the Supplementary Figs 39–41 and 43–66.

## References

1. Steglich, W. & Höfle, G. *N,N*-dimethyl-4-pyridinamine, a very effective acylation catalyst. *Angew. Chem. Int. Ed.* **8,** 981–981 (1969).
2. Spivey, A. C. & Arseniyadis, S. Nucleophilic catalysis by 4-(dialkylamino)pyridines revisited—the search for optimal reactivity and selectivity. *Angew. Chem. Int. Ed.* **43,** 5436–5441 (2004).
3. Xu, S. *et al.* The DMAP-catalyzed acetylation of alcohols—a mechanistic study (DMAP = 4-(dimethylamino)pyridine). *Chem. Eur. J.* **11,** 4751–4757 (2005).
4. Hassner, A., Krepski, L. R. & Alexanian, V. Aminopyridines as acylation catalysts for tertiary alcohols. *Tetrahedron* **34,** 2069–2076 (1978).
5. Höfle, G., Steglich, W. & Vorbrüggen, H. 4-Dialkylaminopyridines as highly active acylation catalysts. *Angew. Chem. Int. Ed.* **17,** 569–583 (1978).
6. De Rycke, N., Couty, F. & David, O. R. P. Increasing the reactivity of nitrogen catalysts. *Chem. Eur. J* **17,** 12852–12871 (2011).
7. Wurz, R. P. Chiral dialkylaminopyridine catalysts in asymmetric synthesis. *Chem. Rev.* **107,** 5570–5595 (2007).
8. Müller, C. E. & Schreiner, P. R. Organocatalytic enantioselective acyl transfer onto racemic as well as *meso* alcohols, amines, and thiols. *Angew. Chem. Int. Ed.* **50,** 6012–6042 (2011).
9. Vedejs, E. & Chen, X. Kinetic resolution of secondary alcohols. Enantioselective acylation mediated by a chiral (dimethylamino)pyridine derivative. *J. Am. Chem. Soc.* **118,** 1809–1810 (1996).
10. Ruble, J. C. & Fu, G. C. Chiral π-complexes of heterocycles with transition metals: a versatile new family of nucleophilic catalysts. *J. Org. Chem.* **61,** 7230–7231 (1996).
11. Shaw, S. A., Aleman, P. & Vedejs, E. Development of chiral nucleophilic pyridine catalysts: applications in asymmetric quaternary carbon synthesis. *J. Am. Chem. Soc.* **125,** 13368–13369 (2003).
12. Spivey, A. C., Zhu, F., Mitchell, M. B., Davey, S. G. & Jarvest, R. L. Concise synthesis, preparative resolution, absolute configuration determination, and applications of an atropisomeric biaryl catalyst for asymmetric acylation. *J. Org. Chem.* **68,** 7379–7385 (2003).
13. Dalaigh, C. O., Hynes, S. J., Maher, D. J. & Connon, S. J. Kinetic resolution of *sec*-alcohols using a new class of readily assembled (*S*)-proline-derived 4-(pyrrolidino)-pyridine analogues. *Org. Biomol. Chem.* **3,** 981–984 (2005).
14. Yamada, S., Misono, T., Iwai, Y., Masumizu, A. & Akiyama, Y. New class of pyridine catalyst having a conformation switch system: asymmetric acylation of various *sec*-alcohols. *J. Org. Chem.* **71,** 6872–6880 (2006).
15. Duffey, T. A., Shaw, S. A. & Vedejs, E. AcOLeDMAP and BnOLeDMAP: conformationally restricted nucleophilic catalysts for enantioselective rearrangement of indolyl acetates and carbonates. *J. Am. Chem. Soc.* **131,** 14–15 (2009).
16. Crittall, M. R., Rzepa, H. S. & Carbery, D. R. Design, synthesis, and evaluation of a helicenoidal DMAP Lewis base catalyst. *Org. Lett.* **13,** 1250–1253 (2011).
17. Mandai, H., Irie, S., Mitsudo, K. & Suga, S. Studies on the synthesis of DMAP derivatives by diastereoselective Ugi reactions. *Molecules* **16,** 8815–8832 (2011).
18. Mandai, H. *et al.* Kinetic resolution of secondary alcohols by chiral DMAP derivatives prepared by the Ugi multicomponent reaction. *Heterocycles* **87,** 329–340 (2013).
19. Mandai, H. *et al.* Enantioselective Steglich rearrangement of oxindole derivatives by easily accessible chiral *N,N*-4-(dimethylamino)pyridine derivatives. *Org. Lett.* **17,** 4436–4439 (2015).
20. Essery, J. M. & Schofield, K. 769. The influence of steric factors on the properties of 4-aminopyridine derivatives. *J. Chem. Soc.* 3939–3953 (1961).
21. Held, I., Villinger, A. & Zipse, H. The stability of acylpyridinium cations and their relation to the catalytic activity of pyridine bases. *Synthesis* 1425–1430 (2005).
22. De Rycke, N., Marrot, J., Couty, F. & David, O. R. P. DMAP-[2.2]paracyclophane: observation of an unusual C–C insertion. *Eur. J. Org. Chem.* **2011,** 1980–1984 (2011).
23. Fu, G. C. Enantioselective nucleophilic catalysis with "planar-chiral" heterocycles. *Acc. Chem. Res.* **33,** 412–420 (2000).
24. Fu, G. C. Asymmetric catalysis with "planar-chiral" derivatives of 4-(dimethylamino)pyridine. *Acc. Chem. Res.* **37,** 542–547 (2004).
25. Wurz, R. P., Lee, E. C., Ruble, J. C. & Fu, G. C. Synthesis and resolution of planar-chiral derivatives of 4-(dimethylamino)pyridine. *Adv. Synth. Catal.* **349,** 2345–2352 (2007).
26. Kawabata, T., Muramatsu, W., Nishio, T., Shibata, T. & Schedel, H. A catalytic one-step process for the chemo- and regioselective acylation of monosaccharides. *J. Am. Chem. Soc.* **129,** 12890–12895 (2007).

27. Kawabata, T., Nagato, M., Takasu, K. & Fuji, K. Nonenzymatic kinetic resolution of racemic alcohols through an "induced fit" process. *J. Am. Chem. Soc.* **119,** 3169–3170 (1997).
28. Priem, G., Pelotier, B., Macdonald, S. J. F., Anson, M. S. & Campbell, I. B. The design of novel *N*-4'-pyridinyl-α-methyl proline derivatives as potent catalysts for the kinetic resolution of alcohols. *J. Org. Chem.* **68,** 3844–3848 (2003).
29. Spivey, A. C., Maddaford, A., Fekner, T., Redgrave, A. J. & Frampton, C. S. Synthesis of $C_2$-symmetric analogues of 4-(pyrrolidino)pyridine: new chiral nucleophilic catalysts. *J. Chem. Soc. Perkin Trans.* **1,** 3460–3468 (2000).
30. De, C. K., Klauber, E. G. & Seidel, D. Merging nucleophilic and hydrogen bonding catalysis: an anion binding approach to the kinetic resolution of amines. *J. Am. Chem. Soc.* **131,** 17060–17061 (2009).
31. Klauber, E. G., De, C. K., Shah, T. K. & Seidel, D. Merging nucleophilic and hydrogen bonding catalysis: an anion binding approach to the kinetic resolution of propargylic amines. *J. Am. Chem. Soc.* **132,** 13624–13626 (2010).
32. Klauber, E. G., Mittal, N., Shah, T. K. & Seidel, D. A dual-catalysis/anion-binding approach to the kinetic resolution of allylic amines. *Org. Lett.* **13,** 2464–2467 (2011).
33. De, C. K. & Seidel, D. Catalytic enantioselective desymmetrization of *meso*-diamines: a dual small-molecule catalysis approach. *J. Am. Chem. Soc.* **133,** 14538–14541 (2011).
34. De, C. K., Mittal, N. & Seidel, D. A dual-catalysis approach to the asymmetric Steglich rearrangement and catalytic enantioselective addition of *O*-acylated azlactones to isoquinolines. *J. Am. Chem. Soc.* **133,** 16802–16805 (2011).
35. Min, C., Mittal, N., De, C. K. & Seidel, D. A dual-catalysis approach to the kinetic resolution of 1,2-diaryl-1,2-diaminoethanes. *Chem. Commun.* **48,** 10853–10855 (2012).
36. Mittal, N., Sun, D. X. & Seidel, D. Kinetic resolution of amines via dual catalysis: remarkable dependence of selectivity on the achiral cocatalyst. *Org. Lett.* **14,** 3084–3087 (2012).
37. Seidel, D. The anion-binding approach to catalytic enantioselective acyl transfer. *Synlett* **25,** 783–794 (2014).
38. Mittal, N. *et al.* A dual-catalysis anion-binding approach to the kinetic resolution of amines: insights into the mechanism via a combined experimental and computational study. *J. Am. Chem. Soc.* **137,** 5748–5758 (2015).
39. Liang, D.-C., Luo, R.-S., Yin, L.-H., Chan, A. S. C. & Lu, G. A new synthetic route for axially chiral secondary amines with binaphthyl backbone and their applications in asymmetric michael reaction of aldehydes to nitroalkenes. *Org. Biomol. Chem.* **10,** 3071–3079 (2012).
40. Ooi, T., Kameda, M. & Maruoka, K. Design of *N*-spiro $C_2$-symmetric chiral quaternary ammonium bromides as novel chiral phase-transfer catalysts: synthesis and application to practical asymmetric synthesis of α-amino acids. *J. Am. Chem. Soc.* **125,** 5139–5151 (2003).
41. Yamada, S., Misono, T. & Tsuzuki, S. Cation – π interactions of a thiocarbonyl group and a carbonyl group with a pyridinium nucleus. *J. Am. Chem. Soc.* **126,** 9862–9872 (2004).
42. Hills, I. D. & Fu, G. C. Catalytic enantioselective synthesis of oxindoles and benzofuranones that bear a quaternary stereocenter. *Angew. Chem. Int. Ed.* **42,** 3921–3924 (2003).
43. Shaw, S. A. *et al.* Enantioselective TADMAP-catalyzed carboxyl migration reactions for the synthesis of stereogenic quaternary carbon. *J. Am. Chem. Soc.* **128,** 925–934 (2006).
44. Ruble, J. C. & Fu, G. C. Enantioselective construction of quaternary stereocenters: rearrangements of *O*-acylated azlactones catalyzed by a planar-chiral derivative of 4-(pyrrolidino)pyridine. *J. Am. Chem. Soc.* **120,** 11532–11533 (1998).
45. Lippert, K. M. *et al.* Hydrogen-bonding thiourea organocatalysts: the privileged 3,5-bis(trifluoromethyl)phenyl group. *Eur. J. Org. Chem.* **2012,** 5919–5927 (2012).
46. Yang, H. & Wong, M. W. Oxyanion hole stabilization by C-H · · · O interaction in a transition state—a three-point interaction model for cinchona alkaloid-catalyzed asymmetric methanolysis of *meso*-cyclic anhydrides. *J. Am. Chem. Soc.* **135,** 5808–5818 (2013).
47. Mizuta, S., Ohtsubo, Y., Tsuzuki, T., Fujimoto, T. & Yamamoto, I. Kinetic resolution of *d,l*-1,2-diols catalyzed by amine-phosphinite bifunctional organocatalysis derived from quinidine. *Tetrahedron Lett.* **47,** 8227–8229 (2006).
48. Mizuta, S., Sadamori, M., Fujimoto, T. & Yamamoto, I. Asymmetric desymmetrization of *meso*-1,2-diols by phosphinite derivatives of cinchona alkaloids. *Angew. Chem. Int. Ed.* **42,** 3383–3385 (2003).
49. Aida, H. *et al.* Enantioselective acylation of 1,2- and 1,3-diols catalyzed by aminophosphinite derivatives of (1*S*,2*R*)-1-amino-2-indanol. *Org. Lett.* **14,** 812–815 (2012).
50. Schedel, H. *et al.* Asymmetric desymmetrization of *meso*-diols by $C_2$-symmetric chiral 4-pyrrolidinopyridines. *Beilstein J. Org. Chem* **8,** 1778–1787 (2012).
51. Oriyama, T., Imai, K., Sano, T. & Hosoya, T. Highly efficient catalytic asymmetric acylation of *meso*-1,2-diols with benzoyl chloride in the presence of a chiral diamine combined with $Et_3N$. *Tetrahedron Lett.* **39,** 3529–3532 (1998).
52. Vedejs, E., Daugulis, O. & Tuttle, N. Desymmetrization of *meso*-hydrobenzoin using chiral, nucleophilic phosphine catalysts. *J. Org. Chem.* **69,** 1389–1392 (2004).
53. Morgan, A. J., Komiya, S., Xu, Y. & Miller, S. J. Unified total syntheses of the inositol polyphosphates: D-I-3,5,6$P_3$, D-I-3,4,5$P_3$, D-I-3,4,6$P_3$, and D-I-3,4,5,6$P_4$ via catalytic enantioselective and site-selective phosphorylation. *J. Org. Chem.* **71,** 6923–6931 (2006).
54. Zhao, Y., Rodrigo, J., Hoveyda, A. H. & Snapper, M. L. Enantioselective silyl protection of alcohols catalysed by an amino-acid-based small molecule. *Nature* **443,** 67–70 (2006).
55. Zhao, Y., Mitra, A. W., Hoveyda, A. H. & Snapper, M. L. Kinetic resolution of 1,2-diols through highly site- and enantioselective catalytic silylation. *Angew. Chem. Int. Ed.* **46,** 8471–8474 (2007).
56. Rodrigo, J. M., Zhao, Y., Hoveyda, A. H. & Snapper, M. L. Regiodivergent reactions through catalytic enantioselective silylation of chiral diols. Synthesis of sapinofuranone A. *Org. Lett.* **13,** 3778–3781 (2011).
57. Manville, N., Alite, H., Haeffner, F., Hoveyda, A. H. & Snapper, M. L. Enantioselective silyl protection of alcohols promoted by a combination of chiral and achiral Lewis basic catalysts. *Nat. Chem.* **5,** 768–774 (2013).
58. Xu, L.-W., Chen, Y. & Lu, Y. Catalytic silylations of alcohols: turning simple protecting-group strategies into powerful enantioselective synthetic methods. *Angew. Chem. Int. Ed.* **54,** 9456–9466 (2015).
59. Matsumura, Y., Maki, T., Murakami, S. & Onomura, O. Copper ion-induced activation and asymmetric benzoylation of 1,2-diols: kinetic chiral molecular recognition. *J. Am. Chem. Soc.* **125,** 2052–2053 (2003).

## Acknowledgements

This research was partially supported by a Grant-in-Aid for Scientific Research on Innovative Areas 'Advanced Molecular Transformations by Organocatalysts' from MEXT (Japan). We thank the Division of Instrumental Analysis, Department of Instrumental Analysis & Cryogenics, Advanced Science Research Center, Okayama University for nuclear magnetic resonance (NMR), high-resolution mass spectrometry measurements (FAB) and X-ray single crystal structural analyses. Some of this work (ESI) was conducted at Institute of Transformative Bio-Molecules (WPI-ITbM) at Nagoya University. The authors are also grateful to Professor Dr Amir, H. Hoveyda (Boston College) for discussions and critical comments on this manuscript.

## Author contributions

H.M. conceived, designed and wrote the manuscript with the assistance of S.S. H.M. and K.F. designed the experiment and analysed the data. K.F. performed all experiments. H.Y. optimized the reaction conditions for the desymmetrization of meso-1,2-diol. K.A. synthesized catalyst **1p**. H.M. and K.M. performed X-ray single-crystal structure analysis. T.K. conducted computational studies and wrote the manuscript involving computational studies. H.M. and S.S. directed the investigations and organized the research.

## Additional information

**Competing financial interests:** The authors declare no competing financial interests.

# 19

# The limits of precision monomer placement in chain growth polymerization

Guillaume Gody[1], Per B. Zetterlund[2], Sébastien Perrier[1,3] & Simon Harrisson[4]

Precise control over the location of monomers in a polymer chain has been described as the 'Holy Grail' of polymer synthesis. Controlled chain growth polymerization techniques have brought this goal closer, allowing the preparation of multiblock copolymers with ordered sequences of functional monomers. Such structures have promising applications ranging from medicine to materials engineering. Here we show, however, that the statistical nature of chain growth polymerization places strong limits on the control that can be obtained. We demonstrate that monomer locations are distributed according to surprisingly simple laws related to the Poisson or beta distributions. The degree of control is quantified in terms of the yield of the desired structure and the standard deviation of the appropriate distribution, allowing comparison between different synthetic techniques. This analysis establishes experimental requirements for the design of polymeric chains with controlled sequence of functionalities, which balance precise control of structure with simplicity of synthesis.

[1] Department of Chemistry, University of Warwick, Warwick CV4 7AL, UK. [2] Centre for Advanced Macromolecular Design, School of Chemical Engineering, University of New South Wales, Sydney, New South Wales 2052, Australia. [3] Faculty of Pharmacy and Pharmaceutical Sciences, Monash University, Parkville, Victoria 3052, Australia. [4] Laboratoire des Interactions Moléculaires et Réactivité Chimique et Photochimique, CNRS UMR5623, Université Paul Sabatier Toulouse III, 118 Route de Narbonne, Toulouse 31062, France. Correspondence and requests for materials should be addressed to S.H. (email: polyharrisson@gmail.com).

How precisely can we control the location of a monomer in a synthetic polymer? This question is crucial to the synthesis of aperiodic copolymers: 'copolymers in which monomer sequence distribution is not regular but follows the same arrangement in all chains'[1]. Included in this definition are polymers that contain an ordered sequence of distinct blocks, and polymers that contain specific monomers at precisely defined locations[2–4]. Such polymers could serve as a medium for information storage, as does DNA, or catalyse complex reactions, as do enzymes[5]. But for these applications to be realized, precise control over structure is essential, and any deviation from the desired structure must be quantified.

If only small quantities of polymer are required, near perfect control over structure can be obtained using stepwise synthesis on a solid[6,7] or soluble polymer[8,9] support, as in peptide synthesis. For the rapid preparation of moderate to large quantities of well-defined polymer, however, we must turn to living or controlled chain growth polymerizations[10–23]. But unlike stepwise techniques, chain growth polymerization is a statistical process that produces a distribution of products. In this paper, we establish a lower bound to the structural variation that exists even under ideal conditions of living polymerization. Our results are applicable to all chain polymerizations, which are carried out under living or near-living conditions, including ring-opening metathesis polymerization (ROMP)[24], the various types of reversible deactivation radical polymerization (RDRP, also known as controlled/'living' radical polymerization)[25], and ionic polymerizations[26,27].

The production of aperiodic copolymers requires control over the arrangement of monomers, which should be replicated in all chains. The degree of control required can be interpreted with varying degrees of severity, corresponding to the sequence-defined, multisite and multiblock copolymers considered by Lutz[1] (Fig. 1): (a) control over the absolute position of a monomer with respect to the chain end, or the absolute separation between two monomers (sequence-defined copolymers); (b) control over the relative position of a monomer with respect to the total chain length, or the relative separation between two monomers (multisite copolymers) or (c) control over the number and order of blocks in a multiblock copolymer, without regard to the distribution of each individual block (multiblock copolymers).

The level of structural control that is actually achieved can be quantified in two ways: in terms of the fraction of chains that correspond to the desired structure (the yield), and in terms of the standard deviation, $\sigma$, of the distribution. High yields and low standard deviations imply high levels of control. In this article, we propose the standard deviation of the monomer distribution as a quantitative measure of precision; derive simple expressions for these distributions under ideal chain growth polymerization conditions and apply these expressions to a number of recent examples of multisite and multiblock copolymers. Our results demonstrate that the level of control over monomer placement is quite limited, even under ideal conditions, and that multiblock copolymers comprised of many short blocks will inevitably contain a large proportion of defective chains.

## Results

**Absolute position of monomers.** Absolute control over the location of an individual monomer is the most demanding goal of polymer synthesis. However, techniques based on chain-growth polymerization lead, even in the best-controlled polymerizations, to a distribution of structures. For very short chains, separation of a single oligomeric structure may be possible, but requires extensive use of chromatography[28,29].

Under ideal conditions of living polymerization, the degree of polymerization obeys a Poisson distribution[30], whose probability mass function is given in Table 1 (full derivations of all results in Table 1 may be found in Supplementary Note 1). In reality, unavoidable side reactions such as termination and chain transfer result in deviations from the Poisson distribution. The ideal is most closely approached by anionic polymerization[31], but this technique must be carried out in dry, nonprotic solvents and is incompatible with most functional groups. RDRP techniques[32–39] tolerate a wide range of functionalities, but lead to broader distributions because of termination reactions and the addition of multiple monomers per activation/deactivation cycle.

The highest level of control over monomer placement achievable using chain growth polymerization techniques is

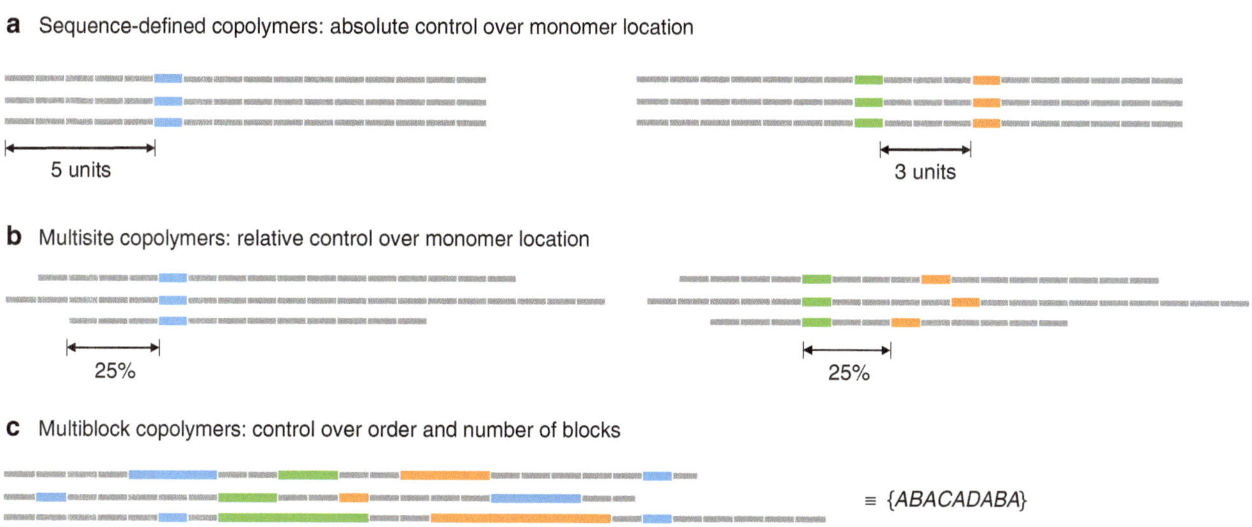


**Figure 1 | Some aims of precision polymer synthesis.** (**a**) Absolute control of monomer position and/or separation in a sequence-defined copolymer. (**b**) Relative control of monomer position and/or separation in a multisite copolymer. (**c**) Control of number and order of blocks in a multiblock copolymer. Throughout this paper, the letters *A*, *B*, *C* and *D* refer to different types of monomer. A collection of polymer chains containing on average *n* units of *A* followed by *m* units of *B* is represented as $\{A_nB_m\}$, whereas a single polymer chain containing exactly *N* units of *A* followed by *M* units of *B* is represented as $A_NB_M$.

**Table 1 | Probability distributions of various properties of Poisson-distributed block copolymers*.**

| Property | Distribution | p.m.f./p.d.f.† | Expected value | Standard deviation |
|---|---|---|---|---|
| Number of monomers, $N$, per chain | Poisson ($L$) | $P(N=k)=\frac{L^k}{k!}e^{-L}$ | $L$ | $\sqrt{L}$ |
| Absolute locations, $k$, contained in $i$th block | Difference of cumulative Poisson distributions | $P(M_k \in block\ i) = \frac{\Gamma(k,L_{i-1})-\Gamma(k,L_i)}{\Gamma(k)}$ | $1+\frac{L_i+L_{i-1}}{2}$ | $\sqrt{\frac{L_i+L_{i-1}}{2}+\frac{l_i^2}{12}}$ |
| Relative position, $r$, of junction between $i$th and $(i+1)$th blocks | Beta ($L_i$, $L_{total}-L_i$) | $f(r)=\frac{r^{L_i-1}(1-r)^{L_{total}-L_i-1}}{B(L_i,L_{total}-L_i)}$ | $\frac{L_i}{L_{total}}$ | $\frac{1}{L_{total}}\sqrt{\frac{L_i(L_{total}-L_i)}{L_{total}+1}}$ |
| Relative locations, $r$, contained in $i$th block | Difference of cumulative beta distributions | $P(M_r \in block\ i) = I_r(L_i, L_{total}-L_i) - I_r(L_{i-1}, L_{total}-L_{i-1})$ | $\frac{L_i+L_{i-1}+1}{2(L_{total}+1)}$ | $\approx \frac{1}{L_{total}}\sqrt{\frac{H(L_{i-1},L_{total}-L_i)}{2}+\frac{(l_i+3)^2-4}{12}}$ |
| Number of blocks in an $N$-block copolymer (all blocks of equal length, $l$) | Binomial | $P(n=k) = \binom{N}{k}\cdot(1-e^{-l})^k \cdot e^{-(N-k)l}$ | $N(1\text{-}e^{-l})$ | $\sqrt{Ne^{-l}(1-e^{-l})}$ |

*Complete derivations of these results are given in Supplementary Note 1. Results for relative locations, $r$, are approximate. The functions $\Gamma(x)$, $\Gamma(x,y)$, $B(x,y)$ and $I_z(x,y)$, the binomial coefficient $\binom{x}{y}$ and the harmonic mean $H(x,y)$ have their usual meanings and are defined in Supplementary Table 1. $L_i$ is the sum of the average lengths of the first $i$ blocks of a copolymer of total average length $L_{total}$. The average length of the $i$th block is denoted $l_i$.
†Probability mass function (for discrete distributions)/probability density function (for continuous distributions).

obtained by addition of a non-homopolymerizing monomer ($B$), such as maleic anhydride, to a polymer of average length $n$ ($\{A_n\}$), followed by subsequent chain extension[10–13]. As $B$ cannot homopolymerize, only one monomer of $B$ will be added to each chain. Use of an excess of $B$ ensures that one monomer of $B$ will be added to every chain, giving the average structure $\{A_n\}B_1$. In a variation on this technique, addition of B may occur by a step-growth reaction, such as via azide-alkyne coupling, followed by chain extension from a suitable functional group[40,41].

Numbering the monomers from the start of the chain, the expected position of $B$ is $n+1$. The probability that $B$ will be found in this position is approximately $\frac{1}{\sqrt{2\pi n}}$ and the standard deviation of the distribution is $\sqrt{n}$. For example, in a polymer with average structure $\{A_{10}\}B_1\{C_{10}\}$, only 12.5% of the chains will contain monomer $B$ as the 11th unit, while in 26.6% of the chains, $B$ will appear before position 8 or after position 14.

In aperiodic copolymers prepared by sequential polymerizations of monomers of similar reactivity[14–20], or by addition of alternating[21,22] or very rapidly polymerizing[23] monomers at intervals during the polymerization, all blocks are Poisson-distributed, at best. Thus, an $\{A_{10}B_1C_{10}\}$ polymer will include chains with no $B$ units, as well as chains that contain 2 or more[5,42]. The proportion of chains that contain a unit of $B$ in position 11 (the midpoint) can be calculated by excluding all chains that contain 11 or more units of $A$ (41.7% of chains), as well as all chains in which the sum of units of $A$ and $B$ is $<11$ (46.0% of chains). The 12.3% of chains that remain are not significantly fewer than the corresponding proportion of $\{A_{10}\}B_1\{C_{10}\}$. In $\{A_{10}\}B_1\{C_{10}\}$, however, each polymer chain contained a single copy of $B$. In $\{A_{10}B_1C_{10}\}$, only 4.6% of chains contain one and only one unit of $B$ in position 11.

In general, the probability that the $k$th monomer of a multiblock copolymer is part of the $n$th block is given by the difference between two Poisson cumulative distribution functions (Table 1). As the average position of a block moves further along the polymer chain, its location becomes more diffuse. Shorter blocks are affected to a greater extent than longer blocks, as shown by Fig. 2a,c, which show the location distributions of each block of two *ABACADACAB* decablock copolymers. Those of the short blocks broaden rapidly and overlap significantly. The distributions of the longer blocks broaden more slowly and overlap less, but fewer changes in composition are possible in a given length of polymer. The standard deviation of the location of the block (Table 1) provides a quantitative measure of the attainable precision: in $\{A_{10}B_1C_{10}\}$, the expected location of $B$ is 11.5, with a standard deviation of 3.3; $B$ occurs before position 8 or after position 14 in 29.2% of chains.

**Relative position of monomers.** Although absolute control over monomer placement is the ideal, it would often be acceptable to control only the relative position of a monomer with respect to the total chain length. In a diblock copolymer, $\{A_nB_m\}$, comprising two Poisson-distributed blocks, the relative length of the $A$ block with respect to the entire polymer chain is closely approximated by a Beta($n$,$m$) distribution (see Supplementary Figs 1 and 9 and Supplementary Note 2). The beta distribution is one of the most widely used distributions in statistics, and is often used to model random variables that are limited to a finite range, such as world cloud cover[43] or the distribution of genetic variation[44].

In the $\{A_{10}\}B_1\{C_{10}\}$ copolymer, the distribution of $B$ can be approximated as the distribution of the interface between $A$ and $C$ blocks in an $\{A_{10}C_{10}\}$ diblock copolymer: a Beta(10,10) distribution with expected value 0.50 and standard deviation 0.11. Thus, in $\sim$30% of the chains, we expect to find $B$ outside the relative length range of 0.39–0.61—that is, outside the central fifth of the polymer. In general, the distribution of the relative location of a single inserted monomer narrows as the total length of the polymer increases, and is broadest when the monomer is inserted at the midpoint of the chain (Supplementary Fig. 2).

In an $\{A_nB_mC_p\}$ copolymer whose inserted block $\{B_m\}$ is also Poisson distributed, the probability that a monomer of type $B$ will be found in a given relative position is given by the difference between two cumulative beta distribution functions with parameters ($n$, $m+p$) and ($n+m$, $p$), respectively (Table 1). Thus, in the $\{A_{10}B_1C_{10}\}$ copolymer, the expected relative position of the $B$ block is 0.50, and its standard deviation is 0.11—almost identical to that of the $\{A_{10}\}B_1\{C_{10}\}$ copolymer. Only 17.6% of chains contain a monomer of type $B$ at the midpoint. Doubling

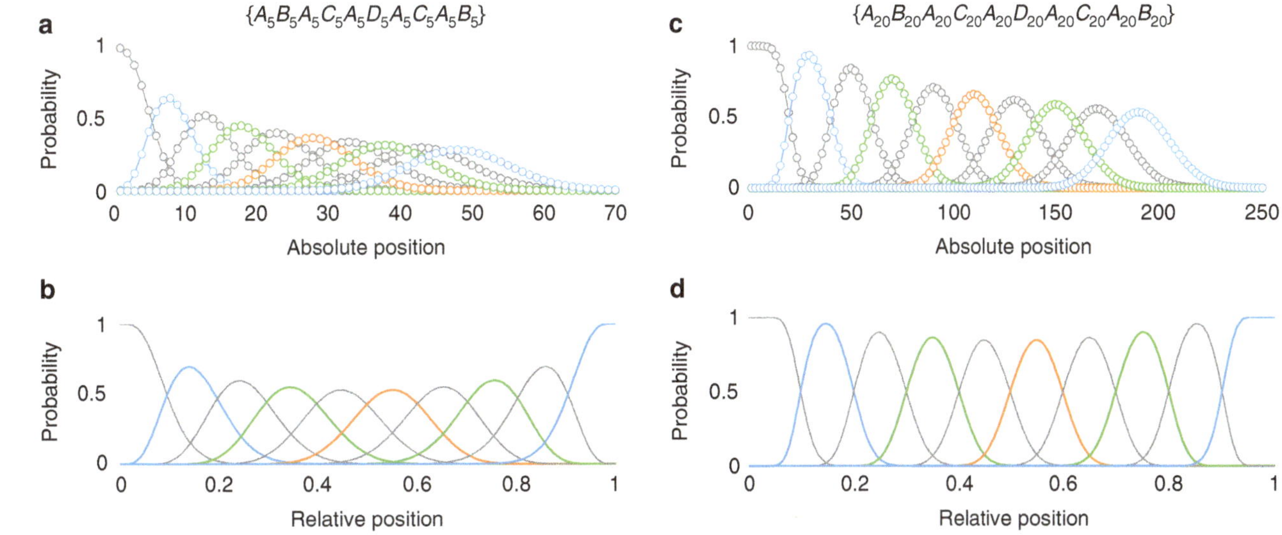


**Figure 2 | Monomer distributions.** Probability of finding monomers from each block of an $\{A_5B_5A_5C_5A_5D_5A_5C_5A_5B_5\}$ (**a**,**b**) or $\{A_{20}B_{20}A_{20}C_{20}A_{20}D_{20}A_{20}C_{20}A_{20}B_{20}\}$ (**c**,**d**) decablock copolymer as a function of absolute position (**a**,**c**) and relative position (**b**,**d**) along the polymer chain.

the average length of the $B$ block ($\{A_{10}B_2C_{10}\}$) increases the proportion of chains that contain a monomer of type $B$ at the midpoint to 33.6%, but the length must be increased to 11 ($\{A_{10}B_{11}C_{10}\}$) to ensure that 95% of the chains contain a monomer of type $B$ at the midpoint. The effects of average block length and position on relative position distributions are illustrated in Fig. 2 for two decablock copolymers.

The effect of Poisson broadening on the control that can be achieved over monomer location can be demonstrated using two recently reported multisite polymers[21]. The polymers were prepared by successive additions of the non-homopolymerizing monomers *N*-benzyl maleimide (Bz-MI), *N*-propyl maleimide (Pr-MI) and *N*-pentafluorophenyl maleimide (PFP-MI) to a single electron transfer living radical polymerization (SET-LRP) of styrene with targeted $\bar{X}_n$ of 65 (ref. 21). For the first polymer, maleimide solutions were added at intervals during the polymerization corresponding to 28, 46 and 70% styrene conversion, leading to the insertion of short segments of copolymer containing on average 1 unit of Bz-MI, Pr-MI or PFP-MI. The second polymer was prepared using a lower initial styrene to initiator ratio and delaying the addition of each type of maleimide until the styrene had reached high conversion. After each maleimide had been fully incorporated, additional styrene was added to generate the intervening polystyrene segments. The effect was to halve the average length of the segments containing Bz-MI and Pr-MI, from 6 in the first synthesis to 3 in the second synthesis. The average length of the PFP-MI block was 3 in both syntheses.

The distribution of each maleimide in the polymer was then calculated, making use of the following simplifying assumptions: (i) all segments were assumed to follow a Poisson distribution; (ii) the maleimides were assumed to be evenly distributed within their respective segments. In this way, the approximate distribution of each maleimide could be obtained by calculating the distribution of each maleimide-containing segment, then normalizing with respect to the segment's maleimide content. Differences in homo- and cross-propagation rate constants of styrene and maleimide were not explicitly considered, as it was assumed that in all cases, the rate of reversible deactivation was much greater than that of propagation. If this assumption is invalid (for example, as a result of very rapid cross-propagation occurring at a comparable rate to deactivation) then broader distributions are obtained.

Despite the much shorter segments obtained in the second synthesis, the maleimide distributions are nearly identical (Fig. 3 and Supplementary Table 2). Their standard deviations are determined primarily by the length of the surrounding blocks, with only a small contribution from the length of the block itself. The calculated distributions underestimate the true degree of structural variation, as a Poisson-distributed copolymer with $\bar{X}_n$ of 53 has a dispersity of 1.02, whereas the measured dispersities of the copolymers were 1.16 (sequential maleimide additions) and 1.27 (alternating maleimide and styrene additions).

We have applied similar analyses to an icosablock copolymer prepared by sequential reversible addition-fragmentation chain transfer (RAFT) polymerizations[14] (Supplementary Fig. 3), a multisite copolymer prepared by sequential additions of *exo*-functionalized norbornene monomers to the ROMP polymerization of an *endo*-functional norbornene monomer (Supplementary Fig. 4)[23], and two undecablock copolymers prepared by sequential anionic polymerizations (Supplementary Fig. 5)[15,16]. In the RAFT and ROMP copolymers, the use of multiple short blocks (typical length 3 and 2, respectively) results in poorly defined structures. The much longer blocks obtained using anionic copolymerization (minimum block lengths 37 (ref. 15) and 130 (ref. 16)) allow greater control over the relative position of a block within the chain.

**Existence of blocks in multiblock copolymers.** The least restrictive goal of aperiodic polymer synthesis is simply to prepare a multiblock copolymer in which each polymer chain contains the desired number of blocks, in the correct order. If, for example, a block of monomer A encoded 1, a block of monomer B encoded 0 and monomer C served to separate consecutive blocks of A or B, a binary string could be represented using a multiblock copolymer, even without specifying the length of each block. The only condition to be fulfilled is that every chain must contain at least one monomer unit from each block.

The probability that a Poisson-distributed block of average length $l$ contains at least one monomer is $1-e^{-l}$. For a polymer composed of $n$ blocks, the proportion of chains that contain all

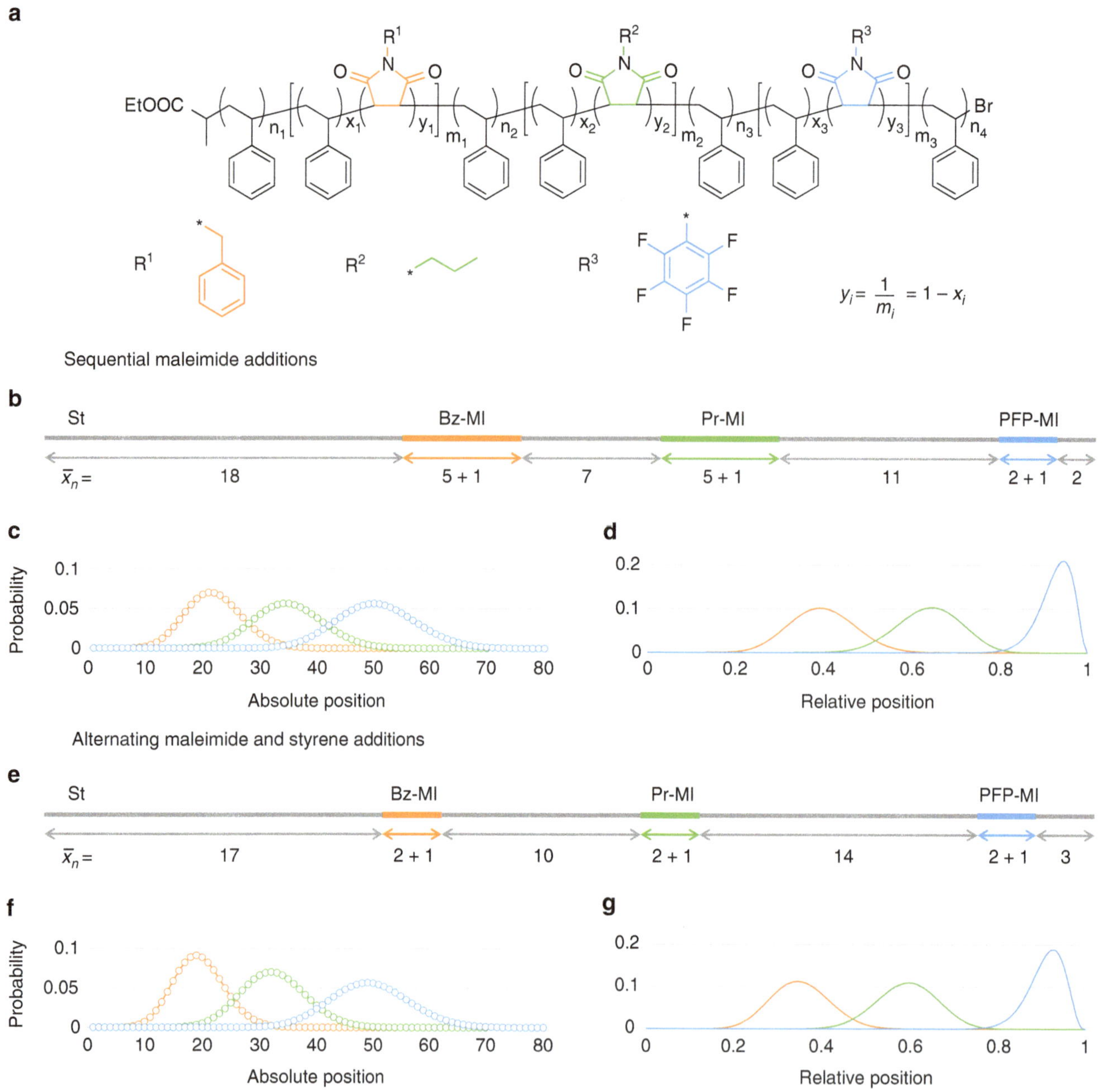


**Figure 3 | Comparison of multisite copolymers.** (**a**) containing benzyl maleimide (Bz-MI), propyl maleimide (Pr-MI) and pentafluorophenyl maleimide (PFP-MI) insertions, prepared using maleimide additions only (**b**-**d**) or by alternating additions of maleimide and styrene (**e**-**g**) to an SET-LRP polymerization of styrene[21]. The expected structures of each copolymer (**a**,**b**,**e**) consist of short regions containing a single maleimide unit, dispersed along a polystyrene chain. The distribution of each type of maleimide is shown in terms of absolute position (**c**,**f**) and position relative to the total chain length (**d**,**g**).

blocks is obtained by multiplying the probabilities for each block. If all blocks are nominally of equal length, the minimum block length, $l_{min}$, required for a given proportion, $\alpha$, of chains to contain all blocks is proportional to $\ln(n)$.

$$l_{min} \geq -\ln\left(1-\alpha^{1/n}\right) \approx \ln(n) - \ln(1-\alpha) \qquad (1)$$

Thus, polymers consisting of many short blocks will contain a significant proportion of defective chains (see Supplementary Figs 3 and 4 for details). Attempts to insert a block of average length 3 will fail in ~5% of chains; in a copolymer containing 18 such blocks, the majority of chains will be defective (Supplementary Fig. 3). Doubling the targeted block length to 6 results in an exponential decrease in the proportion of missing blocks, which would allow a 20-block copolymer to be prepared with only 5% of defective chains (see Supplementary Fig. 6 for details). These theoretical results were confirmed in a real system by quantifying the residual level of unreacted RAFT agent, corresponding to chains of length zero, after preparation of a series of oligomers of 4-acryloyl morpholine of average degree of polymerization 3–6 (see Supplementary Fig. 7 for details).

The validity of the Poisson distribution as a model for the number of units incorporated per chain was further demonstrated by adding a bismaleimide, bis(4-maleimidophenyl)methane, to the early stages of a RAFT polymerization of styrene (Fig. 4). The ratio of bismaleimide to RAFT agent was 1:2, such that on average one maleimide unit was added per chain. The polymerization was allowed to continue for 1 h, during which time the maleimide was entirely consumed. If exactly one maleimide had been added per chain, only the structure H

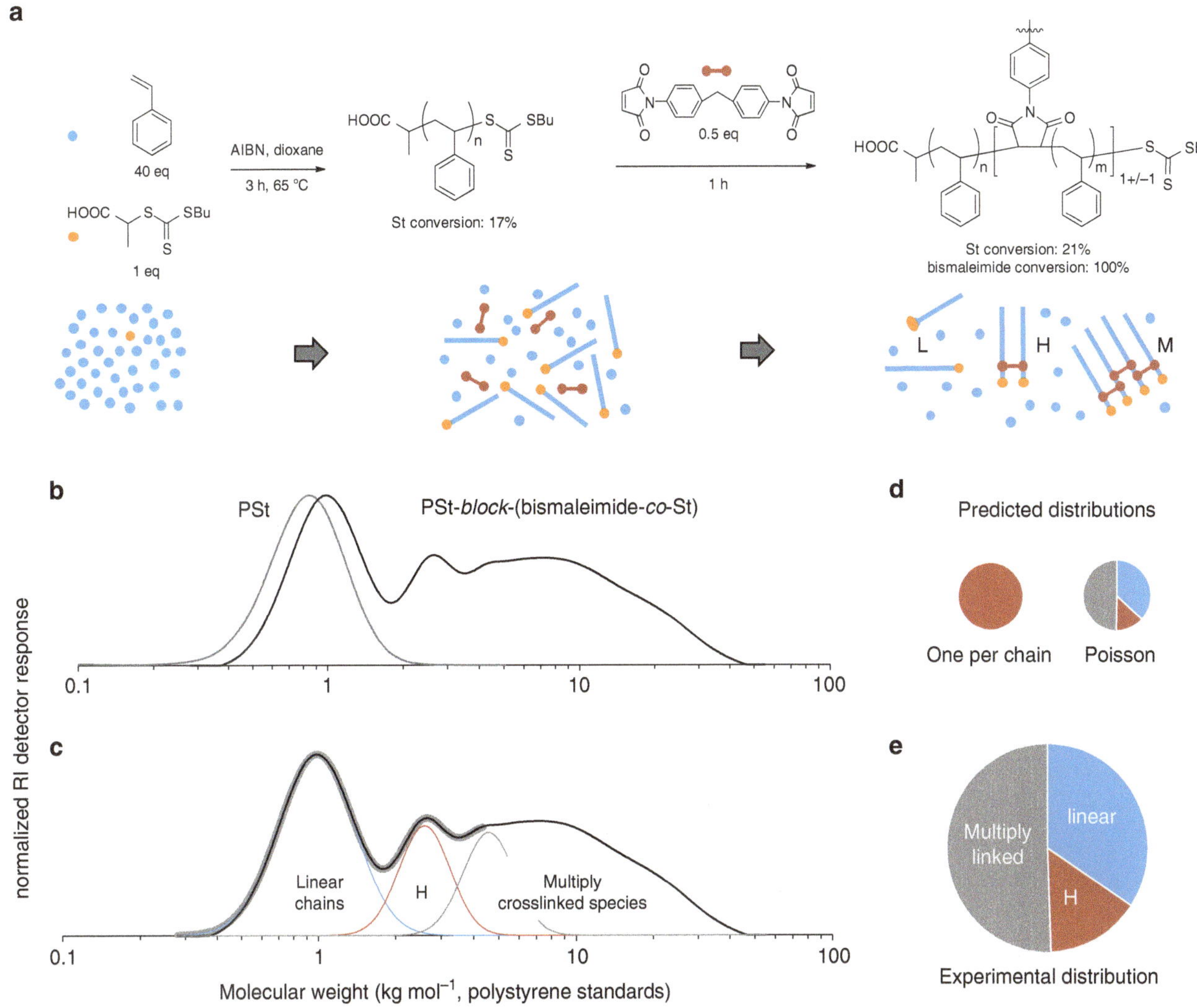


**Figure 4 | Addition of bismaleimide to a RAFT polymerization of styrene (St).** (**a**) After addition of the bismaleimide (0.5 equiv), polymerization was allowed to continue for 1 h, during which time the conversion of St increased from 17 to 21% and the bismaleimide was completely consumed. The resulting copolymer contained, on average, 1 maleimide unit per polystyrene chain (with an estimated standard deviation of 1). The distribution of maleimide units resulted in the formation of a mixture of linear chains (L), double chains (H) and multiply linked structures (M). (**b**) Distribution of chain lengths obtained after addition of 0.5 equiv. of bismaleimide. Linear chains and multiply linked structures are present in addition to H. The distribution immediately before bismaleimide addition is also shown. Deconvolution (**c**) of the low molar mass portion of the peak using Gaussian components (thick grey line represents best fit) allows estimation of the relative mass fractions of each structure (**e**), which are in good agreement with the predicted distribution assuming that the maleimide content of each chain follows a Poisson distribution (**d**).

shown in Fig. 4 would be obtained, with twice the molar mass of its constituent polystyrene chains. In fact, size exclusion chromatography (SEC) of the resulting polymer revealed single polymer chains (L), the linked structure (H) and aggregates of multiple chains (M). Deconvolution of the low molar mass portion of the chromatogram allowed the proportions of L and H species to be estimated at 34.4% and 15.1%, respectively. The expected proportions for a Poisson-distributed number of maleimides per chain are 36.8% L and 13.5% H. We have assumed that no looped structures are formed—a more detailed analysis taking this possibility into account can be found in Supplementary Note 3, but leads to a similar level of non-functionalized chains.

This result conflicts with previous reports that only 10% of chains contain no maleimide when an average of 1 maleimide is incorporated per chain[42,45]. That value was obtained from a matrix-assisted laser desorption/ionization-time of flight mass spectrum in which polystyrene species containing 1 or 2 units of Bz-MI formed the most intense peaks[45], while species containing zero or more than 3 units of Bz-MI accounted for 10% and 15% of the sequence distribution, respectively[42]. This implies that the average number of maleimides per chain is greater than 1, however, and so it seems likely that the frequency of zero-maleimide chains was underestimated. The incorporation of a polar maleimide unit may affect the ability of the polymer to form a charged complex, which can be observed by mass spectroscopy, biasing the results.

These results, applied to the SET-LRP multisite polymers of Fig. 3 (ref. 21), suggest that the yield of copolymer chains that contain at least one of each type of maleimide is only 25%, whereas 5% of chains contain no maleimide at all. Better results are obtained for the ROMP multisite polymer[23] (Supplementary Fig. 4): the use of, on average, two functional monomers per site yields 56% of chains with all four types of functional monomer (the yield of chains with all nine blocks, including spacer blocks, is only 20%). A further increase in average block length to 3, as has been recommended by Lutz and co-workers[42,46], raises the yield of chains containing all four functionalities to over 81%, whereas an average block length of 6 allows the formation of 20-block copolymers with 95% fidelity (Supplementary Fig. 6).

## Discussion

The results presented in this communication assume the ideal case of Poisson-distributed polymer segments, an approximation that has been validated for anionic polymerization, but which does not hold for all controlled chain-growth polymerization techniques. Side reactions such as radical coupling and chain transfer will lead to broader distributions, as will the addition of more than one monomer during each activation–deactivation cycle in RDRP. In such cases, more realistic distributions could be obtained through the use of simulations, or using alternate distribution functions such as the negative binomial distribution which allow for greater variability. An example is given in Supplementary Note 4 for a multiblock polymer with overall dispersity of 1.1 (see also Supplementary Figs 10–14, Supplementary Table 3). The idealized results presented here are valuable because they show the maximum level of precision that can be obtained using chain polymerization, and provide indications of how that precision can be maximized.

We have quantified the maximum level of control over monomer placement that can be obtained using controlled chain growth polymerization techniques. The most ambitious goal of aperiodic polymer synthesis is precise control over the absolute position or separation of introduced monomers. We have shown that, regardless of the technique used, chain growth polymerization is poorly suited to this task, as uncertainty in the position of a given monomer grows in proportion to the square root of the length of the polymer chain. The less ambitious goal of controlling relative position with respect to the total chain length is more achievable, with improved control achieved for longer chains, and as the desired functionality approaches either chain end. Finally, if one desires simply that all blocks of a multiblock copolymer be present in a large majority of chains, this is possible so long as the individual blocks are of sufficient length. In all cases, we have given simple formulae for the probability distributions, expected values and standard deviations of the positions of inserted blocks.

Our research clearly shows the limitations of chain polymerization in the synthesis of very precisely controlled structures—the synthetic analogues of DNA or enzymes. Fortunately, the natural world provides numerous examples of less precisely controlled structures—proteins such as mucins, collagens and elastins—which play vital physiological roles. These proteins are characterized by complex yet variable structures that present multiple functionalities. In many cases, structural complexity enables self-assembly, generating fibrils or gels via a bottom-up process. It is these biomolecules that should provide the inspiration for precision chain polymerization.

Indeed, self-assembly of complex multiblock polymer architectures prepared via chain polymerization has already been used to develop biomimetic stress-stiffening gels[47], and thermoresponsive, flower-like micelles which display multiple functionalities[14]. A recent review by Bates *et al.*[48] highlights the enormous range of phase-separated structures that can be generated by polymers containing a relatively small number of blocks; faced with a bewildering array of possibilities, the challenge is to select the architecture which will give the desired properties. Having done so, it is equally important to know the precision with which that architecture can be prepared. This knowledge will allow the design of a new family of materials that offer an optimal balance of precision in the placement of functionalities and simplicity of production, for a wide range of chemical, biomedical and engineering applications.

## Methods

**Chemicals and reagents.** Tetrahydrofuran (THF, Ajax Finechem, 99%), *N,N*-dimethylformamide (DMF, Merck, HPLC-grade), dimethyl 2,2′-azobis(2-methylpropanoate) (V-601), 2,2′-Azobis[2-(2-imidazolin-2-yl)propane]dihydrochloride (VA-044, Wako), 1,4-dioxane (Sigma-Aldrich, ≥99%) and 4-acryloylmorpholine (NAM, Sigma-Aldrich, 97%) were used without further purification. Styrene (Sigma-Aldrich, ≥99%) was filtered through a basic aluminium oxide (activated, basic, Brockmann I, standard grade, ~150 mesh, 58A) column before use. All polymerizations were carried out under a nitrogen atmosphere. The RAFT agent, 2-(((butylthio)carbonothiolyl)thio)propanoic acid was prepared according to a reported procedure[49].

**NMR spectroscopy.** $^1$H NMR spectra were recorded on Bruker DPX-300 spectrometer using deuterated solvents obtained from Aldrich. Chemical shift values ($\delta$) are reported in p.p.m. The residual proton signal of the solvent was used as internal standard.

**Size exclusion chromatography.** Number-average molar masses ($M_{n,SEC}$) and dispersity values ($Đ$) were determined using THF-SEC performed on an Agilent 390-MDS, comprising of an autosampler and a PLgel 5.0 µm bead-size guard column (50 × 7.5 mm$^2$), followed by two linear 5.0 µm bead-size PLgel Mixed D columns (300 × 7.5 mm$^2$) and a differential refractive index detector using THF (2% (v/v), TEA) as the eluent at 30 °C with a flow rate of 1 ml min$^{-1}$. The SEC system was calibrated with linear PS EasiVial standards (Agilent Ltd.) ranging from 162 to 105 g mol$^{-1}$. All samples were passed through 0.45 µm PTFE filter before SEC analysis. Molar masses were obtained by conventional calibration using ASTRA software.

**RAFT polymerization of styrene with a controlled addition of bismaleimide.** 2-(((Butylthio)carbonothiolyl)thio)propanoic acid (114 mg, 0.480 mmol, 1 equiv.), styrene (2 g, 19.2 mmol, 40 equiv.), V-601 (11 mg, 0.048 mmol, 0.1 equiv.) and DMF (0.540 ml) were placed in a flask equipped with a magnetic stirring bar. The tube was then sealed with a rubber septum, degassed with nitrogen and then immersed in an oil bath thermostated at 65 °C. A degassed solution of bismaleimide (91 mg, 0.240 mmol, 0.5 equiv.) in 0.630 ml of DMF was added with a degassed syringe 3 h after the polymerization started (17% styrene conversion at this stage). After 1 more hour of polymerization, a sample was taken for NMR and THF-SEC analysis to ensure the full incorporation of the bismaleimide in the growing copolymer chain (around 4% of styrene conversion took place in this interval). The THF-SEC chromatogram shows multiple bismaleimide insertion.

**RAFT homopolymerization of NAM.** *For a targeted degree of polymerization ($X_n$) of 3.* RAFT agent (281 mg, 1.181 mmol, 1 equiv.), NAM (0.5 g, 3.541 mmol, 3 equiv.), VA-044 (3.82 mg, 0.012 mmol, 0.01 equiv.), dioxane (0.477 ml) and $H_2O$ (0.257 ml) were placed in a flask equipped with a magnetic stirring bar. The tube was then sealed with a rubber septum, degassed with nitrogen and then immersed in an oil bath thermostated at 70 °C. After 2 h, a sample is withdrawn from the polymerization medium for $^1$H NMR (in dimethylsulphoxide (DMSO-$d_6$)). Conversion was determined by $^1$H NMR to be 94%, corresponding to an $\bar{X}_n$ of 2.82.

*For a targeted $X_n$ of 4.* RAFT agent (211 mg, 0.885 mmol, 1 equiv.), NAM (0.5 g, 3.541 mmol, 4 equiv.), VA-044 (2.20 mg, 0.007 mmol, 0.0077 equiv.), dioxane (0.441 ml) and $H_2O$ (0.294 ml) were placed in a flask equipped with a magnetic stirring bar. The tube was then sealed with a rubber septum, degassed with nitrogen and then immersed in an oil bath thermostated at 70 °C. After 2 h, a sample was withdrawn from the polymerization medium for $^1$H NMR (in DMSO-$d_6$). Conversion was determined by $^1$H NMR to be 97%, corresponding to an $\bar{X}_n$ of 3.89.

*For a targeted $X_n$ of 5.* RAFT agent (169 mg, 0.708 mmol, 1 equiv.), NAM (0.5 g, 3.541 mmol, 5 equiv.), VA-044 (1.53 mg, 0.005 mmol, 0.007 equiv.), dioxane (0.404 ml) and $H_2O$ (0.331 ml) were placed in a flask equipped with a magnetic stirring bar. The tube was then sealed with a rubber septum, degassed with nitrogen and then immersed in an oil bath thermostated at 70 °C. After 2 h, a sample is withdrawn from the polymerization medium for $^1$H NMR (in DMSO-$d_6$). Conversion was determined by $^1$H NMR to be 98%, corresponding to an $\bar{X}_n$ of 4.92.

*For a targeted $X_n$ of 6.* RAFT agent (141 mg, 0.590 mmol, 1 equiv.), NAM (0.5 g, 3.541 mmol, 6 equiv.), VA-044 (0.95 mg, 0.003 mmol, 0.005 equiv.), dioxane (0.367 ml) and $H_2O$ (0.367 ml) were placed in a flask equipped with a magnetic stirring bar. The tube was then sealed with a rubber septum, degassed with nitrogen and then immersed in an oil bath thermostated at 70 °C. After 2 h, a sample is withdrawn from the polymerization medium for $^1$H NMR (in DMSO-$d_6$). Conversion was determined by $^1$H NMR to be 99%, corresponding to an $\bar{X}_n$ of 5.92.

## References


1. Lutz, J.-F. Aperiodic copolymers. *ACS Macro Lett.* **3**, 1020–1023 (2014).
2. Schmidt, B. V. K. J. & Barner-Kowollik, C. Polymer chemistry: macromolecules made to order. *Nat. Chem.* **5**, 990–992 (2013).
3. Lutz, J.-F., Ouchi, M., Liu, D. R. & Sawamoto, M. Sequence-controlled polymers. *Science* **341**, 6146 (2013).
4. Lutz, J.-F. Sequence-controlled polymerizations: the next Holy Grail in polymer science? *Polym. Chem.* **1**, 55–62 (2010).
5. Lutz, J.-F. Writing on polymer chains. *Acc. Chem. Res.* **46**, 2696–2705 (2013).
6. Merrifield, R. B. Solid phase peptide synthesis. I. The synthesis of a tetrapeptide. *J. Am. Chem. Soc.* **85**, 2149–2154 (1963).
7. Merrifield, R. B. Solid phase synthesis (nobel lecture). *Angew. Chem. Int. Ed.* **24**, 799–810 (1985).
8. Pfeifer, S., Zarafshani, Z., Badi, N. & Lutz, J.-F. Liquid-phase synthesis of block copolymers containing sequence-ordered segments. *J. Am. Chem. Soc.* **131**, 9195–9197 (2009).
9. Tong, X., Guo, B.-H. & Huang, Y. Toward the synthesis of sequence-controlled vinyl copolymers. *Chem. Commun.* **47**, 1455–1457 (2011).
10. Feng, X.-S. & Pan, C.-Y. Synthesis of amphiphilic miktoarm ABC star copolymers by RAFT mechanism using maleic anhydride as linking agent. *Macromolecules* **35**, 4888–4893 (2002).
11. Henry, S. M., Convertine, A. J., Benoit, D. S. W., Hoffman, A. S. & Stayton, P. S. End-Functionalized polymers and junction-functionalized diblock copolymers via RAFT chain extension with maleimido monomers. *Bioconj. Chem.* **20**, 1122–1128 (2009).
12. Han, D.-H. & Pan, C.-Y. A novel strategy for synthesis of amphiphilic π-shaped copolymers by RAFT polymerization. *Eur. Polym. J.* **42**, 507–515 (2006).
13. Zhou, C. *et al.* A well-defined amphiphilic polymer co-network from precise control of the end-functional groups of linear RAFT polymers. *RSC Adv* **4**, 8144–8156 (2014).
14. Gody, G., Maschmeyer, T., Zetterlund, P. B. & Perrier, S. Rapid and quantitative one-pot synthesis of sequence-controlled polymers by radical polymerization. *Nat. Commun.* **4**, 3505 (2013).
15. Fleury, G. & Bates, F. S. Structure and properties of hexa- and undecablock terpolymers with hierarchical molecular architectures. *Macromolecules* **42**, 3598–3610 (2009).
16. Nagata, Y. *et al.* Preparation and characterization of a styrene – isoprene undecablock copolymer and its hierarchical microdomain structure in bulk. *Macromolecules* **38**, 10220–10225 (2005).
17. Minoda, M., Sawamoto, M. & Higashimura, T. Sequence-regulated oligomers and polymers by living cationic polymerization. 2. Principle of sequence regulation and synthesis of sequence-regulated oligomers of functional vinyl ethers and styrene derivatives. *Macromolecules* **23**, 4889–4895 (1990).
18. Soeriyadi, A. H., Boyer, C., Nyström, F., Zetterlund, P. B. & Whittaker, M. R. High-order multiblock copolymers via iterative Cu(0)-mediated radical polymerizations (SET-LRP): toward biological precision. *J. Am. Chem. Soc.* **133**, 11128–11131 (2011).
19. Anastasaki, A. *et al.* Photoinduced sequence-control via one pot living radical polymerization of acrylates. *Chem. Sci.* **5**, 3536–3542 (2014).
20. Fetsch, C. & Luxenhofer, R. Highly defined multiblock copolypeptoids: pushing the limits of living nucleophilic ring-opening polymerization. *Macromol. Rapid Commun.* **33**, 1708–1713 (2012).
21. Zamfir, M. & Lutz, J.-F. Ultra-precise insertion of functional monomers in chain-growth polymerizations. *Nat. Commun.* **3**, 1138 (2012).
22. Chan-Seng, D., Zamfir, M. & Lutz, J.-F. Polymer-chain encoding: synthesis of highly complex monomer sequence patterns by using automated protocols. *Angew. Chem. Int. Ed.* **51**, 12254–12257 (2012).
23. Moatsou, D., Hansell, C. F. & O'Reilly, R. K. Precision polymers: a kinetic approach for functional poly(norbornenes). *Chem. Sci.* **5**, 2246–2250 (2014).
24. Grubbs, R. & Tumas, W. Polymer synthesis and organotransition metal chemistry. *Science* **243**, 907–915 (1989).
25. Jenkins, A. D., Jones, R. G. & Moad, G. Terminology for reversible-deactivation radical polymerization previously called 'controlled' radical or 'living' radical polymerization (IUPAC Recommendations 2010). *Pure Appl. Chem.* **82**, 483 (2010).
26. Smid, J., Van Beylen, M. & Hogen-Esch, T. E. Perspectives on the contributions of Michael Szwarc to living polymerization. *Prog. Polym. Sci.* **31**, 1041–1067 (2006).
27. Aoshima, S. & Kanaoka, S. A renaissance in living cationic polymerization. *Chem. Rev.* **109**, 5245–5287 (2009).
28. Houshyar, S. *et al.* The scope for synthesis of macro-RAFT agents by sequential insertion of single monomer units. *Polym. Chem.* **3**, 1879–1889 (2012).
29. Vandenbergh, J., Reekmans, G., Adriaensens, P. & Junkers, T. Synthesis of sequence controlled acrylate oligomers via consecutive RAFT monomer additions. *Chem. Commun.* **49**, 10358–10360 (2013).
30. Flory, P. J. Molecular size distribution in ethylene oxide polymers. *J. Am. Chem. Soc.* **62**, 1561–1565 (1940).
31. Lee, W. *et al.* Molecular Weight Distribution of Polystyrene Made by Anionic Polymerization. *Macromolecules* **33**, 5111–5115 (2000).
32. Matyjaszewski, K. & Xia, J. Atom transfer radical polymerization. *Chem. Rev.* **101**, 2921–2990 (2001).
33. Rosen, B. M. & Percec, V. Single-electron transfer and single-electron transfer degenerative chain transfer living radical polymerization. *Chem. Rev.* **109**, 5069–5119 (2009).
34. Konkolewicz, D. *et al.* Reversible-deactivation radical polymerization in the presence of metallic copper. a critical assessment of the SARA ATRP and SET-LRP mechanisms. *Macromolecules* **46**, 8749–8772 (2013).
35. Keddie, D. J. A guide to the synthesis of block copolymers using reversible-addition fragmentation chain transfer (RAFT) polymerization. *Chem. Soc. Rev.* **43**, 496–505 (2014).
36. Hurtgen, M., Detrembleur, C., Jerome, C. & Debuigne, A. Insight into organometallic-mediated radical polymerization. *Polym. Rev.* **51**, 188–213 (2011).
37. Yamago, S. Precision polymer synthesis by degenerative transfer controlled/living radical polymerization using organotellurium, organostibine, and organobismuthine chain-transfer agents. *Chem. Rev.* **109**, 5051–5068 (2009).
38. Kamigaito, M., Ando, T. & Sawamoto, M. Metal-catalyzed living radical polymerization. *Chem. Rev.* **101**, 3689–3746 (2001).
39. Hawker, C. J., Bosman, A. W. & Harth, E. New polymer synthesis by nitroxide mediated living radical polymerizations. *Chem. Rev.* **101**, 3661–3688 (2001).
40. Altintas, O., Krolla-Sidenstein, P., Gliemann, H. & Barner-Kowollik, C. Single-chain folding of diblock copolymers driven by orthogonal h-donor and acceptor units. *Macromolecules* **47**, 5877–5888 (2014).
41. Altintas, O., Lejeune, E., Gerstel, P. & Barner-Kowollik, C. Bioinspired dual self-folding of single polymer chains via reversible hydrogen bonding. *Polym. Chem.* **3**, 640–651 (2012).
42. Lutz, J.-F., Schmidt, B. V. K. J. & Pfeifer, S. Tailored polymer microstructures prepared by atom transfer radical copolymerization of styrene and N-substituted maleimides. *Macromol. Rapid Commun.* **32**, 127–135 (2011).
43. Falls, L. W. The beta distribution: a statistical model for world cloud cover. *J. Geophys. Res.* **79**, 1261–1264 (1974).
44. Wright, S. Evolution in mendelian populations. *Genetics* **16**, 97–159 (1931).
45. Pfeifer, S. & Lutz, J.-F. Development of a library of N-substituted maleimides for the local functionalization of linear polymer chains. *Chem. Eur. J* **14**, 10949–10957 (2008).
46. Pfeifer, S. & Lutz, J.-F. A facile procedure for controlling monomer sequence distribution in radical chain polymerizations. *J. Am. Chem. Soc.* **129**, 9542–9543 (2007).
47. Kouwer, P. H. J. *et al.* Responsive biomimetic networks from polyisocyanopeptide hydrogels. *Nature* **493**, 651–655 (2013).
48. Bates, F. S. *et al.* Multiblock polymers: Panacea or Pandora's Box? *Science* **336**, 434–440 (2012).
49. Ferguson, C. J. *et al.* *Ab initio* emulsion polymerization by RAFT-controlled self-assembly. *Macromolecules* **38**, 2191–2204 (2005).


## Acknowledgements

We thank M. Destarac for his comments on an early version of the manuscript. The Centre National de Recherche Scientifique (S.H.), the Royal Society Wolfson Merit Award (WM130055; S.P.), the Monash-Warwick Alliance (G.G., S.P.) and the Australian Research Council (Future Fellowship, P.B.Z.) are acknowledged for financial support.

## Author contributions

S.H. developed the model and wrote the manuscript. G.G. and S.P. designed the experiments, which were conducted by G.G. Interpretation of the results was carried out by G.G., S.P., P.B.Z. and S.H.

## Additional information

**Competing financial interests:** The authors declare no competing financial interests.

# 20

# Luminescent platinum(II) complexes with functionalized N-heterocyclic carbene or diphosphine selectively probe mismatched and abasic DNA

Sin Ki Fung[1,*], Taotao Zou[1,2,*], Bei Cao[1], Tianfeng Chen[3], Wai-Pong To[1], Chen Yang[1,2], Chun-Nam Lok[1] & Chi-Ming Che[1,2]

The selective targeting of mismatched DNA overexpressed in cancer cells is an appealing strategy in designing cancer diagnosis and therapy protocols. Few luminescent probes that specifically detect intracellular mismatched DNA have been reported. Here we used Pt(II) complexes with luminescence sensitive to subtle changes in the local environment and report several Pt(II) complexes that selectively bind to and identify DNA mismatches. We evaluated the complexes' DNA-binding characteristics by ultraviolet/visible absorption titration, isothermal titration calorimetry, nuclear magnetic resonance and quantum mechanics/ molecular mechanics calculations. These Pt(II) complexes show up to 15-fold higher emission intensities upon binding to mismatched DNA over matched DNA and can be utilized for both detecting DNA abasic sites and identifying cancer cells and human tissue samples with different levels of mismatch repair. Our work highlights the potential of luminescent Pt(II) complexes to differentiate between normal cells and cancer cells which generally possess more aberrant DNA structures.

[1] State Key Laboratory of Synthetic Chemistry, Institute of Molecular Functional Materials, Chemical Biology Centre, and Department of Chemistry, The University of Hong Kong, Pokfulam Road, Hong Kong, China. [2] HKU Shenzhen Institute of Research and Innovation Shenzhen, Shenzhen 518053, China. [3] Department of Chemistry, Jinan University, Guangzhou 510632, China. * These authors contributed equally to this work. Correspondence and requests for materials should be addressed to C.-M.C. (email: cmche@hku.hk).

The binding interactions between metal complexes and nucleic acids have been the subject of numerous studies in recent decades[1–4]. These interactions are likely to lead to therapeutic (for example, anti-cancer) effects and/or be used for diagnostic (for example, luminescent probes) purposes[5–11]. To date, the selective targeting of nucleic acids in cancer cells by chemotherapeutic metal complexes in a manner that minimizes off-target bindings and hence diminishes side effects remains a great challenge[12,13]. Endeavours in this research area have led to luminescent metal complexes that show emission responses with specificity for specific nucleic acid structures, such as G-quadruplex DNA[14,15], double-stranded RNA[16,17] and mismatched DNA[13,18,19]. The recognition of mismatched DNA is of importance for cancer diagnosis and therapy, because DNA mismatches are associated with oncogenic transformation[20–22]. Furthermore, cancer cells, such as those of colorectal cancer, show a high frequency of DNA mismatches as a result of their deficiency in mismatch repair (MMR)[13,23–26]. Pioneering work by Barton and co-workers has shown that the octahedral $d^6$ metal complexes $[Rh(bpy)_2chrysi]^{3+}$ (bpy = 2-phenylpyridine, chrysi = 5,6-chrysenequinone diimine)[18,27,28] and Δ-$[Ru(bpy)_2dppz]^{2+}$ (dppz = dipyrido[3,2-*a*:2′,3′-c] phenazine)[19,29,30] are able to selectively bind to mismatched DNA. Other researchers have reported that octahedral cobalt(III) complexes[31], a six-coordinated silicon complex containing a sandwich ruthenium moiety[32], some non-metal compounds[33–41] and others[13] can recognize DNA mismatches with specificity.

The $d^8$ pincer Pt(II) complexes have been well documented to intercalate into matched DNA[1,2,4,11,42]. With intrinsic emission properties that are highly sensitive to the microenvironment, these types of Pt(II) complexes have been developed as luminescent DNA probes and have been identified as potent therapeutic agents for cancer treatment[14,17,43–45]. However, metal complexes with bare planar coordination geometries likely exhibit non-discriminative intercalative binding interactions with DNA. In the crystal structures of mismatched DNA bound to Rh(III)[18,28] or Ru(II)[19] complexes, the planar extended π-conjugated ligands (with a width of up to 11.3 Å and a length of up to 9.4 Å) favour strong π-stacking interactions with nearby base pairs at mismatched sites, concomitantly with the ejection of mismatched base pairs, whereas the other auxiliary bipyridine ligands promote a groove binding interaction. We therefore hypothesized that a pincer Pt(II) complex coordinated with an out-of-plane bulky ancillary ligand may exhibit selective binding affinity towards mismatched DNA, as discussed in the next section. The bulky ancillary ligand can be designed to hamper the intercalation of the platinum complex with well-matched DNA while exhibiting high binding affinity with mismatched DNA. This difference is possible because the mismatched sites possess larger binding pockets. We have previously described a luminescent $[Pt^{II}(C\wedge N\wedge N)(NHC^{2C4})]^+$ ($HC\wedge N\wedge N$ = 6-phenyl-2,2′-bipyridine, NHC = *N*-heterocyclic carbene, **1b** in the current report) complex with an NHC ligand almost perpendicular to the C-deprotonated C^N^N plane[46]. This complex exhibits a weak binding affinity to matched DNA. Thus, varying the *N*-alkyl/aryl substitution(s) of NHC, $[Pt^{II}(C\wedge N\wedge N)(NHC)]^+$ may generate a candidate scaffold to target mismatched DNA via cooperative π-stacking and groove binding interactions.

Here we describe several luminescent, mismatched DNA probes based on the $[Pt^{II}(C\wedge N\wedge N)(NHC)]^+$, $[Pt^{II}(N\wedge C\wedge N)(NHC)]^+$ and $[Pt^{II}_2(C\wedge N\wedge N)_2(\mu\text{-dcpm})]^{2+}$ complexes. The mononuclear Pt(II) complex selectively detects DNA containing CC mismatches, and the dinuclear Pt(II) complex exhibits high selectivity towards several types of DNA mismatches. These luminescent Pt(II) complexes also selectively bind thermodynamically unstable abasic DNA, are able to differentiate cancer cells that have different levels of MMR capacity and can differentiate colon tumour tissue from normal tissue.

## Results

**Structure and emission responses to CC mismatched DNA.** We synthesized and characterized a series of $[Pt^{II}(C\wedge N\wedge N)(NHC)]^+$ complexes with different alkyl chains, aromatic groups and hydrophilic alcohol moieties on NHC and/or on C-deprotonated C^N^N ligands (Fig. 1; details in Supplementary Information). In addition, we examined the emission responses of complexes that bind mismatched and matched DNA (Supplementary Fig. 1). The most unstable CC mismatched DNA was chosen in the primary screen because metalloinsertors display the highest binding to this type of mismatched DNA[29]. We found that as the steric hindrance of the NHC ligand increased, the binding of the

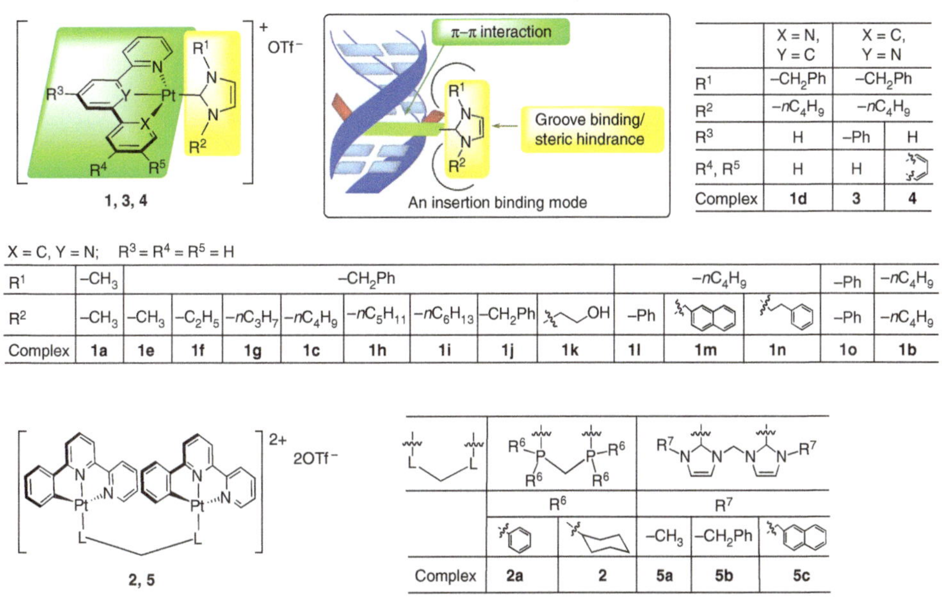


**Figure 1 | Chemical structures and proposed binding modes.** Platinum(II) complexes **1a–1o**, **2**, **2a**, **3**, **4** and **5a–5c**. The counteranions for all complexes are $CF_3SO_3^-$. The inset shows the molecular design of Pt(II) complexes tested for binding mismatched DNA.

Pt(II) complexes to matched DNA decreased, whereas the binding with CC mismatched DNA remained high for certain bulky NHC ligands (Fig. 2; the DNA sequences in all the experiments are depicted in the Supplementary Table 2). For example, for the $[Pt(C^\wedge N^\wedge N)(NHC)]^+$ type complexes of **1e** ($-CH_3$), **1f** ($-C_2H_5$), **1g** ($-C_3H_7$), **1c** ($-C_4H_9$), **1h** ($-C_5H_{11}$), **1i** ($-C_6H_{13}$) and **1j** ($-CH_2Ph$), which contain the same $N^3$-benzyl group but different $N^1$-substituents on the NHC ligands (Table 1), an increase in the length of the $N^1$-alkyl was accompanied by a marked decrease in emission enhancement towards matched DNA ($I_M$), from 38.9-fold to <3-fold. Although increasing the chain length from the $N^1$-methyl of **1e** to the $N^1$-ethyl of **1f** led to a decrease in emission enhancement in the case of mismatched DNA ($I_{MM}$), the emission intensity of **1g**, which bears $N^1$-propyl, was much stronger than that of **1f**, suggesting that $N^1$-propyl favours the groove binding of the Pt(II) complex to DNA. This may enhance the tightness of the insertion binding mode, as proposed. Further lengthening the alkyl chain to *n*-butyl, *n*-pentyl and *n*-hexyl gradually decreased the binding with mismatched DNA. Comparing the $I_{MM}/I_M$ ratios, **1c**, containing $N^1$-(*n*-butyl) and $N^3$-benzyl substitutions on the NHC (Fig. 1) showed the most significantly different emission responses between CC mismatched DNA and matched DNA. We characterized the structure of this complex by $^1H$–$^1H$ correlation spectroscopy (COSY) and nuclear Overhauser effect spectroscopy (NOESY) nuclear magnetic resonance (NMR; Supplementary Fig. 2). As shown in Fig. 3a,c, the emission intensity of **1c** ($\lambda_{max} = 535$ nm) in a Tris-buffered solution increased by 26.4-fold in the presence of 1 equivalent of a hairpin DNA oligomer with a CC mismatch[29] (Fig. 3c). However, the emission of **1c** was enhanced only 2.6-fold upon the addition of the same amount of the hairpin DNA, having the cognate sequence with the mismatched bases replaced by matched GC pairs. In another mismatched DNA model using a 17-mer double-stranded DNA (dsDNA) formed by two complementary oligomers[47], the emission intensity of **1c** displayed 27.5-fold and 2.8-fold increases in emission intensity upon the addition of CC mismatched 17-mer dsDNA and matched 17-mer dsDNA, respectively (Supplementary Fig. 3). In the literature, a rhodium-Oregon Green conjugate has been reported to exhibit a ~3.2-fold higher emission intensity in the presence of CC mismatched DNA compared with matched DNA[47]. In a control experiment, the classical DNA intercalator, ethidium bromide (EB), showed increases in emission intensity with hairpin mismatched DNA and matched DNA of 12.7-fold

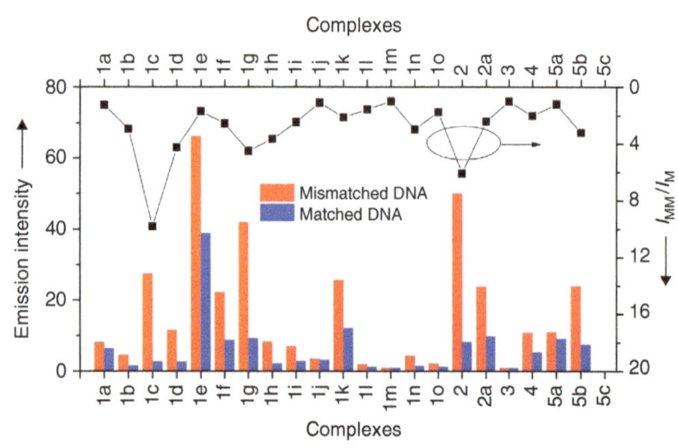


**Figure 2 | Emission responses of Pt(II) complexes to matched and mismatched DNA.** A summary of emission intensities (at near level off) of different Pt(II) complexes towards mismatched and matched DNA and the emission intensity ratios ($I_{MM}/I_M$).

and 14.1-fold, respectively, revealing a 0.9-fold difference (Supplementary Fig. 4). Under similar conditions, **1a**, containing a less bulky *N*-methyl substitution on the NHC, displayed similar 8.5-fold and 7.1-fold emission enhancements for hairpin CC mismatched DNA and matched DNA, respectively (1.2-fold difference; Supplementary Fig. 5). This result can be explained by the fact that the less bulky *N*-methyl substituted NHC does not impose enough steric hindrance to diminish the binding of the Pt complex with matched DNA, resulting in similarly high enhancement in emission intensity for both matched and mismatched DNA. Conversely, the more bulky $N^1$-(*n*-butyl) and $N^3$-benzyl groups diminished the binding of **1c** with matched DNA, while a much stronger binding affinity with mismatched DNA was still maintained.

**Characterizations of bindings to CC mismatched DNA.** The following experiments were performed to shed light on the DNA-binding interactions of the novel complexes. In ultraviolet/visible absorption titration experiments, upon the addition of dsDNA containing a CC mismatch to **1c** in a Tris-buffered solution, a hypochromicity of ~34% at $\lambda = 335$ nm, together with isosbestic spectral changes, was observed (Supplementary Fig. 6a); however, only a minor hypochromicity of ~12% was detected at the same wavelength for **1c** when matched dsDNA with a similar sequence was used instead (Supplementary Fig. 6b). Isothermal titration calorimetry (ITC) experiments (Fig. 3b) revealed that the titration of CC mismatched DNA with **1c** resulted in a significant exothermic reaction, revealing binding between the two components, with a dissociation constant ($K_d$) of $51.5 \pm 8.0$ μM. In contrast, the reaction between **1c** and matched DNA was a very weak exothermic reaction, with a $K_d$ estimated to be >600 μM. In a similar experiment, the $K_d$ values of the binding of **1a** to mismatched and matched DNA were $66.2 \pm 15.7$ and $56.4 \pm 13.6$ μM, respectively (Supplementary Fig. 7a,b). The $K_d$ values of the binding of EB to mismatched DNA and matched DNA were $1.32 \pm 0.24$ and $1.60 \pm 0.28$ μM, respectively (Supplementary Fig. 8a,b). $^1H$ NMR and two-dimensional (2D) NOESY, total correlation spectroscopy (TOCSY) and COSY NMR experiments were performed using a self-complementary DNA oligonucleotide, 5′-$C_1G_2G_3A_4\underline{C}_5T_6C_7C_8G_9$-3′, containing CC mismatches in $D_2O$ (phosphate buffer, pH 6.1; Fig. 4a–d; Supplementary Figs 9 and 10)[48]. In the range of 6.8–8.0 p.p.m., the aromatic $^1H$ signals of C5 and A4 show substantial shifts, whereas those of the other bases remain nearly unshifted in the presence of **1c** ($^1H$ NMR in Fig. 4a). In addition, in the 2D H1′ × aromatic NOESY spectrum of **1c** with DNA at a 1:1 ratio, a sequential NOESY walk (Fig. 4d), which showed a marked shift at A4 and C5 compared with DNA only (Fig. 4b), was found in combination with cytosine H6/H5 TOCSY (Supplementary Fig. 9) and H1′ × H2′–H2″ TOCSY (Fig. 4c) NMR. At [**1c**]:[DNA] ratios of 0.25:1 and 0.5:1 (Supplementary Fig. 10), a gradual disappearance of C5 and A4 aromatic protons of complex-free DNA and the subsequent appearance of new C5 and A4 aromatic protons of **1c**-bound DNA in the NOESY spectra were found. Thus, both one-dimensional and 2D NMR results indicate the possible binding of **1c** at the mismatched site. Collectively, the more significant the hypochromicity in ultraviolet/visible spectral changes, the higher the amount of heat generation, and the marked shifts of NMR signals at the mismatched site together revealed that **1c** preferentially binds to DNA containing CC mismatches. In the emission quenching experiments using $[Cu(phen)_2]^{2+}$ (phen = 1,10-phenanthroline), a minor-groove-specific quencher, the emission intensity of **1c** in the presence of CC mismatched DNA gradually decreased upon increasing the amount of $[Cu(phen)_2]^{2+}$ (Supplementary

**Table 1 | A comparison of structure-emission response relationships of [Pt(C^N^N)(NHC)]$CF_3SO_3$ complexes containing *N*-benzyl groups towards matched and mismatched DNA.**

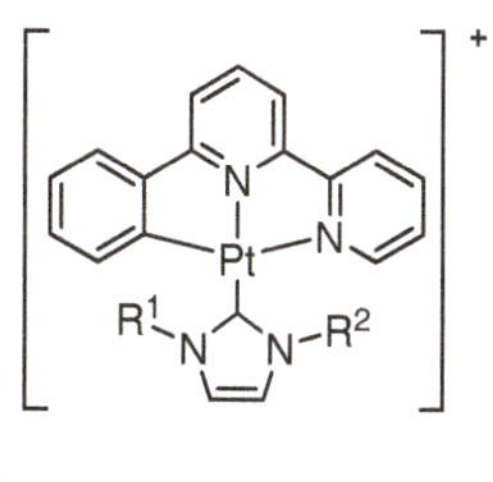


| $R^1$ | $-CH_2Ph$ | | | | | | |
|---|---|---|---|---|---|---|---|
| $R^2$ | $-CH_3$ | $-C_2H_5$ | $-nC_3H_7$ | $-nC_4H_9$ | $-nC_5H_{11}$ | $-nC_6H_{13}$ | $-CH_2Ph$ |
| Complex | **1e** | **1f** | **1g** | **1c** | **1h** | **1i** | **1j** |
| $I_M$* | 38.9 | 8.7 | 9.3 | 2.6 | 2.3 | 2.8 | 3.2 |
| $I_{MM}$* | 66.2 | 22.2 | 42.0 | 26.4 | 8.2 | 7.0 | 3.5 |

*$I_M$ and $I_{MM}$ denote the relative emission intensities of the Pt(II) complexes towards matched and mismatched DNA, respectively.

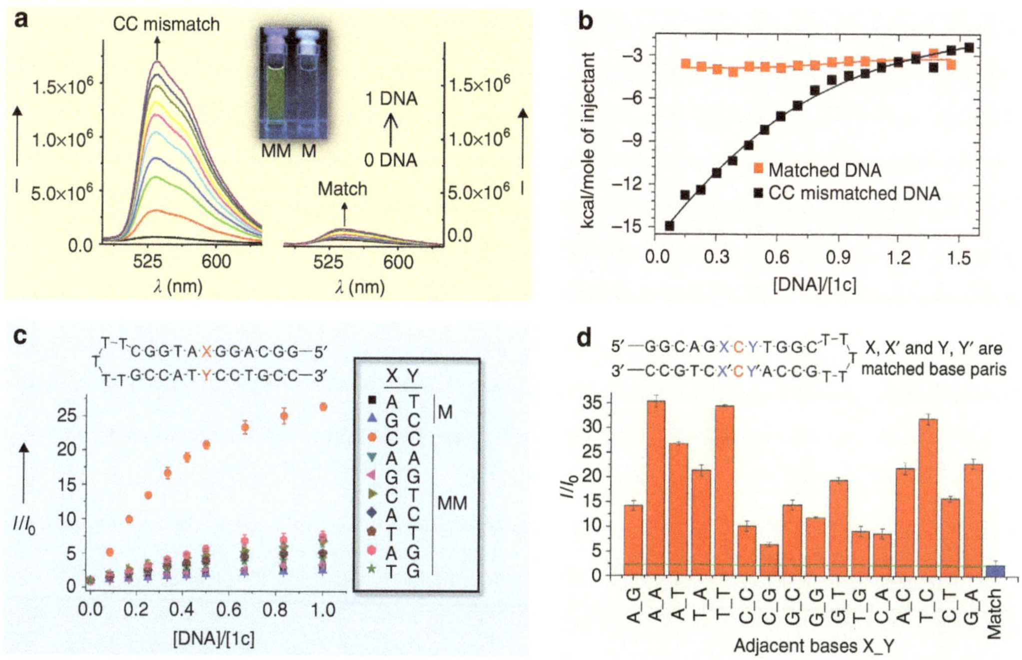


**Figure 3 | Binding of 1c with mismatched DNA.** (**a**) The emission spectra of complex **1c** (5 μM) in a Tris buffer solution (50 mM NaCl and 2 mM Tris, pH 7.5) after binding to different concentrations of CC mismatched DNA (MM) and matched DNA (M). (**b**) Plot of integrated ITC data for the exothermic interaction between CC mismatched DNA (0.75 mM) and **1c** (0.1 mM). (**c**) Changes in emission intensity (± s.e.m.) at 535 nm of complex **1c** (5 μM) in the Tris buffer solution upon the addition of different types of DNA. M, match; MM, mismatch. (**d**) Relative emission enhancement of **1c** towards CC MM DNA with different adjacent base pairs.

Fig. 11), indicating the binding of the complex through the minor groove[29]. Indeed, a favourable binding mode between **1c** and CC mismatched DNA has also been identified in molecular docking experiments (Supplementary Fig. 12), consistently with the insertion of the Pt(C^N^N) plane between base pairs and the favourable minor-groove binding interaction of the $N^1$-(*n*-butyl) and $N^3$-benzyl moieties. Complex **1d**, which contains an N^C^N instead of a C^N^N ligand (Fig. 1), exhibited 4.5-fold higher emission responses towards 1 equivalent of CC mismatched DNA than matched DNA (Supplementary Fig. 13).

The emission responses of **1c** towards all other types of DNA mismatches, including CT, AC, TT, AG, TG, AA and GG mismatches, all of which are more thermodynamically stable than CC mismatches[49], were tested (Fig. 3c). However, no significant difference in emission responses between these mismatched DNA and matched DNA was observed. We also examined the emission responses of **1c** to CC mismatched DNA having 16 different adjacent base pairs (Fig. 3d). Complex **1c** displayed higher emission responses to all the CC mismatched DNA sequences than the cognate matched DNA, with adjacent 5′-ACA-3′ revealing the highest emission enhancement (15.0-fold higher than matched DNA). These experiments suggest that **1c** is a selective probe for CC mismatched DNA.

**Targeting thermodynamically more stable DNA mismatches.** For recognition of the more stable DNA mismatches, an increase in the binding affinity of the Pt(II) complex with DNA is required. The dinuclear analogues of **1c**, which contain an additional positive charge and a larger molecular dimension, are expected to have higher DNA-binding affinities as a result of stronger ionic bonding interactions and possibly a better geometry to fit into the pocket at the mismatched site. Therefore, we examined the emission responses of a series of doubly charged dinuclear platinum complexes with two $Pt^{II}$(C^N^N) moieties linked by bridging bis(NHC) or diphosphine ligands (Fig. 1) in the presence of hairpin mismatched and matched DNA. Among the dinuclear Pt(II) complexes, $[Pt^{II}_2(C^\wedge N^\wedge N)_2(\mu\text{-dcpm})]^{2+}$ **2** (Fig. 1; the X-ray crystallographic structure is depicted in

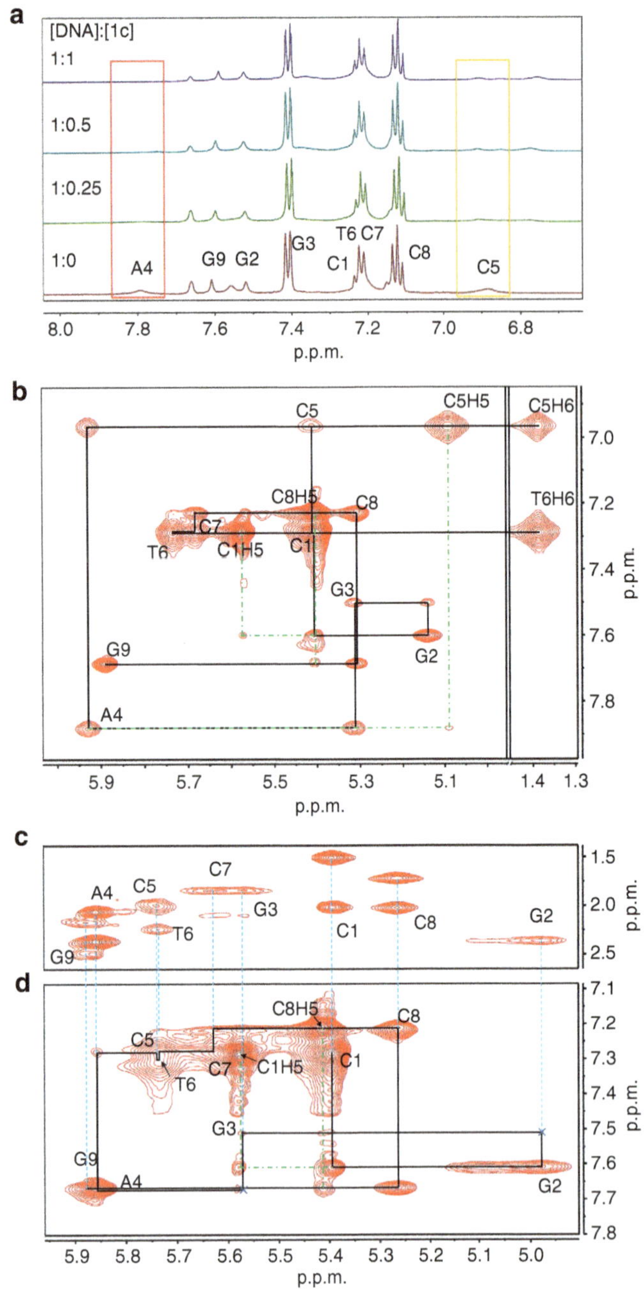


**Figure 4 | NMR analysis of the binding of 1c with mismatched DNA.** (**a**) $^1$H NMR (phosphate buffer = 50 mM, pH 6.1, NaCl = 20 mM, in $D_2O$, 10 °C) spectra of a DNA oligonucleotide in the presence of different amounts of **1c**. (**b**) 2D H1′ × aromatic (left) and T(Me) × aromatic (right) NOESY spectra of a DNA oligonucleotide in phosphate buffer (50 mM in $D_2O$, containing NaCl = 20 mM, 10 °C); a sequential NOESY walk along the full strand was identified. (**c**) TOCSY spectrum of H1′ × H2′-H2″ and (**d**) NOESY spectrum of H1′ × aromatic of a DNA oligonucleotide in the presence of an equimolar concentration of **1c** in phosphate buffer (50 mM in $D_2O$, containing NaCl = 20 mM, 10 °C). We also found a sequential NOESY walk that was significantly different at A4 and C5 in comparison with DNA only in **b**, suggestive of interactions of **1c** at a mismatched site. Note that the interactions of C1H5 with G2H8 and C8H5 with G9H8 at the terminal sites in **b** and **d**, and the interaction of C5H5 with A4H8 at the mismatched site in **b** were identified.

Supplementary Fig. 14 and described in Supplementary Table 1, showing an intramolecular Pt–Pt distance of 3.201 Å) was observed to display higher selectivity towards hairpin CC mismatched DNA (Fig. 2). Figure 5a and Supplementary Fig. 15 illustrate that the emission intensity of **2** at $\lambda = 634$ nm was elevated 50-fold upon the addition of only 0.5 equivalents of hairpin DNA containing a CC mismatch but enhanced only 8.2-fold in the presence of the same amount of matched DNA, that is, a 6.1-fold difference in emission responses.

The binding affinity of **2** with DNA was examined by ITC. The interaction of **2** was much more exothermic with CC mismatched DNA than with matched DNA (Supplementary Fig. 16). Notably, the $K_d$ of the binding of **2** to CC mismatched DNA was found to be 7.0 ± 2.1 μM, which is 7.4-fold higher than that of **1c**. The $K_d$ of the binding of **2** with matched DNA was negligible. In addition, the melting temperature ($T_m$), as determined by the change in absorbance ($OD_{260nm}$) of DNA was increased by 6.8 °C for mismatched DNA in the presence of **2** at an equimolar concentration (Supplementary Fig. 17a). However, only a 0.5 °C increase in melting temperature was found for matched DNA in the presence of **2** (Supplementary Fig. 17b).

Complex **2** also exhibited higher emission responses to CC mismatched DNA containing 16 different adjacent sequences (Fig. 5b), with a nearby 5′-TCT-3′ sequence generating the highest fold increase in emission (13.5-fold higher than matched DNA). In addition to its response to CC mismatches, **2** also showed higher emission responses to several other types of mismatched DNA over matched DNA, and the emission increased 4.9-fold for CT, 3.8-fold for AC and 3.4-fold for TT compared with matched DNA but no higher for other mismatches (AG, 1.8-fold; GT, 1.4-fold; AA, 1.4-fold; GG, 0.8-fold; Fig. 5a). Such differences in the emission responses between mismatched and matched DNA partially correlated with the thermodynamic stability of the DNA base pairs, CC ≤ AC ≤ CT < AA ≤ TT < GA ≤ GT < GG[49], suggesting an insertion binding mode[50]. The emission response to AC was slightly lower than that to CT, and the response to AA was lower than that to TT; these observations are not fully consistent with the trend of thermodynamic stability of mismatched DNA. The lower emission response towards A-containing AC and AA mismatches may have been caused by the weak coordination of the N7 of purine in adenine at the mismatch site to the Pt complex, leading to decrease of emission intensity (note that the pyrimidine N3 in C and T is not accessible to the Pt(II) complexes due to steric hindrance). We performed hybrid quantum mechanics/molecular mechanics (QM/MM) calculations to study the mode of binding between **2** and mismatched DNA. As shown in Fig. 5c, **2** bound to mismatched DNA through an insertion binding mode at the minor groove (this result is also consistent with the results of a $[Cu(phen)_2]^{2+}$ quenching experiment, Supplementary Fig. 18); the C^N^N planes interacted with nearby GC and AT base pairs by π-stacking interactions (inter-planar distance between the C^N^N plane and the base pairs of ~3.2 Å). The QM/MM simulation also indicated that the dinuclear structure better fit the mismatched sites, which have larger pockets (the inter-planar separation between nearby base pairs at mismatched sites can be > 9 Å), but not the matched site, which contains less space between base pairs.

We also tested the emission responses of **2** to dsDNA of poly(dC)-poly(dG), poly(dC-dG)$_2$, poly(dA)-poly(dT) and poly(dA-dT)$_2$; single-stranded DNA of oligo(dT)$_{20}$ and oligo(dC)$_{20}$; dsRNA and bovine serum albumin, all of which were observed to have at least 4.5-fold lower emission intensity than that of CC mismatched DNA (Fig. 5d). Notably, the emission intensity of **2** with RNA containing a CC mismatch was 3.3-fold higher than that with matched RNA (Fig. 5d).

**Photophysical properties upon binding to mismatched DNA.** To better understand the increase in emission upon binding with

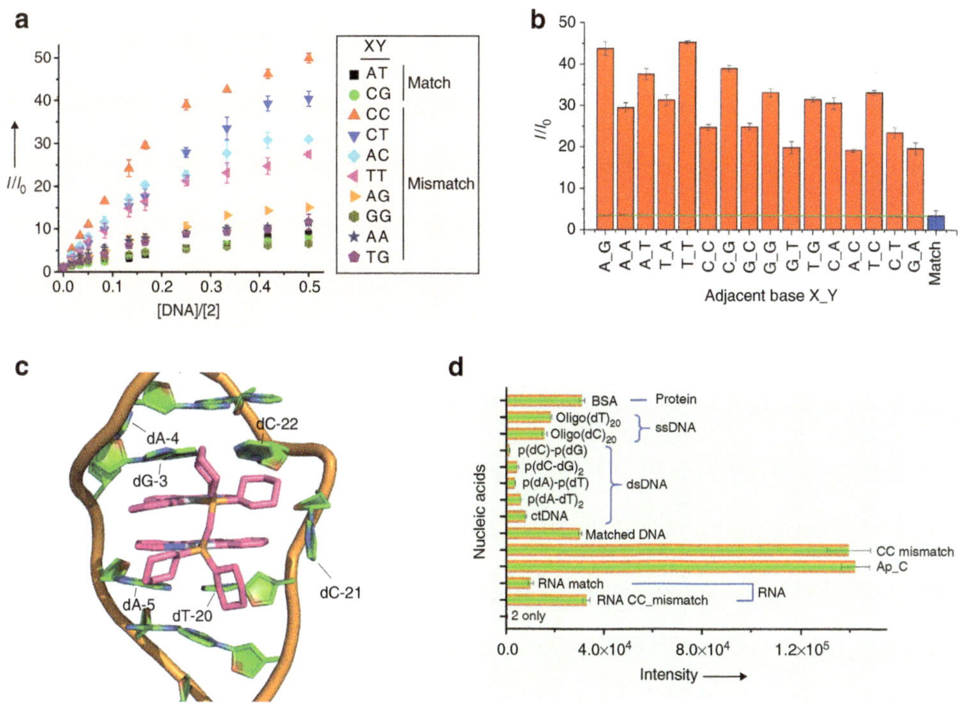


**Figure 5 | Binding of 2 with mismatched DNA.** (**a**) Changes in emission intensity (±s.e.m.) at 634 nm of **2** (5 μM) in Tris buffer solution (50 mM NaCl, 2 mM Tris, pH 7.5) after the addition of different concentrations of various DNA. (**b**) Relative emission enhancement of **2** towards CC mismatched DNA with different adjacent base pairs. (**c**) QM/MM optimized structure of mismatched DNA bound to **2**. (**d**) Emission responses (intensity ± s.e.m.) of **2** towards different analytes.

mismatched DNA, we measured the emission spectra of the Pt(II) complex (**1c** or **2**) in buffer solutions containing CC mismatched DNA under air or argon. Under Ar, no significant changes in emission intensity were found compared with that under air conditions (Supplementary Fig. 19). The two Pt(II) complexes were weakly emissive in buffer solutions, but their emission intensities were significantly higher in acetonitrile, even under air (11-fold for **1c**, 9-fold for **2**, Supplementary Fig. 20a,b; the emission intensity was found to increase with decreasing donor strength of various solvents, as shown in Supplementary Fig. 20c). Thus, the increased emission could not be accounted for by the shielding of oxygen quenching upon mismatched DNA binding to the Pt(II) complexes. In addition, variations of pH from 5 to 11 did not evoke a significant change in emission intensity (Supplementary Fig. 21). We propose that the shielding of quenching by a buffer solution and/or a more ordered structure of the Pt(II) complex after mismatched DNA binding could account for the emission enhancement[51,52].

We used time-resolved emission spectroscopy to examine the binding between the Pt complexes and DNA. As shown in Supplementary Fig. 22, the emission of **2** in the presence of CC mismatches had a lifetime of 2.3 μs, whereas in the presence of matched DNA, the lifetime is composed of a major portion of short decay (<100 ns) and a small amount of 1.2-μs decay. The difference in emission intensity, $I_{mismatch}/I_{match}$, was observed to increase with increasing decay time, from 2.3-fold at $t=0$ to 11.8-fold at $t=10$ μs (Supplementary Fig. 23). Therefore, time-resolved emission spectroscopy may be a useful strategy to achieve higher sensitivity than steady-state emission spectroscopy for the detection of mismatched DNA using luminescent Pt(II) complexes.

**Targeting abasic DNA.** The abasic (AP) site, in which a purine or a pyrimidine base is missing in a nucleotide, represents another type of thermodynamically unstable DNA defect[50]. In the presence of a 27-mer dsDNA oligomer containing an abasic site[50] (AP_C with a complementary base of C), the emission intensity of an equimolar amount of **1c** at $\lambda_{max}=535$ nm exhibited a 32.2-fold increase (Supplementary Fig. 24). This elevation in emission was even higher than that observed for the binding of **1c** with CC mismatched DNA with similar sequences (26.4-fold). For comparison, only a 4.7-fold increase in emission was detected when **1c** was incubated with normal 27-mer dsDNA with a cognate sequence. In the presence of other abasic sites of AP_A, AP_T and AP_G, the emission intensities of **1c** were increased by 14.8, 21.1 and 14.4 times, respectively. Complex **2** showed 17.0–29.3-fold increases in emission intensity in the presence of different abasic DNA sequences, but only a 6.1-fold increase in emission intensity with normal DNA (Supplementary Fig. 25). We also measured the effects of **2** on the $T_m$ of DNA bearing an abasic site (Supplementary Fig. 26). The melting temperature of abasic DNA in the presence of **2** was increased by 5.3 °C, whereas the $T_m$ of the similar DNA sequence without an abasic site was not affected, indicating a high selectivity of **2** towards abasic DNA over matched DNA. Thus, these Pt(II) complexes could be used to probe DNA abasic sites. Such a property is also found in Rh(III)-based metalloinsertors, in which the metal complex binds to thermodynamically unstable DNA structures through an insertion binding mode[50].

**Detection of genomic DNA base pair defects.** The human colorectal carcinoma cell line HCT116 is known to be MMR deficient, owing to mutations in the MMR-associated *hMLH1* gene, whereas HCT116 cells containing a transferred, normal chromosome 3 (denoted as HCT116N) restore MMR activity[53]. We found that **2** showed significantly stronger emission at $\lambda_{max}=634$ nm in the presence of genomic DNA isolated from HCT116 cells compared with that from HCT116N cells ($P<0.01$, $n=3$; Fig. 6a; Supplementary Fig. 27). We then measured and

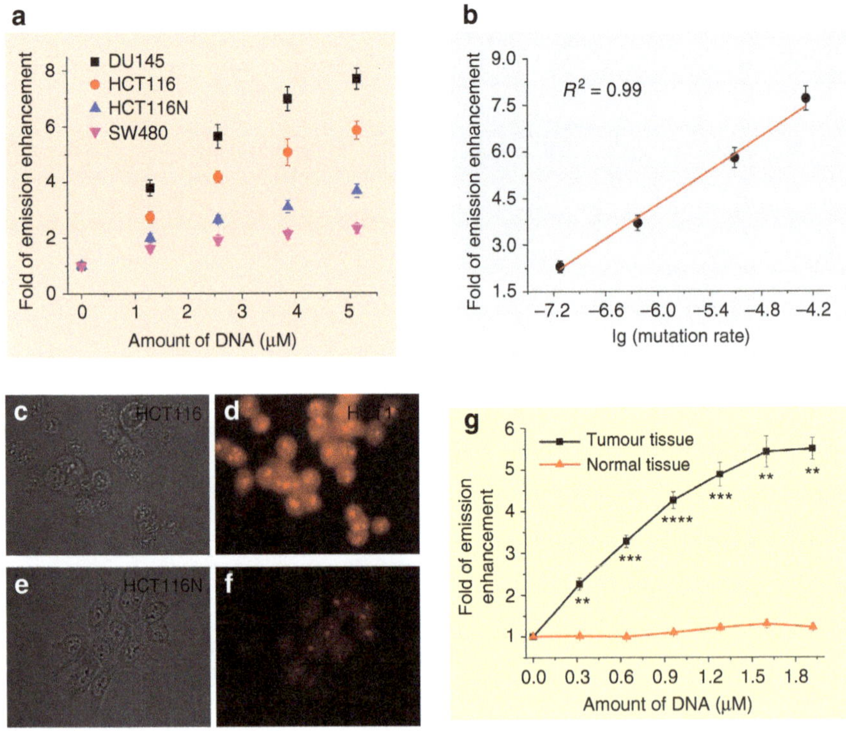


**Figure 6 | Detection of mismatched DNA in cancer cells.** (**a**) Changes in emission intensity (±s.e.m.) of **2** (5 μM) at 634 nm in an aqueous buffer solution (50 mM NaCl and 2 mM Tris, pH 7.5) after binding to genomic DNA extracted from different cancer cell lines. (**b**) Plot of fold of increased emission versus reported mutation rates in different cancer cell lines, lg (mutation rate); a linear fit is shown. Fluorescence microscopy studies of HCT116 (**d**) and HCT116N (**f**) cells pretreated with digitonin and stained with 10 μM **2**; (**c**,**e**) bright-field images of HCT116 and HCT116N cells, respectively. (**g**) Plot of emission responses (fold of increase in emission ± s.e.m.) of **2** versus different concentrations of DNA extracted from colon tumour tissue or adjacent normal tissue. All the concentrations of genomic DNA are shown as base pair concentrations.

compared the emission responses of **2** with several types of cancer cells with different mutation rates. The human cancer cell lines DU145, HCT116, HCT116N and SW480 have mutation rates (mutations per cell per generation) in the *hprt* gene locus of $529 \times 10^{-7}$, $77.5 \times 10^{-7}$, $5.9 \times 10^{-7}$ and $0.75 \times 10^{-7}$, respectively[54], and the relative emission intensities of **2** (5 μM) at 5.1 μM of DNA are 7.7, 5.8, 3.7 and 2.3, respectively (Fig. 6a shows an increase of emission intensity at 0–5.1 μM of genomic DNA from different cancer cells). Notably, the plot of the fold of emission increase versus the lg (mutation rate) yields a linear fit with $R^2 = 0.99$, as shown in Fig. 6b, indicating a strong correlation between emission response and mutation rate.

We next tested whether **2** could be used to selectively detect mismatched DNA in permeabilized cell samples by fluorescence microscopy. Because the Pt(II) complexes are lipophilic and tend to accumulate in cytoplasmic compartments, the HCT116 and HCT116N cancer cells were permeabilized by digitonin and then washed to minimize the fluorescence background due to interactions of the Pt(II) complexes with cytoplasmic contents. As shown in Fig. 6c–f, after labelling with **2**, the emission signal of the HCT116 cells was significantly stronger than that of the HCT116N cells. In addition, complex **2** emitted significantly stronger signals in cancerous HCT116 cells compared with those in the immortalized normal human colon mucosal epithelial cell line NCM460 (ref. 55; Supplementary Fig. 28); complex **2** also consistently exhibited significantly lower emission responses in the presence of the DNA isolated from NCM460 cells compared with that from HCT116 cells (Supplementary Fig. 27). The differences in emission intensity were not caused by variations in cellular uptake of Pt(II) complexes as shown by inductively coupled plasma mass spectrometer measurements (Supplementary Fig. 29). For comparison, no significant differences in the emission properties were found between the two cell lines labelled with the general DNA intercalator EB (Supplementary Fig. 30).

We further examined the ability of the Pt(II) complex to detect mismatched DNA of primary human tumours of colon cancer which is frequently characterized by MMR deficiency. DNA samples were extracted from colorectal adenocarcinoma (well to moderately differentiated) and the surrounding normal tissues of a patient (see details in the Methods section). The results of the emission responses of **2** (5 μM) to the DNA from the tumour and normal tissues are shown in Fig. 6g. In the presence of 0.3–1.9 μM of DNA extracted from normal colon tissues, weak emission responses of up to 1.3-fold of the initial emission intensity were detected; in contrast, the emission intensity increased in a concentration-dependent manner in the presence of 0–1.6 μM of DNA extracted from colon tumour tissues, showing up to a 5.5-fold increase in emission at 1.9 μM. Thus, there was a difference between the emission responses of **2** to the DNA of human colon tumour tissue and the normal adjacent tissue.

In conclusion, we have identified two classes of luminescent Pt(II) complexes that can be used for the detection of DNA mismatches and abasic DNA sites. These Pt(II) complexes are able to differentiate between cells with different levels of MMR activity. Importantly, the differential emission responses of the Pt(II) complexes to DNA from human colon cancer tissues and normal colon tissue suggest that the Pt(II) complexes are promising candidates for tumour diagnosis via simple emission spectroscopy measurements of DNA extracts. The selective targeting of DNA mismatches also lends the complexes therapeutic potential for cancer treatment[21]; and they may be used as scaffolds in the design of new anti-cancer agents for the treatment of MMR-deficient cancers.

## Methods

**General.** The synthesis and characterization of the platinum complexes and the experimental details of X-ray crystallographic analysis, emission titration experiments, ultraviolet/visible absorption titration experiments, ITC, $^{1}H$ NMR titration experiments, melting temperature experiments and molecular docking model experiments are provided in the Supplementary Methods. The nucleic acid sequences for the different experiments are listed in the Supplementary Information.

**Quantum mechanics/molecular mechanics.** The structures of mismatched DNA and complex **2** were taken from the literature (PDB ID: 2O11[56], unit Å) and the X-ray crystal analysis (CCDC 1061095), respectively. In general, molecular docking studies were performed using ICM-pro software to obtain the initial structure of DNA–complex **2** adducts[57]; subsequently, the binding between the DNA and complex was further optimized using the QM/MM approach and NWChem software[58]. The quantum part (QM) included complex **2** (total 133 atoms), and the rest of the system was modelled at the MM level. The QM region was treated at the density functional theory level using the M06L method[59]. The 6-31G* Pople basis set[60] was employed for C, H, N and P atoms, and for the Pt atom, we applied the LANL2DZ[61] basis set with effective core potential. The mismatched DNA was described with the AMBER parm99 force field[62]. Van der Waals parameters for the Pt atom were taken from the literature[63]. The docked structure was solvated in a 65-Å cubic box with 8,830 classical SPC/E water molecules. Twenty-two-positive ($Na^{+}$) counterions were added to neutralize the charges in the system. The structure was optimized using the BFGS algorithm for the QM part and the steepest descent algorithm for the MM part. The optimization of these two regions (QM and MM) was alternated until self-consistency was reached.

**DNA extraction from cancer cells or tumour tissues.** HCT116 and HCT116N cell lines were generously provided by Thomas A. Kunkel (NIH, USA). The human tissue samples were obtained from Jinan University, Guangzhou, China, and the experimental uses were authorized by the Human Experimentation Ethics Committee of Jinan University. Genomic DNA from cells or tissues was isolated using DNAzol Reagent (Thermo Fisher ) according to the manufacturer's protocol. For DNA extraction from cell lines, 1 ml of DNAzol Reagent was added to $1–3 \times 10^{7}$ cells grown on 10-$cm^{2}$ culture plates. Then, the cells were lysed by agitating the culture plate and gently pipetting the lysate into a microfuge tube. For the tissue sample, 1 ml of DNAzol Reagent was added to $\sim 50$ mg of tissues. Then, the tissues were homogenized by applying three strokes in the homogenizer. The homogenate was sedimented for 10 min at 10,000*g* at 4 °C. Following centrifugation, the resulting viscous supernatant was transferred to a fresh tube, mixed with 0.5 ml of 100% ethanol by inversion and stored at room temperature for 1–3 min. Then, the DNA precipitate was washed twice with 1.0 ml of 75% ethanol. The DNA was air-dried and dissolved in 200 µl of 8 mM NaOH, and the pH was adjusted to 7.5 by the addition of HEPES. The emission spectrum of a solution of complex **1c** or **2** (5 µM) in buffer solution (50 mM NaCl and 2 mM Tris, pH 7.5) was recorded and titrated with increasing amounts of the isolated DNA. The emission spectra were recorded after equilibration for 3 min in each addition.

**Fluorescence microscopy examination.** Cells (HCT116 or NCM460) were seeded in glass bottom culture dishes (35 mm) at a density of $1.0 \times 10^{5}\,ml^{-1}$. After 12 h, the dishes were placed on ice, washed twice with 1 ml of cold PBS and then incubated in 1 ml of permeabilization buffer (20 mM HEPES, pH 7.5, 110 mM potassium acetate, 5 mM magnesium sulphate and 250 mM sucrose) containing 40 $\mu g\,ml^{-1}$ digitonin for 30 min. After removal of the digitonin solution, the dishes were washed three times with cold permeabilization buffer for 1, 5 and 10 min. The cells were then incubated with complex 2 (10 µM) in transport buffer (20 mM HEPES, pH 7.3, 110 mM potassium acetate, 5 mM sodium acetate, 2 mM magnesium sulphate and 250 mM sucrose) for 15 min. Cells were directly subjected to fluorescence imaging without removing the transport buffer. For co-culture experiments, HCT116 cells were first seeded in culture dishes for 12 h. Next, the dishes were treated with a non-toxic concentration of Hoechst 33342 (1 $\mu g\,ml^{-1}$) for 30 min and washed twice with PBS. Then, NCM460 cells were seeded in the same dishes for another 12 h. Finally, the co-cultured cells were incubated in permeabilization buffer, stained with complex **2** and subjected to fluorescence imaging as aforementioned.

## References


1. Erkkila, K. E., Odom, D. T. & Barton, J. K. Recognition and reaction of metallointercalators with DNA. *Chem. Rev.* **99,** 2777–2796 (1999).
2. Metcalfe, C. & Thomas Jim, A. Kinetically inert transition metal complexes that reversibly bind to DNA. *Chem. Soc. Rev.* **32,** 215–224 (2003).
3. Haas, K. L. & Franz, K. J. Application of metal coordination chemistry to explore and manipulate cell biology. *Chem. Rev.* **109,** 4921–4960 (2009).
4. Liu, H.-K. & Sadler, P. J. Metal complexes as DNA intercalators. *Acc. Chem. Res.* **44,** 349–359 (2011).
5. Demas, J. N. & DeGraff, B. A. Applications of luminescent transition platinum group metal complexes to sensor technology and molecular probes. *Coord. Chem. Rev.* **211,** 317–351 (2001).
6. Ma, D.-L., He, H.-Z., Leung, K.-H., Chen, D. S.-H. & Leung, C.-H. Bioactive Luminescent Transition-Metal Complexes for Biomedical Applications. *Angew. Chem. Int. Ed.* **52,** 7666–7682 (2013).
7. Zeglis, B. M., Pierre, V. C. & Barton, J. K. Metallo-intercalators and metallo-insertors. *Chem. Commun.* 4565–4579 (2007).
8. Richards, A. D. & Rodger, A. Synthetic metallomolecules as agents for the control of DNA structure. *Chem. Soc. Rev.* **36,** 471–483 (2007).
9. Lo, K. K. W. in *Photofunctional Transition Metals Complexes.* Vol. 123 (ed. Yam, V. W. W.) 204–245 (Springer-Verlag, 2007).
10. Gao, F., Chao, H. & Ji, L.-N. DNA binding, photocleavage, and topoisomerase inhibition of functionalized ruthenium(II)-polypyridine complexes. *Chem. Biodivers.* **5,** 1962–1979 (2008).
11. Cummings, S. D. Platinum complexes of terpyridine: interaction and reactivity with biomolecules. *Coord. Chem. Rev.* **253,** 1495–1516 (2009).
12. Komor, A. C. & Barton, J. K. The path for metal complexes to a DNA target. *Chem. Commun.* **49,** 3617–3630 (2013).
13. Granzhan, A., Kotera, N. & Teulade-Fichou, M.-P. Finding needles in a basestack: recognition of mismatched base pairs in DNA by small molecules. *Chem. Soc. Rev.* **43,** 3630–3665 (2014).
14. Ma, D.-L., Che, C.-M. & Yan, S.-C. Platinum(II) complexes with dipyridophenazine ligands as human telomerase inhibitors and luminescent probes for G-quadruplex DNA. *J. Am. Chem. Soc.* **131,** 1835–1846 (2009).
15. Murat, P., Singh, Y. & Defrancq, E. Methods for investigating G-quadruplex DNA/ligand interactions. *Chem. Soc. Rev.* **40,** 5293–5307 (2011).
16. Carlson, C. B., Stephens, O. M. & Beal, P. A. Recognition of double stranded RNA by proteins and small molecules. *Biopolymers* **70,** 86–102 (2003).
17. Zou, T. *et al.* Luminescent cyclometalated platinum(II) complex forms emissive intercalating adducts with double-stranded DNA and RNA: differential emissions and anticancer activities. *Angew. Chem. Int. Ed.* **53,** 10119–10123 (2014).
18. Pierre, V. C., Kaiser, J. T. & Barton, J. K. Insights into finding a mismatch through the structure of a mispaired DNA bound by a rhodium intercalator. *Proc. Natl Acad. Sci. USA* **104,** 429–434 (2007).
19. Song, H., Kaiser, J. T. & Barton, J. K. Crystal structure of Δ-$[Ru(bpy)_2dppz]^{2+}$ bound to mismatched DNA reveals side-by-side metalloinsertion and intercalation. *Nat. Chem.* **4,** 615–620 (2012).
20. Kinsella, T. J. Coordination of DNA mismatch repair and base excision repair processing of chemotherapy and radiation damage for targeting resistant cancers. *Clin. Cancer Res.* **15,** 1853–1859 (2009).
21. Martin, S. A., Lord, C. J. & Ashworth, A. Therapeutic targeting of the DNA mismatch repair pathway. *Clin. Cancer Res.* **16,** 5107–5113 (2010).
22. Bridge, G., Rashid, S. & Martin, S. A. DNA mismatch repair and oxidative DNA damage: implications for cancer biology and treatment. *Cancers* **6,** 1597–1614 (2014).
23. Lothe, R. A. Microsatellite instability in human solid tumors. *Mol. Med. Today* **3,** 61–68 (1997).
24. Hoeijmakers, J. H. J. Genome maintenance mechanisms for preventing cancer. *Nature* **411,** 366–374 (2001).
25. Loeb, L. A., Loeb, K. R. & Anderson, J. P. Multiple mutations and cancer. *Proc. Natl Acad. Sci. USA* **100,** 776–781 (2003).
26. Iyer, R. R., Pluciennik, A., Burdett, V. & Modrich, P. L. DNA mismatch repair: functions and mechanisms. *Chem. Rev.* **106,** 302–323 (2006).
27. Jackson, B. A., Alekseyev, V. Y. & Barton, J. K. A versatile mismatch recognition agent: specific cleavage of a plasmid DNA at a single base mispair. *Biochemistry* **38,** 4655–4662 (1999).
28. Zeglis, B. M., Pierre, V. C., Kaiser, J. T. & Barton, J. K. A bulky rhodium complex bound to an adenosine-adenosine DNA mismatch: general architecture of the metalloinsertion binding mode. *Biochemistry* **48,** 4247–4253 (2009).
29. Lim, M. H., Song, H., Olmon, E. D., Dervan, E. E. & Barton, J. K. Sensitivity of $Ru(bpy)_2dppz^{2+}$ luminescence to DNA defects. *Inorg. Chem.* **48,** 5392–5397 (2009).
30. McConnell, A. J. *et al.* Luminescent properties of ruthenium(II) complexes with sterically expansive ligands bound to DNA defects. *Inorg. Chem.* **51,** 12511–12520 (2012).
31. Chen, H., Yang, P., Yuan, C. & Pu, X. Study on the binding of base-mismatched oligonucleotide $d(GCGAGC)_2$ by cobalt(III) complexes. *Eur. J. Inorg. Chem.* **2005,** 3141–3148 (2005).
32. Fu, C., Harms, K. & Zhang, L. Meggers, E. DNA mismatch recognition by a hexacoordinate silicon sandwich-ruthenium hybrid complex. *Organometallics* **33,** 3219–3222 (2014).
33. Yang, X. L. *et al.* Imidazole-imidazole pair as a minor groove recognition motif for T:G mismatched base pairs. *Nucleic Acids Res.* **27,** 4183–4190 (1999).
34. Nakatani, K., Sando, S., Kumasawa, H., Kikuchi, J. & Saito, I. Recognition of guanine-guanine mismatches by the dimeric form of 2-amino-1,8-naphthyridine. *J. Am. Chem. Soc.* **123,** 12650–12657 (2001).

35. David, A. *et al.* DNA mismatch-specific base flipping by a bisacridine macrocycle. *Chembiochem.* **4,** 1326–1331 (2003).
36. Chin, K. H., Chen, F. M. & Chou, S. H. Solution structure of the ActD–5′-CCGTT$_3$GTGG-3′ complex: drug interaction with tandem G · T mismatches and hairpin loop backbone. *Nucleic Acids Res.* **31,** 2622–2629 (2003).
37. Nakatani, K. *et al.* Small-molecule ligand induces nucleotide flipping in (CAG)n trinucleotide repeats. *Nat. Chem. Biol.* **1,** 39–43 (2005).
38. Ono, A. *et al.* Specific interactions between silver(I) ions and cytosine-cytosine pairs in DNA duplexes. *Chem. Commun.* 4825–4827 (2008).
39. Jourdan, M., Granzhan, A., Guillot, R., Dumy, P. & Teulade-Fichou, M.-P. Double threading through DNA: NMR structural study of a bis-naphthalene macrocycle bound to a thymine-thymine mismatch. *Nucleic Acids Res.* **40,** 5115–5128 (2012).
40. Takada, T., Ashida, A., Nakamura, M. & Yamana, K. Cationic perylenediimide as a specific fluorescent binder to mismatch containing DNA. *Biorg. Med. Chem.* **21,** 6011–6014 (2013).
41. Lo, Y.-S., Tseng, W.-H., Chuang, C.-Y. & Hou, M.-H. The structural basis of actinomycin D–binding induces nucleotide flipping out, a sharp bend and a left-handed twist in CGG triplet repeats. *Nucleic Acids Res.* **41,** 4284–4294 (2013).
42. Jennette, K. W., Lippard, S. J., Vassiliades, G. A. & Bauer, W. R. Metallointercalation reagents. 2-Hydroxyethanethiolato(2,2′,2″-terpyridine)-platinum(II) monocation binds strongly to DNA by intercalation. *Proc. Natl Acad. Sci. USA* **71,** 3839–3843 (1974).
43. Che, C.-M., Yang, M., Wong, K.-H., Chan, H.-L. & Lam, W. Platinum(II) complexes of dipyridophenazine as metallointercalators for DNA and potent cytotoxic agents against carcinoma cell lines. *Chem. Eur. J.* **5,** 3350–3356 (1999).
44. Wang, P. *et al.* Specific blocking of CREB/DNA binding by cyclometalated platinum(II) complexes. *Angew. Chem. Int. Ed.* **50,** 2554–2558 (2011).
45. Frezza, M. *et al.* *In vitro* and *in vivo* antitumor activities and DNA binding mode of five coordinated cyclometalated organoplatinum(II) complexes containing biphosphine ligands. *J. Med. Chem.* **54,** 6166–6176 (2011).
46. Sun, R. W.-Y. *et al.* Luminescent cyclometalated platinum(II) complexes containing N-Heterocyclic carbene ligands with potent *in vitro* and *in vivo* anti-cancer properties accumulate in cytoplasmic structures of cancer cells. *Chem. Sci.* **2,** 728–736 (2011).
47. Zeglis, B. M. & Barton, J. K. A mismatch-selective bifunctional rhodium-oregon green conjugate: a fluorescent probe for mismatched DNA. *J. Am. Chem. Soc.* **128,** 5654–5655 (2006).
48. Cordier, C., Pierre, V. C. & Barton, J. K. Insertion of a bulky rhodium complex into a DNA cytosine-cytosine mismatch: an NMR solution study. *J. Am. Chem. Soc.* **129,** 12287–12295 (2007).
49. SantaLucia, J. & Hicks, D. The thermodynamics of DNA structural motifs. *Annu. Rev. Biophys. Biomol. Struct.* **33,** 415–440 (2004).
50. Zeglis, B. M., Boland, J. A. & Barton, J. K. Targeting abasic sites and single base bulges in DNA with metalloinsertors. *J. Am. Chem. Soc.* **130,** 7530–7531 (2008).
51. Tong, G. S.-M. & Che, C.-M. Emissive or nonemissive? A theoretical analysis of the phosphorescence efficiencies of cyclometalated platinum(II) complexes. *Chem. Eur. J.* **15,** 7225–7237 (2009).
52. Lo, K. K.-W., Choi, A. W.-T. & Law, W. H.-T. Applications of luminescent inorganic and organometallic transition metal complexes as biomolecular and cellular probes. *Dalton Trans.* **41,** 6021–6047 (2012).
53. Koi, M. *et al.* Human chromosome 3 corrects mismatch repair deficiency and microsatellite instability and reduces *N*-Methyl-*N*′-nitro-*N*-nitrosoguanidine tolerance in colon tumor cells with homozygous *hMLH1* mutation. *Cancer Res.* **54,** 4308–4312 (1994).
54. Glaab, W. E. & Tindall, K. R. Mutation rate at the hprt locus in human cancer cell lines with specific mismatch repair-gene defects. *Carcinogenesis* **18,** 1–8 (1997).
55. Moyer, M., Manzano, L., Merriman, R., Stauffer, J. & Tanzer, L. NCM460, a normal human colon mucosal epithelial cell line. *In Vitro Cell. Dev. Biol. Anim.* **32,** 315–317 (1996).
56. Pierre, V. C., Kaiser, J. T. & Barton, J. K. Insights into finding a mismatch through the structure of a mispaired DNA bound by a rhodium intercalator. *Proc. Natl Acad. Sci. USA* **104,** 429–434 (2007).
57. Neves, M. A. C., Totrov, M. & Abagyan, R. Docking and scoring with ICM: the benchmarking results and strategies for improvement. *J. Comput. Aided Mol. Des.* **26,** 675–686 (2012).
58. Valiev, M. *et al.* NWChem: A comprehensive and scalable open-source solution for large scale molecular simulations. *Comput. Phys. Commun.* **181,** 1477–1489 (2010).
59. Zhao, Y. & Truhlar, D. G. A new local density functional for main-group thermochemistry, transition metal bonding, thermochemical kinetics, and noncovalent interactions. *J. Chem. Phys.* **125,** 194101 (2006).
60. Hariharan, P. C. & Pople, J. A. The influence of polarization functions on molecular orbital hydrogenation energies. *Theor. Chim. Acta* **28,** 213–222 (1973).
61. Hay, P. J. & Wadt, W. R. Ab initio effective core potentials for molecular calculations. Potentials for K to Au including the outermost core orbitals. *J. Chem. Phys.* **82,** 299–310 (1985).
62. Cornell, W. D. *et al.* A second generation force field for the simulation of proteins, nucleic acids, and organic molecules. *J. Am. Chem. Soc.* **117,** 5179–5197 (1995).
63. Hambley, T. W. van der Waals radii of Pt(II) and Pd(II) in molecular mechanics models and an analysis of their relevance to the description of axial M · · · H( − C), M · · · H( − N), M · · · S, and M · · · M (M = Pd(II) or Pt(II)) interactions. *Inorg. Chem.* **37,** 3767–3774 (1998).

## Acknowledgements

This work was supported by the University Grants Committee of the HKSAR of China (Area of Excellence Scheme AoE/P-03/08), the National Key Basic Research Program of China (2013CB834802), the Innovation and Technology Fund (ITF-Tier 2 ITS/130/14FP) and Special Equipment Grant of UGC (SEG_HKU02) for MS analysis. We thank Thomas A. Kunkel (NIH, USA) for providing the colon cancer cell lines, and we are grateful to Kong Hung Sze and Siu-Cheong Yan for help with 2D NMR experiments.

## Author contributions

T.Z. and C.-M.C. designed the research. S.K.F., B.C., T.C., W.-P.T. and C.Y. performed the research. S.K.F., T.Z., C.-N.L and C.-M.C. analysed the data, and S.K.F., T.Z. and C.-M.C. wrote the paper.

## Additional information

**Competing financial interests:** The authors declare no competing financial interests.

# 21

# A competitive and reversible deactivation approach to catalysis-based quantitative assays

Kazunori Koide[1], Matthew P. Tracey[1], Xiaodong Bu[2], Junyong Jo[2], Michael J. Williams[2] & Christopher J. Welch[2]

Catalysis-based signal amplification makes optical assays highly sensitive and widely useful in chemical and biochemical research. However, assays must be fine-tuned to avoid signal saturation, substrate depletion and nonlinear performance. Furthermore, once stopped, such assays cannot be restarted, limiting the dynamic range to two orders of magnitude with respect to analyte concentrations. In addition, abundant analytes are difficult to quantify under catalytic conditions due to rapid signal saturation. Herein, we report an approach in which a catalytic reaction competes with a concomitant inactivation of the catalyst or consumption of a reagent required for signal generation. As such, signal generation proceeds for a limited time, then autonomously and reversibly stalls. In two catalysis-based assays, we demonstrate restarting autonomously stalled reactions, enabling accurate measurement over five orders of magnitude, including analyte levels above substrate concentration. This indicates that the dynamic range of catalysis-based assays can be significantly broadened through competitive and reversible deactivation.

[1] Department of Chemistry, University of Pittsburgh, 219 Parkman Avenue, Pittsburgh, Pennsylvania 15260, USA. [2] Process and Analytical Chemistry, Merck Research Laboratories, 126 East Lincoln Avenue, Rahway, New Jersey 07065, USA. Correspondence and requests for materials should be addressed to K.K. (email: koide@pitt.edu) or to C.J.W. (email: christopher_welch@merck.com).

Development of optical assays for facile quantification of trace analytes is an ever-expanding field. Target analytes range from trace metals[1] and biological signalling agents[2,3] to chemical weapons[4]. Tailored chemosensors interact specifically with an analyte to produce an optically decoded signal, which can manifest itself as a wavelength shift or intensity change in either absorbance or emission. These signals are measured by simple instrumentation, such as a plate reader, or visualized with the naked eye.

Quantitative optical assays exploit either a catalytic or a non-catalytic reaction. Non-catalytic assays rely on a single turnover from the analyte for a chemical conversion or a reversible binding and have the benefit of time-independence; in other words, the signal does not change over time once the reaction or binding event is complete. A major drawback of these systems is the higher limits of quantification due to this limited turnover, rendering these non-catalytic assays undesirable for detection of trace analytes. A more sensitive approach for trace analytes uses catalysis-based assays, where the substrate continues to react over time, amplifying signals.

The continuity of catalysis-based signal amplification presents some practical challenges to assay development. In metal catalysis-based assays, once the metal has entered into the catalytic cycle, the resulting fluorescence signal is dependent on the concentration of the analyte as well as the time elapsed, with the reaction continuing until the fluorogenic substrate is consumed. In enzyme and enzyme-linked immunosorbent assays[5], the reaction continues until the substrate is consumed or a terminating reagent is added[6]. In either case, if an analyte is abundant, the assay substrate will be rapidly consumed, preventing accurate quantitation. In addition, if a reaction with a low concentration of analyte is allowed to continue unchecked, the signal can increase to the point where the detector becomes saturated, again preventing accurate quantification. Finally, when a catalysis-based assay is externally stopped, it cannot be restarted[6] and premature termination requires the assay to be repeated to obtain quantitative data. As such, a significant drawback associated with catalysis-based assays is the far narrower dynamic range (one to two orders of magnitude) compared with more labour-intensive methods, such as inductively coupled plasma mass spectrometry (ICP-MS), which has a detection range is up to five orders of magnitude. New methodologies that overcome these limitations to enable controlled activity of catalytic assays would be broadly useful in chemical and biochemical research.

Herein, we present a new approach to catalysis-based assays in which a catalytic chromogenic reaction competes with the deactivation of the catalyst or depletion of an essential reagent. Under these conditions, a signal-producing reaction proceeds for a limited time, then autonomously stalls, but can be reactivated by reagent addition, generating a graph reminiscent of a staircase function in mathematics. This approach is exemplified by both a new colorimetric method for quantifying palladium (Pd) and a horseradish peroxidase assay system. In the analysis of Pd, multiple cycles of reaction stalling and restarting allow accurate measurement with a detection range of over five orders of magnitude. Moreover, analyte levels significantly above the substrate concentration can be quantified.

## Results

**Resorufin allyl ether as a chemodosimeter for palladium.** We previously reported a fluorescence method for quantifying Pd in pharmaceuticals based on the Pd-catalysed fluorogenic conversion of allyl Pittsburgh Green ether (APE) to Pittsburgh Green (Fig. 1a)[7–10]. Although this method showed excellent sensitivity and an ability to accurately quantify low-level Pd in real-world samples, we realized that a colorimetric version of the assay could allow even simpler, instrument-free access to low-level Pd measurements, a goal previously attempted by several other researchers with limited success[11,12].

Investigation of a number of candidate chromogenic substrates led to the preparation of yellow-coloured resorufin allyl ether (RAE) in one step in 85% yield from commercially available purple-coloured resorufin (Fig. 1b,c, Supplementary Figs 13 and 14). Attempts at Pd-catalysed deallylation of RAE using the optimized conditions for APE (tris(2-furyl)phosphine (TFP)), $NaBH_4$, dimethyl sulphoxide (DMSO)/1.23 M phosphate pH 7 buffer (1:9)) were unsuccessful. However, screening a variety of commercially available phosphines and additives (Supplementary Table S1 and Supplementary Figs 1–3) led to the identification of suitable conditions for carrying out the transformation. Optimized conditions for the Pd-dependent deallylation of RAE used TFP, $NH_4OAc$ and $NaBH_4$ in an EtOH solvent. Further optimization of RAE as a substrate can be found in the Supplementary Methods and Supplementary Figs 2–4.

We found that RAE was selectively responsive to Pd over other metals tested (Ag, Au, Cd, Co, Cr, Fe, Hg, Mn, Ni. Pt, Rh, Ru, Zn, Sr, Ir, Cu; Supplementary Fig. 4a,b) and could detect Pd without interference from these metals, with the exception of Hg, where a small level of interference was observed (Supplementary Fig. 4c). When selectivity was tested by absorbance, higher values were observed in the presence of Au, Ag, and Hg owing to turbidity of the solution, although fluorescence measurement revealed that these were merely false positives (that is, these metals did not convert RAE to resorufin; Supplementary Fig. 4a,b). When exposed to Pd, the fluorescence signal increased linearly with respect to Pd concentration (Fig. 1d), indicating a first-order relationship suitable for convenient quantification.

The Pd-catalysed deallylation of APE in phosphate buffer was more effective in the presence of $NaBH_4$, which reduces Pd(II) and Pd(IV) to catalytically active Pd(0), but did not require this reducing agent as a critical component[9]. In contrast, Pd(II) species did not catalyse the deallylation of RAE in $NH_4OAc$-containing EtOH without the reducing agent, with the amount of $NaBH_4$ dictating the duration of reaction (Fig. 2a). This novel $NaBH_4$-dependence boded well with our aim at competitively and reversibly deactivating catalysis-based assays, as detailed below. Lower concentrations of $NaBH_4$, ranging from 5–25 mM, led to stalling of the colour-forming reaction within 30 s, presumably because of rapid consumption of the reductant, $NaBH_4$, combined with ongoing air-oxidation of catalytically active Pd(0) to higher valent, inactive Pd species. In contrast, $NaBH_4$ concentrations in excess of 50 mM allowed the reaction to continue for several minutes. Importantly, the addition of more $NaBH_4$ could restart a stalled deallylation reaction (Fig. 2b), affording a convenient way to trigger signal generation on demand.

Subsequently, we sought to gain insights into the reaction stalling to rationally expand this developing methodology. The Pd-catalysed deallylation of APE stalled in the presence of $NH_4OAc$ but continued in a phosphate buffer (Supplementary Fig. 6). With 200, 400, 600 and 800 mM $NH_4OAc$ followed by pH adjustment, the reactions stalled nearly at the same time (Supplementary Fig. 7). The Pd-catalysed deallylation reaction of RAE under a nitrogen atmosphere were found to stall more slowly than those carried out in open air (Supplementary Fig. 8), suggesting that aerobic oxidation of Pd(0) to higher order Pd species may account for the observed reaction stalling.

**Demonstration of stop-and-go methodology with RAE.** In an effort to develop a simple, user-friendly colorimetric Pd

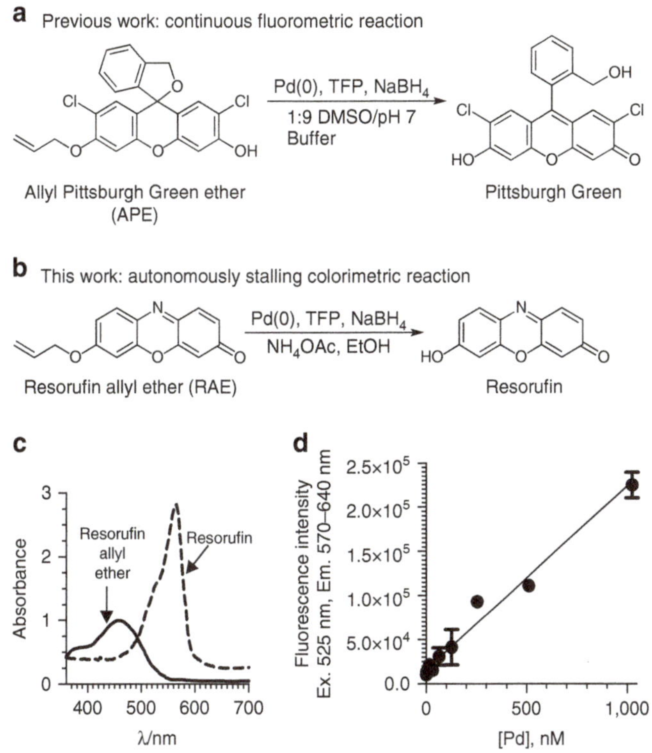


**Figure 1 | Palladium chemodosimeters based on change in optical properties.** (**a**) Structures of previously developed fluorogenic chemodosimeter APE and its conversion to Pittsburgh Green. (**b**) Chromogenic chemodosimeter RAE and its conversion to resorufin. (**c**) Absorption spectra of resorufin and RAE in 800 mM $NH_4OAc$ in EtOH. The data are normalized to 20 µM of each compound. (**d**) Correlation between Pd concentrations and fluorescence signal using RAE. $r^2 = 0.97$, $y = (208 \pm 7.38)x + (15{,}910 \pm 2{,}765)$. Conditions: 29 µM RAE, 0, 8, 16, 32, 64, 128, 256, 512, 1,024 nM Pd(II). 50 mM $NaBH_4$, 200 µM TFP, 800 mM $NH_4OAc$, EtOH, 24 °C, 60 min; $n = 3$.

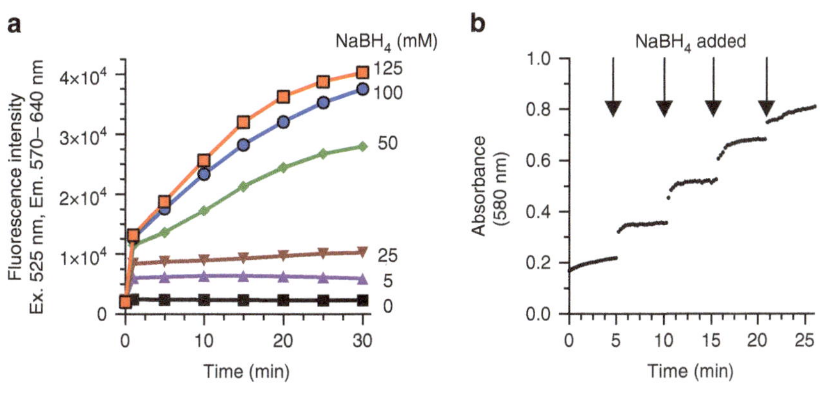


**Figure 2 | Influence of $NaBH_4$ on deallylation of RAE.** (**a**) Reaction lifetime dictated by $NaBH_4$ concentration. Conditions: 29 µM RAE, 10 p.p.b. Pd(II), 200 µM TFP, 800 mM $NH_4OAc$, 0–125 mM $NaBH_4$, EtOH, 25 °C. (**b**) Stalled deallylation reaction can be restarted by $NaBH_4$ addition. Conditions: 29 µM RAE, 0.3 p.p.m. Pd(II), 200 µM TFP, 800 mM $NH_4OAc$, 0. 0.6, 1.2, 1.8, 2.4 mM $NaBH_4$, added as 2.5 M aliquots at indicated time points.

quantification assay, we prepared a reagent cocktail combining all reaction components except $NaBH_4$ in a single solution. This cocktail, which is stable for over 2 weeks when stored at 5 °C, can be dispensed as needed, simplifying application of the colorimetric method. The addition of either 20 µl of a solution or 2–5 mg of a solid sample containing trace Pd to 1 ml of the reaction cocktail, followed by the addition of a $NaBH_4$ solution, generated colour and fluorescence within 1 min. The colour intensity was linearly correlated with Pd concentration, and the dynamic range and reaction time of the assay were tailored by adjusting the $NaBH_4$ concentration.

The power of this method is shown in Fig. 3. Known concentrations of Pd afford widely different colours with a single concentration of $NaBH_4$, with the colour persisting for 24 h (Fig. 3a). If a sample contains 1 p.p.b. Pd, then 100 mM $NaBH_4$ is required to observe a colour change (Fig. 3b). If a sample contains 10 p.p.m. Pd, then no $NaBH_4$ is added to observe a colour change. Thus, Pd concentrations ranging from 1 p.p.b. to 10 p.p.m. (five orders of magnitude) can be distinguished in one reaction solution with $NaBH_4$ titration. Alternatively, a user may prepare multiple wells with variable $NaBH_4$ amounts and count a number of coloured wells to estimate the Pd concentrations.

To confirm that the stop-and-go assay approach is providing quantitative data, we analysed real-world samples. We first tested intermediates used in the preparation of active pharmaceutical

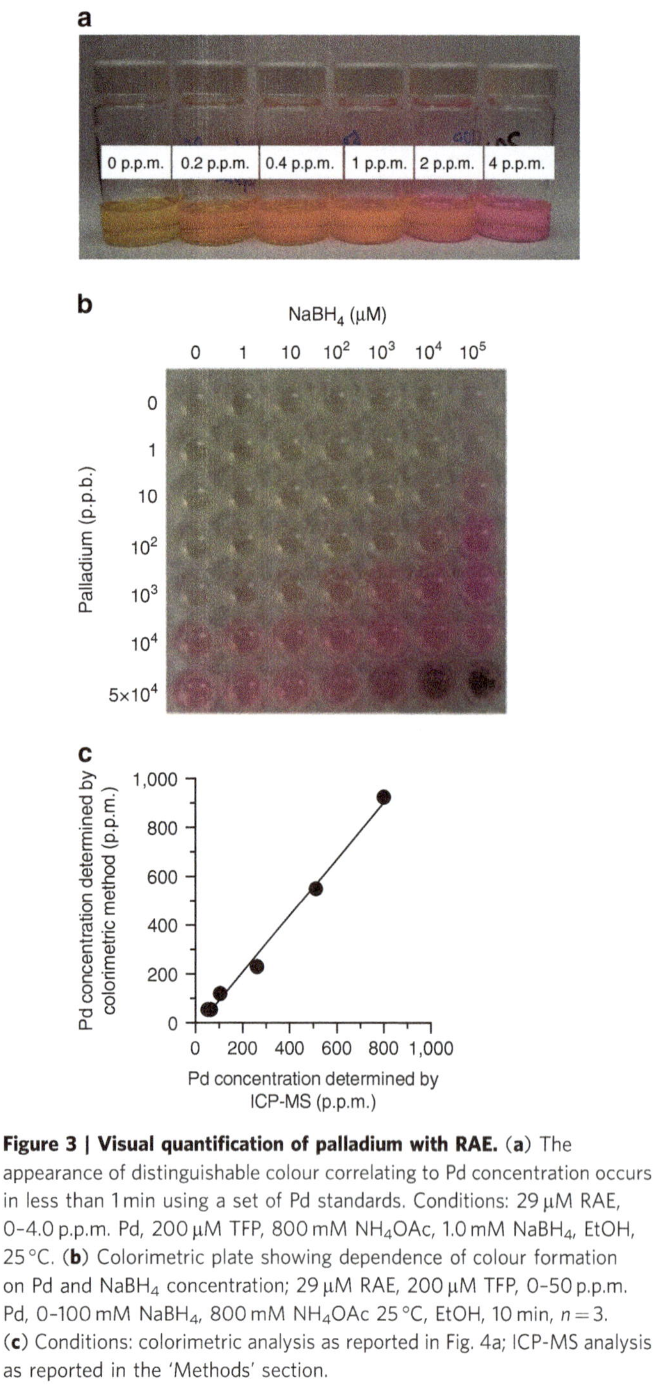


**Figure 3 | Visual quantification of palladium with RAE.** (**a**) The appearance of distinguishable colour correlating to Pd concentration occurs in less than 1 min using a set of Pd standards. Conditions: 29 μM RAE, 0–4.0 p.p.m. Pd, 200 μM TFP, 800 mM $NH_4OAc$, 1.0 mM $NaBH_4$, EtOH, 25 °C. (**b**) Colorimetric plate showing dependence of colour formation on Pd and $NaBH_4$ concentration; 29 μM RAE, 200 μM TFP, 0–50 p.p.m. Pd, 0–100 mM $NaBH_4$, 800 mM $NH_4OAc$ 25 °C, EtOH, 10 min, $n = 3$. (**c**) Conditions: colorimetric analysis as reported in Fig. 4a; ICP-MS analysis as reported in the 'Methods' section.

ingredients. In pharmaceutical synthesis, reactions may leave behind residual Pd in the products, which is often difficult to remove[13]. Various samples were tested from active projects in the Process and Analytical Chemistry Department at Merck Research Laboratories in which residual Pd removal has proven difficult. Quantification of Pd was initially performed by ICP-MS followed by analysis using RAE. Compared with the ICP-MS analysis, the stop-and-go approach with RAE provided accuracy from 70 to 120%, with residual Pd concentrations ranging from 62 to 800 p.p.m. (Fig. 3c). These results were satisfactory for this assay approach to be used for screening dozens of routine Pd remediation protocols.

Microscale screening of process adsorbents is often used to identify resins or activated carbons that can be used for selective adsorption of metal impurities in pharmaceutical process research and development[14,15]. Traditionally, this approach requires close coordination with ICP-MS specialists to allow for quick turnaround time. However, often because of instrument calibration, the vast number of samples, and preparation time, this can be time consuming. As such, the pharmaceutical industry has been interested in a faster technology for trace metal analysis[16].

The application of the colorimetric method enables rapid determination of Pd concentrations 'on the spot', in the same laboratory where the process development studies are being carried out. Figure 4 shows the results of a high throughput screen of Pd impurity remediation treatments of a pharmaceutical intermediate with 48 metal-scavenging adsorbents, using the stop-and-go assay with RAE to visualize relative Pd levels. An aliquot from each well is treated with the reaction cocktail (Fig. 4a), then with $NaBH_4$. In less than 5 min, gross differences in Pd concentration are readily apparent to the naked eye by distinguishable colours (Fig. 4b). At this point, the reaction had stalled, and too many hits were identified. Accordingly, more $NaBH_4$ was added to restart the reaction, accentuating the differences between wells and enabling rapid determination of the potential most-effective treatments for residual Pd remediation (Fig. 4c). A high-throughput mapping of relative Pd concentration was obtained by plotting the ratio of absorbance at 580 and 460 nm using a ultraviolet–visible plate reader (read time for 48 samples <30 s; Fig. 4d). These results quantitatively confirm the most effective Pd removal treatments to be wells A5, C4, E6, F3 and F5. Spot-checking several adsorbent treatment samples using conventional ICP-MS showed a good correlation with the colorimetric method, with the selection of the most-effective adsorbent treatments (A5, E6) being identical in both cases. These results demonstrate the utility of a stop-and-go approach in trace metal quantification, providing an important

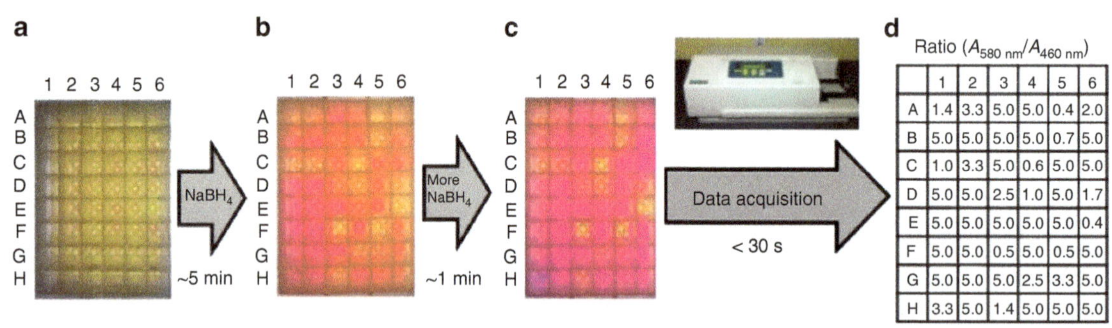


| | 1 | 2 | 3 | 4 | 5 | 6 |
|---|---|---|---|---|---|---|
| A | 1.4 | 3.3 | 5.0 | 5.0 | 0.4 | 2.0 |
| B | 5.0 | 5.0 | 5.0 | 5.0 | 0.7 | 5.0 |
| C | 1.0 | 3.3 | 5.0 | 0.6 | 5.0 | 5.0 |
| D | 5.0 | 5.0 | 2.5 | 1.0 | 5.0 | 1.7 |
| E | 5.0 | 5.0 | 5.0 | 5.0 | 5.0 | 0.4 |
| F | 5.0 | 5.0 | 0.5 | 5.0 | 0.5 | 5.0 |
| G | 5.0 | 5.0 | 5.0 | 2.5 | 3.3 | 5.0 |
| H | 3.3 | 5.0 | 1.4 | 5.0 | 5.0 | 5.0 |

**Figure 4 | Case study of a streamlined process combining adsorbent screening for Pd removal with high-throughput colorimetric Pd detection.** (**a**) Screening kits containing 48 commercial adsorbents[15] are exposed to a solution of a Pd-containing intermediate. (**b**) Aliquots from screening kits are evaluated for Pd content using the colorimetric method, as described in protocol. (**c**) Finding the best potential hits visually by adding more $NaBH_4$. (**d**) High-throughput mapping of relative Pd concentration by measurement of ultraviolet-visible 570 nm/460 nm using ultraviolet-visible plate reader.

advance for process chemists dealing with remediation of Pd impurity problems using point-of-use high-throughput analysis.

We previously applied APE for quantifying Pd in ore samples without requiring acidic sample digestion[17], a significant improvement over standard analytical methods such as ICP-MS, but still requiring the use of a blue light source to check fluorescence intensity. This requirement associated with the fluorometric method was not ideal for turbid samples such as ores. Therefore, we applied our colorimetric method for more convenient visualization. We found that the reaction was stalled before effective Pd extraction from the ores, leading to a need to add a large excess of $NaBH_4$. However, a 1-min pre-incubation in a DMSO solution of TFP with sonication, followed by the addition of RAE, $NH_4OAc$, EtOH and $NaBH_4$ afforded subpar semi-quantitative data. A second addition of $NaBH_4$ after the reaction had stalled provided good colorimetric agreement with previous semi-quantitative analysis[17] (Supplementary Fig. 9) within 20 min.

Trace Pd is also a significant concern in materials science[18,19]. Our colorimetric method could be used to successfully detect trace Pd in polymer materials, as shown in Supplementary Figs 10–12, Supplementary Table 2 and Supplementary Notes 1 and 2.

**Competitive and reversible deactivation in enzymatic assay.** Horseradish peroxidase (HRP) is a common enzyme for detection and quantification in biological assays[20,21]. This enzyme catalytically converts Amplex Red (10-acetyl-3,7-dihydroxyphenoxazine) and $H_2O_2$ to resorufin, acetic acid and $H_2O$. (Fig. 5a)[22]. The reaction continues indefinitely until either Amplex Red or $H_2O_2$ is consumed, at which point, signal generation is stopped.

To illustrate the discontinuous catalysis approach in a different assay, we designed a system in which $PhB(OH)_2$ would competitively reduce $H_2O_2$ (Fig. 5a 'Deactivation of reagent') while the $H_2O_2$-mediated oxidation of Amplex Red occurs ('Conventional HRP assay'). Figure 5b shows that $PhB(OH)_2$ was able to do so in a concentration-dependent manner, affording lower signals. Reactions halted by consumption of $H_2O_2$ could be restarted by an addition of a fresh aliquot of $H_2O_2$ (Fig. 5c). With a further addition of $H_2O_2$, signal saturation occurred (Fig. 5d). With the inclusion of the competitive scavenger, $PhB(OH)_2$ to remove $H_2O_2$ from the system, the discontinuous catalysis alleviated the problem of overshooting signals, as well as allowed us to restart the reaction without problematic increases in fluorescence (Fig. 5d). Although the protocol has not been fully optimized in an HRP system, these data indicate a great potential for the applications of discontinuous catalysis in other enzyme assays.

## Discussion

We have developed a competitive and reversible deactivation approach for catalytic quantification assays (Fig. 6). Conversion of RAE to resorufin via a Pd-catalysed Tsuji–Trost reaction is autonomously stalled by the oxidation of reactive Pd(0) to non-reactive species. The addition of $NaBH_4$ as a reducing agent is able to restart the reaction, enabling accurate measurements over five orders of magnitude. Notably, even in cases where the

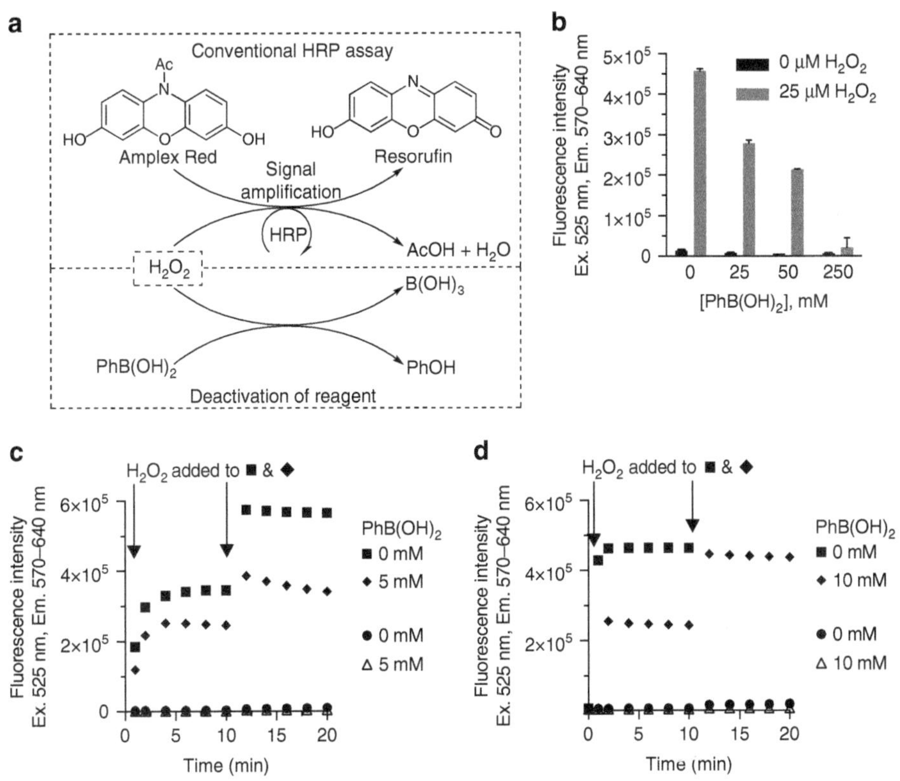


**Figure 5 | Stop-and-go approach in a horseradish peroxidase system.** (**a**) Conversion of Amplex Red to resorufin. (**b**) Effect of $PhB(OH)_2$ on horseradish peroxidase assay. Conditions: 50 µM Amplex Red, 0.1 U ml$^{-1}$ horseradish peroxidase, 25 µM $H_2O_2$, 0, 25, 50, 250 mM $PhB(OH)_2$, PBS pH 7.4. (**c**) Restarting a stopped enzymatic reaction in the presence of an inhibitor. Conditions: 50 µM Amplex Red, 0.05 U ml$^{-1}$ horseradish peroxidase, 0 µM $H_2O_2$ (0–20 min) for the circle and triangle. For others, 10 µM $H_2O_2$ at 0 min, 20 µM $H_2O_2$ at 10 min, PBS pH 7.4. (**d**) Restarting a stopped enzymatic reaction in the presence of an inhibitor with uninhibited saturation. Conditions: 50 µM Amplex Red, 1 U ml$^{-1}$ horseradish peroxidase, 0 µM $H_2O_2$ (0–20 min) for the circle and triangle. For others, 10 µM $H_2O_2$ at 0 min, 30 µM $H_2O_2$ at 10 min, PBS pH 7.4. After the addition of $H_2O_2$ at 10 min, the $PhB(OH)_2$-free sample (square) showed a signal above the upper limit of the instrument (above $2 \times 10^6$ units).

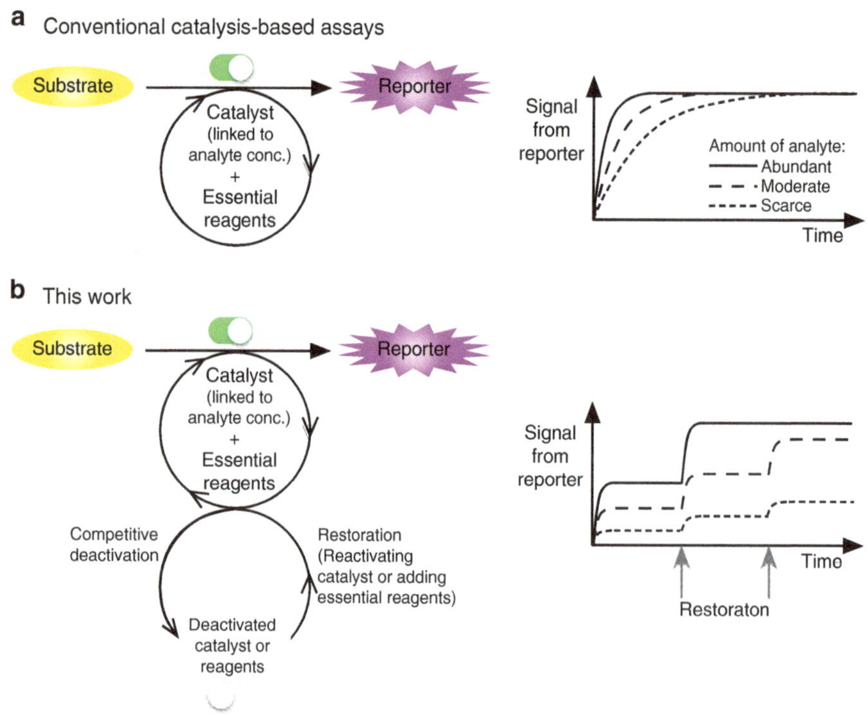


**Figure 6 | Continuous reaction and competitively and reversibly deactivated reaction.** (**a**) Widely used catalytic assays continuously convert a substrate to a reporter molecule. (**b**) This work presents a stop-and-go paradigm, in which there is a competition between the catalytic reaction and autonomous deactivation of the catalyst (that is, analyte) or the essential reagent. The addition of an activator or a reactant restores the system.

amount of Pd far exceeds the amount of RAE, the data remain quantitative. We have also demonstrated the utility of the same concept in a widely used HRP assay system, where competitive destruction of $H_2O_2$ by $PhB(OH)_2$ leads to reaction stalling, broadening the dynamic range of the assay. These approaches should be compatible with automation and may find further applicable arenas to broaden the dynamic ranges of catalysis-based assays.

## Methods

**Ultraviolet-visible spectroscopy.** The ultraviolet–visible spectra of RAE and resorufin solutions were acquired using a diode array spectrophotometer (Agilent Technologies, Santa Clara, CA) in a quartz cuvette. Other absorbance measurements were recorded in either a 96-well plate using a Modulus II Microplate Multimode reader (Promega, Madison, WI) measuring absorbance at 560 nm or in a clear, round bottom 96-well plates on a Spectra Max M5 spectrometer (Molecular Devices, Sunnyvale, CA) under the control of a Windows-based PC running software pro V5. The samples were analysed at $\lambda = 580$ nm for the resorufin, and at $\lambda = 525$ nm for RAE.

**Fluorescence measurement.** Fluorescence measurements were read on a Modulus II Microplate Multimode Reader (excitation 525 nm, emission 580–640 nm) or using a HoribaMax Fluorometer (excitation 578 nm, emission 350–700 nm).

**Metal analysis by ICP-MS.** The samples were either diluted or suspended directly in concentrated nitric acid or evaporated with a rotary evaporator first and then re-dissolved in concentrated nitric acid for ICP-MS analysis. Depending on the concentration range of the element, either a Perkin-Elmer Elan 6000 quadrupole ICP-MS spectrometer (Perkin-Elmer, Norwalk, CT) or a Thermo Finnigan Element 2 high-resolution ICP-MS spectrometer (Finnigan, Bremen, Germany) was used for the analysis.

**General protocol for deallylation of RAE.** A reaction cocktail was prepared by mixing 800 mM $NH_4OAc$ in EtOH (10 ml) with 800 μM RAE in EtOH (400 μl) and 3 mM TFP in DMSO, with 250 p.p.m. BHT (800 μl). The reaction cocktail (1 ml) was added to individual 2-ml Eppendorf tubes. To half of the samples was added 5% TraceMetal $HNO_3$ (20 μl) as a control. To the other half of the samples was added a $Pd^{2+}$ solution in 5% TraceMetal $HNO_3$ (20 μl). To all the samples was added $NaBH_4$ in 10 N NaOH (20 μl). The samples were mixed and transferred (200 μl) to a 96-well black fluorescence well plate. Fluorescence (excitation 525 nm, emission 570–640 nm) was measured every 2 min for 60 min using a Modulus II Microplate Multimode Reader.

## References


1. Carter, K. P., Young, A. M. & Palmer, A. E. Fluorescent sensors for measuring metal ions in living systems. *Chem. Rev.* **114,** 4564–4601 (2014).
2. Namiki, S., Sakamoto, H., Iinuma, S., Iino, M. & Hirose, K. Optical glutamate sensor for spatiotemporal analysis of synaptic transmission. *Eur. J. Neurosci.* **25,** 2249–2259 (2007).
3. Hettie, K. S., Liu, X., Gillis, K. D. & Glass, T. E. Selective catecholamine recognition with NeuroSensor 521: a fluorescent sensor for the visualization of norepinephrine in fixed and live cells. *ACS Chem. Neurosci.* **4,** 918–923 (2013).
4. Zhang, S. W. & Swager, T. M. Fluorescent detection of chemical warfare agents: functional group specific ratiometric chemosensors. *J. Am. Chem. Soc.* **125,** 3420–3421 (2003).
5. Goldberg, M. E. & Djavadiohaniance, L. Methods for measurement of antibody antigen affinity based on ELISA and RIA. *Curr. Opin. Immunol.* **5,** 278–281 (1993).
6. Demontellano, P. R. O., David, S. K., Ator, M. A. & Tew, D. Mechanism-based inactivation of horseradish-peroxidase by sodium azide. Formation of *meso*-azidoprotoporphyrin IX. *Biochemistry* **27,** 5470–5476 (1988).
7. Song, F. L., Garner, A. L. & Koide, K. A highly sensitive fluorescent sensor for palladium based on the allylic oxidative insertion mechanism. *J. Am. Chem. Soc.* **129,** 12354–12355 (2007).
8. Garner, A. L., Song, F. L. & Koide, K. Enhancement of a catalysis-based fluorometric detection method for palladium through rational fine-tuning of the palladium species. *J. Am. Chem. Soc.* **131,** 5163–5171 (2009).
9. Song, F. L., Carder, E. J., Kohler, C. C. & Koide, K. Studies toward an ideal fluorescence method to measure palladium in functionalized organic molecules: effects of sodium borohydride, temperature, phosphine ligand, and phosphate ions on kinetics. *Chemistry* **16,** 13500–13508 (2010).
10. Bu, X. D., Koide, K., Carder, E. J. & Welch, C. J. Rapid analysis of residual palladium in pharmaceutical development using a catalysis-based fluorometric method. *Org. Process Res. Dev.* **17,** 108–113 (2013).
11. Liu, W. *et al.* Water-soluble colorimetric and ratiometric fluorescent probe for selective imaging of palladium species in living cells. *Inorg. Chem.* **53,** 12590–12594 (2014).

12. Yang, M. P. *et al.* A novel selective fluorescent and colorimetric chemosensor for the visual detection of $Pd^{2+}$ and application of imaging in living cells. *Inorg. Chem. Commun.* **46,** 310–314 (2014).
13. Bullock, K. M., Mitchell, M. B. & Toczko, J. F. Optimization and scale-up of a Suzuki-Miyaura coupling reaction: development of an efficient palladium removal technique. *Org. Process Res. Dev.* **12,** 896–899 (2008).
14. Welch, C. J. *et al.* Adsorbent screening for metal impurity removal in pharmaceutical process research. *Org. Process Res. Dev.* **9,** 198–205 (2005).
15. Wang, L. *et al.* Screening binary systems of chelating agents combined with carbon or silica gel adsorbents: the development of a cost-effective method to remove palladium from pharmaceutical intermediates and APIs. *Org. Process Res. Dev.* **15,** 1371–1376 (2011).
16. Thayer, A. M. Trace metals debate. *Chem. Eng. News* **19,** 10–13 (2013).
17. Williams, J. M. & Koide, K. A high-throughput method to detect palladium in ores. *Ind. Eng. Chem. Res.* **52,** 8612–8615 (2013).
18. Krebs, F. C., Nyberg, R. B. & Jorgensen, M. Influence of residual catalyst on the properties of conjugated polyphenylenevinylene materials: palladium nanoparticles and poor electrical performance. *Chem. Mater.* **16,** 1313–1318 (2004).
19. Nielsen, K. T., Bechgaard, K. & Krebs, F. C. Removal of palladium nanoparticles from polymer materials. *Macromolecules* **38,** 658–659 (2005).
20. Engvall, E., Jonsson, K. & Perlmann, P. Enzyme-linked immunosorbent assay. II. Quantitative assay of protein antigen, immunoglobulin G, by means of enzyme-labelled antigen and antibody-coated tubes. *Biochim. Biophys. Acta* **251,** 427–434 (1971).
21. Engvall, E. & Perlmann, P. Enzyme-linked immunosorbent assay (ELISA) quantitative assay of immunoglobulin-G. *Immunochemistry* **8,** 871–874 (1971).
22. Zhou, M. J., Diwu, Z. J., PanchukVoloshina, N. & Haugland, R. P. A stable nonfluorescent derivative of resorufin for the fluorometric determination of trace hydrogen peroxide: applications in detecting the activity of phagocyte NADPH oxidase and other oxidases. *Anal. Biochem.* **253,** 162–168 (1997).

## Acknowledgements

We thank Stillwater Mining Co. for the ore samples. We thank Professor Frederik Krebs for providing the polymer samples. We are grateful to the Merck Research Laboratories New Technologies Review & Licensing Committee (NT-RLC) and to the US National Science Foundation (CHE-0911092 and CHE-1506942) for funding.

## Author contributions

All the authors were involved in designing the experiments and analysing the data. M.P.T., X.B., J.J. and M.J.W. performed the experiments. All the authors contributed towards writing the manuscript.

## Additional information

**Competing financial interests:** The authors declare no competing financial interests.

# 22

# Homochiral $D_4$-symmetric metal-organic cages from stereogenic Ru(II) metalloligands for effective enantioseparation of atropisomeric molecules

Kai Wu[1,*], Kang Li[1,*], Ya-Jun Hou[1], Mei Pan[1], Lu-Yin Zhang[1], Ling Chen[1] & Cheng-Yong Su[1,2]

Absolute chiral environments are rare in regular polyhedral and prismatic architectures, but are achievable from self-assembly of metal-organic cages/containers (MOCs), which endow us with a promising ability to imitate natural organization systems to accomplish stereochemical recognition, catalysis and separation. Here we report a general assembly approach to homochiral MOCs with robust chemical viability suitable for various practical applications. A stepwise process for assembly of enantiopure $\Delta\Delta\Delta\Delta\Delta\Delta\Delta\Delta$- and $\Lambda\Lambda\Lambda\Lambda\Lambda\Lambda\Lambda\Lambda$-$Pd_6(RuL_3)_8$ MOCs is accomplished by pre-resolution of the $\Delta/\Lambda$-Ru-metalloligand precursors. The obtained Pd–Ru bimetallic MOCs feature in large $D_4$-symmetric chiral space imposed by the predetermined Ru(II)-octahedral stereoconfigurations, which are substitutionally inert, stable, water-soluble and are capable of encapsulating a dozen guests per cage. Chiral resolution tests reveal diverse host-guest stereoselectivity towards different chiral molecules, which demonstrate enantioseparation ability for atropisomeric compounds with $C_2$ symmetry. NMR studies indicate a distinctive resolution process depending on guest exchange dynamics, which is differentiable between host-guest diastereomers.

[1] MOE Laboratory of Bioinorganic and Synthetic Chemistry, State Key Laboratory of Optoelectronic Materials and Technologies, Lehn Institute of Functional Materials, School of Chemistry and Chemical Engineering, Sun Yat-Sen University, Guangzhou 510275, China. [2] State Key Laboratory of Applied Organic Chemistry, Lanzhou University, Lanzhou 730000, China. * These authors contributed equally to this work. Correspondence and requests for materials should be addressed to C.-Y.S. (email: cesscy@mail.sysu.edu.cn) or to M.P. (email: panm@mail.sysu.edu.cn).

The design and synthesis of discrete nanoscale metal–organic cages/containers (MOCs) with specific configurations and cavities applying directional bridging ligands and geometrically prefixed metals is emerging as an appealing topic in recent supramolecular coordination chemistry[1–3]. Among this, the controlled assembly of enantiopure chiral cages is of special importance because of their potential applications in stereoselective recognition, catalysis and enzyme mimics[4–11]. Since the chiral space in regular polyhedra only rarely presents in snub dodecahedron and snub cube (all other Platonic, Archimedean, prismatic and antiprismatic solids are achiral)[12,13], assembly of chiral polyhedral MOCs is usually achieved by introducing stereogenic centres into the faces, edges or vertices of a polyhedron to remove inversion and mirror symmetries. In this way, a number of homochiral MOCs of *T*-symmetry[14–19] have been constructed, whereas the chiral MOCs of *O*-symmetry or higher were proved to be more formidable because of more possible stereoisomers and the demand to transmit single chirality from more subcomponents[20,21]. In principle, the chirality of an MOC can be generated either by the organic stereocentres (such as chiral tetrahedral C*) or the metal stereogenic centres. The latter strategy provides a versatile platform for stereochemistry of MOCs because the plentiful metal coordination geometries can afford innumerable stereogenic metal centres for assembly of chiral structures even from achiral components in a supramolecular sense[22–24]. The overall MOC symmetry can be restricted or reduced by the stereochemical coupling between metal centres. For example, transfer of stereoconfiguration information between vertices of a tetrahedron enables absolute assembly[24] of homoconfigurational *ΔΔΔΔ*- or *ΛΛΛΛ*-cages based on the stereogenic tris-chelate metal centres[15–17]. However, the lability of metal–ligand exchange often causes enantiomerization between opposite enantiomers[25], and racemic mixture cannot be prevented during the assembly process. Resolution of the enantiopure product usually has to be accomplished with the aid of chiral auxiliaries to form diastereomers, and stabilization of the dynamic metal centre often needs synergistic effect[15–17].

An alternative way to construct stable and robust homochiral MOCs based on the stereogenic metal centres is to design a metalloligand[26] containing a stereoconfigurationally inert metal centre in lieu of the C* stereocentre in organic ligand. Formation of MOCs by virtue of various metalloligands has been achieved in many excellent lines of works[27–32], in which spontaneous resolution and geometric isomerism were observed[33–35], yet construction of enantiopure MOCs from predetermined chiral metalloligands remains unexplored. On the basis of the well-known stereochemistry of $D_3$-symmetric $[Ru(bpy)_3]^{2+}$- or $[Ru(phen)_3]^{2+}$-type compounds, which are widely explored in DNA interactions, asymmetric catalysis and supramolecular chiral assemblies[36–40], we initiated the design of $[Ru(phen)_3]^{2+}$-type metalloligand for homochiral MOC self-assembly[26]. Since the stereoconfiguration of such tri-chelate Ru-octahedral centres is substitutionally inert and stable in solution assembly and crystallization process, we expect that the predetermined chirality of the Ru metalloligands can direct the assembly of homochiral MOCs with sufficient stability in practical applications. Although stereoselective recognition and catalysis using chiral hosts has been well established[14–19,41,42], enantioseparation of racemic guest molecules by means of homochiral coordination cages remains a challenge. Only a few examples are known to achieve moderate to good diastereoselectivity[43–46], thus urging an extensive study to solve the common problems in this field; for example, (a) efficient resolution of enantiopure cages, (b) effective stabilization of cage stereochemistry and (c) high guest inclusion capacity (more than three guests per host). Herein we report a general approach to assemble homochiral MOCs without post resolution based on the pre-resolved stereogenic Ru-octahedral centres, offering huge cages capable of large amounts of guest encapsulation (>10 guests per host). Specifically, stereoselective separation of atropisomeric molecules rather than C*-based chiral compounds is achieved, and a dynamic resolution process based on differentiable guest exchange by formation of diastereomers is proposed.

## Results

**Assembly of enantiopure MOCs.** We have previously assembled heteronuclear *Δ*/*Λ*-$Pd_6(RuL_3)_8$ MOCs racemate (hereafter assigned as ***rac-Δ/Λ*-MOCs-16**, Fig. 1) from the racemic $RuL_3$ metalloligands (***rac-Δ/Λ*-3**), which show the shape of an octahedron (defined by $Pd_6$ centres) or a rhombic dodecahedron (defined by $Pd_6Ru_8$ centres)[26]. It was noted that the cage assembly proceeded in a homochiral manner, with each individual **MOC-16** integrating the same handed ***Δ*- or *Λ*-3** enantiomers to display either *ΔΔΔΔΔΔΔΔ* or *ΛΛΛΛΛΛΛΛ*

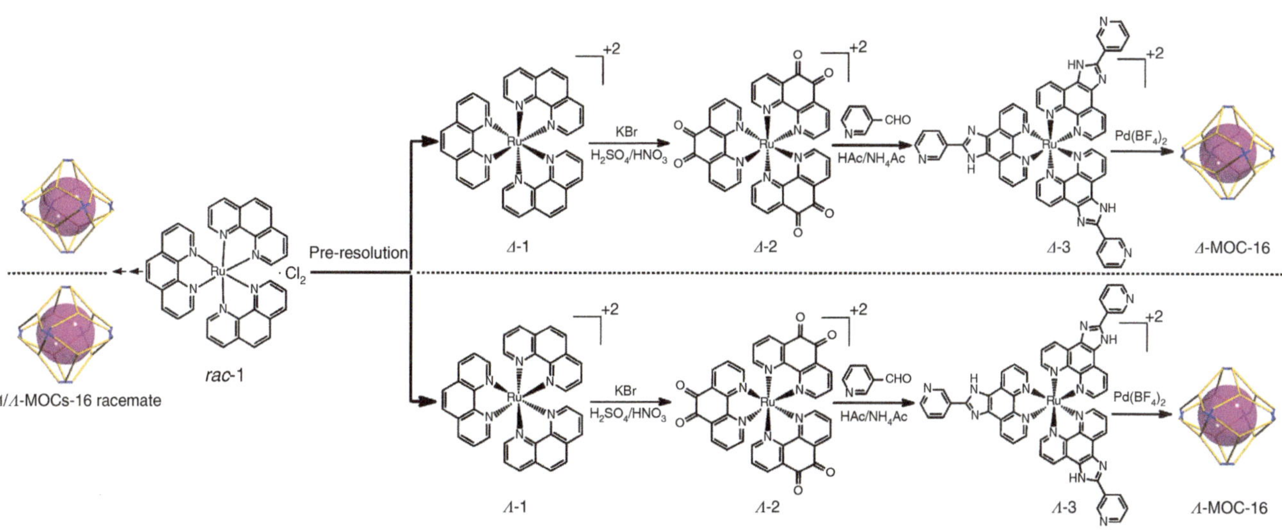


**Figure 1 | Assembly procedures.** Formation of racemic *Δ*/*Λ*-$Pd_6(RuL_3)_8$ cages (*rac-Δ/Λ*-MOCs-16) from mixed precursors, and stepwise syntheses of enantiopure *Δ*- and *Λ*-$Pd_6(RuL_3)_8$ cages (*Δ*-/*Λ*-MOCs-16) from pre-resolved *Δ*-3 and *Λ*-3 metalloligands.

homoconfigurations, indicative of strong cooperative stereochemical coupling between the metal centres[14–19,22–24] to direct the absolute self-organization[24] and exclusive formation of single homochiral *Δ*- or**Λ-MOC-16**. However, thus assembled chiral *Δ*- and**Λ-MOCs** co-crystallize simultaneously to give racemic products that are not ready for practical applications.

To make use of these homochiral cages, we started chiral resolution from well-established pre-resolution of *Δ*/*Λ*-$[Ru(phen)_3]^{2+}$ precursors and developed a pair of enantiomeric triangular metalloligands incorporating fixed chiral octahedral Ru(II) centres and pyridyl (Py) terminals ready for assembly of enantiopure *Δ*- and *Λ*-$Pd_6(RuL_3)_8$ MOCs separately. As shown in Fig. 1 and described in detail in Supplementary Figs 1–10, racemic *Δ*/*Λ*-$[Ru(phen)_3]^{2+}$ was first resolved into a pair of enantiomers (**Δ**- and **Λ-1**) in good yields using $K_2[Sb_2\{(+)\text{-tartrate}\}_2]\cdot 3H_2O$ as chiral induction agent, and then oxidized into *Δ*- and *Λ*-$[Ru(Phendione)_3]^{2+}$ (**Δ**- and **Λ-2**). With the aid of chiral shift reagent $Eu((+)tfc)_3$, the enantiopurity was tested to be 94.8% for **Δ-1** and 95.3% for **Λ-1** (ref. 47). The absolute configurations of the two pairs of **Δ-/Λ-1** and **Δ-/Λ-2** enantiomers have been well established by the single crystal structural analyses (Supplementary Figs 1 and 2), which are in excellent agreement with the experimental resolution and syntheses. The phase purity of the bulk products of **Δ-/Λ-1** and **Δ-/Λ-2** enantiomers has also been verified using the powder X-ray diffraction measurements (Supplementary Figs 1 and 2). Further reaction of **Δ**- and **Λ-2** with 3-pyridinecarboxaldehyde afforded a pair of stereogenic bulky *Δ*- and *Λ*-$RuL_3$ metalloligands (**Δ**- and **Λ-3**), and, finally, the coordination assembly of **Δ**- and **Λ-3** enantiomers with $Pd^{2+}$ ions unambiguously resulted in a pair of homochiral *Δ*- and *Λ*-$Pd_6(RuL_3)_8$ cages, namely **Δ-MOC-16** and **Λ-MOC-16**, respectively. $^1$H NMR spectra of two optically pure **Δ-/Λ-MOCs-16** enantiomers give well-resolved proton patterns basically identical to previously reported racemic ***rac-Δ/Λ*-MOCs-16** (Supplementary Fig. 8), showing distinguishable H resonance between the protons inside and outside cage (Supplementary Fig. 9). The $^1$H-$^1$H-COSY and high-resolution mass spectrometry (HR-ESI-TOF-MS) have also been performed to verify formation of $Pd_6(RuL_3)_8$ cage structures (Supplementary Fig. 10).

The absolute configurational arrangement of the **Δ**- or **Λ-3** metalloligands in **Δ-MOC-16** or **Λ-MOC-16**, respectively, has been undoubtedly established by the single-crystal analyses (Supplementary Fig. 3). The single crystals of **Δ-MOC-16** and **Λ-MOC-16** were grown from their MeCN solutions in the presence of *S*-BINOL and *R*-BINOL, respectively, as absolute structural reference compounds for further authentication of the crystal chirality. Both **Δ-MOC-16** and**Λ-MOC-16** crystallize in the chiral space groups *I*422. In **Δ-MOC-16**, eight **Δ-3** metalloligands are assembled by six square-coordinative $Pd^{2+}$ ions to form $Pd_6(RuL_3)_8$ cage with the *ΔΔΔΔΔΔΔΔ* homoconfigurations (Fig. 2a). The crystal is packed by the identical **Δ-MOC-16** cages in together with *S*-BINOL molecules, giving rise to enantiopure product with the absolute chirality exactly according to the chiral **Δ-3** metalloligands and reference *S*-BINOL used in syntheses and crystal growth. In contrast, **Λ-MOC-16** integrates eight **Λ-3** metalloligands and six $Pd^{2+}$ ions to form $Pd_6(RuL_3)_8$ cage with the *ΛΛΛΛΛΛΛΛ* homoconfigurations (Fig. 2a), and co-crystallizes with *R*-BINOLs to result in enantiopure crystals. For both **Δ-MOC-16** and **Λ-MOC-16**, the powder X-ray diffraction patterns of the bulk samples well match those of the single-crystal simulations, indicating satisfactory phase purity (Supplementary Fig. 3).

Careful examination of the crystal structures of **Δ-/Λ-MOCs-16** enantiomers reveals that the cage molecule possesses crystallographically imposed $D_4$ symmetry (Fig. 2 and Supplementary Fig. 3). If regarding the cage as a pseudo-octahedron, the $C_4$ axis passes two vertices occupied by Pd1 ions, while two pairs of Pd2 ions are located on the $C_2$ axes. Therefore, the cage symmetry may be considered to degrade from chiral *O*-symmetry owing to disposition of the same handed Ru-stereocentres on eight faces of octahedron, or, on eight $C_3$ vertices of rhombic dodecahedron to impose *ΔΔΔΔΔΔΔΔ* or *ΛΛΛΛΛΛΛΛ* homoconfigurations in **Δ-MOC-16** and **Λ-MOC-16**, respectively. In another word, the assembly of the homochiral **Δ-/Λ-MOCs-16** enantiomers proceeds in a way of octahedral face-control or rhombic dodecahedral vertex-control, thus removing inversion *i* and mirror *σ* symmetries to turn an achiral $O_h$ group into a chiral $D_4$ group (Fig. 2c). Furthermore, the stereoconfigurations around six $Pd^{2+}$ vertices are also induced by the fixed Ru-stereocentres. In **Δ-MOC-16**, six Pd-$Py_4$ subcomponents are all in *Λ*-configurations, with the four Py rings showing anticlock fan-like arrangement and *vice versa* in **Λ-MOC-16.** In contrast to other completely labile coordination cages[14–24], in the present cases, the stereoconfiguration around the $Pd^{2+}$ corner is fixed by inserting Ru-stereocentres and cage integrity; therefore, enantiomerization through labile Pd-ligand exchange is inhibited for the whole cage. It is worth mentioning that eight *S*-BINOLs are captured by a **Δ-MOC-16**, or reversely, eight *R*-BINOLs by a **Λ-MOC-16**, on its window pockets but not completely into its cavity (Fig. 2b and Supplementary Figs 3a,b) probably because the crystallization takes place in the MeCN solution where hydrophobic effect is absent and the host–guest inclusion behaviour is different from that in aqueous medium discussed below (*vide infra*).

The circular dichroism (CD) spectra were also employed to monitor the whole synthetic and assembly processes to confirm that the absolute chirality of the starting *Δ*-/*Λ*-$[Ru(phen)_3]^{2+}$ precursors were well preserved all the way down (Fig. 3 and Supplementary Fig. 11). In MeCN solution, *Δ*-/*Λ*-$[Ru(phen)_3]^{2+}$ mainly presents three absorption peaks at 225, 265 and 450 nm (Fig. 3a), with the first two corresponding to the $n$–$\pi^*$ and $\pi$–$\pi^*$ transitions of phen groups, while the last one originating from the metal-to-ligand charge transfer (MLCT) transition between $Ru^{2+}$ and phen ligands. All three absorption bands are reflected in the corresponding CD spectra of the resolved **Δ-1** and **Λ-1** with the middle peak at 265 nm, giving the most prominent CD signal. Taking **Δ-1** as an example, the same tendency of first negative and second positive Cotton effect from longer to shorter wavelength in the three CD bands is in accordance with the *Δ*-type octahedral chirality established for the $Ru^{2+}$ coordination centre[48] and *vice versa* for the **Λ-1** compound. For the rest three pairs of enantiomers, because of the cutoff effect of the solvents (dimethylsulphoxide (DMSO) or $H_2O$), the CD signals corresponding to the absorption at 225 nm were not fully presented; however, the other bands were clearly detected in the whole synthetic process, preserving the same chirality attributes for the same series of enantiomers (**Δ-1, 2, 3, MOC-16** versus *Λ*-**1, 2, 3, MOC-16**, respectively). Furthermore, because of the accumulation effect (eightfold in **Δ**- or **Λ-MOC-16** compared with **Δ**- or **Λ-3**) of multiple chiral Ru centres in one entity in the final enantiopure **Δ-/Λ-MOCs-16**, a remarkable increase in CD signal intensities was observed for **Δ**- or **Λ-MOC-16** ($\Delta\varepsilon = \sim 720\,M^{-1}cm^{-1}$) in comparison with **Δ**- or **Λ-3** ($\Delta\varepsilon = \sim 120\,M^{-1}cm^{-1}$). Optical rotation tests also manifested the absolute configurations in **Δ**- and **Λ-MOCs-16** (*Δ*, $[\alpha]^{20}_D = -266°$; **Λ**, $[\alpha]^{20}D = 272°$, $c = 0.5$, $H_2O$). From these CD studies we see that the stereochemistry of octahedral Ru centres is robust enough to survive all reaction conditions, exactly in agreement with the observations of chirality preservation in crystallographic study. The stereochemical stability of **Δ**- and **Λ-MOCs-16** has also been testified against heating and longtime stay in solution (Fig. 3d), confirming that the absolute

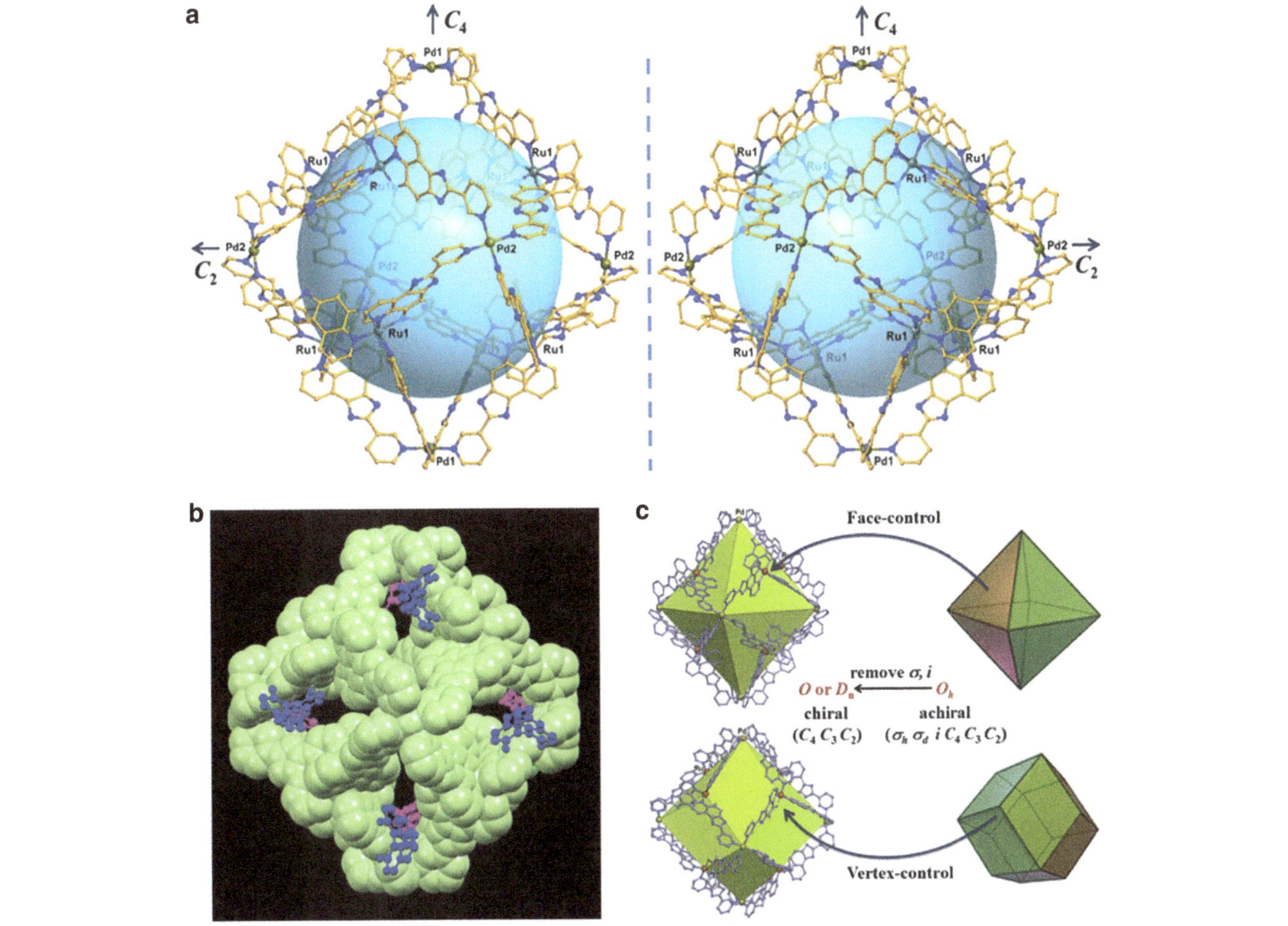


**Figure 2 | Crystal structures.** (**a**) A pair of $D_4$-symmetric homochiral *Δ*- and *Λ*-MOCs-16 showing *ΔΔΔΔΔΔΔΔ* and *ΛΛΛΛΛΛΛΛ* configurations of eight $RuL_3$ metalloligands whereas *ΛΛΛΛΛΛ*- and *ΔΔΔΔΔΔ*-configurations of six Pd-$Py_4$ subcomponents. (**b**) A *Δ*-MOC-16 cage (in space-filling mode) capturing eight *S*-BINOL guests (in ball-and-stick mode) on the windows pockets. (**c**) The demonstration how to form $D_4$ *Λ*-MOC-16 from $O_h$ regular polyhedra by introducing eight *Λ*-3 metalloligands on the faces of an octahedron, or, on the $C_3$ vertices of a rhombic dodecahedron, to reduce molecular symmetry, and further direct *Δ*-arrangement of four pyridyl rings around six vertices of Pd centres.

chirality of each enantiomeric *Δ*- and ***Λ*-MOC-16** is well retained on heating to 373 K and staying in solution for 50 days. Such a stable and substitutionally inert nature of stereogenic Ru centres plays a key role in fixing absolute chirality of *Δ*- and ***Λ*-MOCs-16**, despite intrinsic dynamics of $Pd^{2+}$ centres subject to metal–ligand exchange, thereof paving the way for utilization of these enantiopure cages in, for example, stereoselective catalysis and separation.

**Stereoselective separation of racemic guests.** In an attempt to test enantioseparation ability of ***Δ*-/*Λ*-MOC-16** cages, we selected two types of racemic organic molecules, one carrying a chiral C* centre and the other characteristic of $C_2$-symmetric chirality (Table 1). The host–guest inclusion examined by $^1$H NMR in the $D_2O$ system revealed that all chiral molecules can be well encapsulated by the **MOC-16** host owing to hydrophobic effect (Supplementary Figs 12–16), showing typical upfield shift of guest protons and further splitting of cage protons[26]. Moreover, the host–guest stereochemical relationship between enantiomeric ***Δ*-/*Λ*-MOCs-16** and *R*-/*S*-BINOLs has been examined using $^1$H NMR enantiodifferentiation experiments, where two pairs of host–guest diastereomers, namely *S*-BINOL ⊂ ***Δ*-MOC-16**, *R*-BINOL ⊂ ***Δ*-MOC-16** and *S*-BINOL ⊂ ***Λ*-MOC-16**, *R*-BINOL ⊂ ***Λ*-MOC-16**, and two pairs of host–guest enantiomers, namely *S*-BINOL ⊂ ***Δ*-MOC-16**, *R*-BINOL ⊂ ***Λ*-MOC-16** and *R*-BINOL ⊂ ***Δ*-MOC-16**, *S*-BINOL ⊂ ***Λ*-MOC-16**, are formed. As shown in Fig. 4, the solution dynamics is obviously distinguishable between the diastereomeric pairs, while that between the enantiomeric pairs is similar[41]. This means the homochiral *Δ*- and ***Λ*-MOC-16** cages are able to recognize and differentiate *R*- and *S*-BINOL enantiomeric guests in solution because of their diastereomeric host–guest relationship. As a consequence, the chiral resolution of racemic molecules was carried out by *Δ*- and ***Λ*-MOCs-16** separately in pure $D_2O$ solution based on either a homogeneous or a heterogeneous method (see details in Methods or Supplementary Methods). The resolved guests were determined using high-performance liquid chromatography (HPLC) with enantiomeric excess (*ee*) averaged from three parallel experiments (Table 1 and Supplementary Figs 17–24).

The resolution results unveil that the homochiral *Δ*- or ***Λ*-MOCs-16** have rather poor stereoselectivity towards chiral compounds containing C* stereocentres. As seen in Table 1, no obvious resolution effect can be detected for naproxen, 1-(1-naphthyl)ethanol and benzoin, despite $^1$H NMR-proved inclusion of these guests by the host **MOC-16** (Supplementary Figs 14–16). However, through the same separation process, a pair of *R*-/*S*-BINOL atropisomers was successfully resolved, with the *ee* values reaching 34% or more by ***Δ*-/*Λ*-MOCs-16.** Relatively low enantioseparation results were obtained for *R*-/*S*-3-Br-BINOL

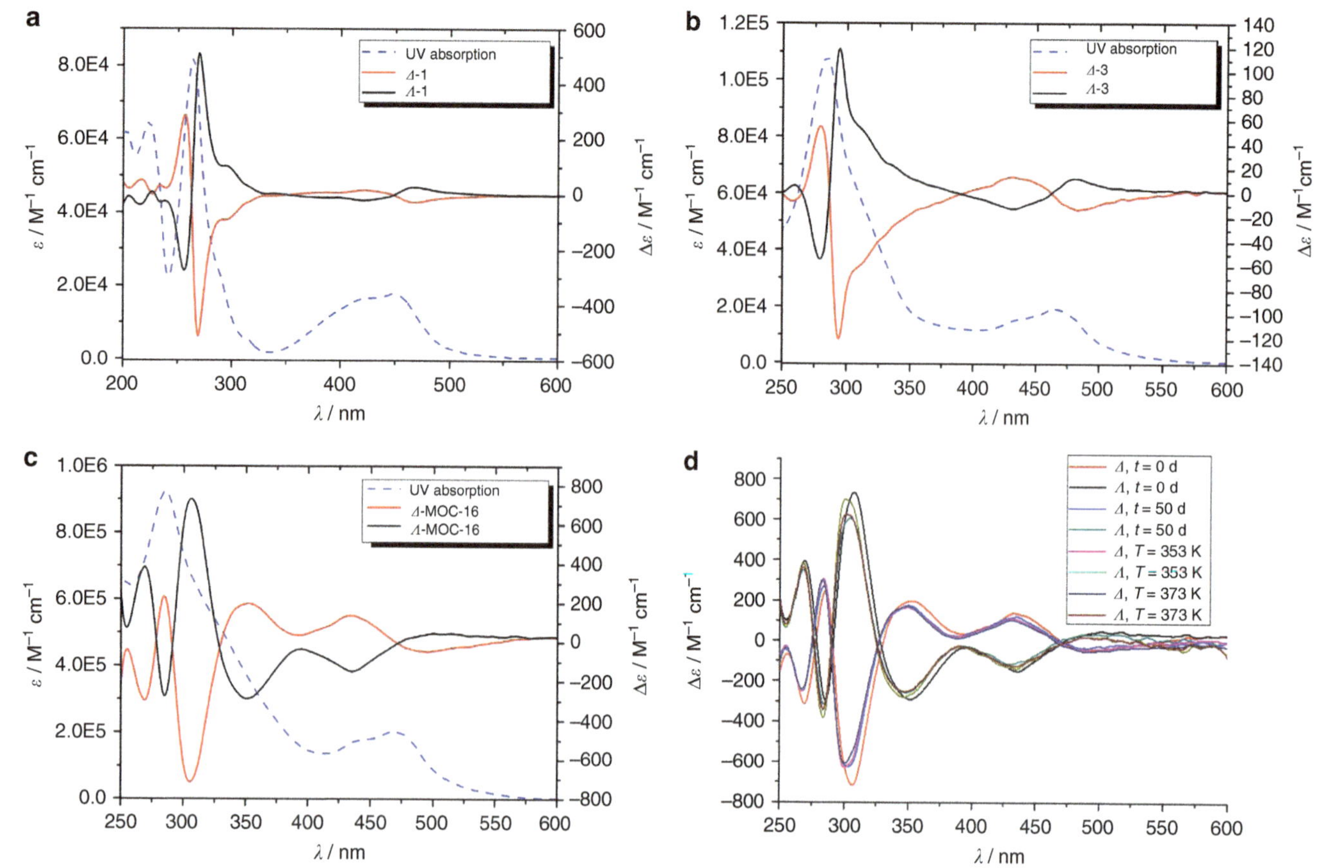


**Figure 3 | CD and UV spectra.** CD (solid lines) and ultraviolet (dotted lines). (**a**) Δ- and Λ-1 in MeCN, (**b**) Δ- and Λ-3 in DMSO, (**c**) Δ- and Λ-MOCs-16 in $H_2O$ and (**d**) the stereochemical stability depending on time and temperature.

**Table 1 | Enantioselective resolution of chiral organic molecules*.**

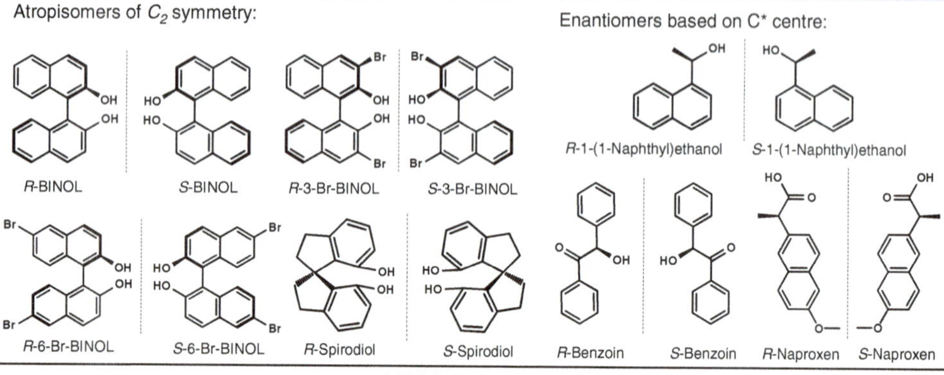


| Guests | Δ-MOC-16 | | Λ-MOC-16 | |
|---|---|---|---|---|
| | **R,S ratio** | **ee (%)** | **R,S ratio** | **ee (%)** |
| BINOL | 67:33 | 34 | 32:68 | 36 |
| 3-Br-BINOL | 54:46 | 8 | 46:54 | 8 |
| 6-Br-BINOL | 77:23 | 54 | 19:81 | 62 |
| 1,1′-spirobiindane-7,7′-diol | 67:33 | 34 | 28:72 | 44 |
| Naproxen | 50:50 | 0 | 50:50 | 0 |
| 1-(1-Naphthyl)ethanol | 51:50 | 1 | 50:50 | 0 |
| Benzoin | 50:50 | 0 | 50:50 | 0 |

ee, enantiomeric excess; MOC, metal-organic cage/container.
*The estimated uncertainty is about ±2% as calculated for the averaged values from the results of three repeating resolution processes.

racemate; however, the resolution effect was greatly improved for the chiral discrimination of *R*-/*S*-6-Br-BINOL enantiomers. By applying ***Δ*-MOC-16**, the resolved product contains 77% of *R*-isomer and 23% of *S*-isomer, giving an *ee* value of 54%. Surprisingly, an *ee* value up to 62% was obtained by ***Λ*-MOC-16** with the product dominant in *S*-isomer. Similar

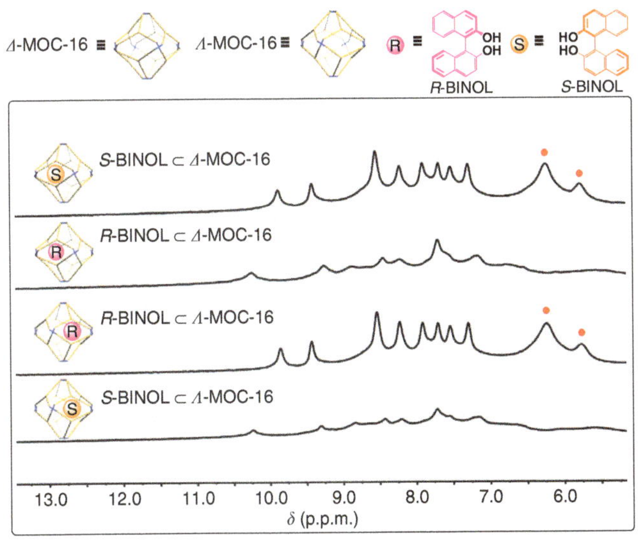


**Figure 4 | $^1$H NMR enantiodifferentiation experiments.** Sequestration of enantiomeric *R*- or *S*-BINOL guests by homochiral Δ- or Λ-MOCs-16 ($d_6$-DMSO/$D_2O$ = 1/5, 298 K). Red circles denote signals of encapsulated guests.

enantioseparation ability of **Δ-/Λ-MOCs-16** was able to extend to another kind of atropisomeric compound *R*-/*S*-spirodiol, exhibiting the same host–guest stereoselectivity. The **Δ-MOC-16** got 34% predominance of *R*-isomer, while a higher *ee* value of 44% was obtained for *S*-isomer by **Λ-MOC-16**. In general, **Δ-MOC-16** shows a preferable stereoselectivity towards *R*-isomer, while **Λ-MOC-16** prefers *S*-isomer for all chiral guests of $C_2$ symmetry. The higher-resolution effect from *S*-isomer ⊂ **Λ-MOC-16** inclusion than from *R*-isomer ⊂ **Δ-MOC-16** inclusion is unexpected, probably owing to the slight difference in optical attribute based on their optical rotation tests. To the best of our knowledge, such a preferable enantiorecognition of chiral guests with $C_2$ symmetry has not been observed before for cage compounds, and the enantioseparation ability of **Δ-/Λ-MOCs-16** reaches high level within the known chiral organic and coordination cages[43–46,49]. In addition, the adequate solubility of **MOCs-16** in water (2.6 g per 100 ml) makes it convenient to implement enantioseparation either in a homogeneous two-phase way (*Method I:* organic-water transfer as shown in Supplementary Fig. 25) or simply in a heterogeneous suspension way (*Method II:* solid-solution transfer as described in Methods). In comparison with the normally insoluble metal–organic frameworks for chiral separation, the water solubility of **MOCs-16** offers advantages by using the hydrophobic effect to transfer water-insoluble organic guests into the aqueous phase, and the guest transformation between the organic-water phases is easy to accomplish. Extraction of the resolved chiral guests from the water phase of **Δ-/Λ-MOCs-16** readily leads to recovery of the empty cages, which can be reused for the next runs of chiral resolution without a significant loss of the enantioseparation ability as tested by four cycles of *R*/*S*-6-Br-BINOL resolution with **Δ-MOCs-16** (*ee* 51–57%, Supplementary Table 1). On the other hand, chiral resolution test of *R*/*S*-6-Br-BINOL racemate within **Δ-MOC-16** and **Λ-MOC-16** using 10 times the amount of host and guest indicates that the enantioselectivity is retained almost the same for the scaling up separation (*ee* 55 and 60%).

**Resolution process study**. To further understand the host–guest interactions for insight into the resolution mechanism, $^1$H NMR titration was performed in an attempt to acquire association constants[42] for the pairs of host–guest diastereomers. However, the experimental results obviously reveal intricate host–guest solution dynamics (Fig. 5 and Supplementary Figs 26–28). Since the **MOC-16** cage has a huge molecular size (3.3 × 3.3 nm) and cavity (2895 Å$^3$ based on VOIDOO calculations) where a large amount of guests could be hosted (for example, 18 Phen guests per host)[26], the proton signals of both cage and guests are generally broadened and poorly resolved because of slow rotational diffusion and dynamics typical of large molecules, thereof preventing us from quantitative study with regard to thermodynamic or kinetic details by using the known methods for relatively simple host–guest systems (usually more than three guests per host)[50–53]. Nevertheless, it is evident that titration of enantiopure **Λ-MOC-16** (or **Δ-MOC-16**) with *R*- and *S*-BINOL guests of $C_2$ symmetry undergoes remarkably different host–guest interaction processes, showing distinguishable guest inclusion behaviours for *R*- and *S*-BINOL atropisomers as demonstrated in Fig. 5a,b. This is in agreement with the observation from above-mentioned $^1$H NMR enantiodifferentiation experiments because of the formation of a pair of host–guest diastereomers *R*-BINOL ⊂ **Λ-MOC-16** and *S*-BINOL ⊂ **Λ-MOC-16**. In contrast, titration of Δ- and **Λ-MOC-16** cages with the same C*-chiral *S*-1-(1-naphthyl)ethanol guest, which should also give a pair of diastereomers *S*-1-(1-naphthyl)ethanol ⊂ **Δ-MOC-16** and *S*-1-(1-naphthyl)ethanol ⊂ **Λ-MOC-16**, just results in rather similar $^1$H NMR chemical shift patterns (Fig. 5d,e), indicating that the homochiral **Λ-MOC-16** (or **Δ-MOC-16**) cage exhibits the same guest inclusion behaviours for *R*- and *S*-enantiomeric guests carrying C* stereocentres.

As demonstrated in Fig. 5, stepwise inclusion of *R*-BINOLs by **Λ-MOC-16** at 298 K causes inverse chemical shifts of cage protons, with those inside the cage moving upfield while those outside the cage moving downfield (Fig. 5a). The guest protons appear as severely broadened doublet and remain almost unmoved up to 12 guest inclusion. Addition of more than 12 equivalent *R*-BINOLs shows little influence on cage protons, but leads to downfield shift and further broadening of guest protons. These results suggest that at least 12 *R*-BINOLs are encapsulated inside the cage, and further guest uptake may speed up dynamic exchange. Inclusion of guests inside the cage is also supported with $^1$H-$^1$H-COSY and NOESY measurements (Supplementary Fig. 26). For comparison, inclusion of *S*-BINOLs at 298 K does not lead to distinct bidirectional shifts of cage protons, while the resonance of guests is even broadened and becomes poorly visible together with the host protons on inclusion of more than 10 *S*-BINOLs (Fig. 5b). To observe guest signals more clearly, titration at a higher temperature 353 K was performed (Fig. 5c), which presents better resolved guest resonances but basically same overall chemical shift patterns as observed at 298. Therefore, similar host–guest interacting manners may be expected at these temperatures (*vide infra*). It is notable that the guest signals display a continuously downfield shift, characteristic of fast guest exchange. These NMR observations imply more dynamic host–guest interactions for *S*-BINOL ⊂ **Λ-MOC-16** inclusion compared with *R*-BINOL ⊂ **Λ-MOC-16** inclusion at the room temperature. Broadening of H resonance is indicative of slow and restricted molecular rotation and tumbling[54,55] as well as of a comparable guest exchange rate with the NMR timescale. For the *S*-BINOL ⊂ **Λ-MOC-16** system, faster guest exchange dynamics may present, showing averaged influence on host protons either inside or outside. When the cage cavity is getting fulfilled (~12 guest per cage), the overall host–guest dynamics is slowed down to make NMR unable to discriminate resonating frequency. In contrast, guest exchange in the *R*-BINOL ⊂ **Λ-MOC-16** system is slow enough at room temperature, thus showing distinguishable impact on host protons inside and outside. Such a guest dynamic difference between two host–guest diastereomers

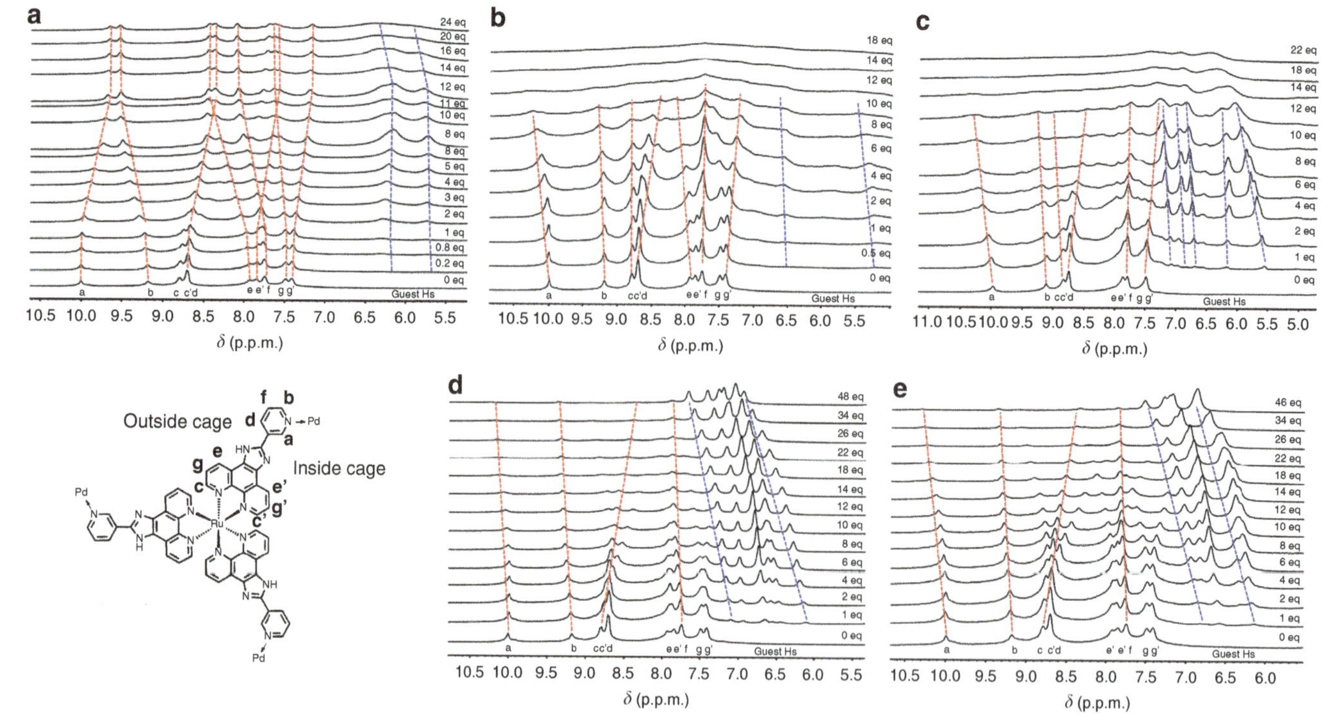


**Figure 5 | $^1$H NMR titration in DMSO-$d_6$/$D_2O$ = 1/5.** (**a**) *R*-BINOL ⊂ *Λ*-MOC-16 at 298 K, (**b**) *S*-BINOL ⊂ *Λ*-MOC-16 at 298 K, (**c**) *S*-BINOL ⊂ *Λ*-MOC-16 at 353 K, (**d**) *S*-1-(1-naphthyl)ethanol ⊂ *Λ*-MOC-16 at 298 K and (**e**) *S*-1-(1-naphthyl)ethanol ⊂ *Δ*-MOC-16 at 298 K.

may account for the intrinsic factor that determines the enantioseparation ability of homochiral *Δ*- or ***Λ*-MOCs-16** towards racemic *R*/*S*-BINOLs. On the contrary, titration of *Δ*- and ***Λ*-MOCs-16** with *S*-1-(1-naphthyl)ethanol guests shows a similar guest exchange dynamics (Fig. 5d,e), where fast guest exchange is evident for both host–guest diastereomers. This may explain why the homochiral ***Δ*-/*Λ*-MOCs-16s** are unable to discriminate *R*/*S*-stereomers carrying opposite C* stereocentres. Although a confinement effect of a cage is usually expected to enhance the intrinsic chirality of the C* guests, discriminable stereoselectivity was not observed for the present C* molecules because of fast guest exchange.

Variable-temperature $^1$H NMR study has been carried out to testify the above-proposed resolution process (Fig. 6 and Supplementary Fig. 29). It is clear that, for both *R*-BINOL ⊂ ***Λ*-MOC-16** and *S*-BINOL ⊂ ***Λ*-MOC-16** diastereomers, heating boosts guest dynamics and accelerates guest exchange, with two broadened signals getting better resolved and moving constantly downfield to approach free guests. This kind of host–guest solution dynamics might be comparable to the NMR-observable molecular dynamics[53–55]. However, the accelation of the guest exchange dynamics from the sufficiently slow state to the fast state may not undergo a normal peak coalescence, but display a turning point where guest H resonances start to resolve apparently owing to NMR-observable freedom of guests from cage restriction. If taking the resonance frequency separation between two slowly restricted guest signals and the turning point of guest signal shifts for Eyring analysis, the guest exchange rates and energy barriers might be estimated at 1,021, 488 $s^{-1}$ and 55.7, 60.5 kJ $mol^{-1}$, for *S*-BINOL ⊂ ***Λ*-MOC-16** and *R*-BINOL ⊂ ***Λ*-MOC-16** diastereomers, respectively. This means that ***Λ*-MOC-16** can capture *S*-BINOL faster at lower energy cost than *R*-BINOL to accomplish a host–guest inclusion process.

## Discussion

The host–guest dynamics and guest exchange mechanism have been vigorously explored for the insight of fundamental host–guest interactions and more complex encapsulation system design, in which both thermodynamics and kinetics play important roles in guest binding[50–55]. On the basis of above enantioseparation and NMR studies, we may speculate that the resolution process of homochiral *Δ*- or ***Λ*-MOCs-16** towards chiral molecules of $C_2$ symmetry is mainly controlled by guest exchange dynamics, in comparison with the more popular thermodynamic resolution of racemic guests by chiral cages[43]. As demonstrated in Fig. 7, encapsulation of racemic *R*/*S* stereomers by, for example, ***Λ*-MOC** may proceed in a dynamic way depending on host–guest interactions and *R*/*S*-guest competition. If inclusion of *S*-stereomers is faster than *R*-stereomers via a lower guest exchange energy barrier, preferable resolution of *S*-stereomers over *R*-stereomers is achievable. It should be noted that such a dynamic resolution based on guest exchange dynamics might be comparable but inherently different from the well-known 'kinetic resolution' based on different reaction rates between a chiral catalyst and enantiomers[41,42,56]. Guest exchange and displacement process in a host–guest system is often sensitive to the synergistic effect from both thermodynamic and kinetic contributions[50–55]. Elongating resolution time may not influence *ee* results so much as by the catalytic kinetic resolution. We have tested the time-dependent chiral resolution of *R*/*S*-BINOL racemate by ***Λ*-MOC-16**. The results indicate an increase in the *ee* value within first 2 h, but remaining nearly unchanged afterwards (Supplementary Tables 2 and 3). We believe that the distinctive host–guest dynamics between *R*-BINOL ⊂ ***Λ*-MOC-16** and *S*-BINOL ⊂ ***Λ*-MOC-16** diastereomers should originate from the stereoconfigurations of the octahedral Ru centres. The twisted arrangement of three Phen motifs around Ru centres in helical sense may not be able to differentiate inclusion behaviour of configurationally free racemic guests carrying C* stereocentres, but significantly affect the interactions between ***Δ*-/*Λ*-MOCs-16** and atropisomeric guests bearing $C_2$ symmetry owing to their intrinsic helical configurations. Formation of adaptive or mismatched host–guest diastereomers through dynamic guest exchange may be more

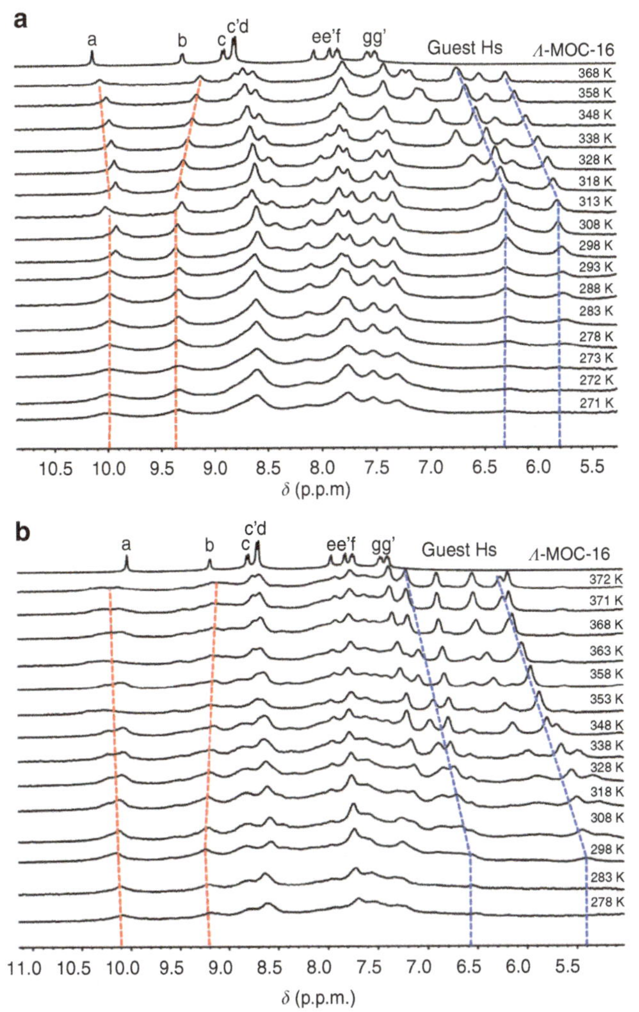


**Figure 6 | VT $^1$H NMR study of guest inclusion dynamics in solvent DMSO-d$_6$/D$_2$O = 1/5. (a) *R*-BINOL** ⊂ Λ-MOC-16. **(b) *S*-BINOL** ⊂ Λ-MOC-16. Red lines show host protons, while blue lines show guest protons.

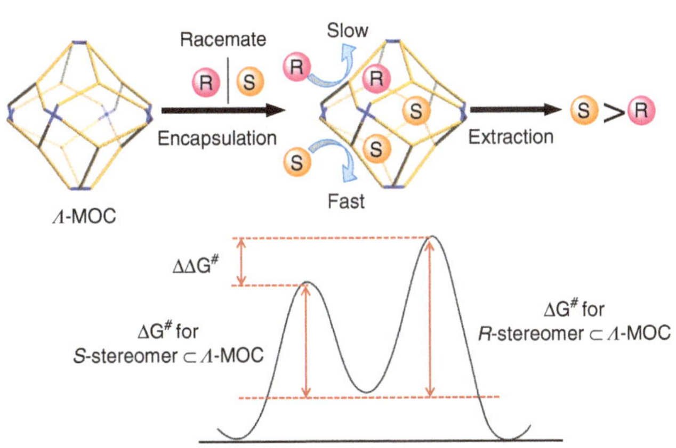


**Figure 7 | Enantioseparation mechanism.** A possible resolution process relying on guest exchange dynamics of atropisomers with homochiral Λ-MOC-16. ΔG$^{\#}$ represents guest exchange energy barrier.

dominated by stereochemical compatibility than by binding constant. Therefore, such stereoconfigurationally predetermined **MOCs** could afford better adaptive inclusion of one atropisomer over the opposite one, thus resulting in stereoselective separation.

In conclusion, pre-resolution of a pair of enantomeric Δ-/Λ-Ru metalloligands has been successfully implemented based on the stereogenic octahedral Ru centres in Δ-/Λ-$[Ru(phen)_3]^{2+}$ precursors, giving rise to the assembly of enantiopure $D_4$-symmetric **Δ-** and **Λ-MOC-16** cages separately, which feature in high guest inclusion capacity and substantial stereochemical stability. The single-crystal diffraction analyses of the individual **Δ-** and **Λ-MOC-16** cages verified the formation of absolute **ΔΔΔΔΔΔΔΔ**- and **ΛΛΛΛΛΛΛΛ** homoconfigurations, respectively, in corresponding $Pd_6(RuL_3)_8$ cages, and the crystallization of optically pure cage products. The stereoselective inclusion of chiral molecules has been tested for two kinds of organic racemates, that is, classic chiral compounds having C* centres and atropisomeric compounds characteristic of $C_2$ symmetry, with the phase transformation resolution processes. Successful enantioseparation of atropisomers has been accomplished by the use of these homochiral **Δ-** and **Λ-MOCs-16**, manifesting an unprecedented dynamic resolution process based on the kinetically driven guest exchange. The possible resolution mechanism has been investigated by the means of $^1$H NMR titration, $^1$H NMR enantiodifferentiation experiments as well as variable-temperature $^1$H NMR study. In general, this kind of assembly process may provide a new platform to study the stereochemical transmission of optically stable metal centres to versatile homochiral entities in coordination chemistry, and the dynamic resolution behaviour imposed by stereoconfiguration of metal centres might be useful in various chiral resolution of synthetic and industrial significance.

## Methods

**Materials and measurements.** Unless otherwise stated, all commercial reagents and solvents were used as commercially purchased without additional purification. The NMR spectra were recorded on Bruker AVANCE III 400 (400 MHz). Circular dichroism spectra and ultraviolet–visible absorption spectra were measured with a JASCO J-810 spectropolarimeter. Specific rotations were recorded on ADP440 + B + S. HR-ESI-TOF mass spectra were tested on Bruker Maxis 4G, and data analyses were processed with the Bruker Data Analysis software. HPLC spectra were measured on Agilent-2000. Diffraction data for the single crystals were collected on an Agilent SuperNova X-ray diffractometer using micro-focus dual X-ray sources (Supplementary Data 1). Syntheses and characterization details for all compounds are given in Supplementary Methods. Selected bond lengths (Å) and bond angles (°) are listed in Supplementary Tables 4–9.

**Crystal data for {Δ-[$Pd_6(RuL_3)_8$](*S*-BINOL)$_4$} · anion · solvent (Δ-MOC-16).** *I*422 space group, $a = 32.2284(4)$ Å, $c = 38.2801(7)$ Å, $V = 39{,}760.5(13)$ Å$^3$, $Mr = 9{,}727.82$, $Dx = 0.813\ g\ cm^{-3}$, $Z = 2$, $\mu = 2.599\ mm^{-1}$, 16,396 independent reflections, of which 9,021 observed ($I > 2\sigma(I)$), $R_1 = 0.0711$, $wR_2 = 0.2375$, $S = 1.008$, Flack parameter = 0.131(13).

**Crystal data for {Λ-[$Pd_6(RuL_3)_8$](*R*-BINOL)$_4$} · anion · solvent (Λ-MOC-16).** *I*422 space group, $a = 32.5722(6)$ Å, $c = 38.7258(9)$ Å, $V = 41{,}086.1(19)$ Å$^3$, $Mr = 9{,}727.82$, $Dx = 0.786\ g\ cm^{-3}$, $Z = 2$, $\mu = 2.515\ mm^{-1}$, 17,111 independent reflections, of which 7,718 observed ($I > 2\sigma(I)$), $R_1 = 0.0913$, $wR_2 = 0.2680$, $S = 1.038$, Flack parameter = 0.139(16).

**Crystal data for {Δ-[$Ru(Phen)_3$]($PF_6$)$_2$}$_2$ · ($C_6H_5CH_3$) · ($CH_3CN$)$_2$ (Δ-1).** $P4_1$ space group, $a = 25.5619(2)$ Å, $c = 12.5769(2)$ Å, $V = 8{,}217.88(18)$ Å$^3$, $Mr = 2{,}037.48$, $Dx = 1.647\ g\ cm^{-3}$, $Z = 4$, $\mu = 0.556\ mm^{-1}$, 19,736 independent reflections, of which 17,221 observed ($I > 2\sigma(I)$), $R_1 = 0.0678$, $wR_2 = 0.1759$, $S = 1.050$, Flack parameter = 0.00(4).

**Crystal data for {Λ-[$Ru(Phen)_3$]($PF_6$)$_2$}$_2$ · ($C_6H_5CH_3$) · ($CH_3CN$)$_2$ (Λ-1).** $P4_3$ space group, $a = 25.5802(1)$ Å, $c = 12.5709(1)$ Å, $V = 8{,}225.73(9)$ Å$^3$, $Mr = 2{,}037.48$, $Dx = 1.645\ g\ cm^{-3}$, $Z = 4$, $\mu = 0.556\ mm^{-1}$, 20,142 independent reflections, of which 17,755 observed ($I > 2\sigma(I)$), $R_1 = 0.0408$, $wR_2 = 0.1071$, $S = 1.025$, Flack parameter = −0.033(8).

**Crystal data for Δ-[Ru(Phendione)$_3$]($ClO_4$)$_2$ · ($H_2O$) · ($CH_3CN$)$_2$ (Δ-2).** $P2_12_12_1$ space group, $a = 13.8114(2)$ Å, $b = 14.0525(2)$ Å, $c = 20.7957(3)$ Å, $V = 4{,}036.13(10)$ Å$^3$, $Mr = 1{,}028.64$, $Dx = 1.693\ g\ cm^{-3}$, $Z = 4$, $\mu = 5.107\ mm^{-1}$, 7,890 independent reflections, of which 7,536 observed ($I > 2\sigma(I)$), $R_1 = 0.0438$, $wR_2 = 0.1195$, $S = 1.025$, Flack parameter = −0.015(4).

**Crystal data for Λ-[Ru(Phendione)$_3$](ClO$_4$)$_2$·(H$_2$O)·(CH$_3$CN)$_2$ (Λ-2).** $P2_12_12_1$ space group, $a = 13.7734(2)$ Å, $b = 14.0148(2)$ Å, $c = 20.7100(3)$ Å, $V = 3{,}997.68(10)$ Å$^3$, $Mr = 1{,}022.59$, $Dx = 1.699$ g cm$^{-3}$, $Z = 4$, $\mu = 5.156$ mm$^{-1}$, 7,980 independent reflections, of which 7,737 observed ($I > 2\sigma(I)$), $R_1 = 0.0409$, $wR_2 = 0.1084$, $S = 1.032$, Flack parameter $= -0.011(3)$.

**General chiral resolution of racemic guests by enantiopure Δ/Λ-MOCs-16.** Two kinds of methods were used to resolve racemic guests depending on whether the guest inclusion leads to precipitation. For racemic *R/S*-BINOL, *R/S*-3-Br-BINOL, *R/S*-6-Br-BINOL and *R/S*-naproxen molecules, *Method I* based on a solution–solution transfer was applied to avoid host–guest precipitation (Supplementary Fig. 25). An aqueous solution of Δ- or **Λ-MOC-16** and an ethereal solution of racemic guest were mixed and stirred vigorously at room temperature, and then the bottom layer was taken out and extracted with CHCl$_3$. The extractants were combined and the solvent was removed using rotary evaporator to afford white solid as resolved guests by the homochiral MOC host. The solid was redissolved in isopropanol and the *ee* of guest molecules was determined using HPLC. For racemic *R/S*-spirodiol, *R/S*-1-(1-naphthyl)ethanol and *R/S*-benzoin molecules, Method II based on a solid-solution transfer was applied directly. The powder of guest racemate was suspended in the aqueous solution of Δ-or **Λ-MOC-16**. The mixture was stirred vigorously at room temperature. After centrifugation, the filtrate was collected and extracted with CHCl$_3$. The extractants were combined and the solvent was removed by rotary evaporator to afford white solid as resolved guest. The *ee* analysis is the same as in Method I.

## References


1. Amouri, H., Desmarets, C. & Moussa, J. Confined nanospaces in metallocages: guest molecules, weakly encapsulated anions, and catalyst sequestration. *Chem. Rev.* **112,** 2015–2041 (2012).
2. Cook, T. R., Zheng, Y.-R. & Stang, P. J. Metal–organic frameworks and self-assembled supramolecular coordination complexes: comparing and contrasting the design, synthesis, and functionality of metal–organic materials. *Chem. Rev.* **113,** 734–777 (2013).
3. Meng, W. *et al.* A self-assembled $M_8L_6$ cubic cage that selectively encapsulates large aromatic guests. *Angew. Chem. Int. Ed.* **50,** 3479–3483 (2011).
4. Smulders, M. M. J., Riddell, I. A., Browne, C. & Nitschke, J. R. Building on architectural principles for three-dimensional metallosupramolecular construction. *Chem. Soc. Rev.* **42,** 1728–1754 (2013).
5. Pluth, M. D., Bergman, R. G. & Raymond, K. N. Proton-mediated chemistry and catalysis in a self-assembled supramolecular host. *Acc. Chem. Res.* **42,** 1650–1659 (2009).
6. Nishioka, Y., Yamaguchi, T., Kawano, M. & Fujita, M. Asymmetric [2+ 2] olefin cross photoaddition in a self-assembled host with remote chiral auxiliaries. *J. Am. Chem. Soc.* **130,** 8160–8161 (2008).
7. Argent, S. P., Riis-Johannessen, T., Jeffery, J. C., Harding, L. P. & Ward, M. Diastereoselective formation and optical activity of an $M_4L_6$ cage complex. *Chem. Commun.* 4647–4649 (2005).
8. Ousaka, N. *et al.* Efficient long-range stereochemical communication and cooperative effects in self-assembled $Fe_4L_6$ cages. *J. Am. Chem. Soc.* **134,** 15528–15537 (2012).
9. Dong, J., Zhou, Y., Zhang, F. & Cui, Y. A highly fluorescent metallosalalen-based chiral cage for enantioselective recognition and sensing. *Chem. Eur. J.* **20,** 6455–6461 (2014).
10. Wang, J., He, C., Wu, P. Y., Wang, J. & Duan, C. Y. An amide-containing metal–organic tetrahedron responding to a spin-trapping reaction in a fluorescent enhancement manner for biological imaging of NO in living cells. *J. Am. Chem. Soc.* **133,** 12402–12405 (2011).
11. Samanta, D. & Mukherjee, S. P. Component selection in the self-assembly of palladium(II) nanocages and cage-to-cage transformations. *Chem. Eur. J.* **20,** 12483–12492 (2014).
12. MacGillivray, L. R. & Atwood, J. L. Structural classification and general principles for the design of spherical molecular hosts. *Angew. Chem. Int. Ed.* **38,** 1018–1033 (1999).
13. Hamilton, T. D. & MacGillivray, L. R. Enclosed chiral environments from self-assembled metal-organic polyhedral. *Cryst. Growth Des.* **4,** 419–430 (2004).
14. Stang, P. J. & Olenyuk, B. Transition-metal-mediated rational design and self-assembly of chiral, nanoscale supramolecular polyhedra with unique *T* symmetry. *Organometallics* **16,** 3094–3096 (1997).
15. Ousaka, N., Clegg, J. K. & Nitschke, J. R. Nonlinear enhancement of chiroptical response through subcomponent substitution in $M_4L_6$ cages. *Angew. Chem. Int. Ed.* **51,** 1464–1468 (2012).
16. Davis, A. V., Fiedler, D., Ziegler, M., Terpin, A. & Raymond, K. N. Resolution of chiral, tetrahedral $M_4L_6$ metal-ligand hosts. *J. Am. Chem. Soc.* **129,** 15354–15363 (2007).
17. Terpin, A. J., Ziegler, M., Johnson, D. W. & Raymond, K. N. Resolution and kinetic stability of a chiral supramolecular assembly made of labile components. *Angew. Chem. Int. Ed.* **40,** 157–160 (2001).
18. Meng, W., Clegg, J. K., Thoburn, J. D. & Nitschke, J. R. Controlling the transmission of stereochemical information through space in terphenyl-edged $Fe_4L_6$ cages. *J. Am. Chem. Soc.* **133,** 13652–13660 (2011).
19. Schweiger, M., Seidel, S. R., Schmitz, M. & Stang, P. J. Rational design of chiral nanoscale adamantanoids. *Org. Lett.* **2,** 1255–1257 (2000).
20. Gütz, C. *et al.* Enantiomerically pure [$M_6L_{12}$] or [$M_{12}L_{24}$] polyhedra from flexible bis(pyridine) ligands. *Angew. Chem. Int. Ed.* **53,** 1693–1698 (2014).
21. Yang, Y. *et al.* Diastereoselective synthesis of *O* symmetric heterometallic cubic cages. *Chem. Commun.* **51,** 3804–3807 (2015).
22. Knof, U. & von Zelewsky, A. Predetermined chirality at metal centers. *Angew. Chem. Int. Ed.* **38,** 302–322 (1999).
23. Castilla, A. M., Ramsay, W. J. & Nitschke, J. R. Stereochemistry in subcomponent self-assembly. *Acc. Chem. Res.* **47,** 2063–2073 (2014).
24. Northrop, B. H., Zheng, Y.-R., Chi, K.-W. & Stang, P. J. Self-organization in coordination-driven self-assembly. *Acc. Chem. Res.* **42,** 1554–1563 (2009).
25. Saalfrank, R. W. *et al.* Enantiomerisation of tetrahedral homochiral [$M_4L_6$] clusters: synchronised four bailar twists and six atropenantiomerisation processes monitored by temperature-dependent dynamic $^1$H NMR spectroscopy. *Chem. Eur. J.* **8,** 2679–2683 (2002).
26. Li, K. *et al.* Stepwise assembly of $Pd_6(RuL_3)_8$ nanoscale rhombododecahedral metal–organic cages via metalloligand strategy for guest trapping and protection. *J. Am. Chem. Soc.* **136,** 4456–4459 (2014).
27. Garrison, J. C. *et al.* Synthesis and characterization of a trigonal bipyramidal supramolecular cage based upon rhodium and platinum metal centers. *Chem. Commun.* 4644–4646 (2006).
28. Smulders, M. M. J., Jiménez, A. & Nitschke, J. R. Integrative self-sorting synthesis of a $Fe_8Pt_6L_{24}$ cubic cage. *Angew. Chem. Int. Ed.* **51,** 6681–6685 (2012).
29. Hiraoka, S., Sakata, Y. & Shionoya, M. Ti(IV)-centered dynamic interconversion between Pd(II), Ti(IV)-containing ring and cage molecules. *J. Am. Chem. Soc.* **130,** 10058–10059 (2008).
30. Sakata, Y., Hiraoka, S. & Shionoya, M. Site-selective ligand exchange on a heteroleptic $Ti^{IV}$ complex towards stepwise multicomponent self-assembly. *Chem. Eur. J.* **16,** 3318–3325 (2010).
31. Kryschenko, Y. K., Seidel, S. R., Arif, A. M. & Stang, P. J. Coordination-driven self-assembly of predesigned supramolecular triangles. *J. Am. Chem. Soc.* **125,** 5193–5198 (2003).
32. Li, H., Yao, Z.-J., Liu, D. & Jin, G.-X. Multi-component coordination-driven self-assembly toward heterometallic macrocycles and cages. *Coord. Chem. Rev.* **293,** 139–157 (2015).
33. Wu, H.-B. & Wang, Q.-M. Construction of heterometallic cages with tripodal metalloligands. *Angew. Chem. Int. Ed.* **48,** 7343–7345 (2009).
34. Metherell, A. J. & Ward, M. D. Stepwise assembly of an adamantoid $Ru_4Ag_6$ cage by control of metal coordination geometry at specific sites. *Chem. Commun.* **50,** 10979–10982 (2014).
35. Zhou, X.-P. *et al.* A high-symmetry coordination cage from 38- or 62-component self-assembly. *J. Am. Chem. Soc.* **134,** 8042–8045 (2012).
36. Carlucci, L., Ciani, G., Maggini, S., Proserpio, D. M. & Visconti, M. Heterometallic modular metal-organic 3D frameworks assembled via new tris-beta-diketonate metalloligands: nanoporous materials for anion exchange and scaffolding of selected anionic guests. *Chem. Eur. J.* **16,** 12328–12341 (2010).
37. Hamelin, O., Rimboud, M., Pecaut, J. & Fontecave, M. Chiral-at-metal ruthenium complex as a metalloligand for asymmetric catalysis. *Inorg. Chem.* **46,** 5354–5360 (2007).
38. Hiort, C., Lincoln, P. & Norden, B. DNA binding of Δ- and Λ-[Ru(phen)$_2$DPPZ]$^{2+}$. *J. Am. Chem. Soc.* **115,** 3448–3454 (1993).
39. Smith, J. A., Collins, J. G., Patterson, B. T. & Keene, F. R. Total enantioselectivity in the DNA binding of the dinuclear ruthenium (II) complex [{Ru(Me$_2$bpy)$_2$}$_2$(μ-bpm)]$^{4+}$ {bpm = 2,2′-bipyrimidine. Me$_2$bpy = 4,4′-dimethyl-2,2′-bipyridine}. *Dalton Trans.* **9,** 1277–1283 (2004).
40. MacDonnell, F. M., Kim, M. J. & Bodige, S. Substitutionally inert complexes as chiral synthons for stereospecific supramolecular syntheses. *Coord. Chem. Rev.* **185,** 535–549 (1999).
41. Zhao, C. *et al.* Chiral amide directed assembly of a diastereo- and enantiopure supramolecular host and its application to enantioselective catalysis of neutral substrates. *J. Am. Chem. Soc.* **135,** 18802–18805 (2013).
42. Kida, T., Iwamoto, T., Asahara, H., Hinoue, T. & Akashi, M. Chiral recognition and kinetic resolution of aromatic amines via supramolecular chiral nanocapsules in nonpolar solvent. *J. Am. Chem. Soc.* **135,** 3371–3374 (2013).
43. Fiedler, D., Leung, D. H., Bergman, R. G. & Raymond, K. N. Enantioselective guest binding and dynamic resolution of cationic ruthenium complexes by a chiral metal-ligand assembly. *J. Am. Chem. Soc.* **126,** 3674–3675 (2004).
44. Hastings, C. J., Pluth, M. D., Biros, S. M., Bergman, R. G. & Raymond, K. N. Simultaneously bound guests and chiral recognition: a chiral self-assembled supramolecular host encapsulates hydrophobic guests. *Tetrahedron* **64,** 8362–8367 (2008).

45. Sawada, T., Matsumoto, A. & Fujita, M. Coordination-driven folding and assembly of a short peptide into a protein-like two-nanometer-sized channel. *Angew. Chem. Int. Ed.* **53,** 7228–7232 (2014).
46. Liu, T. F., Liu, Y., Xuan, W. M. & Cui, Y. Chiral nanoscale metal–organic tetrahedral cages: diastereoselective self-assembly and enantioselective separation. *Angew. Chem. Int. Ed.* **49,** 4121–4124 (2010).
47. Torres, A. S., Maloney, D. J., Tate, D., Saad, Y. & MacDonnell, F. M. Retention of optical activity during conversion of $\Lambda$-[Ru(1,10-phenanthroline)$_3$]$^{2+}$ to $\Lambda$-[Ru(1,10-phenanthroline-5,6-dione)$_3$]$^{2+}$ and $\Lambda$-[Ru(dipyrido[a:3,2-h:2′,3′-c-]-phenazine)$_3$]$^{2+}$. *Inorg. Chim. Acta* **293,** 37–43 (1999).
48. McGee, K. A. & Mann, K. R. Inefficient crystal packing in chiral [Ru(phen)$_3$](PF$_6$)$_2$ enables oxygen molecule quenching of the solid-state MLCT emission. *J. Am. Chem. Soc.* **131,** 1896–1902 (2009).
49. Chen, L. *et al.* Separation of rare gases and chiral molecules by selective binding in porous organic cages. *Nat. Mater.* **13,** 954–960 (2014).
50. Smulders, M. M. J., Zarra, S. & Nitschke, J. R. Quantitative understanding of guest binding enables the design of complex host-guest behavior. *J. Am. Chem. Soc.* **135,** 7039–7046 (2013).
51. Zarra, S., Smulders, M. M. J., Lefebvre, Q., Clegg, J. K. & Nitschke, J. R. Guanidinium binding modulates guest exchange within an [$M_4L_6$] capsule. *Angew. Chem. Int. Ed.* **51,** 6882–6885 (2012).
52. Jiang, W., Ajami, D. & Rebek, Jr J. Alkane lengths determine encapsulation rates and equilibria. *J. Am. Chem. Soc.* **134,** 8070–8073 (2012).
53. Davis, A. V. *et al.* Guest exchange dynamics in an $M_4L_6$ tetrahedral host. *J. Am. Chem. Soc.* **128,** 1324–1333 (2006).
54. Mugridge, J. S., Szigethy, G., Bergman, R. G. & Raymond, K. N. Encapsulated guest-host dynamics: guest rotational barriers and tumbling as a probe of host interior cavity space. *J. Am. Chem. Soc.* **132,** 16256–16264 (2010).
55. Wang, B.-Y. *et al.* The entrapment of chiral guests with gated baskets: can a kinetic discrimination of enantiomers be governed through gating? *Chem. Eur. J.* **19,** 4767–4775 (2013).
56. Ma, G., Deng, J. & Sibi, M. P. Fluxionally chiral DMAP catalysts: kinetic resolution of axially chiral biaryl compounds. *Angew. Chem. Int. Ed.* **53,** 11818–11821 (2014).

## Acknowledgements

We gratefully acknowledge the financial support from the 973 Program (2012CB821701), the NSFC Projects (Grants 91222201, 21373276, 21573291) of China, NSFGP (S2013030013474) and the RFDP of Higher Education (20120171130006).

## Author contributions

C.-Y.S. and M.P. designed the research and wrote the paper. K.W. and K.L. carried out most of the syntheses and measurements. Y.-J.H., L.-Y.Z. and L.C. helped in experiments and data analyses. All authors discussed the results and commented on the manuscript.

## Additional information

**Competing financial interests:** The authors declare no competing financial interests.

# 23

# Selective catalytic two-step process for ethylene glycol from carbon monoxide

Kaiwu Dong[1], Saravanakumar Elangovan[1], Rui Sang[1], Anke Spannenberg[1], Ralf Jackstell[1], Kathrin Junge[1], Yuehui Li[1,2] & Matthias Beller[1]

Upgrading C1 chemicals (for example, CO, CO/$H_2$, MeOH and $CO_2$) with C–C bond formation is essential for the synthesis of bulk chemicals. In general, these industrially important processes (for example, Fischer Tropsch) proceed at drastic reaction conditions ($>$250 °C; high pressure) and suffer from low selectivity, which makes high capital investment necessary and requires additional purifications. Here, a different strategy for the preparation of ethylene glycol (EG) via initial oxidative coupling and subsequent reduction is presented. Separating coupling and reduction steps allows for a completely selective formation of EG (99%) from CO. This two-step catalytic procedure makes use of a Pd-catalysed oxycarbonylation of amines to oxamides at room temperature (RT) and subsequent Ru- or Fe-catalysed hydrogenation to EG. Notably, in the first step the required amines can be efficiently reused. The presented stepwise oxamide-mediated coupling provides the basis for a new strategy for selective upgrading of C1 chemicals.

[1] Leibniz-Institut für Katalyse e.V. an der Universität Rostock, Albert-Einstein Straße 29a, 18059 Rostock, Germany. [2] State Key Laboratory for Oxo Synthesis and Selective Oxidation, Suzhou Research Institute of LICP, Lanzhou Institute of Chemical Physics (LICP), Chinese Academy of Sciences, 730000 Lanzhou, P.R. China. Correspondence and requests for materials should be addressed to Y.L. (email: yuehui.li@catalysis.de and yhli@licp.cas.cn) or to M.B. (email: matthias.beller@catalysis.de).

A major source for innovation in chemistry and material sciences is the use of easily available feedstocks in new transformations. Owing to the rising global demand for bulk chemicals and materials, there exists strong research interest to discover novel production processes from alternative resources[1]. As central raw materials for the chemical industry, C1 compounds such as CO, $CH_3OH$, $CO_2$ and HCN can be obtained on large scale including renewables. Specifically, CO, which is a most versatile and cost-efficient chemical building block, is easily produced from natural gas, coal, biomass and so on, and is currently used in the industrial production of methanol, higher alcohols, aldehydes, alkanes and diols[2–4]. Despite all these processes, selective coupling of CO still represents a major challenge in C1 chemistry. Although numerous selective C–C coupling reactions using CO are known with olefins, alkynes and C–X bonds, the selective dimerization and oligomerization of CO are basically unknown, despite the huge commercial interest in this area. The problem to efficiently create C-C bonds via coupling of carbon monoxide relies on the high activation barrier for reductive transformations. Hence, despite the apparent simplicity there exists no direct method to convert CO selectively into valuable C2 chemicals[5]. Among these products, ethylene glycol (EG) is an important bulk chemical with an annual production and consumption of more than 20 million tons[6]. As a prevalent industrial product, this C2 diol finds widespread applications in different fields, for example, as a solvent, anti-freeze agent and precursor for the manufacture of polyester fibres, resins and so on. In fact, the growing demand for PET (polyethylene terephthalate) resins and polyester fibres leads to a constant growth of the global market for high-priced fibre-quality EG, which is valued at $33.1 billion in 2014 and is estimated to reach $46.8 billion by 2019 (ref. 7).

Owing to the problems of CO dimerization, today's main technology for large-scale industrial production of EG still relies on the vapour-phase oxidation of ethylene to ethylene oxide started by Carbide Company in the 1930s (refs 8,9). Compared with ethylene, CO is easily available from renewable feedstock, for example, biogas. In addition, it can be considered economically advantageous, although the price for such bulk feedstocks varies quite significantly (for example, from 2000 to 2012 ethylene prices were in between $400 and $1,800 per ton and around $750 per ton in early 2015; see http://www.duncanseddon.com/images/ethylene-price-trends.gif.). Therefore, an increasing interest in industry and academia exists to use CO for the production of EG. However, known methods suffer from the necessity of harsh conditions and the insufficient selectivity. Specifically, this latter problem is a critical issue for EG production. For example, fibre-grade EG (*ca.* 40% of production capacity in the world) tolerates only small amounts of glycol oligomers, which constitute major byproducts in the conventional route via ethylene oxide hydrolysis.

As discussed above, the direct synthesis of EG starting from syngas under reductive conditions represents an ideal method, but faces significant difficulties to overcome the low reactivity and selectivity. Thus, alternative methods are under development. Accordingly, EG preparation from methanol, formaldehyde and methyl formate are notable, but low selectivity limits their application too[10–13]. Contrary to all these reductive coupling strategies, oxidative dimerization of CO to oxalates might allow for improved efficiency and two main approaches were developed since the 1990s as follows: (1) carbonylation of MeOH with oxygen in the presence of a Pd/V/Ti system[14] and 2) nitric oxide-mediated carbonylation of alcohols to dialkyl oxalates using palladium complexes. Owing to the higher reactivity, the latter system is preferred in industry and has been applied on ton scale[15]. However, the sensitivity of the oxalate intermediate to water and the use of highly corrosive nitric oxide complicate the latter process and require specific quality of starting materials. Therefore, the development of a selective EG production process from CO without the use/formation of unstable reagents still constitutes a major challenge for industry and continues to attract intense interests of both academic and industrial chemists.

Based on our interest in the catalytic hydrogenation of carboxylic acid derivatives[16], we got inspired to use oxamides as key intermediates for a novel EG process. Notably, the key C–C bond formation in this process should be facilitated in the presence of amines compared with alcohols or water[17]. As shown in Fig. 1, this concept is based on oxidative coupling of CO to oxamides and subsequent selective reduction to EG (to the best of our knowledge, such a reaction has been only proposed in a patent by Shell. However, the selectivity for EG was only *ca.* 65%. For details, see[18]). Although the overall transformation looks apparently simple, there exist several challenges for the individual transformations. For successful oxidative carbonylation of CO to oxamides, side reactions such as oxidation of CO to $CO_2$ and the formylation of amines to formamides or oxycarbonylation of amines to ureas are well known *vide infra* and have to be avoided. In addition, decomposition of the active oxidation species in the presence of carbon monoxide has to be carefully controlled. Obviously, the formation of robust catalysts is critical for this oxidative reaction. The other major challenge of this process is how to achieve selective hydrogenation of oxamides to EG, which, to the best of our knowledge, has not been reported yet. Problematic for this transformation is the deactivation effect that might be caused by coordination of the metal centre with the chelating carbonyl groups of the substrate or by product inhibition. Finally, the strong reduction ability of the formed 1, 2-diol facilitates catalyst deactivation.

Despite all these problems, here we disclose that EG can be selectively produced (up to 99% for CO) by sequential Pd-catalysed oxidative carbonylation of piperidine to the corresponding oxamide and subsequent Ru- or Fe-catalysed hydrogenation to EG. The free amine is easily separated from the reaction mixture and recycled.

## Results

**Amines to oxamides.** Although reactions with CO constitute powerful tools for the introduction of carbonyl groups into all kinds of organic molecules[19–21], oxidative carbonylations including the synthesis of heterocycles, carbonates, carbomates, ureas and oxamides are still challenging regarding catalyst productivity and selectivity[22,23]. For example, reactions of amines with CO in the presence of an oxidant lead typically to mixtures with the corresponding urea as the main product[24]. Hence, even though the synthesis of oxamides represents a valid method to couple two CO molecules, it only attracted limited interest for decades. In fact, the state-of-the-art methodology

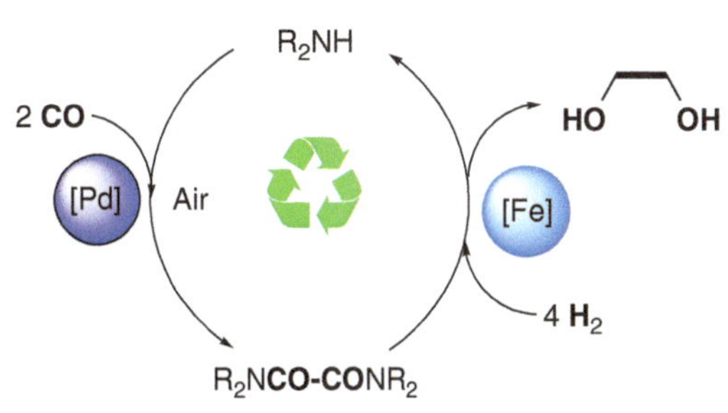


**Figure 1 | Coupling of CO to EG.** Schematic pathway for two-step preparation of EG from CO.

**Table 1 | Towards a practical oxidative carbonylation protocol: influence of ligands and comparison with previously reported catalysts*.**

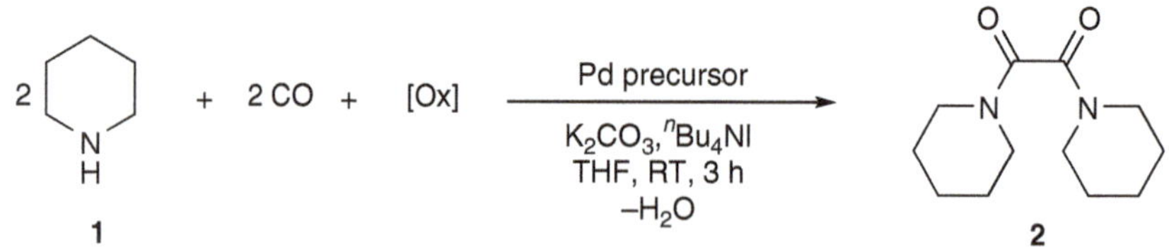


1 2

| Entry | Pd source | Ligand | [Ox] | Yield (%) | TON | TOF ($h^{-1}$) | Selectivity 2 (%) |
|---|---|---|---|---|---|---|---|
| 1 (ref. 25) | $Pd(OAc)_2$ | — | $O_2$ | 38 | 2.8 | 1 | — |
| 2 (ref. 25) | $Pd(OAc)_2$ | $PPh_3$ | $O_2$ | 88 | 6.6 | 2 | — |
| 3 | $Pd(acac)_2$ | 2, 2′-bipyridine | air | 85 | 42 | 14 | 99 |
| 4 | $Pd(acac)_2$ | $H_2IMes$-$CO_2$ | air | 28 | 14 | 5 | 99 |
| 5 | $Pd(acac)_2$ | P(*p*-tol)$_3$ | air | 47 | 24 | 8 | >99 |
| 6 | $Pd(acac)_2$ | Xantphos | air | 99 | 50 | 17 | >99 |
| 7 | $Pd(acac)_2$ | P(*o*-tol)$_3$ | air | 99 | 50 | 17 | >99 |
| 8† | $Pd(acac)_2$ | P(*o*-tol)$_3$ | air | 53 | 26,500 | 158 | 99 |
| 9‡ | $Pd(acac)_2$ | P(*o*-tol)$_3$ | air | 60 | 3,000 | 750 | 90 |

acac, acetylacetonate; $H_2IMes$-$CO_2$, 1, 3-dimesityl-4, 5-dihydro-1*H*-imidazol-3-ium-2-carboxylate; TOF, turnover frequency; TON, turnover number; Xantphos, (9, 9-dimethyl-9*H*-xanthene-4, 5-diyl)bis(diphenylphosphane).
*Reaction conditions for ref. 25: 1.5 mmol **1**, 0.1 mmol $Pd(OAc)_2$, 0.2 mmol $K_2CO_3$, 0.2 mmol $^nBu_4NI$, 2.7 bar CO and 1 bar $O_2$. For this work: 1.0 mmol **1**, 1 mol% $Pd(acac)_2$ (to **1**), 2 mol% monodentate ligand or 1 mol% bidentate ligand, 10 mol% $K_2CO_3$, 5 mol% $^nBu_4NI$, 25 bar CO and 25 bar air. TON = mol of **2** obtained per mol of Pd employed. TOF = mol of **2** obtained per mol of Pd employed per hour.
†0.001 mol% catalyst, 168 h.
‡0.01 mol% catalyst, 120 °C, 4 h.

for this reaction has been disclosed by Pri-Bar and Alper[25], who reported the Pd-catalysed oxycarbonylation of amines to oxamides in the presence of iodide anions as activator. Although the reactions proceeded well under mild conditions using $O_2$ as oxidant, substantial amounts of Pd catalyst (6.7 mol%) are needed to obtain the desired product in high yields (Table 1, entries 1–2).

To bring this reaction closer to practical application, considerable improvements in catalyst efficiency are required (see Supplementary Tables 1–7 for details). To achieve this goal, we envisioned the use of alternative ligands, which should better stabilize the active metal centre. For initial investigations of the ligand effect, the carbonylation of piperidine to 1, 1′-oxalyl dipiperidine (**2**) was performed with air (25 bar) at RT in the presence of various of palladium salts. Indeed, in the presence of nitrogen-based ligands, which are used routinely in palladium-catalysed oxidations, a slight increase in catalyst productivity was observed (Table 1, entry 3). Nevertheless, using nitrogen ligands such as bipyridine did not lead to a stable system and a dark-coloured suspension was formed after the reaction, indicating the decomposition of the active homogeneous catalyst. On the other hand, the use of stronger donating carbene ligand prevented the formation of palladium black and resulted in a minor improvement (Table 1, entry 4). Testing different mono- and bidentate P-based ligands revealed in general a decrease of activity for bidentate phosphines (Table 1, entries 5–6). However, to our surprise, the best result was obtained when using tri(*o*-tolyl)phosphine in combination with $Pd(acac)_2$ as the metal salt (Table 1, entry 7). Here, a colourless suspension was obtained after the reaction, indicating the efficient stabilization effect by this ligand. Interestingly, the corresponding palladacycle showed no reactivity at all, which excludes this species as an active intermediate[26]. Other ligands tested gave only inferior results compared with P(*o*-tol)$_3$ (see Supplementary Table 2 for details).

Obviously, in the absence of $O_2$, this reaction does not occur. Furthermore, no reactivity was observed when using Ni- or Co-based salts under similar conditions. Using 1 mol% of the optimized palladium catalyst, full conversion of piperidine is observed (>99% yield of **2**), demonstrating that the amine is not the limiting reagent under these conditions. Notably, decreasing the metal loading to 0.001 mol% at RT and similar $P_{CO}$ and $P_{air}$ gave 26,500 catalytic turnovers and no byproducts such as ureas or imines were detected (Table 1, entry 8), which demonstrates that this palladium complex is one of the most efficient catalysts for oxidative carbonylations known to date. As shown in Table 1, when reducing the catalyst loading by a factor of 1,000, the activity of the catalyst system (turnover frequency) increased from 17 to 158 $h^{-1}$ (Table 1, entry 7 versus 8). This surprising increase of activity is explained by the lower conversion in the latter case.

Compared with $PPh_3$ as the ligand, the addition of P(*o*-tol)$_3$ improved the stability and efficiency of the corresponding Pd complex under similar conditions. In fact, 60% of $PPh_3$ was oxidized to $P(O)Ph_3$ after 3 h (CO/air = 25/25 bar at RT), whereas < 20% of P(*o*-tol)$_3$ was transformed into the corresponding phosphine oxide after the same time. Notably, at low catalyst loading (0.001 mol% Pd/1.2 mol% P(*o*-tol)$_3$), even after 7 days around 50% of the free phosphine was present as analysed by gas chromatography. Furthermore, the stability of the catalyst system is shown by three consecutive runs on 10-g scale and 30.9 g of **2** was obtained conveniently after crystallization (see Supplementary Methods for details). In addition, this reaction can be performed at higher temperature (up to 120 °C) and the reactivity increased with a turnover frequency up to 750 $h^{-1}$ (Table 1, entry 9). It is important to note that CO is highly selectively converted to oxamide and no $CO_2$ was detected by gas chromatography in the gas phase after the reaction.

**Hydrogenation of oxamides.** Having an efficient procedure for selective preparation of oxamides in hand, we explored its hydrogenation to EG in the presence of Ru catalysts containing hydride phosphine-amine or pincer ligands. Recently, such organometallic complexes were reported for the reduction of carboxylic acid derivatives[27,28]. As an example, John and Bergens[29] described ruthenium complexes with aminophosphine ligands as one of the most efficient homogeneous catalysts for hydrogenation of β-lactams.

**Table 2 | Transition metal catalysed hydrogenation of oxamide 2*.**

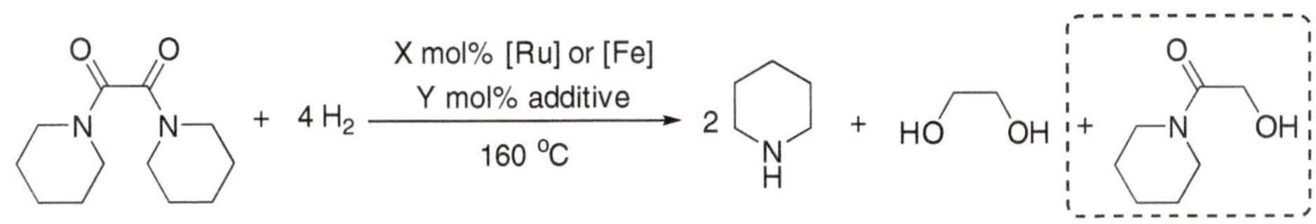


2 1 3 4

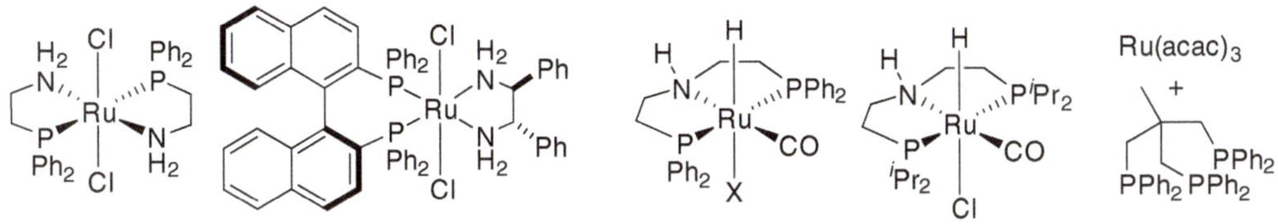


c-1 c-2 c-3: X = $BH_4$ c-4 c-5

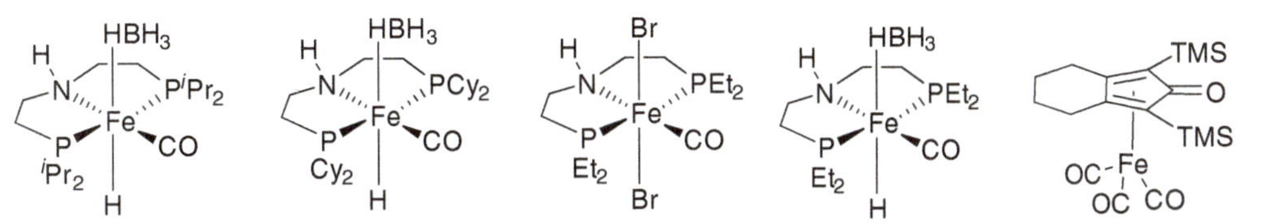


c-6 c-7 c-8 c-9 c-10

| Entry | Catalyst (X) | Additive (Y) | Solvent | Time (h) | Conv. (%) | Yield 3 (%) |
|---|---|---|---|---|---|---|
| 1 | **c-1** (0.5) | $^t$BuOK (2.5) | THF | 6 | 25 | 1 |
| 2 | **c-2** (0.5) | $^t$BuOK (2.5) | THF | 6 | 64 | 8 |
| 3 | **c-3** (0.5) | $^t$BuOK (2.5) | THF | 6 | 20 | 5 |
| 4 | **c-3** (1) | $^t$BuOK (10)/$PhNH_2$ (100) | THF | 24 | 94 | 80 |
| 5 | **c-3** (1) | $^t$BuOK (10) | *i*PrOH | 24 | 85 | 45 |
| 6 | **c-3** (1) | $^t$BuOK (10) | Toluene | 24 | 100 | 95 |
| 7 | **c-3** (1) | NaOH (2) | Toluene | 24 | 100 | 99 |
| 8† | **c-3** (0.1) | $^t$BuOK (2) | Toluene | 24 | 100 | 94 |
| 9† | **c-4** (0.1) | $^t$BuOK (10) | Toluene | 24 | 95 | 70 |
| 10‡ | **c-5** (0.1) | $Al(OTf)_3$ (2) | Toluene | 24 | 100 | 58 |
| 11 | **c-6** (0.2) | $^t$BuOK (1) | Toluene | 6 | 16 | 7 |
| 12 | **c-7** (0.2) | $^t$BuOK (1) | Toluene | 6 | 23 | 10 |
| 13 | **c-8** (0.2) | $^t$BuOK (1.5) | Toluene | 6 | 18 | 8 |
| 14 | **c-9** (0.2) | $^t$BuOK (1) | Toluene | 6 | 53 | 29 |
| 15 | **c-10** (0.2) | $^t$BuOK (1) | Toluene | 6 | 9 | 1 |
| 16 | **c-9** (0.2) | $^t$BuOK (0.2) | Toluene | 6 | 0 | 0 |
| 17 | **c-9** (0.2) | $^t$BuOK (4) | Toluene | 6 | 28 | 11 |
| 18 | **c-9** (0.2) | $^t$BuOK (1.5) | Toluene | 24 | 83 | 46 |
| 19 | **c-9** (2) | KOH (5) | Toluene | 24 | 99 | 77 |

*Reaction conditions: 0.5 mmol **2**, 2 ml solvent; conversion and yield were determined by gas chromatography and $^1H$ NMR, respectively.
†1 mmol **2**, 4 ml toluene.
‡0.1 mol% $Ru(acac)_3$, 0.2 mol% triphos (1,1,1-tri(diphenylphosphinomethyl)ethane), 2 ml toluene and 2 ml $H_2O$.

Unfortunately, **2** was not an eligible substrate under similar conditions. This indicates that oxamides are more difficult to reduce compared with simple carboxamides. Hence, only low-to-moderate reactivity was observed by using catalysts **c-1**, **c-2** and **c-3** in the presence of $^t$BuOK (Table 2, entries 1–3), which are well established for ester hydrogenations. In general, in these processes the added base will accelerate the formation of an active Ru amido complex, which facilitates the heterolytic cleavage of $H_2$ to generate the corresponding active Ru-H species in the catalytic cycle.

Kinetic investigations on the hydrogenation catalysed by Ru-MACHO-BH (**c-3**) showed a strong inhibition of activity after *ca.* 30% conversion due to the formation of coordinating intermediates such as glycol amide **4**. Indeed, this half reduction product was isolated in 24% yield using **c-1** (ref. 30). To solve this problem, different additives and solvents were tested. Gratifyingly, a substantial positive solvent effect was observed by replacing tetrahydrofuran (THF) with isopropanol or toluene, which increased the yield of EG from trace to 45% and 95% yield, respectively (Table 2, entries 3 *versus* 5–6). Finally, we observed the hydrogenation of **2** also works well when inexpensive NaOH was used instead of KO$^t$Bu and full conversion to EG was achieved (Table 2, entry 7). Decreasing the catalyst loading to 0.1 mol% still gave 94% yield of EG (Table 2, entry 8). No byproducts such as tertiary amines that might be easily formed via amination of EG were observed. Notably, under these mild conditions, over-reduction to ethanol was not detected by $^1H$ NMR measurement of the reaction mixture. Interestingly, the

Ru-triphos system **c-5** gave the desired product in 58% yield in the presence of $Al(OTf)_3$ (ref. 31), whereas other catalysts showed no reactivity at all with acid additives (see Supplementary Tables 8–11 for details).

Owing to the obvious advantages of base metals compared with noble metal complexes, we were interested to investigate Fe catalysts for the hydrogenation of **2**. Based on the recent development of Fe pincer complexes (such as **c-6**) for hydrogenation of esters and nitriles[32–34], we prepared a series of complexes **c-7-9** (see Supplementary Fig. 1 for spectra of **c-8**).

Indeed, selective hydrogenation took place and, similar with the reactions using Ru-catalysts, worse reactivity was obtained in THF compared with the reactions in toluene. Moderate reactivity was obtained with Fe pincer complexes, whereas Knölker's Fe complex only produced trace amounts of product (Table 2, entries 11–15)[35]. The concentration of base was critical to achieve higher yields of EG (Table 2, entries 16–17) and the best reactivity was obtained with **c-9** (Table 2, entry 18). Using 2 mol% of **c-9**, full conversion was achieved with **4** as the only byproduct (Table 2, entry 19).

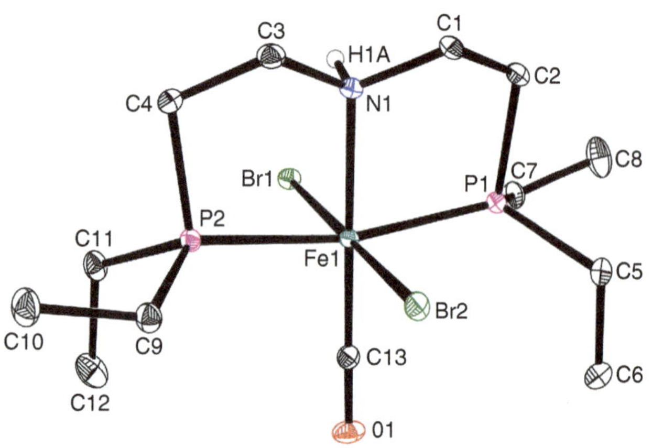


**Figure 2 | Molecular structure of complex c-8.** Displacement ellipsoids are drawn at the 30% probability level. Hydrogen atoms, except of H1A, are omitted for clarity. Selected bond lengths (Å) and angles (°): Fe1-N1 2.0783(10), Fe1-C13 1.7499(13), Fe1-P1 2.26960(4), Fe1-P2 2.2683(4), Fe1-Br1 2.4735(2), Fe1-Br2 2.4562(2), C13-O1 1.1477(16); C13-Fe1-N1 179.05(5), P1-Fe1-P2 167.29 (13), Br1-Fe1-Br2 178.339(8) and N1-Fe1-P1 83.02(3).

The X-ray crystal structure analysis of complex **c-8**, a precursor of **c-9**, reveals a distorted octahedral coordination geometry around the Fe(II) centre, with the CO ligand located *trans* to the nitrogen atom and the two bromine atoms located *trans* with each other (Fig. 2). Apparently, the sterically less demanding $PEt_2$ groups allow for an easier access of the substrate molecule compared with catalyst **c-6**.

Finally, it should be noted that both the iron- and ruthenium-based catalysts can be used for hydrogenation of other oxamides, for example, oxanilide (see Supplementary Tables 12–13 for details), as well as oxalates[36].

## Discussion

The two reaction cycles for the selective synthesis of EG from carbon monoxide are shown in Fig. 3. In the first step, secondary amines such as piperidine or pyrrolidine reacted smoothly to the corresponding oxamides[37–39]. Initially, the piperidine-derived bis(carbamoyl)palladium intermediate[40] is formed from the corresponding monocarbamoyl palladium intermediate at RT highly selectively. Thus, no urea products are detected, which are common side products in such oxidative carbonylations. Next, the corresponding oxamides are formed via reductive elimination giving Pd(0) species. Finally, oxidation with molecular oxygen leads to the regeneration of the active Pd(II) species. Notably, iodide promotes both the generation of the carbamoyl group and the reoxidation of Pd(0)[23]. Besides, the addition of $(o\text{-tol})_3P$ significantly improved the efficiency of the catalyst system despite the well-known sensitivity of phosphine ligands towards oxygen. After the oxidative carbonylation, the solvent (THF) was removed *in vacuo* from the reaction mixture of step 1 followed by addition of toluene. The resulted mixture was quickly filtrated (silica gel, 2–3 cm) and then subjected to the hydrogenation catalyst without further purification (Fig. 4 and see Supplementary Table 14 for details). Subsequent oxamide hydrogenation should proceed via an outer-sphere bifunctional activation mode when using M-PN$^H$P type Ru complex **c-3** or Fe complex **c-9**. In the presence of these catalysts, no formation of water was observed, which suggests the direct hydrogenolysis of the amide bond of intermediate **b'**. From a practical point of view, it is noteworthy that the reaction system is water tolerant, as control experiments with addition of water showed no influence. Based on the two steps shown in Fig. 4, the preparation of EG from CO proceeds with remarkable selectivity (99%). Final purification of the desired

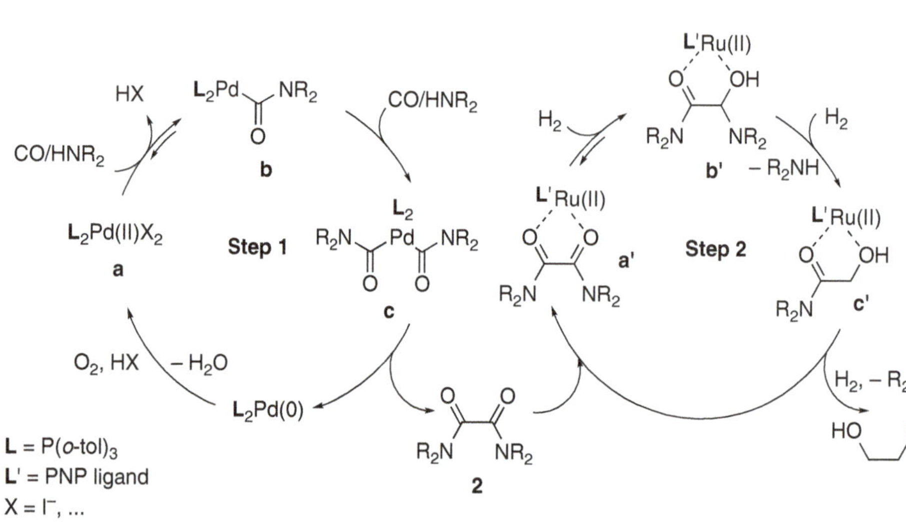


**Figure 3 | Proposed reaction cycles for piperidine-mediated production of EG from CO.** Step 1: Pd-catalysed oxidative carbonylation of piperidine to oxamide **2**. Step 2: Ru-catalysed hydrogenation of oxamide **2** to EG and piperidine.

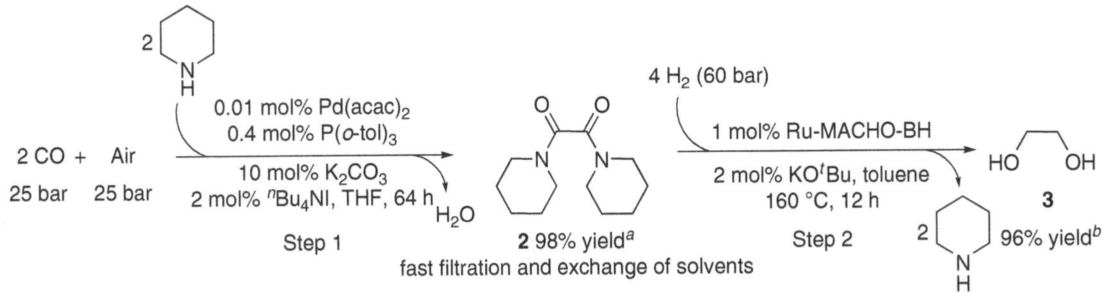


**Figure 4 | Piperidine-mediated production of EG from CO via two steps.** Reaction conditions for step 1: $Pd(acac)_2$ (6.1 mg, 0.02 mmol), $P(o\text{-tol})_3$ (244 mg, 0.8 mmol), $^nBu_4NI$ (1.48 g, 4 mmol), $K_2CO_3$ (2.76 g, 20 mmol), piperidine (20 ml, 200 mmol, in three portions), THF (100 ml), room temperature (RT), 64 h. For step 2: **2** (5.0 mmol in 12 ml toluene), Ru-MACHO-BH (29 mg, 0.05 mmol), $KO^tBu$ (11.2 mg, 0.1 mmol), toluene (8 ml), $H_2$ (60 bar), 160 °C, 12 h. [a]Gas chromatography yield using *iso*-octane as internal standard. [b] $^1H$ NMR yield using *tert*-butanol as internal standard.

product from the solvent (toluene) and the co-product (piperidine) is comparably easy due to the formation of a biphasic mixture. Simple phase separation provides EG with a purity of >95%.

In conclusion, we demonstrated for the first time the selective synthesis of EG from CO, $H_2$ and air. By combination of an efficient room-temperature Pd-catalysed oxidative carbonylation with a selective hydrogenation, it is possible to produce C2 compounds with excellent selectivity (>99% based on CO) and good reactivity. Amines were used and recycled without observable consumption. Key to success of this work is the selective construction of C2 oxamide intermediates. Although the presented catalytic efficiencies are already high for academic standards, clearly a bulk industrial implementation requires further optimization. However, we are confident that the presented methodology has a high potential to become an economical viable process for EG production. At the same time, this work will inspire chemists to discover new concepts on selective transformation of C1 chemicals to C2 or even higher chemicals.

## References


1. Jiang, Q. *et al.* Research progress on solar thermal $H_2O$ and $CO_2$ splitting reactions. *Sci. Sin. Chim.* **44,** 1834–1848 (2014).
2. Hickman, D. A. & Schmidt, L. D. Production of syngas by direct catalytic oxidation of methane. *Science* **259,** 343–346 (1993).
3. Huo, C.-F., Li, Y.-W., Wang, J. & Jiao, H. Insight into $CH_4$ formation in iron-catalyzed Fischer – Tropsch synthesis. *J. Am. Chem. Soc.* **131,** 14713–14721 (2009).
4. Keim, W. (ed) *Catalysis in C1 Chemistry* (D. Reidel Publishing Company, 1983).
5. Yue, H., Ma, X. & Gong, J. An alternative synthetic approach for efficient catalytic conversion of syngas to ethanol. *Acc. Chem. Res.* **47,** 1483–1492 (2014).
6. Yue, H., Zhao, Y., Ma, X. & Gong, J. Ethylene glycol: properties, synthesis, and applications. *Chem. Soc. Rev.* **41,** 4218–4244 (2012).
7. Ethylene Glycol Market Application & Geography - Global Trends & Forecasts to 2019 http://www.researchandmarkets.com/research/jsz7kl/ethylene_glycol April2015.
8. Rebsdat, S. & Mayer, D. in *Ullmann's Encyclopedia of Industrial Chemistry* (Wiley-VCH Verlag GmbH & Co. KGaA, 2000) doi:10.1002/14356007.a10_101.
9. Han, Z. *et al.* Catalytic hydrogenation of cyclic carbonates: a practical approach from $CO_2$ and epoxides to methanol and diols. *Angew. Chem. Int. Ed.* **51,** 13041–13045 (2012).
10. Song, H., Jin, R., Kang, M. & Chen, J. Progress in synthesis of ethylene glycol through C1 chemical industry routes. *Chin. J. Catal.* **34,** 1035–1050 (2013).
11. Drent, E., Mul, W. P. & Ruisch, B. J. Process for the carbonylation of formaldehyde (Shell). US patent 6,376,723 (2002).
12. Celik, F. E., Lawrence, H. & Bell, A. T. Synthesis of precursors to ethylene glycol from formaldehyde and methyl formate catalyzed by heteropoly acids. *J. Mol. Catal. A* **288,** 87–96 (2008).
13. Celik, F. E., Kim, T.-J. & Bell, A. T. Novel vapor-phase carbonylation of dimethoxymethane over H-ZSM-5. *Angew. Chem. Int. Ed.* **48,** 4813–4815 (2009).
14. Gaffney, A. M., Leonard, J. J., Sofranko, J. A. & Sun, H.-N. Heterogeneous catalyst for alcohol oxycarbonylation to dialkyl oxalates. *J. Catal.* **90,** 261–269 (1984).
15. Luo, N., Ji, Y., Mao, Y. & Zhang, B. Syn-gas-based mono ethylene glycol synthesis in Pujing Chemical. *Appl. Petrochem. Res.* **2,** 23–26 (2012).
16. Cabrero-Antonino, J. R. *et al.* Efficient base-free hydrogenation of amides to alcohols and amines catalyzed by well-defined pincer imidazolyl–ruthenium complexes. *ACS Catal.* **6,** 47–54 (2016).
17. Ozaw, F. *et al.* Palladium-catalyzed double carbonylation of aryl halides to give alpha-keto amides. Mechanistic studies. *J. Am. Chem. Soc.* **107,** 3235–3245 (1985).
18. De Boer, E. J. M., Van Der Heijden, H., Meurs, H. H. & Wijnans, S. Process for the production of ethylene glycol and related compounds (Shell). US patent WO 2010130696 A1 (2010).
19. Zuidema, E., Bo, C. & Van Leeuwen, P. W. N. M. Ester versus polyketone formation in the palladium – diphosphine catalyzed carbonylation of ethene. *J. Am. Chem. Soc.* **129,** 3989–4000 (2007).
20. Amoroso, F. *et al.* Catalyst activity or stability: the dilemma in Pd-catalyzed polyketone synthesis. *Dalton Trans.* **42,** 14583–14602 (2013).
21. Gadgea, S. T. & Bhanage, B. M. Recent developments in palladium catalysed carbonylation reactions. *RSC Adv.* **4,** 10367–10389 (2014).
22. Ragaini, F. *et al.* Phenanthroline catalyzed carbonylation of nitroarenes and amines: palladium–carbonyl intermediates and bifunctional effects. *Chem. Eur. J.* **15,** 8064–8077 (2009).
23. Gabriele, B., Mancuso, R. & Salerno, G. Oxidative carbonylation as a powerful tool for the direct synthesis of carbonylated heterocycles. *Eur. J. Org. Chem.* **2012,** 6825–6839 (2012).
24. Della Ca, N. *et al.* Palladium-catalyzed synthesis of symmetrical urea derivatives by oxidative carbonylation of primary amines in carbon dioxide medium. *J. Catal.* **282,** 120–127 (2011).
25. Pri-Bar, I. & Alper, H. Oxidative coupling of amines and carbon monoxide catalyzed by palladium complexes. Mono- and double carbonylation reactions promoted by iodine compounds. *Can. J. Chem.* **68,** 1544–1547 (1990).
26. Beller, M., Fischer, H., Herrmann, W. A., Ofele, K. & Brossmer, C. Palladacycles as efficient catalysts for aryl coupling reactions. *Angew. Chem. Int. Ed.* **34,** 1848–1849 (1995).
27. Korstanje, T. J., Van Der Vlugt, J. I., Elsevier, C. J. & De Bruin, B. Hydrogenation of carboxylic acids with a homogeneous cobalt catalyst. *Science* **350,** 298–302 (2015).
28. Cui, X., Li, Y., Topf, C., Junge, K. & Beller, M. Direct ruthenium-catalyzed hydrogenation of carboxylic acids to alcohols. *Angew. Chem. Int. Ed.* **54,** 10596–10599 (2015).
29. John, J. M. & Bergens, S. H. A highly active catalyst for the hydrogenation of amides to alcohols and amines. *Angew. Chem. Int. Ed.* **50,** 10377–10380 (2011).
30. Zhang, J., Liu, X., Sun, M., Ma, X. & Han, Y. Direct conversion of cellulose to glycolic acid with a phosphomolybdic acid catalyst in a water medium. *ACS Catal.* **2,** 1698–1702 (2012).
31. Li, Y., Topf, C., Cui, X., Junge, K. & Beller, M. Lewis acid promoted ruthenium(II)-catalyzed etherifications by selective hydrogenation of carboxylic acids/esters. *Angew. Chem. Int. Ed.* **54,** 5196–5200 (2015).

32. Bornschein, C. *et al.* Mild and selective hydrogenation of aromatic and aliphatic (di)nitriles with a well-defined iron pincer complex. *Nat. Commun.* **5,** 4111 (2014).
33. Werkmeister, S. *et al.* Hydrogenation of esters to alcohols with a well-defined iron complex. *Angew. Chem. Int. Ed.* **53,** 8722–8726 (2014).
34. Elangovan, S. *et al.* Improved second generation iron pincer complexes for effective ester hydrogenation. *Adv. Synth. Catal.* **358,** 820–825 (2016).
35. Casey, C. P. & Guan, H. An efficient and chemoselective iron catalyst for the hydrogenation of ketones. *J. Am. Chem. Soc.* **129,** 5816–5817 (2007).
36. Zhu, Y. *et al.* The rise of calcination temperature enhances the performance of Cu catalysts: contributions of support. *ACS Catal.* **4,** 3675–3681 (2014).
37. Hiwatari, K. *et al.* Selective oxidative carbonylation of amines to oxamides and ureas catalyzed by pallladium complexes. *Bull. Chem. Soc. Jpn.* **77,** 2237–2250 (2004).
38. De La Fuente, V., Godard, C., Zangrando, E., Claver, C. & Castillón, S. A phosphine-free Pd catalyst for the selective double carbonylation of aryl iodides. *Chem. Commun.* **48,** 1695–1697 (2012).
39. Schranck, J., Wu, X.-F., Tlili, A., Neumann, H. & Beller, M. A novel double carbonylation reaction of aryl halides: selective synthesis of 5-arylfuranones. *Chem. Eur. J.* **19,** 12959–12964 (2013).
40. Aresta, M. *et al.* Synthesis and solid state and solution characterization of mono- and di-($\eta^1$-C) carbamoyl – palladium complexes. New efficient palladium-catalyzed routes to carbamoyl chlorides: key intermediates to isocyanates, carbamic esters, and ureas. *Organometallics* **19,** 3879–3889 (2000).

## Acknowledgements

This work is supported by Chem21, Danish National Research Foundation CADIAC (Carbon Dioxide Activation Centre), the state of Mecklenburg-Vorpommern and the BMBF (Bundesministerium für Bildung und Forschung). We thank Bianca Wendt for her kind support.

## Author contributions

M.B. and Y.L. designed the method. M.B. directed and coordinated the project. Y.L., K.D. and R.S. conducted the experiments and developed the project. S.E. and A.S. prepared and analysed the single crystal of complex **2b**. M.B., Y.L., K.J. and R.J. wrote the manuscript.

## Additional information

**Competing financial interests:** The authors declare no competing financial interests.

# 24

# Orthogonal ring-closing alkyne and olefin metathesis for the synthesis of small GTPase-targeting bicyclic peptides

Philipp M. Cromm[1,2], Sebastian Schaubach[2,3], Jochen Spiegel[1,2], Alois Fürstner[2,3], Tom N. Grossmann[2,4,5] & Herbert Waldmann[1,2]

Bicyclic peptides are promising scaffolds for the development of inhibitors of biological targets that proved intractable by typical small molecules. So far, access to bioactive bicyclic peptide architectures is limited due to a lack of appropriate orthogonal ring-closing reactions. Here, we report chemically orthogonal ring-closing olefin (RCM) and alkyne metathesis (RCAM), which enable an efficient chemo- and regioselective synthesis of complex bicyclic peptide scaffolds with variable macrocycle geometries. We also demonstrate that the formed alkyne macrocycle can be functionalized subsequently. The orthogonal RCM/RCAM system was successfully used to evolve a monocyclic peptide inhibitor of the small GTPase Rab8 into a bicyclic ligand. This modified peptide shows the highest affinity for an activated Rab GTPase that has been reported so far. The RCM/RCAM-based formation of bicyclic peptides provides novel opportunities for the design of bioactive scaffolds suitable for the modulation of challenging protein targets.

[1] Department of Chemical Biology, Max-Planck-Institute of Molecular Physiology, Otto-Hahn-Strasse 11, D-44227 Dortmund, Germany. [2] Technische Universität Dortmund, Fakultät für Chemie and Chemische Biologie, Otto-Hahn-Strasse 6, D-44227 Dortmund, Germany. [3] Max-Planck-Institut für Kohlenforschung, Kaiser-Wilhelm-Platz 1, D-45470 Mülheim/Ruhr, Germany. [4] Chemical Genomics Centre of the Max Planck Society, Otto-Hahn-Strasse 15, D-44227 Dortmund, Germany. [5] Department of Chemistry and Pharmaceutical Sciences, VU University Amsterdam, De Boelelaan 1083, 1081 HV Amsterdam, The Netherlands. Correspondence and requests for materials should be addressed to A.F. (email: fuerstner@kofo.mpg.de) or to T.N.G. (email: tom.grossmann@cgc.mpg.de) or to H.W. (email: herbert.waldmann@mpi-dortmund.mpg.de).

Macrocyclic peptides exhibit unique surface recognition properties and allow the stabilization of bioactive peptide conformations resulting in ligands with increased bioactivity and bioavailability[1–5]. Such scaffolds already proved useful for the modulation of biological targets which are intractable by typical small molecules, such as transcription factors and small GTPases[6–12]. Recently, structurally rigid bicyclic peptides obtained mainly by epitope grafting on disulfide-rich frameworks[13,14] or by phage-display screening[15,16] have emerged as particularly interesting inhibitor types. Given the potential of this new chemical modality as scaffold for next-generation therapeutics, the development of efficient synthetic methods that enable the introduction of non-natural fragments into peptidic bicycles is in high demand. This is particularly true for approaches that allow the design of scaffolds that go beyond the size of small epitopes.

For the synthesis of monocyclic peptides, a variety of methods is available[17–25]. Notably, in such peptides, the crosslink itself can directly contribute to bioactivity[26–28]. In this respect, hydrocarbon crosslinks formed by ruthenium-catalysed ring-closing olefin metathesis (RCM) proved particularly successful owing to the hydrophobic and inert character of those crosslinks[2,29]. Prominent examples involve hydrogen bond surrogates[8,30,31] and hydrocarbon stapling[32–35], which have provided a number of potent inhibitors of protein—protein interactions[2]. In these cases, the synthesis of bicyclic architectures requires the presence of multiple olefins, which causes selectivity problems during cyclization[36,37]. Undesired side reactions can be reduced by the selection of appropriate ring sizes and distances[38], by functional group transformations[39] and by tedious fine tuning of olefin reactivity[39]. Thus, only a small set of scaffolds is accessible by multiple RCM reactions[36–40]. This creates the need for synthesis methods that integrate two consecutive, chemically orthogonal metathesis reactions thereby enabling efficient chemo- and regioselective construction of complex bicyclic peptides. Ideally, such methods would be compatible with solid-phase peptide synthesis (SPPS).

Molybdenum-catalysed ring-closing alkyne metathesis[41] (RCAM) shares many of the advantageous properties of RCM, and is in principle chemically orthogonal to ruthenium-catalysed RCM. RCAM has been applied for peptide macrocyclizations in solution[42–44]. However, the synthesis of bicyclic peptides by orthogonal ring-closing olefin and alkyne metathesis has not been explored so far. Here, we report the solid-phase synthesis of bicyclic peptides by means of orthogonal ring-closing olefin- and alkyne-metathesis reactions. We demonstrate that the alkyne macrocycle can be further functionalized selectively. The orthogonal RCM/RCAM system was successfully used to evolve a monocyclic peptide inhibitor of the small GTPase Rab8 into a bicyclic ligand with increased target affinity.

## Results

**RCAM and functionalization on solid support.** To explore RCAM-based macrocyclization of peptides on solid support (Fig. 1a), two α-methyl-α-alkynyl building blocks (**1–4**) of varying linker length and configuration (Fig. 1b) were introduced into model peptides using Fmoc-based SPPS. Peptide sequences, architectures and relative spacing of non-natural amino acids ($i,i+3$, $i,i+4$ and $i,i+7$) were selected by analogy to previously explored RCM-based peptide macrocyclizations[33]. As proof-of-concept and to test the robustness of the reaction, we designed model peptides that contain all functionalities present among the 20 proteinogenic amino acids and yield macrocyclic peptides **7–9** after RCAM (Fig. 2a). Investigation of various RCAM conditions including the latest generation of stable Mo-complexes[45,46] (Supplementary Table 1) revealed efficient conversions after 3 h at 40 °C in toluene if Tentagel rink amide resin and complex **5** (Fig. 1a) were used (Supplementary Figs 2–5, Supplementary Table 2). Under these conditions, macrocycles were formed for all three architectures (**7–9**) with the best results obtained for an $i,i+4$ geometry and a final crosslink of nine carbon atoms (**8**). Shortening the hydrocarbon bridge from nine to eight carbon atoms reduces the efficiency of the reaction presumably due to increased ring strain (Supplementary Fig. 4).

Selective functionalization of the alkyne linker embedded in the macrocycle was achieved, after treatment of macrocyclic peptide **11** with $CuBr_2$ in dry acetonitrile on solid support to yield dibrominated olefin **12** (Fig. 2b). Notably, the reaction can be performed conveniently with different resin-bound peptides (for all tested architectures: $i,i+3$, $i,i+4$ and $i,i+7$) using standard syringe reactors (Supplementary Figs 6–8, Supplementary Table 2). Full conversion in the dibromination reaction is only achieved after multiple treatments with $CuBr_2$ and for peptides that lack the two N-terminal sulfur containing amino acids (Cys and Met).

**Bicyclic peptide synthesis via orthogonal RCM and RCAM.** To determine whether RCM and RCAM can be performed orthogonally within one peptide sequence (Fig. 3a), peptide **16** was synthesized which embodies two alkyne-functionalized building blocks (**1** and **2**) in $i,i+4$-position at the carboxy (C) terminus, and two olefin-containing amino acids (**6**, Fig. 1b) in $i,i+4$-position at the amino (N) terminus (Fig. 3b). In peptide **16**, an olefin macrocycle can be formed next to an alkyne-bearing macrocycle (Fig. 3b). The treatment of immobilized precursor peptide **13** (blue peak) with either complex **5** or Grubbs first-generation catalyst leads to selective formation of the alkyne (**14**, red peak) and olefin macrocycle (**15**, orange peak), respectively (Fig. 3b, Supplementary Figs 9 and 10). HPLC-MS analyses of the alkyne and olefin crosslinked intermediates (**14** and **15**) reveal highly selective formation of the desired macrocycle without formation of an alternative cyclization product (Supplementary Fig. 10). Both monocycles can be converted into the bicyclic product **16** by means of the second metathesis reaction. This result is remarkable since previous attempts of orthogonal macrocycle formation within peptides failed[44], but were successful only for the assembly of simple building blocks[47]. In an even more demanding set-up, the simultaneous closure of both macrocycles in a one-pot reaction was tested (instead of the previous sequential synthesis). Strikingly, treatment of the open peptide precursor **13** with a mixture of complex **5** and Grubbs first-generation catalyst also yields the desired bicyclic peptide **16** (green peak, Fig. 3b and Supplementary Fig. 10).

In peptide **16**, the two individual macrocycles are sequentially arranged along the amino acid chain, that is, the two individual macrocycles are linked by a linear amino acid sequence. A synthetically more challenging setup involves the synthesis of two entangled macrocyles resulting in more constrained peptide scaffolds (for example, peptide **17**, Fig. 4). In this architecture, an edge-on bimacrocycle structure is generated as opposed to a linear macrocycle arrangement as in peptide **16**. Most notably, entangled bicyclic peptide **17** (Fig. 4) was also efficiently formed in both the sequential as well as the one-pot synthesis (Supplementary Figs 11 and 12).

**Bicyclic ligands of the small GTPase Rab8.** To demonstrate the potential of this robust orthogonal RCM/RCAM macrocyclization, we aimed at the improvement of a monocyclic bioactive peptide targeting a challenging protein. Small GTPases comprise a protein superfamily with clinically highly relevant, yet

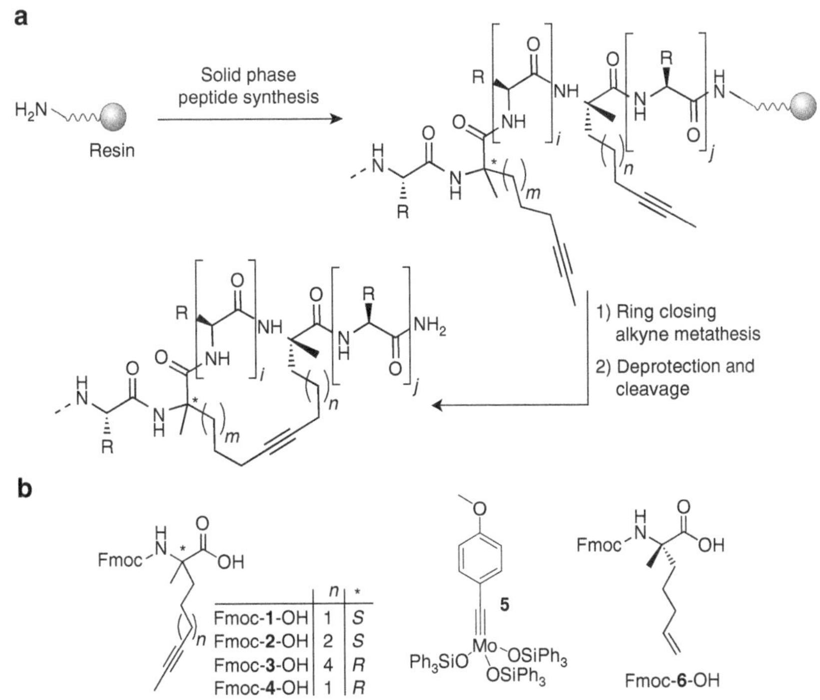


**Figure 1 | Alkyne macrocyclization.** (**a**) The linear peptide is assembled via SPPS including the incorporation of two α-methylated-α-alkynylated building blocks (**1-4**). The C-terminal building block is always (*S*)-configured, the configuration of the N-terminal building block varies between the different architectures. Complex **5** is used to perform the RCAM reaction. ($i = 2, 3, 6$, number of amino acids between non-natural building blocks; $j = 3, 6$; $m = 1, 2, 4$; $n = 1, 2$; R = side chain of a proteinogenic amino acid) (**b**) Fluorenylmethoxycarbonyl (Fmoc) protected non-natural amino acids incorporated into the peptide sequence (alkyne: **1-4**, olefin: **6**). The alkyne building blocks **1-4** are either used in the (*S*)- or (*R*)-configuration depending on the macrocycle architecture. Mo-complex (**5**) used for RCAM.

particularly challenging, drug targets[48–54]. Despite enormous efforts, no efficient inhibitors of small GTPases have reached clinical trials. Importantly, the target affinity of small molecule modulators typically does not exceed the low-to-medium micromolar range[50,51]. Among the small GTPase superfamily, Rab proteins (Ras-related in brain) constitute key regulators of intracellular vesicular transport and trafficking[55,56]. As starting point for the generation of a bicyclic peptide, we selected the hydrocarbon-stapled monocyclic peptide **StRIP3,** which binds the small GTPase Rab8a with moderate affinity (dissociation constant ($K_d$) = 20.7 μM, Table 1, entry 2)[12]. **StRIP3** resembles the only known inhibitor of a Rab protein–protein interaction[12] and is based on the interaction motif of Rab6-interacting protein 1 (**wt-R6IP**, Table 1, entry 1)[57]. Initially, alkyne-bearing macrocycles based on **StRIP3** were explored in which the $i,i+4$ olefin crosslink was replaced by alkynes with varying crosslink length (9–10 carbon atoms, Table 1, entry 3–5). The 10-carbon crosslink requires double incorporation of building block **2** (entry 3). A nine-carbon crosslink is generated by incorporation of two different building blocks (**1** and **2**), which results in two different architectures (entry 4: **1/2**, entry 5: **2/1**). The synthesis of an eight-carbon crosslinked **StRIP3** derivative containing building block **1** twice was not possible, most likely due to high ring strain caused by the linear geometry of the triple bond.

In addition, dibrominated (entry 6–8) and bicyclic peptides (entry 9–14) were synthesized resulting in a total of 12 **StRIP3** derivatives grouped into four subfamilies (Table 1, Supplementary Table 3): (i) alkyne mono-macrocyclic peptides **18–20** (entry 3–5); (ii) dibrominated olefin macrocyclic peptides **21–23** (entry 6–8); (iii) orthogonally macrocyclized peptides **24–26** carrying the original olefin crosslink and an additional alkyne crosslink at the C terminus (entry 9–11); and (iv) bicyclic peptides **27–29** with exchanged positions for the alkyne and the olefin crosslink (entry 12–14). Since the N-terminal part of parent peptide **StRIP3** is already constrained by the olefin macrocycle, we aimed for the introduction of a new macrocycle in the C-terminal part. We reasoned that additional constraint could further stabilize the bioactive peptide conformation. Owing to a lack of structural information, it is not obvious which amino acids are directly involved in Rab-binding. For this reason, we selected two amino acids with hydrophobic side chains (L911 and A915) for macrocycle introduction as their non-polar side chains are potentially mimicked by the hydrocarbon macrocycle. All the peptides were synthesized via SPPS and modified with an N-terminal fluorescein–polyethyleneglycol label (Supplementary Table 3) to enable determination of their binding affinity towards activated Rab8a$_{6-176}$(GppNHp) in a fluorescence polarization (FP) assay (Supplementary Fig. 13). After initial ranking of the peptides by means of relative $K_d$ values (rel. $K_d$, Table 1, Supplementary Table 4), the affinity of the best binders (peptide **21**, **25** and **28**) was determined in an independent FP assay run in triplicates (Table 1, Supplementary Fig. 14). Replacement of the olefin by an alkyne crosslink yields peptides **18–20** with affinities comparable to **StRIP3**. In contrast, the dibrominated olefin derivatives **21–23** show improved affinity towards Rab8a$_{6-176}$ with peptide **21** being the most potent binder within this subfamily ($K_d$ = 10.7 μM, Table 1). Peptide **21** shows a 2-fold increased binding affinity when compared with **StRIP3**. Notably, an even higher improvement in binding affinity to Rab8a$_{6-176}$(GppNHp) is observed for two of the bicyclic peptides **24–29**, namely peptide **25** and **28** (Table 1 and Supplementary Table 4). In both the cases, the nine-carbon alkyne crosslink

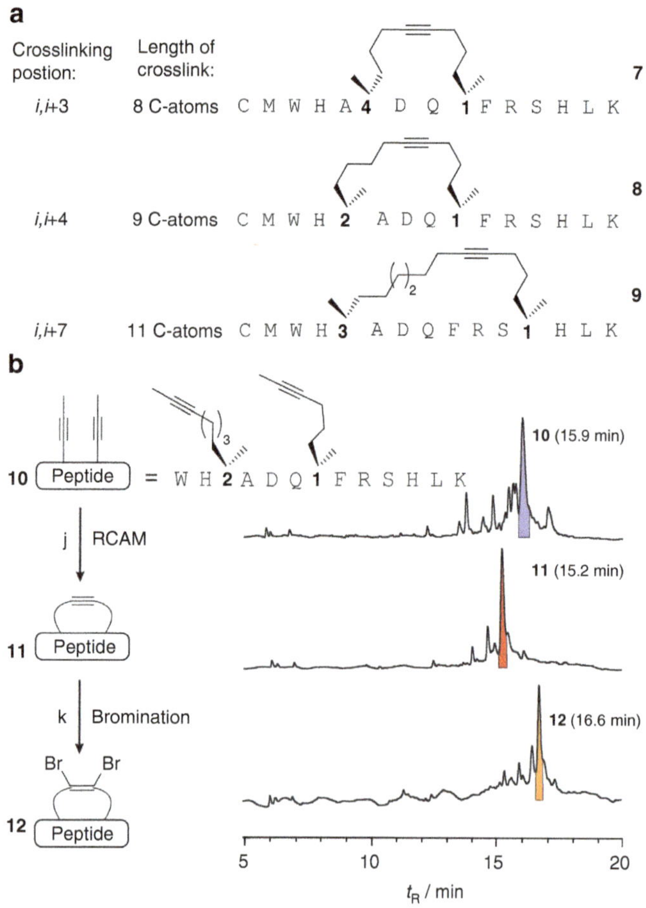

**Figure 2 | Ring-closing alkyne metathesis on solid support.** (**a**) Sequences of peptides **7-9** containing different macrocylic architectures. (**b**) Sequence of peptide **10** with non-natural amino acids **1** and **2** at position *i* and *i* + 4 (nine-carbon crosslink) with corresponding chromatograms of crude reaction mixtures before (**10**, top) and after RCAM (**11**, middle) and after dibromination (**12**, bottom). Corresponding product peaks are highlighted: open (**10**, blue), closed (**11**, red) and dibrominated (**12**, orange) macrocycle. Chromatograms were obtained after deprotection and release of intermediates from the resin. 'j' represents Complex **5**, dry toluene, 40 °C, 2 × 1.5 h; 'k' represents $CuBr_2$, dry MeCN, 3 × 1 h.

(with **1** at N-terminal and **2** at C-terminal position within the sequence) provides the most potent architecture resulting in two significantly improved ligands for activated $Rab8a_{6-176}$ ($K_d$[**25**] = 6.6 μM; $K_d$[**28**] = 9.6 μM). Bicyclic peptide **25** (Fig. 5a) is more than three times more potent than the parent hydrocarbon stapled peptide **StRIP3** and displays a more than 15-fold increased binding affinity compared with the unmodified wild-type peptide **wt-R6IP**. Binding affinity of peptide **25** was confirmed in microscale thermophoresis measurements. On the basis of fluorescence intensity, an affinity for $Rab8a_{6-176}$(GppNHp) was observed ($K_d$[**25**] = 11 μM, Supplementary Table 5, Supplementary Figs 15 and 16), which is in the range of our FP measurements ($K_d$[**25**] = 6.6 μM, see above). In addition, FP competition experiments were performed using a complex between labelled peptide **25** and $Rab8a_{6-176}$(GppNHp), which was treated with an excess of acetylated **StRIP3**. In this setup, we observed full displacement of peptide **25** ($IC_{50}$ = 33 μM, red Fig. 5b). As one would expect, the acetylated low-affinity peptide **wt-R6IP** does not compete with peptide **25** (black, Fig. 5b). These results verify reversible binding of peptide **25** to the same site on Rab8 as parent peptide **StRIP3**.

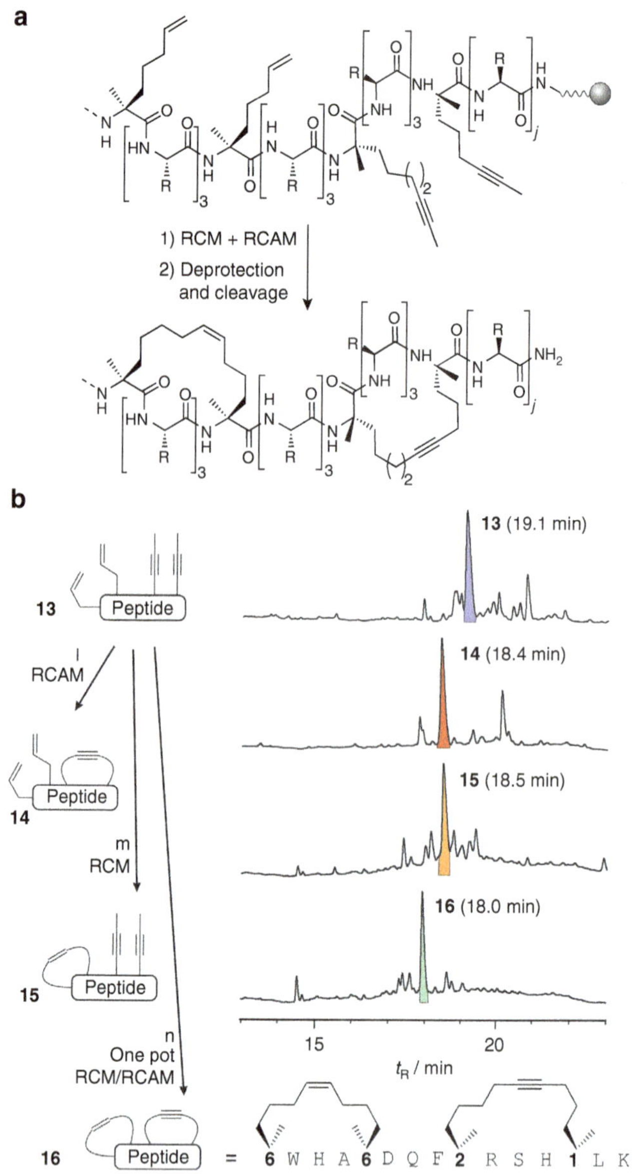

**Figure 3 | Solid phase synthesis of bicyclic peptides by means of the RCM/RCAM method.** (**a**) General scheme for the synthesis of bicyclic peptides obtained by means of RCM and RCAM of the acyclic precursor peptides. (*j* = 2, number of C-terminal amino acids; R = side chain of a proteinogenic amino acid except Cys or Met). (**b**) Sequence of bicyclic test peptide **16** bearing an *i*,*i* + 4 olefin crosslink (eight C-atoms) and an *i*,*i* + 4 alkyne crosslink (nine C-atoms). Chromatograms of crude reaction mixtures of peptide **16** before macrocyclization (**13**, top), after RCAM (**14**, second) and RCM (**15**, third), respectively, and after simultaneous (one-pot) RCM and RCAM (**16**, bottom). Corresponding product peaks are highlighted: fully open (**13**, blue), alkyne monocycle (**14**, red), olefin monocycle (**15**, orange) and bicyclic peptide (**16**, green). Chromatograms were obtained after deprotection and cleavage of resin-bound intermediates. 'l' represents Complex **5**, dry toluene, 40 °C, 2 × 1.5 h; 'm' represents Grubbs first-generation catalyst, DCE, 3 × 2 h; 'n' represents Complex **5**, Grubbs first-generation catalyst, dry toluene, 40 °C, 2 × 1.5 h.

## Discussion

We identified conditions that enable the performance of RCAM reactions in conjunction with SPPS allowing alkyne-based macrocyclization of peptide sequences involving all natural side-chain functionalities. Subsequent functionalization of the alkyne allows further modification of the macrocycle, which opens new perspectives in the design of macrocyclic scaffolds. In addition, we report the chemo- and regioselective synthesis of bicyclic peptides bearing an alkyne as well as an olefin crosslink accessible via orthogonal ring-closing olefin and alkyne metathesis on solid support. This approach allows direct control of the individual macrocyclization reaction and enables the formation of bicyclic peptides with novel architectures combining two different macrocycles within the same peptide sequence. The applicability of such scaffolds for highly challenging targets, such as small GTPases, was demonstrated via identification of the currently most potent binder of an activated Rab GTPase (peptide **25**, $K_d = 6.6\,\mu M$). Since the bioactivity of a peptide is mainly determined by its secondary structure[1,2], alkyne macrocyclization and the orthogonal introduction of bicyclic alkyne/olefin macrocycles within the same peptide sequence give rise to novel constrained peptide architectures with high potential for the targeting of currently intractable proteins.

17 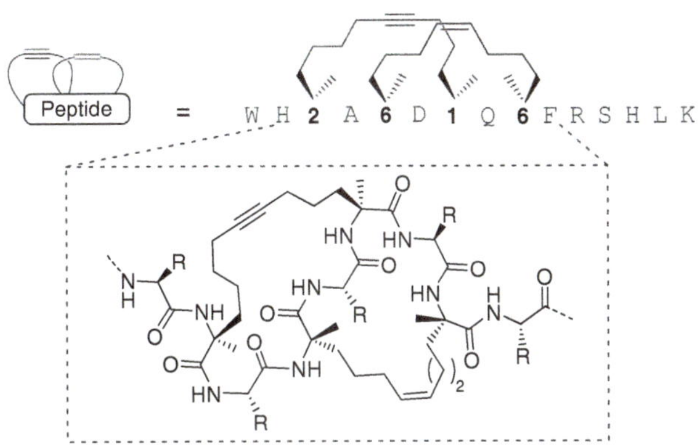


**Figure 4 | Sequence and chemical structure of the engulfed bicyclic peptide 17.** R, side chain of a proteinogenic amino acid except Cys or Met.

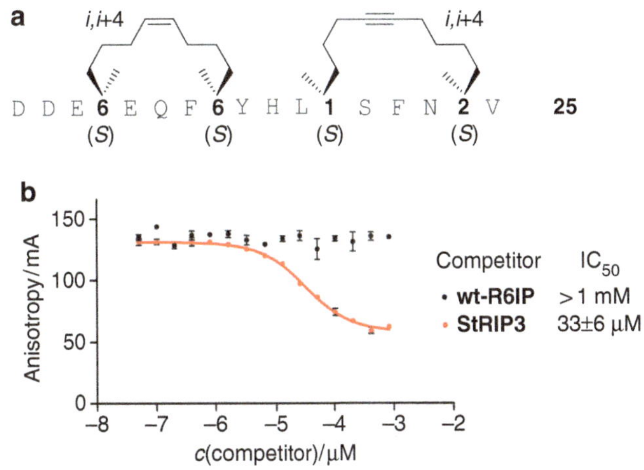


**Figure 5 | Sequence and binding studies of peptide 25.** (**a**) Sequence of bicyclic peptide **25** showing highest affinity for Rab8a$_{6-176}$(GppNHp). (**b**) Competition of fluorescein-labelled peptide **25** (60 nM) bound to Rab8a(GppNHp; 15 μM) with increasing concentrations of acetylated peptides **StRIP3** and **wt-R6IP** (competitors). Errors represent 1σ of triplicates.

## Methods

**General.** For abbreviations and detailed information about the experimental procedures, analytical data and FP binding curves, see Supplementary Figs 1–16, Supplementary Tables 1–5, Supplementary Note 1 and Supplementary Methods.

**Synthesis of building blocks 1-4.** Synthesis of the Fmoc protected building blocks 1–4 was performed according to adapted protocols using Ni(II)-BPB ((*R/S*)-2-[N-(N′-benzylpropyl)amino]benzophenone) complexes[58,59]. For a detailed description of building block synthesis and analytical data, see Supplementary Methods.

**Peptide synthesis.** Peptides were synthesized according to standard Fmoc-chemistry for SPPS using HCTU (*O*-(6-chlorobenzotriazol-1-yl)-*N*,*N*,*N*′,

**Table 1 | Small GTPase targeting peptides.**

900 905 910 915
D D E [K] E Q F [L] Y H L [L] S F N [A] V **wtR6IP**

| Entry | Peptide | aa 903 | aa 907 | aa 911 | aa 915 | Rel. $K_d$ | $K_d$ (μM) |
|---|---|---|---|---|---|---|---|
| 1 | **wt-R6IP** | K | L | L | A | ND | >100 |
| 2 | **StRIP3** | 6 | 6 | L | A | 1.00 | 20.7 ± 0.7 |
| 3 | **18** | 2 | 2 | L | A | 0.92 | ND |
| 4 | **19** | 1 | 2 | L | A | 0.99 | ND |
| 5 | **20** | 2 | 1 | L | A | 0.89 | ND |
| 6 | **21*** | 2 | 2 | L | A | 0.40 | 10.7 ± 0.5*** |
| 7 | **22*** | 1 | 2 | L | A | 0.46 | ND |
| 8 | **23*** | 2 | 1 | L | A | 0.52 | ND |
| 9 | **24** | 6 | 6 | 2 | 2 | 0.43 | ND |
| 10 | **25** | 6 | 6 | 1 | 2 | 0.27 | 6.6 ± 0.4*** |
| 11 | **26** | 6 | 6 | 2 | 1 | 0.44 | ND |
| 12 | **27** | 2 | 2 | 6 | 6 | 0.44 | ND |
| 13 | **28** | 1 | 2 | 6 | 6 | 0.42 | 9.6 ± 0.4*** |
| 14 | **29** | 2 | 1 | 6 | 6 | 0.56 | ND |

ND, not determined. Wild-type sequence of linear starting sequence **wtRIP** (aa 900-916 of Rab6 interacting protein) with table of mono- and bicyclic peptide derivatives. Peptides were synthesized using SPPS, macrocyclized and N-terminally labelled with a fluorescein-polyethylene glycol conjugate. Affinities towards Rab8a$_{6-176}$(GppNHp) were determined via fluorescence polarization (FP) assays (rel. $K_d = K_d$ [Peptide]/$K_d$ [StRIP3] (these $K_d$-values are single point measurements)). Given $K_d$-values are average of triplicate measurements (errors account for 1σ). ***$P<0.001$.
*Dibrominated olefin.

N′-tetramethyluronium hexafluorophosphate) and COMU (1-[(1-(cyano-2-ethoxy-2-oxoethylidenaminooxy)-dimethylaminomorpholino)]-uronium hexafluorophosphate) as coupling reagents (Supplementary Fig. 1). For more detailed information about peptide synthesis, see Supplementary Methods and Supplementary Tables 2 and 3.

**Ring-closing alkyne metathesis.** The dried resin was swollen and shrunken under argon alternating in dry diethyl ether and dry toluene (3 × each). Afterwards 0.5 ml of a solution of the alkyne-metathesis complex **5** (2 mg $ml^{-1}$) in dry toluene was added and the reaction mixture was stirred at 40 °C for 1.5 h. During the reaction time, argon was bubbled through the reaction mixture to evaporate the 2-butyne. After addition of 0.5 ml of fresh complex **5** solution, the mixture was stirred at 40 °C for 1.5 h.

**Ring-closing olefin metathesis.** The dried resin was swollen in 1,2-dichloroethane (DCE) for 15 min. Subsequently, 0.5 ml of a solution of Grubbs first-generation catalyst (2 mg $ml^{-1}$) in DCE was added to the resin and reacted for 2 h at room temperature. During the reaction time, argon was bubbled through the reaction mixture to remove ethene. The procedure was repeated twice.

**One-pot ring-closing alkyne and olefin metathesis.** The dried resin was swollen and shrunken under argon alternating in dry diethyl ether and dry toluene (3 × each). Afterwards 0.5 ml of a solution of the alkyne-metathesis complex **5** (2 mg $ml^{-1}$) and Grubbs first-generation catalyst (2 mg $ml^{-1}$) in dry toluene was added and the reaction mixture stirred at 40 °C for 1.5 h. During the reaction time, argon was bubbled through the reaction mixture to evaporate 2-butyne and ethene. After the addition of 0.5 ml of fresh complex solution (alkyne complex **5** and Grubbs first-generation catalyst), the mixture was stirred at 40 °C for 1.5 h.

**Dibromination of alkyne macrocycles.** The dried resin was swollen in dry MeCN for 15 min and treated with a mixture of $CuBr_2$ in dry MeCN (2 mg $ml^{-1}$) for 2 h. The reaction was performed in a Syringe reactor and the procedure was repeated twice.

**Protein expression and purification.** The expression and purification of $Rab8a_{6-176}$ was performed by analogy to full-length Rab8a according to established protocols[60,61].

**Nucleotide exchange.** Nucleotide exchange was performed according to previously established protocols[60,62]. Briefly, for nucleotide removal, $Mg^{2+}$ was removed by the addition of a 5-fold excess of EDTA and reacted for 1 h at room temperature. The protein solution was desalted using a PD-10 desalting column Sephadex G-25 DNA Grade (GE Healthcare) with elution buffer consisting of 20 mM HEPES (pH 7.5), 50 mM NaCl, 1 mM TCEP. After removal of $Mg^{2+}$, the protein was diluted to 80–100 μM before the addition of $ZnCl_2$ (500 μM) and $(NH_4)_2SO_4$ (200 mM). After the addition of alkaline phosphatase (5 U $mg^{-1}$ Rab protein), the mixture was incubated for 16 h at 4 °C. For nucleotide exchange, the mixture contained a 5-fold excess of GppNHp during alkaline phosphatase incubation. Afterwards, the mixture was desalted using a PD-10 desalting column Sephadex G-25 DNA Grade (GE Healthcare) with elution buffer consisting of 25 mM HEPES (pH 7.5) 150 mM NaCl, 1 mM TCEP, 1 mM $MgCl_2$ and 1 μM GppNHp.

**Fluorescence polarization assay.** $Rab8a_{6-176}$(GppNHp) was serially diluted in a buffer containing 25 mM HEPES (pH 7.5), 150 mM NaCl, 1 mM $MgCl_2$, 1 mM TCEP, 0.01% Tween 20 and 1 μM GppNHp, treated with 66 nM fluorescein-labelled peptides and incubated for 4 h at room temperature. Fluorescence polarization values ($\lambda_{ex} = 470$ nm, $\lambda_{em} = 525$ nm) were determined at room temperature. Initial studies for alkyne macrocyclized peptides were performed as single measurements. Final affinity measurements of a subset of peptides were performed in triplicates. After correction for changes in fluorescence intensity upon binding, the fluorescence anisotoropy data were converted into fraction bound of the FITC-labelled peptide and fitted to a one-site binding model derived from the law of mass action using $K_d$ as the only fitting parameter (for details, see Supplementary Methods)[63]. Nonlinear regression was performed in Prism 5.0 (Graphpad)[64].

**Competition fluorescence polarization assay.** Acetylated peptides were serially diluted and incubated with a mixture of the fluorescein-labelled peptide and $Rab8a_{6-176}$ (GppNHp) at room temperature for 1 h. Fluorescence polarization was determined and $IC_{50}$ values were calculated by nonlinear regression analysis using Prism 5.0 software (GraphPad)[64].

## References


1. Bock, J. E., Gavenonis, J. & Kritzer, J. A. Getting in shape: controlling peptide bioactivity and bioavailability using conformational constraints. *ACS Chem. Biol.* **8,** 488–499 (2013).
2. Pelay-Gimeno, M., Glas, A., Koch, O. & Grossmann, T. N. Structure-based design of inhibitors of protein-protein interactions: mimicking peptide binding epitopes. *Angew. Chem. Int. Ed.* **54,** 8896–8927 (2015).
3. Hill, T. A., Shepherd, N. E., Diness, F. & Fairlie, D. P. Constraining cyclic peptides to mimic protein structure motifs. *Angew. Chem. Int. Ed.* **53,** 13020–13041 (2014).
4. Mas-Moruno, C., Rechenmacher, F. & Kessler, H. Cilengitide: the first anti-angiogenic small molecule drug candidate. design, synthesis and clinical evaluation. *Anticancer Agents Med. Chem.* **10,** 753–768 (2010).
5. Srinivas, N. *et al.* Peptidomimetic antibiotics target outer-membrane biogenesis in *Pseudomonas aeruginosa*. *Science* **327,** 1010–1013 (2010).
6. Leshchiner, E. S. *et al.* Direct inhibition of oncogenic KRAS by hydrocarbon-stapled SOS1 helices. *Proc. Natl Acad. Sci. USA* **112,** 1761–1766 (2015).
7. Moellering, R. E. *et al.* Direct inhibition of the NOTCH transcription factor complex. *Nature* **462,** 182–188 (2009).
8. Patgiri, A., Yadav, K. K., Arora, P. S. & Bar-Sagi, D. An orthosteric inhibitor of the Ras-Sos interaction. *Nat. Chem. Biol.* **7,** 585–587 (2011).
9. Grossmann, T. N. *et al.* Inhibition of oncogenic Wnt signaling through direct targeting of β-catenin. *Proc. Natl Acad. Sci. USA* **109,** 17942–17947 (2012).
10. Upadhyaya, P. *et al.* Inhibition of Ras signaling by blocking Ras-effector interactions with cyclic peptides. *Angew. Chem. Int. Ed.* **54,** 7602–7606 (2015).
11. Giordanetto, F. & Kihlberg, J. Macrocyclic drugs and clinical candidates: what can medicinal chemists learn from their properties? *J. Med. Chem.* **57,** 278–295 (2014).
12. Spiegel, J. *et al.* Direct targeting of Rab-GTPase-effector interactions. *Angew. Chem. Int. Ed.* **53,** 2498–2503 (2014).
13. Northfield, S. E. *et al.* Disulfide-rich macrocyclic peptides as templates in drug design. *Eur. J. Med. Chem.* **77,** 248–257 (2014).
14. Poth, A. G., Chan, L. Y. & Craik, D. J. Cyclotides as grafting frameworks for protein engineering and drug design applications. *Biopolymers* **100,** 480–491 (2013).
15. Heinis, C., Rutherford, T., Freund, S. & Winter, G. Phage-encoded combinatorial chemical libraries based on bicyclic peptides. *Nat. Chem. Biol.* **5,** 502–507 (2009).
16. Baeriswyl, V. & Heinis, C. Polycyclic peptide therapeutics. *Chem. Med. Chem.* **8,** 377–384 (2013).
17. White, C. J. & Yudin, A. K. Contemporary strategies for peptide macrocyclization. *Nat. Chem.* **3,** 509–524 (2011).
18. Montalbetti, C. A. & Falque, V. Amide bond formation and peptide coupling. *Tetrahedron* **61,** 10827–10852 (2005).
19. Góngora-Benítez, M., Tulla-Puche, J. & Albericio, F. Multifaceted roles of disulfide bonds. peptides as therapeutics. *Chem. Rev.* **114,** 901–926 (2014).
20. Assem, N., Ferreira, D. J., Wolan, D. W. & Dawson, P. E. Acetone-linked peptides: a convergent approach for peptide macrocyclization and labeling. *Angew. Chem. Int. Ed.* **54,** 8665–8668 (2015).
21. Wang, Y. & Chou, D. H.-C. A thiol-ene coupling approach to native peptide stapling and macrocyclization. *Angew. Chem. Int. Ed.* **54,** 10931–10934 (2015).
22. Lau, Y. H., Andrade, P., de McKenzie, G. J., Venkitaraman, A. R. & Spring, D. R. Linear aliphatic dialkynes as alternative linkers for double-click stapling of p53-derived peptides. *ChemBioChem.* **15,** 2680–2683 (2014).
23. Lau, Y. H. *et al.* Double strain-promoted macrocyclization for the rapid selection of cell-active stapled peptides. *Angew. Chem. Int. Ed.* **54,** 15410–15413 (2015).
24. Mendive-Tapia, L. *et al.* New peptide architectures through C-H activation stapling between tryptophan-phenylalanine/tyrosine residues. *Nat. Commun.* **6,** 7160 (2015).
25. Lau, Y. H., Andrade, P., de Wu, Y. & Spring, D. R. Peptide stapling techniques based on different macrocyclisation chemistries. *Chem. Soc. Rev.* **44,** 91–102 (2015).
26. Baek, S. *et al.* Structure of the stapled p53 peptide bound to Mdm2. *J. Am. Chem. Soc.* **134,** 103–106 (2012).
27. Phillips, C. *et al.* Design and structure of stapled peptides binding to estrogen receptors. *J. Am. Chem. Soc.* **133,** 9696–9699 (2011).
28. Glas, A. *et al.* Constrained peptides with target-adapted cross-links as inhibitors of a pathogenic protein-protein interaction. *Angew. Chem. Int. Ed.* **53,** 2489–2493 (2014).
29. Pérez de Vega, M. J., García-Aranda, M. I. & González-Muñiz, R. A role for ring-closing metathesis in medicinal chemistry: mimicking secondary architectures in bioactive peptides. *Med. Res. Rev.* **31,** 677–715 (2011).
30. Wang, D., Chen, K., Kulp, J. L. & Arora, P. S. Evaluation of biologically relevant short alpha-helices stabilized by a main-chain hydrogen-bond surrogate. *J. Am. Chem. Soc.* **128,** 9248–9256 (2006).
31. Mahon, A. B. & Arora, P. S. End-capped α-helices as modulators of protein function. *Drug Discov. Today Technol.* **9,** e57–e62 (2012).

32. Cromm, P. M., Spiegel, J. & Grossmann, T. N. Hydrocarbon stapled peptides as modulators of biological function. *ACS Chem. Biol.* **10,** 1362–1375 (2015).
33. Kim, Y.-W., Grossmann, T. N. & Verdine, G. L. Synthesis of all-hydrocarbon stapled α-helical peptides by ring-closing olefin metathesis. *Nat. Protoc.* **6,** 761–771 (2011).
34. Blackwell, H. E. & Grubbs, R. H. Highly efficient synthesis of covalently cross-linked peptide helices by ring-closing metathesis. *Angew. Chem. Int. Ed.* **37,** 3281–3284 (1998).
35. Schafmeister, C. E., Po, J. & Verdine, G. L. An all-hydrocarbon cross-linking system for enhancing the helicity and metabolic stability of peptides. *J. Am. Chem. Soc.* **122,** 5891–5892 (2000).
36. Ghalit, N., Rijkers, D. T. S., Kemmink, J., Versluis, C. & Liskamp, R. M. J. Pre-organization induced synthesis of a crossed alkene-bridged nisin Z DE-ring mimic by ring-closing metathesis. *Chem. Commun.* 192–194 (2005).
37. Slootweg, J. C., Kemmink, J., Liskamp, R. M. J. & Rijkers, D. T. Synthesis and structural characterization of the individual diastereoisomers of a cross-stapled alkene-bridged nisin DE-ring mimic. *Org. Biomol. Chem.* **11,** 7486–7496 (2013).
38. Bird, G. H. *et al.* Hydrocarbon double-stapling remedies the proteolytic instability of a lengthy peptide therapeutic. *Proc. Natl Acad. Sci. USA* **107,** 14093–14098 (2010).
39. Robinson, A. J. *et al.* Regioselective formation of interlocked dicarba bridges in naturally occurring cyclic peptide toxins using olefin metathesis. *Chem. Commun.* **28,** 4293–4295 (2009).
40. Hilinski, G. J. *et al.* Stitched α-helical peptides via bis ring-closing metathesis. *J. Am. Chem. Soc.* **136,** 12314–12322 (2014).
41. Fürstner, A. Alkyne metathesis on the rise. *Angew. Chem. Int. Ed.* **52,** 2794–2819 (2013).
42. Aguilera, B. *et al.* Synthesis of diaminosuberic acid derivatives via ring-closing alkyne metathesis. *J. Org. Chem.* **66,** 3584–3589 (2001).
43. IJsselstijn, M. *et al.* Ring-closing alkyne metathesis mediated synthesis of cyclic β-turn mimetics. *Tetrahedron Lett.* **45,** 4379–4382 (2004).
44. Ghalit, N., Poot, A. J., Fürstner, A., Rijkers, D. T. S. & Liskamp, R. M. J. Ring-closing alkyne metathesis approach toward the synthesis of alkyne mimics of thioether A-, B-, C-, and DE-ring systems of the lantibiotic nisin Z. *Org. Lett.* **7,** 2961–2964 (2005).
45. Heppekausen, J. *et al.* Optimized synthesis, structural investigations, ligand tuning and synthetic evaluation of silyloxy-based alkyne metathesis catalysts. *Chem. Eur. J.* **18,** 10281–10299 (2012).
46. Heppekausen, J., Stade, R., Goddard, R. & Fürstner, A. Practical new silyloxy-based alkyne metathesis catalysts with optimized activity and selectivity profiles. *J. Am. Chem. Soc.* **132,** 11045–11057 (2010).
47. Burnley, J., Jackson, W. R. & Robinson, A. J. One-pot selective homodimerization/hydrogenation strategy for sequential dicarba bridge formation. *J. Org. Chem.* **80,** 9057–9063 (2015).
48. Wennerberg, K., Rossman, K. L. & Der, C. J. The Ras superfamily at a glance. *J. Cell Sci.* **118,** 843–846 (2005).
49. McCormick, F. KRAS as a therapeutic target. *Clin. Cancer Res.* **21,** 1797–1801 (2015).
50. Spiegel, J., Cromm, P. M., Zimmermann, G., Grossmann, T. N. & Waldmann, H. Small-molecule modulation of Ras signaling. *Nat. Chem. Biol.* **10,** 613–622 (2014).
51. Cromm, P. M., Spiegel, J., Grossmann, T. N. & Waldmann, H. Direct modulation of small GTPase activity and function. *Angew. Chem. Int. Ed.* **54,** 13516–13537 (2015).
52. Stephen, A. G., Esposito, D., Bagni, R. K. & McCormick, F. Dragging ras back in the ring. *Cancer Cell* **25,** 272–281 (2014).
53. Ostrem, J. M., Peters, U., Sos, M. L., Wells, J. A. & Shokat, K. M. K-Ras(G12C) inhibitors allosterically control GTP affinity and effector interactions. *Nature* **503,** 548–551 (2013).
54. Agola, J. O. *et al.* A competitive nucleotide binding inhibitor: *in vitro* characterization of Rab7 GTPase inhibition. *ACS Chem. Biol.* **7,** 1095–1108 (2012).
55. Bhuin, T. & Roy, J. K. Rab proteins: the key regulators of intracellular vesicle transport. *Exp. Cell Res.* **328,** 1–19 (2014).
56. Hutagalung, A. H. & Novick, P. J. Role of Rab GTPases in membrane traffic and cell physiology. *Physiol. Rev.* **91,** 119–149 (2011).
57. Recacha, R. *et al.* Structural basis for recruitment of Rab6-interacting protein 1 to Golgi via a RUN domain. *Structure* **17,** 21–30 (2009).
58. Belokon', Y. N., Tararov, V. I., Maleev, V. I., Savel'eva, T. F. & Ryzhov, M. G. Improved procedures for the synthesis of (S)-2-[N-(N'-benzylprolyl)amino] benzophenone (BPB) and Ni(II) complexes of Schiff's bases derived from BPB and amino acids. *Tetrahedron* **9,** 4249–4252 (1998).
59. Bird, G. H., Crannell, W. C. & Walensky, L. D. Chemical synthesis of hydrocarbon-stapled peptides for protein interaction research and therapeutic targeting. *Curr. Protoc. Chem. Biol.* **3,** 99–117 (2011).
60. Hou, X. *et al.* A structural basis for Lowe syndrome caused by mutations in the Rab-binding domain of OCRL1. *EMBO J.* **30,** 1659–1670 (2011).
61. Bleimling, N., Alexandrov, K., Goody, R. & Itzen, A. Chaperone-assisted production of active human Rab8A GTPase in Escherichia coli. *Protein Expr. Purif.* **65,** 190–195 (2009).
62. Simon, I., Zerial, M. & Goody, R. S. Kinetics of interaction of Rab5 and Rab7 with nucleotides and magnesium ions. *J. Biol. Chem.* **271,** 20470–20478 (1996).
63. Huang, X. & Aulabaugh, A. Application of fluorescence polarization in HTS assays. *Methods Mol. Biol.* **565,** 127–143 (2009).
64. Motulsky, H. & Christopoulos, A. *Fitting Models to Biological Data Using Linear and Nonlinear Regression. A Practical Guide to Curve Fitting* (Oxford Univ. Press, 2004).

## Acknowledgements

We thank N. Bleiming for protein expression/purification and technical assistance. P.M.C. is grateful to the Studienstiftung des Deutschen Volkes for a Fellowship. S.S. and J.S. acknowledge financial support by the Fonds der Chemischen Industrie. T.N.G. thanks the Deutsche Forschungsgemeinschaft (Emmy Noether program GR3592/2-1), AstraZeneca, Bayer CropScience, Bayer HealthCare, Boehringer Ingelheim, Merck KGaA and the Max-Planck Society for their support.

## Author contributions

P.M.C. designed and carried out the building block synthesis, peptide synthesis, as well as FP assays and analysed the data. S.S. supported alkyne-metathesis reactions and analysed the data. J.S. supported the design and performance of experiments. A.F., T.N.G. and H.W. designed the experiments and supervised the project. All the authors discussed the results and commented on the manuscript. P.M.C., T.N.G. and H.W. wrote the manuscript.

## Additional information

**Competing financial interests:** The authors declare no competing financial interests.

# 25

# Total synthesis of atropurpuran

Jing Gong[1,*], Huan Chen[1,*], Xiao-Yu Liu[1,*], Zhi-Xiu Wang[1], Wei Nie[1] & Yong Qin[1]

Due to their architectural intricacy and biological significance, the synthesis of polycyclic diterpenes and their biogenetically related alkaloids have been the subject of considerable interest over the last few decades, with progress including the impressive synthesis of several elusive targets. Despite tremendous efforts, conquering the unique structural types of this large natural product family remains a long-term challenge. The arcutane diterpenes and related alkaloids, bearing a congested tetracyclo[5.3.3.$0^{4,9}$.$0^{4,12}$]tridecane unit, are included in these unsolved enigmas. Here we report a concise approach to the construction of the core structure of these molecules and the first total synthesis of (±)-atropurpuran. Pivotal features of the synthesis include an oxidative dearomatization/intramolecular Diels-Alder cycloaddition cascade, sequential aldol and ketyl-olefin cyclizations to assemble the highly caged framework, and a chemoselective and stereoselective reduction to install the requisite allylic hydroxyl group in the target molecule.

[1] Key Laboratory of Drug Targeting and Drug Delivery Systems of the Ministry of Education, Department of Chemistry of Medicinal Natural Products, West China School of Pharmacy, and State Key Laboratory of Biotherapy, Sichuan University, No. 17, Section 3, Renmin Nan Road, Chengdu 610041, China. * These authors contributed equally to this work. Correspondence and requests for materials should be addressed to Y.Q. (email: yongqin@scu.edu.cn).

Three-dimensionally complex and highly cage-like molecules represent some of the amazing natural products that provide both enormous benefits and new challenges to mankind[1]. For instance, the polycyclic diterpenes and their alkaloidal congeners, both featuring difficult chemical structures and possible medicinal applications, fall into such a category[2–4]. From a biogenetic viewpoint, the formation of structurally diverse diterpenes relies on cyclization and oxidation stages, starting from geranylgeranyl diphosphate[5,6], and the introduction of nitrogen atoms into the well-established diterpene skeletons affords the diterpenoid alkaloids[7,8]. While the genera of *Aconitum* and *Delphinium* provided rich sources for various types of these pseudo-alkaloids[7], only a few of their parent diterpene molecules were isolated from nature. As illustrated in Fig. 1a, identification of the atisane-, hetidane- and arcutane-type diterpenes (**1–3**, Fig. 1a) belonged to such exceptional cases[9–11] in which the corresponding natural alkaloids (**4–6**) were isolated[12–16]. All these structural types were biosynthetically derived from atisane diterpenes and are closely related to each other. Specifically, based on the atisane/atisine cores, C14–C20 bond construction afforded hetidanes/hetidines[17] and further bond rearrangement (C10–C20 → C5–C20) gave rise to the arcutane/arcutine frameworks[17,18].

Over the past decade, the synthetic community has made great efforts in pursuit of the total synthesis of atisane diterpenes and their related diterpenoid alkaloids with some notable successes[2–4,19–28]. The teams of Carreira and Liu reported creative solutions for the total synthesis of a 3,4-seco atisane diterpene (crotogoudin, 7, Fig. 1b)[23,26], respectively. Baran and co-workers disclosed a unified access to methyl atisenoate (**1**), isoatisine (**8**) and the hetidine skeleton from a common intermediate[24]. Additionally, the Fukuyama group contributed to the first total synthesis of a denudatine alkaloid (lepenine, **9**)[25], while Gin *et al* realized a refined synthesis of the hetisine-type alkaloid nominine (**10**)[19]. Very recently, an elegant approach to weisaconitine D (**11**) and liljestrandinine (**12**) was developed by Sarpong and co-workers employing a network analysis strategy[27], which has enabled this area to reach new technical heights. Despite these advances, the total synthesis of the unique arcutane/arcutine group (for example, atropurpuran **3** and arcutinine **6**) has been impeded by the complexity of the cage-like tetracyclo[5.3.3.0$^{4,9}$.0$^{4,12}$]tridecane unit (blue bonds shown in Fig. 1a). Aside from the bridged ring systems, the challenging task was further complicated by the presence of multiple stereogenic centres embedded in the molecules, including two all-carbon bridgehead quaternary centres (C5 and C8). These natural products have been the focus of many synthetic studies[29–32] due not only to their alluring structures, but also because their biological functions are unknown. Here, we describe our recent endeavours that led to the first total synthesis of (±)-atropurpuran.

## Results

**Retrosynthetic analysis**. Our retrosynthetic analysis is outlined in Fig. 2. The target molecule atropurpuran (**3**) could be accessed through functional group manipulations from an advanced intermediate **13**. Central to the synthesis of **3** would be the assembly of its core structure (that is, **13**), which can be realized from keto aldehyde **14** via a sequence of aldol addition and ketyl-olefin cyclization[33]. Construction of the C5–C20 bond before the C9–C10 bond would be critical for delivering the right stereochemical outcome at C9. We speculated that the precursor **14** could in turn be obtained by a reductive Knoevenagel

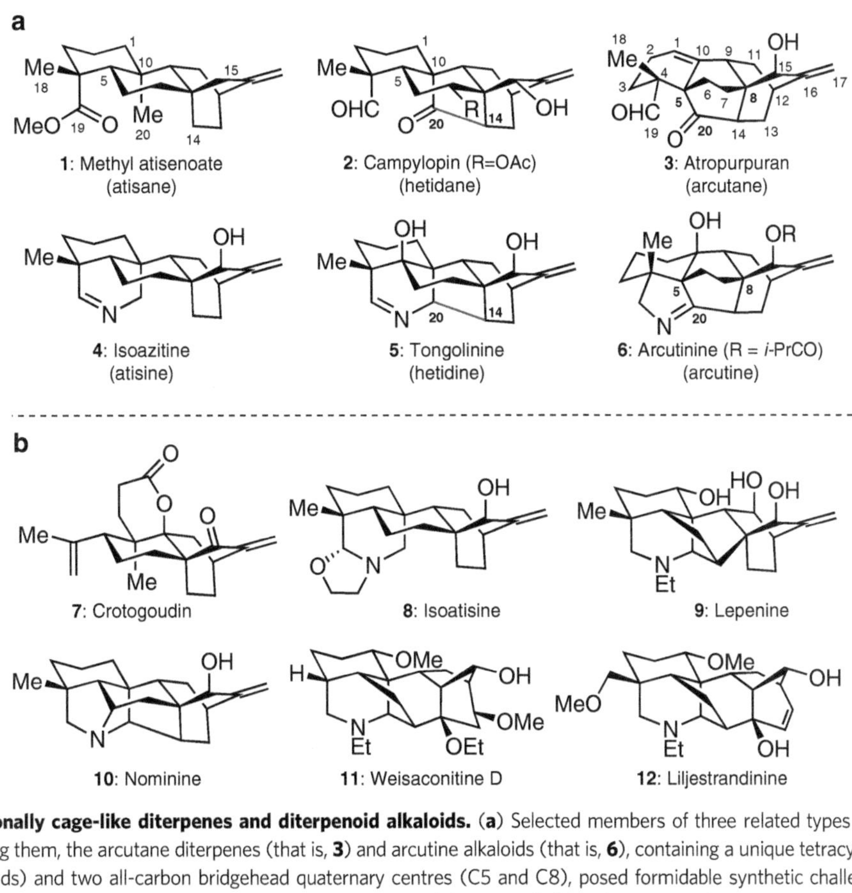


**Figure 1 | Three-dimensionally cage-like diterpenes and diterpenoid alkaloids.** (**a**) Selected members of three related types of diterpenes and their alkaloidal congeners. Among them, the arcutane diterpenes (that is, **3**) and arcutine alkaloids (that is, **6**), containing a unique tetracyclo[5.3.3.0$^{4,9}$.0$^{4,12}$]tridecane moiety (shown in blue bonds) and two all-carbon bridgehead quaternary centres (C5 and C8), posed formidable synthetic challenges. (**b**) Recent accomplishments in the total synthesis of complex atisane-type diterpenes and diterpenoid alkaloids. Ac, acetyl; *i*-Pr, isopropyl.

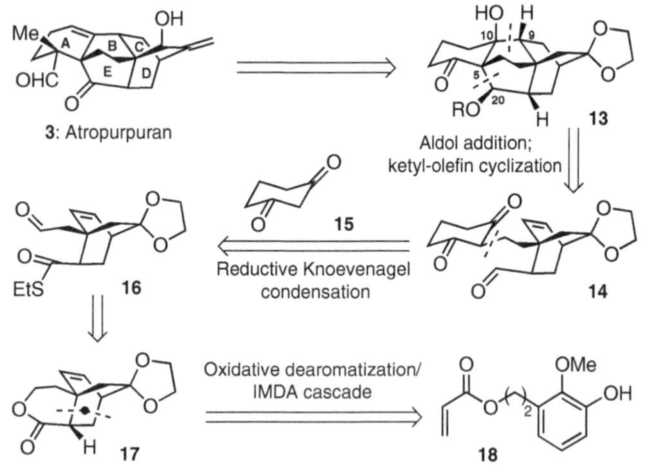


3: Atropurpuran

**Figure 2 | Retrosynthetic analysis of atropurpuran (3).** The key disconnections involve an aldol reaction and a ketyl-olefin cyclization to construct rings E and B, respectively, at a late stage.

condensation[34] between 1,3-cyclohexanedione (**15**) and aldehyde **16**. Generation of tricycle **17** (the precursor of **16**) would rely on an oxidative dearomatization/intramolecular Diels-Alder cycloaddition (IMDA) cascade reaction from the olefinic phenol **18**, which has been proven to be efficient for stereoselective synthesis of bicyclo[2.2.2]octane systems[35].

**Synthesis of the pentacyclic core structure**. The multiple step synthesis begins with preparation of the tricyclic intermediate **17** as shown in Fig. 3a. Homologation of readily available aldehyde **19** (ref. 36) was achieved using the *in situ* generated ylide $Ph_3P=CHOMe$ followed by workup; reduction of the resultant aldehyde gave alcohol **20** in 55% overall yield in two steps. Subjecting **20** to a sequence of esterification with acryloyl chloride and desilylation by TBAF (tetra-*n*-butylammonium fluoride) generated the olefinic phenol **18**. Having a rapid approach to the precursor **18** established, the stage was thus set for the crucial oxidative dearomatization/IMDA cascade reaction. As anticipated, phenol **18** underwent facile oxidation on exposure to PIDA (iodobenzene diacetate) in MeOH. After switching the solvent from methanol to xylene, the ensuing IMDA cycloaddition proceeded smoothly at 150 °C through the *endo* transition state (**21**), securing the desired bicyclo[2.2.2]octane unit (**22**) as a single diastereomer (Supplementary Figs 11–13). Removal of the two methoxy groups in **22** in the presence of $SmI_2$ (ref. 37) followed by subsequent ketone protection furnished **17** in 76% overall yield.

The next step was to assemble the pentacyclic backbone of atropurpuran (Fig. 3a). The cleavage of the lactone in **17** with EtSH, followed by Dess-Martin oxidation of the resultant primary alcohol, led to aldehyde **16** in 65% overall yield. A reductive Knoevenagel condensation between 1,3-cyclohexanedione (**15**) and aldehyde **16** took place smoothly utilizing L-proline as the catalyst and Hantzsch ester as the hydrogen source. Given the limited stability of the resulting product, this intermediate was masked immediately as an enol silane in one pot to give thioester **23** in 82% yield. Conversion of thioester **23** into aldehyde **24** was readily achieved according to Fukuyama's protocol[38]. At this point, we were faced with the construction of two critical C–C bonds (that is, C5–C20 and C9–C10) before arriving at the pentacyclic framework of the target molecule. We reasoned that establishment of the C5–C20 bond first would be beneficial to the installation of the requisite stereocenter at C9. As a result, subjection of the labile silyl enol ether (**24**) to TBAF secured the formation of the E ring to afford tetracyclic diketone **25** and its 20-hydroxyl epimer (91% yield, >20:1 dr).

With the aforementioned strategy in mind, we then focused on investigating the pivotal ketyl-olefin cyclization to set up the B ring. After much experimentation, the cyclization of **25** proved to be challenging in that the starting material remained under most attempted conditions using $SmI_2$ as a radical initiator. This could be explained by the assumption that a chair conformation was adopted for the E ring due to its tendency to release the steric repulsion between the cyclohexanedione and the bridged bicyclo[2.2.2]octane system (**25**, Fig. 3b), which resulted in a significant spatial distance between the generated ketyl radical at C10 and the radical acceptor of $\Delta^{9,11}$ olefin. In addition, the intramolecular H-bonding between the proximal ketone and free hydroxyl group might stabilize this chair conformation. In this context, we hypothesized that a bulky substituent at the C20 position would not only interrupt the H-bonding but also force the formation of a boat conformation for the E ring because of the steric repulsion between the bulky OR group at C20 and the ketone group at C4 (**26**, Fig. 3b). In the boat conformation, the ketone moiety at C10 was in close proximity to the olefin, thus favouring the desired cyclization[33]. Working on this hypothesis, a bulky TBS protecting group was then installed on the C20 hydroxyl group in **25** to give **26**. Gratifyingly, on treatment of intermediate **26** with $SmI_2$/HMPA, the expected cyclization proceeded with remarkable efficiency to yield **27** as a single diastereomer in 95% yield. Its structure was unambiguously confirmed by X-ray crystallographic analysis (see Supplementary Methods). Overall, the 14-step sequence enabled an efficient access to the core structure for both arcutane diterpenes and arcutine alkaloids on a gram scale.

**Total synthesis of atropurpuran**. The final steps of our synthesis of (±)-atropurpuran are illustrated in Fig. 4. Dehydration of the tertiary hydroxyl group in **27** with $SOCl_2$ in pyridine, followed by removal of dioxolane in aqueous acetone, afforded diketone **28** in 67% overall yield. A three-step transformation including selective Wittig methylenation at C16, Corey-Chaykovsky epoxidation[39] at C4, and epoxide opening with $BF_3 \bullet OEt_2$ (ref. 40) enabled access to aldehyde **29** as a single diastereomer for further methylation at C4. The stereochemistry of the aldehyde group in **29** was not determined at this stage. Because the initial efforts for α-methylation of aldehyde **29** failed, probably due to its sterically hindered environment, desilylation was conducted using TBAF, leading to a pair of inseparable hydroxyl aldehydes along with corresponding mixtures of a cyclic hemiacetal. On treating the above crude mixture with Dess-Martin periodinane, clean keto aldehyde **30** was obtained as a pair of separable diastereomers (~1:1 dr). As anticipated, under the conditions of *t*-BuOK/MeI, both diastereomers were converted to the desired product **31** and its C4 methyl epimer in 75% yield (5:1 dr) based on recycling of **30a/b** (Supplementary Methods).

With **31** in hand, we envisioned that an allylic oxidation of the exocyclic alkene might give access to our target. However, the oxidation occurred predominately from the less hindered α face and produced the undesired 15-epi-atropurpuran **32** as the major product, along with the minor product atropurpuran **3**. Interestingly, the yield and diastereomeric ratio for the allylic oxidation were observed to be 45% and 14:1, respectively, when using an old $SeO_2$ stored in our laboratory, while the yield increased to 72% and the dr value dropped to 5:1 with a freshly ordered $SeO_2$, possibly owing to the different reactivity of the two reagents (see Supplementary Methods). After extensive experimentation, inversion of the stereochemistry of the C15 hydroxyl group in **32** proved to be fruitless. In this context, we

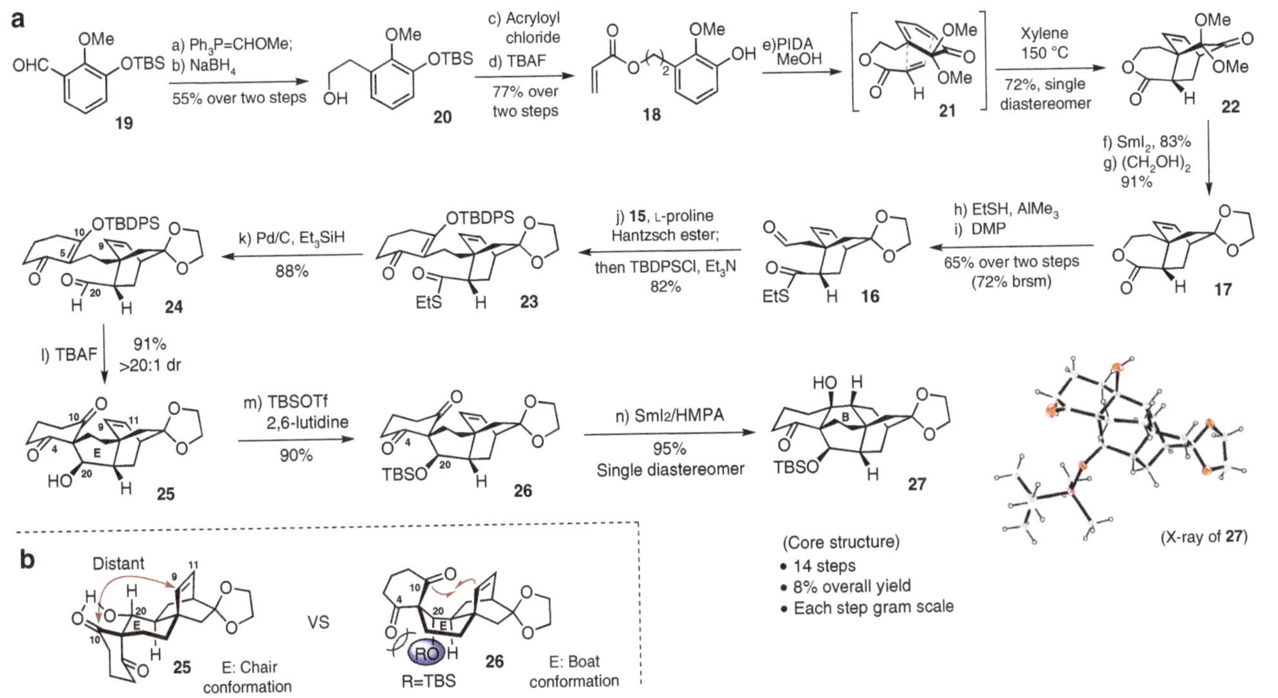


**Figure 3 | Access to the core structure for both arcutane diterpenes and arcutine alkaloids.** (**a**) Conditions: (a) $Ph_3PCH_2OMeCl$, *t*-BuOK, THF, 0 °C; then TMSCl, NaI, MeCN, −18 °C; (b) $NaBH_4$, MeOH, 0 °C, 55% over two steps; (c) acryloyl chloride, $Et_3N$, $CH_2Cl_2$, rt; (d) TBAF, THF, rt, 77% over two steps; (e) PIDA, MeOH, 0 °C; then xylene, BHT, 150 °C; 72%; (f) $SmI_2$, THF/MeOH (20:1), rt, 83%; (g) ethylene glycol, TMSCl, $CH_2Cl_2$, rt, 91%; (h) EtSH, $AlMe_3$, $CH_2Cl_2$, 0 °C to rt; (i) DMP, $CH_2Cl_2$, rt, 65% over two steps (72% brsm); (j) 1,3-cyclohexanedione (**15**), Hantzsch ester, L-proline, $CH_2Cl_2$; then TBDPSCl, $Et_3N$, $CH_2Cl_2$, rt, 82%; (k) Pd/C, $Et_3SiH$, $CH_2Cl_2$, rt, 88%; (l) TBAF, THF, rt, 91% (>20:1 dr); (m) TBSOTf, 2,6-lutidine, $CH_2Cl_2$, -40 °C, 90%; (n) $SmI_2$, HMPA, THF/*t*-BuOH (20:1), rt, 95%. (**b**) Comparison of the conformational preferences of intermediates **25** and **26**. BHT, 2,6-bis(1,1-dimethylethyl)-4-methylphenol; brsm, based on recovered starting material; HMPA, hexamethylphosphoramide; PIDA, iodobenzene diacetate; TBAF, tetra-*n*-butylammonium fluoride; TBDPSCl, *tert*-butyl(chloro)diphenylsilane; TBS, *tert*-butyldimethylsilyl; TBSOTf, *tert*-butyldimethylsilyl trifluoromethanesulfonate; THF, tetrahydrofuran; TMSCl, chlorotrimethylsilane.

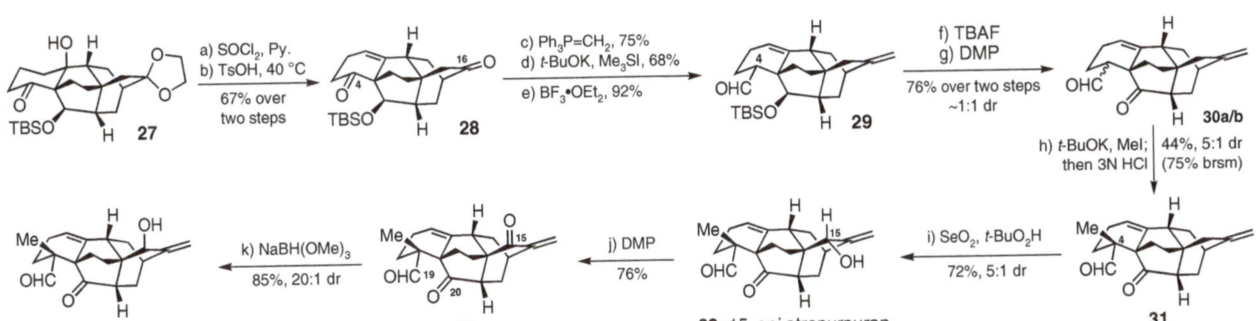


**Figure 4 | Completion of the total synthesis of (±)-atropurpuran (3).** Conditions: (a) $SOCl_2$, Py., 0 °C; (b) TsOH, acetone/$H_2O$ (10:1), 40 °C, 67% over two steps; (c) $Ph_3PCH_3Br$, *t*-BuOK, PhMe, rt, 75%; (d) $Me_3SI$, *t*-BuOK, DMSO, THF, rt, 68%; (e) $BF_3$•$OEt_2$, PhMe, -20 °C, 92%; (f) TBAF, THF, reflux; (g) DMP, $CH_2Cl_2$, rt, 76% over two steps (~1:1 dr); (h) *t*-BuOK, MeI, *t*-BuOH, rt; then 3N HCl, THF, 44% (75% brsm, 5:1 dr); (i) $SeO_2$, *t*-$BuO_2H$, $CH_2Cl_2$, 30 °C, 72% (5:1 dr); (j) DMP, $CH_2Cl_2$, rt, 76%; (k) $NaBH(OMe)_3$, THF/MeOH (10:1), −30 °C, 85% (20:1 dr). DMP, Dess-Martin periodinane; THF, tetrahydrofuran.

turned to an oxidation–reduction sequence for setting up the right stereochemistry at C15, which seemed challenging but possible. On one hand, hydride attack to the ketone at C15 (**33**) from the less hindered α face during the reduction would stereoselectively generate the desired C15 hydroxyl group. Furthermore, a chemoselective reduction could be realized through a substrate-controlled manner, owing to the crowded circumstances of two carbonyl groups at C19 and C20 compared with that of C15. Thus, enone **33** was readily prepared via Dess-Martin oxidation of alcohol **32**, before we screened the reduction conditions in detail. Initially, reductants such as reactive diisobutylaluminium hydride, lithium tri-*sec*-butylborohydride (L-selectride), and $NaBH_4$–$CeCl_3$ resulted in over-reduction even at a low temperature (−78 °C). In contrast, no reaction was observed using the less reactive sodium triacetoxyborohydride, sodium triethylborohydride and potassium tri-*sec*-butylborohydride (K-selectride) at temperatures below −30 °C. Finally, it was found that reduction of enone **33** with $NaBH_4$ at −60 °C in tetrahydrofuran/MeOH (10:1) for 1 h provided atropurpuran **3** and 15-*epi*-atropurpuran **32** in 75% yield and 2:1 dr, without reduction of the carbonyl groups at C19 and C20. The stereoselectivity was greatly improved to 20:1 dr to give atropurpuran **3** with 85% yield when more bulky sodium trimethoxyborohydride was employed as the reductant at −30 °C in tetrahydrofuran/MeOH (10:1) for 16 h. The spectral data of synthetic atropurpuran **3** were identical in all respects to those reported for the natural product[11].

## Discussion

In summary, we have disclosed an expedient and scalable approach to synthesize the core structure for arcutane diterpenes and related alkaloids and achieved the first total synthesis of (±)-atropurpuran. The oxidative dearomatization/IMDA cascade was employed as a key reaction in constructing the bicyclo[2.2.2]octane ring. Sequential aldol and ketyl-olefin cyclizations were used to secure the highly bridged tetracyclo[5.3.3.$0^{4,9}$.$0^{4,12}$]tridecane unit with excellent stereoselectivity. Of particular note, a chemoselective[41] and stereoselective reduction allowed us to install the requisite allylic hydroxyl group in the target atropurpuran. Having had success with this synthetic approach, our future efforts will focus on pursuit of an asymmetric version, as well as an extension to the total synthesis of corresponding diterpenoid alkaloids.

## Methods

**General.** All reactions that require anhydrous conditions were performed in flame-dried glassware under Ar atmosphere and all reagents were purchased from commercial suppliers. Solvent purification was conducted according to Purification of Laboratory Chemicals[42]. The products were purified by flash column chromatography on silica gel (200–300 meshes).

The reactions were monitored by thin layer chromatography. Visualization was accomplished with ultraviolet light, exposure to iodine, staining with an ethanolic solution of phosphomolybdic acid or basic solution of $KMnO_4$. $^1H$ nuclear magnetic resonance (NMR) and $^{13}C$ NMR spectra were recorded on Varian INOVA-400/54 and Agilent DD2-600/54 instruments and calibrated by using residual undeuterated chloroform (δ, $^1H$ NMR = 7.260, $^{13}C$ NMR = 77.00). The following abbreviations were used to explain the multiplicities: br, broad; d, doublet; dt, double triplet; m, multiplet; s, singlet; t, triplet; td, triple doublet; q = quartet and coupling constants (*J*) are reported in Hertz (Hz). Infrared (IR) spectra were recorded on a Perkin Elmer Spectrum Two fourier transform infrared spectrometer. High-resolution mass spectra were recorded on Bruker Apex IV FTMS or Thermo Scientific LTQ Orbitrap XL ESI mass spectrometers.

**Experimental data.** For NMR spectra of synthetic intermediates, see Supplementary Figs 1–52. For the comparisons of $^1H$ and $^{13}C$ NMR spectra of the natural and synthetic atropurpuran, see Supplementary Figs 53–58. For the comparisons of $^1H$ and $^{13}C$ NMR spectroscopic data of the natural and synthetic atropurpuran, see Supplementary Tables 1 and 2. For the experimental procedures and spectroscopic and physical data of compounds and the crystallographic data of compound **27**, see Supplementary Methods.

## References


1. Nicolaou, K. C. & Montagnon, T. *Molecules that changed the world* (Wiley, 2008).
2. Hamlin, A. M., Kisunzu, J. K. & Sarpong, R. Synthetic strategies toward hetidine and hetisine-type diterpenoid alkaloids. *Org. Biomol. Chem.* **12,** 1846–1860 (2014).
3. Liu, X.-Y. & Qin, Y. Ongoing pursuit of diterpenoid alkaloids: a synthetic view. *Asian J. Org. Chem.* **4,** 1010–1019 (2015).
4. Zhu, G., Liu, R. & Liu, B. Total synthesis of atisane-type diterpenoids and related diterpenoid alkaloids. *Synthesis. (Mass)* **47,** 2691–2708 (2015).
5. Dewick, P. M. *Medicinal Natural Products: A Biosynthetic Approach* 3rd edn (Wiley, 2009).
6. Chen, K. & Baran, P. S. Total synthesis of eudesmane terpenes by site-selective C–H oxidations. *Nature* **459,** 824–828 (2009).
7. Wang, F.-P., Chen, Q.-H. & Liu, X.-Y. Diterpenoid alkaloids. *Nat. Prod. Rep.* **27,** 529–570 (2010).
8. Cherney, E. C. & Baran, P. S. Terpenoid-alkaloids: their biosynthetic twist of fate and total synthesis. *Isr. J. Chem.* **51,** 391–405 (2011).
9. Bohlmann, F., Abraham, W.-R. & Sheldrick, W. S. Weitere diterpene mit helifulvan-gerüst und andere inhaltsstoffe aus *Helichrysum chionosphaerum*. *Phytochemistry* **19,** 869–871 (1980).
10. Wang, F.-P. & Yan, L.-P. Campylopin from *Delphinium campylocentrum*, the first hetidane $C_{20}$-diterpene, suggests a new alkaloid biogenetic pathway. *Tetrahedron* **63,** 1417–1420 (2007).
11. Tang, P., Chen, Q.-H. & Wang, F.-P. Atropurpuran, a novel diterpene with an unprecedented pentacyclic cage skeleton, from *Aconitum hemsleyanum* var. *atropurpureum*. *Tetrahedron Lett.* **50,** 460–462 (2009).
12. Pelletier, S. W. & Jacobs, W. A. The aconite alkaloids. XXVII. The structure of atisine. *J. Am. Chem. Soc.* **76,** 4496–4497 (1954).
13. Díaz, J. G., Ruiz, J. G. & de la Fuente, G. Alkaloids from *Delphinium staphisagria*. *J. Nat. Prod.* **63,** 1136–1139 (2000).
14. He, L., Chen, Y.-Z., Ding, L.-S. & Li, B.-G. New alkaloids tongoline and tongolinine from *Delphinium tongolense*. *Chin. Chem. Lett.* **7,** 557–560 (1996).
15. Tashkhodzhaev, B., Saidkhodzhaeva, S. A., Bessonova, I. A. & Antipin, M. Y. Arcutin – a new type of diterpene alkaloids. *Chem. Nat. Compd.* **36,** 79–83 (2000).
16. Saidkhodzhaeva, S. A., Bessonova, I. A. & Abdullaev, N. D. Arcutinine, a new alkaloid from *Aconitum arcuatum*. *Chem. Nat. Compd.* **37,** 466–469 (2001).
17. Wang, F.-P. & Liang, X.-T. in *The Alkaloids: Chemistry and Biology* Vol. 59 (ed. Cordell, G. A.) 1–280 (Elsevier, 2002).
18. Weber, M., Owens, K. & Sarpong, R. Atropurpuran – missing biosynthetic link leading to the hetidine and arcutine $C_{20}$-diterpenoid alkaloids or an oxidative degradation product? *Tetrahedron Lett.* **56,** 3600–3603 (2015).
19. Peese, K. M. & Gin, D. Y. Efficient synthetic access to the hetisine $C_{20}$-diterpenoid alkaloids. a concise synthesis of nominine via oxidoisoquinolinium-1,3-dipolar and dienamine-Diels-Alder cycloadditions. *J. Am. Chem. Soc.* **128,** 8734–8735 (2006).
20. Liu, X.-Y. *et al.* Oxidative dearomatization/intramolecular Diels-Alder cycloaddition cascade for the syntheses of (±)-atisine and (±)-isoazitine. *Org. Biomol. Chem.* **10,** 1411–1417 (2012).
21. Shi, Y., Wilmot, J. T., Nordstrøm, L. U., Tan, D. S. & Gin, D. Y. Total synthesis, relay synthesis, and structural confirmation of the $C_{18}$-norditerpenoid alkaloid neofinaconitine. *J. Am. Chem. Soc.* **135,** 14313–14320 (2013).
22. Hamlin, A. M., de Jesus Cortez, F., Lapointe, D. & Sarpong, R. Gallium(III)-catalyzed cycloisomerization approach to the diterpenoid alkaloids: construction of the core structure for the hetidines and hetisines. *Angew. Chem. Int. Ed.* **52,** 4854–4857 (2013).
23. Breitler, S. & Carreira, E. M. Total synthesis of (+)-crotogoudin. *Angew. Chem. Int. Ed.* **52,** 11168–11171 (2013).
24. Cherney, E. C., Lopchuk, J. M., Green, J. C. & Baran, P. S. A unified approach to *ent*-atisane diterpenes and related alkaloids: synthesis of (−)-methyl atisenoate, (−)-isoatisine, and the hetidine skeleton. *J. Am. Chem. Soc.* **136,** 12592–12595 (2014).
25. Nishiyama, Y., Han-ya, Y., Yokoshima, S. & Fukuyama, T. Total synthesis of (−)-lepenine. *J. Am. Chem. Soc.* **136,** 6598–6601 (2014).
26. Song, L. *et al.* Total synthesis of atisane-type diterpenoids: application of Diels – Alder cycloadditions of podocarpane-type unmasked *ortho*-benzoquinones. *J. Am. Chem. Soc.* **137,** 13706–13714 (2015).
27. Marth, C. J. *et al.* Network-analysis-guided synthesis of weisaconitine D and liljestrandinine. *Nature* **528,** 493–498 (2015).
28. Cheng, H. *et al.* Collective total syntheses of atisane-type diterpenes and atisine-type diterpenoid alkaloids: (±)-spiramilactone B, (±)-spiraminol, (±)-dihydroajaconine, and (±)-spiramines C and D. *Angew. Chem. Int. Ed.* **55,** 392–396 (2016).
29. Suzuki, T., Sasaki, A., Egashira, N. & Kobayashi, S. A synthetic study of atropurpuran: construction of a pentacyclic framework by an intramolecular reverse-electron-demand Diels–Alder reaction. *Angew. Chem. Int. Ed.* **50,** 9177–9179 (2011).
30. Hayashi, R., Ma, Z. X. & Hsung, R. P. A tandem 1,3-H-shift-6π-electrocyclization–cyclic 2-amido-diene intramolecular Diels–Alder cycloaddition approach to BCD-ring of atropurpuran. *Org. Lett.* **14,** 252–255 (2012).
31. Chen, H., Zhang, D., Xue, F. & Qin, Y. Synthesis of the atropurpuran A-ring via an organocatalytic asymmetric intramolecular Michael addition. *Tetrahedron* **69,** 3141–3148 (2013).
32. Chen, H. *et al.* Synthetic approach to the functionalized tricyclic core of atropurpuran. *Tetrahedron* **72,** 347–353 (2016).
33. Szostak, M., Fazakerley, N. J., Parmar, D. & Procter, D. J. Cross-coupling reactions using samarium(II) iodide. *Chem. Rev.* **114,** 5959–6039 (2014).
34. Elamparuthi, E., Fellay, C., Neuburger, M. & Gademann, K. Total synthesis of cyrneine A. *Angew. Chem. Int. Ed.* **51,** 4071–4073 (2012).
35. Roche, S. P. & Porco, Jr J. A. Dearomatization strategies in the synthesis of complex natural products. *Angew. Chem. Int. Ed.* **50,** 4068–4093 (2011).

36. Harayama, T., Sato, T., Nakano, Y., Abe, H. & Takeuchi, Y. Concise synthesis of fagaridine and decarine, phenolic benzo[c]phenanthridine alkaloids, using the palladium-assisted biaryl coupling reaction. *Heterocycles* **59,** 293–301 (2003).
37. Hwang, J.-T. & Liao, C.-C. Synthesis of angularly and linearly fused triquinanes via the common intramolecular Diels-Alder adduct of a masked *o*-benzoquinone. *Tetrahedron Lett.* **32,** 6583–6586 (1991).
38. Fukuyama, T., Lin, S.-C. & Li, L. Facile reduction of ethyl thiol esters to aldehydes: application to a total synthesis of ( + )-neothramycin A methyl ether. *J. Am. Chem. Soc.* **112,** 7050–7051 (1990).
39. Corey, E. J. & Chaykovsky, M. Dimethyloxosulfonium methylide ($(CH_3)_2SOCH_2$) and dimethylsulfonium methylide ($(CH_3)_2SCH_2$). formation and application to organic synthesis. *J. Am. Chem. Soc.* **87,** 1353–1364 (1965).
40. Toyota, M., Yokota, M. & Ihara, M. Remarkable control of radical cyclization processes of cyclic enyne: total syntheses of ( ± )-methyl gummiferolate, ( ± )-methyl 7β-hydroxykaurenoate, and ( ± )-methyl 7-oxokaurenoate and formal synthesis of ( ± )-gibberellin $A_{12}$ from a common synthetic precursor. *J. Am. Chem. Soc.* **123,** 1856–1861 (2001).
41. Shenvi, R. A., O'Malley, D. P. & Baran, P. S. Chemoselectivity: the mother of invention in total synthesis. *Acc. Chem. Res.* **42,** 530–541 (2009).
42. Perrin, D. D., Armarego, W. L. F. & Perrin, D. R. *Purification of Laboratory Chemicals* 2nd edn (Pergamon Press, Oxford, 1980).

## Acknowledgements

We are grateful to the National Natural Science Foundation of China (21321061, 21132006, 21572139 and 21502126) for financial support of this research.

## Author contributions

Y.Q. conceived and designed the project and wrote the paper with assistance from X.-Y.L.; J.G., H.C., X.-Y.L., Z.-X.W. and W.N. performed the experiments. All authors discussed the results and commented on the manuscript. J.G., H.C. and X.-Y.L. contributed equally to this research.

## Additional information

**Competing financial interests:** The authors declare no competing financial interests.

# Permissions

All chapters in this book were first published in NC, by Nature Publishing Group; hereby published with permission under the Creative Commons Attribution License or equivalent. Every chapter published in this book has been scrutinized by our experts. Their significance has been extensively debated. The topics covered herein carry significant findings which will fuel the growth of the discipline. They may even be implemented as practical applications or may be referred to as a beginning point for another development.

The contributors of this book come from diverse backgrounds, making this book a truly international effort. This book will bring forth new frontiers with its revolutionizing research information and detailed analysis of the nascent developments around the world.

We would like to thank all the contributing authors for lending their expertise to make the book truly unique. They have played a crucial role in the development of this book. Without their invaluable contributions this book wouldn't have been possible. They have made vital efforts to compile up to date information on the varied aspects of this subject to make this book a valuable addition to the collection of many professionals and students.

This book was conceptualized with the vision of imparting up-to-date information and advanced data in this field. To ensure the same, a matchless editorial board was set up. Every individual on the board went through rigorous rounds of assessment to prove their worth. After which they invested a large part of their time researching and compiling the most relevant data for our readers.

The editorial board has been involved in producing this book since its inception. They have spent rigorous hours researching and exploring the diverse topics which have resulted in the successful publishing of this book. They have passed on their knowledge of decades through this book. To expedite this challenging task, the publisher supported the team at every step. A small team of assistant editors was also appointed to further simplify the editing procedure and attain best results for the readers.

Apart from the editorial board, the designing team has also invested a significant amount of their time in understanding the subject and creating the most relevant covers. They scrutinized every image to scout for the most suitable representation of the subject and create an appropriate cover for the book.

The publishing team has been an ardent support to the editorial, designing and production team. Their endless efforts to recruit the best for this project, has resulted in the accomplishment of this book. They are a veteran in the field of academics and their pool of knowledge is as vast as their experience in printing. Their expertise and guidance has proved useful at every step. Their uncompromising quality standards have made this book an exceptional effort. Their encouragement from time to time has been an inspiration for everyone.

The publisher and the editorial board hope that this book will prove to be a valuable piece of knowledge for researchers, students, practitioners and scholars across the globe.

# List of Contributors

**Dan Zhang, Jun Zhou, Fei Xia, Zhenghui Kang and Wenhao Hu**
Shanghai Engineering Research Center of Molecular Therapeutics and New Drug Development, and Department of Chemistry, East China Normal University, Shanghai 200062, China

**Masayuki Naruto and Susumu Saito**
Graduate School of Science, Nagoya University, Chikusa, Nagoya 464-8602, Japan

**Susumu Saito**
Institute for Advanced Research, Nagoya University, Chikusa, Nagoya 464-8601, Japan.
3ACT-C, JST, Chikusa, Nagoya 464-8602, Japan

**Kuan Sun, Zeyun Xiao, Shirong Lu, Jonathan M. White, Jegadesan Subbiah, Andrew B. Holmes, Wallace W.H. Wong and David J. Jones**
School of Chemistry, Bio21 Institute, The University of Melbourne, 30 Flemington Road, Parkville, Victoria 3010, Australia

**Jianyong Ouyang and Kuan Sun**
Department of Materials Science and Engineering, National University of Singapore, 7 Engineering Drive 1, Singapore 117574, Singapore

**Kuan Sun**
Department of Renewable Energy, School of Power Engineering, Chongqing University, 174 Shazhengjie, Shapingba, Chongqing 400044, China

**Wojciech Zajaczkowski and Wojciech Pisula**
Max Planck Institute for Polymer Research, Ackermannweg 10, 55128 Mainz, Germany

**Eric Hanssen**
Advanced Microscopy Facility, Bio21 Institute, The University of Melbourne, 30 Flemington Road, Parkville, Victoria 3010, Australia

**Rachel M. Williamson**
MX Beamlines, Australian Synchrotron, 800 Blackburn Road, Clayton, Victoria 3168, Australia

**Kyohei Ozaki, Katsuaki Kawasumi, Mari Shibata and Hideto Ito**
Institute of Transformative Bio-Molecules (WPI-ITbM) and Graduate School of Science, Nagoya University, Chikusa, Nagoya 464-8602, Japan

**Kenichiro Itami**
Institute of Transformative Bio-Molecules (WPI-ITbM) and Graduate School of Science, Nagoya University, Chikusa, Nagoya 464-8602, Japan
JST-ERATO, Itami Molecular Nanocarbon Project, Nagoya University, Chikusa, Nagoya 464-8602, Japan

**Huimin Yang and Xingchao Dai**
State Key Laboratory for Oxo Synthesis and Selective Oxidation, Centre for Green Chemistry and Catalysis, Lanzhou Institute of Chemical Physics, Chinese Academy of Sciences, No.18, Tianshui Middle Road, Lanzhou 730000, China
University of Chinese Academy of Sciences, No. 19A, Yuquanlu, Beijing 100049, China

**Xinjiang Cui, Youquan Deng and Feng Shi**
State Key Laboratory for Oxo Synthesis and Selective Oxidation, Centre for Green Chemistry and Catalysis, Lanzhou Institute of Chemical Physics, Chinese Academy of Sciences, No.18, Tianshui Middle Road, Lanzhou 730000, China

**Qiang Liu, Lipeng Wu, Ralf Jackstell and Matthias Beller**
Leibniz-Institut für Katalyse an der Universität Rostock, Albert-Einstein-Strasse 29a, 18059 Rostock, Germany

**R.N. Straker, Q. Peng, A. Mekareeya, R.S. Paton and E.A. Anderson**
Chemistry Research Laboratory, University of Oxford, 12 Mansfield Road, Oxford OX1 3TA, UK

**Shuhei Kusumoto and Kyoko Nozaki**
Department of Chemistry and Biotechnology, Graduate School of Engineering, The University of Tokyo, 7-3-1 Hongo, Bunkyo-ku, Tokyo 113-8656, Japan

**Jasper H. M. van der Velde, Jochem H. Smit, Atieh Aminian Jazi Giorgos Guoridis and Thorben Cordes**
Molecular Microscopy Research Group, Zernike Institute for Advanced Materials, University of Groningen, Nijenborgh 4, 9747 AG Groningen, The Netherlands

**Jens Oelerich and Gerard Roelfes**
Stratingh Institute for Chemistry, University of Groningen, Nijenborgh 4, 9747 AG Groningen, The Netherlands

**Jingyi Huang and Andreas Herrmann**
Department of Polymer Chemistry, Zernike Institute for Advanced Materials, University of roningen, Nijenborgh 4, 9747 AG Groningen, The Netherlands

**Silvia Galiani and Christian Eggeling**
MRC Human Immunology Unit, Weatherall Institute of Molecular Medicine, University of Oxford, Headley Way, Oxford OX3 9DS, UK

**Kirill Kolmakov**
Department NanoBiophotonics, Max-Planck-Institute of Molecular Medicine, Am Fassberg 1, 37077 Goettingen, Germany

**Michal Amar, Sukanta Bar, Hila Toledo, Boris Tumanskii, Mark Botoshansky, Natalia Fridman and Alex M. Szpilman**
Schulich Faculty of Chemistry, Technion-Israel Institute of Technology, 32000 Haifa, Israel

**Mark A. Iron and Linda J.W. Shimon**
Department of Chemical Research Support,Weizmann Institute of Science, 76100 Rehovot, Israel.

**Xiaheng Zhang and Biao Yu**
State Key Laboratory of Bioorganic and Natural Products Chemistry, Shanghai Institute of Organic Chemistry, Chinese Academy of Sciences, 345 Lingling Road, Shanghai 200032, China

**Yu Zhou and Jianping Zuo**
State Key Laboratory of New Drug Research, Shanghai Institute of Materia Medica, Chinese Academy of Sciences, Shanghai 201203, China

**Zuyong Deng, Jin-Hong Lin and Ji-Chang Xiao**
Key Laboratory of Organofluorine Chemistry, Shanghai Institute of Organic Chemistry, Chinese Academy of Sciences, 345 Lingling Road, Shanghai 200032, China

**Zhanchao Meng, Hao Chen, Ming Wan, Peng Yang and Haixin Yu**
State Key Laboratory of Bioorganic and Natural Products Chemistry, Shanghai Institute of Organic Chemistry, Chinese Academy of Sciences, 345 Lingling Road, Shanghai 200032, China

**Li Li, Wanyin Tao and Jin Zhong**
Unit of Viral Hepatitis, Key Laboratory of Molecular Virology and Immunology, Institut Pasteur of Shanghai, Shanghai Institutes for Biological Sciences, Chinese Academy of Sciences, Shanghai 200032, China

**David J. Edmonds**
Pfizer Worldwide Research and Development, Cambridge, Massachusetts 02139, USA

**Ang Li**
Collaborative Innovation Center of Chemistry for Life Sciences, 22 Hankou Road, Nanjing, Jiangsu 210093, China

**Jing He, Lawrence G. Hamann and Rohan E.J. Beckwith**
Department of Global Discovery Chemistry, Novartis Institutes for BioMedical Research, 250 Massachusetts Avenue, Cambridge, Massachusetts 02139, USA

**Huw M.L. Davie**
Department of Chemistry, Emory University, 1515 Dickey Drive, Atlanta, Georgia 30322, USA

**Depeng Zhao, Thomas M. Neubauer and Ben L. Feringa**
Centre for System Chemistry, Stratingh Institute for Chemistry, Faculty of Mathematics and Natural Sciences, University of Groningen, Nijenborgh 4, 9747AG Groningen, The Netherlands

**Yivan Jiang, Jessica L. Freyer, Spencer D. Brucks, Jeffrey S. Bandar, Christopher Torsitano, Tristan H. Lambert and Luis M. Campos**
Department of Chemistry, Columbia University, New York, New York 10027, USA

**Pepa Cotanda**
Joint Center for Artificial Photosynthesis, Lawrence Berkeley National Laboratory, Berkeley, California 94720, USA
Department of Chemical and Biomolecular Engineering, University of California, Berkeley, California 94720, USA

**Kato L. Killops**
Edgewood Chemical Biological Center, Aberdeen Proving Ground, Aberdeen, Maryland 21010, USA

**Nitash P. Balsara**
Department of Chemical and Biomolecular Engineering, University of California, Berkeley, California 94720, USA
Materials Sciences Division, Environmental Energy Technologies Division and Joint Center for Energy Storage, Lawrence Berkeley National Laboratory, Berkeley, California 94720, USA

**Shangda Li, Lei Cai, Huafang Ji and Long Yang**
State Key Laboratory of Structural Chemistry, Fujian Institute of Research on the Structure of Matter, Chinese Academy of Sciences, Fujian 350002, China

**Gang Li**
State Key Laboratory of Structural Chemistry, Fujian Institute of Research on the Structure of Matter, Chinese Academy of Sciences, Fujian 350002, China
Key Laboratory of Coal to Ethylene Glycol and Its Related Technology, CAS, Fujian 350002, China

**Hiroki Mandai, Kazuki Fujii, Hiroshi Yasuhara, Kenko Abe and Koichi Mitsudo**
Division of Applied Chemistry, Graduate School of Natural Science and Technology, Okayama University, 3-1-1 Tsushima-naka, Kita-ku, Okayama 700-8530, Japan

**Toshinobu Korenaga**
Faculty of Engineering, Department of Chemistry and Bioengineering, Iwate University, Morioka, Iwate 020-8551, Japan

**Seiji Suga**
Division of Applied Chemistry, Graduate School of Natural Science and Technology, Okayama University, 3-1-1 Tsushima-naka, Kita-ku, Okayama 700-8530, Japan
Research Center of New Functional Materials for Energy Production, Storage and Transport Okayama University, 3-1-1 Tsushima-naka, Kita-ku, Okayama 700-8530, Japan
Japan Science and Technology Agency, ACT-C, 4-1-8 Honcho, Kawaguchi, Saitama 332-0012, Japan

**Guillaume Gody**
Department of Chemistry, University of Warwick, Warwick CV4 7AL, UK

**Per B. Zetterlund**
Centre for Advanced Macromolecular Design, School of Chemical Engineering, University of New SouthWales, Sydney, New SouthWales 2052, Australia

**Sébastien Perrier**
Department of Chemistry, University of Warwick, Warwick CV4 7AL, UK
Faculty of Pharmacy and Pharmaceutical Sciences, Monash University, Parkville, Victoria 3052, Australia

**Simon Harrisson**
Laboratoire des Interactions Moléculaires et Réactivité Chimique et Photochimique, CNRS UMR5623, Université Paul Sabatier Toulouse III, 118 Route de Narbonne, Toulouse 31062, France

**Sin Ki Fung, Bei Cao, Wai-Pong To and Chun-Nam Lok**
State Key Laboratory of Synthetic Chemistry, Institute of Molecular Functional Materials, Chemical Biology Centre, and Department of Chemistry, The University of Hong Kong, Pokfulam Road, Hong Kong, China

**Taotao Zou, Chen Yang and Chi-Ming Che**
State Key Laboratory of Synthetic Chemistry, Institute of Molecular Functional Materials, Chemical Biology Centre, and Department of Chemistry, The University of Hong Kong, Pokfulam Road, Hong Kong, China
HKU Shenzhen Institute of Research and Innovation Shenzhen, Shenzhen 518053, China

**Tianfeng Chen**
Department of Chemistry, Jinan University, Guangzhou 510632, China

**Kazunori Koide and Matthew P. Tracey**
Department of Chemistry, University of Pittsburgh, 219 Parkman Avenue, Pittsburgh, Pennsylvania 15260, USA

**Xiaodong Bu, Junyong Jo, Michael J. Williams and Christopher J. Welch**
Process and Analytical Chemistry, Merck Research Laboratories, 126 East Lincoln Avenue, Rahway, New Jersey 07065, USA

**Kai Wu, Kang Li, Ya-Jun Hou, Mei Pan, Lu-Yin Zhang and Ling Chen**
MOE Laboratory of Bioinorganic and Synthetic Chemistry, State Key Laboratory of Optoelectronic Materials and Technologies, Lehn Institute of Functional Materials, School of Chemistry and Chemical Engineering, Sun Yat-Sen University, Guangzhou 510275, China

**Cheng-Yong Su**
MOE Laboratory of Bioinorganic and Synthetic Chemistry, State Key Laboratory of Optoelectronic Materials and Technologies, Lehn Institute of Functional Materials, School of Chemistry and Chemical Engineering, Sun Yat-Sen University, Guangzhou 510275, China
State Key Laboratory of Applied Organic Chemistry, Lanzhou University, Lanzhou 730000, China

**Kaiwu Dong, Saravanakumar Elangovan, Rui Sang, Anke Spannenberg, Ralf Jackstell, Kathrin Junge and Matthias Beller**
Leibniz-Institut für Katalyse e.V. an der Universität Rostock, Albert-Einstein Stra_e 29a, 18059 Rostock, Germany

**Yuehui Li**
Leibniz-Institut für Katalyse e.V. an der Universität Rostock, Albert-Einstein Stra_e 29a, 18059 Rostock, Germany
State Key Laboratory for Oxo Synthesis and Selective Oxidation, Suzhou Research Institute of LICP, Lanzhou Institute of Chemical Physics (LICP), Chinese Academy of Sciences, 730000 Lanzhou, P.R. China

**Philipp M. Cromm, Jochen Spiegel and Herbert Waldmann**
Department of Chemical Biology, Max-Planck-Institute of Molecular Physiology, Otto-Hahn-Strasse 11, D-44227 Dortmund, Germany
Technische Universität Dortmund, Fakultät für Chemie and Chemische Biologie, Otto-Hahn-Strasse 6, D-44227 Dortmund, Germany

**Sebastian Schaubach and Alois Fürstner**
Technische Universität Dortmund, Fakultät für Chemie and Chemische Biologie, Otto-Hahn-Strasse 6, D-44227 Dortmund, Germany
Max-Planck-Institut für Kohlenforschung, Kaiser-Wilhelm-Platz 1, D-45470 Mülheim/Ruhr, Germany

**Tom N. Grossmann**
Technische Universität Dortmund, Fakultät für Chemie and Chemische Biologie, Otto-Hahn-Strasse 6, D-44227 Dortmund, Germany
Chemical Genomics Centre of the Max Planck Society, Otto-Hahn-Strasse 15, D-44227 Dortmund, Germany
Department of Chemistry and Pharmaceutical Sciences, VU University Amsterdam, De Boelelaan 1083, 1081 HV Amsterdam, The Netherlands

**Jing Gong, Huan Chen, Xiao-Yu Liu, Zhi-Xiu Wang, Wei Nie and Yong Qin**
Key Laboratory of Drug Targeting and Drug Delivery Systems of the Ministry of Education, Department of Chemistry of Medicinal Natural Products, West China School of Pharmacy, and State Key Laboratory of Biotherapy, Sichuan University, No. 17, Section 3, Renmin Nan Road, Chengdu 610041, China

# Index


**A**

Achiral Catalyst, 60
Acyl Transfer Catalysis, 157
Antiviral Activity, 118, 123
Architectural Intricacy, 216
Aromatic Cyclopropenium Ion, 142
Aryl Methyl Ethers, 69-74
Arylmetallic Reagents, 110, 112
Atropisomeric Molecules, 192-193
Atropurpuran, 216, 220

**B**

Benzodithiophene Terthiophene Rhodanine (btr), 18-19
Benzoic Acid, 10, 12, 149, 151, 154
Bioactive Scaffolds, 209
Bond Cleavage, 1-2, 5-6, 51, 74

**C**

C-h Functionalization, 58, 124, 126-127, 129-130, 133-134, 149, 155-156
Carbenoid Insertion, 126
Carbon Dioxide, 16, 45-47, 49, 54, 56-59, 75, 114-115, 207-208
Carbon Monoxide, 54, 58, 202-203, 206-207
Carboxylic Acids, 9-10, 16, 54, 58, 62, 98, 109, 155, 207
Catalyst Prototypes, 9-10
Chain Growth Polymerization, 168-169
Chemical Transformations, 76, 86, 136
Chiral Phosphoric Acid, 1, 7
Chirality, 29, 33, 64, 135-136, 159, 193-195, 198, 200
Common Catalytic Motifs, 157
Complex Alkaloids, 126-127, 129
Computational Ligand Design, 60, 68
Cyclic Nitroxides, 92, 99
Cycloisomerization, 60-68, 220
Cyclopropenium Ions, 142, 148

**D**

Diastereoselective Ynamide, 60-61
Diels-alder Cycloaddition, 33, 216
Diphosphine, 17, 176, 179, 207
Drug Molecules, 126-127, 130, 132, 149
Dynamic Control, 135-136

**E**

Electrostatic Interactions, 109, 157, 160
Enantiomeric Ratio (er), 2, 157, 160
Enantioselective Catalysis, 135, 138, 141, 200
Enantioseparation, 192-193, 195, 197-199
Ethylene Glycol, 56, 149, 202, 207

**F**

Fluorescence Microscopy, 76-78, 85, 89, 182-183
Fluorophore Derivatives, 76-78, 85, 87, 89
Fragment Modification, 1-2
Functional Polyelectrolytes, 142
Functionalization, 10, 16, 27-29, 31-33, 46, 57-58, 124-127, 129-130, 132-134, 149, 155-156, 175, 210, 212

**H**

Hydrogenation, 9-17, 34, 38, 41, 43, 46, 54, 57-58, 69-70, 74-75, 141, 184, 202-208, 215
Hydrogenation Reaction, 34, 41
Hydroxycyclopentadienyl Iridium, 69, 71, 74

**I**

Immunosuppressive Effects, 100-101
Indolosesquiterpenoids, 118
Intermolecular Rhodium, 126
Intramolecular Photostabilization, 76-78, 80, 85-89
Isothermal Titration Calorimetry, 176

**K**

K-region Selectivity, 27, 29, 31-32
Ketone Reduction, 34, 38-39, 42, 122-123

**L**

Liquid Crystalline Material, 18
Luminescent Platinum, 176

**M**

Macrocycle Geometries, 209
Metal-organic Cages, 192, 200
Metalloligands, 192-195, 199-200
Monomer Placement, 168-169, 174
Monomeric Cyclopropenium, 142
Multiblock Copolymers, 168-169, 171, 175

**N**

Nanographene Synthesis, 27-29
Natural Products, 33, 100, 108, 118, 122-124, 126-127, 130, 132-133, 149, 216, 220
Nitroxides, 91-93, 95-99
Nitroxyl Radicals, 91, 93, 99
Nucleophilic Arylation, 110, 115

## O
Olefin Metathesis, 209-210, 214-215
Olefination Reactions, 151
Organic Phosphonium Salts, 110
Organic Photovoltaics, 18, 25-26
Organic Synthesis, 10, 28, 45-47, 56-57, 116, 124, 133, 149, 221
Organometallic Chemistry, 45-47
Oridamycin, 118, 122-123
Oxidative Dearomatization, 216, 220

## P
Periploside A, 100-101, 103-109
Phosphine Ligands, 92, 135-136, 206
Polycyclic Aromatic Hydrocarbons, 27, 33
Polymeric Chains, 168
Power Conversion Efficiencies (pces), 18
Pregnane Hexasaccharide, 100, 105, 108

## Q
Quantitative Assays, 185

## R
Reductive Hydrogen Atom, 34, 38, 41-42
Resource Management, 45
Reversible Deactivation, 169, 171, 185, 189
Reversible Deactivation Approach, 185, 189
Ring-closing Alkyne, 209-210, 212, 214-215
Room Temperature (rt), 103, 202, 207
Ruthenium Carboxylates, 9-10

## S
Selective Hydrogenolysis, 69, 71, 74
Selective Transformation, 207
Signal Amplification, 185-186
Stereoelectronic Effects, 61, 91, 98
Stereogenic Ru, 192-193, 195
Stereoselective Reduction, 216
Synthetic Chemistry, 56, 110, 126, 133, 176, 192

## T
Tetraarylphosphonium Salts, 110, 112, 115
Three-component Reactions, 1, 7
Total Synthesis, 68, 100-101, 103, 106, 108-109, 118, 120, 123-125, 133, 141, 216, 220-221
Transition Metal Catalysts, 34
Triple Deoxygenation, 69

## V
Valorization, 45-46, 54, 56-57, 69-70, 74

## X
Xiamycin, 118, 120, 122-124